D1544567

Methods in Enzymology

Volume 353
REDOX CELL BIOLOGY AND GENETICS
Part B

METHODS IN ENZYMOLOGY

EDITORS-IN-CHIEF

John N. Abelson   Melvin I. Simon

DIVISION OF BIOLOGY
CALIFORNIA INSTITUTE OF TECHNOLOGY
PASADENA, CALIFORNIA

FOUNDING EDITORS

Sidney P. Colowick and Nathan O. Kaplan

Methods in Enzymology

Volume 353

Redox Cell Biology and Genetics

Part B

EDITED BY

Chandan K. Sen

LABORATORY OF MOLECULAR MEDICINE
DEPARTMENTS OF SURGERY AND MOLECULAR AND CELLULAR BIOCHEMISTRY
DAVIS HEART AND LUNG RESEARCH INSTITUTE
THE OHIO STATE UNIVERSITY MEDICAL CENTER
COLUMBUS, OHIO

Lester Packer

DEPARTMENT OF MOLECULAR PHARMACOLOGY AND TOXICOLOGY
UNIVERSITY OF SOUTHERN CALIFORNIA
LOS ANGELES, CALIFORNIA

EDITORIAL ADVISORY BOARD

John F. Engelhardt
Pascal J. Goldschmidt-Clermont
Rajiv R. Ratan
Seppo Ylä-Herttuala
Jay L. Zweier

ACADEMIC PRESS

An imprint of Elsevier Science

Amsterdam Boston London New York Oxford Paris
San Diego San Francisco Singapore Sydney Tokyo

Magale Library
Southern Arkansas University
Magnolia, Arkansas 71753
DISCARDED

This book is printed on acid-free paper. ∞

Copyright © 2002, Elsevier Science (USA).

All Rights Reserved.
No part of this publication may be reproduced or transmitted in any form or by any means, electronic or mechanical, including photocopy, recording, or any information storage and retrieval system, without permission in writing from the Publisher.

The appearance of the code at the bottom of the first page of a chapter in this book indicates the Publisher's consent that copies of the chapter may be made for personal or internal use of specific clients. This consent is given on the condition, however, that the copier pay the stated per copy fee through the Copyright Clearance Center, Inc. (222 Rosewood Drive, Danvers, Massachusetts 01923), for copying beyond that permitted by Sections 107 or 108 of the U.S. Copyright Law. This consent does not extend to other kinds of copying, such as copying for general distribution, for advertising or promotional purposes, for creating new collective works, or for resale. Copy fees for pre-2002 chapters are as shown on the title pages. If no fee code appears on the title page, the copy fee is the same as for current chapters. 0076-6879/2002 $35.00

Explicit permission from Academic Press is not required to reproduce a maximum of two figures or tables from an Academic Press chapter in another scientific or research publication provided that the material has not been credited to another source and that full credit to the Academic Press chapter is given.

Academic Press
An imprint of Elsevier Science.
525 B Street, Suite 1900, San Diego, California 92101-4495, USA
http://www.academicpress.com

Academic Press
84 Theobalds Road, London WC1X 8RR, UK
http://www.academicpress.com

International Standard Book Number: 0-12-182256-7

PRINTED IN THE UNITED STATES OF AMERICA
02   03   04   05   06   07   SB   9   8   7   6   5   4   3   2   1

Table of Contents

Section I. Protein Structure and Function

Section II. Nucleic Acids and Genes

Contributors to Volume 353

Article numbers are in parentheses following the names of contributors.
Affiliations listed are current.

SHIN AIZAWA (8), *Department of Anatomy, Nihon University School of Medicine, Itabashi, Tokyo 173-8610, Japan*

LYNN M. ALMLI (33), *Department of Ecology and Evolutionary Biology, University of Tennessee, Knoxville, Tennessee 37916*

RICHARD G. W. ANDERSON (13), *Department of Cell Biology, University of Texas Southwestern Medical Center-Dallas, Dallas, Texas 75390*

FRANÇOISE AUCH'ERE (14), *Section of Hematology Research, Department of Biochemistry and Molecular Biology, Mayo Clinic and Foundation, Rochester, Minnesota 55905*

JOSE AVALOS (26), *Department of Biophysics and Biophysical Chemistry, Johns Hopkins University School of Medicine, Baltimore, Maryland 21205*

MICHAEL A. BALDWIN (6), *Mass Spectrometry Facility, Department of Pharmaceutical Chemistry, University of California, San Francisco, California 94143*

JACQUELINE K. BARTON (43), *Division of Chemistry and Chemical Engineering, California Institute of Technology, Pasadena, California 91125*

NICOLE BEC (11), *French National Institute for Health and Medical Research, U 128, IFR 24, 34293 Montpellier, France*

PHIL BEFFREY (41), *Laboratory of Molecular Medicine, Departments of Surgery and Molecular and Cellular Biochemistry, Davis Heart and Lung Research Institute, Ohio State University Medical Center, Columbus, Ohio 43210*

KIMBERLY BENTLEY (41), *Laboratory of Molecular Medicine, Departments of Surgery and Molecular and Cellular Biochemistry, Davis Heart and Lung Research Institute, Ohio State University Medical Center, Columbus, Ohio 43210*

CHRISTOPHER C. BENZ (6), *Buck Institute for Age Research, Novato, California 94945*

GUNNAR I. BERGLUND (27), *Department of Biochemistry, Uppsala University, S-751 23 Uppsala, Sweden*

ERIK A. BEY (36), *Department of Cell Biology, Lerner Research Institute, Cleveland Clinic Foundation, Cleveland, Ohio 44195*

JEF D. BOEKE (26), *Department of Molecular Biology and Genetics, Johns Hopkins University School of Medicine, Baltimore, Maryland 21205*

CELIA BONAVENTURA (18), *Marine Laboratory, Nicholas School of the Environment, Duke University, Beaufort, North Carolina 28516*

ELIZABETH M. BOON (43), *Division of Chemistry and Chemical Engineering, California Institute of Technology, Pasadena, California 91125*

GREGORY G. BORISENKO (25), *Department of Environmental and Occupational Health, Graduate School of Public Health, University of Pittsburgh, Pittsburgh, Pennsylvania 15260*

MARIE-CHRISTINE BROILLET (19), *Institute of Pharmacology and Toxicology, University of Lausanne, CH-1005 Lausanne, Switzerland*

ix

ANNE-LAURE BULTEAU (23), *Laboratory of Biological and Biochemical Cellular Aging, Université Denis Diderot-Paris 7, 75251 Paris Cedex 05, France*

JANUARIO CABRAL-NETO (45), *Carlos Chagas Filho Institute of Biophysics, Universidade Federal do Rio de Janeiro, 21949-900 Rio de Janeiro, Brazil*

GUNILLA H. CARLSSON (27), *Department of Biochemistry, Uppsala University, S-751 23 Uppsala, Sweden*

MARTHA K. CATHCART (36), *Department of Cell Biology, Lerner Research Institute, Cleveland Clinic Foundation, Cleveland, Ohio 44195*

IVANA CELIC (26), *Department of Molecular Biology and Genetics, Johns Hopkins University School of Medicine, Baltimore, Maryland 21205*

CHING K. CHOW (34), *Graduate Center for Nutritional Sciences, University of Kentucky, Lexington, Kentucky 40506*

FENG CHU (9), *Department of Cancer Biology, University of Texas M. D. Anderson Cancer Center, Houston, Texas 77030*

FUNG-LUNG CHUNG (44), *Division of Carcinogenesis and Molecular Epidemiology, American Health Foundation, Valhalla, New York 10595*

AL CLAIBORNE (5), *Department of Biochemistry, Bowman Gray School of Medicine, Wake Forest University, Winston-Salem, North Carolina 27157*

MARIANGELA CONCONI (23), *Laboratory of Biological and Biochemical Cellular Aging, Université Denis Diderot-Paris 7, 75251 Paris Cedex 05, France*

PRISCILLA K. COOPER (45), *Life Sciences Division, Lawrence Berkeley National Laboratory, Berkeley, California 94720*

DAVID L. COX (14), *Division of Sexually Transmitted Diseases Laboratory Research, Centers for Disease Control and Prevention, Atlanta, Georgia 30341*

ALVIN L. CRUMBLISS (18), *Department of Chemistry, Duke University, Durham, North Carolina 27708*

DIPAK K. DAS (30), *Department of Surgery, Cardiovascular Research Center, University of Connecticut School of Medicine, Farmington, Connecticut 06030*

VICTOR L. DAVIDSON (12), *Department of Biochemistry, University of Mississippi Medical Center, Jackson, Mississippi 39216*

WOLFGANG DILLMANN (30), *Department of Medicine, University of California at San Diego, La Jolla, California 92093*

CHRISTINA EKERFELT (3), *Division of Clinical Immunology, Department of Health and Environment, University of Linköping, SE-581 85 Linköping, Sweden*

JOHN F. ENGELHARDT (28), *Department of Anatomy and Cell Biology, University of Iowa, Iowa City, Iowa 52242*

JAMES P. FABISIAK (25), *Department of Environmental and Occupational Health, Graduate School of Public Health, University of Pittsburgh, Pittsburgh, Pennsylvania 15260*

KLAUS FELIX (37), *Laboratory of Genetics, Center for Cancer Research, National Cancer Institute, National Institutes of Health, Bethesda, Maryland 20892*

WEI FENG (22), *Department of Molecular Biosciences, University of California, Davis, California 95616*

DONNA M. FERRIERO (33), *Department of Neurology, University of California, San Francisco, California 94143*

TOREN FINKEL (10), *Laboratory of Molecular Biology, National Heart, Lung, and Blood Institute, National Institutes of Health, Bethesda, Maryland 20892*

ANDRÁS FISER (2), *Laboratory of Molecular Biophysics, The Rockefeller University, New York, New York 10021*

MARCO W. FRAAIJE (17), *Laboratory of Biochemistry, Groningen Biomolecular Sciences and Biotechnology Institute, University of Groningen, 9747 AG Groningen, The Netherlands*

BERTRAND FRIGUET (23), *Laboratory of Biological and Biochemical Cellular Aging, Université Denis Diderot-Paris 7, 75251 Paris Cedex 05, France*

KALPANA GHOSHAL (40), *Department of Molecular and Cellular Biochemistry, Ohio State University, Columbus, Ohio 43210*

BERND R. GLOSS (30), *Department of Medicine, University of California at San Diego, La Jolla, California 92093*

ANTONIUS C. F. GORREN (11), *Institute for Pharmacology and Toxicology, Karl-Franzens-Universität Graz, A-8010 Graz, Austria*

PRABHAT C. GOSWAMI (38), *Free Radical and Radiation Biology Program, University of Iowa, Iowa City, Iowa 52242*

KATHY K. GRIENDLING (20), *Division of Cardiology, Emory University, Atlanta, Georgia 30322*

MARK W. GRINSTAFF (46), *Department of Chemistry, Duke University, Durham, North Carolina 27708*

JANOS HAJDU (27), *Department of Biochemistry, Uppsala University, S-751 23 Uppsala, Sweden*

HIROSHI HANDA (8), *Department of Biological Information, Graduate School of Bioscience and Biotechnology and Frontier Collaborative Research Center, Tokyo Institute of Technology, Midori-ku, Yokohama 226-8503, Japan*

MAMORU HATAKEYAMA (8), *Frontier Collaborative Research Center, Tokyo Institute of Technology, Midori-ku, Yokohama 226-8503; Faculty of Science and Technology, Keio University, Kouhoku-ku, Yokohama 223-8522, Japan*

KARSTEN R. O. HAZLETT (14), *Center for Microbial Pathogenesis, University of Connecticut Health Center, Farmington, Connecticut 06030*

RYUJI HIGASHIKUBO (38), *Radiation Oncology Center, Mallinckrodt Institute of Radiology, Washington University Medical School, St. Louis, Missouri 63110*

MASAKI HIRAMOTO (8), *Department of Biological Information, Graduate School of Bioscience and Biotechnology, Tokyo Institute of Technology, Midori-ku, Yokohama 226-8501; Department of Anatomy, Nihon University School of Medicine, Itabashi, Tokyo 173-8610, Japan*

YE-SHIH HO (30), *Institute of Toxicology, Wayne State University, Detroit, Michigan 48202*

CHANTAL HOUÉE-LEVIN (4), *Laboratory of Physical Chemistry, Interdisciplinary Research Laboratory 8000, Centre Universitaire, F-91405 Orsay Cedex, France*

XI HU (46), *Department of Chemistry, Duke University, Durham, North Carolina 27708*

WISSAM IBRAHIM (34), *Graduate Center for Nutritional Sciences, University of Kentucky, Lexington, Kentucky 40506*

NOBUYA ISHIBASHI (39), *Department of Surgery, Kurume University School of Medicine, Kurume, Fukuoka 830-0011, Japan*

SAMSON T. JACOB (40), *Department of Molecular and Cellular Biochemistry, Ohio State University, Columbus, Ohio 43210*

SIEGFRIED JANZ (37), *Laboratory of Genetics, Center for Cancer Research, National Cancer Institute, National Institutes of Health, Bethesda, Maryland 20892*

CAROLYN J. JOHNSON (42), *Health Science Center, University of Colorado, Denver, Colorado 80262*

VALERIAN E. KAGAN (25), *Department of Environmental and Occupational Health, Graduate School of Public Health, University of Pittsburgh, Pittsburgh, Pennsylvania 15260*

HARUMA KAWAGUCHI (8), *Faculty of Science and Technology, Keio University, Kouhoku-ku, Yokohama 223-8522, Japan*

SAVITA KHANNA (41), *Laboratory of Molecular Medicine, Departments of Surgery and Molecular and Cellular Biochemistry, Davis Heart and Lung Research Institute, Ohio State University Medical Center, Columbus, Ohio 43210*

JENNIFER L. KISKO (43), *Division of Chemistry and Chemical Engineering, California Institute of Technology, Pasadena, California 91125*

DAVID LANDO (1), *Department of Molecular BioSciences (Biochemistry), Adelaide University, Adelaide 5005, South Australia*

FRANCE LANDRY (15), *Cancer Research Center, Burnham Institute, La Jolla, California 92037*

REINHARD LANGE (11), *French National Institute for Health and Medical Research U 128, IFR 24, 34293 Montpellier, France*

NILS-GÖRAN LARSSON (35), *Department of Medical Nutrition and Biosciences, Karolinska Institute, NOVUM, Huddinge Hospital, S-141 86 Huddinge, Sweden*

JOHANNA LAUKKANEN (29), *Department of Molecular Medicine, A.I. Virtanen Institute, University of Kuopio, 70211 Kuopio, Finland*

MATTHEW D. LAYNE (16), *Department of Medicine, Brigham and Women's Hospital, Boston, Massachusetts 02115*

STEPHEN J. LEE (46), *Department of Chemistry, Duke University, Durham, North Carolina 27708*

FLORENCE LE PAGE (45), *Division of Life Sciences, CEA, 92265 Fontenay aux Roses, France*

RODNEY L. LEVINE (10), *Laboratory of Biochemistry, National Heart, Lung, and Blood Institute, National Institutes of Health, Bethesda, Maryland 20892*

KURT M. LIN (30), *Department of Medicine, University of California at San Diego, La Jolla, California 92093*

CHRISTOPHER D. LINK (42), *Institute for Behavioral Genetics, University of Colorado, Boulder, Colorado 80309*

SHANG-XI LIU (25), *Department of Environmental and Occupational Health, Graduate School of Public Health, University of Pittsburgh, Pittsburgh, Pennsylvania 15260*

CHRISTIAN R. LOMBARDO (15), *Cancer Research Center, Burnham Institute, La Jolla, California 92037*

CHARLES J. LOWENSTEIN (21, 24), *Division of Cardiology, Department of Medicine, Johns Hopkins University School of Medicine, Baltimore, Maryland 21205*

MAHIN D. MAINES (32), *Department of Biochemistry and Biophysics, University of Rochester Medical Center, Rochester, New York 14642*

SARMILA MAJUMDER (40), *Department of Molecular and Cellular Biochemistry, Ohio State University, Columbus, Ohio 43210*

BERND MAYER (11), *Institute for Pharmacology and Toxicology, Karl-Franzens-Universität Graz, A-8010 Graz, Austria*

LUIS G. MELO (16), *Department of Medicine, Brigham and Women's Hospital, Boston, Massachusetts 02115*

FRANCIS J. MILLER, JR. (20), *Department of Internal Medicine, University of Iowa, Iowa City, Iowa 52242*

OLEG MIROCHNITCHENKO (39), *Department of Biochemistry, Robert Wood Johnson Medical School, University of Medicine and Dentistry of New Jersey, Piscataway, New Jersey 08854*

SHABAZZ MUHAMMAD (26), *Department of Biophysics and Biophysical Chemistry, Johns Hopkins University School of Medicine, Baltimore, Maryland 21205*

ARNOLD MUNNICH (47), *French National Institute for Health and Medical Research U393, Hôpital Necker-Enfants Malades, 75015 Paris, France*

TAKEYUKI NISHI (8), *Department of Biological Information, Graduate School of Bioscience and Biotechnology, Tokyo Institute of Technology, Midori-ku, Yokohama 226-8501, Japan*

CATHERINE A. O'BRIAN (9), *Department of Cancer Biology, University of Texas M. D. Anderson Cancer Center, Houston, Texas 77030*

STAFFAN PAULIE (3), *Immunology Unit, Wenner-Gren Institute, Stockholm University, SE-106 91 Stockholm, Sweden*

DANIEL J. PEET (1), *Department of Molecular BioSciences (Biochemistry), Adelaide University, Adelaide 5005, South Australia*

MARK A. PERRELLA (16), *Department of Medicine, Brigham and Women's Hospital, Boston, Massachusetts 02115*

ISAAC N. PESSAH (22), *Department of Molecular Biosciences, University of California, Davis, Davis, California 95616*

ISABELLE PETROPOULOS (23), *Laboratory of Biological and Biochemical Cellular Aging, Université Denis Diderot-Paris 7, 75251 Paris Cedex 05, France*

BRUCE R. PITT (25), *Department of Environmental and Occupational Health, Graduate School of Public Health, University of Pittsburgh, Pittsburgh, Pennsylvania 15260*

INGEMAR PONGRATZ (1), *Department of Cell and Molecular Biology, Medical Nobel Institute, Karolinska Institutet, S-171 77 Stockholm, Sweden*

JUSTIN D. RADOLF (14), *Center for Microbial Pathogenesis, University of Connecticut Health Center, Farmington, Connecticut 06030*

TIFFANY A. REITER (7), *Section of Hematology Research, Department of Biochemistry and Molecular Biology, Mayo Clinic and Foundation, Rochester, Minnesota 55905*

TERESA C. RITCHIE (deceased) (28), *Department of Anatomy and Cell Biology, University of Iowa, Iowa City, Iowa 52242*

LYNNE D. ROCKWOOD (37), *Laboratory of Genetics, Center for Cancer Research, National Cancer Institute, National Institutes of Health, Bethesda, Maryland 20892*

ANDERS ROSÉN (3), *Division of Cell Biology, Department of Biomedicine and Surgery, University of Linköping, SE-581 85 Linköping, Sweden*

AGNÈS RÖTIG (47), *French National Institute for Health and Medical Research U393, Hôpital Necker-Enfants Malades, 75015 Paris, France*

SASHWATI ROY (41), *Laboratory of Molecular Medicine, Departments of Surgery and Molecular and Cellular Biochemistry, Davis Heart and Lung Research Institute, Ohio State University Medical Center, Columbus, Ohio 43210*

FRANK RUSNAK (7, 14), *Section of Hematology Research, Department of Biochemistry and Molecular Biology, Mayo Clinic and Foundation, Rochester, Minnesota 55905*

PIERRE RUSTIN (47), *French National Institute for Health and Medical Research U393, Hôpital Necker-Enfants Malades, 75015 Paris, France*

BITA SAHAF (3), *Herzenberg Laboratory, Beckman Center, Department of Genetics, Stanford University School of Medicine, Stanford, California 94305*

DARET K. ST. CLAIR (34), *Graduate Center for Toxicology, University of Kentucky, Lexington, Kentucky 40506*

ALAIN SARASIN (45), *Laboratory of Genetic Instability and Cancer, UPR 2169 Centre National de la Recherche Scientifique, 94800 Villejuif, France*

JOSÉ SEGOVIA (31), *Department of Physiology, Biophysics and Neuroscience, Centro de Investigación y de Estudios Avanzados del IPN, Mexico, 07300, D. F.*

CHANDAN K. SEN (41), *Laboratory of Molecular Medicine, Departments of Surgery and Molecular and Cellular Biochemistry, Davis Heart and Lung Research Institute, Ohio State University Medical Center, Columbus, Ohio 43210*

R. ANN SHELDON (33), *Department of Neurology, University of California, San Francisco, California 94143*

DAISUKE SHIMA (8), *Department of Biological Information, Graduate School of Bioscience and Biotechnology, Tokyo Institute of Technology, Midori-ku, Yokohama 226-8501, Japan*

NORIAKI SHIMIZU (8), *Frontier Collaborative Research Center, Tokyo Institute of Technology, Midori-ku, Yokohama 226-8503, Japan*

ROBERT A. SIKKINK (14), *Section of Hematology, Department of Biochemistry and Molecular Biology, Mayo Clinic and Foundation, Rochester, Minnesota 55905*

ISTVÁN SIMON (2), *Institute of Enzymology, Biological Research Center, Hungarian Academy of Sciences, H-1113 Budapest, Hungary*

TOVE SJÖGREN (27), *Department of Biochemistry, Uppsala University, S-751 23 Uppsala, Sweden*

ERIC J. SMART (13), *Department of Physiology, University of Kentucky Medical School, Lexington, Kentucky 40536*

JEFFREY S. SMITH (26), *Department of Biochemistry and Molecular Genetics, University of Virginia Health Sciences Center, Charlottesville, Virginia 22908*

JEFFREY W. SMITH (15), *Cancer Research Center, Burnham Institute, La Jolla, California 92037*

ANITA SÖDERBERG (3), *Division of Cell Biology, Department of Biomedicine and Surgery, University of Linköping, SE-581 85 Linköping, Sweden*

DOUGLAS R. SPITZ (38), *Free Radical and Radiation Biology Program, University of Iowa, Iowa City, Iowa 52242*

DANIEL M. SULLIVAN (10), *Laboratory of Molecular Biology, National Heart, Lung, and Blood Institute, National Institutes of Health, Bethesda, Maryland 20892*

DAPENG SUN (12), *Department of Biochemistry, University of Mississippi Medical Center, Jackson, Mississippi 39216*

CÉLINE H. TABOY (18), *Department of Chemistry, Duke University, Durham, North Carolina 27708*

HIROTOSHI TANAKA (8), *Institute of Medical Science, University of Tokyo, Minato-ku, Tokyo 108-8639, Japan*

ALEKSANDRA TRIFUNOVIC (35), *Department of Medical Nutrition, Karolinska Institute, NOVUM, Huddinge Hospital, S-14186 Huddinge, Sweden*

VLADIMIR A. TYURIN (25), *Department of Environmental and Occupational Health, Graduate School of Public Health, University of Pittsburgh, Pittsburgh, Pennsylvania 15260*

WILLEM J. H. VAN BERKEL (17), *Department of Agrotechnology and Food Sciences, Laboratory of Biochemistry, Wageningen University, 6703 HA Wageningen, The Netherlands*

ROBERT H. H. VAN DEN HEUVEL (17), *Department of Microbiology and Genetics, University of Pavia, 27100 Pavia, Italy*

NANCY E. WARD (9), *Department of Cancer Biology, University of Texas M. D. Anderson Cancer Center, Houston, Texas 77030*

MURRAY L. WHITELAW (1), *Department of Molecular BioSciences (Biochemistry), Adelaide University, Adelaide 5005, South Australia*

CARRIE M. WILMOT (27), *Department of Biochemistry, Molecular Biology and Biophysics, University of Minnesota, Minneapolis, Minnesota 55455*

CYNTHIA WOLBERGER (26), *Department of Biophysics and Biophysical Chemistry, Johns Hopkins University School of Medicine, Baltimore, Maryland 21205*

BOXU YAN (15), *Program on Cell Adhesion, Cancer Research Center, Burnham Institute, La Jolla, California 92037*

JUSAN YANG (28), *Department of Anatomy and Cell Biology, The University of Iowa, Iowa City, Iowa 52242*

JOANNE I. YEH (5), *Department of Molecular Biology, Cell Biology, and Biochemistry, and Department of Chemistry, Brown University, Providence, Rhode Island 02912*

HSIU-CHUAN YEN (34), *School of Medical Technology, Chang Gung University, Tao-Yuan, Taiwan*

SHAW-FANG YET (16), *Pulmonary and Critical Care Division, Department of Medicine, Brigham and Women's Hospital, Boston, Massachusetts 02115*

SEPPO YLÄ-HERTTUALA (29), *Department of Molecular Medicine, A. I. Virtanen Institute, University of Kuopio, 70211 Kuopio, Finland*

LEI ZHANG (44), *Division of Carcinogenesis and Molecular Epidemiology, American Health Foundation, Valhalla, New York 10595*

Preface

Oxidants may serve as cellular messengers. Changes in oxidoreductive or redox status in the cell regulate several signal transduction pathways. Redox-sensitive changes in signal transduction processes translate to functional changes at the cellular, tissue, as well as organ levels.

Redox changes in biological cells, tissues, and organs are often transient. For years, investigators have been challenged by the lack of reliable techniques to assess such changes in intact biological samples. Only recently have novel cell biology and genetic techniques to visualize and document redox changes in intact cells become available. Unlike biochemical methods that rely on the study of biological extracts, these cell biology- and genetics-related techniques arrest transient redox changes in the intact cell, tissues, and even organs. Technologies dependent on laser illumination, advanced spectroscopy, DNA microarray, and related approaches allow visualization of redox changes in the intact biological sample. Such approaches, including but not limited to redox imaging of intact organs, gene therapy, gene screening, flow cytometry, and advanced microscopy, represent the "cutting-edge" technology currently available to only select laboratories.

Our objective was to compile detailed protocols describing and critiquing essential methods in the field of redox cell biology and genetics. Redox Cell Biology and Genetics, Parts A and B, Volumes 352 and 353 of *Methods in Enzymology,* feature a diverse collection of novel cell biology and genetic protocols authored by highly recognized leaders in the field. Part A covers cellular responses and tissues and organs; Part B covers structure and functions of proteins and nucleic acids and genes.

The excellent editorial assistance of Dr. Savita Khanna and the outstanding contributions of the authors are gratefully acknowledged. We hope that this volume will contribute to the further development of this important field of biomedical research.

CHANDAN K. SEN
LESTER PACKER

METHODS IN ENZYMOLOGY

Section I

Protein Structure and Function

[1] Mammalian Two-Hybrid Assay Showing Redox Control of HIF-Like Factor

By DAVID LANDO, DANIEL J. PEET, INGEMAR PONGRATZ, and MURRAY L. WHITELAW

Introduction

A growing number of transcription factors including activating protein-1 (AP-1), Myb, activating transcription factor/cAMP response element-binding protein (ATF/CREB), nuclear factor-κB (NF-κB), and p53 have been shown to have their DNA-binding activities modulated by changes in redox status (reviewed by Morel and Barouki[1]). For the Fos–Jun heterodimer that constitutes the AP-1 complex, the nuclear redox protein Ref-1 imparts DNA-binding activity by reducing specific cysteine sulfhydryl groups in the basic DNA-binding regions of Fos and Jun. Although the precise mechanism by which Ref-1 reduces Fos and Jun is not known, chemical cross-linking studies with bacterially expressed proteins have suggested that a direct cysteine-mediated interaction may occur between Ref-1 and Jun.[2] However, *in vivo* cellular interactions between Ref-1 and Fos or Jun have not been demonstrated, nor have interactions been shown in cell extracts by standard biochemical techniques such as coimmunoprecipitation. This is most likely due to the transient nature of such interactions.

The mammalian hypoxia-inducible factors HIF-1α and HIF-like factor (HLF) are two highly related basic helix–loop–helix/Per–Arnt–Sim homology (bHLH/PAS) transcription factors that are rapidly activated by oxygen deprivation to induce a network of genes responsible for maintaining oxygen homeostasis (reviewed by Semenza[3]). The molecular mechanisms of oxygen sensing and signaling that lead to the activation of HIF-1α and HLF are poorly understood. However, evidence suggests the involvement of an oxygen-regulated redox signal that may activate a kinase cascade or modify the HIFs directly.[4] In support of this, overexpression of redox factors Ref-1 and thioredoxin enhances the hypoxic response of HIF-1α and HLF.[5–7] We have shown that Ref-1 can specifically enhance the

[1] Y. Morel and R. Barouki, *Biochem. J.* **342**, 481 (1999).

[2] S. Xanthoudakis, G. G. Miao, and T. Curran, *Proc. Natl. Acad. Sci. U.S.A.* **91**, 23 (1994).

[3] G. L. Semenza, *Annu. Rev. Cell Dev. Biol.* **15**, 551 (1999).

[4] N. S. Chandel, D. S. McClintock, C. E. Feliciano, T. M. Wood, J. A. Melendez, A. M. Rodriguez, and P. T. Schumacker, *J. Biol. Chem.* **275**, 25130 (2000).

[5] L. E. Huang, Z. Arany, D. M. Livingston, and H. F. Bunn, *J. Biol. Chem.* **271**, 32253 (1996).

[6] M. Ema, K. Hirota, J. Mimura, H. Abe, J. Yodoi, K. Sogawa, L. Poellinger, and Y. Fujii-Kuriyama, *EMBO J.* **18**, 1905 (1999).

[7] P. Carrero, K. Okamoto, P. Coumailleau, S. O'Brien, H. Tanaka, and L. Poellinger, *Mol. Cell. Biol.* **20**, 402 (2000).

Copyright 2002, Elsevier Science (USA).
All rights reserved.
0076-6879/02 $35.00

in vitro DNA-binding activity of HLF by reducing a cysteine residue in the basic DNA-binding domain of HLF.[8]

Here we describe a cell-based mammalian two-hybrid assay that is capable of detecting the intracellular interaction of Ref-1 with the basic DNA-binding domain of HLF. By using this assay we are able to demonstrate that the interaction between Ref-1 and HLF is dependent on a redox-sensitive cysteine in the basic region of HLF.

Methods and Procedures

Principle

The mammalian two-hybrid system is a simple and inexpensive, yet powerful technique capable of assaying for protein–protein interactions within a cellular environment. Like the yeast two-hybrid system originally described by Fields and Song,[9] this is an *in vivo* assay based on the functional reconstitution of a transcription activator. Typically, in this system, one protein of interest is expressed as a fusion protein with the yeast Gal4 DNA-binding domain (DBD), and the other protein is expressed as a fusion to the activation domain of the VP16 protein of the herpes simplex virus. The vectors that encode these fusion proteins are then cotransfected along with a luciferase reporter gene into a mammalian cell line. The expression of the luciferase gene within the reporter plasmid is under the control of Gal4-binding elements. If the two fusion proteins interact, this will bring together the Gal DBD and VP16 domains, forming a hybrid transcription activator capable of binding the Gal4 response elements and increasing the expression of the luciferase reporter gene. The increase in luciferase can then be quantitated with a luminometer.

Plasmids

The mammalian expression vectors we use to analyze Ref-1 interaction with HLF and HIF-1α were constructed from pCMX/GalDBD and pCMX/VP16 vectors described previously[8,10] and schematically outlined in Fig. 1. A description of general molecular biology techniques such as cloning, mutagenesis, and plasmid propagation and isolation can be found elsewhere.[11] Briefly, the Gal4DBD chimeras contain the yeast DBD of Gal4 (amino acids 1–147) fused in frame with either full-length Ref-1 (GalDBD/Ref-1), HLF amino acids 1–265 (GalDBD/HLF

[8] D. Lando, I. Pongratz, L. Poellinger, and M. L. Whitelaw, *J. Biol. Chem.* **275,** 4618 (2000).

[9] S. Fields and O. Song, *Nature (London)* **340,** 245 (1989).

[10] I. Pongratz, C. Antonsson, M. L. Whitelaw, and L. Poellinger, *Mol. Cell. Biol.* **18,** 4079 (1998).

[11] F. M. Ausubel, R. Brent, R. E. Kingston, D. D. Moore, J. G. Seidman, J. A. Smith, and K. Struhl, "Current Protocols in Molecular Biology," Vols. 1–3. John Wiley & Sons, New York, 1997.

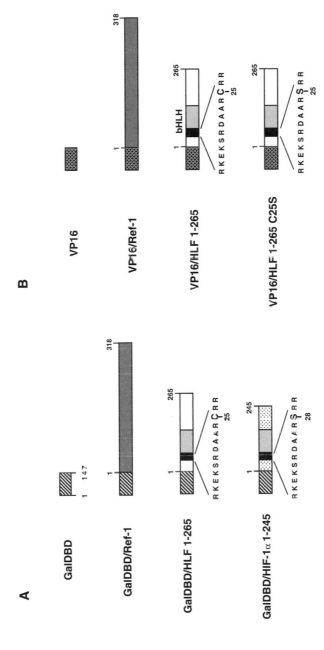

FIG. 1. Schematic representation of GalDBD (A) and VP16 (B) fusion constructs used to analyze the interaction of Ref-1 with HLF and HIF-1α. The amino acid compositions of the basic regions for HLF and HIF-1α are included. bHLH, basic helix–loop–helix domain

1–265), or HIF-1α amino acids 1–245 (GalDBD/HIF-1α 1–245). The VP16 chimeras contain the 78-amino acid activation domain from the herpes simplex virus fused in frame with either full-length Ref-1 (VP16/Ref-1), HLF amino acids 1–265 (VP16/HLF 1–265), or HLF 1–265 cysteine (amino acid residue 25)-to-serine point mutant (VP16/HLF 1–265 C25S). The luciferase reporter construct G5E1b-Luc contains five copies of the Gal4-binding site upstream of the minimal E1b promoter-driven firefly luciferase gene.

Two-Hybrid Assay

Several methods to transfect plasmid DNA into mammalian cells are available, including calcium phosphate, DEAE-dextran, and liposome-mediated transfection. To analyze the *in vivo* interaction of Ref-1 with HLF we use the monkey kidney COS-1 cell line in conjunction with the liposomal transfection reagent LipofectAMINE 2000 (Life Technologies, Rockville, MD) and the Dual-Luciferase reporter assay system (Promega, Madison, WI). Although we have obtained similar results with other transfection reagents (1,2-dioleoyl-3-trimethylammonium-propane, DOTAP; Boehringer Mannheim, Germany) and cell lines (human embryonic kidney, HEK293T), in our hands COS-1 cells and the LipofectAMINE 2000 reagent routinely give high transfection efficiencies, resulting in consistent two-hybrid results. A description of general mammalian cell tissue culture techniques is beyond the scope of this chapter; readers who require further information on these techniques are therefore referred to methods and protocols given elsewhere.[11]

For two-hybrid assays COS-1 cells are cultured in a humidified incubator (95% air/5% CO_2) at 37° in Dulbecco's modified Eagle's medium (DMEM, GIBCO-BRL, Gaithersburg, MD) supplemented with 10% (v/v) fetal calf serum (FCS; GIBCO-BRL). Before transfection, cells are trypsinized and plated onto 24-well tissue culture plates (Falcon; Becton Dickinson Labware, Lincoln Park, NJ) at a density of 5×10^4 viable cells per well in a final volume of 450 μl of DMEM–FCS. After 24 hr (50–60% cell confluency), transfections are carried out with LipofectAMINE 2000 reagent as follows.

1. For each transfection (or well of cells), 425 ng of plasmid DNA consisting of 300 ng of G5E1b-Luc reporter, 50 ng of GalDBD or GalDBD chimera, 50 ng of VP16 or VP16 chimera, and 25 ng of *Renilla* internal control pRL-TK plasmid (Promega) is diluted into DMEM without FCS to a final volume of 25 μl. The pRL-TK control plasmid provides constitutive expression of *Renilla* luciferase from the herpes simplex virus thymidine kinase (TK) promoter and its addition serves as an internal control of transfection efficiency.

2. For each well of cells to be transfected, dilute 1 μl of LipofectAMINE 2000 reagent into 25 μl of DMEM without FCS.

3. Combine the 25 μl of plasmid–DMEM mix (step 1) with 25 μl of diluted LipofectAMINE 2000 reagent (step 2). Incubate the mixture at room temperature

for 20 min to allow plasmid DNA–LipofectAMINE 2000 reagent complexes to form. *Note:* Diluted LipofectAMINE 2000 from step 2 must be combined with plasmid DNA mix within 10 min of preparation.

4. After incubation add the plasmid DNA–LipofectAMINE 2000 reagent complexes (50 μl) directly to each well. To obtain consistency and high transfection efficiencies it is important to disperse the plasmid DNA–LipofectAMINE 2000 reagent complexes thoroughly and evenly throughout each well. This is best done by gently rocking the plate back and forth three or four times.

5. Return the transfected cells to the incubator and incubate for 24 to 48 hr. We have found that it is not necessary to remove the complexes or change the medium during the assay, as prolonged cell exposure to LipofectAMINE 2000 does not affect the transfection activity.

6. After transfection wash all wells once with 500 μl of phosphate-buffered saline solution and harvest with Dual-Luciferase lysis buffer (100 μl/well; Promega). Lysed extracts (3 μl) are then analyzed for luciferase activity in a Turner Designs (Sunnyvale, CA) T20/20 luminometer, using the Dual-Luciferase reporter assay system, and measured G5E1b-Luc reporter activity is normalized to the activity of the internal control (pRL-TK).

Results and Discussion

Typical results of a two-hybrid assay showing Ref-1 interaction with HLF are presented in Fig. 2. A fusion protein containing the N-terminal domain of HLF encompassing the basic DNA-binding region attached to the activation domain of VP16 (VP16/HLF 1–265) is assayed for interaction with GalDBD/Ref-1 fusion protein in COS-1 cells. Coexpression of GalDBD/Ref-1 with VP16/HLF 1–265 results in an approximately 4-fold increase in reporter gene activity over that seen when GalDBD alone and VP16/HLF 1–265 are coexpressed, suggesting that GalDBD/Ref-1 can interact with VP16/HLF 1–265 via the Ref-1 portion of the chimera. Because it is possible that the Ref-1 fusion may interact with the activation domain of VP16, it is also important to assay the GalDBD fusion protein with nonchimeric VP16. Therefore, Fig. 2 also shows reporter gene activity when GalDBD/Ref-1 or GalDBD are coexpressed with nonchimeric VP16. Clearly, when GalDBD/Ref-1 is coexpressed with nonchimeric VP16 the activity remains at the low basal level seen when GalDBD alone is coexpressed with VP16, and thus the interaction of Ref-1 with the VP16/HLF 1–265 fusion is specific for the HLF 1–265 portion of the fusion construct.

Oxidoreductase factors such as Ref-1 are thought to interact by forming transient intermediate disulfide linkages with cysteine residues in target proteins.[2,12] In analogy with the mechanism proposed for the interaction of Ref-1 with Fos and Jun, we have previously suggested that Ref-1 enhances HLF DNA-binding

[12] A. Holmgren, *Annu. Rev. Biochem.* **54,** 237 (1985).

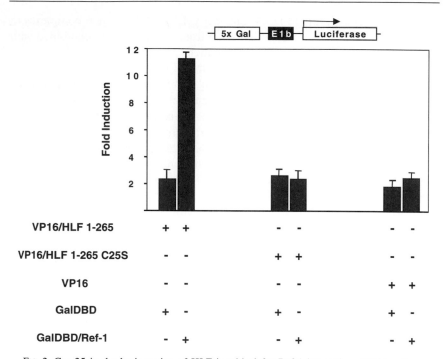

FIG. 2. Cys-25 in the basic region of HLF is critical for Ref-1 interaction. COS-1 cells were cotransfected with 300 ng of Gal4-luciferase reporter G5E1b-Luc, 50 ng of expression vector chimeric for the activation domain of VP16 and HLF 1–265 or the HLF cysteine-to-serine point mutant HLF 1–265 C25S, and 50 ng of GalDBD/Ref-1 or GalDBD as indicated. After 24 hr, cells were harvested and luciferase activity was normalized relative to cotransfected pRL-TK control reporter. Results are from two independent experiments performed in triplicate and are represented as fold induction relative to reporter gene activity of nonchimeric VP16 with GalDBD. Error bars represent the standard deviation.

activity by reducing Cys-25 located in the basic DNA-binding region of HLF.[8] As shown in Fig. 2, when HLF Cys-25 is mutated to a serine residue (VP16/HLF 1–265 C25S) interaction with GalDBD/Ref-1 is abolished. This suggests that the interaction between Ref-1 and HLF may depend on the formation of transient disulfide linkages, and that the two-hybrid system is capable of detecting this type of weak interaction.

A possible caveat of the mammalian two-hybrid system is that artifactual interactions between protein fragments may be invoked by spurious structures created by the chimeric fusion proteins. Although such false-positive interactions are not commonly observed, they can be guarded against by creating reciprocal chimeras to verify interactions in a new protein context. Thus, to demonstrate that the interaction of Ref-1 with HLF observed in Fig. 2 was not peculiar to the GalDBD and VP16 constructs utilized, we tested (Fig. 3) reciprocal chimeras in which

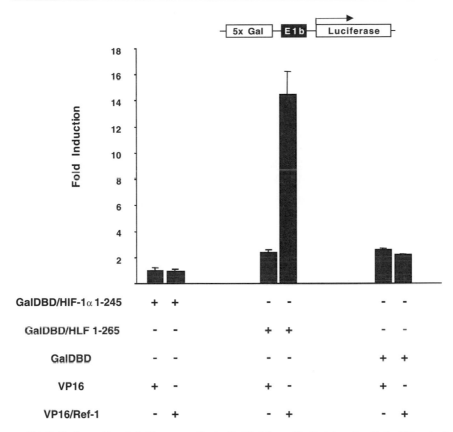

FIG. 3. Reciprocal two-hybrid assay confirming that Ref-1 specifically interacts with the N-terminal region of HLF but not with HIF-1α. COS-1 cells were cotransfected with 300 ng of Gal4-luciferase reporter G5E1b-Luc, 50 ng of expression vector chimeric for the DNA-binding domain of Gal4 and HLF 1–265 or HIF-1α 1–245, and 50 ng of VP16/Ref-1 or VP16 as indicated. After 40 hr cells were assayed for luciferase activity as outlined in Fig. 2. Results are from two independent experiments performed in triplicate and are represented as fold induction relative to reporter gene activity of nonchimeric GalDBD with VP16. Error bars represent the standard deviation.

Ref-1 was fused to the activation domain of VP16 and then assayed for interaction with a GalDBD chimera containing HLF. As in Fig. 2, coexpression of Ref-1 (VP16/Ref-1) and HLF (GalDBD/HLF 1–265) fusion proteins resulted in an increase in reporter gene activity when compared with coexpression of VP16/Ref-1 and GalDBD alone, further confirming that Ref-1 interacts with the N-terminus of HLF. We have previously shown that whereas HIF-1α can recognize the same DNA response element as HLF, its DNA-binding activity is not regulated by Ref-1.[8] This is due to the basic DNA-binding region of HIF-1α containing a serine rather than a cysteine at the critical position 28 (see Fig. 1). In agreement with the DNA-binding experiments, coexpression of VP16/Ref-1 with an N-terminal HIF-1α fusion (GalDBD/HIF-1α 1–245) resulted in no increase in reporter gene activity.

We believe the major advantages of analyzing redox protein interactions with the mammalian two-hybrid system are that (1) it is a cell-based *in vivo* interaction assay, which is more physiologically relevant than an *in vitro* interaction assay; and (2) it is often more sensitive than other assays such as coimmunoprecipitation, this being important because most redox interactions tend to be weak and transient in nature. Although other methods to observe transient interactions between redox proteins exist, for example, solution nuclear magnetic resonance analysis of the interaction between thioredoxin and Ref-1,[13] the mammalian two-hybrid approach is by comparison simple, quick, and inexpensive, and requires no specialized expertise.

Acknowledgments

We are grateful to Professor Fujii-Kuriyama for HLF cDNA and to Dr. T. Curran for Ref-1 cDNA.

[13] J. Qin, G. M. Clore, W. M. Kennedy, J. Kuszewski, and A. M. Gronenborn, *Structure* **4,** 613 (1996).

[2] Predicting Redox State of Cysteines in Proteins

By ANDRÁS FISER and ISTVÁN SIMON

Introduction

Protein sequences are usually determined by sequencing the corresponding cDNA. Although this approach is efficient, it is unable to account for posttranslational covalent modifications such as the oxidation of cysteines forming disulfide bridges. In the era of genome-wide sequencing projects the insufficiency of this type of information is becoming more apparent. Statistics on recently sequenced organisms indicate that the function of 22–60% of putative reading frames is unknown.[1,2] To reveal the full covalent structure of a protein, either the expressed or extracted protein must be analyzed experimentally or its structure needs to

[1] M. A. Martí-Renom, A. Stuart, A. Fiser, R. Sánchez, F. Melo, and A. Šali, *Annu. Rev. Biophys. Biomol. Struct.* **29,** 291 (2000).

[2] J. Cedano, P. Aloy, J. A. Pérez-Pons, and E. Querol, *J. Mol. Biol.* **266,** 594 (1997).

Copyright 2002, Elsevier Science (USA).
All rights reserved.
0076-6879/02 $35.00

be determined. Both of these approaches are time-consuming.[3–5] The sequence databases are approximately 100-fold larger than the three-dimensional protein databases and the gap is growing rapidly.

The oxidation state of cysteine plays an important role in protein structure and function. In its thiol form, cysteine is the most reactive amino acid under physiological conditions, and is often used for adding fluorescent groups and spin labels.[6] In oxidized forms, cysteines form disulfide bonds, which are the primary covalent cross-links found in proteins and which stabilize the native conformation of a protein. Thus accurate predictions of the oxidation state of cysteines would have numerous applications, for example, in engineering stabilizing cystines or reactive thiol groups,[7–11] in locating key reactive thiol groups in enzymatic reactions,[12] and in determining topologies to aid in three-dimensional structure predictions.[13]

In the early 1990s two methods predicting disulfide bond-forming cysteine residues were published.[14,15] Both methods used sequence information hidden in the specific sequence environment of cysteines and half-cystines. One method employed a neural network (NN) to recognize disulfide-bonded cysteine,[14] whereas the other method performed a statistical analysis of the amino acid frequencies in the sequence environment of cysteine.[15] The NN method, tested on an independent data set, achieved 81% accuracy, whereas the statistical method performed at 71% prediction accuracy as tested by a jack-knife procedure.

An important development in sequence-based prediction methods was the incorporation of evolutionary information (see, e.g., Fariselli et al.[16]). This important factor is exploited by two methods to predict the bonding state of cysteine residues. The method of Fariselli et al. uses a neural network approach, as does the method of Muskal et al.,[14] but incorporates evolutionary information by feeding the neural network multiple sequence alignments.[16] This method has achieved a slightly

[3] A. Kremser and I. Rasched, Biochemistry 33, 13954 (1994).

[4] H. R. Morris and P. Pucci, Biochem. Biophys. Res. Commun. 126, 1122 (1985).

[5] T. W. Tannhauser, Y. Konishi, and H. A. Scheraga, Anal. Biochem. 138, 181 (1984).

[6] T. E. Creighton, "Proteins: Structures and Molecular Properties," 2nd Ed., p. 162. W. H. Freeman, New York, 1993.

[7] M. Matsumura and B. W. Matthews, Methods Enzymol. 202, 336 (1991).

[8] J. Clarke and A. R. Fersht, Biochemistry 32, 4322 (1993).

[9] N. E. Zhou, C. M. Kay, and R. S. Hodges, Biochemistry 32, 3178 (1993).

[10] J. Eder and M. Wilmanns, Biochemistry 31, 4437 (1992).

[11] A. Yokota, K. Izutani, M. Takai, Y. Kubo, Y. Noda, Y. Koumoto, H. Tachibana, and S. Segawa, J. Mol. Biol. 295, 1275 (2000).

[12] V. B. Ritov, R. Goldman, D. A. Stoyanovsky, E. V. Menshikova, and V. E. Kagan, Arch. Biochem. Biophys. 321, 140 (1995).

[13] I. Simon, L. Glasser, and H. A. Scheraga, Proc. Natl. Acad. Sci. U.S.A. 88, 3661 (1991).

[14] S. M. Muskal, S. R. Holbrook, and S. H. Kim, Protein Eng. 3, 667 (1990).

[15] A. Fiser, M. Cserzö, E. Tüdös, and I. Simon, FEBS Lett. 302, 117 (1992).

[16] P. Fariselli, P. Riccobelli, and R. Casadio, Proteins 36, 340 (1999).

higher accuracy than either of the two methods published previously. They also illustrate that the achieved higher accuracy is primarily due to the incorporation of evolutionary information and is only partly due to the fact that the databases have increased more than 10-fold during the last decade. The importance of evolutionary information is illustrated in our more recent work,[17] which shows that a relatively simple conservation analysis of multiple sequence alignments can make a sensitive distinction between chemically different cysteine residues, achieving a prediction accuracy above 82%. Another conclusion of the analysis is that the natural border-line between differently conserved cysteines lies in between the different oxidation states rather than between cysteines forming or not forming disulfide bridges. In other words, oxidized cysteines such as those bound to cofactors or to ligands or cysteines participating in binding sites are as conserved as cysteines participating in disulfide bonds. From the viewpoint of conservation it does not seem to matter whether certain cysteine residues are important for structural or for functional reasons.

Here we briefly review the differences between oxidized and reduced cysteines on the surface and in the interior of proteins, the differences in the surrounding microenvironments, and the occurrence of different oxidation states of cysteine in secondary structural elements. We also analyze the types of secondary structural elements linked by a disulfide bridge, the most dominant form of the oxidized state, and the correlation between the cellular location of a protein and the oxidation state of its cysteines. Next, a method is explained for predicting the covalent state of cysteine by analyzing residue conservation of multiple sequence alignments. The results demonstrate that the analysis of multiple sequence alignments is an efficient tool for distinguishing oxidized cysteines from those with reactive sulfhydryl groups.

Applied Databases and Methods

In our more recent work two data sets were used for analysis and for calibrating the prediction method.[17] First, a database of 81 representative protein alignments was created in such a way that in each alignment at least one sequence had its structure solved. The protein structures were selected from the Protein Data Bank (PDB)[18] by a two-step procedure. First, the sequences of all PDB proteins longer than 50 residues and having a crystallographic resolution better than 2.5 Å were compared by calculating the correlation coefficient of dipeptide frequencies. A set of 101 proteins remained after requiring that any pair should have a dipeptide

[17] A. Fiser and I. Simon, *Bioinformatics* **47**, 251 (2000).

[18] E. E. Abola, F. C. Bernstein, S. H. Bryant, T. F. Koetzle, and J. Weng, Protein Data Bank, *in* "Crystallographic Databases—Information, Content, Software Systems, Scientific Applications" (F. H. Allen, G. Bergerhoff, and R. Sievers, eds.), p. 107. Data Commission of the International Union of Crystallography, Bonn, Germany, 1987.

frequency correlation smaller than 0.4. In the second step, every pair of proteins in the filtered set was compared by a rigorous sequence comparison method[19,20] followed by cluster analysis, which yielded the final 81 proteins. The four-letter PDB codes and chain identifiers are as follows: 155C, 1ACX, 1ALC, 1BBPA, 1CC5, 1ECA, 1FKF, 1FNF, 1FNR, 1GCR, 1GPLA, 1HDSB, 1HIP, 1HOE, 1LRD4, 1PAZ, 1PCY, 1PHH, 1PRCC, 1RBP, 1RHD, 1RNH, 1SN3, 1TGS, 1TPKA, 1WSYB, 256BA, 2ALP, 2AZAA, 2CAB, 2CD4, 2CDV, 2CPP, 2FXB, 2GN5, 2LH7, 2LIV, 2LTNA, 2ORLL, 2PABA, 2RNT, 2RSPA, 2SECI, 2SNLE, 2SNS, 2SODB, 2SSI, 2STV, 2TSL, 2UTGA, 3ADK, 3B5C, 3CLA, 3FXC, 3GAPB, 3LZM, 3SGB1, 451C, 4BP2, 4FDL, 4FXN, 4HHBA, 4PEP, 4PFK, 4PTP, 4TNC, 5CTS, 5CYTR, 5EDX, 5RUBA, 5RXN, 6LDH, 6TMNE, 7PTI, 8ADH, 8ATCB, 8CATA, 8DFR, 9PAP, 9RSAA, and 9WGAA. Of the protein set, 51% contained only free cysteines, 27% contained only half-cystines, 5% contained both forms of cysteine, and 15% contained neither form. By chance the number of half-cystines (plus liganded ones) and free cysteines in the set turned out to be equal: 148 half-cystines (plus 9 liganded cysteines) and 157 free cysteines. Each of these sequences was compared with the Protein Information Resource (PIR) database[21] by the program SCANPS (http://www2.ebi.ac.uk/scanps/). Sequences that gave a probability lower than 10^{-6} were used to produce the multiple sequence alignments by the method of Barton and Sternberg.[22] The number of sequences in the alignment varied between 3 and 499, with a median of 28. Another, less strictly selected data set of proteins was used (including lower resolution X ray structures with a crystallographic R factor less than 25%) to confirm some observations made with the smaller set. This larger set contained 233 proteins: 161 (69.1%) had only free cysteines, 24 (10.3%) had only half-cystines, 33 (14.2%) had neither, and 15 (6.4%) contained both forms of cysteine.

Accessible surface areas were calculated by the program DSSP[23] and converted to relative accessibilities by dividing by the accessibility of the residue in a Gly-X-Gly tripeptide.[24] Two relative accessibility classes were considered: buried ($A \leq 0.25$) and exposed ($A > 0.25$).

Conservation scores based on the physicochemical properties of the amino acids were calculated for each position in each alignment according to Livingstone and Barton.[25] Such conservation scores range from 0 to 10 and count the number of the properties shared at a position, where the properties are as follows: Hydrophobic, Positive, Negative, Polar, Charged, Small, Tiny, Aliphatic, Aromatic, and Proline, and their negation (e.g., *not* Hydrophobic). For each position

[19] T. F. Smith and M. S. Waterman, *J. Mol. Biol.* **147**, 195 (1981).

[20] G. J. Barton, *Protein Eng.* **6**, 37 (1993).

[21] D. G. George, W. C. Barker, and L. T. Hunt, *Nucleic Acids Res.* **14**, 11 (1986).

[22] G. J. Barton and M. J. E. Sternberg, *J. Mol. Biol.* **198**, 327 (1987).

[23] W. Kabsch and C. Sander, *Biopolymers* **22**, 2577 (1983).

[24] G. Rose, A. Geselowitz, G. Lesser, R. Lee, and M. Zehfus, *Science* **229**, 834 (1985).

[25] C. D. Livingstone and G. J. Barton, *Comput. Appl. Biosci.* **9**, 745 (1993).

in each protein this score was then divided by the average conservation of the protein to give a relative conservation score C_r. We refer to a position as "conserved" if $C_r > 1$, that is, the conservation of the given position is higher than the average conservation of the sequence.

Characterizing Conservation, Microenvironment, Occurrence, and Distribution of Cysteine Residues

The average relative conservation score (C_r) and average relative accessibility are shown for the 20 residues (Fig. 1). Cysteine is by far the most conserved residue, both on the surface and in the interior, which reflects its crucial role in structure stabilization and biochemical functions. The difference in the relative conservation score between the bonded (i.e., if one groups liganded cysteines and half-cystines into a new category of "bonded" cysteines) and free forms of cysteine is significant. The standard deviations of the averages are about the difference, as follows: bonded, 1.56 ($\delta = 0.53$); free, 1.13 ($\delta = 0.44$). Functionally important cysteine residues covalently bonded to prosthetic groups or active site residues appear to correlate better with disulfide-bonded cysteines than with free cysteines. In other words, from the viewpoint of residue conservation functionally or structurally conserved residues behave similarly, and therefore the natural borderline lies between

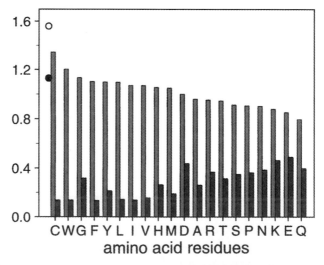

FIG. 1. Average relative conservation and accessibility of the 20 amino acids (see Prediction Method). Lighter columns represent the average relative conservation score; darker columns correspond to the average relative accessibilities of each of the 20 residues. In the case of cysteine residues an open circle and a filled circle represent the average relative conservation of bonded and nonbonded cysteines, respectively. The amino acids are indicated underneath by their one-letter code.

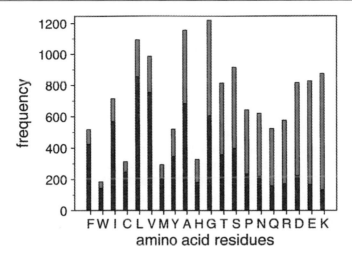

FIG. 2. Distribution of the 20 amino acids between exposed and buried positions in the 81 protein structures analyzed. Dark-shaded portions of the columns correspond to the number of buried residues, whereas the light-shaded portions correspond to the exposed residue. The length of the columns corresponds to the total number of occurrences of the residues in the overall composition. The residues are ranked from left to right, the leftmost one (phenylalanine) being the most buried, the second (tryptophan) the second most buried, etc. The amino acid one-letter codes are below the graph.

the different oxidation states of cysteine rather than between half-cystines and cysteines. It is suggested that this division between oxidized and reduced cysteine should be taken into account in order to incorporate evolutionary information into prediction methods efficiently.

Figure 2 shows the distribution and occurrence of residues: Cysteine is among the most buried residues (in order, right after phenylalanine, tryptophan, and isoleucine), presumably because it has the most reactive side chain.[6] The distribution of half-cystines and cysteines between the surface and the interior is almost identical (84 and 80% are buried, respectively). This result is slightly modified if we use the bonded and nonbonded categories instead of the half-cystine and cysteine categories: 79 and 80% of the bonded cysteines and "free" (i.e., nonbonded) cysteines are buried, respectively.

Free cysteines have a tendency to interact with apolar residues.[14,17] This observation was confirmed in a detailed study of Petersen *et al.,* who concluded that free cysteines are slightly more buried than cystines and accordingly they prefer a more hydrophobic environment.[26] In addition, cysteines are often surrounded by histidines and methionines whereas cystines are more likely to make contact with tryptophans and tyrosines according to the study by Bagley and Altman.[27]

[26] M. T. Petersen, P. H. Jonson, and S. B. Petersen, *Protein Eng.* **12,** 535 (1999).
[27] S. C. Bagley and R. B. Altman, *Protein Sci.* **4,** 622 (1995).

Cysteines are the most frequent active site residues, followed by histidine,[28] which might explain why they prefer one another's surroundings.

The original observation that glycine is abundant around half-cystines[14,17] was not confirmed by a more recent study,[26] whereas another did, but only at a low significance level.[27] However, the original observation referred to sequence neighbors, whereas the latter referred to structural neighbors. An even more recent study by Abkevich and Shakhnovich[29] illustrated a strong preference for glycine in the structural neighborhood of disulfide bonds, but it was shown that this effect becomes insignificant if the disulfide bond bridged cysteines were closer than eight residues away in the sequence. The most frequent sequential separation between two half-cystines forming a disulfide bond peaks at 11 and 16 residues.[26] A higher frequency of glycine could be explained by entropic reasons because a glycine-rich chain segment has high flexibility in the unfolded state.

Various studies agree that polar residues are more abundant around disulfide bridges, but the possible explanations are different.[14,15,17,26,29] Fiser and Simon[17] and Dosztányi et al.[30] suggest that a disulfide bridge with its surrounding network of hydrogen bond-forming polar residues often forms a special case of *stabilization centers* in protein structures,[17,30] whereas Abkevich and Shakhnovich hypothesize that disulfide bonds are necessary to compensate the destabilizing, buried polar environment.[29] One should not exclude the obvious effect of the different amino acid compositions between extra- and intracellular proteins, which strongly correlate with the presence of disulfide bridges.[2,31] Moreover, smaller proteins, which are generally extracellular, have a larger surface-to-volume ratio, that is, they contain more polar residues than do large proteins. These small proteins are often stabilized by disulfide bridges, and free cysteines were not observed at all in proteins with fewer than 50 residues.[26] The observed differences between the amino acid compositions of extra- and intracellular proteins are also partly a consequence of the fact that extracellular proteins have a higher β-sheet content, whereas intracellular proteins have higher helical content.[32]

Exceptions: Proteins Containing Both Cysteines and Half-Cystines

In the few proteins that have both cysteine and cystine residues, the cysteines are usually bonded, for example, 1CC5 (155C) cytochromes, in which two cysteines are bonded to the heme group, whereas the other two occur in disulfide bonds. Other examples include 2AZA, the electron transport protein azurin, in which the "free" cysteine is in the active site, forming a ligand to copper ion together with two histidines and a methionine, and 9PAP, the sulfhydryl proteinase papain, in which

[28] A. Fiser, I. Simon, and G. J. Barton, *FEBS Lett.* **47**, 45 (1996).

[29] V. I. Abkevich and E. I. Shakhnovich, *J. Mol. Biol.* **300**, 975 (2000).

[30] Z. Dosztányi, A. Fiser, and I. Simon, *J. Mol. Biol.* **272**, 597 (1997).

[31] H. Nakashima and K. Nishikawa, *J. Mol. Biol.* **238**, 54 (1994).

[32] Z. Gugolya, Z. Dosztányi, and I. Simon, *Proteins* **27**, 360 (1997).

cysteine can be found in the active site. The only protein in our set that contains both the bonded and free forms of cysteine is 2SODB, which is an oxidoreductase (Cu,Zn-superoxide dismutase).[33] In this molecule there are three cysteines, two of which Cys-144 and Cys-55 form a disulfide bond, whereas the third (Cys-6) is free.

The statistical correlations appear more pronounced if cysteines are grouped as bonded versus free rather than as cysteine versus cystine. This hypothesis was checked on a larger, less strictly selected data set (including lower resolution X-ray structures with a crystallographic R factor less than 25%). This larger set contained 233 proteins: 161 (69.1%) had only free cysteines, 24 (10.3%) had only half-cystines, 33 (14.2%) had neither, and 15 (6.4%) contained both forms of cysteine. By investigating this latter group, we found that at least half the proteins containing both cysteines and half-cystines occur in the same oxidation state, as in the cytochromes, sulfhydryl proteinases, and electron transport proteins cited above. This is true also for protein complexes such as endodeoxyribonuclease complex with actin (e.g., 1ATN). The actin forms a 1 : 1 complex with DNase I, with the actin having four sulfhydryl groups and the DNase having two disulfide bridges, but that situation corresponds to two separately folded molecules.

In the two data sets analyzed only a small percentage (2–4%) of the proteins contain both redox types of cysteines in the same molecule, but a specific role is often suspected for the free cysteines such as interdomain links, heavy atom-binding sites, and active sites. The number of cysteines in these proteins is not even.

Correlation between Disulfide-Bonding State of Cysteine and Subcellular Location

Thornton[34] mentions that free thiols are unstable outside the cell, that is, cysteines predominantly occur in disulfide bridges (see also Fahey et al.[35]). In the intracellular environment the thiols are kept reduced by glutathione, but once outside they are reactive and may even cause polymerizations. In our survey we grouped the proteins into three subgroups (intracellular, extracellular, and periplasmic), according to their cellular locations, and checked the occurrence of different types of cysteine in the subgroups (Table I). The cellular location shows a high correlation with the oxidation state of the cysteines, but this correlation is not exclusive. In our set, among the intra- and extracellular proteins only one contains both types of cysteine (Table I) whereas 10% of extracellular proteins contain free cysteine. Inside the cell, none of the proteins contain disulfide bridges, according to our data set. Only a few periplasmic proteins are present in the data set, but they

[33] D. Bordo, K. Dijnovic, and M. Bolognesi, *J. Mol. Biol.* **238**, 366 (1994).
[34] J. M. Thornton, *J. Mol. Biol.* **151**, 261 (1981).
[35] R. C. Fahey, J. S. Hunt, and G. C. Windham, *J. Mol. Evol.* **10**, 155 (1977).

TABLE I
CYSTEINE IN DIFFERENT SUBCELLULAR LOCATIONS[a]

	Protein location [% (no.) of cysteines]		
Cysteine state	Intracellular (46)	Extracellular (30)	Periplasmic (5)
Free	78 (36)	10 (3)	40 (2)
SS	0 (0)	70 (21)	20 (1)
Bonded	4 (2)	0 (0)	0 (0)
—	13 (6)	17 (5)	20 (1)
Free + SS	2 (1)	0 (0)	0 (0)
Bonded + SS	2 (1)	3 (1)	20 (1)
Free + bonded	0 (0)	0 (0)	0 (0)

[a] The distribution among different cellular locations (extracellular, intracellular, and periplasmic) of different oxidation states of cysteines in 81 analyzed proteins. *Free,* Cysteines, free thiols; *SS,* cystines, that is, cysteines in disulfide bridges; *bonded,* cysteines liganded to a prosthetic group or in an active center; —, proteins without any cysteine; and combinations of the subgroups (*free + SS; bonded + SS; free + bonded*).

already show the most variation in the redox state of cysteine. There is a high (but not perfect) correlation between the oxidation state of cysteine and the location of the protein within the cell; intracellular proteins all have reduced cysteines, whereas one-tenth of extracellular proteins also have reduced cysteines.

Occurrence in Secondary Structures, Linked Structures

The frequencies of oxidized and reduced cysteines in secondary structure elements are similar. Cysteines occur most often in coil structures, and these cysteines are also the most conserved, but conserved bonded cysteines also occur in helixes (Table II). The high frequency of half-cystines in coil structures is not connected with a distortion of the regular structure by disulfide links. Such a situation occurs

TABLE II
CYSTEINE CONSERVATION AND OCCURRENCE IN SECONDARY STRUCTURES[a]

	All residues	Nonbonded cysteine		Bonded cysteine	
Secondary structure	frequency (%)	Frequency (%)	Conservation	Frequency (%)	Conservation
Helix	32	31	1.137	25.0	1.770
Sheet	23	26	1.033	28.7	1.444
Coil	45	42	1.350	47.0	1.587

[a] Residue conservation and frequency of bonded and nonbonded cysteines in secondary structural elements. The second column refers to overall database statistics considering all types of residues.

TABLE III
FREQUENCY OF SECONDARY STRUCTURAL ELEMENTS LINKED
BY DISULFIDE BONDS

Type of linked secondary structure	Frequency	Percentage
Helix–helix	5	6.6
Helix–sheet	15	19.7
Helix–coil	13	17.1
Sheet–sheet	5	6.6
Sheet–coil	19	25.0
Coil–coil	19	25.0
	76	100

only twice among our examples, where the preceding and following residues occur in the same secondary structural elements, while the central cysteine is in coil structure according to the DSSP definition.[23]

Among the secondary structural elements that are linked by disulfide bonds, the most predominant is the coil–coil linkage followed, in order, by the "regular structure"–coil linkage (sheet–coil and helix–coil) and the "regular structure"– "different regular structure" linkage (e.g., sheet–helix), whereas the sheet–sheet or helix–helix linkage (i.e., the "regular structure"–"similar regular structure") is rare (6.6%) (Table III). This latter linkage occurs only five times, and three of them are in the same molecule: bovine pancreatic prophospholipase (4BP2). It seems that the high frequency of disulfide bridges between linked coils or between a coil and a regular structure is more general and important for the three-dimensional structures of the proteins. This result agrees with the experimental observation of Matsumura and Matthews,[7,36] who introduced nonnative disulfide bonds into T4 lysozyme. If the disulfide bond was located in a regular–α-helical–structural region, which was the most rigid part of the protein, there was no observable increase in stability, whereas in case of flexible regions the engineered disulfide bond increased the thermostability of the molecule. The disulfide links in these structural elements appear to be as important as conserved salt bridges with aspartate residues, which occur mainly in coil structures.[28]

Prediction Method

In light of the results of the conservation analyses of multiple sequence alignment we can set up a simple and efficient prediction approach. As discussed, the covalent state of cysteine is determined almost exclusively by the location of the proteins. In our representative set there is hardly a single protein in which oxidized and reduced cysteines occur together, except when a cysteine is covalently

[36] M. Matsumura and B. W. Matthews, *Science* **243,** 792 (1989).

bonded to heteroatoms, prosthetic groups, or other amino acids in active sites, etc., so that the cysteine is also oxidized. Therefore, we use the criterion that if a larger fraction of the predicted cysteines belongs to one oxidation state (with the higher conservation score to the group of oxidized cysteines, and with the lower conservation score to the group of reduced cysteines) then the other cysteines in the same molecule can be assumed to be in the same oxidation state.

If the predicted number n of reduced and oxidized cysteines is equal in a protein, then the relative conservation score must be taken into account. To compare them, take the average of the relative conservation scores for the predicted bonded cystines and predicted free cysteines, and then take the logarithm of these averages and compare their absolute value. Mathematically, if

$$\ln\left[\text{abs}\left(\frac{\sum_{i=1}^{n} C_{r/\text{oxidized}/}^{i}}{n} - \text{mean}\right)\right] - \ln\left[\text{abs}\left(\frac{\sum_{i=1}^{n} C_{r/\text{reduced}/}^{i}}{n} - \text{mean}\right)\right] > 0$$

then the cysteines are predicted to be oxidized; otherwise they are reduced. The overall mean for the relative conservation score is 1.27. This normalized score is not sensitive to the number of sequences or to their similarities in the alignment.

The efficiency of the prediction was tested by the jack-knife procedure. One alignment was removed in each step and the averages of the relative conservation of half-cystines and cysteines were calculated from the remaining alignments. The average of the two gave the threshold; if the larger fraction of the cysteines in the tested protein fell below the threshold, every cysteine was predicted to be in the reduced state; otherwise every cysteine was considered to be oxidized. In the case of an equal number of predicted oxidized and reduced cysteines, we considered the absolute average deviation from the actually determined threshold, which was determined from the remaining part of the set, as described above. In this case it is possible to define the covalent state of 75.8% of cysteines (119 good vs 38 bad predictions and 89.8% of half-cystines (141 good vs 16 bad predictions) in a jack-knife test (overall average, 82.8%). If we used a constant threshold (1.27) throughout the test the overall efficiency rose above 84%.

Three types of mispredictions can occur. First, some free cysteines are strongly conserved for functional reasons as in the case of the bilin-binding protein (1BBPA), plastocyanin (1PCY), and ferredoxin (3FXC). If we exclude from the prediction those free cysteines that are conserved for functional reasons, the prediction accuracy for cysteines increases by nearly 8%, which means an overall increase in prediction accuracy of 4%. However, in this case 17% of the cysteine residues are excluded from the prediction. Second, the method fails if both forms of cysteine occur in the same protein, for example, oxidoreductase (2SODB). Third, misprediction can occur when the normalized conservation of the cysteines is near the prediction threshold, as in cytochrome c (155C), serine protease (2ALP), and azurin (2AZAA). If the C_r values of cysteines and cystines are close to one another (near the mean), this usually results from an uninformative alignment that has either too few sequences or the sequences are similar and C_r does not vary enough along the sequence.

Server

A Web server has been set up to calculate the conservation of cysteines for a given protein sequence. Because of advances in sequence search methods the server applies a somewhat different approach to prepare the input sequence alignment than has been described in the original paper and used to calibrate the system (see Applied Databases and Methods). The submitted sequence is searched against a frequently updated non redundant (NR) database of sequences.[37] The NR database currently contains more than 700,000 sequence entries. The sequence database search method uses PSI-BLAST[38] with a strict (0.00001) e-value cutoff to extract significant hits. This default cutoff can be changed by the user if more remote homologs need to be included in the analysis and especially if no or only a few sequences have been found using the default values.

Once the PSI-BLAST search returns a set of homologous sequences and fragments, all the full-length sequences are extracted from the sequence database and aligned by CLUSTAL W.[39] After the alignments are prepared the absolute and relative conservations of each position are calculated as described in Methods, using the approach of Livingstone and Barton as described in the AMAS program.[25] The user can set the fraction of gaps that can be ignored at each position before the conservation scores are calculated. The Results page allows optional download of the full PSI-BLAST search. However, for practical reasons the output shows only a fraction of the used alignment, with the calculated absolute conservation values underneath. The absolute and relative conservation values for the cysteines are reported at the end of the Results page.

The Web server implementing the prediction method is accessible at http://guitar.rockefeller.edu/~andras/cysredox.html.

Acknowledgments

This work has been supported by OTKA Grant T030566. A.F. is a Burroughs Wellcome Fellow. The authors are grateful to Dr. Eva Fodor for discussions of this manuscript.

[37] A. Bairoch and R. Apweiler, *Nucleic Acids Res.* **27**, 49 (1999).
[38] S. F. Altschul, T. L. Madden, A. A. Schaffer, J. Zhang Zhang, W. Miller, and D. J. Lipman, *Nucleic Acids Res.* **25**, 3389 (1997).
[39] J. D. Thompson, D. G. Higgins, and T. J. Gibson, *Nucleic Acids Res.* **22**, 4673 (1994).

[3] Enzyme-Linked Immunospot Assay for Detection of Thioredoxin and Thioredoxin Reductase Secretion from Cells

By BITA SAHAF, ANITA SÖDERBERG, CHRISTINA EKERFELT,
STAFFAN PAULIE, and ANDERS ROSÉN

Oxidative stress response was determined in this study by enzyme-linked immunospot (ELISpot) assays for thioredoxin (Trx) and Trx reductase (TrxR). On exposure to oxidative stress, cells can launch a variety of defense mechanisms, including release of antioxidant proteins. The Trx system, consisting of Trx, TrxR, and NADPH, constitutes one of these cellular defense systems for maintenance of a healthy reduction–oxidation (redox) balance. Trx and TrxR are rapidly upregulated and released from monocytes, lymphocytes, and other normal and neoplastic cells on exposure. Secreted Trx and TrxR have proved to be eminent indicators of oxidative stress. Trx is a small, 12-kDa protein released through a leaderless pathway, whereas TrxR, which is a 116-kDa selenoprotein and required for regeneration of Trx, is secreted through the Golgi pathway. In this chapter we present a detailed laboratory bench protocol for enumeration of single cells secreting redox-active Trx and TrxR after oxidative stress exposure. Physiological stimuli (such as interferon γ, lipopolysaccharide, interleukin 1, and CD23 ligation; and phorbol 12-myristate 13-acetate and ionophore) as well as UV light and hydrogen peroxide were used to generate oxidative stress, and some are presented in detail. The protocol includes a description of cell isolation, preparation, handling, and development of ELISpot plates, troubleshooting notes, presentation of results, statistical evaluation, and comments on alternative sources of materials and manufacturer Web addresses. We conclude that the ELISpot assay is a useful method for detection of single cells secreting the redox-active proteins Trx and TrxR after oxidative stress exposure.

Introduction

Cell growth and proliferation, programmed cell death, and cytoprotection are cellular events that involve signal transduction pathways, which depend on the generation of oxidants and their scavenging by antioxidants. For maintenance of a healthy reduction–oxidation (redox) balance, the thioredoxin (Trx) system, consisting of Trx, Trx reductase (TrxR), and NADPH, has been shown to play a pivotal role.[1,2]

[1] B. Sahaf, Thioredoxin System in Normal and Transformed Human Cells. Medical Dissertation no. 642. Linköping University, Linköping, Sweden, 2000.

[2] H. Nakamura, K. Nakamura, and J. Yodoi, *Annu. Rev. Immunol.* **15,** 351 (1997).

Copyright 2002, Elsevier Science (USA).
All rights reserved.
0076-6879/02 $35.00

Thioredoxin constitutes a family of small redox-active proteins with multiple functions, including cytoprotective antioxidant,[3,4] cytokine,[5] and chemokine[6] activities. Trx is secreted by monocytes, lymphocytes, and other normal and neoplastic cells[7] through a leaderless pathway.[8] Mammalian cells have at least three different types of Trx: (1) cytoplasmic,[9] (2) mitochondrial,[10] and (3) plasma membrane bound.[11] The cytoplasmic form has 104 amino acids and a molecular mass of 12 kDa, so-called full-length Trx. It can be cleaved into a 10-kDa form consisting of 80 to 84 amino acids, termed truncated Trx. Both forms of Trx can be secreted from cells.[12]

Thioredoxin reductase is required for redox regeneration of Trx. TrxR is a 116-kDa flavoprotein consisting of two homodimers of M_r 58,000. It was found to have a catalytically active COOH-terminal elongation containing the amino acid selenocysteine.[13] Expression of constitutive Trx and TrxR has been observed in several cell types of the mammalian body, including keratinocytes of the skin, placental cells, hepatocytes, secretory cells and leukocytes.[14,15] Physiological stimuli, including UV light, hydrogen peroxide, and mitogens, can induce the expression of Trx and TrxR, suggesting an important role in protection against oxidative stress.[2,16] We found that TrxR was upregulated and released from human monocytes after mitogenic stimulation that generated oxidative stress.[16]

Dysregulation and overexpression of Trx and TrxR have also been found in a number of human primary cancers, such as malignant melanoma,[11,17] adult T cell leukemia, lung, colon, cervical, and liver carcinoma.[18] Previously, the lack

[3] H. Nakamura, H. Masutani, Y. Tagaya, A. Yamauchi, T. Inamoto, Y. Nanbu, S. Fujii, K. Ozawa, and J. Yodoi, *Immunol. Lett.* **42**, 75 (1994).

[4] J. Nilsson, O. Söderberg, K. Nilsson, and A. Rosén, *Blood* **95**, 1440 (2000).

[5] A. Rosén, P. Lundman, M. Carlsson, K. Bhavani, B. R. Srinivasa, G. Kjellström, K. Nilsson, and A. Holmgren, *Int. Immunol.* **7**, 625 (1995).

[6] R. Bertini, O. M. Z. Howard, H.-F. Dong, J. J. Oppenheim, C. Bizzarri, R. Sergi, G. Caselli, S. Pagliei, B. Romines, J. A. Wilshire, M. Mengozzi, H. Nakamura, J. Yodoi, K. Pekkari, R. Guranath, A. Holmgren, L. A. Herzenberg, L. A. Herzenberg, and P. Ghezzi, *J. Exp. Med.* **189**, 1 (1999).

[7] M. Ericson, J. Hörling, V. Wendel-Hansen, A. Holmgren, and A. Rosén, *Lymphokine Cytokine Res.* **11**, 201 (1992).

[8] A. Rubartelli, A. Bajetto, G. Allavena, and R. Sitia, *J. Biol. Chem.* **267**, 24161 (1992).

[9] Y. Tagaya, Y. Maeda, A. Mitsui, N. Kondo, H. Matsui, J. Hamura, N. Brown, K. Arai, T. Yokota, and H. Wakasugi, *EMBO J.* **8**, 757 (1989).

[10] G. Spyrou, E. Enmark, A. Miranda-Vizuette, and J. Gustafsson, *J. Biol. Chem.* **272**, 2936 (1997).

[11] B. Sahaf, A. Söderberg, G. Spyrou, A. M. Barral, K. Pekkari, A. Holmgren, and A. Rosén, *Exp. Cell Res.* **236**, 181 (1997).

[12] B. Sahaf and A. Rosén, *Antioxid. Redox Signal.* **4**, 717 (2000).

[13] V. N. Gladyshev, K. T. Jeang, and T. C. Stadtman, *Proc. Natl. Acad. Sci. U.S.A.* **93**, 6146 (1996).

[14] B. Rozell, H. A. Hansson, M. Luthman, and A. Holmgren, *Eur. J. Cell Biol.* **38**, 79 (1985).

[15] K. Schallreuter, M. R. Pittelkow, and J. M. Wood, *J. Invest. Dermatol.* **87**, 728 (1986).

[16] A. Söderberg, B. Sahaf, and A. Rosén, *Cancer Res.* **60**, 2281 (2000).

[17] A. M. Barral, R. Källström, B. Sander, and A. Rosén, *Melanoma Res.* **10**, 331 (2000).

[18] G. Powis, M. Briehl, and J. Oblong, *Pharmacol. Ther.* **68**, 149 (1995).

of reliable techniques to assess redox changes in intact biological samples has been a challenge to investigators. Technical advances have allowed for significant development. We present a detailed protocol of an immunological technique called the enzyme-linked immunospot (ELISpot) assay. Single cells secreting redox-active proteins Trx and TrxR after the induction of a signal pathway involving oxidative stress can be enumerated. The basic principle of observing the secretion of immunoreactive substances from single cells was first explored in the hemolytic plaque assay by Niels Jerne and co-workers in 1963 (Jerne plaque assay).[19] He was awarded the Nobel prize in 1984.[20] Later, assays were developed for detection of single immunoglobulin (Ig)-secreting cells, using erythrocytes coated with anti-immunoglobulins or staphylococcal protein A.[21,22] The instability of the erythrocytes and reagents in the plaque assay was, however, a drawback. The assays for detection of antibody (Ab)-secreting cells were considerably improved when solid-phase material such as polystyrene plastic surfaces,[23–25] nitrocellulose,[26,27] or polyvinylidene difluoride (PVDF)[28] membranes was used for antigen coupling. Enzyme-linked anti-Ig conjugates and chromogenic substrates were used for visualization of secretory cells. Sedgwick and Holt named the technique "spot" ELISA[23] and Czerkinsky et al. coined the ELISpot[24] assay. The technique has been successfully adopted for enumeration of cytokine-secreting cells,[25,26,29] or cells releasing redox-active proteins after an oxidative burst.[12,16]

Principles

The enzyme-linked immunospot assay is based on incubating live cells, stimulated or unstimulated, in a microtiter plate in which the solid-phase bottom (nitrocellulose, PVDF, or polystyrene) is precoated with a specific monoclonal antibody (MAb), termed the "capture Ab," against the antigen, for example, Trx or TrxR (Fig. 1). After an incubation period of 16 to 48 hr, as a rule, the cells are removed, and the position of the captured antigen is visualized at the bottom of the plate by

[19] N. K. Jerne and A. A. Nordin, *Science* **140**, 405 (1963).
[20] N. K. Jerne, *Scand. J. Immunol.* **38**, 1 (1984).
[21] A. Rosén, P. Gergely, M. Jondal, G. Klein, and S. Britton, *Nature (London)* **267**, 52 (1977).
[22] E. A. Gronowicz, A. Coutinho, and F. Melchers, *Eur. J. Immunol.* **6**, 588 (1976).
[23] J. D. Sedgwick and P. G. Holt, *J. Immunol. Methods* **57**, 301 (1983).
[24] C. C. Czerkinsky, L.-Å. Nilsson, H. Nygren, Ö. Ouchterlony, and A. Tarakowski, *J. Immunol. Methods* **65**, 109 (1983).
[25] J. Rönnelid and L. Klareskog, *J. Immunol. Methods* **200**, 17 (1997).
[26] C. C. Czerkinsky, G. Andersson, H.-P. Ekre, L.-Å. Nilsson, L. Klareskog, and Ö. Ouchterlony, *J. Immunol. Methods* **110**, 29 (1988).
[27] M. Steinitz, A. Rosén, and G. Klein, *J. Immunol. Methods* **136**, 119 (1991).
[28] P. Schielen, W. van Rodijnen, T. Tekstra, R. Albers, and W. Seinen, *J. Immunol. Methods* **188**, 33 (1995).
[29] A. Lalvani, R. Brookes, S. Hambleton, W. J. Britton, A. V. S. Hill, and A. J. McMichael, *J. Exp. Med.* **186**, 859 (1997).

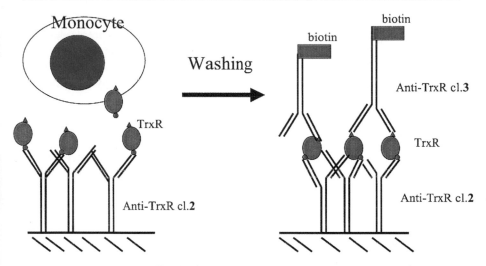

FIG. 1. Principles of the ELISpot assay.

the binding of a biotinylated "indicator Ab" directed against an antigen epitope different from that recognized by the capture Ab. The imprints or immunospots are then developed by an avidin–biotin–enzyme or streptavidin–enzyme conjugate. Alkaline phosphatase with its chromogenic substrates is the enzyme system of choice. One immunospot corresponds to one secretory cell. The total number of spots in a well can be enumerated manually with a stereomicroscope or, preferably, analyzed by an automated imaging system instrument.

Reagents

Multititer plates (96 well): Nitrocellulose-bottomed, MultiScreen MAHAS-4510, or PVDF-bottomed MAIPS4510 or alternatively ELIIP10SSP plates (Millipore, Bedford, MA; www.millipore.com)

Thioredoxin (12 kDa, full length) antibodies[11,12]: (1) Capture Ab, anti-Trx (goat polyclonal IgG) (IMCO, Stockholm, Sweden; www.imcocorp.se); (2) indicator Ab, biotinylated anti-Trx clone 2G11 (BD Pharmingen, San Diego, CA; www.pharmingen.com)

Thioredoxin (10 kDa, truncated) antibodies:[11,12] (1) Capture Ab, anti-Trx (goat polyclonal IgG) (IMCO); (2) indicator Ab, biotinylated anti-Trx clone 7D11 (BD Pharmingen)

Thioredoxin reductase antibodies[30]: (1) Capture Ab, anti-TrxR clone; (2) indicator Ab, biotinylated anti-TrxR clone 3 (Mabtech, Stockholm, Sweden; www.mabtech.com)

30 A. Söderberg, B. Sahaf, A. Holmgren, and A. Rosén, *Biochem. Biophys. Res. Commun.* **249**, 86 (1998).

Note. The two anti-TrxR MAbs are of $IgG_1(\kappa)$ type; they are designated anti-TrxR clone 2 and anti-TrxR clone 3 and are specific for native placenta-derived human 58-kDa TrxR (TrxR1) as tested by Western blot.[30] These MAbs do not react with TrxR2 (65 kDa), as described by Sun *et al.*,[31] or with mitochondrial TrxR, as described by Miranda-Vizuete *et al.*[32] (A. Rosén and G. Spyrou, personal communications, 2000).

Tumor necrosis factor α (TNF-α) control antibodies: (1) Capture MAb, anti-TNF-α (MAB610; R&D Systems, Minneapolis, MN; www.rndsystems. com); (2) indicator MAb, biotinylated anti-TNF-α (BAF210; R&D Systems)

Avidin–biotin complex–alkaline phosphatase conjugate: Vectastain ABC–AP (Vector Laboratories, Burlingame, CA; www.vectorlabs.com) or streptavidin–alkaline phosphatase conjugate (Mabtech, Stockholm, Sweden; www.mabtech.se)

5-Bromo-4-chloro-3-indolyl phosphate/nitroblue tetrazolium (BCIP/NBT) liquid substrate (Bio-Rad, Hercules, CA; www.bio-rad.com) or BCIP/NBT liquid substrate Plus (ICN Pharmaceuticals, Costa Mesa, CA; www.icnpharm.com)

Phosphate-buffered saline (PBS), pH 7.2: For a stock solution that can be kept at room temperature, dissolve 1.0 g of KCl, 1.0 g of KH_2PO_4, 40.0 g of NaCl, and 6.0 g of $Na_2HPO_4 \cdot 2H_2O$ in 1 liter of distilled water. Dilute 1 : 5 in distilled sterile water for use. Sterile filter the buffer through a 0.2-μm pore size filter

PBS-T: PBS with 0.05% (v/v) Tween 20

Tween 20 (Sigma, St. Louis, MO; www.sigma.com)

RPMI 1640 cell culture medium (GIBCO-BRL, Glasgow, UK)

AIM-V medium (GIBCO-BRL)

Fetal calf serum (FCS; GIBCO-BRL): Heat inactivated (30 min at 56°)

Ficoll-Paque (research grade) density gradient medium for isolation of peripheral blood mononuclear cells (PBMCs) and monocytes (Amersham Pharmacia Biotech, Uppsala, Sweden; www.APBiotech.com)

Gelatin, microbiology grade (Merck, Darmstadt, Germany; www.merckeurolab.com)

Escherichia coli lipopolysaccharide (LPS; Sigma)

Interleukin 1 (IL-1; Sigma)

Interferon γ (IFN-γ) (Boehringer Ingelheim, Mannheim, Germany; http://biochem.boehringer-mannheim.com)

---

[31] Q.-A. Sun, Y. Wu, F. Zappacosta, K.-T. Jeang, B. J. Lee, D. L. Hatfield, and V. N. Gladyshev, *J. Biol. Chem.* **274,** 24522 (1999).

[32] A. Miranda-Vizuete, A. E. Damdimopoulos, J. R. Pedrajas, J.-Å. Gustavsson, and G. Spyrou, *Eur. J. Biochem.* **261,** 405 (1999).

Ionomycin (Calbiochem, La Jolla, CA; www.calbiochem.com)

CD23 (EBVCS-2) MAb IgG[33]: Purified from hybridomas grown in our laboratory and used for stimulation of cells via CD23 membrane receptor cross-linking. The hybridomas were a kind gift from Dr. Bill Sugden.

N-Formyl-L-methionyl-1-leucyl-1-phenylalanine (fMLP; Sigma)

Phorbol 12-myristate 13-acetate (PMA; Biomol Research Laboratories, Plymouth Meeting, PA)

Sodium selenite (Sigma)

Diamide (Sigma)

Procedure

The protocol for the TrxR ELISpot assay is described below, and the procedure for detection of truncated Trx and full-length Trx follows the same protocol except for the antibodies used. We have included troubleshooting advice as notes at appropriate points. Please observe that we have used sterile filtered solutions (0.2-μm pore size) throughout the procedure, and that the microtiter plates have been handled in sterile hoods until the step of cell removal

Day 1

Coating of ELISpot Plates

1. Prewet the nitrocellulose-bottomed 96-well microtiter plates (MultiScreen MAHA4510) by addition of 200 μl of distilled water for 30 min. Add 100 μl of anti-TrxR clone 3 (15 μg/ml) or control anti-TNF Ab diluted in PBS to sterile nitrocellulose-bottomed 96-well microtiter plates (MultiScreen MAHA4510). Divide the plate into sections to allow space for three to six replicates of each of the following cell samples (PBMCs or monocytes): (1) unstimulated cells and (2) stimulated cells (mitogen LPS, IL-1, and IFN-γ; PMA and ionomycin; or fMLP; or oxidative stress generators diamide and selenite). Include the following controls: (1) anti-hTNF-α MAbs (15 μg/ml) as a positive control for secretion, (2) replicate wells containing capture and indicator Abs in the absence of cells; and (3) a control for nonviable/necrotic dying cells. Treat the cells with detergent [0.1% (v/v) Tween 20] after adding the cells to the plate.

2. Incubate the plates overnight at 4° in a humidified chamber.

Note. If PVDF plates (MAIPS4510 or ELIIP10SSP) are being used, note that an extra pre-wetting step is required. Add 100 μl of 70% (v/v) ethanol per well, incubate the plates for 5 min, and rinse in 200 μl of sterile distilled H$_2$O (three times) followed by 200 μl of PBS (three times).

---

[33] C. Kintner and B. Sugden, *Nature* **294**, 458 (1981).

Day 2

Preparation of Plates

3. Decant the Ab solution from the plates by flicking them upside down, or by attaching the plates to a vacuum manifold for flowthrough emptying. Wash the plates four times with sterile PBS (200 μl/well). At each washing step, let the plates stand for 5 min in PBS.

4. Block the remaining protein-binding sites on the solid-phase nitrocellulose by addition of 200 μl of RPMI 1640 medium containing 10% (v/v) FCS and incubation for at least 2 hr at 37°.

Isolation of Mononuclear Cells and Monocytes from Peripheral Blood

Peripheral blood mononuclear cells (PBMCs) are isolated from healthy blood donor buffy coat by Ficoll-Paque density gradient centrifugation according to previously described methods.[34] Briefly:

a. Carefully add 15 ml of buffy-coat cell suspension, prediluted 1 : 2 in RPMI 1640 medium, on top of 30 ml of Ficoll-Paque liquid in a 50-ml tube (Falcon; Becton Dickinson Labware, Lincoln Park, NJ). Centrifuge at 400g for 30–40 min at 18–20°.

b. Collect the cells at interphase. Wash the cells in 3 volumes of PBS, twice.

c. Count the cells in a Bürker chamber. These cells constitute the PBMC population. Set aside half the cell number for ELISpot and suspend them in RPMI 1640 medium with 5% (v/v) FCS to a concentration of 10^6 cells/ml. The second half is used to isolate pure monocytes. Suspend these cells in AIM-V medium to a concentration of 5×10^6/ml. Continue as follows.

Monocyte-macrophages are isolated from healthy blood donor buffy coat according to Freundlich and Avdalovic.[35] Briefly:

a. Precoat 75-cm^2 tissue culture flasks by addition of 5 ml of 2% (w/v) gelatin dissolved in distilled water. Incubate the flasks for 2 hr at 37°. Remove the gelatin solution by vacuum aspiration and allow the flasks to dry. Add 10 ml of autologous plasma; incubate for 30–60 min at 37°.

b. Remove the plasma by aspiration, and wash twice with PBS.

c. Add 15 ml of PBMC suspension, prepared as described above (5×10^6/ml), in AIM-V medium to each flask, and then incubate the flasks for 40 min at 37°.

d. Remove nonadherent cells by gently washing three times.

---

[34] A. Bøyum, *Clin. Lab. Invest.* **21**(Suppl.), 77 (1968).
[35] B. Freundlich and N. Avdalovic, *J. Immunol. Methods* **62,** 31 (1983).

e. Detach the remaining adherent monocytes by addition of 10 ml of a mixture containing 5 ml and PBS with 5 mM EDTA and 5 ml of AIM-V medium. Incubate for 15 min. Cells are then collected and pelleted by centrifugation and resuspended in RPMI 1640–5% (v/v) FCS to a density of 1 × 10^6 cells/ml.

Note. The monocyte-macrophage cell purity is >90%. Cell viability is determined by trypan blue dye exclusion; the cells are counted in a Bürker chamber.

Stimulation of Cells from Peripheral Blood

For stimulation of PBMCs or human monocytes, the following reagents are used.

CD23 MAb at a final concentration of 10 μg/ml
IL-1α, *E. coli* LPS, and IFN-γ at final concentrations of 0.5 ng/ml, 1 μg/ml, and 200 U/ml, respectively
fMLP at a final concentration of 10^{-7} M
PMA and ionomycin, 0.5 μM and 300 ng/ml, respectively
Sodium selenite, 1 to 10 μM
Diamide, 25 to 100 μM
RPMI 1640 medium with 5% (v/v) FCS only

Note. Monocytes adhere easily to plastic tube walls, and therefore it is preferable to distribute cells first into microtiter plates, and then to add stimulating reagents, rather than to stimulate cells in tubes before plate distribution.

Plating Cells for Assay

5. Decant the blocking medium.
6. Add 100 μl of the cell suspension in RPMI 1640 with 5% (v/v) FCS containing 10^5 cells to each well. Add the seven stimulating reagents to cells (in 5 μl) in six replicates.
7. Incubate the cells in 5% CO_2 at 37° in humidified air for 20 hr.

Note. Use an isolated incubator in which there is minimal disturbance or movements such as opening and closing doors. This prevents cells from moving, thus avoiding poorly defined spots, streaks, etc.

The incubation time must be optimized for each cell type and stimulation protocol. The suggested time of 20 hr has been optimal for Trx and TrxR. Ten to 40 hr of incubation should be tested.

Day 3

8. Remove cells from the microtiter plates by three washes in PBS and three washes in PBS–0.05% (v/v) Tween 20.

Note. Careful washing is necessary: Cell fragments sticking to the solid phase and disturbing the assay will form if PBS–Tween washing buffer is used before PBS washings. Inspect the solid-phase surface for remaining cells. The surface should appear smooth; if not, further washing may be required.

Development of ELISpot Plates

9. Develop the microtiter plates by addition of 100 μl of biotinylated anti-TrxR clone 2 or biotinylated anti-hTNF, respectively, diluted to 300 ng/ml in PBS-T with 0.1% (w/v) BSA. Incubate the plates for 2 hr at room temperature in a humidified chamber.

10. Wash five times with PBS-T.

11. Add 100 μl of Vectastain ABC–AP or streptavidin–alkaline phosphatase diluted 1:1000 in PBS-T with 0.1% (w/v) BSA. Incubate the plates for 1 hr at room temperature in a humidified chamber.

Note. We recommend sterile filtering the substrate solution through a 0.2-μm pore size filter before use.

12. Wash the plates six times in PBS.

13. Add 100 μl of BCIP/NBT substrate solution to the wells. Incubate the plates for 30 min. Monitor the color development visually; check for staining in negative control wells.

14. Stop the reaction by rinsing the plates three times in distilled water and then remove the underdrain of the plates. Rinse the plates three times in tap water. Empty and dry the plates carefully, first on a paper towel upside down and then overnight at room temperature in the dark.

Day 4

Evaluation and Enumeration of Spots

15. Enumerate the spots in each well, using a stereomicroscope (Nikon [Garden City, NY] model SMZ-1B; ×35 magnification). Calculate the median value for each of the six replicate wells. For objective enumeration of the ELISpot assay, we use an ELISpot reader instrument for automatic imaging analysis and counting (Autoimmun Diagnostika [AID], Strassberg, Germany; www.elispot.com).

Statistical Analysis

Statistical differences between stimulated and nonstimulated control cultures of monocytes or PBMCs are evaluated by the nonparametric Mann–Whitney U test. All statistical evaluations are performed with JMP version 3.2.5 (SAS Institute, Cary, NC; JMPdiscovery.com) software and a Gateway G6-300 PC microcomputer.

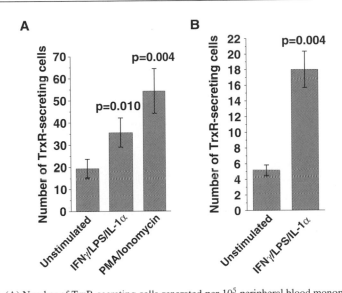

FIG. 2. (A) Number of TrxR-secreting cells generated per 10^5 peripheral blood mononuclear cells (PBMCs) during a 20-hr incubation, as determined by ELISpot. Cells were unstimulated [5% (v/v) FCS in RPMI 1640] or stimulated with IFN-γ (200 U/ml), LPS (1 μg/ml), and IL-1α (0.5 ng/ml) or with PMA (0.5 μM) and ionomycin (300 ng/ml). p Values show significant differences between non-stimulated and stimulated PBLs (Mann–Whitney U test). Error bars represent the SE. (B) Number of TrxR-secreting cells generated per 10^5 peripheral monocytes during a 20-hr incubation, as determined by ELISpot. Cells were unstimulated [5% (v/v) FCS in RPMI] or stimulated with IFN-γ (200 U/ml), LPS (1 μg/ml), and IL-1 (0.5 ng/ml). p Values show significant differences between nonstimulated and stimulated PBLs (Mann–Whitney U test). Error bars represent the SE. [Reproduced with permission from A. Söderberg, B. Sahaf, and A. Rosen, *Cancer Res.* **60,** 2281 (2000).]

The statistical program Shapiro–Wilks is used for analysis of normal distribution. Data presented in Fig. 2 show a normal distribution and data are presented as means $\pm$ standard error.

The reproducibility of the ELISpot assay has been tested. Intraassay variability between the six replicates is four or five spots. Interassay variability has been tested on frozen cell samples from the same blood donor on two separate days, and has been found to be eight or nine spots. The major parameter that influences variability is the condition and viability of cells.

Presentation of Results

Figure 2 shows the results from an ELISpot assay in which PBMCs and monocytes were stimulated via pathways that generate reactive oxygen species. We observed that the signal pathways employed by mitogenic stimulation with LPS, IL-1, and IFN-γ significantly induced the secretion of TrxR, which indicates a prompt antioxidant modulation. Results are plotted as mean values of six replicates $\pm$SE.

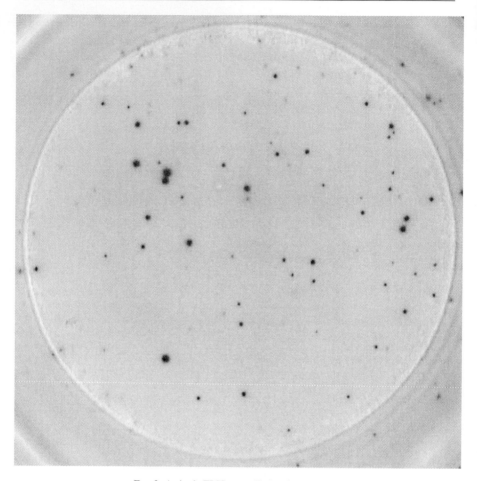

FIG. 3. A single ELISpot well after development.

Figure 3 illustrates the visual appearance of a single ELISpot well after development.

Validation of ELISpot by Three Methods

For a correct validation of the ELISpot technique, immunoprecipitation with metabolic labeling, Western blotting, and flow cytometry are performed. The protocol for stimulation that was used in the analysis of secreted Trx or TrxR is the same as in the ELISpot assay, as it is for the immunoprecipitation, Western blotting, and flow cytometry.

Immunoprecipitation

First, immunoprecipitation is performed. Cells (5×10^6) are resuspended in 5 ml of cysteine- and methionine-free RPMI 1640 medium (Amersham Life Science, Rockville, MD) with 5% (v/v) PBS-dialyzed FCS added. A 50-μCi/ml concentration of L-[^{35}S]methionine and L-[^{35}S]cysteine (>1000 Ci/mmol) (Amersham Life Science) is added. The supernatants are collected after 24 hr. Cells are harvested for extract preparation as described above. The supernatants and cell extracts are preabsorbed with Sepharose–protein A for 1 hr and immunoprecipitated with 1 μg of anti-TxR clone 3 (anti-TrxR-cl.3) mixed into 50 μl of a 50% slurry of Sepharose–protein A for 18 hr at 4°. The precipitates are separated in 10–20% (w/v) polyacrylamide–sodium dodecyl sulfate (SDS) gels and transferred to PVDF membranes (Amersham Life Science). Autoradiography exposure is performed at −70° for 10 days, using Eastman Kodak (Rochester, NY) X-Omatic intensifying screens and Hyperfilm ECL (Amersham Life Science).

Western Blot

All samples are electrophoresed under reducing conditions in 10–20% (w/v) polyacrylamide–SDS gels. After electrophoresis, proteins are transferred to 0.2-μm nitrocellulose membranes (Amersham Pharmacia Biotech). The membranes are blocked with 5% (w/v) skim milk powder (Semper, Stockholm, Sweden) and incubated with anti-TrxR MAbs (50 ng/ml). The bound antibodies are visualized with horseradish peroxidase (HRP)-labeled anti-mouse IgG (diluted 1 : 10,000) and an ECL detection kit purchased from Amersham Pharmacia Biotech).

Flow Cytometry

Flow cytometry is employed to assess whether the different stimulation pathways yielded intracellular expression changes. Cells are harvested and washed in balanced saline solution (BSS) with 1% (v/v) HEPES (BSS–HEPES) and fixed in a 4% (w/v) paraformaldehyde (PFA) solution for 10 min. After another rinse with BSS–HEPES–saponin, 1×10^6 cells are incubated with 100 μl of unlabeled anti-TrxR-cl.3 (10 μg/ml) diluted in BSS–HEPES with 0.1% (w/v) saponin (BSS–HEPES–saponin) for 30 min according to the method of Sander *et al.*[36] Cells are washed in BSS–HEPES–saponin and incubated with 100 μl of fluorescein isothiocyanate (FITC)-conjugated F(ab′)$_2$ fragment of goat anti-mouse Ig (Dako, Glostrup, Denmark) (25 μg/ml) for 30 min. After another rinse with BSS–HEPES–saponin, cells are blocked with 10 μl of normal mouse serum (Dako) for 10 min and incubated with 100 μl of biotinylated anti-Trx (10 μg/ml) for 30 min, followed by two washes and a final incubation with 100 μl of streptavidin-conjugated Cy5-R phycoerythrin; Dako) (10 μg/ml). Cells are analyzed on a FACSCalibur flow

[36] B. Sander, J. Andersson, and U. Andersson, *Immunol. Rev.* **119**, 65 (1991).

cytometer with a secured voltage supply of 220 to 230 V and further analyzed with CellQuest software (Becton Dickinson, Franklin Lakes, NJ).

Comments

Inducible Thioredoxin Reductase Secretion

TrxR secretion was analyzed in cultures from PBMCs and monocytes after 20 hr of incubation at 37° in an ELISpot assay detecting secretory cells at the single-cell level. PBMCs and monocytes were stimulated with IFN-γ, LPS, and IL-1α. We found a significant increase in the number of TrxR-secreting cells, compared with cultures without stimuli. The mean number of TrxR-secretory PBMCs after stimulation was 41 ± 4 (SE) per 10^5 cells compared with control cultures [23 ± 3 (SE)] ($p = 0.010$) (Fig. 2A). For purified monocytes the corresponding values were 18 ± 2 (control, 5 ± 0.7) ($p = 0.001$) (Fig. 2B). Increased TrxR release was also found with PMA/ionomycin (Fig. 2A). We also explored the effect of alternative stimuli previously reported to activate human monocytes. Reactive oxygen species released were determined directly after stimulation in a chemoluminescence assay.[37] Parallel, identically stimulated cultures were analyzed after 20 hr for the number of TrxR-secreting cells in an ELISpot assay. A dose-dependent response was observed: increased chemoluminescence response induced more TrxR-secretory cells (Spearman, $R = 0.85$).

The quality of the ELISpot assay was determined by including several positive and negative controls: (1) The secretory capacity of the cells was determined by analyzing TNF-α secretion in parallel, identically stimulated cell cultures. A doubling of TNF-α-specific spots was found after IFN-γ/LPS/IL-1α stimulation compared with nonstimulated cells; (2) wells containing capture and indicator MAbs with substrate did not give any spots in the absence of cells; (3) only viable cells secreted TrxR or TNF; necrotic cells did not give any spots at all, as determined with detergent-lysed cells [0.1% (v/v) Tween 20]; and (4) the specificity of the two anti-TrxR MAbs was carefully controlled in separate immunoprecipitation experiments showing intracellular TrxR from cell extracts and secreted TrxR in monocytes and PMA-stimulated monocytes; the secreted TrxR appeared as a single 58-kDa protein band with identical molecular weight as compared with purified human placenta TrxR (TrxR1).

Conclusions

Secretion of the redox-active proteins thioredoxin and thioredoxin reductase from peripheral blood mononuclear cells and monocytes was observed at the

---

[37] A. Johansson and C. Dahlgren, *J. Leukoc. Biol.* **45,** 444 (1989).

single-cell level by using a sensitive enzyme-linked immunospot (ELISpot) assay. The release was inducible and physiological stimulation of human monocytes by IFN-γ, LPS, and IL-1α significantly increased the number of Trx- and TrxR-secreting cells. To validate the TrxR secretion, we performed metabolic labeling of cells followed by immunoprecipitation, Western blot, and flow cytometry. Results revealed that Trx and TrxR were actively secreted into the medium for 24 h. We conclude that the ELISpot assay contributes an important, novel method for determination of oxidative stress-induced release of redox-active proteins.

Acknowledgment

This work was supported by a grant from the Swedish Cancer Society (no. 3171).GSD.

[4] Determination of Redox Properties of Protein Disulfide Bonds by Radiolytic Methods

By CHANTAL HOUÉE-LEVIN

Disulfide/dithiol redox systems are involved in the regulation of cell growth and proliferation,[1,2] in human cancer development,[3,4] and in the development of postirradiation effects.[5]

Protein sulfur functions are essential in numerous events related to oxidative stress. In the regulation of all events involving free radicals, such as oxidative stress and irradiation by ionizing radiation, one-electron redox processes take place and sulfur free radicals are created. Their properties play a key role in the reactions leading to cellular effects. The most important sulfur free radicals are thiyl and disulfide radicals. They interconvert into each other (Scheme 1), because the disulfide radical anion undergoes S–S bond opening and the thiyl radical adds to thiol function to give the disulfide radical.

However, this interconversion may be only partial, because carbon-centered free radicals can also be created. For instance, in glutathione, the thiyl radical is in

[1] H. Nakamura, K. Nakamura, and J. Yodoi, *Annu. Rev. Immunol.* **15**, 351 (1997).

[2] G. Powis, J. R. Gasdaska, P. Y. Gasdaska, M. Berggren, D. L. Kirkpatrick, L. Engma, I. A. Cotgreave, M. Angulo, and A. Baker, *Oncol. Res.* **6–7**, 303 (1997).

[3] A. Baker, C. M. Payne, M. M. Brieh, and G. Powis, *Cancer Res.* **57**, 5162 (1997).

[4] A. Gallegos, M. Berggren, J. R. Gasdaska, and G. Powis, *Cancer Res.* **57**, 4965 (1997).

[5] S. Kojima, O. Matsuki, T. Nomura, A. Kubodera, Y. Honda, S. Honda, H. Tanooka, H. Wakasugi, and K. Yamoka, *Biochim. Biophys. Acta* **1381**, 312 (1998).

Copyright 2002, Elsevier Science (USA).
All rights reserved.
0076-6879/02 $35.00

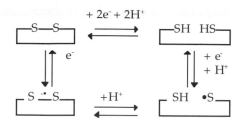

SCHEME 1. One-electron redox intermediates in the disulfide/dithiol couple.

tautomeric equilibrium with a carbon-centered form.[6] In hen egg white lysozyme, reduction of the disulfide bridge between Cys-6 and Cys-127 leads also to cleavage of the polypeptidic backbone,[7] which indicates a similar electron migration.

The redox properties of both disulfide and thiyl sulfur free radicals are believed to be different; disulfide radical is reductant whereas thiyl radical is oxidant. However, to date there is no experimental method for measuring the one-electron reduction potentials of the disulfide/disulfide anion and thiyl/thiol couples in proteins or in small molecules.

Principle of Radiolysis Methods

General Kinetic Scheme

γ radiolysis and pulse radiolysis provide a unique way of testing the redox properties of the sulfur free radicals and the reactivity of disulfide and thiol functions toward reductants and oxidants, respectively. The easiest way is through reduction of disulfide bonds, because the CO_2^- free radical has proved to be able to induce a rapid and specific one-electron reduction of disulfide bonds. Kinetic studies of the reduction process have been investigated in small molecules[8–11] as well as in some proteins.[12–16] The elementary reactions induced during this reduction process are given in Scheme 2.

[6] L. Grierson, K. Hildebrand, and E. Bothe, *Int. J. Radiat. Biol.* **62,** 265 (1992).

[7] J. Bergès, E. Kassab, D. Conte, E. Adjadj, and C. Houée-Levin, *J. Phys. Chem.* **42,** 7809 (1997).

[8] P. C. Chan and B. H. J. Bielski, *J. Am. Chem. Soc.* **95,** 5504 (1973).

[9] A. J. Elliot and F. C. Sopchyshyn, *Radiat. Phys. Chem.* **19,** 417 (1982).

[10] A. J. Elliot, S. Simsons, and F. C. Sopchyshyn, *Radiat. Phys. Chem.* **23,** 377 (1984).

[11] C. Von Sonntag, "The Chemical Basis of Radiation Biology." Taylor & Francis, Philadelphia, NJ, 1987.

[12] M. Faraggi, J. P. Steiner, and M. H. Klapper, *Biochemistry* **24,** 3273 (1985).

[13] V. Favaudon, H. Tourbez, C. Houée-Levin, and J.-M. Lhoste, *Biochemistry* **29,** 10978 (1990).

[14] V. Favaudon, H. Tourbez, C. Houée-Levin, and J.-M. Lhoste, *J. Chim. Phys.* **88,** 993 (1991).

[15] C. J. Koch and J. A. Raleigh, *Arch. Biochem. Biophys.* **287,** 75 (1991).

[16] D. Conte and C. Houée-Levin, *J. Chim. Phys.* **90,** 971 (1993).

(1) $CO_2^{\cdot-} + P/SS$ $\longrightarrow$ $CO_2 + P/SS^{\cdot}$

(2) $P/SS^{\cdot-} + H^+$ $\rightleftharpoons$ $P/SSH^{\cdot}$

(3) $P/SS^{\cdot-} + H^+$ $\rightleftharpoons$ $P/(SH)S^{\cdot}$

(4) $P/SSH^{\cdot}$ $\longrightarrow$ $P/(SH)S^{\cdot}$

(5) $2\,P/SS^{\cdot-}\ (+2H^+)$ $\longrightarrow$ $P/SS + P/(SH)_2$

(6) $P/(SH)S^{\cdot} + HCOO^-$ $\rightleftharpoons$ $P/(SH)_2 + CO_2^{\cdot-}$

(7) $2\,P/(SH)S^{\cdot}$ $\longrightarrow$ $(HS)/P/S...S/P/(SH)$

(8) $P/SS^{\cdot-}$ $\longrightarrow$ $P^{\cdot}/SS$

(9) $P^{\cdot}/SS$ $\longrightarrow$ products

SCHEME 2. Kinetic scheme of the one-electron reduction of protein or peptide disulfide bond reduction by $COO^{\cdot-}$ radicals.

The first step is the formation of the disulfide radical $P/SS^{\cdot-}$ [reaction (1)]. Under acidic conditions it is protonated ($P/SSH^{\cdot}$) [reaction (2)]. The disulfide radical decay may involve several reactions: disproportionation [reaction (5)] and SS bond cleavage leading to the thiyl radical and a thiol or thiolate function [$P/(SH)S^{\cdot}$] [reaction (3)]. This latter reaction is usually much faster with the protonated radical [reaction (4)]. The relative importance of these reactions varies from one protein to another. Electron transfer from disulfide anion to other residues and/or the polypeptidic backbone may take place [reaction (8)]. It was suggested that thiyl radicals are able to oxidize tyrosine residues in an inter- or intramolecular process,[17,18] but there is no experimental evidence for this reaction, and thus it is not shown in Scheme 2. For biological effects, the key reaction is thus the protonation equilibrium of the disulfide radical [reaction (2)], because it allows the transformation of a mild reductant (disulfide radical) to a strong oxidant (thiyl radical).[18] The thiyl radical dimerizes, leading to an intermolecular disulfide bond, that is, a protein dimer [reaction (7)]. In the presence of formate ions, it can also oxidize formate ions [reaction (6)]. The whole process is thus a chain reaction.

Interaction of Ionizing Radiation with Aqueous Solutions
  and Formation of Reductant

The methods of radiolysis are based on use of the interaction of ionizing radiation with an aqueous solution containing appropriate solutes. More details

[17] W. A. Prütz, J. Butler, E. J. Land, and A. J. Swallow, *Free Radic. Res. Commun.* **2,** 69 (1986).
[18] W. A. Prütz, J. Butler, E. J. Land, and A. J. Swallow, *Int. J. Radiat. Biol.* **55,** 539 (1989).

TABLE I
RADIOLYTIC YIELDS[a] OF WATER FREE RADICALS AND OF HYDROGEN
PEROXIDE INDUCED BY IRRADIATION OF DEAERATED WATER BY
HIGH-ENERGY ELECTRONS OR γ RAYS FROM ^{60}Co, pH 3–11[b]

	H	OH	e_{aq}^-	H_2O_2
G value (μmol J^{-1})	0.062	0.28	0.28	0.073

[a]Also called G values.
[b]Data taken from J. W. T. Spinks and R. J. Woods, "Introduction to Radiation Chemistry," 3rd Ed. John Wiley & Sons, New York, 1990.

concerning the interaction of ionizing radiation with matter are given in the literature.[19] Briefly, γ rays from ^{60}Co, or high-energy electrons, interact similarly with an aqueous solution. Because the energy of the radiation is much higher than that of electrons in molecules, both rays extract an electron from the major compound, that is, water. The result is the creation of a homogeneous solution of OH, H, and e_{aq} free radicals in the nanosecond time scale after interaction of ionizing radiation with matter. Some hydrogen peroxide is also created. The absorbed dose is the amount of energy absorbed per unit mass and is expressed in grays (1 Gy $= 1$ J kg^{-1}). The quantities of the radiolytic species are known and are proportional to the dose. Table I gives the yields of radiolytic compounds.

Water free radicals may then react with solutes or with themselves (by recombinations, disproportionation, or dimerization). In an aqueous solution containing formate ions in high quantity (10^{-2} mol liter^{-1} or higher), and in an atmosphere of nitrous oxide, the most probable reaction of OH and H radicals is with formate. Hydrated electrons react with N_2O. The kinetic scheme is the following:

$$OH + HCOO^- \rightarrow H_2O + COO^{\cdot-} \tag{10}$$

$$H + HCOO^- \rightarrow H_2 + COO^{\cdot-} \tag{11}$$

$$e_{aq}^- + N_2O \rightarrow OH + OH^- + N_2 \tag{12}$$

Thus all water free radicals are transformed into the reductant COO$^{\cdot-}$ radical. Under the conditions of concentration described above, and in the pH range 4–10, its yield G(COO$^{\cdot-}$) is thus equal to

$$G(COO^{\cdot-}) = g(H) + g(OH) + g(e_{aq}^-) = 0.62 \, \mu\text{mol J}^{-1}$$

In this expression $g(H)$, $g(OH)$, and $g(e_{aq}^-)$ denote the respective G values given in Table I. The whole process [reactions (10)–(12)] is finished in less than 1 μsec after interaction of the ionizing radiation.

[19] J. W. T. Spinks and R. J. Woods, "Introduction to Radiation Chemistry," 3rd Ed. John Wiley & Sons, New York, 1990.

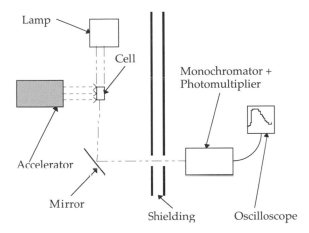

FIG. 1. Schematic drawing of a pulse radiolysis setup. The solution to be irradiated is in the cell. The electron beam (coming from the accelerator) is usually perpendicular to the analyzing light beam.

What Are γ and Pulse Radiolysis?

Pulse radiolysis is a fast kinetic method, whereas γ radiolysis is a steady state method. In pulse radiolysis, pulses of high-energy electrons (2–10 MeV; in our setup ~4 MeV) are delivered to the irradiation cell from a linear accelerator (ours is located at the Curie Institute, Orsay, France). A schematic drawing of such a setup is given in Fig. 1.

The pulse duration is usually 2–200 nsec long. The sample is contained in a quartz cell (100–400 μl in our case). The modifications of the solution are followed by absorption (the most frequent method). A UV–visible beam is delivered by a lamp (Xe–Hg "super-quiet" or Xe arc) perpendicularly to the electron beam. The residual light is analyzed by a conventional system (monochromator, photomultiplier, oscilloscope, and computer). The equipment that we use has been described.[13] The doses per pulse (2–60 Gy) are calibrated by the thiocyanate dosimeter ($10^{-2} M$ sodium thiocyanate, nitrous oxide atmosphere).[19] The data collected are absorption spectra of free radicals and kinetics of their formation and decay. In this work, it is thus possible to distinguish between disulfide radicals in anionic or protonated form[13] and thus to measure the pK_a of this free radical (if it is above ~4). If the lifetime of the thiyl radical is long enough, it can be observed.[13]

In steady state γ radiolysis, the ionizing radiation (from ^{60}Co or ^{137}Cs) is delivered at a low rate compared with that of pulse radiolysis; however, for this system the chemistry is the same. Free radicals can be assumed to reach steady states that last as long as the irradiation. This method is convenient for preparing large quantities of final products and thus allows further analysis. For instance, the number of reducible disulfide bonds can be determined, and it is possible to observe whether products other than reduced protein (with thiol functions) and dimers linked by intermolecular disulfide bonds are created. The chain length, expressed as the ratio of the yield of thiol functions versus that of reductant $G(-SH)/G(COO^{\cdot-})$, can be measured. Its value would depend on the pK_a of the disulfide radical and on the reduction potential of the thiyl/thiol couple (Scheme 2). Thus it allows an evaluation of the one-electron reduction potential of the thiyl/thiol redox couple compared with that of the $CO_2^{\cdot-}/HCOO^-$ couple (1.07 V at pH 7[20]).

Experimental Procedures

Reagents

Radiolysis is sensitive to metal impurities, and thus salts for radiolysis should be as pure as possible. Also, many buffers such as Tris or carbonate react with water free radicals; therefore their use should be avoided except in small concentrations. Whenever possible, phosphate buffer should be used.

Sodium formate and potassium hydrogen phosphate are of the highest quality available (Normatom [Prolabo, Fontenay sous Bois, France] or Suprapure [Merck, Fontenay sous Bois, France]). 5,5′-Dithio-bis(2-nitrobenzoic acid) (DTNB) is provided by Sigma (L'Isle d'Abeau, France). Nitrous oxide is delivered by Air Liquide (Grigny, France). Its purity is higher than 99.99% (>20 ppm O_2). Water is purified by an Elga Maxima system (Plessis-Robinson, Hauts-de-Seine, France) (resistivity, 18.2 MΩ).

All proteins should be as pure as possible. Purification by dialysis is recommended to eliminate organic buffers, conservators, and transition metal cations.

1. Unless otherwise stated, samples to be irradiated are made up in 20 mM phosphate, 100 mM sodium formate buffer adjusted to the required pH with sulfuric acid or sodium hydroxide, and saturated with N_2O. The doses per pulse are $\sim$5–40 Gy ($[CO_2^{\cdot-}] \approx 3$–25 μM). The protein concentration is 10–100 μM. The optical path is 1 or 2 cm. The irradiated volume is 200–400 μl per pulse.

2. Radiolysis results are interpreted with the help of other methods. In addition to radiolysis, we use nuclear magnetic resonance, high-performance liquid chromatography, electrophoresis, etc.

3. The free sulfhydryl group concentration is determined by optical titration with DTNB at pH 8.0 (100 mM Tris-HCl buffer) using $\varepsilon_{410\,nm} = 13.6$ mM^{-1} cm^{-1} for the 3-carboxylato-4-nitrothiophenolate anion[21] (verified by us, using glutathione).

4. Thiol groups in the reduced protein are alkylated before electrophoresis by reaction with 10 mM iodoacetamide (10 min, room temperature).

[20] P. Wardman, *J. Phys. Chem. Ref. Data* **18**, 1637 (1989).
[21] G. L. Ellman, *Arch. Biochem. Biophys.* **74**, 443 (1958).

5. For γ radiolysis, the samples are subjected to irradiation in a panoramic IL60PL (CIS Bio International, Gif sur Yvette, France) ^{60}Co source at a dose rate, determined by Fricke dosimeter,[19] of approximately 1.0 Gy sec^{-1}.

Some Typical Results

In the following sections we emphasize the conclusions that can be drawn from our studies.

Free Radical Formation

As a result of the reaction of COO$^{\cdot-}$ radical with the protein, disulfide free radicals are created. Their absorption spectrum is made of a broad band with no structure, centered at about 420–440 nm (Fig. 2).

In acidic medium, the spectrum is shifted to the UV, indicating the formation of the protonated disulfide radical. The maximum is at about 400 nm and the extinction coefficients are lower.

Under the conditions described above, formation of the free radical is over in ~10–250 μsec. The rate constants of disulfide radical formation [reaction (1) of

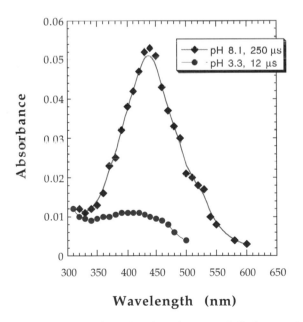

FIG. 2. Absorption spectra of disulfide anion (◆) and protonated (●) immunoglobulin G (IgG). These spectra were taken at the time when the absorbances were maximum. [IgG], 250 μM; [HCOO$^-$], 0.1 M; phosphate, 10 mM; N$_2$O atmosphere; dose, ~25 Gy.

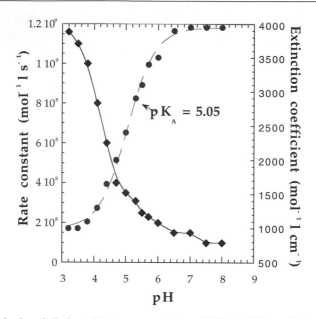

FIG. 3. Reduction of riboflavin-binding apoprotein (apro-RBP) by $COO^{\cdot-}$ radicals. Shown is the variation in the rate constant of reaction (1) (Scheme 2) (♦) (*left scale*), and of the average extinction coefficient of disulfide radical (●) (*right scale*), with pH. Points represent experimental data. For extinction coefficients, the line is the best fit to experimental points, using Eq. (13) (see text), which gives a pK_a of 5.05. [apo-RBP], 100 μM; [$HCOO^-$], 0.1 M; phosphate, 10 mM; N_2O atmosphere; dose, ~25 Gy.

Scheme 2] do not vary much with the protein, but increase strongly with decreasing pH (Fig. 3).

Free Radical Decay

Conversely, the free radical decay varies with the protein, reflecting the relative importance of reactions (5) to (8) (Scheme 2). Also, the final compounds may be different.

The free radical decay becomes much faster in acidic medium: the lifetime of the protonated disulfide radical is ~20–100 μsec, because of the weakness of the protonated disulfide bond[22] and because of the catalysis of reaction (3) by protons. The opening of the radical, which is the main reaction [see reactions (3) and/or (4) of Scheme 2], is first order.

[22] J. Berges, F. Fuster, J.-P. Jacquot, B. Silv, and C. Houée-Levin, *Nukleonika* **45,** 23 (2000).

Determination of Disulfide Radical pK_a

The determination of the free radical pK_a is obtained by plotting the free radical average extinction coefficient ε_T at fixed wavelength (such as 420 nm), as a function of pH (Fig. 3). The extinction coefficients of the basic and of the acidic forms (respectively, ε_A and ε_B) are measured in acidic and basic solutions, respectively, and thus the only adjustable parameter is K_A in Eq. (13).

$$\varepsilon_T = \frac{\varepsilon_A \times 10^{-pH}}{10^{-pH} + K_A} + \frac{\varepsilon_B + K_A}{10^{-pH} + K_A} \tag{13}$$

Typical values of pK_A lie between ~4.5 and 6 for small molecules and for proteins. Thioredoxin behavior is markedly different[23] because its pK_A is less than 3. This abnormal behavior is suppressed by site-directed mutagenesis of Trp-35.

Final Products and Chain Length

Final products can be detected after pulse radiolysis; however, it is often easier to prepare larger amounts of irradiated solutions by steady state γ radiolysis. Both conditions differ mainly by the dose rate, which is several orders of magnitude higher in pulse radiolysis.

Final compounds include reduced protein (with two thiol functions) and dimers linked by intermolecular disulfide bridges. For lysozyme we also found a fragment.[22] For immunoglobulin G (IgG), the combination of reduction–dimerization leads to mixtures of heavy and light chains, with varous kinds of heavy–light combinations.[13] For thioredoxin, only the reduced protein is found, but mutations of Trp-35 or of Asp-30 allow the formation of dimers. It is possible that thioredoxin reduces its own dimer and that mutants are not active enough to do so.

The chain length also varies according to the protein. Typical values are 10–20 in basic medium. It increases with acidity, because of the easier formation of thiyl radicals, and it can reach values as high as 200 for lysozyme or glutathione at pH 4. Hence the low values obtained for thioredoxin (1 in basic and neutral medium, <20 at pH 4.5)[24] reflect (1) the absence of formation of thiyl radicals in neutral and basic medium and (2) a reduction potential and/or accessibility much lower than that of other thiyl radicals in acidic medium.

Specificity of Thioredoxin

Our results show several differences in the behavior of thioredoxin sulfur free radicals compared with that of other sulfur free radicals.[23,24] First, the disulfide radical is much more acidic, and this acidity seems to be controlled by Trp-35. Asp-30 also plays a role, although less clear, in the proton transfer. Second, the thiyl

[23] C. El Hanine, D. Conte, J.-P. Jacquot, and C. Houée-Levin, *Biochemistry* **39**, 9295 (2000).
[24] C. El Hanine, D. Conte, J.-P. Jacquot, and C. Houée-Levin, *Res. Chem. Intermediates* **25**, 313 (1999).

radical is much less oxidant. The two cysteines responsible for the redox properties of thioredoxin (Cys-36 and Cys-39) have different properties. Cys-36 is solvent accessible,[25] whereas Cys-39 has a pK_a that is at least 4 units higher than physiological pH,[26–28] and is relatively unreactive. Our results would indicate a preferential localization of the thiyl radical on Cys-39, and thus a nonsymmetric bond opening of the disulfide radical. Thiyl radicals are responsible for the pro-oxidant side of sulfur compounds. Although thioredoxin and glutathione play similar roles in antioxidant defense, thioredoxin would be a much better antioxidant because of its weak pro-oxidant properties. It would be interesting to know whether this specificity is also found in other members of the thiol–disulfide oxidoreductase family.

Acknowledgment

I thank the Curie Institute (Dr. V. Favaudon) for the use of the accelerator.

[25] P. T. Chivers and R. T. Raines, *Biochemistry* **36,** 15810 (1997).

[26] P. T. Chivers, K. E. Prehoda, B. F. Volkman, B. M. Kim, J. L. Markley, and R. T. Raines, *Biochemistry* **36,** 14985 (1997).

[27] J. F. Andersen, D. A. Sanders, J. R. Gasdaska, A. Weichsel, G. Powis, and W. R. Montfort, *Biochemistry* **36,** 13979 (1997).

[28] D. M. LeMaster, P. A. Springer, and C. J. Unkefer, *J. Biol. Chem.* **272,** 29998 (1997).

[5] Crystal Structures of Oxidized and Reduced Forms of NADH Peroxidase

By JOANNE I. YEH and AL CLAIBORNE

Introduction

Modulations of protein function by oxidation–reduction reactions through cofactors such as FAD, iron–sulfur centers, heme prosthetic groups, and redox-active disulfides are well known. However, only more recently have cysteine–sulfenic acid (Cys–SOH) derivatives been recognized as novel moieties (cofactors) functioning in enzyme catalysis and redox regulation. A major reason for the delayed recognition of the functional importance of this moiety in proteins is that sulfenic acids are difficult to identify and have long been characterized primarily as transient intermediates because of their highly reactive and unstable chemical nature. There are three major simple organic oxyacids of sulfur: (1) sulfenic acids (RSOH), (2) sulfinic acids (RSO_2H), and (3) sulfonic acids (RSO_3H).[1] Of these, sulfonic acids are by far the most stable; sulfinic acids are somewhat unstable thermally,

[1] J. L. Kice, *Adv. Phys. Org. Chem.* **17,** 65 (1980).

Copyright 2002, Elsevier Science (USA).
All rights reserved.
0076-6879/02 $35.00

and sulfenic acids are for the most part elusive reactive intermediates. In proteins, organic oxyacids are obtained by the oxidation of cysteine residues and, until more recently, the presence of these sulfur oxyacids, primarily as cysteine–sulfinic and cysteine–sulfonic acids, was thought to be nonnative, caused by the oxidation of surface cysteines by ambient oxygen or as experimental artifacts formed, for example, through oxidation by free radicals generated during X-ray data collection. In many enzymes that require sulfhydryls or sulfenic acids for catalysis, the oxidation to a sulfinic or sulfonic acid prohibits the formation of an "activated" form of the sulfur, such as a thiolate (S^-) or sulfenate (SO^-), and irreversibly inhibits the enzyme. However, the crystal structure of NADH peroxidase[2] (Npx) led to an unambiguous identification of a reversibly reducible cysteine–sulfenic acid as a secondary redox center in this enzyme. This, along with substantial biochemical data, established this moiety as an intrinsic functional group in this protein. Npx utilizes an essential Cys–SOH and the reversibility of the Cys–SH ↔ Cys–SOH redox cycle is inherent to its catalytic mechanism. In general, the facile redox potential of this reversible redox partner highlights its utilization in a variety of regulatory functions and other biological reactions. The noted reactivity of sulfenic acid, which until recently prevented its identification as important functional groups in proteins, is also a hallmark as Cys–SOH is considerably more reactive than the corresponding Cys–SS–Cys disulfide.[3] This reactivity is advantageous when utilized in roles such as redox control; the identification of regulation via reversible SOH ↔ SH oxidation substantiates this idea.[4–7] Among the distinct protein systems that utilize cysteine–sulfenic acids for catalysis and/or regulation are FAD-dependent peroxide and disulfide reductases, peroxiredoxins, transcription factors, and transport proteins, among others.[8] These diverse systems illustrate the functional significance of sulfenic acids in proteins and in protein–DNA complexes. Further descriptions of redox regulation can be found in other chapters in this volume and in reviews, which give additional consideration to the topic of redox regulation and catalysis via cysteine–sulfenic acids.[3,8]

Mechanistic Considerations

The defining property of Npx relative to other FAD-dependent disulfide reductases is the absence of a redox-active protein disulfide. Npx contains only one-half

[2] J. I. Yeh, A. Claiborne, and W. G. J. Hol, *Biochemistry* **35**, 9951 (1996).

[3] A. Claiborne, J. I. Yeh, C. Mallett, J. Luba, E. J. Crane III, V. Charrier, and D. Parsonage, *Biochemistry* **38**, 15407 (1999).

[4] M. Zheng and G. Storz, *Science* **279**, 1718 (1998).

[5] S.-R. Lee, K.-S. Kwon, S.-R. Kim, and S. G. Rhee, *J. Biol. Chem.* **273**, 15366 (1998).

[6] S. Veeraraghava, C. C. Mello, E. J. Androphy, and J. D. Baleja, *Biochemistry* **38**, 16115 (1999).

[7] M. A. Gorman, S. Morera, D. G. Rothwell, E. de La Fortelle, C. D. Mol, J. A. Tainer, I. D. Hickson, and P. S. Freemont, *EMBO J.* **16**, 6548 (1997).

[8] A. Claiborne, T. C. Mallett, J. I. Yeh, J. Luba, and D. Parsonage, *Adv. Protein Chem.* **58**, 215 (2001).

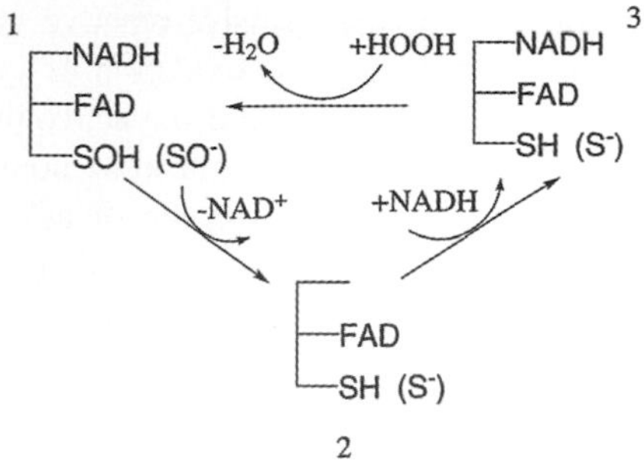

1

```
  ┌─NADH      -H₂O    +HOOH  ┌─NADH    3
  ├─FAD      ◄─────────────  ├─FAD
  └─SOH (SO⁻)                └─SH (S⁻)
       ╲-NAD⁺          +NADH ╱
              ╲        ╱
              ┌─FAD
              └─SH (S⁻)
```

2

SCHEME 1

cystine per subunit and NADH reduction generates one Cys–SH.[9] In Npx, the cycling between the reduced thiol or thiolate and the oxidized sulfenate or sulfenic acid of residue 42 is an intrinsic part of the catalytic cycle (species 1 and 3, Scheme 1) as the formation of the catalytically competent thiolate form (species 2, Scheme 1) precedes the binding of H_2O_2.

The structures of native, oxidized Npx to 2.1 Å resolution and the NADH-bound reduced form at 2.85 Å resolution have yielded insight into how elements of the protein help to stabilize and modulate the redox and catalytic properties of this enzyme. This chapter describes the experimental approaches used to capture the active, sulfenic acid form of Npx as well as the NADH-reduced form, to outline the steps that were taken to minimize oxidation to an inactive form during crystal growth and X-ray data collection.

When working with a protein that is suspected to contain a functional cysteine–sulfenic acid, it should be kept in mind that the presence of stable Cys–SOH derivatives in proteins implies that the respective Cys–SOH must exist in an unusual microenvironment, protected from solvent and preferentially associated with apolar regions of the protein structure (Fig. 1). All known structures of functional Cys–SOH-containing proteins have the Cys–SOH group effectively immersed within a cavity and isolated from other reactive groups.[2,10,11] A point worth emphasizing is that the presence of Cys–SOH in proteins occurs when the surface topology of the protein prevents the formation of interchain disulfide bonds and there are no sulfhydryl groups in the vicinity to form intramolecular cystine bonds. This provides the environment for the formation of a sulfenic acid on reaction of a cysteinyl side chain with an oxidant.[12]

---

[9] L. B. Poole and A. Claiborne, *J. Biol. Chem.* **261**, 14525 (1986).

[10] K. Becker, S. N. Savvides, M. Keese, R. H. Schirmer, and P. A. Karplus, *Nat. Struct. Biol.* **5**, 267 (1998).

[11] H.-J. Choi, S. W. Kang, C.-H. Yang, S. G. Rhee, and S.-E. Ryu, *Nat. Struct. Biol.* **5**, 400 (1998).

[12] W. S. Allison, *Accounts Chem. Res.* **9**, 293 (1976).

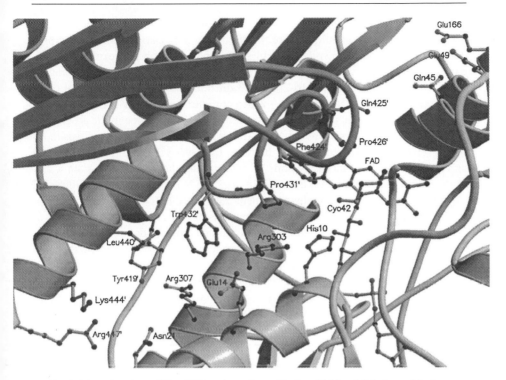

FIG. 1. Dimer interface of Npx. FAD and cysteine–sulfenic acid (Cys–SOH, denoted in figure as Cyo-42) residue indicate the active site of Npx, which is formed at the dimer interface. Residues from each monomer subunit at the interface contribute to the tight packing; numbered residues from one monomer subunit are differentiated from the second by primes. The tight packing shown limits solvent accessibility, stabilizing the labile cysteine–sulfenic acid (Cyo-42) by burying the group in an apolar environment of the protein. In addition, the hydrogen-bonding chain formed by Arg-307, Glu-14, Arg-303, His-10, and Cyo-42 stabilizes the sulfenate form (SO⁻) of the cysteine–sulfenic acid, further enhancing the stability of this reactive group.

For a protein suspected of utilizing a cysteine–sulfenic acid, several approaches can be taken to determine the oxidation state of the suspected group. After examination of the sequence to determine the number of cysteines and whether cystine bonds are possible, based on a known structure or prediction of the fold, site-directed mutagenesis can be done to systematically map the function of each cysteine. If a cysteine is determined not to be in a disulfide bond, owing to the absence of a partner or because it is the lone cysteine in the protein, then modification of the cysteine can be done to determine whether the oxidation state of the residue affects protein function. In all the proteins that utilize Cys–SOH, the reversibility of Cys–SH $\leftrightarrow$ Cys–SOH is obligatory for activity. Consequently, the addition of an equivalent of an oxidizing reagent, such as H_2O_2, should have measurable effects on activity whereas addition of a reducing agent, such as dithiothreitol (DTT), should reverse the effects, as long as oxidation has not proceeded to the sulfinic or sulfonic acid state. One approach that exploits the intrinsic nucleophilicity of the Cys–SO$^-$ anion toward the amine and thiol reagent NBD-Cl (7-chloro-4-nitrobenzo-2-oxa-1,3-diazole) results in a protocol for Cys–SOH modification[13] that can be analyzed in conjunction with UV–visible, fluorescence, and electrospray interface-mass spectrometric data. Even in the absence of chemical modification such as the NBD-Cl approach, spectroscopic analysis monitoring charge-transfer interactions between the Cys–SOH and a cofactor, such as FAD, can provide another means of determining the presence of such a catalytic pair if such interaction exists, as it does for Npx.[14] Similarly, site-directed mutagenesis,[15] ^{13}C-based nuclear magnetic resonance (^{13}C-NMR) analysis of labeled cysteines, and mass spectrometric analysis of reduced and oxidized forms of the protein under study can be done to ascertain whether oxidation occurs via 1,2- or 3-oxygen addition to a sulfhydryl group when an oxidizing agent is added, where the presence of these oxidized forms can be correlated to functional activity.[9,14]

An understanding of the redox mechanism can be helpful in monitoring the state of the enzyme or protein before structural analysis. For Npx, a simple spectrophotometric assay[9] to determine the activity of the enzyme allowed us to assess the oxidation state of the crystals before data collection in most cases or to confirm the oxidation state of the enzyme in the crystals after structural analysis.

General Considerations

Although protein sulfenates are generally unstable, reactivities of sulfenic acids in proteins can span a wide range. For some proteins, severe limitation of exposure to oxygen and persistent presence of reducing agents are necessary to prevent significant oxidation whereas for others, oxidation/inactivation by ambient oxygen is a slow process and these proteins can be stored under ambient conditions for hours to days before a substantial population of the protein molecules is oxidized. Hence, it may be necessary to empirically determine the sensitivity of a protein through various means, as indicated in the preceding section. For Npx, we were able to use a spectrophotometric activity assay to ascertain whether freshly expressed and purified protein, stored protein, and protein obtained from dissolved crystals at various time points were active. These measurements gave us some general understanding of the time frame over which oxidation of the enzyme occurred and the expected oxidation state of the protein crystals before X-ray structure analysis.

[13] H. R. Ellis and L. B. Poole, *Biochemistry* **36,** 15013 (1997).
[14] E. J. Crane III, D. Parsonage, and A. Claiborne, *Biochemistry* **35,** 2380 (1996).
[15] D. Parsonage and A. Claiborne, *Biochemistry* **34,** 435 (1995).

A final unambiguous view of the oxidation state of the enzyme was obtained after our structural analysis.

To obtain a native, active, cysteine–sulfenic acid-containing crystal structure of Npx, several precautionary measures were taken. X-ray structure determination of a protein containing a cysteine–sulfenic acid may require additional precautions in order to prevent the formation of oxidation artifacts. With the advent of cryotechniques during data collection, the production of free radicals that result in crystal decay, among other phenomena, has been minimized. These free radicals can also presumably be the cause of oxidation of sulfenic acids as the radicals can readily oxidize the acid to its more stable, higher oxidation states and, consequently, use of cryogenic temperatures during data collection can also minimize these oxidation events. However, in some systems, data collection at cryogenic temperatures alone may not be sufficient to prevent the oxidation of sulfenic acids as these groups are highly labile and could presumably react with any residual molecules, including ambient oxygen, before and during crystal growth. An earlier X-ray crystal structure of the Npx had shown oxidation of what is now known to be the sulfenic acid to a sulfonic acid.[16] To maximize the possibility of crystallizing the active, sulfenic acid form of the enzyme, freshly prepared protein was used for crystallization setups and crystals were stored under argon, as described below.

It should be noted that the structures of other proteins/enzymes have indicated that the above-described precautionary measures were not necessary to limit oxidation of the cysteine–sulfenic acid. In human glutathione reductase, a sulfenate of residue 63 was found under fully aerobic conditions and at ambient temperatures.[10] Here, the sulfenate was generated through oxidation of Cys-63 by S-nitrosoglutathione, a physiological carrier of NO, and the formation of this reversible, inactive form of glutathione reductase may play a role in NO-based redox signaling, where interactions at the active site were speculated to be responsible for the stability of the sulfenic acid. In a human cysteine peroxidase enzyme, a cysteine–sulfenic acid was unexpectedly found at the active site although data were collected at room temperature.[11] The authors propose that the unusual stability of the sulfenic acid was due to interaction with an Mg^{2+} ion, which stabilized the ionized sulfenate by charge interactions and blocked the narrow entrance of the active site pocket, inhibiting the access of additional O_2 or H_2O_2. The stability of protein sulfenates can be attributed to specific interactions and factors based on the localized protein environment surrounding the sulfenic acid residue. Criteria that represent factors for stabilization include limited solvent accessibility, an apolar microenvironment, hydrogen bonding and/or ionization of –SOH to the sulfenate (–SO⁻) form, and the absence of vicinal protein thiols.[3] These criteria have been satisfied in all the proteins that have been structurally characterized, including Npx, glutathione reductase, and human cysteine peroxidase enzyme, in which a

[16] T. Stehle, S. A. Ahmed, A. Claiborne, and G. E. Schulz, *J. Mol. Biol.* **221**, 1325 (1991).

Cys–SOH is utilized in catalysis or regulation, highlighting the significance of the local protein environment in stabilization and function of this reactive moiety.

General Protocol

To minimize the possibility of oxidation under ambient conditions, crystals are grown from freshly prepared protein, thereby limiting aging time and exposure of the enzyme to atmospheric oxygen before crystallization. Typically the amount of time from crystallization to data collection is between 2 weeks and 2 months and crystallization trays are stored under inert argon, purged of oxygen. It is not known whether this measure is necessary; it is likely, for freshly grown crystals that are frozen, with data collection at 100 K soon after crystal formation, that this would be unnecessary. For crystals that are to be stored for more than a few weeks, the storing of these crystals under inert atmospheres prevents oxidation. It is known from cryogenic data collection on crystals grown and allowed to age for lengths of time spanning weeks to months under ambient oxygen conditions that this storage time contributes to oxidation. A sulfinic acid form of Npx has been obtained by cryogenic data collection methods,[17] indicating that oxidation from the sulfenic acid occurred during crystal storage and that this is likely a stepwise event, depending on the amount of storage time. Storage of crystals for as long as 7 months before cryogenic data collection results in the sulfinic acid form of residue 42. The formation of a sulfonic acid may be due to long, ambient oxygen exposure of the protein and/or protein crystal, in conjunction with data collection at room temperature, which was the original data collection protocol.[16]

The setup for storage under inert atmospheres can be as sophisticated as a glovebox or as simple as oxygen-resistant bags constructed of polymers such as polyvinyl chloride, which with sufficient thickness can provide reasonable resistance to oxygen permeation. To purge the trays of oxygen, a gentle stream of argon is passed over the sitting drop tray before sealing it with tape. For hanging drop trays, each coverslip is lifted and argon is passed over the reservoir before reseating each drop. These crystallization trays are then placed into an oxygen-resistant bag and the bag is purged with argon. As argon is heavier than air, the argon displaces oxygen from the system. Care is taken to flush out the system on a routine basis to sustain an inert atmosphere and crystals of active Npx are maintained for more than 6 months by this method. A monitoring system is not used for Npx, but for more sensitive proteins oxygen sensors can be used to measure the amount of oxygen in the setup and care can be taken to maintain the oxygen levels at a low value.

Npx crystals can be readily dissolved and analyzed by a simple spectrophotometric method that monitors the decrease in 340-nm absorbance of NADH in the

---

[17] J. I. Yeh *et al.,* manuscript in preparation (2002).

presence of H_2O_2.[9] This assay, which can be directly correlated to the cysteine–sulfenic acid oxidation state, is a useful and expedient means of monitoring oxygen-aging effects. A unit of activity is defined as the amount of enzyme that catalyzes the oxidation of 1 μmol of NADH/min at 25°. This allows the monitoring of the redox state of Npx, because oxidation of the sulfenic acid at residue 42 to sulfinic or sulfonic acid forms can be distinguished in this assay as diminished or abolished activity of the enzyme. During the catalytic cycle, SOH must first be reduced to the thiolate by NADH; this precedes binding of H_2O_2 (see Scheme 1). When the sulfenic acid is oxidized to the sulfinic or sulfonic acid form, the initial "priming step" for the formation of the critical thiolate cannot occur and, hence, hydrogen peroxide cannot bind to the active site. Although NADH can still bind, this assay readily ascertains the oxidation state of the cysteine at position 42, as reductants such as dithionite or NADH can monitor the presence of redox intermediates through UV–visible and fluorescence properties.[18] Reductive titrations of wild-type Npx generate a charge-transfer intermediate characteristic of the two-electron reduced (EH_2) forms of most disulfide reductases.[9] In Npx, this absorbance band is centered at 540 nm and is due to the electronic interaction between the nascent $Cys-42-S^-$ and the oxidized FAD. When the sulfenic acid is oxidized to sulfinic or sulfonic acid forms, reduction to $Cys-42-S^-$ is not possible and hence reduction or absence of the 540-nm absorbance band can be used to monitor the presence of these oxidized species. Similarly, NADH reduction experiments with intact crystals (described below) can be done, where color changes in the crystals can be simply visualized under a microscope to ascertain the oxidation state of the sulfenic acid at position 42.

Structural Analysis

The structural analysis of Npx utilizes conventional structure validation and assessment tools including SA-omit maps and iterative model building correlated to statistical factors and map correctness to confirm the final accuracy of the structure, especially in the assignment of a sulfenic acid at residue 42. For the 2.1 Å native structure, confirmation that the Cys-42–SOH has not been oxidized to the sulfinic Cys-42–SO_2H or sulfonic Cys-42–SO_3H acid is achieved through refinements with several models that differ at residue 42. These models are generated with a Cys-42–SOH model previously refined to 2.8 Å (Yeh et al.[2]) and the C_β substituent at position 42 is changed to include –SH, –SO_2H, and –SO_3H as well as a truncated form (–H; alanine model). Omit difference maps, calculated after simulated annealing to reduce model bias, are then used to ascertain the amount of electron density that is present after omitting a specific region from the model. Refined omit maps are calculated by omitting from the model a region of 8 Å

[18] C. H. Williams, Jr., L. D. Arscott, R. G. Matthews, C. Thorpe, and K. D. Wilkinson, *Methods Enzymol.* **62,** 185 (1979).

around a residue and refining the remaining model. Atoms within a 3-Å shell are restrained in order to avoid artificial movement into the omitted region. Each cycle includes manual adjustment and fitting into difference Fouriers [$(F_{soak} - F_{native})$ $\exp(i\alpha_{calc-native})$] and additive Fouriers [$(2F_{soak} - F_{native})\exp(i\alpha_{calc-native})$] maps.[19] These maps show unambiguously that the resulting difference electron density at position 42 best fits a Cys-42–SOH model (Fig. 2), whereas Cys–SO$_3$H results in two negative density peaks in the maps, indicating that the model contains too many atoms at that site.

To obtain the NADH-reduced structure, NADH is added to intact crystals; this is done with crystals of the cysteine–sulfenic acid and cysteine–sulfinic acid forms. In the case of the active, sulfenic acid form, addition of NADH immediately turns the yellow crystals red, signifying reduction of the sulfenic acid. To obtain this complex of Npx with NADH, a 100 mM solution of NADH is freshly prepared in Tris buffer, pH 6.8, and diluted 1 : 10 with the artificial mother liquor, for a final concentration of 10 mM. Crystals are soaked in this solution for several minutes, during which the crystals turn from yellow to red, signifying reduction of the sulfenic acid. Once the red color is observed, the crystal is flash-frozen to 100 K and data are collected to capture the reduced form. The frozen crystal remains red throughout data collection, indicating a charge-transfer complex between Cys-42–Sγ, and the isoalloxazine of FAD is formed. When this same procedure is performed on crystals that have the inactive, sulfinic acid-oxidized form, the red shift is not observed, even with addition of a molar excess of NADH. The crystal structure of this sulfinic acid form shows that although NADH is bound, reduction of the sulfinic acid cannot occur[17] and, hence, a charge-transfer complex does not form with the isoalloxazine ring of FAD.

Although soaking in NADH causes minor changes in cell parameters and overall structural conformation, considerable changes are found at the active site. The overall extent of structural changes that occurs on soaking is reflected in R_{iso} values, when compared with the apo form, of 19–25%. These values are appreciably higher than R_{merge} values of 5–9% within apo forms or within NADH-bound forms. Structurally, these changes reflect reorientation of several side chain groups immediately at the active site of Npx, to accommodate bound NADH, as well as a shift in the isoalloxazine ring of FAD.

Structural analyses for the NADH-reduced form of Npx are done similarly, as described for the oxidized form of Npx. Models are generated as described above except that the C$_\beta$ substituent at position 42 is changed to include –SH, –SOH, and a truncated form (–H; alanine) for the 2.85-Å NADH-reduced native structure. Models are refined, and after every cycle of refinement maps are calculated to visually inspect the positive and negative residual densities in a σ_A-weighted

[19] P. A. Karplus and G. E. Schulz, *J. Mol. Biol.* **210,** 163 (1989).

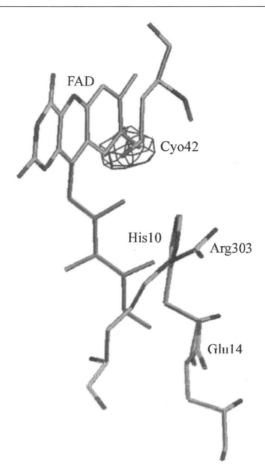

FIG. 2. Residual electron density. A σ_A-weighted $F_o - F_c$ electron density map contoured at 3.5 σ of the active site of Npx. The strong positive electron density identifies the position of the oxygen of the sulfenic acid residue (Cyo42). The map was calculated with a model containing a cysteine residue at position 42. The single, strong peak unambiguously identifies the residue at position 42 as a cysteine–sulfenic acid. Active site residues that are important for stabilizing the sulfenic acid include His-10 and Arg-303, whereas Glu-14 participates in forming the tight dimer interface that limits solvent accessibility, important for maintaining the oxidation state of the sulfenic acid.

difference ($F_o - F_c$) density map. After these first maps are examined, a cysteine is placed at residue 42 and another round of refinement is completed. The difference Fouriers clearly show densities corresponding to NADH and reveal areas of conformational change in the structures. Examination of the difference Fourier and additive Fourier maps allows assignment of the oxidation state of residue 42 to correspond to a sulfhydryl.

Summary

X-ray structural characterization of cysteine–sulfenic acid-containing proteins is one of the most defining approaches to characterizing this rapidly growing class of protein functional groups. Although outside the scope of this chapter, these structural analyses can lead to kinetic measurements in the crystal that allow intermediate states to be trapped, visualized, and studied. An understanding of the biochemistry of these reactive groups can be more fully gained by studying the localized protein environment in which these groups function. Increased perception of how elements of a protein can stabilize and contribute to modulation of function in these systems will allow novel means of enhancing or inhibiting function in important classes of protein molecules, including transcription factors and redox-regulated enzymes.

Acknowledgments

This work was supported by the National Institutes of Health NRSA Fellowship DK-09568 (J.I.Y.) and the National Institutes of Health Grant GM-35394 (A.C.).

[6] Redox Control of Zinc Finger Proteins

By MICHAEL A. BALDWIN and CHRISTOPHER C. BENZ

Introduction

Recognized as an essential trace element for well over a century, zinc is the second most abundant metal ion in mammalian cells, with the adult human body containing up to 3 g of ionic zinc (Zn^{2+}).[1] Unlike transition metal cations such as chromium, nickel, iron, and copper, which can engage in Fenton/Haber–Weiss chemistry intracellularly to generate free radicals and induce potentially carcinogenic oxidative damage,[2] Zn^{2+} produces structure-stabilizing protein cross-links without the undesirable effects of metal ion redox reactivity. Since 1940, it has been known that Zn^{2+} is structurally essential for the catalytic activity of critical enzymes (e.g., carbonic anhydrase, alkaline phosphatase, lactate dehydrogenase, superoxide dismutase, and RNA and DNA polymerases), in which it is often tetravalently coordinated by imidazole nitrogens (from histidines) and carboxylic groups (e.g., glutamic or aspartic acid). In most of these enzymes, zinc catalytically interacts with the substrate molecule undergoing transformation; in some it

[1] J. M. Berg and Y. Shi, *Science* **271,** 1081 (1996).
[2] K. S. Kasprzak, *Cancer Invest.* **13,** 411 (1995).

Copyright 2002, Elsevier Science (USA).
All rights reserved.
0076-6879/02 $35.00

TABLE I
SELECTED ZINC FINGER PROTEIN FAMILIES[a]

Zinc finger family motif	Representative family member		
	Repeats	Protein acronym	Binding target
Cys_4	1	GATA-1	DNA
	2	Nuclear receptor (ER)	DNA
Cys_6	1	GAL4	DNA
Cys_2His_2	1	TFIIIA	RNA
	3	Sp1	DNA
	10	KS1 (KRAB-ZFP)	DNA
Cys_3His_1	2	HIV-1 NCP7	RNA
	2	PKC	Protein
$Cys_3HisCys_4$	1	LIM domain	Protein
$Cys_3HisCys_4$ (RING)	1	BRCA-1	Protein

[a] Within each finger there can be different numbers of cysteine and histidine residues tetrahedrally coordinated to zinc. Individual zinc finger proteins possess different numbers of zinc finger motifs sequentially repeated within a binding domain that targets nucleic acid (RNA or DNA, single or double stranded) or protein.

plays a purely structural role. Less than 20 years ago, the first of many zinc-based motifs termed "zinc fingers" was identified in a eukaryotic transcription factor (TFIIIA), where it was determined to be essential for the DNA-binding ability of the factor. Since then, more than 10 different classes of zinc finger motifs have been discovered and characterized, many for their ability to bind nucleic acids (DNA, RNA, and heteroduplexes of DNA–RNA) in a sequence-specific manner, with others thought to specifically mediate protein–protein interactions (e.g., dimerization box in C-terminal zinc finger of steroid receptors). Unlike the typical mode of Zn^{2+} coordination within the catalytic center of enzymes, tetrahedral coordination of Zn^{2+} within zinc fingers characteristically involves two to four potentially redox-reactive sulfhydryl (cysteine) groups.[1,3–5]

It is estimated that zinc finger proteins constitute up to 1% of all human gene products, each of these proteins containing from 2 to 37 repeats of cysteine ($\pm$ histidine)-containing zinc finger motifs.[1,3] Table I summarizes the known zinc finger motif classes and illustrates each with a representative protein or protein family. X-ray crystallography of many DNA-bound zinc finger domains has shown stabilized folded and helical structures within a single domain that binds to specific

[3] A. Klug and J. W. R. Schwabe, *FASEB J.* **9,** 597 (1995).
[4] K.-D. Kroncke and V. Kolb-Bachofen, *Methods Enzymol.* **269,** 279 (1996).
[5] X. Wu, N. H. Bishopric, D. J. Discher, B. J. Murphy, and K. A. Webster, *Mol. Cell. Biol.* **16,** 1035 (1996).

bases within the major groove of a duplexed DNA target site (DNA response element). This apparent structural commonality belies the fact that most zinc finger proteins are still biochemically and structurally poorly characterized. In fact, a vast but not well-characterized subset of zinc finger proteins, including many implicated in human tumorigenesis, appears to act by binding directly to macromolecules other than nucleic acids (e.g., the $Cys_3HisCys_4$-type LIM domain proteins, $Cys_3HisCys_4$ RING finger proteins such as BRCA1 and BRCA2, and the $Cys_3HisCys_3His$ cysteine-rich subdomain of protein kinase C).[1,3,6,7] A better characterized subset of zinc finger proteins includes transcription factors implicated in both mammalian aging and tumorigenesis, and comprises diverse superfamilies whose major groove DNA binding results in either direct gene upregulation or its repression (e.g., the Cys_4-type zinc finger steroid receptor superfamily or the Cys_2His_2-type zinc finger Sp/Kruppel-like superfamily).[8,9] However, these same DNA-binding transcription factors can also regulate gene expression without directly binding DNA; instead, they use one of their zinc fingers to mediate protein–protein interactions (e.g., dimerization) with other DNA-bound transcription factors.[6,10] Despite the redox insensitivity of the coordinated Zn^{2+} in these transcription factors, a large number of them are known to be redox regulated by intracellular levels of reactive oxygen species (ROS), which can impair their DNA-binding activity and alter their protein–protein interactions with the oxidation of coordinating cysteines and ejection of Zn^{2+} that maintains the critical zinc finger structure.[6,11,12] Apart from hydrogen peroxide and shorter lived superoxide and hydroxyl radicals, other ROS produced intracellularly under both normal and diseased conditions include the free radical product of nitric oxide synthase, NO, which is also known to structurally impair the DNA-binding and gene-regulating function of zinc finger transcription factors.[13]

Redox Sensitivity of Zinc Finger Transcription Factors Such as Estrogen Receptor

It is well established that intracellular oxidative stress and ROS are produced by growth factors as well as by many other exogenous agents and during normal physiologic responses (e.g., hypoxia–reperfusion injury, inflammation).[14] If

[6] L. Zheng, S. Li, T. G. Boyer, and W.-H. Lee, *Oncogene* **19,** 6159 (2000).

[7] S. Kuroda, C. Tokunaga, Y. Kiyohara, O. Higuchi, H. Konishi, K. Mizuno, G. N. Gill, and U. Kikkawa, *J. Biol. Chem.* **271,** 31029 (1996).

[8] C. C. Benz, *Endocr.-Relat. Cancer* **5,** 271 (1998).

[9] G. Suske, *Gene* **238,** 291 (1999).

[10] J. Liden, F. Delaunay, I. Rafter, J.-A. Gustafsson, and S. Okret, *J. Biol. Chem.* **272,** 21467 (1997).

[11] Y. Sun and L. W. Oberley, *Free Radic. Biol. Med.* **21,** 335 (1996).

[12] R. G. Allen and M. Tresini, *Free Radic. Biol. Med.* **28,** 463 (2000).

[13] K.-D. Kroncke and C. Carlberg, *FASEB J.* **13,** 166 (2000).

[14] B. S. Berlett and E. R. Stadtman, *J. Biol. Chem.* **272,** 20313 (1997).

intracellular glutathione (GSH), antioxidant, and/or superoxide dismutase (SOD)-detoxifying systems of the cell are deficient or overwhelmed, increases in intracellular ROS produce various cellular responses ranging from cell proliferation to apoptosis or senescence, and contribute to either transient stress injury or more permanent manifestations such as aging and malignancy.[14–16] Of physiologic interest, natural aging is known to be associated with *in vivo* accumulation of oxyradical tissue damage, resulting in selective loss of Sp1 and glucocorticoid receptor DNA-binding activities but without significant decline in the tissue content of these zinc finger transcription factors.[11] In preclinical models of breast tumorigenesis, growth-associated periodic fluctuations in blood flow through the tumor microvasculature have been found to be sufficient to induce hypoxia–reperfusion injury, with the resulting redox-sensitive gene induction thought to be associated with enhanced tumor aggressiveness and therapeutic resistance.[17]

A critical member of the nuclear and steroid receptor superfamily essential for normal reproductive gland function and whose transcriptional overexpression contributes to the development of most human breast cancers is the zinc finger transcription factor estrogen receptor (ER, α isoform).[8] ER activity is regulated allosterically by ligand binding with either an agonist (estrogen) or antagonist (antiestrogen). On binding to an agonistic ligand, homodimerization and DNA binding by the activated ER complex occurs, producing *trans*-activation of ER-inducible genes. As with many other well-described transcription factors, a number of intracellular proteins and conditions (e.g., ER phosphorylation) are thought to modulate the binding of ligand-activated steroid receptors to cognate response elements (e.g., the estrogen response element, ERE) within the inducible gene promoter. Dimerization of ER, in particular, is necessary for DNA binding; and at least two dimerization domains have been mapped within the full-length receptor, one in the DNA-binding domain near the second zinc finger and the other in the more C-terminal ligand-binding domain of ER. Interference with either of the two Cys_4-type zinc finger structures located in the ER DNA-binding domain diminishes ER dimerization, DNA binding, and *trans* activation of ERE containing target gene promoters.

Structural changes in the zinc finger DNA-binding domain can certainly be induced by amino acid alterations, or by loss or substitutions of the two tetrahedrally coordinated Zn^{2+} cations either by chelation (e.g., treatment with *o*-phenanthroline) or replacement with another coordinating metal ion. In this latter regard, transition metal ions of environmental or physiologic concern include Cu(II), Cd(II), Ni(II), and Cr(II); and of these, Cu(II) is of special interest because of its significant increase in patients with malignant breast disease, its known ability to inhibit

[15] T. Finkel and N. J. Holbrook, *Nature* (*London*) **408,** 239 (2000).

[16] R. A. DePinho, *Nature* (*London*) **408,** 248 (2000).

[17] H. Kimura, R. D. Braun, E. T. Ong, R. Hsu, T. W. Secomb, D. Papahadjopoulos, and K. Hong, *Cancer Res.* **56,** 5522 (1996).

ER function both *in vitro* and *in vivo,* and the ease with which it replaces Zn^{2+} in the ER DNA-binding domain (forming four bicoordinated Cu^{2+} complexes of higher affinity than the Zn^{2+} complexes), leading to an alteration in ER tertiary structure.[18–20] Although the concentration of Zn^{2+} exceeds that of Cu^{2+} in most normal cells, the intracellular and intranuclear levels of Zn^{2+} and Cu^{2+} in different tissues, including estrogen target organs, support the possibility that Cu^{2+} may occupy the ER DNA-binding domain (DBD) in some human breast tumors.[21]

As a Cys$_4$-type zinc finger protein, ER is also expected to be redox sensitive. Interestingly, the *trans*-activating and DNA-binding capacities of ER have previously been shown to be modulated by the redox effector protein, thioredoxin.[22] Moreover, thiol-specific oxidation of ER appears to account at least partially for the fact that otherwise intact (67 kDa) and immunoreactive ER present in about one-third of all ER-overexpressing primary breast tumors appears to be completely unable to bind DNA when assayed *in vitro* in either the presence or absence of exogenous zinc.[23–25] This ER DNA-binding dysfunction is partially reversible under strong thiol-reducing conditions (excess dithiothreitol [DTT]), indicating that mild cysteine oxidation has occurred with tumor growth in some of these samples. However, free radical intermediates formed under stronger oxidant stress can also alkylate cysteine and noncysteine residues in and outside of the ER DNA-binding domain, which would then result in nonreversible ER dysfunction, as found in the majority of breast tumor samples exhibiting loss of ER DNA-binding capacity.

Thus, we have proposed that enhanced intratumor ROS production accompanying aggressive tumor cell proliferation and invasion, in association with inadequate tumor neovascularization and a fluctuating supply of oxygen, potentially produces a subset of ER-overexpressing breast tumors that are not only more aggressive in their clinical behavior but also unresponsive to endocrine therapy (e.g., with the antiestrogen tamoxifen), because of the structural damage induced in their ER zinc finger DNA-binding domains. Clinical studies are pending to prove that this subset of ER-overexpressing breast tumors contain such oxidatively damaged and dysfunctional ER. In the interim, however, mass spectrometric analysis of oxidant-stressed recombinant ER protein and ER-overexpressing human breast cancer cell lines (e.g., MCF-7) is providing important new mechanistic

[18] J. H. Fishman and J. Fishman, *Biochem. Biophys. Res. Commun.* **152,** 783 (1988).

[19] P. F. Predki and B. Sarkar, *J. Biol. Chem.* **267,** 5842 (1992).

[20] T. W. Hutchens, M. H. Allen, C. M. Li, and T.-T. Yip, *FEBS Lett.* **309,** 170 (1992).

[21] J. H. Freedman, R. J. Weiner, and J. Peisach, *J. Biol. Chem.* **261,** 11840 (1986).

[22] S.-I. Hayashi, K. Harjiro-Nakanishi, Y. Makino, H. Eguchi, J. Yodoi, and H. Tanaka, *Nucleic Acids Res.* **25,** 4035 (1997).

[23] G. K. Scott, P. Kushner, J. L. Vigne, and C. C. Benz, *J. Clin. Invest.* **88,** 700 (1991).

[24] P. A. Montgomery, G. K. Scott, M. C. Luce, M. Kaufmann, and C. C. Benz, *Breast Cancer Res. Treat.* **26,** 181 (1993).

[25] X. Liang, B. Lu, G. K. Scott, C.-H. Chang, M. A. Baldwin, and C. C. Benz, *Mol. Cell. Endocrinol.* **146,** 151 (1998).

insights into the structural changes induced by oxidant stress of zinc finger transcription factors.[26]

Monitoring Protein Oxidation by Electrospray Ionization Mass Spectrometry

Most oxidation processes in proteins involve a change in molecular mass, such as the addition of 16 Da for conversion of a methionine residue to methionine sulfoxide or 32 Da for formation of the sulfone. Therefore mass spectrometry (MS) is well suited to monitor protein oxidation. Oxidation of cysteine by disulfide formation may be regarded as a special case as the mass difference between two reduced cysteine residues and the disulfide-linked product cystine is only 2 Da, and only for smaller proteins can such a difference be measured unambiguously. If necessary, chemical alkylation of free thiols may be used to give substantially larger mass differences between the reduced and oxidized states. This provides the additional advantage that, once alkylated, the thiol groups are resistant to further oxidation. Electrospray ionization (ESI)[27] is particularly advantageous as it ionizes directly from solution, is fully compatible with the analysis of involatile biomolecules including peptides and proteins, and can be interfaced with high-performance liquid chromatography (HPLC). Thus, not only may specific cysteine residues that are more or less susceptible to oxidation be identified by proteolytic digestion of a partially oxidized protein and analysis of individual peptides by HPLC-MS, cysteines modified by addition of the oxidizing moiety may also be revealed. ESI-MS carried out under nondenaturing conditions is finding increasing application for monitoring the biophysical properties of proteins and the formation of noncovalent complexes, allowing additional information about the structural consequences of protein oxidation to be derived.[28,29] MS also gives an unambiguous measure of stoichiometry for complex formation, for example, it confirmed that the DBD of vitamin D binds to DNA as a dimer, and that this assembly requires a total of four Zn^{2+} ions.[30]

Several MS studies of the DBDs of ER and other nuclear hormone receptors have concentrated on their metal-binding properties, as these DBDs possess Cys_4-liganded zinc fingers that are essential for normal activity. Thus ESI-MS showed that Cu^{2+} can replace Zn^{2+} in the ER-DBD, possibly with higher affinity and with the higher stoichiometry of two Cu^{2+} ions per zinc finger,[20] and the glucocorticoid receptor DNA-binding domain (GR-DBD) can bind two Cd^{2+} ions as

[26] R. M. Whittal, C. C. Benz, G. Scott, J. Semyonov, A. L. Burlingame, and M. A. Baldwin, *Biochemistry* **39**, 8406 (2000).

[27] J. B. Fenn, M. Mann, C. K. Meng, S. F. Wong, and C. M. Whitehouse, *Science* **246**, 64 (1989).

[28] J. A. Loo, *Mass Spectrom. Rev.* **16**, 1 (1997).

[29] A. M. Last and C. V. Robinson, *Curr. Opin. Chem. Biol.* **3**, 564 (1999).

[30] T. D. Veenstra, L. M. Benson, T. A. Craig, A. J. Tomlinson, R. Kumar, and S. Naylor, *Nat. Biotechnol.* **16**, 262 (1998).

an alternative to Zn^{2+}.[31] The study by Veenstra *et al.,* referred to above, showed that either zinc or cadmium allowed the vitamin D receptor DBD to bind to DNA.[30] More recently, we demonstrated that impaired DNA binding of ER resulting from zinc finger oxidation could be monitored by ESI-MS, and that the susceptibility of the second zinc finger to oxidation was enhanced relative to the first.[26] Other zinc finger proteins studied by mass spectrometry include the HIV type 1 nucleopcapsid protein P7 (NcP7).[32] Unlike the two Cys_4 zinc fingers of ER, NcP7 contains two Cys_3His zinc finger motifs. Nevertheless, there are marked similarities between the two proteins, in that the C-terminal zinc finger demonstrated a higher susceptibility toward oxidation in both cases. When NcP7 was oxidized by 2,2′-dithiopyridine, oxidation intermediates were observed in which either one or two cysteines were linked to thiopyridine, with the corresponding loss of one or two Zn^{2+} ions. By contrast, the more reactive disulfiram gave only a single adduct with the loss of a single Zn^{2+}.[32] Similarly, we have observed an intermediate formed by the addition of *S*-nitrosylglutathione to cysteine in ER-DBD when this material is used as an oxidizing agent.[32a]

Other established analytical methods can provide valuable complementary information. These include electrophoretic mobility shift assays (EMSAs), metal ion titrations, and various spectroscopic techniques including absorption spectroscopy, circular dichroism, and nuclear magnetic resonance. Absorption spectroscopy established that for peptide analogs of a simian NcP, Zn^{2+} is complexed in a zinc finger with tetrahedral geometry; Ni^{2+} also gives a native-like tetrahedral complex, whereas replacement of Zn^{2+} by Co^{2+} or Cd^{2+} results in square planar structures that are more sensitive to oxidation.[33]

Equipment for Electrospray Ionization-Mass Spectrometry and High-Performance Liquid Chromatography-Mass Spectrometry

ESI-MS can be effective with several different types of mass spectrometer. Initial studies of the purity and of the oxidation and reduction of ER-DBD employed a quadrupole mass spectrometer of relatively modest performance. In other studies we used an orthogonal acceleration time-of-flight mass spectrometer (oa-ToFMS) of higher resolving power and higher sensitivity.[26] Magnetic sector instruments have also been employed, but for ESI-MS these are mostly being superseded by quadrupoles, ion traps, and oa-TOFs or Qoa-TOFs. The purity and oxidation state

[31] H. E. Witkowska, C. H. L. Shackleton, K. Dahlman-Wright, J. Y. Kim, and J.-Å. Gustaffson, *J. Am. Chem. Soc.* **117,** 3320 (1995).

[32] Y. Hathout, D. Fabris, M. S. Han, R. C. Sowder II, L. E. Henderson, and C. Fenselau, *Drug Metab. Dispos.* **24,** 1395 (1996).

[32a] J. E. Meza, C. C. Benz, G. K. Scott, and M. A. Baldwin, in preparation (2002).

[33] X. Chen, M. Chu, and D. P. Giedroc, *J. Biol. Inorg. Chem.* **5,** 93 (2000).

of a peptide or protein can be determined by ESI from acidified solution, infusing from a syringe pump a 5–10 μM solution of the protein or peptide in H_2O–methanol (1 : 1, v/v) acidified with 1% (v/v) acetic acid, directly into the ESI source. Alternatively, an ESI-MS instrument is readily interfaced with the HPLC, perhaps using a 1-mm i.d. reversed-phase (RP) column with 300-Å pore size and 5-μm particle size. This allows samples to be desalted by injection onto the HPLC before being transferred to the mass spectrometer. Typically solvent A is water with 0.06% (v/v) trifluoroacetic acid (TFA) to control pH, and solvent B is 80% (v/v) acetonitrile with 0.052% (v/v) TFA at a flow rate of 50 μl/min. Mass spectrometric sensitivity may be adversely affected by TFA, in which case this may be replaced with formic acid, and acetonitrile can be replaced by a 5 : 2 mixture of ethanol and n-propanol.[34] The column effluent should pass through a UV flow cell and then, depending on the optimum flow rate for the ESI source, may be split such that ~10% passes into the ESI source of the mass spectrometer and the balance can be collected in fractions for later studies. Alternatively, a capillary column operating at lower flow rate will allow the entire effluent to be transferred without splitting. Mass spectra should be recorded and accumulated continuously throughout the running of each sample. The UV chromatogram recorded at 215 nm can then be compared with the total ion current (TIC) trace obtained from the mass spectrometer. ESI spectra corresponding to peaks in the UV and TIC traces can be selected manually for averaging and deconvolution, using software provided by the mass spectrometer manufacturer.

The conventional RP-HPLC conditions of low pH are denaturing and do not support the formation or maintenance of noncovalent complexes. To study a metal-loaded holoprotein such as a zinc finger protein at neutral pH, it must be introduced via a syringe pump. The buffer should contain 5–10 μM protein in 2.5 mM ammonium acetate solution (pH 7.4)–10% (v/v) methanol, and varying concentrations of Zn^{2+} diluted from zinc(II) sulfate stock solution. A reducing agent such as 100 μM DTT may be included to inhibit aerobic oxidation. The methanol is needed to enhance the spray formation. Ammonium acetate is favored because it is volatile and will not contaminate the ESI source with salt deposits; ammonium formate or bicarbonate can also be used at either lower or higher pH respectively. Each solution should be infused to establish a steady state, and then the spectra can be recorded and averaged over a period of a few minutes.

Interpretation of Mass Spectra

Under normal ESI-MS conditions, peptides and proteins become protonated at the basic residues, that is, arginine, lysine, and histidine and the amino terminus,

[34] K. F. Medzihradszky, D. A. Maltby, S. C. Hall, C. A. Settineri, and A. L. Burlingame, *J. Am. Soc. Mass Spectrom.* **5**, 350 (1994).

giving a distribution of multiply charged species. This is observed as a series of peaks, separated according to their mass/charge (m/z) ratios. This series can be mathematically deconvoluted to give a single profile for the molecular weight. Because of heavy isotopes such as ^{13}C, ^{15}N, and ^{18}O, each molecule gives a cluster of peaks with individual components separated by 1 Da. These are easily separated for small peptides, the "monoisotopic" molecular weight being defined by the mass of the first peak in the cluster. For larger peptides and proteins these may not be resolved, in which case an "average" mass is derived from the overall profile of the cluster.[35] Quantitative data can be derived from the raw mass spectra by addition of the signal intensities from all of the individual charge states, or from peak intensities in the deconvoluted spectra. For recombinant ER-DBD of $M_r \sim 11.2$ kDa, the isotopic components at adjacent masses could not be resolved by either mass spectrometer, and therefore the measured mass of each peak, representing an average of the distribution of all isotopes present, was compared with the value calculated from average atomic masses. For synthetic peptides and peptides originating from tryptic digestion, the isotopic clusters were resolved by the oa-TOF mass spectrometer to give separate peaks, giving the monoisotopic molecular mass.

Recombinant Estrogen Receptor DNA-Binding Domain

Using well-established techniques, the 84-residue DNA-binding domain of the estrogen receptor (ER-DBD) with an N-terminal histidine tag was expressed in *Escherichia coli* as a protein of 99 amino acids (Fig. 1), and purified by immobilized metal ion affinity chromatography (IMAC). Activity was confirmed by a positive EMSA response for binding to the cognate ERE. Some samples were subjected to chemical treatments such as oxidation, reduction, or alkylation before separation and analysis by HPLC-MS. Other experiments were conducted by direct infusion of samples into the ESI mass spectrometer without HPLC separation. The UV chromatogram and total ion current trace for a typical HPLC separation showed a single chromatographic peak, but ESI-MS revealed that this contained two coeluting components. The measured M_r of the lower mass component (11,232.7 Da) was close to that calculated for fully reduced ER-DBD (11,232.9 Da), whereas the mass of the second component was higher by ~ 14 Da. Digestion with trypsin and HPLC-MS separation identified the adduct as an N-terminal methylation, presumably occurring during protein expression.

To confirm the presence of nine cysteine residues, ER-DBD was reduced with 2 mM DTT and reacted with 6 mM iodoacetic acid (IAA) for 1 hr at 4° to

---

[35] S. A. Carr, A. L. Burlingame, and M. A. Baldwin, *in* "Mass Spectrometry in Biology and Medicine" (A. L. Burlingame, S. A. Carr, and M. A. Baldwin, eds.), p. 553. Humana Press, Totowa, NJ, 2000.

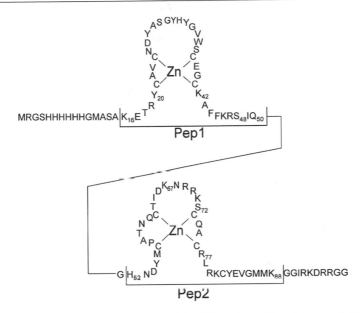

FIG. 1. Recombinant histidine-tagged ER–DBD of 99 amino acids and two synthetic peptides [K16–Q50 (Pep1) and H52–K88 (Pep2)] prepared for these studies. Amino acid residues at the trypsin cleavage sites that give peptides that identify oxidation of the individual fingers are numbered.

carboxymethylate the thiol groups, each of which should increase the molecular mass by 58 Da. A mixture of products was obtained, the most abundant of which corresponded to the alkylation of eight and nine cysteines. A minor degree of over-reaction gave other, weaker peaks for the addition of 10 and 11 carboxymethyl groups. As anticipated, ER-DBD that had been oxidized with hydrogen peroxide and then treated with IAA proved to be resistant to carboxymethylation. Thus it was confirmed that ER-DBD reduced with DTT had the anticipated number of free thiols.

Oxidation of Estrogen Receptor DNA-Binding Domain

Initial experiments on the oxidation of recombinant ER-DBD were carried out with either 5 mM hydrogen peroxide or 5 or 25 mM diamide for periods extending from 25 min to overnight. Diamide [$(CH_3)_2NCON = NCON(CH_3)_2$] is a reagent that specifically oxidizes cysteine to cystine.[36] The extent and nature of the oxidation were evaluated by HPLC-MS of the intact protein without digestion to separate

---

[36] N. S. Kosower, E. M. Kosower, and B. Wertheim, *Biochem. Biophys. Res. Commun.* **37,** 593 (1969).

peptides. ER-DBD contains nine cysteines and five methionines. Treatment of ER-DBD with hydrogen peroxide induced two different types of oxidation: (1) the loss of one or more pairs of hydrogen atoms as disulfide bonds were formed, and (2) the addition of multiple oxygen atoms, almost certainly to form methionine sulfoxide. Peroxide treatment at 5 mM for 25 min removed only four hydrogens and added zero, one, or two oxygens, whereas overnight treatment removed six hydrogens and added one to five oxygens. After oxidation for 25 min, further treatment with 50 mM DTT reduced the disulfide bonds but did not reduce the methionine sulfoxides. Treatment with 5 mM diamide for 25 min partially oxidized ER-DBD by the removal of an average of four hydrogen atoms, whereas 25 mM diamide removed eight hydrogens but also gave rise to a new species corresponding to the addition of a diamide molecule. Treatment with DTT reduced the cystines back to cysteine and removed the diamide adduct. Unlike peroxide, diamide induced little methionine sulfoxide formation.

Subsequent titrations with oxidizing agents were followed by digestion with trypsin so that differential oxidation of different regions of ER could be monitored. Initially the resulting peptides were found to be prone to oxidation during digestion and chromatography, probably because of dissolved air, even when the digestion time was limited to 3 min. To minimize unwanted oxidation, dissolved air was removed from all solutions by vacuum filtration immediately before their use and all solutions were maintained under a positive pressure of helium. Oxidation was carried out by incubating 10 μM reduced ER-DBD (previously stored in 100 μM DTT) in 11.5 mM ammonium acetate, pH 7.4, and either 60 μM zinc(II) sulfate or 100 μM Na$_2$EDTA with various concentrations of either hydrogen peroxide or diamide. The oxidation was allowed to proceed for 15 min before analysis by rapid tryptic digestion. To achieve faster enzymatic digestion of ER-DBD protein than would be obtained in solution, the protein was injected onto a Porozyme packed immobilized trypsin column (PE Biosystems, Framingham, MA) at a flow rate of 50 μl min^{-1}. The digestion buffer was 70 mM ammonium bicarbonate–5% (v/v) acetonitrile, pH 8.0, from which dissolved air had been removed by vacuum filtration followed by helium purging and continuous exposure to helium at 3 lb in^{-2}. The pH was checked and adjusted with acetic acid or ammonium hydroxide before each analysis. The enzyme column was connected in series via a Rheodyne (Rohnert Park, CA) switching valve to a Vydac (Hesperia, CA) 150 × 1 mm C$_{18}$ microbore column. After 3 min the valve was switched to allow analysis of the resulting peptides by HPLC-MS at a flow rate of 50 μl min^{-1}. The enzyme column was rinsed after each analysis with at least 10 ml of digest buffer. For this HPLC analysis, solvent A was 0.1% (v/v) formic acid and solvent B was 5 : 2 (v/v) ethanol–n-propanol with 0.05% (v/v) formic acid; both solvents were vacuum filtered and purged with helium and then continuously exposed to helium at 3 lb in^{-2}. After holding the HPLC at 5% (v/v) solvent B for 2 min a linear gradient was run from 5 to 60% solvent B in 30 min. The eluted peptides were passed through the

UV flow cell with detection at 215 nm and the flow was split to transfer $\sim$10% to the ESI source.

Calculation of Degree of Oxidation

Oxidation of cysteine residues in a zinc finger results in the formation of disulfide bonds with the loss of two protons for each disulfide formed. For cysteine-containing peptides in a tryptic digest, the degree of peptide oxidation can be calculated by determining the amount of oxidized peptide present (M_r lower than that of the reduced peptide by $2n$, where n = the number of disulfides), compared with the amount of reduced peptide present, based on the heights of the peaks in the mass spectrum. If the peptides are not separated chromatographically, the isotopic patterns for the reduced and oxidized forms will overlap. For a peptide containing two cysteines the peak height of the reduced form can be corrected by subtracting the isotopic contribution of the third isotopic peak of the oxidized form. Because ER-DBD has no trypsin cleavage sites in zinc finger 1, the entire four-cysteine structure was observed intact, containing none, one, or two disulfide bonds. Therefore, calculation of the amount of singly oxidized peptide (one disulfide bond) was corrected by subtracting the contribution of the third isotopic peak of the doubly oxidized peptide (two disulfide bonds); the peak height of the reduced peptide was corrected for the contribution of the fifth isotope of the doubly oxidized peptide and the third isotope of the singly oxidized peptide. As a result of incomplete digestion, zinc finger 1 was actually represented by two overlapping peptides corresponding to amino acids 17–47 and 20–42. The results from these two peptides were averaged to give the percent oxidation of zinc finger 1. Baseline levels of oxidation observed in the absence of any added oxidant were subtracted from the data.

The target peptides were identified from an analysis of the multiply charged ions in the HPLC-MS spectra, from which the relative abundances of the reduced and oxidized forms were calculated. For zinc finger 2 of ER-DBD there were four possible trypsin cleavage sites (Fig. 1). Again the digestion was incomplete, giving rise to additional peptides. The data from two peptides spanning amino acids 48–67 and 48–69 were averaged to give the percent oxidation of the first half (ZF2$_{C1-2}$). Although two peptides were also observed representing the second half of zinc finger 2, that is, amino acids 71–77 and 72–77, the peak for amino acids 71–77 was usually weak and in a more noisy region of the spectrum. Thus, only the amino acid 72–77 peptide fragment was used in the calculation of the percent oxidation of the second half of finger 2 (ZF2$_{C3-4}$). Taking account of the isotope corrections, the amount of oxidation occurring under different conditions was calculated for different concentrations of oxidant. This revealed only low levels of oxidation of finger 1 under conditions that caused extensive oxidation of finger 2; furthermore, the oxidation of finger 2 was asymmetric, with a much higher level of disulfide formation between cysteines 3 and 4 (Cys3 and Cys4) than between cysteines 1 and 2 (Cys1 and Cys2). For example, 2 mM peroxide was sufficient to oxidize 100% of ZF2$_{Cys3-4}$, whereas 7.5 mM was not sufficient to fully oxidize any of the other pairs of cysteines. Similar trends were observed with 0–500 μM diamide in the presence of zinc ions. Interestingly, the removal of zinc by EDTA increased the sensitivity of finger 1 to oxidation such that both fingers were oxidized to the same degree. The observation that finger 1 is more resistant to oxidation than finger 2 only when zinc ions are present and not in the absence of zinc is consistent with structural studies by nuclear magnetic resonance (NMR), showing that zinc in finger 2 is less well coordinated than in finger 1.[37]

If the tetrahedral arrangement of the cysteine residues was maintained during oxidation, within a single finger there would be equal cross-linking of peptides by disulfide formation between Cys1–Cys2/Cys3–Cys4, Cys1–Cys3/Cys2–Cys4, and Cys1–Cys4/Cys2–Cys3. Statistically, a tryptic digest of finger 2 should reveal twice the abundance of the cross-linked peptides, such as peptides 48–67/72–77 compared with the separate oxidized peptides 48–67 and 72–77. However, the signal for the cross-linked peptide was of much lower intensity ($\sim$10%) than the separate peptides, suggesting that oxidation disrupts the geometry and expels the Zn^{2+} ion from this finger before the disulfide bonds are formed. It was not possible to make a direct comparison with the behavior of finger 1 as this was not cleaved by trypsin, but HPLC separation of the fully oxidized species gave three peaks of approximately equal abundance, which are likely attributable to the three different disulfide-bonded isomers. This would indicate that the finger structure was maintained more strongly during oxidation of finger 1 than finger 2.

Methionine Oxidation: Absence of Protection Against Thiol Oxidation

It has been proposed that methionine residues at strategic sites in proteins might act as scavengers for ROS and thereby protect cysteine thiols from oxidation.[38] Oxidation of methionine to methionine sulfoxide was monitored in the same experiment as a function of oxidant concentration by observing the formation of peaks corresponding to the addition of one oxygen atom, that is, 16 Da. Three methionine residues are located in or adjacent to zinc finger 2 as follows: Met-56, immediately adjacent to the first cysteine of finger 2, and Met-86 and Met-87 situated just after the C-terminal helix of this finger. A peptide spanning Ser-48 to

---

[37] A. Wilkström, H. Berglund, C. Hambraeus, S. van den Berg, and T. Härd, *J. Mol. Biol.* **290**, 96 (1999).

[38] R. L. Levine, L. Mosoni, B. S. Berlett, and E. R. Stadtman, *Proc. Natl. Acad. Sci. U.S.A.* **93**, 15036 (1996).

Lys-67 revealed a background level of ~3–4% oxidation for Met-56, which did not increase at all on exposure to up to 7.5 mM hydrogen peroxide or 500 mM diamide, even though these amounts were sufficient to completely disrupt and oxidize finger 2. Furthermore, the other methionine residues (Met-86 and Met-87), situated just outside the C terminus of the finger 2 helix, showed limited oxidation totaling less than 10% above baseline on exposure to up to 7.5 mM peroxide and no significant oxidation on exposure to diamide. Thus, contrary to the suggestion that methionine might protect against oxidation, the zinc finger cysteines are more susceptible to oxidation than any of the three methionine residues contained in the ER-DBD.

Stoichiometry of Zinc Binding to Estrogen Receptor DNA-Binding Domain Monitored by Electrospray Ionization-Mass Spectrometry

Zinc binding by recombinant ER-DBD was monitored by ESI-MS in a weak buffer at pH 7.4, conditions designed to allow observation of intact noncovalent complexes and reveal the stoichiometry of any complex formed between the protein and zinc ions. The strongest peak representing approximately two-thirds of total protein molecules was free of zinc, whereas the remaining third added a single Zn^{2+} cation. There was no observable addition of two Zn^{2+} ions. No zinc had been added at any stage in the expression or purification, and therefore it was concluded that the protein had scavenged the zinc from the cells in which it was expressed. The presence of less than the stoichiometric amount of zinc was consistent with a circular dichroic (CD) spectrum showing a lower than normal degree of α- helicity compared with fully zinc-loaded ER-DBD, which shifted to higher helix content on addition of zinc. The addition to the buffer of 2 mol of Zn^{2+} per mole of reduced ER-DBD showed strong formation of the complex ER-DBD · Zn^{2+} and a lesser amount of ER-DBD · $2Zn^{2+}$. The CD spectrum confirmed that this amount of Zn^{2+} was sufficient to restore the α-helicity to its normal level. Exposure of ER-DBD to 5 mol of Zn^{2+} per mole of protein caused the ER-DBD · $2Zn^{2+}$ peak to become the most intense in the mass spectrum. Weak binding of a third Zn^{2+} ion was also seen to occur, possibly by binding to the histidine tag, or simply as a nonspecific effect of ESI-MS. In contrast, ER-DBD oxidized with either peroxide or diamide could bind only low levels of Zn^{2+} in a nonspecific manner, again perhaps involving the histidine tag.

The precision of the ESI-MS mass measurements for ER-DBD · $2Zn^{2+}$ was sufficient to establish that the binding of two Zn^{2+} ions involved the elimination of four protons from ER-DBD, that is, on average each Zn^{2+} ion added only 63.5 Da to the mass, even though the average atomic mass of zinc is 65.5 Da. Four cysteine residues participate in stabilizing each of the two zinc fingers, and thus it can be concluded that each finger involves two thiolate anions forming ionic bonds and two thiols forming coordinate bonds. A similar observation by ESI-MS was also reported for a Cys_3His zinc finger.[39]

Cooperative Action of Two Zinc Fingers in Stabilizing Structure

To probe the behavior of each zinc finger of ER-DBD in isolation, both were synthesized as separate peptides (Pep1 and Pep2; Fig. 1) and probed by EMSA, ESI-MS, and CD. Although we determined by EMSA that neither peptide was able to bind to ERE, ESI-MS showed that with a peptide: Zn^{2+} molar ratio of 1 : 1, each peptide strongly bound a single Zn^{2+} ion. The effect of Zn^{2+} on the secondary structures of these peptides was monitored by CD spectroscopy and compared with the behavior of ER-DBD. We had previously established that salt at near-physiological concentration was necessary for ER-DBD to maintain its structure and therefore the same was assumed to be true of the individual peptides. Buffers containing 100 mM sodium fluoride were used for all CD experiments, rather than sodium chloride, as the chloride ion absorbs UV radiation below 200 nm. The CD spectrum of a typical random coil structure should show a minimum at $\sim$195 nm and relatively little signal at $\sim$222 nm, whereas α-helical proteins generally show a maximum at $\sim$195 nm and two minima at 208 and 222 nm that correspond to $\pi-\pi^*$ and $n-\pi^*$ electronic transitions, respectively. In the absence of Zn^{2+} both synthetic peptides showed minima strongly blue-shifted from the α-helix value to $\sim$198 nm, and weak signals at 222 nm. This was indicative of low α-helix content, which was confirmed by secondary structural analysis, giving only 5% α-helix for Pep1 and 2% for Pep2. Furthermore, there were only weak increases in α-helical structure on the addition of one molar equivalent of Zn^{2+}, particularly for Pep2, for which the α-helix increased to only 6% despite the relatively strong Zn^{2+}-binding properties of both peptides demonstrated by ESI-MS. For neither peptide was the α-helical content comparable to the 23% determined for ER-DBD itself.

Further Experimental Considerations

Quantitative studies such as those described here involve a comparison between peptides that contained cysteines that are either oxidized or reduced. It was found that if peptides containing free cysteines are kept in solution for any length of time, such a procedure may be problematical as dissolved air during trypsin digestion or HPLC purification may cause additional oxidation. An alternative is to alkylate the free thiols with a reagent such as IAA, iodoacetamide, or N-ethylmaleimide before any processing or manipulation, which will prevent further reaction. A number of authors have adopted a strategy of alkylating free thiols with one reagent and then reducing the oxidized cysteines and alkylating with a second agent, giving

[39] D. Fabris, J. Zaia, Y, Hathout, and C. Fenselau, *J. Am. Chem. Soc.* **118,** 12242 (1996).

two groups of differently alkylated cysteines that can be distinguished by their characteristic masses. An affinity tag may be attached to the first alkylating agent to selectively purify cysteine-containing peptides from a tryptic digest.[40] However, reduction and alkylation would destroy other disulfide-bonded species, such as the S-nitrosylglutathione adducts that we have observed as intermediates from oxidation by S-nitrosoglutathione. Nevertheless, in studying a zinc finger protein extracted from a biological material such as a human tumor, the risk of unwanted oxidation is much greater and an initial alkylation step should be regarded as mandatory. It is likely that a successful purification of a specific target protein will involve an immunoaffinity purification; therefore it is important to establish that alkylation does not diminish the affinity of the antibody for the protein.

The products of other oxidation processes may be sought, and it should be noted that although disulfide formation is the common product of cysteine oxidation, at least five other more extensively oxidized products could potentially be formed, although not all occur in biological samples. Xu and Wilcox have described three products of oxidation of a single cysteine formed by introduction of one, two, or three oxygen atoms: cysteine sulfenic acid, sulfinic acid, and sulfonic acid (cysteic acid), respectively. Also, a disulfide can add one oxygen to give thiosulfinate or two oxygens to give thiosulfonate.[41] Other oxidative processes can proceed through nitric oxide and S-nitrosoglutathione, and these may play a role in the regulation of gene expression.[42] Consequently it may be relevant to search for corresponding cysteine derivatives in the zinc fingers of other transcription factors.

Acknowledgments

We thank Drs. J. Meza and R. M. Whittal for valuable contributions. Mass spectrometry was carried out in the UCSF Mass Spectrometry Facility, supported by NIH NCRR 01614. This work was also supported in part by NIH Grant CA71468 as well as by Hazel P. Munroe (Buck Institute) and Janet Landfear (Mt. Zion Health Systems) memorial funds.

[40] T.-Y. Yen, R. K. Joshi, H. Yan, N. O. L. Seto, M. M. Palcic, and B. A Macher, *J. Mass Spectrom.* **35,** 990 (2000).
[41] H. E. Marshall, K. Merchant, and J. S. Stamler, *FASEB J.* **14,** 1889 (2000).
[42] Y. Xu and D. E. Wilcox, *J. Am. Chem. Soc.* **120,** 7375 (1998).

[7] Quantification of Intracellular Calcineurin Activity and H_2O_2-Induced Oxidative Stress

By TIFFANY A. REITER and FRANK RUSNAK

Introduction

Oxidative stress has been loosely defined as the exposure of cells to increased concentrations of reactive oxygen or reactive nitrogen species (ROS and RNS, respectively). Exposure of cells to ROS and RNS can lead to change in the activities of redox-sensitive proteins. Cells protect themselves from oxidative stress by several mechanisms. These include the maintenance of a redox-buffered intracellular environment[1]; the production of detoxifying enzymes such as catalase,[2] superoxide dismutase,[3,4] and superoxide reductase[5]; and the presence of specific ROS and RNS sensors that can activate various defense mechanisms allowing the organism to survive and adapt to the stress condition.[6] In prokaryotes, many redox-sensitive proteins are transcription factors that regulate genes involved in anaerobic and aerobic metabolism.

One of the best strategies to quantitatively measure oxidative stress brought on by specific ROS and RNS is an activity measurement of a redox-sensitive enzyme. For example, the iron–sulfur protein aconitase is inactivated by nitric oxide, peroxynitrite, and superoxide and is used to quantify production of these RNS and ROS.[7] Quantitative measurements of oxidative stress in the intact cell, tissue, and organism, however, are lacking due to the dearth of specific assays for various ROS and RNS species *in vivo*. In this chapter, we describe a sensitive method to quantify the effects of hydrogen peroxide-induced oxidative stress in an intact cell, using a cell culture system and a luciferase reporter plasmid as an indirect measure of intracellular calcineurin activity.

---

[1] F. Rusnak and T. Reiter, *Trends Biomed. Sci.* **25,** 527 (2000).

[2] M. R. Murthy, T. J. Reid III, A. Sicignano, N. Tanaka, and M. G. Rossmann, *J. Mol. Biol.* **152,** 465 (1981).

[3] J. Selverstone-Valentine, L. M. Ellerby, J. A. Graden, C. R. Nishida, and E. B. Gralla, in "Bioinorganic Chemistry: An Inorganic Perspective of Life" (D. P. Kessissoglou, ed.), p. 77. Kluwer, Dordrecht, 1995.

[4] I. Fridovich, *Annu. Rev. Biochem.* **64,** 97 (1995).

[5] F. E. Jenney, Jr., M. F. Verhagen, X. Cui, and M. W. Adams, *Science* **286,** 306 (1999).

[6] C. E. Bauer, S. Elsen, and T. H. Bird, *Annu. Rev. Microbiol.* **53,** 495 (1999).

[7] J. C. Drapier and J. B. Hibbs, Jr., *J. Clin. Invest.* **78,** 790 (1986).

METHODS IN ENZYMOLOGY, VOL. 353

Copyright 2002, Elsevier Science (USA).
All rights reserved.
0076-6879/02 $35.00

Calcineurin is a serine/threonine protein phosphatase found in most eukaryotic organisms including yeast, protozoa, and mammals.[8–10] Calcineurin consists of a 59-kDa catalytic subunit (calcineurin A) and a 19-kDa regulatory subunit (calcineurin B). Calcineurin is activated by calmodulin in the presence of Ca^{2+}. In resting cells, in which the intracellular $[Ca^{2+}]$ is low, calcineurin exists in an inactive state. On activation of signal transduction pathways that lead to a sustained rise in intracellular Ca^{2+}, Ca^{2+} binds to calmodulin, which in turn binds to calcineurin, resulting in a 20-fold increase in phosphatase activity.

This model of calcineurin regulation has existed for several years but data from several groups have led to the observation that calcineurin may be regulated by H_2O_2-induced oxidative stress. Yu et al. first demonstrated the redox sensitivity of calcineurin activity in vitro.[11,12] Subsequent experiments showed that calcineurin lost activity on aerobic exposure to Ca^{2+}/calmodulin in vitro and that calcineurin activity in crude cell lysate was inhibited by hydrogen peroxide, dithionite, and high concentrations of dithiothreitol (DTT).[13–16] Although there is good evidence for redox regulation of calcineurin in these in vitro systems, the study of redox-regulated enzymes such as calcineurin outside the cell is fraught with difficulties. The reducing environment of the cytosol has an estimated redox potential of -230 to -255 mV[17–20] and is buffered by redox-active species such as protein thiols, glutathione, pyridine nucleotides, and other redox-active metabolites. Cell lysis disrupts this reducing environment. As a result, lysis buffers rarely mimic the intracellular redox potential, thereby exposing redox-sensitive proteins and enzymes to oxygen and a different redox environment. In an attempt to circumvent this, we have developed an assay that provides an indirect measure of calcineurin activity in intact cells. Using this assay, we have demonstrated that H_2O_2, but not other physiological oxidants, leads to a dose-dependent loss of calcineurin activity. Thus, this assay also provides a sensitive and quantitative measure of H_2O_2-induced oxidative stress.

[8] F. Rusnak and P. Mertz, *Physiol. Rev.* **80**, 1483 (2000).

[9] J. Aramburu, A. Rao, and C. B. Klee, *Curr. Topics Cell. Regul.* **36**, 237 (2000).

[10] S. Shenolikar and A. C. Nairn, *Adv. Second Messenger Phosphoprotein Res.* **23**, 3 (1991).

[11] L. Yu, J. Golbeck, J. Yao, and F. Rusnak, *Biochemistry* **36**, 10727 (1997).

[12] L. Yu, A. Haddy, and F. Rusnak, *J. Am. Chem. Soc.* **117**, 10147 (1995).

[13] X. Wang, V. C. Culotta, and C. B. Klee, *Nature (London)* **383**, 434 (1996).

[14] M. Carballo, G. Márquez, M. Conde, J. Martin-Nieto, J. Monteseirin, J. Conde, E. Pintado, and F. Sobrino, *J. Biol. Chem.* **274**, 93 (1999).

[15] T. A. Reiter, R. T. Abraham, M. Choi, and F. Rusnak, *J. Biol. Inorg. Chem.* **4**, 632 (1999).

[16] A. Ferri, R. Gabbianelli, A. Casciati, E. Paolucci, G. Rotilio, and T. Carrï, *J. Neurochem.* **75**, 606 (2000).

[17] D. Williamson, P. Lund, and H. Krebs, *Biochem. J.* **103**, 514 (1967).

[18] R. L. Veech, R. Guynn, and D. Veloso, *Biochem. J.* **127**, 387 (1972).

[19] H. F. Gilbert, *Adv. Enzym. Relat. Areas Mol. Biol.* **63**, 69 (1990).

[20] C. Hwang, A. Sinskey, and H. Lodish, *Science* **257**, 1496 (1992).

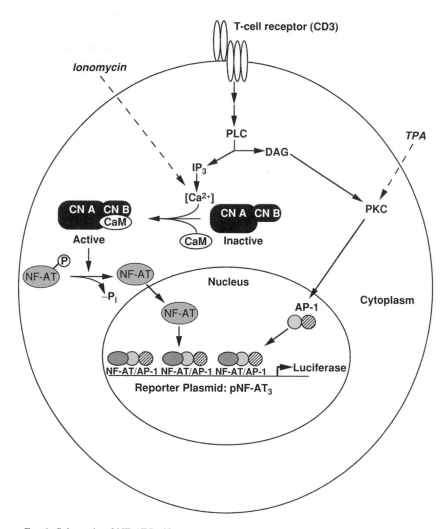

FIG. 1. Schematic of NF-AT/luciferase reporter plasmid assay. T cell activation results in an increase in two secondary messengers, diacylglycerol (DAG) and inositol 1,4,5-trisphosphate (IP_3). IP_3 produces a rise in intracellular Ca^{2+}, which binds to calmodulin to activate calcineurin phosphatase activity. Calcineurin dephosphorylates the cytoplasmic subunit of NF-AT, which undergoes nuclear translocation. DAG stimulates the activity of protein kinase C (PKC), which results in new synthesis and processing of the transcription factor AP-1. Treatment with ionomycin and TPA can be used to mimic the upstream signaling events that normally activate calcineurin and PKC. The transcription factors NF-AT and AP-1 bind to the promoter and drive transcription of luciferase. Luciferase activity, therefore, provides an indirect and quantitative measure of calcineurin phosphatase activity inside intact transfected T cells. CN A, calcineurin A; CN B, calcineurin B; CaM, calmodulin; NF-AT, nuclear factor of activated T cells; PLC, phospholipase C; AP-1, Fos–Jun heterodimer; TPA, tetradecanoylphorbol acetate; pNF-AT$_3$, luciferase reporter plasmid pNF-AT$_3$. [Reprinted with permission from T. A. Reiter, R. T. Abraham, M. Choi, and F. Rusnak, *J. Biol. Inorg. Chem.* **4**, 632 (1999), copyright © 1999, Society of Biological Inorganic Chemistry.]

Assay Methods

Principle

In T lymphocytes, calcineurin dephosphorylates the transcription factor NF-AT (nuclear factor of activated T lymphocytes) (Fig. 1). On dephosphorylation, NF-AT undergoes nuclear translocation, where it binds to specific promoter sequences and activates gene transcription. A reporter plasmid (Fig. 2) containing three tandem copies of the murine interleukin 2 (IL-2) NF-AT/AP-1 promoter

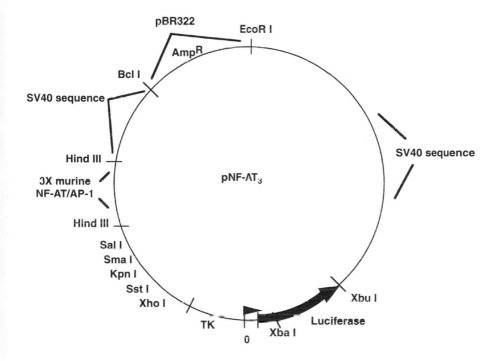

Murine NF-AT/AP-1

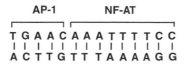

```
    AP-1           NF-AT
 ┌───────┐   ┌───────────────┐
 T G A A C A A A T T T T C C
 | | | | | | | | | | | | | |
 A C T T G T T T A A A A G G
```

FIG. 2. Plasmid map of pNF-AT$_3$. The plasmid pNF-AT$_3$ contains the luciferase gene upstream of three tandem NF-AT/AP-1 murine IL-2 promoter sequences. Restriction endonuclease recognition sites, the transcription start site (arrowhead), the ampicillin resistance gene (Amp^R), the luciferase open reading frame, and pBR322- and simian virus 40 (SV40)-derived sequences are labeled. The murine NF-AT/AP-1 promoter sequence used in pNF-AT$_3$ is shown below the plasmid map.

sequence upstream of the luciferase gene is transfected into the human T lymphocyte cell line Jurkat, a cell line that expresses both endogenous calcineurin and NF-AT. Treatment of the transfected cells with tetradecanoylphorbol acetate (TPA, a phorbol ester) activates AP-1 (Fos–Jun heterodimer) whereas ionomycin, a calcium ionophore, mobilizes Ca^{2+}, thereby activating calcineurin (Fig. 1). Activated cells can be treated with a redox reagent and harvested for luciferase measurements to provide a specific and sensitive assay for calcineurin activity in intact Jurkat cells.

Reagents

Plasmids

The luciferase reporter plasmid pNF-AT3-luc (pNF-AT$_3$) contains three copies of the murine NF-AT/AP-1 mixed promoter sequence at approximately -287 bp of the murine IL-2 promoter upstream of the luciferase gene.[21] The reporter plasmid pAP1D3-luc (pAP-luc) contains three tandem copies of the distal AP-1 sequence at approximately -180 bp of the murine IL-2 promoter placed upstream of the luciferase gene. Both luciferase reporter plasmids are constructed from pT81Luc (ATCC 37584; American Type Culture Collection, Rockville, MD), a derivative of p232AL-Δ5'.[22] The plasmid pcDNA3.1/His is purchased from Invitrogen (Carlsbad, CA). All plasmids are prepared by CsCl density centrifugation at 340,000g in a Beckman (Fullerton, CA) NVT65 rotor.

Other Materials

2-Mercaptoethanol (2-ME), TPA, ionomycin, and hydrogen peroxide (H_2O_2) are purchased from Sigma (St. Louis, MO). Fetal calf serum (FCS) is purchased from HyClone Laboratories (Logan, UT). RPMI medium and L-glutamine are purchased from GIBCO-BRL Life Technologies (Frederick, MD). The Promega luciferase assay system is purchased from Promega (Madison, WI).

Buffers and Media

Phosphate-buffered saline (PBS): Dissolve 8 g of NaCl, 0.2 g of KCl, 1.44 g of Na_2PO_4, and 0.24 g of KH_2PO_4 in 800 ml of H_2O. Adjust to pH 7.4 with HCl and bring the final volume to 1 liter

Tris–EDTA (TE) buffer: Prepare 10 mM Tris–1 mM EDTA buffer and adjust to pH 7.4 with hydrochloric acid

Sodium acetate (3 M): Dissolve solid sodium acetate in H_2O and adjust to pH 4.8 with glacial acetic acid. Bring to final volume with distilled H_2O

[21] K. E. Hedin, M. P. Bell, K. R. Kalli, C. J. Huntoon, B. M. Sharp, and D. J. McKean, *J. Immunol.* **159,** 5431 (1997).

[22] J. R. de Wet, K. V. Wood, M. DeLuca, D. R. Helinski, and S. Subramani, *Mol. Cell. Biol.* **7,** 725 (1987).

RPMI–10% (v/v) FCS: Sterile fetal calf serum (FCS, 55 ml) is heated at 55° for 30 min and added to 500 ml of sterile RPMI medium (without glutamine) under sterile conditions in a biosafety hood. Using sterile pipettes, 5 ml of 1 M HEPES, 200 mM L-glutamine, and 2 μl of 14.3 M 2-ME are added. The medium is filtered through sterile 0.2-μm pore size filters and stored at 4°

RPMI–0.5% (v/v) FCS: RPMI–10% FCS is diluted 20-fold in sterile RPMI medium to create a final FCS concentration of 0.5% (v/v). The medium is stored at 4°

Trypan blue cell viability dye: Dissolve trypan blue in 0.15 M NaCl to give a final concentration of 0.2% (w/v). Store the dye solution at room temperature

Ionomycin stock solution: A 2.64 mM ionomycin stock solution is prepared by dissolving ionomycin in ethanol–dimethyl sulfoxide (DMSO) (1 : 1, v/v). Additions to culture medium are made so that ethanol–DMSO $\leq$ 0.1% (v/v) of the final volume. The stock solution is stored at −20°

TPA stock solution: A 100-μg/ml stock solution of TPA is prepared in ethanol and stored at −20°. Dilutions of TPA are made in RPMI–10% (v/v) FCS before addition to cell culture

Procedure

Cell Culture

The human T lymphocyte cell line Jurkat is cultured in RPMI–10% (v/v) FCS at 37° and 5% CO$_2$ in a humidified incubator. Cells are passaged every 48 hr and are not allowed to exceed a cell density of 1×10^6 cells/ml. Cells are counted periodically with a cytometer and viability is determined by staining with trypan blue. To count cells, 1 ml of medium is removed from a culture flask and briefly vortexed, and 100 μl is removed and diluted with an equal volume of trypan blue dye. After another brief vortex, 10 μl of the mixture is placed on a cytometer and the cells are counted under a light microscope ($\times$100 magnification). Viable cells are clear but stain dark blue if nonviable.

Transfection

In a single transfection, 10^7 Jurkat cells are centrifuged at 100g for 5 min, the supernatant is discarded, and the cell pellet is resuspended in 250 μl of RPMI–10% (v/v) FCS (final concentration of 40×10^6 cells/ml). The cells are kept at room temperature while the plasmids are prepared. A total of 30 μg of plasmid DNA is prepared for each transfection. In a single transfection, 10 μg of either pNF-AT$_3$ or pAP-luc and 20 μg of pcDNA3.1/His are aliquoted into sterile microcentrifuge tubes and brought to a 200-μl final volume with TE buffer. The plasmid DNA is precipitated by addition of 20 μl of 3 M sodium acetate (pH 4.8) and 500 μl of cold ethanol and centrifuged at 16,000g at 4°. The supernatant is removed by gentle aspiration and the DNA pellets are washed once with 100 μl of 70% (v/v) ethanol. The microcentrifuge tubes are then placed in the hood and opened for 5 min to allow complete evaporation of ethanol from the pellets. The pellets are resuspended in 50 μl of serum-free RPMI and placed in a 37° water bath for 10 min. After incubation the tubes are vortexed briefly and 250 μl of the cell suspension prepared above (10 × 10^6 cells) is added to each. The plasmid DNA and cells are allowed to incubate at room temperature for 10 min with no agitation, after which the cell–DNA suspensions are transferred to sterile 4-mm gap cuvettes and electroporated at 300 V for 10 msec (Electro Square Porator T820; BTX, San Diego, CA). After electroporation, the cells are allowed to rest for 10 min in the cuvettes, transferred to 21 ml of RPMI–10% (v/v) FCS (for pNF-AT$_3$ transfections) or RPMI–0.5% (v/v) FCS (for pAP-luc transfections), and incubated overnight (16–18 hr) at 37° and 5% CO_2. Cells are then aliquoted into six-well tissue culture plates (5 ml per well) and stimulated by the addition of ionomycin (2 μM final concentration) and TPA (2-ng/ml final concentration) for pNF-AT$_3$ transfections or with TPA (2 ng/ml) alone for pAP-luc transfections. The transfected cells are treated with H_2O_2 or a preferred reagent coincident with ionomycin and/or TPA addition. Cells are incubated for a total of 6 hr (at 37° and 5% CO_2) and harvested for luciferase assays as described below.

Luciferase Assays

The content (5 ml) of each well is removed and centrifuged in a 15-ml Falcon tube (Becton Dickinson Labware, Lincoln Park, NJ) at 100g for 5 min at room temperature. The supernatant is carefully aspirated and discarded. Each pellet is resuspended in 1 ml of cold PBS and transferred to a new microcentrifuge tube and centrifuged at 100g for 5 min at room temperature. The supernatant is removed and 50 μl of 1× lysis buffer [25 mM phosphate (pH 7.8), 2 mM DTT, 2 mM EDTA, 10% (v/v) glycerol, 1% (v/v) Triton X-100] from the Promega luciferase assay system is added to each cell pellet. The suspensions are incubated at room temperature for 10 min and subsequently centrifuged at 16,000g for 5 min at 4°. The supernatant (50 μl) of each sample is then carefully removed and placed in a luminometer tube (Starstedt, Newton, NC), and the light emission is measured with an EG&G Berthold (Pforzheim, Germany) LB 9507 luminometer after automated addition of 100 μl of Promega luciferin substrate.

Discussion and Example Results

Our reporter plasmid assay of calcineurin activity is based on the T cell receptor signal transduction pathway that regulates interleukin 2 (IL-2) gene expression

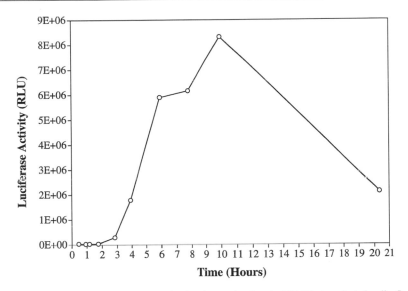

FIG. 3. Time course of luciferase production from stimulated pNF-AT$_3$-transfected cells. Jurkat cells transfected with pNF-AT$_3$ were stimulated at $t = 0$ hr with TPA (2 ng/ml) and 2 μM ionomycin. Cells were harvested at various time points after stimulation (40 min to 20 hr) and cell lysate was measured for luciferase activity. Data are represented in units of luciferase activity (RLU). [Reprinted with permission from T. A. Reiter, R. T. Abraham, M. Choi, and F. Rusnak, *J. Biol. Inorg. Chem.* **4**, 632 (1999), copyright © 1999, Society of Biological Inorganic Chemistry.]

(Fig. 1). Although in this pathway several other proteins are involved upstream of calcineurin and protein kinase C (PKC) activation, treatment with ionomycin and TPA bypasses these upstream signaling components, resulting in calcineurin and PKC activation. Calcineurin dephosphorylates NF-AT, which migrates to the nucleus and activates IL-2 transcription. Nuclear translocation of NF-AT is blocked when calcineurin is inhibited with the drugs cyclosporin A and FK506. Thus, luciferase activity, which is under the control of NF-AT in pNF-AT3-luc, provides an indirect measurement of calcineurin activation in intact cells.

Positive and negative controls must be included for each experiment performed. The negative control measures luciferase activity in unstimulated (by TPA and ionomycin), transfected cells. The positive control measures luciferase activity in transfected cells stimulated by ionomycin and TPA for 6 hr. Transfected cells are harvested after 6 hr of ionomycin–TPA treatment because luciferase production is maximized after this time point (Fig. 3). Relative light unit (RLU) readings from the luminometer for negetive and positive controls, using a single plasmid preparation, are 2.55($\pm$0.63) × 10^4 RLU and 1.92($\pm$0.04) × 10^7 RLU, respectively, representing a 590-fold increase in NF-AT activity on treatment of the cells

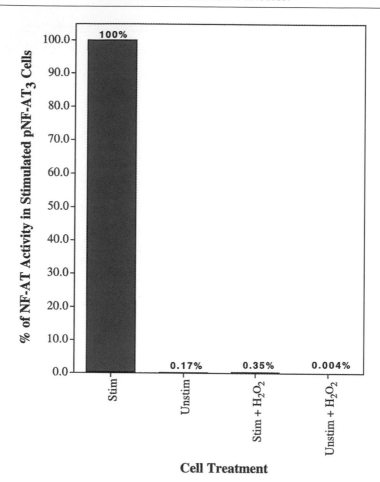

FIG. 4. Effect of H_2O_2 on calcineurin-dependent NF-AT activation of luciferase. pNF-AT$_3$-transfected Jurkat cells were either stimulated with ionomycin (2 μM) and TPA (2 ng/ml) (Stim), left unstimulated (Unstim), stimulated in the presence of 200 μM H_2O_2 (Stim + H_2O_2), or left unstimulated and treated with 200 μM H_2O_2 (Unstim + H_2O_2) for 6 hr. Cells were harvested for luciferase assays as described. NF-AT activity is reported as the percentage of relative light units produced by pNF-AT$_3$-transfected Jurkat cells stimulated with ionomycin and TPA alone. Data represent the means of two experiments. [Adapted with permission from T. A. Reiter, R. T. Abraham, M. Choi, and F. Rusnak, *J. Biol. Inorg. Chem.* **4,** 632 (1999), copyright © 1999, Society of Biological Inorganic Chemistry.]

with ionomycin and TPA (Fig. 4). RLU readings for the positive control vary with each plasmid preparation.

Using this assay, H_2O_2 was tested for its effect on calcineurin activity *in vivo.* Transfected cells were treated with ionomycin and TPA and varying concentrations of H_2O_2 for 6 hr and then harvested for luciferase assays. The activity of calcineurin

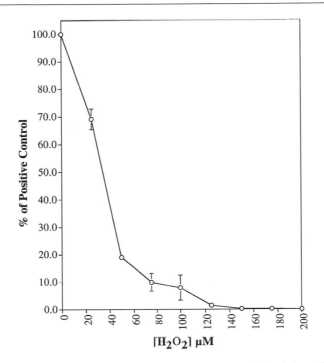

FIG. 5. Concentration dependence of H_2O_2 inhibition of NF-AT activity. Jurkat cells transfected with pNF-AT$_3$ were stimulated with 2 μM ionomycin, TPA (2 ng/ml), and various concentrations of H_2O_2 for 6 hr. Cells were harvested and the luciferase activity in the lysate was measured. NF-AT activity is shown as the percentage of luciferase activity found in ionomycin/TPA-stimulated, H_2O_2-untreated cells. Data represent the means of two individual experiments. [Adapted with permission from T. A. Reiter, R. T. Abraham, M. Choi, and F. Rusnak, *J. Biol. Inorg. Chem.* **4,** 632 (1999), copyright © 1999, Society of Biological Inorganic Chemistry.]

can be represented by the luciferase activity in the cell lysate. Using this assay, it can be shown that NF-AT activity is sensitive to exogenous H_2O_2 (Fig. 5). The results indicate that H_2O_2 inhibits calcineurin-dependent signaling, with a 50% inhibitory concentration (IC$_{50}$) of ~30–40 μM. Addition of exogenous paraquat, which would generate superoxide as well as H_2O_2 via endogenous mechanisms, did not cause a significant decrease in luciferase production.[23] We hypothesize that exogenous H_2O_2 inactivates calcineurin by altering the redox state of active site iron ion(s).

The pNF-AT$_3$ plasmid is driven by a promoter activated by both NF-AT and AP-1 proteins. Therefore, the inhibitory effect of H_2O_2 on pNF-AT$_3$ luciferase production may also be caused by a loss of AP-1 activity in the cell. A second reporter plasmid, pAP-luc, was used to determine whether AP-1 activity was affected

[23] T. A. Reiter and F. Rusnak, *J. Biol. Inorg. Chem.,* in press, 2002.

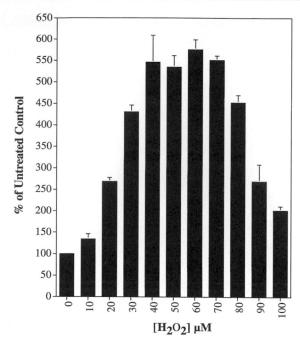

FIG. 6. Effect of H_2O_2 on AP-1 activity. Jurkat cells transfected with pAP-luc were stimulated with TPA (2 ng/ml) and varying doses of H_2O_2 (0 to 100 μM) for 6 hr. Cells were harvested and luciferase activity in the lysate was measured. Data are represented as the percentage of luciferase activity found in TPA-stimulated, H_2O_2-untreated cells. Data represent the means of two individual experiments. [Reprinted with permission from T. A. Reiter, R. T. Abraham, M. Choi, and F. Rusnak, *J. Biol. Inorg. Chem.* **4,** 632 (1999), copyright © 1999, Society of Biological Inorganic Chemistry.]

by H_2O_2 treatment. pAP-luc contains three tandem copies of the AP-1 promoter sequence upstream of the luciferase gene. After pAP-luc transfection, cells were incubated in low serum to prevent activation of AP-1 by serum. Jurkat cells were transfected with this plasmid and treated with TPA and various concentrations of H_2O_2 for 6 hr. This control experiment shows that AP-1 activity is stimulated by all H_2O_2 concentrations tested (Fig. 6).

Cell viability experiments should also be performed when treating cells with any potentially cytotoxic compound. An easy viability assay can be performed with trypan blue dye. The viability assay showed minimal cell death after cell exposure to H_2O_2 for 6 hr (data not shown).

The reporter plasmid assay herein described is an indirect, yet sensitive assay for calcineurin activity *in vivo*. Calcineurin is a redox-sensitive enzyme and therefore we propose that this assay can be used as a qualitative measure of H_2O_2-induced oxidative stress. The assay, however, does have limitations. The assay will not work in cell lines that fail to express either calcineurin or NF-AT. These problems can be circumvented by cotransfection with NF-AT and/or calcineurin expression plasmids. Electroporation has proved to be an easy and effective means of transfecting Jurkat cells with plasmid DNA, whereas other cell lines may require a different means of transfection (liposomes, retrovirus, $CaCl_2$, etc.). This assay could be adapted to yield significant information about calcineurin activity in different human cell lines and provide a readout of H_2O_2-induced oxidative stress.

[8] High-Performance Affinity Beads for Identifying Anti-NF-κB Drug Receptors

By Masaki Hiramoto, Noriaki Shimizu, Takeyuki Nishi, Daisuke Shima, Shin Aizawa, Hirotoshi Tanaka, Mamoru Hatakeyama, Haruma Kawaguchi, and Hiroshi Handa

Introduction

Affinity purification is an established technique used to identify ligand-binding proteins[1,2]; however, its widespread use has been limited by the inefficiency and instability of conventional matrices. The unstable nature of conventional matrices narrows the spectrum of ligands that can be used. Moreover, nonspecific binding of proteins to the solid support has complicated identification of target proteins. Another difficulty is frequent low purification efficiency, which requires partial purification of the source material before affinity chromatography to improve results.

This chapter describes the preparation and use of affinity beads for identification of drug receptors. The drugs of interest are immobilized to the matrix, which consists of latex beads covalently coupled with spacers. The latex beads are composed of glycidylmethacrylate (GMA)-covered GMA–styrene (St) copolymer cores (SG beads) that were developed originally for the affinity purification of DNA-binding proteins.[3-5] To reduce steric hindrance, divalent epoxide, that is, ethyleneglycol diglycidyl ether (EGDE), molecules are introduced as spacers after

[1] P. Cuatrecasas, M. Wilchek, and C. B. Anfinsen, *Proc. Natl. Acad. Sci. U.S.A.* **61**, 636 (1968).

[2] P. Cuatrecasas and C. B. Anfinsen, *Annu. Rev. Biochem.* **40**, 259 (1971).

[3] H. Kawaguchi, A. Asai, Y. Ohtsuka, H. Watanabe, T. Wada, and H. Handa, *Nucleic Acids Res.* **17**, 6229 (1989).

[4] Y. Inomata, H. Kawaguchi, M. Hiramoto, T. Wada, and H. Handa, *Anal. Biochem.* **206**, 109 (1992).

[5] T. Wada, H. Watanabe, H. Kawaguchi, and H. Handa, *Methods Enzymol.* **254**, 595 (1995).

Copyright 2002, Elsevier Science (USA).
All rights reserved.
0076-6879/02 $35.00

A

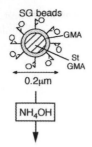

SG beads

GMA

St GMA

0.2μm

NH₄OH

SGN beads

NH₂

OH

excess EGDE

$CH_2CHCH_2-O-(CH_2)_2-OCH_2CHCH_2$

SGNEGDE beads

EGDE

NH_2—Drug NH_2—Drug

NH_2—Drug

Drug-fixed beads

Drug

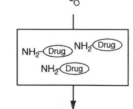

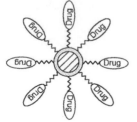

B

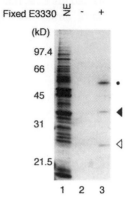

Fixed E3330   NE   -   +

(kD)

97.4

66

45

31

21.5

1   2   3

C

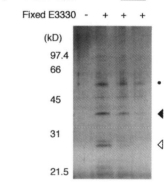

Free E3330   -   -

Fixed E3330   -   +   +   +

(kD)

97.4

66

45

31

21.5

1   2   3   4

D

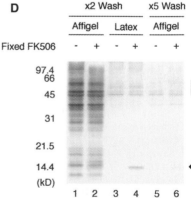

x2 Wash   x5 Wash

Affigel   Latex   Affigel

Fixed FK506   - +   - +   - +

97.4
66

45

31

21.5

14.4

(kD)

1   2   3   4   5   6

aminolysis of epoxy groups (SGN beads) on the surfaces of the beads (SGNEGDE beads; Fig. 1A).[6]

Compared with commonly used supports such as agarose,[7] the new beads offer several distinct advantages: (1) The lack of pores on the beads results in efficient removal of residual proteins during the critical washing steps, as well as easy access for target proteins to the fixed ligand, irrespective of their molecular mass; (2) the extremely small diameter of the beads (0.2 μm) provides a large surface area (1 g of beads has a surface area of 20 m^2), giving the beads a relatively high capacity; (3) the presence of GMA and the EGDE spacer arm produces a hydrophilic surface, which minimizes nonspecific protein binding; (4) the chemical and physical stability of the beads permits coupling of ligands in the presence of organic solvents; and (5) activation of the beads is not required for ligand coupling, because the bead surface contains chemically active epoxy groups derived from EGDE, allowing drug amino derivatives to be coupled to the beads easily by simple mixing. These advantages have enabled us to identify drug receptors directly from crude cell extracts within a few hours.

A brief description of the procedure is as follows. Latex beads composed of GMA–St cores, GMA surfaces, and EGDE spacers are prepared. Drug amino derivatives are then covalently coupled to the epoxy groups on the latex beads by simple mixing. Crude cell extracts are incubated with the affinity beads. The desired drug-binding proteins are then purified in a batchwise manner. The desired drug-binding proteins bind to the drug on the affinity beads, while other proteins flow through the beads. The drug-binding proteins are eluted from the affinity beads with a high-salt solution after brief centrifugation. A typical target protein is

[6] N. Shimizu, K. Sugimoto, J. Tang, T. Nishi, I. Sato, M. Hiramoto, S. Aizawa, M. Hatakeyama, R. Ohba, H. Hatori, T. Yoshikawa, F. Suzuki, A. Oomori, H. Tanaka, H. Kawaguchi, H. Watanabe, and H. Handa, *Nat. Biotechnol.* **18,** 877 (2000).

[7] M. W. Harding, A. Galat, D. E. Uehling, and S. L. Schreiber, *Nature (London)* **341,** 758 (1989).

FIG. 1. Purification of anti-NF-κB drug receptors. (A) Preparation of drug-fixed latex beads. Epoxy groups on SG beads were aminolysed by NH$_4$OH and coupled to the EGDE spacer to produce SGNEGDE beads. Amino derivatives of drugs were then fixed to the epoxy groups of EGDE. (B) Affinity purification with E3330-fixed beads. Jurkat cell nuclear extracts (lane 1) were subjected to affinity purification with E3330-fixed latex beads (lane 3). As a control, latex beads alone were used (lane 2). Arrowheads and a dot indicate the specific bands eluted from the E3330-fixed latex beads. (C) Competition analysis of E3330-binding proteins. To confirm the binding specificity of the E3330-binding proteins, an equal molar amount (lane 3) or a 10-fold molar excess (lane 4) of NH$_2$-E3330 was added to the extracts before purification. The fraction obtained by using beads alone is also shown in lane 1. (D) Comparison of affinity latex beads with affinity agarose resins (AffiGel). Purified fractions using FK506-fixed agarose (lanes 2 and 6) and FK506-fixed latex beads (lane 4) are shown. As a control, each matrix alone was also included (lanes 1, 3, and 5). In lanes 5 and 6, the matrices were washed three additional times. The vertical bar indicates artifacts in silver staining. [Adapted with permission from N. Shimizu, K. Sugimoto, J. Tang, T. Nishi, I. Sato, M. Hiramoto, S. Aizawa, M. Hatakeyama, R. Ohba, H. Hatori, T. Yoshikawa, F. Suzuki, A. Oomori, H. Tanaka, H. Kawaguchi, H. Watanabe, and H. Handa, *Nat. Biotechnol.* **18,** 877 (2000).]

purified more than a 1000-fold with a yield of 70%. If necessary, the batchwise step can be repeated, enabling further purification. This procedure is not only effective, but is also simple and straightforward to perform. Using the affinity beads, we have identified a redox-related factor, Ref-1, as a target protein of the anti-NF-κB drug, (E)-3-[2-(5,6-dimethoxy-3-methyl-1,4-benzoquinonyl)]-2-nonyl propenoic acid (E3330).[6]

Materials and Methods

Preparation of Latex Beads

Latex beads are prepared as described.[4] Briefly, GMA, St, and divinylbenzene (DVB) are used as monomers. To prepare the latex beads, 1.8 g of GMA, 1.2 g of St, and 0.04 g of DVB are mixed in 110 ml of distilled water in a 200-ml three-necked round-bottom flask equipped with a stirrer, a nitrogen gas inlet, and a condenser. To purge oxygen from the mixture, nitrogen gas is passed through the mixture, which is then kept at 70° in a water bath. Distilled water (10 ml) containing 0.06 g of azobisamidinopropane dihydrochloride is then added to initiate soap-free emulsion polymerization. Because the resulting GMA–St copolymer has a partially hydrophobic surface due to exposed polystyrene microdomains, 0.3 g of GMA is added to the mixture 2 hr after the initiation of polymerization. The mixture is then allowed to stand for 24 hr until the whole surface of the copolymer is covered with poly (GMA). The beads, containing epoxy groups on their surfaces derived from GMA, are collected by centrifugation, and washed three times with distilled water before being suspended in distilled water at a final concentration of 10% (v/v). The beads can be stored at 4°. Determination of the concentration of epoxy groups on the beads is performed by back titration of hydrogen ions. The concentration of hydrogen ions is measured after a known excess of hydrochloric acid is allowed to react with the epoxy groups. Specifically, known quantities (0.1, 0.2, and 0.4 g) of latex beads are mixed with 3.6 ml of 1 M HCl-CaCl$_2$ and 60 ml of distilled water in a 100-ml beaker. The mixture is covered with Parafilm and shaken for 3 hr at 40° to open the epoxy rings. After the reaction, 14.4 ml of 0.5 M KOH is added to the mixture, which is kept standing for 15 min at room temperature. Conductometric titration of the latex beads is then carried out with 0.1 M HCl.

Introduction of Spacers onto Latex Beads

To reduce steric hindrance, spacers are introduced onto the surface of the latex beads as follows. The first step is introduction of amino groups into the epoxy groups on the surfaces of SG beads. One gram of SG beads is suspended in 100 ml of 3 M NH$_4$OH (pH 11), incubated at 70° for 24 hr, and washed with distilled water (SGN beads). The concentration of amino groups on the bead surfaces is calculated as follows. Ten milligrams of SGN beads is mixed with 60 ml of distilled water in a 100-ml beaker. The pH of the mixture is adjusted to pH 11 with 0.1 M NaOH, and the dispersion is titrated with 0.01 M HCl by using a TOA (South San Francisco, CA) conductometer AUT-3000/FUT-3040/ABT-1010. The concentration of amino groups on the bead surfaces is calculated on the basis of inflection points derived from the conductometric titration curve. The standard concentration of amino groups obtained is 0.4–0.6 mmol/g beads. The second step is the coupling of EGDE with the amino groups on the surfaces of the SGN beads. The SGN beads are incubated with 100 mmol of EGDE at 30° for 24 hr at pH 11, and then washed with distilled water (SGNEGDE beads). In this reaction, excess amounts of EGDE are used in order to prevent them from coupling at both their ends. The concentration of epoxy groups at the uncoupled ends of spacers is calculated as described above. The SGNEGDE beads are suspended in 20 volumes of distilled water and stored at 4° until use.

Coupling of Drug Amino Derivatives to SGNEGDE Beads

Transfer an aliquot containing 10 mg of beads with a 200-μl tip to a 2.0-ml flat-bottomed microcentrifuge tube with a screw cap and collect the beads by centrifugation at 15,000 rpm (18,000g) for 5 min at room temperature. Add 1 ml of the solvent used in the coupling reaction, for example, distilled water, phosphate buffer, or 1,4-dioxane, and vortex for at least 2 min until the beads are completely resuspended. Centrifuge as described above and discard the cleared supernatant. Repeat this washing process twice.

In the case of the E3330 amino derivative, (E)-3-[2 (5,6-dimethoxy-3-methyl-1,4-benzoquinonyl)]-2-(9-aminononyl) propenoic acid (NH$_2$-E3330), incubate 10 mg of SGNEGDE beads with 10 μmol of NH$_2$-E3330 in 0.5 ml of 1,4-dioxane at 37° for 24 hr.[6] Occasional mixing during the incubation is likely to increase the drug-coupling efficiency. We usually put a micro stirrer bar into the reaction tube, and keep the reaction mixture agitated by using a magnetic stirrer. Collect the beads bearing the drug by centrifugation at 15,000 rpm for a few minutes at room temperature and wash them three times with 200 μl of 1,4-dioxane. After this, resuspend the beads in 1 ml of 1 M Tris-HCl (pH 7.4), and incubate at 4° for 24 hr to inactivate the unreacted epoxy groups on the beads. This inactivating step is not always necessary. Generally, approximately 15% of the NH$_2$-E3330 is immobilized. The beads can be stored at 4°.

Purification of Drug-Binding Protein (Drug Receptor)

The main advantage of using the drug affinity beads is the ability to perform one-step purification of drug-binding proteins from crude cell extracts. However, optimal binding conditions should be carefully established. The effects of the composition of binding buffer (pH, concentration and valency of salt, and concentration of detergent), temperature, amount of immobilized drugs, concentration of crude cell extract, and incubation time are some of the variables that should be optimized by a small-scale purification.

Here we present a typical purification procedure for the isolation of anti-NF-κB drug receptors.[6] Transfer 1-mg volumes of beads bearing the drug to 1.5-ml microcentrifuge tubes with screw caps, using 200-μl tips, and centrifuge at 15,000 rpm for 5 min at 4°. As a negative control experiment, we usually use, in parallel, equal amounts of SGNEGDE beads with inactivated epoxy groups. After discarding the supernatants, wash the beads three times with 400 μl of buffer D [20 mM HEPES (pH 7.9), 10% (v/v) glycerol, 0.1 M KCl, 0.2 mM EDTA, 1 mM dithiothreitol (DTT)] and keep them on ice until required. Add 200 μl of Jurkat cell nuclear extract[8] (dialyzed in buffer D) to the beads in the 1.5-ml microcentrifuge tubes and mix well. Allow the mixtures to stand at 4° for 30 min with occasional agitation. Centrifuge at 15,000 rpm for 5 min at 4° to clarify the supernatant. The length of the centrifugation time depends on the buffer and the concentration of proteins. Transfer the supernatants to fresh microcentrifuge tubes, quickly freeze with liquid nitrogen, and store at −80°. Wash the beads three times with 500 μl of buffer D by vortexing. Remove as much of the last wash as possible. As some buffer will adhere to the sides of the tubes, a brief centrifugation to remove the last few microliters may lower backgrounds for critical applications. Then, add 50 μl of buffer D containing 1.0 M KCl. Vortex well and collect the beads by brief centrifugation. Leave the beads on ice for a few minutes, centrifuge them for 5 min, and collect the supernatant. Quickly freeze the supernatant with liquid nitrogen, and store it at −80° until the characterization step. Affinity-purified proteins can be characterized by sodium dodecyl sulfate–polyacrylamide gel electrophoresis (SDS–PAGE). Usually, the protein concentration is so low that silver staining is necessary to identify the purified proteins. Three polypeptides (55, 38, and 27 kDa) are specifically eluted from the E3330-fixed latex beads (Fig. 1B). Binding between the E3330-fixed latex beads and the putative target proteins is specific, because these proteins cannot be eluted from control SGNEGDE beads. Furthermore, the binding reaction is inhibited by excess amounts of NH$_2$-E3330 in the binding buffer, confirming that these three proteins bind to the latex beads in a ligand-specific manner (Fig. 1C). This suggests that excess amounts of drug instead of a high concentration of salt may be used to elute the specifically binding proteins in some cases. The protein with a molecular weight of 38,000 has been further examined and has been identified as the redox-related factor Ref-1.[6]

Comparison of Latex Beads with Agarose Resins

To compare the purification efficiency of the latex beads with that of conventional agarose resins[7] for drug receptor isolation, we used the immunosuppressive drug FK506 as a model ligand. After incubation of FK506-immobilized latex beads with a Jurkat cell cytoplasmic fraction, the beads were washed and the bound

[8] J. D. Dignam, R. M. Lebovitz, and R. G. Roeder, *Nucleic Acids Res.* **11,** 1475 (1983).

protein analyzed by SDS–PAGE. A major protein with a molecular weight of 12,000, which was previously identified as FK506-binding protein (FKBP), was purified with high specificity directly from the crude cytoplasmic fraction (Fig. 1D). In contrast, when FK506-fixed agarose resin was used in the same manner as the latex system, increased background levels of protein staining were observed on the SDS–polyacrylamide gel and recovery of the specific binding protein was minimal. Although extensive washing reduced the background binding to the agarose resins, the yield of FKBP was also reduced. In other words, FKBP was present in the agarose resin-purified fraction but at a reduced level. These results demonstrate that latex beads have high binding efficiency and low nonspecific adsorption properties, indicating that they are superior to agarose resins for identifying drug receptors.

Troubleshooting

Low Coupling Efficiency of Drug

Drugs are coupled to the latex beads through the reaction of the amino groups on the drugs with epoxy groups on the beads. This reaction is easily inhibited by the presence of other amino groups. To couple drugs to the latex beads efficiently, it is necessary to eliminate other amino groups. Furthermore, it is useful to try a variety of solvents for the coupling reaction. However, use of solvents containing amino groups, hydroxyl groups, carboxyl groups, or sulfonic groups should be avoided in the coupling reaction because these functional groups display variable binding activity for epoxy groups. In the case of E3330, we tried a variety of solvents such as 10 mM phosphate buffer, dimethyl sulfoxide, N,N-dimethyl formamide, and 1,4-dioxane. We found that the use of 1,4-dioxane as the solvent in the coupling reaction produced the best coupling efficiency with the drug, and high recovery rates of pure protein. Because the optimal pH in the coupling reaction between epoxy groups and amino groups is about pH 11, the reaction should be performed at this pH, or as near as possible to this pH depending on the stability of the drug. Similarly, it may be possible to raise the reaction temperature and to extend the incubation period. An important point to note is that the beads bearing high concentrations of drug are not necessarily the most effective for purification of target proteins. These beads tend to aggregate, and occasionally diminish the quality of purified proteins. We therefore recommend optimizing the concentration of drugs used in the coupling reaction.

Low Recovery of Purified Protein

There are three main reasons for failure to recover drug-binding activity.

1. Improper binding conditions: This may be resolved by determining again the optimal binding conditions. The binding conditions can be affected by factors

such as the composition of the binding buffer (pH, concentration and valency of salt, concentration of detergent), temperature, amount of immobilized drugs, concentration of crude cell extract, and incubation time.

2. Nonspecific adsorption to tips and tubes: This can be reduced by the addition of 0.01–0.1% (v/v) detergent such as Nonidet P-40, Triton X-100, or Tween 20 to the buffer. Siliconization of plastic ware is also effective, although it is not always necessary.

3. Inefficient coupling reactions or improper storage of the drug-fixed latex beads: Optimization of the coupling reaction was mentioned above. The drug-fixed latex beads can be stored at 4° for more than 1 year; however, this will depend on the stability of the drug in the storage buffer.

Low Purity of Isolated Protein

Sometimes multiple bands are observed when the purified fraction is subjected to SDS–PAGE. The main reason for this is insufficient washing of the affinity beads. Because the beads have a tendency to stick together, latex beads must be resuspended well in the washing buffer with vigorous agitation. A middle washing buffer, containing 0.2 to 0.3 M NaCl or KCl, is also effective in reducing background protein levels. Use of this wash is dependent on the characteristics of the drug–drug-binding factor interaction and, therefore, it should be confirmed that the drug-binding protein is not eluted by the middle washing buffer. The best washing condition can be determined by performing small-scale purifications. Prepare five tubes, each containing 100 μl of cell extract, mix the extracts with the affinity beads, and wash three times with buffer D. Wash each sample three more times with buffer D containing various concentrations of salt (usually from 0.1 to 0.5 M), respectively, and then elute the bound proteins with buffer D containing 1.0 M KCl. Sometimes elution buffer containing a lower salt concentration enables isolation of a protein of higher purity. The optimal salt concentration of the elution buffer can be determined by a small-scale purification experiment as just described. To improve the quality of purified proteins, use of an elution buffer containing an excess amount of free drug instead of a high concentration of salt is often effective. Nonspecific adsorption of proteins to the affinity beads can be reduced by adding to the buffer one of the detergents mentioned previously (0.01–0.1%, v/v). Another way to reduce nonspecific adsorption to the latex beads is as follows. Before mixing with drug-fixed latex beads, cell extracts are added to a 1.5-ml microcentrifuge tube containing SGNEGDE beads with inactivated epoxy groups. After mixing and incubation on ice for 30 min with occasional agitation, the suspensions are centrifuged at 15,000 rpm for 5 min to clarify the supernatant. The supernatant is then transferred to a 1.5-ml tube containing the drug-fixed latex beads and the remainder of the regular purification procedure is carried out as described above.

[9] Regulation of Protein Kinase C Isozyme Activity by S-Glutathiolation

By NANCY E. WARD, FENG CHU, and CATHERINE A. O'BRIAN

I. Introduction

Protein S-thiolation is a posttranslational modification of select proteins produced by a nonenzymatic, oxidative mechanism, which entails the formation of disulfide linkages between thiols of reactive cysteine residues and low molecular weight (LMW) thiols.[1] Common endogenous S-thiolating species are glutathione (GSH), cysteine, and cystamine.[2] Mechanisms of protein S-glutathiolation reactions include thiol–disulfide exchange reactions between protein thiols and GSH disulfide (GSSG) and reactions between the protein thiols and GSH that involve the oxidant-produced sulfenic acid intermediates GSOH or PrSOH, or the thiyl radical intermediate PrS·.

Oxidative stimuli that produce protein S-thiolation in cells include physiological stimuli that induce intracellular hydrogen peroxide production and bolus treatment with either the thiol-specific oxidant diamide or hydrogen peroxide.[3,4] Protein S-thiolation serves as a line of defense against oxidative damage to cells by reversibly blocking protein thiols that might otherwise be irreversibly oxidized to sulfinic or sulfonic acid under conditions of oxidative stress.[3,5] The existence of enzymatic protein dethiolation mechanisms, which are catalyzed by glutaredoxin (thioltransferase) and thioredoxin, provides evidence that protein S-thiolation serves as a regulatory mechanism for certain enzymes and binding proteins.[6] Indeed, the demonstrated ability of S-glutathiolation to dramatically and reversibly alter the catalytic activity of some enzymes strongly implicates protein S-glutathiolation as a mechanism of regulation of select redox-sensitive enzymes.[1,4,5]

In this chapter, we describe methods that have been used to demonstrate that the isozyme protein kinase Cα (PKCα) is exquisitely sensitive to oxidant-induced S-glutathiolation, which fully and reversibly inactivates the kinase activity.[4] The methods are likely to be broadly applicable to other PKC isozymes, as preliminary observations indicate that several other PKC isozymes are inactivated by analogous

[1] J. A. Thomas, B. Poland, and R. Honzatko, *Arch. Biochem. Biophys.* **319**, 1 (1995).

[2] J. A. Thomas, W. Zhao, S. Hendrich, and P. Haddock, *Methods Enzymol.* **251**, 423 (1995).

[3] D. M. Sullivan, N. B. Wehr, M. M. Fergusson, R. L. Levine, and T. Finkel, *Biochemistry* **39**, 11121 (2000).

[4] N. E. Ward, J. R. Stewart, C. G. Ioannides, and C. A. O'Brian, *Biochemistry* **39**, 10319 (2000).

[5] E. Cabiscol and R. L. Levine, *Proc. Natl. Acad. Sci. U.S.A.* **93**, 4170 (1996).

[6] S. A. Gravina and J. J. Mieyal, *Biochemistry* **32**, 3368 (1993).

METHODS IN ENZYMOLOGY, VOL. 353
Copyright 2002, Elsevier Science (USA).
All rights reserved.
0076-6879/02 $35.00

GSH-dependent, oxidative mechanisms.[6a] The methods may also be applied to other protein kinases that are potential candidates for regulation by S-thiolation, for example, the cAMP-dependent protein kinase (PKA), which harbors a highly reactive cysteine residue in its active site.[7]

The methodology for the study of PKC S-glutathiolation is likely to be relevant in investigations of the mechanism of GSH antagonism of tumor promotion,[4] and the potential role of PKC S-glutathiolation as a negative feedback loop for PKC-mediated oxidant production in cells.[8,9] The methods described will also be valuable in defining the relationship between the opposing effects on PKC activity that have been produced by oxidant treatment of mammalian cells, that is, a GSH-independent PKC activation mechanism that involves PKC phosphorylation at tyrosine[10] and the GSH-dependent inactivation mechanism of PKC S-glutathiolation.[4]

II. Methods for Analyzing the Inactivation of Purified Protein Kinase C Isozymes by S-Glutathiolation

A. Refreshing Protein Kinase C Isozyme Thiols

1. Principle. Refreshing PKC isozyme thiols to their native, reduced state is a necessary preliminary step for studies of the oxidative regulation of purified PKC. Although PKC isozymes are typically stored in the presence of 5–15 mM 2-mercaptoethanol, over long-term storage, the 2-mercaptoethanol slowly oxidizes and may eventually fail to fully protect PKC thiols from oxidation. In the case of commercial purified PKC isozymes supplied in volumes of 30–200 μl [Calbiochem (La Jolla, CA), PanVera (Madison, WI), Biomol (Hamburg, Germany)], air oxidation of the 2-mercaptoethanol in solution may occur rapidly and cannot be directly measured because of the expense of the material. Instead, it is more practical and cost-effective to routinely refresh PKC thiols before analysis of oxidant effects on PKC activity. Human PKC isozymes contain 16 to 23 cysteine residues, for example, human PKCα contains 20 cysteine residues.[4] We have found that omission of the reductive thiol preconditioning step can sometimes result in the production of aberrant and unreproducible effects on PKC activity by a given oxidative insult, for example, if PKC thiols are not refreshed, diamide may occasionally appear to activate a purified rat brain PKC isozyme mixture described in Ward *et al.*[4] We attribute those unreproducible effects to partial oxidation of

---

[6a] F. Chu, N. E. Ward, and C. A. O'Brian, *Carcinogenesis* **22**, 1221 (2001).

[7] H. N. Bramson, N. Thomas, R. Matsueda, N. C. Nelson, S. S. Taylor, and E. T. Kaiser, *J. Biol. Chem.* **257**, 10575 (1982).

[8] H. M. Korchak, M. W. Rossi, and L. E. Kilpatrick, *J. Biol. Chem.* **273**, 27292 (1998).

[9] M. A. Stevenson, S. S. Pollock, C. N. Coleman, and S. K. Calderwood, *Cancer Res.* **54**, 12 (1994).

[10] H. Konishi, M. Tanaka, Y. Takemura, H. Matsuzaki, Y. Ono, U. Kikkawa, and Y. Nishizuka, *Proc. Natl. Acad. Sci. U.S.A.* **94**, 11233 (1997).

PKC thiols that may occur to variable extents in storage. In addition, we have found that if PKC thiols are not refreshed before analysis, reducing agents such as dithiothreitol (DTT) may activate PKC by as much as 2-fold in the system of analysis. Refreshing PKC thiols before analysis of oxidative regulation of the enzyme achieves a reliable baseline of PKC activity, which remains unchanged ($\pm 10\%$) when the enzyme is subsequently incubated with millimolar DTT in the analysis.

2. *Method.* The method entails incubation of PKC with DTT followed by chromatography of the PKC sample on a small desalting gel-filtration column to remove excess DTT from the enzyme.[4] First, a 2-ml G-25 column (bed height, $\sim$4 cm) is equilibrated in equilibration buffer [20 mM Tris-HCl (pH 7.5), 1 mM EDTA, 1 mM EGTA, soybean trypsin inhibitor (20 μg/ml), leupeptin (10 μg/ml), 250 μM phenylmethylsulfonyl fluoride (PMSF)] at 4°. Next, in the case of commercial purified human PKCα (available from PanVera, Calbiochem, and Biomol), 5 μg of PKCα (this is about 50 μl of the commercial stock) is diluted into Equilibration buffer containing 2 mM DTT (final volume, 500 μl) and then incubated for 20 or 30 min at 4° to refresh the thiols. To remove the excess DTT from the PKCα sample, the sample is then gel filtered on the G-25 column at 4°. The column may be washed and reused numerous times, but before its initial use, the column must be calibrated to identify the elution positions of PKCα and DTT. This can be done by measuring PKC activity and the reactivity of DTT with 5,5′-dithiobis(2-nitrobenzoic acid) (DTNB, Ellman's reagent) (yellow product formation, A_{max} at 412 nm) in fractions of 200–250 μl eluted from the column. Typically, pooling the fractions that span 750 to 1500 μl (this includes the eluted volume of the sample) achieves excellent recovery of PKCα activity ($>$95%) and $>$90% removal of DTT. PKCα should be routinely recovered from the calibrated column in a single vial that can be capped and that has minimal air space after PKCα collection to minimize air oxidation of thiols. Because only residual levels of the sulfhydryl protectant DTT remain, it is best to use the gel-filtered PKCα sample for analysis of oxidative regulation of the isozyme within a relatively short period of time, that is, $<$2 hr.

B. Inducing Inactivation of Purified PKCα by S-Glutathiolation

1. *Principle.* The oxidant activity of diamide is thiol specific and restricted to disulfide bridge formation.[11] The defined oxidant activity of diamide facilitates the elucidation of PKC inactivation mechanisms that involve thiol oxidation. Measurement of the DTT reversibility of diamide-induced effects on PKCα activity is a useful control to verify that the effects result from disulfide bridge formation.[11] Because diamide only weakly inactivates PKCα,[4] diamide can be used as an inducing agent for investigations of the effects of PKCα S-glutathiolation on the activity of the isozyme.

[11] N. S. Kosower and E. M. Kosower, *Methods Enzymol.* **251**, 123 (1995).

TABLE I

PREINCUBATION CONDITIONS FOR ANALYSIS OF PKCα INACTIVATION
BY DIAMIDE-INDUCED S-GLUTATHIOLATION

Condition	Purpose
1. PKCα alone	Positive control that represents 100% PKCα activity
2. With diamide	Control to measure weak or no PKCα inactivation by diamide alone
3. With GSH	Control to demonstrate lack of effect of 100 μM GSH on PKCα activity
4. With GSH + diamide	Conditions that support potent PKCα inactivation by S-glutathiolation

The potent induction of GSH-dependent, DTT-reversible inactivation of PKCα by diamide, coupled with the direct demonstration of the S-glutathiolating modification of the isozyme in association with the inactivation (which is described in Section II,C), is indicative of PKCα inactivation by S-glutathiolation. This is because the only other inactivation mechanisms possible, intramolecular disulfide bridge formation within PKCα and intermolecular disulfide bridge formation between PKCα molecules, are excluded by the inability of diamide alone to potently inactivate the isozyme.[4] The experimental conditions and procedures described below produce potent PKCα inactivation by S-glutathiolation, which is measured as GSH-dependent, diamide-induced isozyme inactivation.[4] In an analysis of recombinant human PKCα, these conditions resulted in a loss of >80% of the activity of the isozyme by a GSH-dependent and DTT-reversible, diamide-induced mechanism across a broad range of diamide concentrations (0.1–5.0 mM). Diamide alone inactivated the isozyme weakly or not at all, depending on the diamide concentration, and GSH alone (100 μM GSH) had no effect on PKCα activity.[4] It is worthwhile to note that millimolar GSH concentrations should be avoided, as non-redox-inhibitory effects against purified PKC isozymes have been observed at GSH concentrations in the range of 1–10 mM.[12] The wide range of diamide concentrations achieving near-full PKCα inactivation by an S-glutathiolation mechanism indicates the flexibility of the system and thus its potential for detecting PKCα inactivation by S-glutathiolation when the conditions are modified to tailor the experiments to specific applications.

2. *Method*. To demonstrate diamide-induced PKCα inactivation by S-glutathiolation, a PKCα sample with refreshed thiols (see Section II,A) is preincubated under the four conditions specified in Table I. Preincubation mixtures containing the components indicated in Table II (105 μl in capped 0.2-ml tubes) are vortexed (briefly, to minimize aeration), preincubated for 5 min at 30° to support

[12] N. E. Ward, D. S. Pierce, S. E. Chung, K. R. Gravitt, and C. A. O'Brian, *J. Biol. Chem.* **273,** 12558 (1998).

TABLE II
COMPOSITION OF PREINCUBATION MIXTURES FOR ANALYSIS OF PKCα INACTIVATION
BY DIAMIDE-INDUCED S-GLUTATHIOLATION

Component (final concentration)[a]	Method of addition
1. 0.1–5.0 mM diamide (250 μM is recommended for routine use)	Added as 5 μl of a diamide stock solution freshly made from the solid in 20 mM Tris-HCl, pH 7.5
2. 100 μM GSH	Added as 5 μl of a 2.1 mM GSH stock solution in 20 mM Tris-HCl, pH 7.5
3. 20 mM Tris-HCl, pH 7.5	To 105 μl (final volume)
4. 800 ng of human recombinant PKCα	Present in equilibration buffer (95 μl)

[a] Components presented in order of addition.

the PKCα S-glutathiolation reaction, placed on ice, and then assayed for PKCα activity, as described below.

To analyze the DTT reversibility of PKCα inactivation, 30 mM DTT (final concentration) is added to the preincubation mixtures at the end of the 5-min preincubation period at 30°; the preincubation mixture volumes are adjusted to 110 μl by the addition of 5 μl of either 20 mM Tris-HCl, pH 7.5 (−DTT samples), or 660 mM DTT in Tris buffer (+DTT samples) at 30°. The isozyme is then further preincubated at 30° for 5–20 min, placed on ice, and assayed for PKCα activity. An incubation period of at least 20 min in the presence of 30 mM DTT is generally necessary to fully reverse S-glutathiolation-mediated PKCα inactivation.[4]

Numerous assay procedures, including commercial assay kits, are available to measure PKC activity and can be used for the analysis of PKCα inactivation by S-glutathiolation. We have employed an assay that measures the transfer of ^{32}P from $[\gamma$-^{32}P]ATP to a synthetic peptide substrate. The assay mixture components are shown in Table III. Addition of $[\gamma$-^{32}P]ATP to the assay mixtures initiates a 5- to 15-min phosphotransferase reaction at 30° with linear kinetics, and the reaction is terminated by the widely used practice of pipetting an aliquot of the assay mixture (40 μl) onto an approximately 1-in^2 piece of P-81 phosphocellulose paper (Fisher Scientific, Pittsburgh, PA), which binds the $[^{32}$P]phosphopeptide product but not $[\gamma$-^{32}P]ATP. After the reacted mixture is pipetted onto each filter paper, the paper (numbered in pencil) is dropped into a 1-liter beaker filled with deionized water. The papers are collectively washed by decanting and refilling the beaker three additional times, and then counted in 5 ml of scintillation fluid.

C. Demonstrating S-Glutathiolation of Purified PKCα

1. Principle. The demonstration that diamide induces GSH-dependent and DTT-reversible PKCα inactivation (see Section II,B) points to PKCα S-gluta-thiolation as the inactivation mechanism for the following reasons. The involvement

TABLE III
COMPOSITION OF PKCα ASSAY MIXTURES

Component (final concentration)[a]	Method of addition
Phosphatidylserine (PS), 30 μg/ml	10 μl of a sonicated dispersion of 0.36-mg/ml bovine brain PS (>98% pure) in water; not present in assays of basal activity
20 mM Tris-HCl (pH 7.5, 20 mM), 10 mM MgCl$_2$, 0.2 mM CaCl$_2$ (or 1 mM EGTA)	20 μl of a 6× stock solution; EGTA is present in assay tubes that measure basal activity
10 μM [Ser-25]PKC(19–31) peptide substrate	10 μl of a 120 μM peptide stock solution (peptide is purchased from Peninsula or BACHEM Bioscience)
Water	To adjust final volume to 120 μl
PKCα, 80 ng	10 μl of preincubated PKCα
6 μM [γ-^{32}P]ATP (5000–8000 cpm/pmol)	10 μl of 72 μM [γ-^{32}P]ATP is added to initiate the reaction. Reactions are initiated in tubes at 30° at 15-sec intervals and terminated 5–15 min later at 15-sec intervals

[a] Components presented in order of addition.

of the oxidant activity of diamide in the inactivation mechanism and the reversal of inactivation by DTT both indicate a role for disulfide bridge formation, and a requirement for GSH is also evident. The only simple inactivation mechanism that can account for these mechanistic features is PKCα S-glutathiolation [Eq. (1)].

$$\text{PKC}\alpha + \text{GSH} \underset{\text{DTT}}{\overset{\text{Diamide}}{\rightleftharpoons}} \text{PKC}\alpha\text{-SG (inactive)} \qquad (1)$$

The close association that has been observed between GSH-dependent, diamide-induced PKCα inactivation and PKCα S-glutathiolation indicates that PKCα S-glutathiolation is the cause of the inactivation.[4] PKCα S-glutathiolation can be detected by nonreducing sodium dodecyl sulfate–polyacrylamide gel electrophoresis (SDS–PAGE) as the diamide-induced and DTT-reversible covalent incorporation of [^{35}S]GSH into the isozyme.[4] This is a broadly applicable approach that can be used to investigate whether any PKC isozyme or other purified protein of interest is subject to S-glutathiolation. In some cases, S-glutathiolation of a protein may produce a shift in its migration position in nonreducing SDS–PAGE; this is true of recombinant human PKCα, which shifts from an apparent molecular weight of 82,000 to 88,000.[4] If a measurable shift in apparent molecular weight is detected for an ^{35}S-glutathiolated protein by nonreducing SDS–PAGE, then nonreducing Western analysis can be used as a nonradioactive method for routine monitoring of the S-glutathiolation modification. However, because a comparable shift in the migration position could, in principle, be produced by intramolecular disulfide bridge formation, it is important to first establish through [^{35}S]GSH labeling

TABLE IV
PREINCUBATION CONDITIONS REQUIRED TO DEMONSTRATE PKCα ^{35}S-GLUTATHIOLATION
BY NONREDUCING SDS–PAGE

Condition	Purpose
1. PKCα + [35S]GSH	To establish that PKCα is not covalently labeled by [35S]GSH in the absence of an oxidant. Corresponds to noninactivating conditions
2. PKCα + [35S]GSH + diamide	To demonstrate oxidant-induced covalent labeling of PKCα by [35S]GSH. Corresponds to inactivating conditions
	To establish a close correlation between PKCα inactivation and ^{35}S-glutathiolation, a range of diamide concentrations should be employed
3. PKCα + [35S]GSH + diamide + DTT	To demonstrate that the covalent [35S]GSH labeling of PKCα in step 2 is due to PKCα S-glutathiolation, based on the DTT reversibility of the labeling. PKCα is preincubated with [35S]GSH + diamide followed by further incubation with DTT. Corresponds to noninactivating conditions

experiments that the shift is indeed due to protein S-glutathiolation. To measure the stoichiometry of the covalent incorporation of [35S]GSH into PKCα, [35S]GSH-modified PKCα can be precipitated with trichloroacetic acid (TCA) and captured on filter paper[4] as described below.

2. *Method.* To detect PKCα ^{35}S-glutathiolation by nonreducing SDS–PAGE and autoradiography, the preincubation mixtures employed in the analysis of PKCα inactivation (Table II) are modified as follows. A 100 μM [35S]GSH ($\sim$125 mCi/mmol; DuPont-NEN, Boston, MA) concentration is substituted for 100 μM GSH, and the DTT concentration is reduced to 10 mM, because higher DTT concentrations may produce dark backgrounds in the autoradiograms. PKCα preincubation mixtures that correspond to the three sets of conditions outlined in Table IV are required to demonstrate PKCα ^{35}S-glutathiolation. At the end of the 5-min preincubation period at 30°, an equal volume of 2× nonreducing SDS–PAGE sample buffer [0.125 M Tris-HCl (pH 6.8), 4.0% (w/v) SDS, 20% (v/v) glycerol, 0.002% (w/v) bromphenol blue] is added to each preincubation mixture, and the samples are boiled for 3 min. The samples are analyzed by nonreducing 10% (w/v) SDS–PAGE (80–120 ng of PKCα per lane). [Gels with 7.5% (w/v) polyacrylamide are recommended for the alternative approach of monitoring the migration position shift of PKCα by Western analysis.] The gel is stained with Coomassie dye to detect molecular weight markers for subsequent determination of the migration positions of bands detected on the autoradiogram. Next, the gel is washed with deionized water and impregnated with Amplify fluorographic reagent (Amersham Pharmacia Biotech, Piscataway, NJ) by shaking the gel for 15 min at room temperature in the presence of sufficient Amplify solution to cover the gel. By converting

TABLE V
PREINCUBATION MIXTURES EMPLOYED TO MEASURE STOICHIOMETRY
OF DIAMIDE-INDUCED PKCα ^{35}S-GLUTATHIOLATION

Component (final concentration/amount)[a] (in order of addition)	Method of addition
1. 0.01–1.0 mM diamide	Added as 10 μl of a diamide stock solution in 20 mM Tris-HCl, pH 7.5
2. 100 μM [^{35}S]GSH (12 nmol)	Added as 10 μl of a 1.2 mM [^{35}S]GSH stock solution. The stock solution is prepared by adding 50 μl of [^{35}S]GSH (0.5 mCi; specific activity, ~404 Ci/mmol) (DuPont-NEN) per 1 ml of 1.26 mM cold GSH in 20 mM Tris-HCl, pH 7.5
3. 20 mM Tris-HCl, pH 7.5	To 120 μl (final volume)
4. 5–10 pmol of PKCα	Present in 100 μl of equilibration buffer

[a] Components are presented in order of addition.

weak β emissions to light, Amplify increases the efficiency of ^{35}S-glutathiolated PKCα detection and reduces the exposure time needed for detection to 6–48 hr. Next, the gel is vacuum dried onto filter paper. The filter paper is then spotted at its corners with a few drops of ^{35}S-labeled ink, to allow orientation of the autoradiogram to the gel for molecular weight calculations based on Coomassie-stained molecular weight markers. The gel is exposed to Amersham Hyperfilm MP, using an intensifying screen in a film cassette at $-70°$.

To determine the stoichiometry of the covalent incorporation of [^{35}S]GSH into PKCα that is sufficient to induce GSH-dependent, oxidative PKCα inactivation, PKCα (with refreshed thiols; see Section II,A) is preincubated for 5 min at 30° with 100 μM [^{35}S]GSH and 0.01–1.0 mM diamide, as described in Table V; either the full range of diamide concentrations or a single diamide concentration sufficient for >90% GSH-dependent PKCα inactivation may be employed in the analysis. A preincubation mixture containing 100 μM [^{35}S]GSH and PKCα alone (no diamide) is required to measure background radioactivity. Preincubations are terminated by adding an equal volume (120 μl) of ice-cold TCA–PP$_i$ solution [20% (w/v) trichloroacetic acid–1% (w/v) pyrophosphate] to the samples, which are then vortexed briefly and placed on ice. Nitrocellulose filter circles (25 mm, 0.2 μm GSWP; Millipore, Bedford, MA) are placed on a vacuum-filtration manifold (Pharmacia/Hoeffer) and premoistened with TCA–PP$_i$ solution. The TCA-precipitated samples (240 μl) are pipetted onto the filters with a Pipetman, and the filters are washed with TCA–PP$_i$ solution (three times, 5 ml each) by vacuum filtration and counted in 5 ml of scintillation fluid. To calculate the stoichiometry of the covalent binding of [^{35}S]GSH to PKCα (mole per mole), the background counts per minute are subtracted from the total filter-bound counts per minute. The counts per minute-to-picomole conversion can be made by simultaneously counting an [^{35}S]GSH standard, for example, 1 nmol of [^{35}S]GSH from the 1.2 mM [^{35}S]GSH stock solution. By this method, we have determined that an S-glutathiolation stoichiometry of approximately 1 mol of GSH per mole of PKCα is sufficient for >90% inactivation of human recombinant PKCα.[4]

III. Methods for Analyzing PKCα Inactivation by S-thiolation in Mammalian Cells

A. Demonstrating Dithiothreitol-Reversible, Oxidative PKCα Inactivation in Mammalian Cells

1. Principle. To focus on oxidant-induced effects on PKCα activity in mammalian cells that may involve PKCα S-glutathiolation, DTT-reversible changes in PKCα activity produced by diamide treatment of cells are measured. The focus on DTT-reversible changes in PKCα activity distinguishes effects of oxidant-induced PKCα S-glutathiolation on kinase activity[4] from effects stemming from oxidant-mediated stabilization of phosphotyrosine residues in the catalytic domain of PKCα, which induces PKC isozyme activation.[10] Furthermore, diamide treatment of cells limits the types of oxidative modifications that may be introduced into PKC isozymes to those involving disulfide linkages and thus simplifies the parallel analysis of the oxidative modification of cellular PKCα. NIH 3T3 cells are convenient for the analysis of oxidative PKCα regulation, because PKCα activity can be directly assayed in DEAE-Sepharose-extracted cell lysates owing to the lack of expression of other Ca^{2+}-dependent PKC isozymes in the cells.[4] Diamide treatment of NIH 3T3 cells offers an excellent model of potent PKCα inactivation by S-thiolation.[4] Methods used to demonstrate DTT-reversible inactivation of PKCα by diamide treatment of NIH 3T3 cells, which occurs in association with PKCα S-thiolation[4] (see Section III,B), are described below.

2. Method. NIH 3T3 fibroblasts (60–80% confluent) cultured at 37° under standard conditions [Dulbecco's modified Eagle's medium (DMEM) plus 10% (v/v) bovine serum] (Life Technologies, Rockville, MD) are incubated with cycloheximide (50 μg/ml) for 5 hr at 37°, to match conditions used to analyze cellular PKCα ^{35}S-thiolation in parallel (see Section III,B). The cells are washed with 10 ml of Hanks' balanced salt solution (HBSS; Life Technologies) and then incubated with diamide (0.1–5.0 mM) under serum-free conditions for 10 min at 37° (8–12 $\times$ 10^6 cells per treatment group, e.g., one 100-mm-diameter dish at 75% confluency); potent DTT-reversible PKCα inactivation is achieved by $\geq$2.5 mM diamide.[4] At the end of the treatment period, the cells are washed with ice-cold phosphate-buffered saline (PBS; Life Technologies), and lysed with 1% Triton X-100 in equilibration buffer (defined in Section II,A) (1.5 ml per treatment group) by stirring in 1.5-ml capped tubes for 15 min at 4°. The samples are spun in a microcentrifuge at 14,000g for 2 min at 4° to remove debris and then loaded onto 0.5-ml DEAE-Sepharose columns equilibrated in equilibration buffer (without Triton). After the columns are washed with 2 ml of equilibration buffer, equilibration buffer with 0.3 M NaCl (1 ml) is used to elute PKCα activity, and the protein concentration of the eluted sample is determined. To measure DTT-reversible PKCα inactivation, the eluted PKC is divided into two portions, which are preincubated with/without 30 mM DTT for 15 min at 30° in capped tubes, placed on ice, and assayed immediately. Assays are done as delineated in Table III, except that 5 μg of partially purified PKCα is included per assay mixture, and the assay mixtures that measure background kinase activity include Ca^{2+} and the PKC inhibitor Go6976 (100 nM) (Calbiochem), which selectively inhibits Ca^{2+}-dependent PKC isozymes and, thus, PKCα in NIH 3T3 cells. PKCα activity is calculated as total minus background kinase activity. The DTT-reversible PKCα inactivation produced by diamide treatment is calculated as the percent inactivation observed with DEAE-extracted PKCα samples preincubated without DTT minus the percent inactivation observed when the samples are preincubated with DTT. Because DTT reversal of disulfide bridge formation is time and concentration dependent, the ratio of DTT-reversible to DTT-irreversible inactivation observed by this method may be increased by prolonging the preincubation with DTT or by increasing the DTT concentration. Thus, disulfide bridge formation cannot be ruled out as the mechanism for the minor DTT-irreversible component of inactivation that is observed in the analysis.

B. Demonstrating Oxidant-Induced PKCα ^{35}S-Thiolation in Mammalian Cells

1. Principle. Because diamide by itself only weakly inactivates purified PKCα across a broad range of diamide concentrations but supports potent inactivation of the purified isozyme by S-glutathiolation,[4] demonstration that potent diamide-induced inactivation of PKCα in NIH 3T3 cells (see Section III,A) is associated with induction of cellular PKCα S-thiolation and is not associated with the formation of disulfide-linked complexes between PKCα and other cellular proteins provides compelling evidence that the inactivation mechanism is PKCα S-thiolation.[4] Therefore, NIH 3T3 cells are analyzed for diamide-induced, DTT-reversible PKCα inactivation (see Section III,A) and PKCα ^{35}S-thiolation in parallel, under conditions that achieve potent PKCα inactivation. Complex formation between PKCα and other proteins in the cells is analyzed on the basis of the migration position of PKCα in nonreducing SDS–polyacrylamide gels as measured by Western analysis; full recovery of cellular PKCα at its normal migration position is indicative of negligible complex formation with other proteins.[4] Measurement of PKCα ^{35}S-thiolation in NIH 3T3 cells entails selective, metabolic ^{35}S labeling of cellular GSH and other cysteine-derived LMW thiols, oxidative induction of PKCα ^{35}S-thiolation by diamide treatment of the metabolically labeled cells, immunoprecipitation of ^{35}S-thiolated PKCα from the cell lysates, and analysis of ^{35}S-thiolated PKCα (±DTT) by nonreducing SDS–PAGE/autoradiography.[4]

The general method of ^{35}S metabolic labeling of LMW thiols for analysis of oxidative induction of protein ^{35}S-thiolation in cells is described in Thomas *et al.*[2]

The method entails culturing the cells in the absence of sulfur-containing amino acids to deplete GSH and other LMW thiols, followed by labeling of the cells with [^{35}S]cysteine under conditions in which protein synthesis is inhibited, so that the predominant ^{35}S-labeled species are cysteine, cystamine, and GSH. The ^{35}S-labeled cells are then exposed to an oxidative stimulus and lysed under two sets of conditions: one that preserves the oxidatively produced ^{35}S-thiolated protein species present in the cells and one that dethiolates the proteins.[2] Cellular protein ^{35}S-thiolation is preserved by lysing cells in the presence of the thiol-modifying agent N-ethylmaleimide (NEM), which blocks free thiols and thus prevents the migration of the ^{35}S-thiolating species to other protein thiols during cell lysis. Cellular protein ^{35}S-thiolation is reversed by lysis of the cells in the presence of DTT.[2] Protein S-thiolation is operationally defined in this system as DTT-reversible ^{35}S labeling of cellular proteins detected by nonreducing SDS–PAGE/autoradiography. Because the method of sample preparation for SDS–PAGE/autoradiography entails boiling samples in the presence of SDS, ≥ 10 mM DTT fully reverses protein ^{35}S-thiolation in this system. Any protein labeling observed in the DTT-treated samples by SDS–PAGE/autoradiography is generally due to residual protein backbone labeling and may be minimized by more stringent protein synthesis inhibition.

2. *Method.* To analyze PKCα ^{35}S-thiolation in diamide-treated NIH 3T3 cells under conditions that produce potent oxidative inactivation of PKCα, the cells are cultured under the conditions employed in the analysis of PKCα inactivation (see Section III,A), except that LMW thiols are depleted by culturing the cells with dialyzed serum and DMEM lacking sulfur-containing amino acids for 16 hr at 37°, before the addition of the protein synthesis inhibitor cycloheximide (50 μg/ml). After the cells are incubated with cycloheximide for 1 hr at 37°, [^{35}S]cysteine (Amersham, Arlington Heights, IL) is added to the medium (30–50 μCi/ml medium), and the cells are further incubated for 4 hr at 37° so that selective metabolic labeling of GSH and other LMW thiols can ensue. Next, the radiolabeled cells are treated with diamide, and cell lysates are prepared as described in Section III,A, with the following modification. For each treatment condition, two plates ($\sim 10^7$ cells per plate) are needed for the analysis; one is for cell lysis in the presence of NEM, and the other is for cell lysis in the presence of DTT. Equilibration buffer (defined in Section II,A) with 1% (v/v) Triton X-100 and either 50 mM NEM or 25 mM DTT is employed to lyse the cells (1 ml/plate).

A fraction of each cell lysate (about 50 μl) should be reserved for analysis of whether diamide induces disulfide-linked complexes between PKCα and other proteins in the cells. This is done by Western analysis with a PKCα monoclonal antibody (mAb) (Transduction Laboratories, Lexington, KY) under nonreducing conditions, that is, by nonreducing SDS–PAGE. Either the appearance of DTT-sensitive PKCα-immunoreactive bands at retarded migration positions and/or the loss of PKCα immunoreactivity at its normal migration position will occur if a substantial amount of PKCα forms complexes with other proteins; we have observed neither phenomenon.[4]

To detect PKCα ^{35}S-thiolation, PKCα is immunoprecipitated from the radio-labeled cell lysates with a polyclonal PKCα antibody available from Santa Cruz Biotechnology (Santa Cruz, CA); PKCα S-thiolation does not interfere with the binding interactions between PKCα and this antibody.[4] Because the analysis entails immunoprecipitation of PKCα, it is necessary to add 50 mM NEM to the DTT-treated cell lysates before proceeding, to prevent reduction of the antibody to its subunits. Cell lysates are precleared with a 50% (v/v) slurry of protein A–Sepharose (100 μl of slurry/ml cell lysate) by end-over-end rotation for 30 min at 4°, followed by a brief spin in a microcentrifuge (10 min, 4°) to recover the supernatant. Sample protein concentrations are normalized by adjusting sample volumes with equilibration buffer, followed by incubation of the samples with equal volumes of the polyclonal PKCα antibody (300–500 μg of lysate protein per 5 μg of antibody; the recommended total volume is 1 ml). The antibody is prepared in 2× IP buffer [IP buffer includes 10 mM Tris-HCl (pH 7.5), 1 mM EDTA, 1 mM EGTA, 150 mM NaCl, 1% (v/v) Nonidet P-40 (NP-40), 10% (v/v) glycerol, leupeptin (10 μg/ml), and 0.2 mM PMSF]. After incubating the precleared cell lysates with the PKCα antibody for 2 hr at 4° with end-over-end rotation, the incubation is continued for at least 30 min to 1 hr in the presence of protein A–Sepharose [100 μl of a 50% (v/v) slurry], and the Sepharose beads are washed three times with IP buffer. The washed beads are resuspended with 100 μl of 2× nonreducing SDS–PAGE sample buffer (defined in Section II,C) and boiled for 5 min. The supernatants are analyzed for PKCα ^{35}S-thiolation, as described in Section II,C. To demonstrate diamide-induced PKCα ^{35}S-thiolation, the analysis must include samples corresponding to (1) untreated cells/NEM–lysate; (2) diamide-treated cells/NEM–lysate, (3) untreated cells/DTT–lysate; and (4) diamide-treated cells/DTT–lysate. Western analysis of the PKCα samples is done in parallel to verify equivalent recovery of immunoprecipitated PKCα from the treatment groups analyzed. Demonstration of diamide-dependent, DTT-sensitive ^{35}S labeling of PKCα by autoradiography and confirmation of equivalent recovery of each PKCα sample by Western analysis are indicative of PKCα ^{35}S-thiolation.

Acknowledgments

This work was supported by NCI Grant CA74831 and by Robert A. Welch Foundation Award G-1141.

[10] Detection and Affinity Purification of Oxidant-Sensitive Proteins Using Biotinylated Glutathione Ethyl Ester

By DANIEL M. SULLIVAN, RODNEY L. LEVINE, and TOREN FINKEL

Introduction

Effects of oxidative stress in biological systems have long been the subject of intense scientific scrutiny. This work has established at least an ancillary role for oxidants in an astonishingly wide range of human diseases and, more recently, a role in normal signal transduction. Many of these studies have focused on toxicological effects of highly reactive species such as peroxyl, alkoxyl, and hydroxyl radicals, which are capable of modifying a wide range of cellular constituents. In proteins, potent oxidants such as the ones listed above can attack a variety of amino acid side chains and the polypeptide backbone,[1,2] inflicting irreversible damage. Such modifications are likely important in pathological processes, such as reperfusion injury and inflammation, in which cells are exposed to high levels of oxidative stress. However, a variety of antioxidant systems are employed by cells to prevent accumulation of highly reactive species in association with normal redox metabolism. Therefore processes associated with lower levels of oxidative stress, such as redox-dependent signal transduction, are more likely mediated by the less reactive species nitric oxide (NO), hydrogen peroxide (H_2O_2), and superoxide (O_2^-). The cellular constituents susceptible to attack by these radicals are more limited, and in proteins the sulfur-containing amino acids cysteine and methionine are likely to be the predominant sites of modification.

Among the most oxidant-sensitive proteins are those containing ionized cysteine thiols.[3] These thiolate anions, also referred to as reactive cysteines, are a feature of a variety of proteins, including a number of proteins involved in signal transduction. For example, thiolates have been described in transcription factors,[4,5] kinases,[6,7]

[1] R. T. Dean, S. Fu, R. Stocker, and M. J. Davies, *Biochem. J.* **324,** 1 (1997).

[2] E. R. Stadtman and R. L. Levine, *Ann. N.Y. Acad. Sci.* **899,** 191 (2000).

[3] G. H. Snyder, M. J. Cennerazzo, A. J. Karalis, and D. Field, *Biochemistry* **20,** 6509 (1981).

[4] Y. Sun and L. W. Oberley, *Free Radic. Biol. Med.* **21,** 335 (1996).

[5] M. Zheng and G. Storz, *Biochem. Pharmacol.* **59,** 1 (2000).

[6] R. Gopalakrishna and S. Jaken, *Free Radic. Biol. Med.* **28,** 1349 (2000).

[7] N. E. Ward, J. R. Stewart, C. G. Ioannides, and C. A. O'Brian, *Biochemistry* **39,** 10319 (2000).

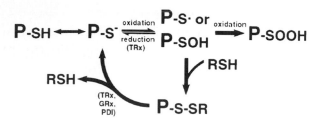

FIG. 1. Scheme for reversible redox modification of reactive cysteines. A reactive cysteine is a protein thiol (P–SH) that is ionized to the relatively nucleophilic thiolate anion (P–S$^-$) at physiological pH. The reactive cysteine can be readily oxidized to a thiyl radical (P–S$^\cdot$) or sulfenic acid (P–SOH) under mild oxidative stress. These oxidized species are frequently unstable under physiological conditions and, if not enzymatically reduced, will react to form more stable intermediates. Further oxidation yields a sulfinic acid (P–SOOH), which is thought to be irreversible in biological systems. Alternatively, the thiyl radical or sulfenic acid can react with a low molecular weight thiol (RSH) to form a stable mixed disulfide (P–S–SR). The mixed disulfide can then be reduced by one of three known enzymes systems [thioredoxin (TRx), glutaredoxin (GRx), or protein disulfide-isomerase (PDI)] to restore the cysteine to its fully reduced state.

phosphatases[8–10] and small GTP-binding proteins.[11] In most, if not all cases, thiolates play an important role in the normal functioning of the protein. Therefore their modification would be expected to have functional consequences. Oxidation of a thiolate by superoxide or hydrogen peroxide produces a thiyl radical or sulfenic acid (Fig. 1). Under physiological conditions, these oxidation products are unstable and, unless enzymatically reduced back to the thiol, will rapidly react with other molecules to form more stable products. Further oxidation of a thiyl radical or sulfenic acid results in irreversible oxidation of the cysteine. Alternatively, these species can react with other thiols to form stable disulfides. In some instances the second thiol is contributed by another cysteine contained within the same protein or an associated protein, resulting in the formation of an intra- or intermolecular protein disulfide. If there are no protein cysteinyl thiols in the vicinity of the thiyl radical or sulfenic acid, they might react with one of several low molecular weight thiols that are present in the cell to form a mixed disulfide. Because of the relatively high concentration of glutathione (GSH) in the cell (1 to 10 mM),[12] it is expected that the majority of protein mixed disulfides formed as a consequence of oxidative stress contain GSH. Indeed, the transient incorporation of glutathione into cellular protein is a well-established

[8] G. Zhou, J. M. Denu, L. Wu, and J. E. Dixon, *J. Biol. Chem.* **269,** 28084 (1994).
[9] G. H. Peters, T. M. Frimurer, and O. H. Olsen, *Biochemistry* **37,** 5383 (1998).
[10] J. M. Denu and K. G. Tanner, *Biochemistry* **37,** 5633 (1998).
[11] H. M. Lander, D. P. Hajjar, B. L. Hempstead, U. A. Mirza, B. T. Chait, S. Campbell, and L. A. Quilliam, *J. Biol. Chem.* **272,** 4323 (1997).
[12] F. Tietze, *Anal. Biochem.* **27,** 502 (1969).

response to oxidative challenge of intact tissues or cells in culture (reviewed in Refs. 13–15).

This process of protein thiolate oxidation and glutathiolation is intriguing in that it represents the reversible covalent modification of a protein attribute with functional importance. As such, the glutathiolation state of cellular proteins could serve to gauge the redox status of the intracellular environment, with the altered functional state of some modified proteins serving to transduce the oxidative stress into a biological response. Therefore a thorough accounting of proteins modified in this way and the functional consequences of modification could provide insight into the ways in which cells sense and respond to oxidative stress.

A number of techniques have been developed to study protein glutathiolation[16–18] or redox-dependent modification of reactive cysteines.[19,20] However, although these protocols can provide information regarding changes in the global glutathiolation or oxidation status of protein thiols, or can be used to study the glutathiolation state of individual proteins, it remains difficult to identify new glutathiolated proteins by these technologies. In addition, some of the available technologies are prone to artifacts arising from the need to inhibit protein synthesis while labeling the intracellular GSH pool, or the need to label proteins after cell lysis. We have described a novel reagent, biotinylated glutathione ethyl ester (BioGEE), and protocols that allows for the rapid purification of proteins that are oxidatively modified at reactive cysteine thiols in situ. This approach has a number of advantages over existing technologies. Because the tracer molecule can be incorporated into protein only as a consequence of cysteinyl thiol oxidation, the cell can be loaded without the need to inhibit protein synthesis. In addition, excess label is scavenged from the system before and during cell lysis so that incorporation of the label accurately reflects the glutathiolation state of the proteins before cell lysis. Finally, the use of biotin as a label allows sensitive nonradioactive detection and rapid affinity purification of labeled proteins with streptavidin conjugates.

By using BioGEE, we were able to demonstrate oxidative modification of several proteins in conjunction with tumor necrosis factor α (TNF-α)-stimulated apoptosis, and identified two proteins that had not previously been shown to be

[13] P. Klatt and S. Lamas, *Eur. J. Biochem.* **267**, 4928 (2000).

[14] J. A. Thomas, B. Poland, and R. Honzatko, *Arch. Biochem. Biophys.* **319**, 1 (1995).

[15] I. A. Cotgreave and R. G. Gerdes, *Biochem. Biophys. Res. Commun.* **242**, 1 (1998).

[16] Y. C. Chai, S. Hendrich, and J. A. Thomas, *Arch. Biochem. Biophys.* **310**, 264 (1994).

[17] Y. C. Chai, S. S. Ashraf, K. Rokutan, R. B. Johnston, Jr., and J. A. Thomas, *Arch. Biochem. Biophys.* **310**, 273 (1994).

[18] J. A. Thomas, W. Zhao, S. Hendrich, and P. Haddock, *Methods Enzymol.* **251**, 423 (1995).

[19] Y. Wu, K. S. Kwon, and S. G. Rhee, *FEBS Lett.* **440**, 111 (1998).

[20] J. R. Kim, H. W. Yoon, K. S. Kwon, S. R. Lee, and S. G. Rhee, *Anal. Biochem.* **283**, 214 (2000).

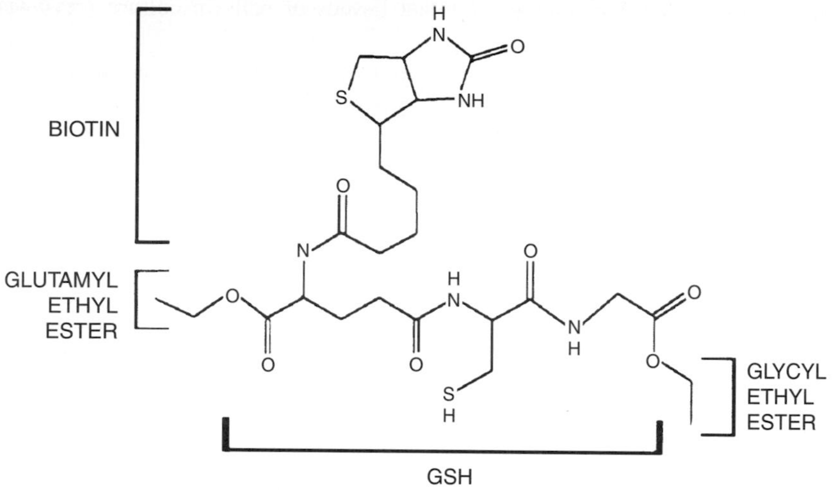

FIG. 2. Structure of BioGEE. The reagent BioGEE is composed of the glutathione tripeptide with a molecule of biotin incorporated at the γ-glutamyl amine and ethyl esters at one (glycyl) or both of the carboxyl groups.

glutathiolated.[21] In this chapter we present detailed protocols for the synthesis, purification, and use of BioGEE.

Preparation of Biotinylated Glutathione Ethyl Ester

BioGEE (Fig. 2) consists of the glutathione tripeptide modified to include ethyl esters at one or both of the carboxylates, and a molecule of biotin attached at the γ-glutamyl amine. The use of ethyl ester derivatives of glutathione is based on data indicating that the GSH mono(glycyl) or diethyl esters cross the plasma membrane much more efficiently than does GSH.[22–24] Incorporation of biotin into the molecule allows for simple and highly specific streptavidin-based detection and purification of proteins into which BioGEE is incorporated.

In our initial publication, establishing the utility of BioGEE, synthesis was accomplished by simply biotinylating a commercially available preparation of glutathione ethyl ester.[21] The crude biotinylation reaction was then added directly to cells in culture. Although this approach is simple and ultimately proved effective,

[21] D. M. Sullivan, N. B. Wehr, M. M. Fergusson, R. L. Levine, and T. Finkel, *Biochemistry* **39,** 11121 (2000).
[22] E. J. Levy, M. E. Anderson, and A. Meister, *Methods Enzymol.* **234,** 499 (1994).
[23] A. Meister, *in* "Glutathione: Chemical Biochemical and Medical Aspects" (D. Dolphin, R. Poulson, and O. Auramovic, eds.), p. 22. John Wiley & Sons, New York, 1989.
[24] M. E. Anderson, E. J. Levy, and A. Meister, *Methods Enzymol.* **234,** 492 (1994).

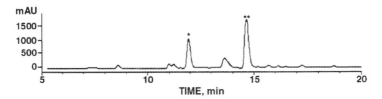

FIG. 3. BioGEE synthesis product. An aliquot of BioGEE prepared as described in text was solubilized in 0.05% (w/v) TFA and separated by reversed phase HPLC with both spectrophotometric and mass spectrometric detection (model 1100; Hewlett-Packard, Palo Alto, CA), using a Vydac narrow-bore C_{18} column (218TP5205; Vydac, Hesperia, CA). The initial solvent was 0.05% (w/v) trifluoroacctic acid and the other solvent was acetonitrile–0.05% (w/v) trifluoroacetic acid. The acetonitrile gradient was developed at 2%/min, with a flow rate of 0.2 ml/min. Mass analysis of the major peaks indicates that they are monoethyl (∗) and diethyl (∗∗) esters of biotinylated glutathione, with measured masses of 561.24 and 589.29, respectively. The expected masses were 561.6 and 589.6.

it has a number of drawbacks. First, it is necessary to limit both the concentration of the biotinylating reagent and the length of the biotinylation reaction in order to prevent incorporation of biotin at the glutathione –SH group. Second, only the monoethyl ester derivative of glutathione, which is taken up more slowly by cells than is the diethyl ester form,[25] is presently commercially available. Finally, the presence of unlabeled GSH ethyl ester in the crude reaction mixture might decrease the overall sensitivity of the protocol by competing with the biotinylated reagent.

To address these concerns, we devised the synthesis and purification scheme described below. First, the biotinylation reaction is carried out with glutathione disulfide. Because the sulfur moiety is protected in the oxidized form, the biotinylation reaction can go to completion without concern for incorporation of biotin at the sulfur. The biotinylated product is then reduced and purified by ion-exchange chromatography. The first purification step, using a strong cation exchanger, separates biotinylated from unbiotinylated GSH on the basis of the conversion of the GSH primary amine to a secondary amine in the biotinylated GSH. The second purification step uses a strong anion exchanger to separate biotinylated glutathione from excess reducing agent. The eluate from the anion-exchange column is then incubated in acidic ethanol to esterify the carboxylic acids. The products of this scheme consist primarily of mono- and diethyl esters of biotinylated glutathione (Fig. 3). As mentioned above, both the mono- and diethyl esters of glutathione are capable of crossing the plasma membrane, although the diethyl ester appears to be taken up more quickly. If desired, the yield of diethyl ester can be increased by extending the incubation time in acidic ethanol.

[25] E. J. Levy, M. E. Anderson, and A. Meister, *Proc. Natl. Acad. Sci. U.S.A.* **90,** 9171 (1993).

Materials

NHS–biotin, Ellman's reagent [5,5′-dithio-bis(2-nitrobenzoic acid)], *N,N*-dimethylformamide, GSH, oxidized GSH (GSSG), and HCl in anhydrous ethanol can be obtained from Sigma (St. Louis, MO), and the ion-exchange resins AG-50 (hydrogen form) and AG-1 (chloride form) can be obtained from Bio-Rad (Hercules, CA). The AG-50 resin should be swelled in water and stored as a slurry. AG-1 should be equilibrated with 50 m*M* sodium acetate (pH 5.2) and can be stored as a slurry or in columns. The necessary equipment is a spectrophotometer or plate reader equipped for absorbance measurements at 405 nm and a vacuum centrifuge.

Method for Determination of Free –SH Using Ellman's Reagent

At several points in the BioGEE synthesis and purification scheme it is advantageous to determine the concentration of GSH, or other free thiols, in solution. This can be accomplished quickly and easily with Ellman's reagent and the protocol below.

1. Set up a GSH standard curve. We do a 2-fold serial dilution starting at 10 m*M* and going down to 0.312 m*M*.

2. Dissolve Ellman's reagent at 200 μM in 0.1 m*M* NaPO$_4$, pH 8.0.

3. Mix the standards and unknowns with the Ellman's reagent solution at 1 : 40 (v/v). Incubate the reactions for 5 min at room temperature and determine the absorbance of the solutions at 405 nm.

Method for Synthesis and Purification of Biotinylated Glutathione Ethyl Ester

1. Make up a solution containing 40 m*M* GSSG in 50 m*M* Na$_2$HPO$_4$ and adjust the pH to between pH 7.5 and 8.0 with NaOH.

2. Dissolve NHS–biotin to 0.1 *M* in *N,N*-dimethylformamide and add it, dropwise, to the GSSG solution with gentle mixing to a final concentration of 20 m*M*. Incubate the biotinylation reaction overnight at room temperature with mild agitation (shaking too vigorously will cause precipitation of the NHS–biotin).

3. Add dithiothreitol (DTT) to the biotinylation reaction to a final concentration of 80 m*M* and incubate for 1 hr at 60°.

4. Add the reaction mixture to a bed of AG-50 resin (0.5 volume of AG-50 per volume reaction mixture) and check that the pH of the slurry is pH 5.0 or less, using pH paper. Incubate the slurry for 1 hr at room temperature with shaking.

5. Centrifuge briefly to sediment the resin and transfer the supernatant to a fresh tube.

6. Adjust the pH of the supernatant to approximately pH 5.0 and load it onto an AG-1 column equilibrated with sodium acetate (pH 5.2). We typically use a 5-ml bed volume of AG-1 for 40 ml of AG-50-extracted biotinylation reaction.

7. Wash the column with 6 bed volumes of sodium acetate (pH 5.2), and then with 6 volumes of 90% (v/v) ethanol. If desired, the concentration of free thiol in the washes can be monitored with Ellman's reagent and the protocol described above. A large thiol peak, representing excess DTT, should come off the column in the wash. The thiol concentration should be nearly 0 by the end of the ethanol wash.

8. Elute glutathione from the anion-exchange column with 0.5 N HCl in 90% (v/v) ethanol. Collect the eluate in 1-ml fractions and determine the pH and thiol content of each. Glutathione should elute when the pH of the effluent falls below pH 3.

9. Combine the thiol-containing fractions, neutralize them with NaOH, and remove the precipitate by either centrifuging or filtering the sample.

10. Dry the eluate in a vacuum centrifuge and resuspend the pellet in an approximately one-half elution volume of 1.25 M HCl in anhydrous ethanol and incubate for at least 16 hr at room temperature with mild agitation.

11. Determine the thiol content of the esterification reaction with Ellman's reagent as described above. By 16 hr, free thiol in the reaction mixture represents predominantly mono- or diethyl esters of biotinylated glutathione.

12. On the basis of the thiol content and intended use, divide the reaction into single-use aliquots and dry the aliquots in a vacuum centrifuge. Store the aliquots at $-80°$.

Assay of Biotinylated Glutathione Ethyl Ester Incorporation into Cellular Protein

The experiments shown in Fig. 4 demonstrate oxidant-induced incorporation of BioGEE into cellular proteins and, consistent with the mechanism described above, the apparent selectivity of the reagent for reactive cysteines. These data along with our previously published work indicate that BioGEE is a sensitive and selective marker for redox-dependent modification of thiolates. Determination of BioGEE incorporation into total cellular protein or into selected proteins is quite simple and requires only standard western-blotting equipment and reagents.

Materials

The sulfhydryl-alkylating reagents N-ethylmaleimide (NEM) and iodoacetamide (IAM) can be obtained from Sigma. We use the Complete Mini protease inhibitor cocktail from Roche Molecular Biochemicals (Indianapolis, IN), nonfat dry milk from Bio-Rad, and the bicinchoninic acid (BCA) protein assay reagent, HRP–streptavidin, and Supersignal chemiluminescent substrate from Pierce (Rockville, IL). The following solutions are needed: phosphate-buffered saline (PBS; 150 mM NaCl, 6 mM NaPO$_4$; pH 7.4), PBS-T [PBS, 0.1% (v/v) Tween 20], radioimmunoprecipitation assay (RIPA) buffer [1% (v/v) Nonidet P-40

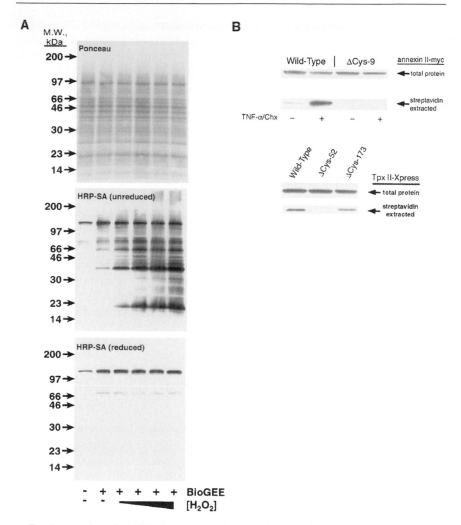

FIG. 4. Detection of BioGEE incorporation into protein, and selectivity for reactive cysteines. (A) HeLa cells were loaded with 0.5 mM BioGEE for 1 hr and then exposed to H_2O_2 (0, 0.12, 0.25, 0.5, and 1 mM) for 15 min. A soluble protein extract was obtained as described in text and 40 μg of protein from each extract was resolved by SDS–PAGE under either nonreducing or reducing conditions. The proteins were transferred to nitrocellulose and the blots were stained with Ponceau S. The blots were then blocked with milk and probed with horseradish peroxidase-conjugated streptavidin (HRP–SA). (B) HeLa cells were transfected with cDNAs encoding epitope-tagged wild-type or mutant annexin II and thioredoxin peroxidase II (Tpx II). The annexin II mutant contained a glycine substitution for the reactive cysteine at amino acid 9 (see Ref. 21), and the Tpx II mutants contained serine substitutions at the active site cysteines Cys-52 (reactive, see Ref. 26) and Cys-173 (not reactive). The cells were exposed to the indicated oxidative stress (TNF-α/cycloheximide 3-hr stimulation; H_2O_2, 15-min exposure), and a soluble protein extract was prepared as described in text. Biotin-containing proteins were affinity purified, blotted to nitrocellulose, and probed with antibodies against the indicated epitope tags.

(NP-40), 0.1% (w/v) sodium dodecyl sulfate (SDS), sodium deoxycholate (0.5 mg/ml), 150 mM NaCl, and 50 mM Tris-HCl (pH 7.5)], and hypotonic lysis buffer [1 mM EDTA, 1 mM EGTA, 50 mM Tris-HCl (pH 7.0)]. Standard equipment, solutions, and materials for SDS–polyacrylamide gel electrophoresis (PAGE) and transfer of proteins to nitrocellulose are also needed.

Method for Loading Cells with Biotinylated Glutathione Ethyl Ester

To load cells in culture we typically resolubilize BioGEE in culture medium at a concentration of approximately 0.5 mM, filter the medium, and then add it directly to cells. To estimate the concentration of BioGEE, one pellet from each synthesis lot should be dissolved in PBS and the concentration determined with Ellman's reagent as described above. In our experience, good results can be obtained by loading cells from 1 hr to overnight and in the absence or presence of up to 10% (v/v) fetal bovine serum (FBS). Although there is some basal incorporation of BioGEE into protein in unchallenged cells, the effect of oxidative stress on incorporation is obvious even after 16 to 18 hr of loading. Basal incorporation of BioGEE is probably a steady state phenomenon as oxidative stress appears to increase labeling of many of the same bands detected in the basal state. In addition, the extent of BioGEE incorporation in the basal state appears to be protein specific as we cannot detect basal incorporation of BioGEE into glyceraldehyde-3-phosphate dehydrogenase (GAPDH),[21] a protein for which incorporation is readily detectable after oxidative stress.

Because BioGEE is likely to enter the cell by passive diffusion, we anticipate that it will be possible to load a wide variety of cell types and tissues with the reagent. The efficiency of loading might vary, however, depending on how BioGEE is metabolized by the individual cell type. For example, results of one study indicate that the diethyl ester of glutathione crosses the plasma membrane in both directions more rapidly than the monoethyl ester.[25] In some human cell types the diethyl ester is rapidly metabolized to the monoester, resulting in a partial trapping and concentration of the reagent inside of the cell. However, the rate of conversion of the monoethyl ester to GSH is slow in these same cells. In our experience oxidant-induced incorporation of BioGEE declines rapidly once the reagent is washed out of the culture medium, probably because of efflux of incompletely deesterified label from the cell. Therefore we typically leave the reagent in the culture medium during stimulation. The need to do this might, however, vary from one cell type to the next, depending on how rapidly the cell deesterifies the second carboxylate.

---

[26] S. Hirotsu, Y. Abe, K. Okada, N. Nagahara, H. Hori, T. Nishino, and T. Hakoshima, *Proc. Natl. Acad. Sci. U.S.A.* **96,** 12333 (1999).

Method for Harvesting Proteins: For Both Blotting and Affinity Purification

1. Wash the cells twice with ice-cold PBS and once with ice-cold PBS containing 50 mM N-ethylmaleimide (NEM) to scavenge any remaining BioGEE.

2. Harvest cells by scraping in RIPA or hypotonic lysis buffer containing NEM or iodoacetamide (IAM) at 50 mM concentration and protease inhibitors. We have obtained good results with both NEM and IAM. However, it is important to note that we do not know of a peptide-mapping resource that includes NEM-modified cysteines, and therefore IAM should be used if the goal of the experiment is to identify a protein by peptide mapping.

3. For cells harvested in RIPA buffer, homogenize by passing the lysate several times through a 22-gauge needle. For cells harvested in hypotonic buffer, lyse by freezing in a dry ice–ethanol bath and thawing at room temperature.

4. Centrifuge the lysates for 10 min at 12,000g and 4° to pellet insoluble material and transfer the supernatant to a fresh tube.

5. Determine the protein concentration in the supernatant, using BCA or other standard methods.

Method for Detection of Biotinylated Glutathione Ethyl Ester Incorporation into Protein

Standard laboratory Western blotting and immunoprecipitation protocols can be easily adapted for assaying BioGEE incorporation into total or selected proteins. It is, of course, important to keep in mind that BioGEE is incorporated via a disulfide bond and therefore reducing agents must be omitted from buffers and solutions unless the intention is to remove the label. We have obtained good results with both freshly prepared and frozen (−80°) lysates, and from as little as 5 μg of soluble protein. For detection of biotin in proteins blotted to nitrocellulose membranes, we typically block the membranes with 5% (w/v) nonfat dry milk in PBS-T for 1 hr at room temperature. The membranes are then probed with HRP-conjugated streptavidin at 1 μg/ml for 1 hr at room temperature. The blots are washed five times with PBS-T and developed with HRP chemiluminescent substrate according to the manufacturer protocol. It is important to be aware that biotin-containing proteins such as carboxyltransferases might also be detected by this protocol. However, because incorporation of biotin via a disulfide is unique to BioGEE, labeling with BioGEE can be confirmed by running a duplicate blot under reducing conditions (Fig. 4A).

BioGEE incorporation into individual proteins can be determined by probing blots of immunoprecipitates with streptavidin, or by probing blots of affinity-purified proteins (described below) with antibodies. In our experience the former approach is both more sensitive and more reliable than the latter.

Affinity Purification of Biotinylated Glutathione Ethyl Ester-Labeled Proteins

Perhaps the greatest strength of the method described here is that it provides a means to affinity purify proteins on the basis of their propensity to undergo redox modification at select amino acid residues. We have used the simple two-step scheme described below to successfully identify, by mass spectrometric peptide mapping and peptide sequencing, proteins purified from as little as 5 mg of soluble protein (Fig. 5). In the protocol described below proteins covalently bound to biotin are extracted in batch. The protocol can be readily adapted to a column if so desired.

Materials

Streptavidin–agarose can be obtained from Sigma and Centricon-10 ultrafiltration units can be obtained from Amicon (Beverly, MA). Biotin-blocked streptavidin is prepared by suspending 1 ml of streptavidin–agarose in 10 ml of (D-biotin 3 mg/ml in PBS), incubating the mixture for at least 1 hr at room temperature, and then washing the beads extensively with PBS. For in-gel protein detection we use the SilverXpress silver staining kit (Invitrogen, Carlsbad, CA) and GelCode Blue Coomassie staining reagent (Pierce).

Method

1. Obtain protein extracts as described above.
2. Add biotin-blocked streptavidin–agarose (50 μl/mg of soluble protein) to the extract and incubate for 30 min at 4° with gentle shaking.
3. Pellet the beads by brief centrifugation and transfer the supernatant to a fresh tube.
4. Add 100 μl of agarose-conjugated streptavidin per milligram of protein and incubate for 2 hr at 4° with gentle shaking.
5. Pellet the beads by brief centrifugation and remove the supernatant.
6. Wash the beads five times with approximately 10 volumes of ice-cold RIPA buffer.
7. Wash the beads two times with 10 volumes of room temperature PBS containing 0.1% (w/v) SDS.
8. Resuspend the agarose pellet in 1 volume of PBS–0.1% (w/v) SDS and incubate for 30 min at room temperature with gentle shaking.
9. Pellet the agarose and save the supernatant to be run in parallel as a −DTT control.
10. Resuspend the agarose pellet in 1 volume of PBS–0.1% (w/v) SDS containing 10 mM DTT and incubate for 30 min at room temperature with gentle shaking.

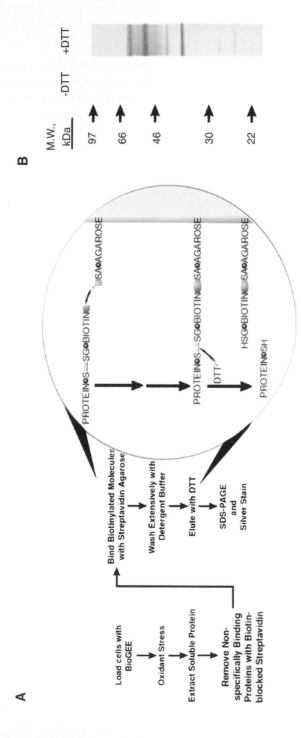

FIG. 5. Affinity purification of proteins labeled with BioGEE. (A) Diagram of the protocol devised for labeling and purification of S-glutathiolated proteins. Abbreviations: S, Sulfur; SG, glutathione; SA, streptavidin; DTT, dithiothreitol. (B) A soluble protein extract was obtained, as outlined in (A) and text, from HeLa cells treated with H_2O_2. Biotin-containing proteins were extracted from 0.5 mg of soluble protein with streptavidin–agarose. The beads were washed and eluted sequentially, first with PBS containing 0.1% (w/v) SDS (−DTT eluate) and then with PBS containing 0.1% (w/v) SDS and 10 mM DTT (+DTT eluate). The eluates were resolved by SDS–PAGE and proteins were detected by silver staining the gel.

11. Pellet the agarose and transfer the supernatant to a fresh tube.

12. Wash the beads with 1 volume of PBS–0.1% (w/v) SDS and combine the wash with the DTT eluate.

13. Add DTT to the −DTT eluate to a final concentration of 10 mM and transfer both the −DTT and DTT eluates to Centricon-10 ultrafiltration units. Concentrate proteins according to the manufacturer protocol.

14. Add SDS–PAGE sample buffer to the concentrated proteins and separate by SDS–PAGE.

15. Detect proteins by Coomassie or silver staining the gel, using standard protocols.

Interpretation of Data Obtained with Biotinylated Glutathione Ethyl Ester

The dynamic incorporation of BioGEE into protein in association with oxidative stress, its structural similarity to glutathione and apparent selectivity for reactive cysteines make BioGEE an excellent marker for reversible redox modification of proteins. However, as mentioned above, many details regarding the metabolism of BioGEE remain to be established. In particular, it is not known whether the enzyme systems involved in removing GSH from protein are effective at removing BioGEE. Even if BioGEE is completely deesterified to biotinylated GSH, it is conceivable that the remaining biotinyl moiety could affect the ability of enzymes, such as glutaredoxin, to use biotinylated GSH as a substrate. If that were the case, BioGEE might be expected to have a slower off rate and therefore accumulate in protein to a greater extent than endogenous glutathione. Likewise, the extent of BioGEE incorporation into protein is dependent on the unknown ratio of BioGEE to endogenous glutathione in the cell. It should therefore be stressed that the extent to which BioGEE is incorporated into a given protein cannot at this point be directly correlated with the extent to which that protein is glutathiolated. A finding that BioGEE is incorporated into a protein should therefore be taken as evidence of the existence of a redox-active cysteine that is sufficiently exposed to accept a molecule of glutathione. Such information is a crucial first step in understanding the effects of oxidants on the modified protein and ultimately on the cell.

[11] Redox Role for Tetrahydrobiopterin in Nitric Oxide Synthase Catalysis: Low-Temperature Optical Absorption Spectral Detection

By Antonius C. F. Gorren, Nicole Bec, Reinhard Lange, and Bernd Mayer

Introduction

Nitric Oxide Synthase

The biosynthesis of nitric oxide (NO) is catalyzed by nitric-oxide synthases (NOS, EC 1.14.13.39; for reviews see Refs. 1–3). These enzymes are homodimers, in which each monomer consists of an oxygenase domain and a reductase domain that is homologous to cytochrome P-450 reductase. NOS catalyzes the transformation of L-arginine to L-citrulline and NO in two discrete steps, with N^G-hydroxy-L-arginine (NHA) as intermediate. Both reactions take place at a P-450-type heme in the oxygenase domain and consume 1 equivalent of O_2. Both reactions also require electrons (two in the first step, and one in the second step), which are supplied by NADPH via two flavin moieties in the reductase domain. Reaction mechanisms proposed for the first step, the conversion of arginine to NHA, are based on the reaction cycle that is thought to describe P-450 catalysis. According to this model, ferric heme is reduced, binds O_2, and accepts a second electron. This provides the energy to achieve O–O bond scission, expulsion of H_2O, and formation of oxyferryl heme. This compound hydroxylates arginine bound to the distal heme pocket, resulting in formation of NHA and regeneration of ferric heme. Because of the different electronic requirements of the second step (net consumption of only one NADPH-derived electron), it has generally been assumed that hydroxylation of L-arginine and conversion of NHA to L-citrulline and NO must follow entirely different pathways.

Tetrahydrobiopterin

Tetrahydrobiopterin (BH$_4$) is an essential cofactor for NO synthesis. Although the main functions of the other cofactors and prosthetic groups (heme, FAD, FMN, Ca^{2+}/calmodulin, and Zn^{2+}) are known, the role of BH$_4$ proved to be more contentious, with the argument focusing on whether BH$_4$ is redox active

[1] O. W. Griffith and D. J. Stuehr, *Annu. Rev. Physiol.* **57,** 707 (1995).
[2] A. C. F. Gorren and B. Mayer, *Biochemistry (Moscow)* **63,** 870 (1998).
[3] S. Pfeiffer, B. Mayer, and B. Hemmens, *Angew. Chem. Int. Ed.* **38,** 1714 (1999).

Copyright 2002, Elsevier Science (USA).
All rights reserved.
0076-6879/02 $35.00

in NOS catalysis (see Refs. 2–5 for reviews). In the absence of BH$_4$, oxidation of NADPH becomes uncoupled from NO synthesis, and reduction of O$_2$ results in formation of superoxide anion (O$_2^-$). The well-established structural and allosteric effects of BH$_4$ on NOS do not account for these observations. Moreover, studies with a range of BH$_4$ analogs illustrated the importance of the redox properties of BH$_4$ for NOS catalysis. In the aromatic amino acid hydroxylases, which are the only other enzymes known to use BH$_4$ as a cofactor, BH$_4$ is directly involved in binding and reduction of O$_2$, and it exhibits two-electron redox shuttling between the tetrahydro and quinonoid dihydro species. However, a similar function of BH$_4$ in NOS catalysis can be ruled out, because 5-methyl-BH$_4$, which is unable to undergo reversible two-electron oxidation, is an activator of NO synthesis.[6]

Cryoenzymology

We decided to use UV–visible (UV–vis) spectroscopy at subzero temperatures to study the reaction of reduced NOS with O$_2$, about which little was known when we started our investigations. Cryoenzymology, the study of enzyme reactions at low temperatures, can provide vital information regarding reaction intermediates that are kinetically or thermodynamically inaccessible by other methods (see Refs. 7–12 for reviews). Preliminary experiments had shown that the reaction of NOS with O$_2$ proceeds too fast for detection of intermediates by standard absorption spectroscopy at ambient temperature. We intended to slow the reaction sufficiently by performing experiments at $-30°$ in a mixed hydro-organic solvent. This technique has been successfully applied to identify intermediates in the reaction of cytochrome P-450 with O$_2$.[13–17]

[4] D. J. Stuehr, *Biochim. Biophys. Acta* **1411**, 217 (1999).

[5] B. Mayer and E. R. Werner, *Naunyn-Schmiedebergs Arch. Pharmacol.* **351**, 453 (1995).

[6] C. Riethmüller, A. C. F. Gorren, E. Pitters, B. Hemmens, H.-J. Habisch, S. J. R. Heales, K. Schmidt, E. R. Werner, and B. Mayer, *J. Biol. Chem.* **274**, 16047 (1999).

[7] T. E. Barman and F. Travers, *Methods Biochem. Anal.* **31**, 1 (1985).

[8] P. Douzou, *Methods Biochem. Anal.* **22**, 401 (1974).

[9] P. Douzou, "Cryobiochemistry—An Introduction." Academic Press, London, 1977.

[10] A. L. Fink and M. A. Geeves, *Methods Enzymol.* **63**, 336 (1979).

[11] P. Douzou, *Adv. Enzymol. Relat. Areas Mol. Biol.* **51**, 1 (1980).

[12] F. Travers and T. Barman, *Biochimie* **77**, 937 (1995).

[13] L. Eisenstein, P. Debey, and P. Douzou, *Biochem. Biophys. Res. Commun.* **77**, 1377 (1977).

[14] C. Bonfils, P. Debey, and P. Maurel, *Biochem. Biophys. Res. Commun.* **88**, 1301 (1979).

[15] C. Larroque, R. Lange, L. Maurin, A. Bienvenue, and J. E. van Lier, *Arch. Biochem. Biophys.* **282**, 198 (1990).

[16] N. Bec, P. Anzenbacher, E. Anzenbacherová, A. C. F. Gorren, A. W. Munro, and R. Lange, *Biochem. Biophys. Res. Commun.* **266**, 187 (1999).

[17] I. Schlichting, J. Berendzen, K. Chu, A. M. Stock, S. A. Maves, D. E. Benson, R. M. Sweet, D. Ringe, G. A. Petsko, and S. G. Sligar, *Science* **287**, 1615 (2000).

Experimental Procedures

Equipment

We perform experiments with a Cary 3E (Varian, Palo Alto, CA) spectrophotometer that is adapted for low-temperature studies in a similar way as described in Maurel et al.[18] These adaptations, which have been carried out in-house, also enable flushing the cuvette holder with dry nitrogen to prevent condensation and formation of ice on cuvette windows and spectrophotometer lenses. For our studies, it has the additional advantage of keeping the cuvette holder anaerobic, and thus of preventing premature oxidation of the samples. The home-built sample compartment contains an aluminum block that serves as a cuvette holder for sample and reference. The temperature of the block is regulated with circulating ethanol that is thermostatted with a Thermo Haake (Karlsruhe, Germany) F3-Q bath with a lower limit of $-50°$. The temperature of the sample holder is monitored with a thermocouple connected to an AOIP (Evry, France) voltmeter. For thermal insulation the walls of the sample compartment are made of polyvinylchloride, equipped with double quartz windows.

Cryosolvent

An important aspect of cryoenzymology is the choice of a suitable solvent. We have opted for a mixed solvent consisting of aqueous buffer and ethylene glycol in a $1 : 1$ (v/v) ratio. The physicochemical properties of such cryosolvents have been reported.[9,19] In the case of phosphate buffer, the pH varies only slightly as a function of temperature. Preliminary studies have shown that this solvent does not affect the spectroscopic properties of NOS at ambient temperature, and the enzyme remains active.[20] However, detailed measurements of pterin binding and enzyme activity have revealed that ethylene glycol decreases the affinity of NOS for BH_4, such that the dimer appears to bind only one equivalent of BH_4.[21]

Sample Preparation

Samples are prepared in Teflon-capped 3-ml quartz cuvettes at final NOS concentrations of 2–4 μM in total volumes of 1–2 ml, containing 50 mM potassium phosphate (pH 7.4), 1 mM 3-[(3-cholamidopropyl)dimethylammonio]-1-propanesulfonate (CHAPS), 1 mM 2-mercaptoethanol, 0.5 mM EDTA, and 50% ethylene glycol. Each sample is gently flushed with argon for 30 min via two

[18] P. Maurel, F. Travers, and P. Douzou, Anal. Biochem. **57**, 555 (1974).

[19] P. Douzou, G. Hui Bon Hoa, P. Maurel, and F. Travers, in "Handbook of Chemistry and Molecular Biology" (G. D. Fasman, ed.), p. 520. CRC Press, Cleveland, OH, 1976.

[20] N. Bec, A. C. F. Gorren, C. Völker, B. Mayer, and R. Lange, J. Biol. Chem. **273**, 13502 (1998).

[21] A. C. F. Gorren, N. Bec, A. Schrammel, E. R. Werner, R. Lange, and B. Mayer, Biochemistry **39**, 11763 (2000).

holes in the Teflon cap. After anaerobiosis, 230 μM sodium dithionite (10–20 μl of a freshly prepared anaerobic 23 mM stock solution) is added with a Hamilton syringe and the sample is incubated under an atmosphere of argon for another 30 min. The cuvette is then placed in the cuvette holder at ambient temperature, and an absorbance spectrum is measured to check whether reduction is complete. If so, the temperature is lowered to $-30°$, which takes about 1 hr. Meanwhile, a 30-ml syringe is filled with O$_2$ and stored at $-70°$ until use. When the temperature of the sample reaches $-30°$, another spectrum is taken. Subsequently, 2–5 ml of O$_2$ is carefully administered with the precooled syringe. Spectra are measured immediately after O$_2$ addition, and then at certain intervals (usually 2 min) for several hours or until oxidation is complete.

Data Acquisition

Spectra are measured in double-beam mode between 350 and 700 nm, with 1.5-nm slit width. Data acquisition usually is in steps of 0.5 or 1.0 nm with an integration time of 0.5 sec/data point. All spectra are baseline corrected.

Results

The primary goal of our investigations is the identification of reaction intermediate(s) in the oxidation of reduced NOS by O$_2$, particularly of the oxyferrous complex Fe^{II}O$_2$, which accumulates under similar conditions in the case of cytochrome P-450. Indeed, we observed formation (within 2 min) of a compound exhibiting a red-shifted absorption maximum at 416/7 nm that we ascribed to oxyferrous (Fe^{II}O$_2$) heme.[20] Most importantly, this intermediate did not accumulate with BH$_4$-containing NOS in the presence of arginine, which implied that under those conditions the reaction cycle continued beyond the Fe^{II}O$_2$ state. This was unexpected, because the next step in the cycle requires reduction of the oxyferrous complex, and under our reaction conditions no obvious reductant was present. Partly on the basis of this observation we proposed that BH$_4$ can reduce the oxyferrous complex, most likely as a one-electron donor, and that this capacity of enzyme-bound BH$_4$ provides the explanation for the absolute dependence of NO synthesis on the pterin cofactor.[20] This hypothesis was confirmed by the fact that under single-turnover conditions NHA was formed in substantial yields only when arginine and BH$_4$ were both present.[20]

In a subsequent study we made similar observations with NHA instead of arginine.[21] With BH$_4$-free NOS the same 417-nm intermediate was found (Fig. 1), whereas with BH$_4$-containing NOS the reaction ran to completion immediately, resulting in the formation (within 2 min) of an Fe$^{III} \cdot$ NO complex (Fig. 2). This suggests that BH$_4$ has the same function in both reaction cycles. In the presence of the inhibitory redox-inactive pteridine 7,8-BH$_2$ the reaction cycle was halted at the stage of the Fe^{II}O$_2$ complex with either arginine or NHA, whereas

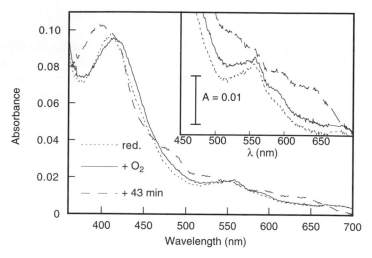

FIG. 1. Oxidation of BH$_4$-free neuronal deoxyferrous NOS by O$_2$ at $-30°$ in the presence of 1 mM NHA. Shown are the spectra of dithionite-reduced BH$_4$-free NOS (dotted line), as well as the spectra recorded immediately (within 2 min, continuous line) and 43 min (dashed line) after O$_2$ addition. *Inset:* The visible part of the spectra at greater magnification and vertically offset for clarity. Experimental conditions: 1.8 μM NOS, 1 mM NHA, 50 mM KP$_i$ (pH 7.2), 1 mM CHAPS, 0.5 mM EDTA, 1 mM 2-mercaptoethanol, 0.23 mM sodium dithionite, and 50% (v/v) ethylene glycol. See Experimental Procedures for further details.

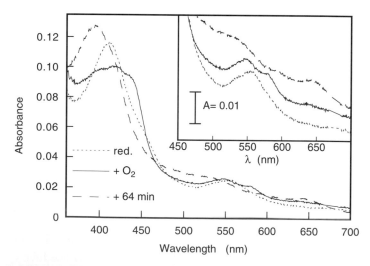

FIG. 2. Oxidation of BH$_4$-containing (as isolated) neuronal deoxyferrous NOS at $-30°$ in the presence of 1 mM NHA. Shown are the spectra of dithionite-reduced BH$_4$-containing NOS (dotted line), as well as the spectra recorded immediately (within 2 min, continuous line) and 64 min (dashed line) after O$_2$ addition. *Inset:* The visible part of the spectra at greater magnification and vertically offset for clarity. Experimental conditions: 2.1 μM NOS, 1 mM NHA, 50 mM KP$_i$ (pH 7.2), 1 mM CHAPS, 0.5 mM EDTA, 1 mM 2-mercaptoethanol, 0.23 mM sodium dithionite, and 50% (v/v) ethylene glycol. See Experimental Procedures for further details.

5-methyl-BH_4 mimicked the effects of BH_4. The latter observation is particularly significant, because the structure of this BH_4 analog, which supports NO synthesis,[6] precludes reversible two-electron oxidation, while allowing one-electron oxidation. From these two studies we concluded that BH_4 functions as an obligate one-electron donor to the oxyferrous complex during NO synthesis.[21] The $BH3^{\cdot}$ radical that is formed in the process has now been demonstrated directly by electron spin resonance spectroscopy.[22–24]

Comparison with Published Results

There have been several studies applying the same technique to other P-450-type enzymes. In most cases formation of an oxyferrous complex with a red-shifted absorbance maximum between 416 and 423 nm was reported,[13–16,25] similar to what we found with NOS. Interestingly, with cytochrome P-450$_{cam}$ and P-450$_{scc}$ blue-shifted intermediates (λ_{max} 405 nm) were observed that were ascribed to the oxyferryl state ($Fe^{IV}=O$).[15,26,27] The spectrum with λ_{max} at 404 nm that we observed immediately after addition of oxygen to BH_4-containing NOS in the presence of arginine was tentatively attributed to the same species.[20] Definitive evidence supporting this assignment is still lacking.

In line with our observations, Ledbetter *et al.* independently reported formation of an $Fe^{II}O_2$ complex, absorbing at 419 nm, using the same technique and almost identical conditions.[28] Curiously, Ledbetter *et al.* achieved optimal complex formation in the presence of N^G-methyl-L-arginine (NMA), whereas we observed instantaneous reoxidation without any detectable intermediate in the presence of that substrate analog (our unpublished observations, 1997). The cause of these and other subtle differences remains to be determined. Interestingly, NMA stimulates O_2^- formation by inducible NOS.[29] A similar stimulation of the uncoupled reaction cycle could explain the rapid reoxidation we observed in the presence of NMA.

[22] A. R. Hurshman, C. Krebs, D. E. Edmondson, B. H. Huynh, and M. A. Marletta, *Biochemistry* **38,** 15689 (1999).

[23] N. Bec, A. C. F. Gorren, B. Mayer, P. P. Schmidt, K. K. Andersson, and R. Lange, *J. Inorg. Biochem.* **81,** 207 (2000).

[24] P. P. Schmidt, R. Lange, A. C. F. Gorren, E. R. Werner, B. Mayer, and K. K. Andersson, *J. Biol. Inorg. Chem.* **6,** 151 (2001).

[25] R. C. Tuckey and H. Kamin, *J. Biol. Chem.* **257,** 9309 (1982).

[26] R. Lange, C. Larroque, and J. E. van Lier, *Biochem. Life Sci. Adv.* **7,** 137 (1988).

[27] R. Lange, G. Hui Bon Hoa, C. Larroque, and I. C. Gunsalus, *in* "Cytochrome P450: Biochemistry and Biophysics" (I. Schuster, ed.), p. 272. Taylor & Francis, London, 1989.

[28] A. P. Ledbetter, K. McMillan, L. J. Roman, B. S. S. Masters, J. H. Dawson, and M. Sono, *Biochemistry* **38,** 8014 (1999).

[29] H. M. Abu-Soud, P. L. Feldman, P. Clark, and D. J. Stuehr, *J. Biol. Chem.* **269,** 32318 (1994).

The oxyferrous complex of NOS has also been observed by stopped-flow/ rapid-scan optical spectroscopy.[30–33] With this method the absorbance maximum appeared at significantly higher wavelengths (427–430 nm), although one group reported a maximum at 420 nm.[31] This has caused some controversy regarding the true nature of the intermediates observed by us and others. In fact, most of the differences may be artifactual. Manual addition and mixing of O_2, followed by scanning of the first spectrum, puts a lower limit in the range of minutes to our method. Because several of the intermediates decayed on a similar time scale ($t_{1/2}$ 3–8 min), the absorbance maxima observed by us represent mixtures of the initial state (deoxyferrous NOS, absorbing at 410–412 nm), the final state (ferric NOS, absorbing at 394 or 418 nm depending on the heme spin state), and the oxyferrous complex, for which we estimate an absorption maximum between 420 and 425 nm. Some variation appears to be due to the presence of different substrates and pterin analogs. Similar subtle variations in peak position have been reported for cytochrome P-450 in the presence of different substrates.[25]

Advantages and Drawbacks

The main general advantage of subzero optical absorption spectroscopy is that the application of low temperatures combined with the obligatory use of a cryosolvent enable detection and characterization of reaction intermediates that are inaccessible to research by other methods. On the other hand, for the same reason the physiological significance of such intermediates is not always clear. For our studies we chose a 1 : 1 (v/v) mixture of ethylene glycol and aqueous buffer, because this did not affect the spectra of ferric and ferrous NOS, while the enzyme remained active.[20,28] Nevertheless, the substantially diminished activity and the decrease in the affinity of NOS for BH_4 indicate that the cryosolvent is not entirely innocuous.

As already mentioned, NOS binds only one equivalent of BH_4 per dimer in 50% ethylene glycol. Because the enzyme species with one BH_4 per dimer may be significant physiologically, this property of ethylene glycol is in some respects fortunate. However, as it prevents us from studying the pterin-saturated enzyme, we are currently exploring other cryosolvents as well.

A particularly useful aspect of our experimental procedure is that it provides a convenient method to generate ferrous NOS for single-turnover studies. Reduction of the enzyme with a small excess of sodium dithionite at room temperature, and

[30] H. M. Abu-Soud, R. Gachhui, F. M. Raushel, and D. J. Stuehr, *J. Biol. Chem.* **272**, 17349 (1997).

[31] H. Sato, I. Sagami, S. Daff, and T. Shimizu, *Biochem. Biophys. Res. Commun.* **253**, 845 (1998).

[32] M. Couture, D. J. Stuehr, and D. L. Rousseau, *J. Biol. Chem.* **275**, 3201 (2000).

[33] S. Boggs, L. Huang, and D. J. Stuehr, *Biochemistry* **39**, 2332 (2000).

subsequent addition of O_2 at $-30°$, circumvents complications by the remaining reductant, because at that temperature reduction of NOS by sodium dithionite is extremely slow ($t_{1/2}$ on the order of weeks; our unpublished observations, 1998). It is this simple and effective way to achieve a situation in which only the electron on the heme is available for oxygen reduction that enabled us to conclude that BH_4 must serve as the donor of the second electron in the reaction cycle with arginine. Furthermore, as the sample chamber is continuously flushed with nitrogen, anaerobic conditions are guaranteed. One can apply the same experimental procedure to generate samples for other spectroscopic techniques, such as electron paramagnetic resonance (EPR) spectroscopy.[23]

One disadvantage of the method is that about 2 min must elapse before the first spectrum after O_2 addition can be measured. It is mainly because of this limitation that, for some intermediates, spectral decomposition, and hence determination of the absorption spectra, could not be achieved. This can and will be redressed by application of low-temperature stopped-flow spectroscopy in future experiments.

Acknowledgments

This work was supported by Grant 13013-MED of the Fonds zur Förderung der Wissenschaftlichen Forschung in Österreich and the Human Frontier Science Program (RGP 0026/2001-M).

[12] Lysozyme–Osmotic Shock Methods for Localization of Periplasmic Redox Proteins in Bacteria

By VICTOR L. DAVIDSON and DAPENG SUN

Introduction

The cell envelope of gram-negative bacteria is composed of a cytoplasmic membrane (plasma membrane), a murein peptidoglycan layer, and an outer membrane that is linked by lipoproteins to the murein layer. The compartment between the cytoplasmic and outer membrane is called the periplasm, or periplasmic space.[1] Several proteins are specifically localized in the periplasm. These have long been known to include binding proteins for nutrients, and hydrolytic and degradative enzymes. More recently it has become clear that the periplasm is also home to a wide variety of redox enzymes and electron transfer

---

[1] J. W. Costerton, J. M. Ingram, and K.-J. Cheng, *Bacteriol. Rev.* **39,** 87 (1974).

Copyright 2002, Elsevier Science (USA).
All rights reserved.
0076-6879/02 $35.00

proteins.[2-5] These include c-type cytochromes, redox-active metalloproteins, thiol : disulfide oxidoreductases, and several $NAD(P)^+$-independent dehydrogenases and reductases. In particular, a disproportionate number of soluble electron transfer proteins of gram-negative bacteria are localized in the periplasm relative to the cytoplasm. Unlike eukaryotic cells, bacteria lack internal organelles and a complex intracellular membrane system. The membrane-bound respiratory chain of bacteria is localized in the cytoplasmic membrane. From a bioenergetic perspective, the periplasm is analogous to the mitochondrial intermembrane space of eukaryotic cells.

To determine whether a bacterial protein is periplasmic or cytoplasmic, a method must be available to cleanly separate the periplasmic and cytoplasmic contents of the cell. Furthermore, to purify a periplasmic protein from the cell, it would be desirable to purify it from the isolated periplasmic fraction rather than from a whole cell extract. The most common methods for disrupting bacterial cells are ultrasonic disruption (sonication) or passage through a French press. Neither of these methods is appropriate for fractionation of bacterial cells and the localization of periplasmic proteins. These methods not only disrupt the cell wall and outer membrane, but also the plasma membrane. The resulting cell extract will contain both periplasmic and cytoplasmic proteins. These procedures will also release DNA and some membrane lipids, which will interfere with the further processing of the cell extract. Furthermore, sonication is a relatively harsh treatment that may physically damage proteins.

For *Escherichia coli,* several methods for the release of periplasmic proteins have been described. These include osmotic shock,[6] magnesium chloride treatment,[7] chloroform treatment,[8] and polymyxin treatment[9] of intact cells. The most widely used technique for the release of periplasmic proteins from *E. coli* combines treatment of cells with lysozyme and exposure to a mild osmotic shock.[10] These methods for the release of the periplasmic proteins from *E. coli* are not, however, universally applicable to all gram-negative bacteria.

In this chapter we describe two different procedures that use the combination of lysozyme and osmotic shock for the fractionation of bacterial cells and selective

---

[2] C. Anthony (ed.), "Bacterial Energy Transduction." Academic Press, San Diego, CA, 1988.

[3] V. L. Davidson (ed.), "Principles and Applications of Quinoproteins." Marcel Dekker, New York, 1993.

[4] T. E. Meyer and M. A. Cusanovich, *Biochim. Biophys. Acta* **975**, 1 (1989).

[5] R. A. Fabianek, H. Hennecke, and L. Thony-Meyer, *FEMS Microbiol. Rev.* **24**, 303 (2000).

[6] N. G. Nossal and L. A. Heppel, *J. Biol. Chem.* **241**, 3055 (1966).

[7] K. J. Cheng, J. M. Ingram, and J. W. Costerton, *J. Bacteriol.* **104**, 748 (1970).

[8] G. F.-L. Ames, C. Prody, and S. Kustu, *J. Bacteriol.* **160**, 1181 (1984).

[9] Y. Kimura, H. Matsinaga, and M. Vaarga, *J. Antibiot. (Tokyo)* **45**, 742 (1992).

[10] B. Witholt, M. Boekhout, M. Brock, J. Kingma, H. van Heerikhuizen, and L. de Leij, *Anal. Biochem.* **74**, 160 (1976).

release of periplasmic proteins. Procedure 1 has been used in our laboratory to fractionate *E. coli*,[11] *Paracoccus denitrificans*,[12] and *Rhodobacter sphaeroides*.[13] Procedure 2 was used with *Alcaligenes faecalis*,[14] which could not be fractionated by procedure 1. The development of procedure 2 required significant modification of the generally applicable technique that was developed originally for *E. coli*. It is likely that these procedures will require further modificaiton for optimal effectiveness with other bacteria. As such, a general protocol is also described for optimizing conditions for the fractionation of other bacterial cells that are not efficiently fractionated by either of these two procedures.

Assay of Cytoplasmic and Periplasmic Marker Proteins

To evaluate the efficiency of a particular technique for the selective release of periplasmic proteins, it is necessary to monitor the release of general markers for periplasmic and cytoplasmic proteins. The latter will also be released from cells in which the cytoplasmic membrane is inadvertently ruptured during spheroplast formation. Spheroplasts are vesicles with an intact cytoplasmic membrane that remains after removal of the cell wall and outer membrane and release of the periplasm. Some percentage of spheroplasts is likely to rupture during any fractionation procedure. The goal of the fractionation procedure is to maximize release of the periplasmic contents while minimizing release of the cytoplasmic contents. It should be noted that not all periplasmic proteins will be released to the same extent during fractionation procedures. The extent of release will depend on the size of the protein and the degree of its association with the plasma membrane. Specific examples of cytoplasmic and periplasmic marker proteins are given below with methods by which they may be quantitated. Assay of such marker proteins provides a reasonable estimate of the efficiency of fractionation procedures.

The bacterial cytoplasm is the site of most metabolic pathways. As such, enzymes that participate in glycolysis and the citric acid cycle are reasonable cytoplasmic markers. Because NADH dehydrogenase faces the cytoplasmic side of the plasma membrane, any NAD^+-dependent enzyme may be presumed to be cytoplasmic. The NAD^+-dependent enzyme, malate dehydrogenase, is common to a wide range of bacteria. It is commonly used as a marker to gauge the release of cytoplasmic proteins. A method by which to assay its activity[15] is given below.

[11] V. L. Davidson, L. H. Jones, M. E. Graichen, F. S. Mathews, and J. P. Hosler, *Biochemistry* **36,** 12733 (1997).

[12] V. L. Davidson, *Methods Enzymol.* **188,** 241 (1990).

[13] M. E. Graichen, L. H. Jones, B. Sharma, R. J. M. van Spanning, J. P. Hosler, and V. L. Davidson, *J. Bacteriol.* **181,** 4216 (1999).

[14] Z. Zhu, D. Sun, and V. L. Davidson, *J. Bacteriol.* **181,** 6540 (1999).

[15] P. R. Alefounder and S. J. Ferguson, *Biochem. J.* **192,** 231 (1980).

Assays of several non-redox-active periplasmic marker enzymes have been described. Commonly used enzymes include acid phosphatase[16] and alkaline phosphatase,[17] and kits with which to assay their activities are commercially available. Soluble c-type cytochromes are widely distributed in gram-negative bacteria.[4] These cytochromes are localized exclusively in the periplasm, or on the periplasmic face of the plasma membrane. Cytochrome c is not an enzyme and therefore has no specific activity that may be assayed. However, cytochromes c may be readily monitored by a method described below, which allows quantitative detection of cytochromes c after sodium dodecyl sulfate–polyacrylamide gel electrophoresis (SDS–PAGE) of crude cell extracts.[18]

Assay of Malate Dehydrogenase: Cytoplasmic Marker

Reagents

Potassium phosphate (pH 7.5), 0.1 M
Oxaloacetate
β-Nicotinamide adenine dinucleotide, reduced form (NADH)

Procedure. Malate dehydrogenase catalyzes the reversible reaction shown below.

$$\text{Oxaloacetate} + \text{NADH} + \text{H}^+ \leftrightarrow \text{malate} + \text{NAD}^+$$

The spectrophotometric assay of malate dehydrogenase is performed as follows. The reaction mixture contains 0.1 M potassium phosphate, pH 7.5, with 0.2 mM oxaloacetate and 0.27 mM NADH. The reaction is initiated by the addition of an aliquot of the fractionated cell extract. The conversion of NADH to NAD$^+$ is monitored by the decrease in absorbance at 340 nm. An assay without oxaloacetate acid should be performed to correct for background oxidation of NADH. Preparations of malate dehydrogenase are commercially available and may be used as a positive control. This assay may also be applied to any other NADH-dependent enzyme that does not require additional cofactors by simply substituting the appropriate substrate for the enzyme in place of oxaloacetate.

Detection of Cytochromes c: Periplasmic Markers

Reagents

Dimethylbenzidine (also called o-dianisidine)
Sodium citrate (pH 4.4), 0.5 M

---

[16] L. A. Hepple, D. R. Harkness, and R. J. Hilmoe, *J. Biol. Chem.* **237,** 841 (1962).
[17] H. F. Dvorak, R. W. Brockman, and L. A. Hepple, *Biochemistry* **6,** 1743 (1967).
[18] R. T. Francis and R. B. Becker, *Anal. Biochem.* **136,** 509 (1984).

H$_2$O$_2$, 30% (v/v)
Trichloroacetic acid, 12% (w/v)

Procedure. Fractionated cell extracts are subjected to denaturing SDS–PAGE by standard methods. Sets of samples are run in duplicate on different portions of the gel or on two identical gels. One gel is stained for total protein, using standard reagents such as Coomassie blue or commercially available reagents such as Gelcode blue stain reagent (Pierce, Rockford, IL). The total protein stain is performed to ensure that the gel was run properly with the expected amount of protein present. The other set of samples is specifically stained for heme, using the protocol described below. This stain is relatively specific for *c*-type cytochromes. In contrast to most heme-containing proteins, the heme is covalently bound in cytochromes *c*, and is retained during denaturation and SDS–PAGE.[18]

To prepare the staining solution, 100 mg of dimethylbenzidine is dissolved in 90 ml of H$_2$O. Immediately before incubating with the gel to be stained, 10 ml of sodium citrate plus 0.2 ml of H$_2$O$_2$ are added to the solution. Immediately after SDS–PAGE, the gel to be stained is placed in 12% (w/v) trichloroacetic acid and incubated for 30 min. The gel is then rinsed with H$_2$O and incubated in the staining solution until covalent heme-containing proteins appear as green bands (15–60 min). The gel is then rinsed with H$_2$O to reduce background. Relative amounts of heme-stained proteins present in the gel may be quantitated by densitometry and related to the percentage of the total extract that was loaded on the gel. This will provide a reasonable estimate of the percent release of periplasmic cytochrome *c*. Any cytochrome *c* that appears in the cytoplasmic fraction reflects the percentage of the periplasmic protein that was not released by the procedure. As a positive control for the heme stain a sample of commercially available horse heart cytochrome *c* may be included on the gel to be stained.

Lysozyme–Osmotic Shock Procedures

The two procedures described below and summarized in Fig. 1 share the common features of treatment of cells with lysozyme plus EDTA in conjunction with the induction of a mild osmotic shock. The rationale for this method[19] is as follows. Lysozyme will digest, at least partially, the murein layer of the cell wall. Divalent cations are believed to play a role in maintaining the integrity of the cell wall and outer membrane. EDTA will modify the outer cell membrane of the cell wall to make it more permeable to lysozyme and agents that are used to adjust the osmotic strength. In this way the combined actions of lysozyme and EDTA make the

---

[19] D. C. Birdsell and E. H. Cota-Robles, *J. Bacteriol.* **93,** 427 (1967).

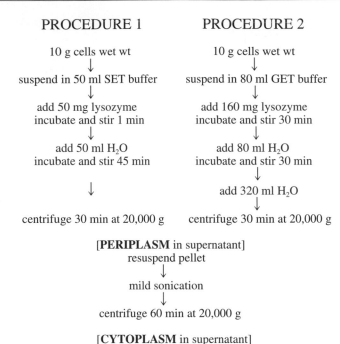

FIG. 1. Flow chart for fractionation of gram-negative bacteria by lysozyme–osmotic shock procedures. These examples show the general scheme for the fractionation of a sample size of 10 g of cells (wet weight). The details for each procedure are described in text.

cell wall and outer membrane more vulnerable to the mechanic disruption caused by the mild osmotic shock, while the inner membrane remains intact, causing selective release of the periplasmic contents.

Procedure 1

  Reagents

  SET buffer: 0.5 M sucrose and 0.5 mM EDTA in 0.2 M Tris-HCl, pH 7.4
  DNase I (solid)
  Lysozyme (10 mg/ml) in H_2O: Prepare immediately before use
  $MgCl_2$, 1 M in H_2O
  Phenylmethylsulfonyl fluoride (PMSF, 100 mM) dissolved in dry isopropanol

  Procedure. Cells are resuspended immediately after harvesting to a concentration of 1 g wet weight per 5 ml in SET buffer at 30°. Lysozyme that has been previously dissolved in H_2O is added to the cell suspension to a final concentration of 1 mg/ml. The following are then added with the amounts listed per milliliter of this suspension: 0.01 mg of DNase I, 1 μl of MgCl$_2$, and 0.5 μl of PMSF. The suspension is stirred with a glass or plastic rod for approximately 1 min. An equal volume of H$_2$O, which has been warmed to 30°, is added to the suspension to induce a mild osmotic shock. This suspension is incubated for 45 min at 30° while being stirred. The suspension is centrifuged for 30 min at 20,000g to remove spheroplasts and any unbroken cells from the periplasmic fraction. DNase I was added to reduce viscosity caused by the release of DNA from any spheroplasts that lyse during the procedure. PMSF is a protease inhibitor that was added to prevent degradation of proteins of interest. Other protease inhibitors may be added as well if the stability of the protein of interest is a concern.

Procedure 2

Reagents

GET buffer: 0.5 M glucose and 1.0 mM EDTA in 0.2 M Tris-HCl, pH 7.4
DNase I (solid)
Lysozyme (10 mg/ml) in H$_2$O: Prepare immediately before use
MgCl$_2$, 1 M in H$_2$O
PMSF (100 mM), dissolved in dry isopropanol

Procedure. Cells are resuspended immediately after harvesting to a concentration of 1 g wet weight per 8 ml of GET buffer at 30°. Lysozyme that has been previously dissolved in H$_2$O is added to the cell suspension to a final concentration of 2 mg/ml. The following are also added with the amounts listed per milliliter of this suspension: 0.01 mg of DNase, 1 μl of MgCl$_2$, and 0.5 μl of PMSF. This suspension is incubated for 30 min with stirring. An equal volume of H$_2$O, which has been warmed to 30°, is added to the suspension to induce a mild osmotic shock. The suspension is incubated for another 30 min at 30° while being stirred. To this suspension is then added 2 volumes of H$_2$O, which has been warmed to 30°, to induce a second mild osmotic shock. The suspension is centrifuged for 30 min at 20,000g to remove spheroplasts and any unbroken cells from the periplasmic fraction.

Disruption of Spheroplasts to Release Cytoplasmic Proteins

To fully assess the accuracy of the localization of proteins as cytoplasmic or periplasmic, and to assess the efficiency of the release of periplasmic proteins by the procedures described above, it is necessary to release and quantitate the cytoplasmic proteins from spheroplasts. This is most easily accomplished by mild sonication (30 W of power for 10 min) of the resuspended spheroplasts. The cytoplasmic components may then be separated from the disrupted cell membranes by centrifugation. Analysis of the putative periplasmic and cytoplasmic fractions for proteins of interest and cytoplasmic and periplasmic marker proteins will provide an estimate of the efficiency of the fractionation procedure that is used.

Protocol for Optimization of Lysozyme–Osmotic Procedures for Use with Different Bacteria

Our experience has taught us that no single standard protocol for lysozyme–osmotic shock can be generally applied to every gram-negative bacterium. Procedure 1 is effective for the fractionation of *E. coli*,[11] *P. denitrificans*,[12] and *R. sphaeroides*.[13] However, when we applied this method to *A. faecalis*, no detectable release of periplasmic proteins was observed.[14] This necessitated the development of procedure 2. When attempting to fractionate a bacterium that has not previously been studied in this manner, we suggest that procedure 1 be tried first. Determination of the extent of release of target proteins and selected marker proteins will provide an estimate of the efficiency of procedure 1. If little release of periplasmic proteins is observed with procedure 1, then procedure 2 should be tried. For bacteria that are not efficiently fractionated by either procedure 1 or 2, a protocol is described below for the optimization of the lysozyme–osmotic shock method for use with a new bacterium.

Choice of Appropriate Solute with Which to Adjust Osmotic Strength

Sucrose is the most commonly used solute for adjusting osmotic strength during lysozyme–osmotic shock procedures. However, when used with *A. faecalis* it was completely ineffective.[11] We hypothesized that the inability of sucrose to act as an osmotic agent was because the exclusion size of the outer membrane porin of *A. faecalis* had been shown to be too small to allow easy diffusion of sucrose.[20] Glucose, glycine, and NaCl were tested as alternative osmotic agents, and glucose was found to be the most effective. When little or no release of periplasmic proteins is observed with procedure 1, smaller alternative osmotic agents may be substituted for sucrose to see whether an improvement in the extent of release of periplasmic proteins is observed.

Lysozyme Sensitivity

Different bacteria may exhibit different sensitivities to lysozyme. Concentrations of lysozyme as low as 60 μg/ml of cell suspension have been reported for use with *E. coli*.[10] For *A. faecalis* we found that a concentration of approximately

[20] J. Ishii and N. Taiji, *Antimicrob. Agents Chemother.* **32,** 378 (1988).

2 mg/ml, before the osmotic shock, was needed to obtain maximum release of the periplasm. Furthermore, for bacteria treated by procedure 1, only a brief incubation with lysozyme before the osmotic shock was necessary. With *A. faecalis* a 30-min incubation with lysozyme before administration of the osmotic shock was needed for optimum results. Thus, the lysozyme–osmotic shock procedure should be optimized for both the concentration of lysozyme that is used as well as the length of time of incubation. We have also found that the effectiveness of lysozyme in this procedure will vary with the grade of lysozyme that is used and its freshness. For reproducible results it is important to use a comparable batch of lysozyme for each application.

Osmotic Shock Conditions

For the bacteria that were treated by procedure 1, a single osmotic shock with a 1 : 1 volume of water was sufficient to selectively release most of the periplasmic contents from the cell. This was not true for *A. faecalis,* which necessitated the modifications to procedure 1 seen in procedure 2. The efficiency of the fractionation procedure is dependent on both the number of times the cells are subjected to the osmotic shock, and the volume of water used to create the osmotic shock. We found when monitoring the release of a periplasmic dehydrogenase from *A. faecalis* that only 3% of the enzyme was released after a 1-hr incubation with lysozyme with no osmotic shock. After a single osmotic shock, about 15% of the enzyme was released. When a second osmotic shock was performed, as in procedure 2, approximately 75% of the enzyme was released. The efficiency of the osmotic shock also depended on the relative volume used to shock the cells. With *A. faecalis,* significant improvement was observed when a 2 : 1 ratio of water was used for the second osmotic shock rather than a 1 : 1 ratio. We examined the effect of increasing this ratio as high as 10 : 1. Increasing the ratio above 2 : 1 did increase the percentage of periplasm released, however, the extent of release of cytoplasmic markers also increased. This was likely due to rupture of the plasma membrane, which was caused by the more severe osmotic shock. If release of periplasm with a 1 : 1 ratio of water is not effective, then the ratio that maximizes the release of periplasmic proteins while minimizing the release of cytoplasmic proteins should be determined.

Potential Disadvantages of Lysozyme–Osmotic Shock Methods

It should be noted that the lysozyme–osmotic procedure does pose some potential problems depending on the particular application. These problems are primarily due to the requirements for large sample volumes and relatively long incubations. If large-scale cultures of cells are being processed, the volume of the supernatant that contains the released periplamic proteins may be quite large.

With procedure 1, the volume of the suspension after osmotic shock is 10 ml/g wet weight of cells. With procedure 2, this volume is 48 ml/g wet weight of cells. Thus, extensive concentration of the periplasmic extract will be required for further analysis or processing. These procedures also require relatively long incubation of cells in the presence of lysozyme at 30°. If the proteins of interest are relatively unstable, then they may be degraded during the procedure. As discussed above, inclusion of protease inhibitors during the incubation can decrease protein degradation. When applying these procedures, particularly for the purpose of purifying periplasmic proteins, the stability of the protein to be purified should be considered.

It should also be acknowledged that it may not be possible to use the lysozyme–osmotic shock method to fractionate some bacteria, even with the suggested protocol for optimizing these procedures. We have found at least one bacterium, *Methylophilus methylotrophus* sp. W3A1, that could not be fractionated by these procedures.[12] In that case, the problem was that significant release of cytoplasmic proteins occurred under the most gentle conditions that could be used to release periplasmic proteins.

General Relevance of Lysozyme–Osmotic Shock Methods

The ability to efficiently fractionate bacterial cells and localize enzymes and proteins is important for understanding the physiologic roles of these proteins. The application of relatively gentle procedures for disrupting and fractionating cells is also an important first step for the purification of proteins, as it will improve yields and prevent damage or artifactual modification of proteins. Application of the methods described here to other bacteria that are difficult to disrupt and to fractionate will help us to better characterize novel redox proteins and understand their physiologic functions.

Acknowledgment

Work from this laboratory was supported by the National Institutes of Health Grant GM-41574.

[13] Alterations in Membrane Cholesterol That Affect Structure and Function of Caveolae

By ERIC J. SMART and RICHARD G. W. ANDERSON

Introduction

A widely acknowledged function of cellular membrane is to promote biological reactions by creating a two-dimensional space that facilitates interactions among resident molecules. Less well appreciated is that cell membranes are subdivided into specific domains that compartmentalize a variety of essential activities. Three examples of membrane domains are clathrin-coated pits, focal adhesion sites, and caveolae. Each requires specific cellular machinery for their existence and is constructed from specific membrane proteins and lipids. In the case of caveolae, cholesterol is an essential lipid. Several studies have demonstrated that it functions as a structural molecule for this domain. In this chapter we describe several methods for modifying the cholesterol composition of caveolae and discuss how these changes alter the structure and function of this membrane domain.

Caveolae were originally described as small, noncoated plasma membrane invaginations.[1] There is now clear evidence, however, that caveolae can be either invaginated,[2] flat within in the plane of the membrane,[2,3] or detached vesicles.[4,5] In 1992, Rothberg *et al.*[2] demonstrated that a 22-kDa protein called caveolin-1 was associated with a filamentous coat that decorates the cytoplasmic side of morphologically identifiable caveolae. Caveolin-1 has since been shown to be a key player in the structure and function of caveolae. Although caveolin-1 is not exclusively localized to caveolae, it is considered by many to be the defining marker for invaginated caveolae. Many but not all cells express caveolin. Interestingly, cells not expressing caveolin-1 contain plasma membrane domains that have biochemical properties similar to those of caveolae.[6,7] The relationship between caveolin-containing membranes and caveolin-deficient caveola-like membrane domains remains to be determined. For purposes of this chapter, caveolae are defined

[1] G. E. Palade, *J. Appl. Phys.* **24,** 1424 (1953).

[2] K. G. Rothberg, J. E. Heuser, W. C. Donzell, Y. Ying, J. R. Glenney, and R. G. W. Anderson, *Cell* **68,** 673 (1992).

[3] E. J. Smart, D. C. Foster, Y. S. Ying, B. A. Kamen, and R. G. Anderson, *J. Cell Biol.* **124,** 307 (1994).

[4] J. R. Henley, E. W. Krueger, B. J. Oswald, and M. A. McNiven, *J. Cell Biol.* **141,** 85 (1998).

[5] P. Oh, D. P. McIntosh, and J. E. Schnitzer, *J. Cell Biol.* **141,** 101 (1998).

[6] R. G. W. Anderson, *Annu. Rev. Biochem.* **67,** 199 (1998).

[7] E. J. Smart, G. A. Graf, M. A. McNiven, W. C. Sessa, J. A. Engelman, P. E. Scherer, T. Okamoto, and M. P. Lisanti, *Mol. Cell. Biol.* **19,** 7289 (1999).

Copyright 2002, Elsevier Science (USA).
All rights reserved.
0076-6879/02 $35.00

as cholesterol/sphingolipid-rich membrane domains that contain caveolin-1 and are capable of internalizing molecules.

Caveolae, Caveolin, and Cholesterol

The morphological and functional behavior of caveolae depends on specific lipids and proteins. A key molecule is cholesterol. Indeed, most of the current methods for modulating caveolar function depend on perturbing the cholesterol composition of this domain. Examples of how cholesterol affects caveolar structure and function are outlined below. Specific methods for modulating cholesterol in caveolae are described in the following sections.

Invaginated caveolae are easily seen in thin-section electron microscopy (EM) images of cells. Rapid-freeze deep-etch EM has shown that these invaginated caveolae are decorated with a filamentous coat structure.[2] These images also provided strong evidence that caveolar shape is variable, ranging from flat to deeply invaginated. The first indication that cholesterol was an essential molecule of caveolae came from the discovery that depletion of membrane cholesterol inhibits the clustering of glycosylphosphatidylinositol (GPI)-anchored membrane proteins in caveolae.[8,9] These studies also showed that cholesterol depletion causes the loss of invaginated caveolae. Subsequent studies demonstrated that pharmacological reagents (see below) that bind cholesterol flatten caveolae and cause disassembly of the filamentous coat structure. Biochemical assays have confirmed that caveolae from control cells have a 5-fold higher cholesterol-to-protein ratio than caveolae from cholesterol-depleted cells.[10]

Cholesterol depletion also affects caveolar function. Caveolae in the kidney epithelial cell line MA104 are involved in the selective uptake of 5-methyltetrahydrofolic acid.[11,12] The uptake is selective because 5-methyltetrahydrofolic acid is delivered to the cytoplasm but the receptor remains associated with the cell membrane. This process is called potocytosis.[12] Internalization involves the binding of the vitamin to the folate receptor, which is a glycosylphosphatidylinositol (GPI)-anchored receptor enriched in caveolae.[9,11,13] In these cells, caveolae cycle between an open or extracellular exposed conformation and a closed or sequestered conformation once every 60 min. When caveolae close, a bafilomycin-sensitive proton pump transports protons into the caveolae lumen, thereby decreasing the

[8] W. J. Chang, K. G. Rothberg, B. A. Kamen, and R. G. Anderson, *J. Cell Biol.* **118,** 63 (1992).

[9] K. G. Rothberg, Y. Ying, B. A. Kamen, and R. G. W. Anderson, *J. Cell Biol.* **111,** 2931 (1990).

[10] E. J. Smart, Y. S. Ying, P. A. Conrad, and R. G. Anderson, *J. Cell Biol.* **127,** 1185 (1994).

[11] K. G. Rothberg, Y. Ying, J. F. Kolhouse, B. A. Kamen, and R. G. W. Anderson, *J. Cell Biol.* **110,** 637 (1990).

[12] R. G. Anderson, B. A. Kamen, K. G. Rothberg, and S. W. Lacey, *Science* **255,** 410 (1992).

[13] W.-J. Chang, K. G. Rothberg, B. A. Kamen, and R. G. W. Anderson, *J. Cell Biol.* **118,** 63 (1992).

pH. The ligand dissociates from the receptor and translocates to the cytosol through an anion transporter. The caveolae then reopen and the receptors are free to repeat the cycle. A critical feature of this mechanism is the ability of caveolae to completely invaginate and form a sequestered compartment, a process that may involve pinching off from the membrane to form a vesicle. The potocytosis of 5-methyltetrahydrofolic acid is inhibited when caveolae are depleted of cholesterol but resumes on sterol repletion.[9,13]

In addition to being the vehicle for potocytosis, caveolae also compartmentalize a variety of signaling activities that take place at the cell surface.[6,7,14] Caveolar cholesterol is essential for signal transduction that originates in this domain. Many signaling molecules enriched in caveolae are either acylated or prenylated and cholesterol depletion causes their mislocalization.[6,7,14] This leads to an alteration in signal transduction by these molecules. For example, the localization of endothelial nitric-oxide synthase (eNOS) to caveolae requires that it be both myristoylated and palmitoylated.[15] Depletion of caveolae cholesterol in endothelial cells, using either oxidized low-density lipoprotein or cyclodextrin (see below), causes the relocalization of eNOS to an intracellular compartment without a corresponding alteration in the acylation state of the enzyme.[16,17] Relocated eNOS no longer is activated by physiological stimuli such as acetylcholine.[16,17] Cholesterol is also required for the appropriate activity of many signaling molecules that reside in caveolae. Depletion of caveolar cholesterol in fibroblasts, for example, causes spontaneous activation of Extracellular signal-regulated kinase-1/2 (ERK1/2) and the addition of epithelial growth factor (EGF) causes further activation of this crucial regulatory kinase.[18]

Caveolin-1 is thought to play a role in maintaining the proper cholesterol level of caveolae. Exactly how it does this is not clear. The caveolae in cells that express caveolin-1 have a higher cholesterol-to-protein ratio than the surrounding membrane. A cytosolic pool of caveolin-1 in a complex with several heat shock proteins is an intermediate in transport of cholesterol from the endoplasmic reticulum (ER) to caveolae.[19] Palmitoylation of caveolin-1 appears to be necessary to promote association of cholesterol with caveolin-1 and the formation of cholesterol-rich caveolae.[20] Caveolin-1 has also been shown to function

[14] P. W. Shaul and R. G. Anderson, *Am. J. Physiol.* **275,** L843 (1998).

[15] P. W. Shaul, E. J. Smart, L. J. Robinson, Z. German, I. S. Yuhanna, Y. Ying, R. G. Anderson, and T. Michel, *J. Biol. Chem.* **271,** 6518 (1996).

[16] A. Uittenbogaard, P. W. Shaul, I. S. Yuhanna, A. Blair, and E. J. Smart, *J. Biol. Chem.* **275,** 11278 (2000).

[17] A. Blair, P. W. Shaul, I. S. Yuhanna, P. A. Conrad, and E. J. Smart, *J. Biol. Chem.* **274,** 32512 (1999).

[18] T. Furuchi and R. G. Anderson, *J. Biol. Chem.* **273,** 21099 (1998).

[19] A. Uittenbogaard, Y. Ying, and E. J. Smart, *J. Biol. Chem.* **273,** 6525 (1998).

[20] A. Uittenbogaard and E. J. Smart, *J. Biol. Chem.* **275,** 25595 (2000).

as an apolipoprotein in certain cells and to be secreted in lipid particles that have the characteristics of high-density lipoprotein (HDL).[21] Caveolin-1 may also function as a scaffolding protein for multiple signaling molecules. It contains a short amino acid sequence that has been termed the scaffolding domain.[22] Numerous *in vitro* binding studies have shown that this region of the molecule interacts with proteins that have a characteristic amino acid sequence called the caveolin-1-binding domain.[22–25] Thus, the ability of caveolin-1 to form a coat structure may position the scaffolding domain so that it directly binds to signaling molecules containing the caveolin-1-binding motif and maintain the protein in an inactive state.[7,14] Many methods that alter caveolar cholesterol levels cause caveolin-1 to move to other cellular locations. Regardless of its function, however, depletion of caveolin-1 from caveolae will perturb signal transduction from this domain.

Cholesterol-Binding Drugs

The first pharmacological reagents identified that disrupts the structure of caveolae were cholesterol-binding drugs such as filipin and nystatin.[9] Filipin is a polyene macrolide antibiotic that binds to cholesterol and disrupts the organization of the surrounding membrane.[26,27] de Kruijff and Demel proposed that plasma membrane cholesterol associates with filipin to generate bulky complexes that deform the plasma membrane.[28] Despite the possibility of nonspecific effects, filipin will disrupt the caveolar coat structure, cause caveolae to flatten, and prevent internalization of caveolae.[9] Another important feature of filipin is that it does not inhibit clathrin-mediated endocytosis, which makes it potentially useful for distinguishing between these two endocytic pathways.[9]

Method

1. Prepare a fresh 10-mg/ml stock solution of filipin in dimethyl sulfoxide (DMSO) (filipin complex is from Sigma, St. Louis, MO).
2. Wash the cells once with phosphate-buffered saline (PBS).
3. Add fresh culture medium without serum to the cells.

[21] P. Liu, W. P. Li, T. Machleidt, and R. G. Anderson, *Nat. Cell Biol.* **1,** 369 (1999).
[22] S. Li, J. Couet, and M. P. Lisanti, *J. Biol. Chem.* **271,** 29182 (1996).
[23] J. Couet, S. Li, T. Okamoto, T. Ikezu, and M. P. Lisanti, *J. Biol. Chem.* **272,** 6525 (1997).
[24] J. A. Engelman, C. Chu, A. Lin, H. Jo, T. Ikezu, T. Okamoto, D. S. Kohtz, and M. P. Lisanti, *FEBS Lett.* **428,** 205 (1998).
[25] G. García-Cardeña, P. Martasek, B. S. Masters, P. M. Skidd, J. Couet, S. Li, M. P. Lisanti, and W. C. Sessa, *J. Biol. Chem.* **272,** 25437 (1997).
[26] J. Bolard, *Biochim. Biophys. Acta* **864,** 257 (1986).
[27] J. Milhaud, *Biochim. Biophys. Acta* **1105,** 307 (1992).
[28] B. de Kruijff and R. A. Demel, *Biochim. Biophys. Acta* **26,** 57 (1974).

4. Add filipin to achieve a final concentration of 5–10 μg/ml [DMSO <0.05% (v/v)].

5. Incubate the cells at 37° for 15–60 min.

Potential Problems. The ability of filipin to permeabilize membranes makes it necessary to determine the concentration that will affect caveolar function without damaging the cell. At low concentrations, filipin will disrupt caveolar function with minimum side effects. Too high a concentration, however, will dramatically affect cell morphology, cell permeability, and cell viability. Trypan Blue exclusion, lactate dehydrogenase release, and transferrin uptake are standard tests of cell viability that should be used to establish how each cell type responds to the drug. As with any reagent, filipin can have side effects and should not be considered a specific inhibitor of caveolar function. Nevertheless, when combined with other tests, filipin can be useful for identifying events in the cell that depend on caveolae.

Cholesterol Oxidase

Cholesterol oxidase is a bacterial enzyme that converts cholesterol to cholest-4-en-3-one.[29] The activity of the enzyme on membrane cholesterol is greatly influenced by the local lipid environment within the membrane. The lateral surface pressure of the membrane, the phospholipid composition, and the amount of cholesterol are all critical factors that influence the function of cholesterol oxidase.[29,30] One of the first uses of cholesterol oxidase was for detecting the distribution of cellular cholesterol in glutaraldehyde-fixed cells.[30] Nevertheless, cholesterol oxidase will work on live cells, where it preferentially oxidizes caveolar cholesterol and causes caveolin-1 to accumulate in the Golgi apparatus.[10] This property of cholesterol oxidase has found use in studying signal transduction from caveolae as well as the transport of cholesterol to caveolae.

Method

1. Cholesterol oxidase is from Boehringer Mannheim (Indianapolis, IN) (*Rhodococcus erythropolis,* 25 units/mg).

2. Wash cells once with PBS.

3. Add fresh cell culture medium without serum to the cells.

4. Add cholesterol oxidase to achieve a final concentration of 0.5 units/ml.

5. Incubate the cells at 37° for 60 min.

---

[29] J. MacLachlan, A. T. L. Wotherspoon, R. O. Ansell, and C. J. W. Brooks, *J. Steroid Biochem. Mol. Biol.* **72,** 169 (2000).

[30] Y. Lange, *J. Lipid Res.* **33,** 315 (1992).

Potential Problems. The main difficulty with this method is that not all cells respond to cholesterol oxidase.[29,30] Therefore, the activity of the enzyme needs to be tested on each cell type by assaying for the conversion of caveolar cholesterol to cholest-4-en-3-one. Cells that do respond need to be grown and handled according to a standard format. For example, the cell density can affect the ability of cholesterol oxidase to access the sterol. The presence of other cholesterol sources such as serum lipoproteins also dramatically affects the activity of the enzyme. Serum-free conditions should always be used. The temperature of the reaction is another variable that needs to be controlled. The enzyme is less active at reduced temperatures. Thus, cholesterol oxidase is an effective tool for studying the function of caveolae and caveolin-1 when the conditions are optimized and the activity of the enzyme is determined for each experiment.

Cyclodextrin

Cyclodextrins are gaining wide acceptance as a tool for modifying the structure and function of caveolae. β-Cyclodextrins are cyclic heptasaccharides consisting of $\beta(1-4)$-glucopyranose units.[31] Numerous cyclodextrins exist but the one most effective at removing membrane cholesterol is methyl-β-cyclodextrin.[32] This cyclodextrin is water soluble and contains a hydrophobic core that has a specific affinity for cholesterol. Cholesterol bound to methyl-β-cyclodextrin is soluble but can exchange with plasma membrane cholesterol. Moreover, the addition of methyl-β-cyclodextrin to cells effectively removes cholesterol from living cells. The efficiency of extraction varies with cell type.[33] Cyclodextrins remove cholesterol from all parts of the membrane, not just caveolae. Work by Haynes *et al.* suggests that for each cell type conditions can be found wherein methyl-β-cyclodextrin will selectively extract cholesterol from different membrane pools of the sterol.[34] Selective removal of cholesterol from caveolae may be favored at low concentrations of methyl-β-cyclodextrin because of the high cholesterol content of this membrane domain.

Method

1. Prepare a fresh 500 mM stock solution of methyl-β-cyclodextrin in water.
2. Wash cells once with PBS.
3. Add fresh cell culture medium without serum to the cells.
4. Add methyl-β-cyclodextrin to a final concentration of 5 mM.
5. Incubate the cells at 37° for 1–3 hr.

[31] J. Pitha, T. Irie, P. B. Sklar, and J. S. Nye, *Life Sci.* **43,** 493 (1988).
[32] T. Irie, K. Fukunaga, and J. Pitha, *J. Pharm. Sci.* **81,** 521 (1992).
[33] E. P. Kilsdonk, P. G. Yancey, G. W. Stoudt, F. W. Bangerter, W. J. Johnson, M. C. Phillips, and G. H. Rothblat, *J. Biol. Chem.* **270,** 17250 (1995).
[34] M. P. Haynes, M. C. Phillips, and G. H. Rothblat, *Biochemistry* **39,** 4508 (2000).

Potential Problems. There are several factors that influence the ability of methyl-β-cyclodextrin to remove cholesterol. Because methyl-β-cyclodextrin can promote both efflux and influx of cholesterol, the culture medium cannot contain a source of cholesterol such as serum. Long-term incubation of cells in the presence of methyl-β-cyclodextrin (>6 hr) will reduce cellular cholesterol and upregulate genes that control endogenous cholesterol levels. It can also affect cell viability. Each experimental paradigm needs to be tested to ensure that methyl-β-cyclodextrin is removing caveolar cholesterol without causing massive changes in the total plasma membrane cholesterol. Therefore, tests should be performed to determine the amount of cholesterol in the caveolar fraction compared with the total plasma membrane after methyl-β-cyclodextrin treatment. Finally, if methyl-β-cyclodextrin is observed to have an affect, the addition of cholesterol to the medium should block this affect.

Oxidized Low-Density Lipoprotein

Oxidized low-density lipoprotein (LDL) also alters the level of caveolar cholesterol and the localization of caveolin-1 in a receptor-dependent manner.[16,17] Oxidized LDL bound to CD36 receptors located in caveolae promotes the efflux of caveolar cholesterol. In addition, we have demonstrated that HDL, in a scavenger receptor class B, type I (SR-BI)-dependent manner, delivers sterol to caveolae and counteracts the effects of oxidized LDL.[16,17] Cells must express CD36 in caveolae for this method to work. Immunoblotting of isolated caveolar fractions and immunofluorescence are the recommended methods for determining whether CD36 is localized to caveolae. The preparation of LDL and oxidized LDL has been described elsewhere.[35–37]

Method

1. Prepare fresh oxidized LDL [5–15 TBARs (thiobarbituric acid-reactive substances)] as described Buege and Aust.[37]
2. Wash cells once with PBS.
3. Add fresh cell culture medium without serum to the cells.
4. Add oxidized LDL to a final concentration of 10–50 μg/ml.
5. Incubate the cells at 37° for 1–3 hr.

Potential Problems. Cells should not be exposed to oxidized LDL for more than a few hours. Lipoproteins potentially can affect the level of caveolin-1 in cells.

---

[35] W. R. Fisher and V. N. Schumaker, *Methods Enzymol.* **128,** 247 (1986).
[36] J. L. Kelley and A. W. Kruski, *Methods Enzymol.* **128,** 170 (1986).
[37] J. A. Buege and S. D. Aust, *Methods Enzymol.* **52,** 302 (1978).

In addition, LDL bound to CD36 may activate peroxisome proliferator-activated receptor γ (PPARγ)-regulated genes. As with all the experimental protocols, the state of cholesterol in the caveolar fraction needs to be determined.

Quantification of Cholesterol

Because it is necessary to determine the amount of cholesterol associated with caveolae and plasma membranes, we have included two common methods for quantifying cholesterol: radioisotope labeling and direct mass measurement. Both these methods are reliable and widely used. The radioisotope labeling method is useful for pulse–chase experiments and distinguishing newly synthesized cholesterol from existing cholesterol pools. The direct mass measurement method circumvents potential problems with exchange and isotope dilution but is less sensitive than isotope labeling.

Method: Radioisotope Labeling

See Smart et al.[38] for details.

1. Wash cells once with an acetate-free buffer such as PBS.
2. Add medium and lipoprotein-deficient serum to the cells.
3. Add unlabeled acetate to a final concentration of 10 μM and add 500 μCi of [^{3}H]acetate (New England Nuclear, Boston, MA) to one 150-mm plate of cells.
4. Incubate the cells for 16–24 hr at 37°.
5. Process the cells as required (fractionation, lysate, etc.).
6. Adjust the volume of each sample to 1 ml with water and transfer the sample to a 13 × 100 mm glass tube.
7. Add 1.2 ml of 2% (v/v) acetic acid in methanol and mix thoroughly.
8. Add 1.2 ml of chloroform and mix thoroughly.
9. Centrifuge at 1000g for 15 min at room temperature.
10. Take the bottom layer (chloroform) and transfer it to a microcentrifuge tube.
11. Add 5 μl of cholesterol (10 mg/ml in chloroform) to each sample to aid in the identification of the appropriate spots on the thin-layer chromatography plates.
12. Dry the samples under nitrogen.
13. Suspend samples in the thin-layer chromatography solvent system [petroleum ether–ethyl ether–acetic acid, 80 : 20 : 1 (v/v/v)].
14. Spot the samples and standards onto a silica gel plate [Si250-PA (19C); VWR, Scientific, San Francisco, CA].
15. Place the plate in the solvent chamber and allow the solvent to migrate to within 1 in. of the top of the plate.

[38] E. J. Smart, Y.-S. Ying, W. C. Donzell, and R. G. Anderson, *J. Biol. Chem.* **271**, 29427 (1996).

16. Remove the plate and air dry.

17. Lightly spray the plate with a fresh solution of 5% (v/v) sulfuric acid made in ethanol.

18. Let the plate completely air dry and then bake for 10 min at 170°.

19. Scrape the appropriate spots into scintillation fluid and count.

Method: Direct Mass Determination

See Uittenbogaard et al.[16] for details.

1. Process the cells as required (fractionation, lysate, etc.).

2. Adjust the volume of each sample to 1 ml with water and transfer the sample to a 13 × 100 mm glass tube.

3. Add 3 ml of chloroform–acetic acid (1 : 2, v/v) and mix.

4. Add 1 ml of chloroform and mix.

5. Add 1 ml of water and mix.

6. Centrifuge at 1000g of 10 min at room temperature.

7. Take the bottom layer (chloroform) and transfer to a 10 × 75 mm glass tube.

8. Completely dry the samples and standards [1, 2, 3, 5, 10, and 20 μg of standard solution from a Wako (Tokyo, Japan) cholesterol determination kit] with nitrogen.

9. Add 1 ml of 1% (v/v) Triton X-100 made in chloroform to each sample.

10. Dry with nitrogen. The sample will coat the tube.

11. Add 0.5 ml of water to dissolve the Triton X-100 and sample.

12. Add 0.5 ml of 2× enzyme mix to samples and standards. The enzyme mix is from a commercial cholesterol determination kit (Wako). The enzyme mix must be diluted fresh each time.

13. Incubate the samples and standards at 37° for 15 min.

14. Measure the absorbance at 505 nm.

Summary

Most of the available methods for modifying caveolae structure and function depend on altering the cholesterol content of caveolae. The most important aspect of each method is to ensure the reagents are working in the cells that are being studied. The idiosyncrasies of each method are such that they cannot be universally applied without carefully optimizing the conditions. When used correctly, these methods are accepted as a specific way to perturb the structure and function of caveolae.

[14] Contribution of Neelaredoxin to Oxygen Tolerance by *Treponema pallidum*

By KARSTEN R. O. HAZLETT, DAVID L. COX, ROBERT A. SIKKINK,
FRANÇOISE AUCH'ERE, FRANK RUSNAK, and JUSTIN D. RADOLF

Introduction

Treponema pallidum, the syphilis spirochete, remains one of the few major bacterial pathogens of humans that cannot be cultivated *in vitro,* although limited replication has been achieved by coculture with rabbit epithelial cells.[1–4] During the course of infection, the bacterium encounters ambient oxygen tensions ranging from the relatively low partial pressures within peripheral tissues (approximately 40 mmHg) to the much higher partial pressures of arterial blood (100 mmHg).[5] Moreover, infection of the cerebrospinal fluid (CSF), a body fluid presumed to have a relatively high level of oxygenation, occurs in a substantial proportion of syphilis patients.[6,7] That *T. pallidum* readily disseminates hematogeneously and can survive within well-oxygenated host environments presents something of a paradox when viewed against *in vitro* studies demonstrating that oxygen concentrations above 5% are inhibitory, and those above 12% are lethal.[8–12] Assuming that this conception of the *T. pallidum*'s exposure to oxygen and reactive oxygen species (ROS) *in vivo* is correct, it follows that *T. pallidum* is protected from ROS during infection by mechanisms that are not accurately reproduced by current *in vitro* cultivation systems. Identification of these factors could provide us with a much better understanding of how *T. pallidum* survives within its obligate human host

[1] D. L. Cox, R. A. Moeckli, and A. H. Fieldsteel, *In Vitro* **20,** 879 (1984).

[2] A. H. Fieldsteel, D. L. Cox, and R. A. Moeckli, *Infect. Immun.* **32,** 908 (1981).

[3] A. H. Fieldsteel, D. L. Cox, and R. A. Moeckli, *Infect. Immun.* **35,** 449 (1982).

[4] S. J. Norris, *Infect. Immun.* **36,** 437 (1982).

[5] G. J. Tortora and S. R. Grabowski, "Principles of Anatomy and Physiology." John Wiley & Sons, New York, 2000.

[6] R. T. Rolfs, M. R. Joesoef, A. M. Rompalo, M. H. Augenbraun, M. Chiu, G. Bolan, S. G. Johnson, P. French, E. Steen, J. D. Radolf, and S. A. Larsen for the Syphilis and HIV Study Group, *N. Engl. J. Med.* **337,** 307 (1997).

[7] S. A. Lukehart, E. W. Hook, S. A. Baker-Zander, A. C. Collier, C. W. Critchlow, and H. H. Handsfield, *Ann. Intern. Med.* **109,** 855 (1988).

[8] H. Noguchi, *J. Exp. Med.* **14,** 98 (1911).

[9] J. B. Baseman and N. S. Hayes, *Infect. Immun.* **10,** 1350 (1974).

[10] W. H. Cover, S. J. Norris, and J. N. Miller, *Sex. Transm. Dis.* **7,** 1 (1982).

[11] D. L. Cox, B. Riley, P. Chang, S. Sayahtaheri, S. Tassell, and J. Hevelone, *Appl. Environ. Microbiol.* **56,** 3063 (1990).

[12] S. J. Norris, J. N. Miller, J. A. Sykes, and T. J. Fitzgerald, *Infect. Immun.* **22,** 689 (1978).

Copyright 2002, Elsevier Science (USA).
All rights reserved.
0076-6879/02 $35.00

and could be instrumental for achieving continuous *in vitro* propagation, a goal that has eluded syphilis researchers for nearly a century.

Resolving the apparent contradiction between the oxygen sensitivity of the syphilis spirochete and its ability to withstand host environments with high redox potential will require (1) delineation of the pathways in this bacterium for oxygen detoxification and (2) a comparative analysis of their function *in vivo* and *in vitro*. The complete genomic sequence[13] represents a powerful new tool with which to identify the oxidative defense enzymes of the spirochete, a line of investigation now undergoing a resurgence among syphilis researchers.[14,15] *Treponema pallidum* lacks the oxidative defense enzymes superoxide dismutase (SOD), catalase, and peroxidase[13] typically found in aerotolerant microorganisms. It does, however, express neelaredoxin, a superoxide reductase (SOR) that catalyzes the reduction of superoxide to hydrogen peroxide, which, presumably, compensates for the lack of SOD.[14,15] In addition, *T. pallidum* appears also to possess homologs for the additional constituents of a primitive, SOR-dependent, oxygen detoxification pathway (i.e., NADH oxidase, rubredoxin, thioredoxin reductase, thioredoxin, and the C subunit of alkyl hydroperoxide reductase).[13] Consistent with the microaerophilic nature of the spirochete *in vitro,* SORs have, thus far, been identified only in anaerobic sulfate-reducing archaea and microaerophilic sulfate-reducing bacteria.[16-21]

Although serial passage has yet to be achieved, the current *in vitro* cocultivation system allows investigators to monitor treponemal responses to various oxygen tensions. As such, syphilis researchers are now poised to compare both the transcriptional and enzymatic activities of these putative oxygen detoxification

[13] C. M. Fraser, S. J. Norris, G. M. Weinstock, O. White, G. C. Sutton, R. Dodson, M. Gwinn, E. K. Hickey, R. Clayton, K. A. Ketchum, E. Sodergren, J. M. Hardham, M. P. McLeod, S. Salzberg, J. Peterson, H. Khalak, D. Richardson, J. K. Howell, M. Chidambaram, T. Utterback, L. McDonald, P. Artiach, C. Bowman, M. D. Cotton, C. Fujii, S. Garland, B. Hatch, K. Horst, K. Roberts, M. Sandusky, J. Weidman, H. O. Smith, and J. C. Venter, *Science* **281,** 375 (1998).

[14] M. Lombard, D. Touati, M. Fontecave, and V. Nivière, *J. Biol. Chem.* **275,** 27021 (2000).

[15] T. Jovanovíc, C. Ascenso, K. R. O. Hazlett, R. Sikkink, C. Krebs, R. Litwiller, L. M. Benson, I. Moura, J. J. G. Moura, J. D. Radolf, B. H. Huynh, and F. Rusnak, *J. Biol. Chem.* **275,** 28439 (2000).

[16] F. E. Jenney, Jr., M. F. J. M. Verhagen, X. Cui, and M. W. W. Adams, *Science* **286,** 306 (1999).

[17] M. Lombard, M. Fontecave, D. Touati, and V. Nivière, *J. Biol. Chem.* **275,** 115 (2000).

[18] H. L. Lumppio, N. V. Shenvi, A. O. Summers, G. Voordouw, and D. M. Kurtz, Jr., *J. Bacteriol.* **183,** 101 (2000).

[19] I. A. Abreu, L. M. Saraiva, J. Carita, H. Huber, K. O. Stetter, D. Cabelli, and M. Teixeira, *Mol. Microbiol.* **38,** 322 (2000).

[20] C. V. Romão, M. Y. Liu, J. Le Gall, C. M. Gomes, V. Braga, I. Pacheco, A. V. Xavier, and M. Teixeira, *Eur. J. Biochem.* **261,** 438 (1999).

[21] C. Ascenso, F. Rusnak, I. Cabrito, M. J. Lima, S. Naylor, I. Moura, and J. J. G. Moura, *J. Biol. Inorg. Chem.* **5,** 720 (2000).

genes under controlled oxygen tensions. Conversely, applying these methodologies to treponemes harvested from rabbit tissues, blood, and cerebrospinal fluid could yield clues to the ability of the treponeme to survive hematogeneous dissemination and CSF invasion. Presumably, comparing transcriptional profiles of *in vitro*- and *in vivo*-grown treponemes would shed light on the critical "missing ingredients" of the current *in vitro* cultivation techniques.

This chapter summarizes the methods of treponemal propagation both *in vivo* and *in vitro,* and the techniques employed to analyze the expression and enzymatic activity of neelaredoxin. A similar conceptual and methodological synthesis could be engaged to empirically confirm the identity of the remaining constituents of the oxygen detoxification pathway in this bacterium.

Propagation of *Treponema pallidum* and Effects of Oxidative Stress

Soon after the isolation of *Treponema pallidum* as the etiologic agent of syphilis, Noguchi[8] reported in 1911 that anaerobiosis was beneficial to its survival *in vitro*. Over the next 70 years, many attempts were made to culture *T. pallidum,* the best of which prolonged viability a few days, but failed to promote treponemal replication. Starting in the mid-1970s, results from pure culture methods began to refine our understanding of treponemal oxygen tolerance. Norris *et al.* demonstrated that anaerobic or near anaerobic ($<0.5\%$ O_2) conditions were actually detrimental to spirochete viability, and that O_2 above 12.5%, even in the presence of reducing agents, caused rapid loss of viability.[12] Baseman and Hayes[9] reported that low oxygen tensions stimulated treponemal protein synthesis and glucose utilization. Subsequently, it was reported that *T. pallidum* lacked endogenous catalase and superoxide dismutase activities,[22] was sensitive to superoxide,[23] and was 10 times more sensitive to hydrogen peroxide than *Escherichia coli*.[24] Results from early work with coculture systems supported these findings in that dissolved oxygen tensions, from 2 to 5%, were found to be most beneficial for long-term (5- to 21-day) viability of the treponemes.[2,25,26]

In 1980, by incorporating a reduced oxygen partial pressure, the potent reducing agent dithiothreitol (DTT), a slow-growing rabbit cell line (SflEp), and prescreened lots of fetal bovine serum (FBS), Fieldsteel and co-workers established the first *in vitro* coculture system that supported treponemal replication.[2] DTT was used in place of cysteine and glutathione as these reducing agents are

[22] F. E. Austin, J. T. Barbieri, R. E. Corin, K. E. Grigas, and C. D. Cox, *Infect. Immun.* **33,** 372 (1981).

[23] B. M. Steiner, G. H. W. Wong, and S. R. Graves, *Br. J. Vener. Dis.* **60,** 14 (1984).

[24] B. M. Steiner, G. H. W. Wong, P. Sutrave, and S. R. Graves, *Can. J. Microbiol.* **30,** 1467 (1983).

[25] T. J. Fitzgerald, R. C. Johnson, J. A. Sykes, and J. N. Miller, *Infect. Immun.* **15,** 444 (1977).

[26] A. H. Fieldsteel, F. A. Becker, and J. G. Stout, *Infect. Immun.* **18,** 173 (1977).

readily inactivated in the presence of oxygen. By screening multiple lots of fetal bovine serum, it was noted that a low FBS iron content was critical for spirochetal growth. Although useful as an enzyme cofactor, iron is well recognized as a redox-active element that can generate reactive oxygen species (ROS) and participate in Fenton chemistry. Although the reason(s) that the Sf1Ep cell line supports treponemal growth better than other cell lines is unknown, it has been postulated that the slow growth of these cells results in decreased generation of ROS. During the next 10 years, further improvements in the tissue culture system came from efforts to neutralize ROS. The addition of exogenous antioxidants such as SOD (25 U/ml), catalase (10 U/ml), mannitol (550 μM), histidine (230 μM), and vitamin E (16 nM) was reported to further increase the growth of *T. pallidum in vitro*.[11] The rationale for adding multiple antioxidants was that each of the compounds acted on a distinct ROS. Moreover, the chemical antioxidants were hypothesized to cross the treponemal outer membrane to protect cellular constituents.

Although serial passage *in vitro* has yet to be achieved, the similarities between the *in vitro* cultivation conditions and those *in vivo* suggest that the cocultivation system is suitable to molecular analysis of treponemal oxygen tolerance, such as oxygen- and/or ROS-dependent transcriptional responses. *in vivo,* the generation time of *T. pallidum* is estimated to be between 30 and 33 hr[27]; the generation time *in vitro* is between 35 and 40 hr. The dissolved oxygen (dO_2) in both the tissue culture flasks and human tissues is near 5%.[5] The following section summarizes both the rabbit model of treponemal propagation and the current *in vitro* coculture system that has been used to study the oxygen requirements and tolerances of *T. pallidum*. With minor modification, these systems should prove to be valuable tools in the molecular analysis of *T. pallidum* oxygen detoxicification pathways.

In Vivo Propagation of Treponema pallidum in Rabbit Testis

Preparation of Medium for Extraction of Treponema pallidum from Rabbit Testis. Usually 50 ml of *T. pallidum* culture medium base (TpCM base; Table I) without DTT is prepared the day before treponemes are to be harvested. The pH of the base is adjusted to pH 7.4, filter sterilized, and placed in a sterile 125-ml side-arm flask and sealed. The flask is alternately evacuated and gassed with a 5% CO_2–95% (v/v) N_2 mixture three times and stored. The next day a fresh solution of DTT is prepared (5 mg in 2 ml of base), filter sterilized, and added to the flask. The flask is regassed as described above and stored for the extraction of treponemes. It must be noted that the lot of FBS used for extraction and cultivation must first be screened for its ability to support treponemal growth; many lots of FBS are toxic. Several samples can then be obtained from various serum companies and compared with a reference lot. In our experience, about one out of every four or

[27] M. C. Cumberland and T. B. Turner, *Am. J. Syph.* **33**, 201 (1949).

TABLE I
FORMULATION OF *Treponema pallidum* CULTURE MEDIUM

Component	Amount
1. Base medium	
Earle's salt solution, 10×	100 ml
Essential amino acids (EAA),[a] 50×	10 ml
Non-EAA,[a] 100×	10 ml
Vitamins,[a] 100×	10 ml
Glucose	2.5 g
L-Glutamine, 200 mM	10 ml
NaHCO$_3$, 7.5% (w/v)	27 ml
MOPS buffer, 1 M (pH 7.4)	25 ml
Sodium pyruvate	100 mg
Dithiothreitol	150 mg
FBS[b]	100–200 ml
2. Antioxidants	
CoCl$_2$	5 μg
Cocarboxylase	2 μg
Mannitol	100 mg
Histidine	50 mg
Catalase	10,000 U
Superoxide dismutase	25,000 U
3. Ultrapure water	to 1 liter

[a] Flow Laboratories (Rockville, MD).
[b] Depends on the particular lot; some lots support growth better at 10% (v/v).

five is suitable for cultivation. All lots used in our experiments support growth that is 90% or greater than that of the reference lot.

Propagation of Treponema pallidum in Vivo and Harvesting Treponemes for in Vitro Cultivation. New Zealand White male rabbits (10–12 lb) are used for *in vivo* growth of *T. pallidum* and as a source of treponemes for inoculation of Sf1Ep tissue culture cells. For passage, 10^8 treponemes in 0.25 ml of TpCM are inoculated into rabbit testis. This inoculum originates from a 50% (v/v) glycerol frozen stock of treponemes (4×10^8/ml) from a previous rabbit harvest. Ten days after infection, the rabbits are killed and the testes are aseptically removed. Treponemes are extracted from the testis in a laminar flow hood, using universal precautions. The testes are trimmed of any fatty tissue, minced in a sterile petri dish, and placed in a sterile 125-ml Erlenmeyer flask containing 12 ml of fresh TpCM base (see Table I). The flask is gassed with a mixture of 92% N$_2$–5% CO$_2$–3% O$_2$ (v/v) for 30 sec, sealed with a sterile silicone stopper, and placed on a orbital shaker operating at 120 oscillations per minute for 30 min. The TpCM base containing treponemes and testis extract is removed from the minced testicular tissue with a pipette and transferred into a 50-ml polypropylene conical centrifuge tube. The tube is briefly gassed with the gas mixture described above and sealed. The extract is then centrifuged at $500g$ for 10 min at $4°$ to remove gross debris. Dilutions of the supernatant (1 : 10 and 1 : 100) are prepared and the treponemes are counted by dark-field microscopy, using a Petroff–Hausser counting chamber. The remaining treponemal suspension is gassed and sealed for later use. This suspension is used as a source of inoculum and also for the preparation of testis extract.

In Vitro Cocultivation of Treponema pallidum with Sf1Ep Cells

Cultivation of Tissue Culture Cells. The slow-growing Sf1Ep cells are grown in Eagle's minimal essential medium (EMEM; GIBCO, Grand Island, NY) containing 10% (v/v) FBS unless otherwise noted. It is critical that the medium be free of any antibiotics because they will prevent infection of the monolayers with *T. pallidum.* Approximately 4×10^6 cells in 25 ml of EMEM are seeded into a 150-cm^2 tissue culture flask. The medium is removed and replaced with fresh EMEM after 5 and 9 days of incubation. The cells are harvested and passaged every 2 weeks. The Sf1Ep monolayers are usually set up at a confluency of 25% two days before infection with *T. pallidum.*

Preparation of Treponema pallidum Culture Medium for Cocultivation. TpCM is made from the base medium by adding the appropriate amounts of antioxidants listed in Table I and 3.33 ml of testis extract per 100 ml of TpCM. Testis extract is prepared by removing most of the treponemes from a portion of the treponemal suspension by centrifugation at $12,000g$ for 10 min. The supernatant is removed with a pipette and placed into a 50-ml polypropylene tube, gassed with the CO_2–N_2 mixture, and sealed. The extract is heat inactivated at $56°$ for 30 min. Any precipitate is removed by centrifugation at $12,000g$ for 10 min. The complete TpCM is alternately evacuated and gassed with a 5% CO_2–95% N_2 mixture three times. The appropriate number of treponemes is added to the TpCM before infection of the tissue cultures.

Infection of Sf1Ep Monolayers with Treponema pallidum. To infect tissue cultures, the EMEM is removed with a pipette and replaced with the appropriate volume of TpCM containing the treponemes. If the tissue culture vessel is a flask, it is briefly gassed with 5% CO_2 and 95% N_2 and sealed. In contrast, tissue culture plates are placed in a vacuum chamber (Coy Laboratory Products, Ann Arbor, MI), evacuated (to -10 lb/in^2), and gassed with a 5% CO_2/95% N_2 mixture three times. When all the cultures have been prepared, the caps on tissue culture flasks are loosened and the tissue culture plates are removed from the vacuum chamber. All cultures are incubated at $34°$ in a Tri-Gas incubator (Forma Scientific, Marietta, OH), where a microaerophilic environment (3.5% O_2, 5% CO_2, and 91.5% N_2) is maintained throughout the cultivation period.

Quantitation of Treponemal Growth. Cultures are routinely counted after 5, 7, 10, 12, 14, and 17 days of incubation to obtain a complete growth curve. Treponemal growth is routinely monitored by harvesting the cultures with a mixture of 0.05% (w/v) trypsin and 0.53 mM EDTA in phosphate-buffered saline (PBS). To harvest the cultures, the TpCM is removed and placed into a 15-ml conical centrifuge tube. The cultures are then rinsed with 2 ml of PBS, which is removed and placed in the conical centrifuge tube with the TpCM. Three milliliters of trypsin–EDTA solution is added and the cultures are briefly gassed with 5% CO_2–95% N_2. The centrifuge tubes are also gassed. After the cultures are incubated at 34° for 5 min, the culture monolayers are repeatedly triterated to dislodge all the tissue culture cells. This suspension is added to the TpCM in the conical centrifuge tube. If the sample is stored for more than 5 min before counting, it is gassed with CO_2–N_2. The concentration of treponemes in the TpCM suspension is determined by placing 10 μl in a Petroff–Hausser counting chamber followed by enumeration by dark-field microscopy.

Analysis of Effect of Atmospheric Oxygen Concentration on Growth of Treponema pallidum. To study the effects of oxygen tension on the growth of *T. pallidum* in coculture, multiple flasks of Sp1Ep cells are infected with treponemes in TpCM lacking antioxidants. When the cocultures have been prepared, the tissue culture flasks are gassed with 5% CO_2, O_2 varying from 0.3 to 12.5%, and nitrogen. The flasks are then sealed and incubated at 34°. Treponemal growth is measured every 2 days over 16 days by harvesting and enumerating spirochetes from one or two flasks per oxygen concentration. By using this method we have found that treponemal growth in coculture occurs with O_2 concentrations between 0.5 and 10%, with optimal growth between 1.5 and 5% (Fig. 1). Interestingly, this oxygen concentration (5%) is the same as that of many human tissues.[5]

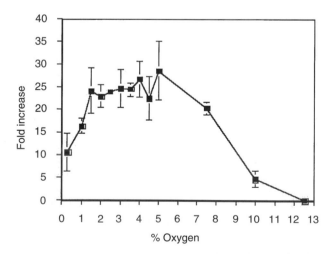

FIG. 1. The effect of atmospheric oxygen concentration on the growth of *T. pallidum* cocultured with Sf1Ep cells. Treponemes were grown in culture with Sf1Ep cells in TpCM without antioxidants (base medium). The flasks were gassed with 5% CO_2, O_2 varying from 0.3 to 12.5%, with the remainder nitrogen. Growth was monitored over 12 days and maximum growth was recorded.

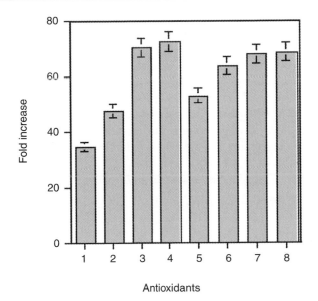

FIG. 2. Effects of antioxidants on the growth of *T. pallidum*. Treponemes were grown in culture with Sf1Ep cells in TpCM base medium with 5% oxygen and various antioxidants as listed in Table I. Column 1, no antioxidants; column 2, catalase; column 3, catalase and superoxide dismutase; column 4, mannitol, histidine, CoCl$_2$, and cocarboxylase; column 5, histidine, CoCl$_2$, and cocarboxylase; column 6, mannitol, CoCl$_2$, and cocarboxylase; column 7, mannitol, histidine, and cocarboxylase; column 8, mannitol, histidine, and CoCl$_2$.

Analysis of Effect of Various Antioxidants on Growth of Treponema pallidum. To study the effects of antioxidants on the growth of *T. pallidum* in coculture, multiple flasks of Sp1Ep cells are infected with treponemes in TpCM containing various antioxidants. When the cocultures have been prepared, the caps on tissue culture flasks are loosened. All cultures are incubated at 34° in a Tri-Gas incubator (Forma Scientific), where a microaerophilic environment (5% O$_2$, 5% CO$_2$, and 90% N$_2$) is maintained throughout the cultivation period. Treponemal growth is measured every 2 days over 16 days by harvesting and enumerating spirochetes from one or two flasks per antioxidant condition. We have found that treponemal growth in coculture is markedly enhanced by the addition of superoxide dismutase, catalase, histidine, and mannitol (Fig. 2), whereas CoCl$_2$ and cocarboxylase make smaller contributions.

Reactive Oxygen Species Detoxification in *Treponema pallidum*: Novel Mechanism for Elimination of Superoxide

The molecular basis for the susceptibility of *T. pallidum* to oxygen has been clarified, in part, by the finding that it lacks genes encoding SOD, catalase, and peroxidase.[13] However, a complete lack of oxygen detoxification systems is clearly at odds with the treponemal requirement for oxygen noted previously.[10,11] This paradox is potentially resolved by the observation that *T. pallidum* contains a gene (TP0823) encoding neelaredoxin (TpNlr), a mononuclear iron-containing enzyme. Orthologs of Nlr, found in microaerophilic and anaerobic organisms, have been shown to complement *E. coli* SOD mutants,[28,29] and more recently, to catalyze the reduction of superoxide to hydrogen peroxide without the production of molecular oxygen characteristic of superoxide dismutation.[16–18] We and others have employed the following techniques to characterize the SOR activity of TpNlr *in vitro,* and the expression of TpNlr by *T. pallidum.*[14,15]

In Vitro Analysis of Treponema Neelaredoxin Enzymatic Activity

Purification of Recombinant Neelaredoxin. Whereas quantities of native super-oxide reductases sufficient for enzymatic analysis have been purified from hundreds of liters of cultivable organisms,[16,19,30,31] the inability to continuously cultivate *T. pallidum* precludes this approach with the treponemal homolog, TpNlr. To obtain TpNlr for enzymatic analysis, the *nlr* gene is overexpressed in the heterologous host *E. coli.*

1. The *nlr* gene is amplified from *T. pallidum* chromosomal DNA, using primers 5′-CCATGG<u>CATATG</u>GGACGGGAGTTGTCG-3′ and 5′-TTAAGCTT <u>GGATCC</u>CTACTTACCTGACCACAC-3′, which contain engineered *Nde*I and *Bam*HI sites (underlined), respectively. The *Nde*I- and *Bam*HI-digested polymerase chain reaction (PCR) product is ligated to similarly digested plasmid pT7-7 and transformed into *E. coli* DH5-α, which are plated on Luria–Bertani (LB) agar containing ampicillin (100 μg/ml) for overnight growth at 37°. Plasmids purified from 5-ml broth cultures of individual colonies are screened for *nlr* by both restriction enzyme digestion and nucleic acid sequence analysis.

2. Competent *E. coli* BL21(DE3) are transformed with a pT7-7 : *nlr* plasmid and plated on LB agar containing ampicillin (100 μg/ml). After growth at 37°, an individual colony is used to inoculate 10 ml of LB broth containing ampicillin (100 μg/ml) for aerated growth; this is subsequently used to inoculate 1.5 liters of LB broth. When the OD_{595} of the 1.5-liter culture reaches 0.8, high-level expression of Nlr is induced by the addition of isopropyl-β-D-thiogalactopyranoside (IPTG) to 1 mM followed by 6 hr of continued aerated growth.

3. *Escherichia coli* are harvested by centrifugation at 2500g at 4°, resuspended in 50 mM Tris-HCl (pH 7.8), and lysed by three passages through a French press

[28] M. J. Pianzzola, M. Soubes, and D. Touati, *J. Bacteriol.* **178,** 6736 (1996).

[29] S. I. Liochev and I. Fridovich, *J. Biol. Chem.* **272,** 25573 (1997).

[30] I. Moura, P. Tavares, J. J. G. Moura, N. Ravi, B. H. Huynh, M. Y. Liu, and J. Legall, *J. Biol. Chem.* **265,** 21596 (1990).

[31] L. Chen, P. Sharma, A. M. Mariano, M. Teixeira, and A. V. Xavier, *Eur. J. Biochem.* **226,** 613 (1994).

operating at 15,000 lb/in^2. The lysate is clarified by 1 hr of centrifugation at 39,100g, after which the supernatant is loaded on a column (2.6 × 17.5 cm) containing DEAE-Sepharose CL6B anion-exchange resin equilibrated with 20 mM Tris-HCl, pH 7.8.

4. After washes with the same buffer, a 0 to 1.0 M NaCl linear gradient in 20 mM Tris-HCl, pH 7.8, is applied to the column. TpNlr usually elutes as a blue fraction at an NaCl concentration of ~0.1 M. The blue color results from the Fe^{3+} oxidation state. Depending on the redox potential of the column fractions as well as the presence of other redox-active species, it is possible that fractions containing TpNlr will appear colorless (Fe^{2+} form of the protein). In either case, it is best to assay fractions for the presence of TpNlr by either sodium dodecyl sulfate–polyacrylamide gel electrophoresis (SDS–PAGE) and/or superoxide reductase activity as described below. The best fractions in terms of purity and activity are pooled and concentrated to ~10 ml, using an Amicon (Danvers, MA) concentrator equipped with a PM10 membrane, and applied to a column containing Sephadex G-75 gel-filtration resin equilibrated with 50 mM Tris-HCl–0.3 M NaCl (pH 7.8). The protein is eluted from this column with the same buffer.

Enzymatic Analysis of Treponema Neelaredoxin. Although orthologs of TpNlr were originally characterized as SODs,[20,32] Adams and colleagues were the first to clarify the fundamental differences between the reaction catalyzed by SODs and SORs.[16] Indeed, although *Pyrococcus furiosus* SOR exhibited a high SOD activity (~4000 U/mg) in the classic cytochrome c reduction assay, the SOD activity of SOR was significantly lower when acetylated cytochrome c was used or in three other SOD assays, demonstrating that SORs have little, if any, SOD activity. Similar results have been demonstrated for homologs from *Desulfovibrio desulfuricans*,[20] *Desulfoarculus baarsii*,[17] *T. pallidum*,[14,15] *Archaeoglobus fulgidis*,[19] and *Desulfovibrio vulgaris*,[21] all of which have measurable but catalytically inefficient SOD activities of 10–70 U/mg.

Superoxide Reduction/Superoxide Reductase Oxidation. In the classic cytochrome c reduction assay, a steady state concentration of superoxide generated by xanthine–xanthine oxidase reduces cytochrome c, thereby leading to an increase in absorbance at 550 nm.

1. The assay is performed at 25° in a 1-ml quartz optical cuvette containing 50 mM potassium phosphate, pH 7.8, containing 100 μM EDTA, 10 μM horse heart cytochrome c, and 7.5 mM xanthine.

2. An amount of xanthine oxidase that gives an initial rate of ΔA_{550} of ~0.025 absorbance units/min is added. Various amounts of TpNlr are added and the

[32] G. Silva, S. Oliveira, C. M. Gomes, I. Pacheco, M. Y. Liu, A. V. Xavier, M. Teixeira, J. Legall, and C. Rodrigues-Pousada, *Eur. J. Biochem.* **259**, 235 (1999).

absorbance at 550 nm is monitored over time. One unit of SOD activity is defined as the amount of protein that inhibits the rate of cytochrome c reduction by 50%.

In the presence of SOR, which usually exists as a mixture of reduced (Fe^{2+}) and oxidized (Fe^{3+}) species, the reduction of cytochrome c by superoxide occurs in two distinct phases (Fig. 3). In the initial lag phase, in which no increase in A_{550} occurs, the superoxide generated by xanthine–xanthine oxidase is immediately reduced by the Fe^{2+} form of SOR present in solution. The length of this initial phase is proportional to the amount of SOR present in solution (Fig. 3, inset), as expected for this bimolecular reaction. When all the SOR is oxidized, the second phase commences, corresponding roughly to a linear rate of increase in A_{550} as superoxide reduces cytochrome c. At high enough concentrations of SOR (15 μg), its low but measurable SOD activity results in a slight inhibition of cytochrome c reduction, as evidenced by a decrease in slope of the A_{550}-versus-time graph. From this observed change in slope of these curves, the SOD activity of TpNlr is calculated to be 10 $\pm$ 5 U/mg.[15]

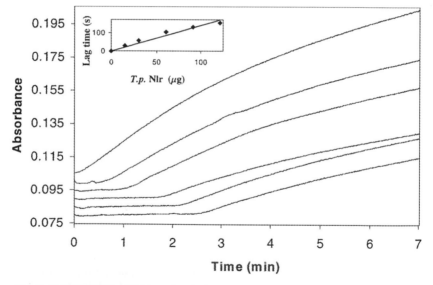

FIG. 3. The effect of *T. pallidum* neelaredoxin (*T.p.* Nlr) on the rate of cytochrome c reduction by superoxide. The reduction of cytochrome c is monitored at 550 nm as a function of time in the presence of increasing amounts of neelaredoxin. At $t = 0$, xanthine oxidase is added to initiate the reaction. Each curve has been offset to facilitate comparisons between curves. From top to bottom, 0, 15, 30, 61, 91, and 121 μg of neelaredoxin has been included in each assay. Inset: A plot of the lag time of the initial phase as a function of the neelaredoxin concentration. [Figure used with permission from T. Jovanovíc, C. Ascenso, K. R. O. Hazlett, R. Sikkink, C. Krebs, R. Litwiller, L. M. Benson, I. Moura, J. J. G. Moura, J. D. Radolf, B. H. Huynh, and F. Rusnak, *J. Biol. Chem.* **275**, 28439 (2000). Copyright © 2000, The American Society for Biochemistry and Molecular Biology.]

Additional studies have shown that superoxide is capable of oxidizing TpNlr to the Fe^{3+} state, but not the opposite.[14,15] The second-order rate constant for the SOR activity of TpNlr with superoxide has been measured by monitoring the increase in absorbance due to the ferric form of the active site as a function of time in the presence or absence of SODs, yielding values of $10^8-10^9 M^{-1} sec^{-1}$,[14] indicating that TpNlr is a potent antioxidant enzyme.

Reduction of Treponema Neelaredoxin. As an SOR reduces superoxide the protein becomes oxidized and therefore catalytically inactive; to restore activity, SORs must subsequently be reduced.[15]

1. Reduction of SORs can be observed by monitoring the $A_{644-656}$, which decreases as the iron center of the SOR is reduced.[14,15,17] To assay *E. coli* for SOR reductase activity, Lombard *et al.* have added 5–20 μg of cell extract to 110 μM oxidized recombinant SOR in 50 mM Tris-HCl, pH 7.6, containing 600 μM NAD(P)H under anaerobic conditions. Using this method, TpNlr reduction by *E. coli* extracts was shown to be dependent on either NADH or NADPH, with optimal TpNlr reduction activity found in cytoplasmic preparations.[14]

2. Alternatively, to evaluate individual electron donors for SOR, such as rubredoxin (Rd), the A_{495} of reduced Rd is measured in the presence of SOR and superoxide.[16] As reduced Rd donates electrons to oxidized TpNlr during the SOR cycle, the A_{495} increases due to the formation of oxidized (Fe^{3+}) Rd.

The assay is performed at 25° in a 1-ml quartz optical cuvette containing 17 μM rubredoxin, 50 mM potassium phosphate (pH 7.8), and 100 μM EDTA. Sodium dithionite is added to reduce Rd, which is reflected by a decrease in the A_{495} value (Fig. 4). Reduced Rd is allowed to slowly reoxidize (due to oxygen in the assay buffer) for ~100 sec, after which xanthine (0.4 mM) and xanthine oxidase (0.15 U) are added to generate superoxide. After the addition of 50 nM TpNlr, the rate of Rd oxidation increases as Rd reduces TpNlr (Fig. 4). The specific activity of Rd oxidation is derived by comparing the rate of Rd oxidation in the presence and absence of TpNlr. Using this method, we have determined that recombinant *D. vulgaris* Rd can serve as an electron donor for TpNlr with a specific activity of 3–7 μmol of Rd oxidized per minute per milligram of TpNlr. It should be noted that the reaction rate was not always proportional to the neelaredoxin concentration, possibly because of some other factor in the assay becoming rate determining. Significantly, *T. pallidum* encodes an Rd homolog, genome designation TP0991.[13]

Analysis of Treponema Neelaredoxin Activity and Expression in Vivo

Whereas classic genetic techniques such as knockout and complementation analysis can be used in other organisms to study SOR activity *in vivo*,[18] no system currently exists for the genetic manipulation of *T. pallidum*. For this reason, analysis

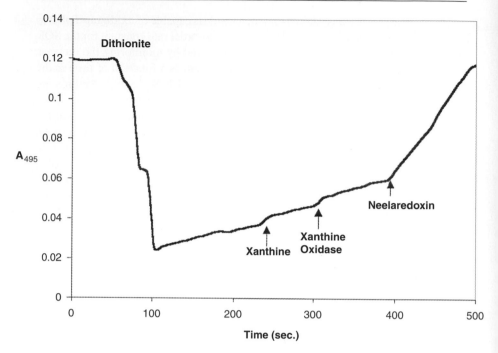

FIG. 4. Rubredoxin can serve as an electron donor to neelaredoxin/superoxide reductase. At $t = 0$, 17 μM recombinant *D. vulgaris* rubredoxin was added to 50 mM potassium phosphate (pH 7.8)–0.1 mM EDTA in a quartz optical cuvette and the absorbance due to the oxidized form was monitored at 495 nm. Sodium dithionite was added at ~80 sec, resulting in reduction of rubredoxin and the corresponding decrease in the A_{495} value. This was followed by a gradual increase in the A_{495} value due to reoxidation from oxygen in the assay buffer. Xanthine (0.4 mM) and xanthine oxidase (0.15 U) were added at 250 and 300 sec, respectively. At $t = 400$ sec, TpNlr was added to 50 nM. The increased rate of rubredoxin oxidation (ΔA_{495}) compared with the rate before TpNlr addition was used to calculate a specific activity of 7 μmol of rubredoxin oxidized per minute per milligram of TpNlr.

of TpNlr *in vivo* relies on heterologous complementation and analysis of TpNlr expression by *T. pallidum.*

1. Pianzzola *et al.* were the first to report that the aerobic growth defect of superoxide dismutase-deficient *E. coli* could be corrected by complementation with a *D. baarsii* Nlr homolog.[28] Also using complementation, Liochev and Fridovich proposed, for the first time, that the *D. baarsii* Nlr homolog catalyzed reduction, rather than dismutation, of superoxide.[29] Similar analyses have subsequently shown that Nlr homologs of *D. vulgaris* and *T. pallidum* complement *E. coli* SOD mutants.[14,28] Expression of TpNlr in *E. coli* sodA,B, recA triple mutants almost fully restored aerobic growth.[14]

2. The expression of TpNlr has been examined in *T. pallidum* by both reverse transcriptase (RT)-PCR and Western blot. For RT-PCR analysis of TpNlr transcript, RNA is isolated from freshly harvested *T. pallidum* with the Ultraspec RNA isolation reagent (Biotecx, Houston, TX) according to the manufacturer recommendations. Isolated RNA is treated with RNase-free DNase, phenol–chloroform extracted, ethanol precipitated, and resuspended in diethyl pyrocarbonate-treated water along with an RNase inhibitor. RT-PCR analysis is carried out with the Titan one-tube RT-PCR system (Roche Molecular Biochemicals, Indianapolis, IN) and primers specific for neelaredoxin (5′-AGGCAGTAGTGTCGCGTGCGG-3′ and 5′-AAAGGTCACCTCAGGCGCTCC-3′) and *flaA* (5′-TGAATTATCCTCATGG TTTGTACGTG-3′ and 5′-TCAGCACCGCCTTATCATAGATAATC-3′). For each primer set, four reactions are performed: (1) RT-PCR with 50 ng of RNA template, (2) PCR with 50 ng of RNA template, (3) RT-PCR with water only, and (4) PCR with 3 ng of DNA. In reactions without reverse transcription, the RT–DNA polymerase mixture is replaced by the Expand High Fidelity DNA polymerase (Roche). After the RT reaction (45° for 30 min) PCR is performed, using the following parameters: 98° for 2 min followed by 40 cycles of 98° for 10 sec, 60° for 10 sec, and 68° for 30 sec followed by a single terminal extension for 2 min at 68°. One-fifth (5 μl) of each reaction is electrophoresed through a 2% (w/v) agarose gel containing ethidium bromide before photography. Using this technique, we were able to detect transcript for TpNlr in freshly harvested *T. pallidum*, albeit at lower levels than the abundantly expressed flagellin subunit (Fig. 5A).

Before immunoblot analysis of *T. pallidum*, antisera against TpNlr are raised by priming rats intraperitoneally with 20 μg of purified TpNlr emulsified in a 1 : 1 mixture with Freund's complete adjuvant and PBS, pH 7.4. Booster immunizations consisting of 20 μg of protein in a 1 : 1 mixture with Freund's incomplete adjuvant and PBS are administered by the same route on weeks 4 and 6. Anesthetized rats are exsanguinated on week 8 by cardiac puncture. The coagulated blood is centrifuged for 10 min at 2000g, after which the sera are collected.

For immunoblot analysis of TpNlr, samples of whole *T. pallidum* (5×10^7) or dilutions of purified TpNlr are diluted 1 : 2 in Tricine sample buffer and boiled for 5 min before electrophoresis through 16.5% (w/v) polyacrylamide Tris–Tricine gels (Bio-Rad, Hercules, CA). Resolved proteins are electrophoretically transferred to 0.2-mm pore size nitrocellulose and blocked in PBS containing 5% (w/v) skim milk, 5% (v/v) fetal calf serum, and 0.05% (v/v) Tween 20 for 1 hr before immunoblotting. Immunoblots are incubated with 1 : 1000 dilutions of antisera in blocking buffer for 1 hr, after which the membranes are extensively washed with PBS containing 0.05% (v/v) Tween 20 and then incubated with a 1 : 75,000 dilution of goat anti-rat horseradish peroxidase conjugate (Southern Biotechnology Associates, Birmingham, AL) in blocking buffer for 1 hr. Immunoblots are developed with SuperSignal Femto chemiluminescence substrate (Pierce, Rockville, IL)

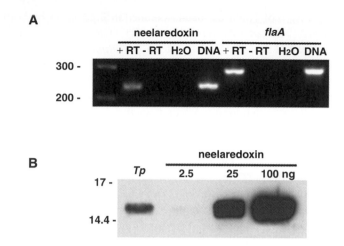

FIG. 5. Expression of neelaredoxin in *T. pallidum* (*Tp*). (A) Detection of transcripts for neelaredoxin and *flaA* by RT-PCR (+RT). Controls consisted of PCR performed on 50 ng of RNA template (–RT), RT-PCR of water (H₂O), and PCR of 3 ng of DNA. (B) Detection of protein expression by immunoblot analysis of *T. pallidum* (5×10^7) and purified recombinant protein, using rat antiserum generated against purified recombinant protein. Numbers above the neelaredoxin lanes indicate the amount of protein (in nanograms). Numbers to the left of (A) and (B) indicate standards in base pairs (A) and kilodaltons (B). [Figure used with permission from T. Jovanovíc, C. Ascenso, K. R. O. Hazlett, R. Sikkink, C. Krebs, R. Litwiller, L. M. Benson, I. Moura, J. J. G. Moura, J. D. Radolf, B. H. Huynh, and F. Rusnak, *J. Biol. Chem.* **275**, 28439 (2000). Copyright © 2000, The American Society for Biochemistry and Molecular Biology.]

followed by exposure to chemiluminescence film. Using these parameters and densitometric analysis of immunoblots, we detected 11 ng of TpNlr in lysates of 5×10^7 freshly harvested testicular *T. pallidum* (Fig. 5B). By using these detection methods in concert with the *in vitro* cultivation system described above, it should be possible to analyze the effects of oxygen tension on the regulation of TpNlr in *T. pallidum*.

Summary and Future Directions

The syphilis spirochete expresses an active SOR, suggesting that *T. pallidum* has retained a primitive mechanism for coping with oxidative stress that appears to be unique among human pathogens.[14,15] The findings that Rd can reduce TpNlr to restore SOR activity, and that *T. pallidum* has an Rd homolog, suggests that the spirochete encodes enzymes for the reduction of superoxide to hydrogen peroxide. By scanning the genome, we have identified treponemal homologs for the alkyl hydroperoxide reductase protein subunit C, thioredoxin, and thioredoxin reductase. Although the reduction of hydrogen peroxide by AhpC has been most frequently

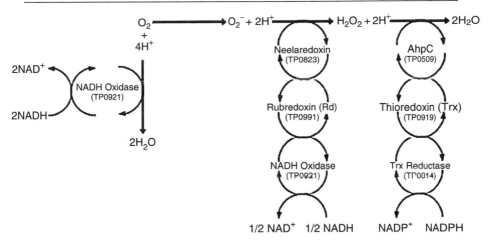

FIG. 6. Tentative model for oxygen detoxification by *T. pallidum*. The annotation and activities of neelaredoxin and rubredoxin are based on empirically derived data reviewed in this work. The putative annotations and activities proposed for NADH oxidase, alkyl hydroperoxide reductase C (AhpC), thioredoxin, and thioredoxin reductase are based on homology with enzymes in other organisms and their empirically determined activities. The numerical designations (TP#) below each enzyme are in accordance with the nomenclature of The Institute for Genomic Research (www.tigr.org).

associated with either AhpF or a Nox-1-type NADH oxidase, thioredoxin and thioredoxin reductase have been shown to interact with AhpC in *Helicobacter pylori* to reduce hydrogen peroxide to water.[33] Thus, we speculate that *T. pallidum* encodes all the activities required for the complete reduction of superoxide to water (Fig. 6). In addition, *T. pallidum* encodes a Nox-2-like NADH oxidase that may i) lower the potential for superoxide generation by directly reducing molecular oxygen to water, and ii) provide reducing potential to rubredoxin. Isogenic mutants of Nox-2 in other organisms, including the spirochetal pathogen *Brachyspira hyodysenteriae*, have been shown to be hypersensitive to oxygen and less virulent in animal models of infection.[34–36]

Testing this scheme for oxygen detoxification will require a multifaceted approach including biochemical analysis of recombinant proteins, complementation of appropriate heterologous mutants, and transcriptional analysis of candidate treponemal ROS detoxification genes. Ultimately, *in vitro* reconstitution of the

[33] L. M. Baker, A. Raudonikiene, P. S. Hoffman, and L. B. Poole, *J. Bacteriol.* **183,** 1961 (2001).

[34] J. Yu, A. P. Bryant, A. Marra, M. A. Lonetto, K. A. Ingraham, A. F. Chalker, D. J. Holmes, D. Holden, M. Rosenberg, and D. McDevitt, *Microbiology* **147,** 431 (2001).

[35] C. M. Gibson, T. C. Mallett, A. Claiborne, and M. G. Caparon, *J. Bacteriol.* **182,** 448 (2000).

[36] T. B. Stanton, E. L. Rosey, M. J. Kennedy, N. S. Jensen, and B. T. Bosworth, *Appl. Environ. Microbiol.* **65,** 5028 (1999).

full enzymatic pathway, in concert with detection of these activities in lysates of Percoll-purified spirochetes, will be required to conclusively delineate the components of the treponemal oxidative defense pathway(s). Moreover, comparative functional analysis of these pathways *in vivo* and *in vitro* may reveal a relationship between treponemal oxidative stress responses and syphilis pathogenesis. These lines of investigation hold the potential to yield knowledge essential to the long-sought continuous cultivation of *T. pallidum in vitro*.

Acknowledgments

This research was supported in part by NRSA postdoctoral fellowship award AI-10573 to K.R.O.H., U.S. Public Health Service Grant AI-26756 to J.D.R., and by support from the Mayo Foundation to F.R.

[15] Redox Control of Integrin Adhesion Receptors

By Jeffrey W. Smith, Boxu Yan, France Landry, and Christian R. Lombardo

Reagents

The following is a list of reagents that have been used in our studies of the integrin redox site. The manufacturer of each reagent is provided.

Triton X-100 (Sigma, St. Louis, MO)
Phenylmethylsulfonyl fluoride (PMSF; Sigma)
Leupeptin (Sigma)
Methyl-α-D-mannopyranoside (Sigma)
Guanidine-HCl (Sigma)
Trichloracetic acid (Sigma)
Dimethyl sulfoxide (Sigma)
Glutathione (reduced form; Sigma)
L-Cysteine (Sigma)
Avidin–horseradish peroxidase (HRP) (Sigma)
Concanavalin A (ConA)–Sepharose (Amersham Pharmacia Biotech, Piscataway, NJ)
CNBr–Sepharose (Amersham Pharmacia Biotech)
Aquacide (CalBiochem-Novabiochem, La Jolla, CA)
[^{14}C]Iodoacetamide (New England Nuclear, Boston, MA)
Dimethylformamide (Aldrich, Milwaukee, WI)

Copyright 2002, Elsevier Science (USA).
All rights reserved.
0076-6879/02 $35.00

Integrin Redox Site

Integrins are transmembrane adhesion receptors that play an essential role in normal tissue development and homeostasis. Many integrins bind to their ligands through the Arg-Gly-Asp tripeptide sequence, which is displayed by a number of extracellular matrix proteins, plasma proteins, and even viruses.[1] Unlike many other classes of receptors, the integrins participate in bidirectional signaling across the membrane.[2] "Inside-out" signals activate, and deactivate, the integrin ligand-binding function. "Outside-in" signals are generated when integrin binds to its ligands and regulate a host of cellular processes including cell proliferation and cell death. We identified a redox site within the extracellular domain of the integrin.[3] This redox site is composed of between two and five unpaired cysteines, which appear to reshuffle during conformational transitions in integrin. Therefore, we suggested that this redox site has the properties that might be expected of a conformational switch involved in bidirectional signaling.

The realization that integrins contain a redox site is so recent that only the initial steps toward establishing experimental methodology have been completed. We describe methods for (1) quantifying the number of free sulfhydryls within the integrin redox site, (2) tagging these free sulfhydryls with site-specific modification reagents, and (3) pinpointing the position of the free sulfhydryls that comprise the integrin redox site. Much of our effort has been directed toward the platelet integrin $\alpha_{IIb}\beta_3$, also known as platelet glycoprotein IIb IIIa.[4] Integrin $\alpha_{IIb}\beta_3$ serves as an excellent paradigm for the study of other integrins because it participates in both inside-out and outside-in signaling, and because it can be purified from outdated platelets in milligram quantities. Equally as important, two conformers of this integrin can be obtained. Activation state 1 (AS-1) represents the resting integrin, has low affinity for ligand, and an "oxidized" redox site. In contrast, activation state 2 (AS-2) has high affinity for physiologic ligands and a "reduced" redox site. The purification of AS-1 and AS-2 is described as a starting point.

Purification of Resting and Active Conformers of Platelet Integrin $\alpha_{IIb}\beta_3$

Our strategy for purifying AS-1 and AS-2 is rooted in methods that have been previously reported,[5-7] with some modifications. A key aspect of the purification

[1] E. Ruoslahti, *Annu. Rev. Cell Dev. Biol.* **12,** 697 (1996).

[2] R. O. Hynes, *Cell* **69,** 11 (1992).

[3] B. Yan and J. W. Smith, *J. Biol. Chem.* **275,** 39964 (2000).

[4] D. R. Phillips, I. F. Charo, and R. M. Scarborough, *Cell* **65,** 359 (1991).

[5] L. A. Fitzgerald, B. Leung, and D. R. Phillips, *Anal. Biochem.* **151,** 169 (1985).

[6] R. Pytela, M. D. Pierschbacher, S. Argraves, S. Suziki, and E. Ruoslahti, *Methods Enzymol.* **144,** 475 (1987).

[7] W. C. Kouns, P. Hadvary, and B. Steiner, *J. Biol. Chem.* **267,** 18844 (1992).

is the separation of AS-1 from AS-2 by ligand affinity chromatography. Outdated human platelets (100 units) are subjected to centrifugation at 5500g for 30 min at 4° in a Beckman (Fullerton, CA) JA-10 rotor (or equivalent). Red blood cells also pellet with the platelets but are removed by gently inverting the centrifugation tube. The platelet pellet is further washed by gentle resuspension in 20 mM Tris-HCl (pH 7.4), 150 mM NaCl, containing 0.1 mM EDTA, followed by centrifugation at 5500g for 30 min at 4°. This wash is repeated one additional time. Platelets are subsequently lysed at 4° by addition of 5 volumes of 20 mM Tris-HCl (pH 7.4), 150 mM NaCl, 1 mM CaCl$_2$ and containing 1% Triton X-100, 5 mM PMSF, and 10^{-5} M luepeptin. This sample is maintained at 4° for 14 hr with gentle rocking. The lysate is cleared of insoluble material by centrifugation at 50,000g in a Beckman JA-25.5 rotor at 4° for 1 hr. The supernatant, containing $\alpha_{IIb}\beta_3$, is collected and used for subsequent chromatography. We have found it convenient to store the supernatant from the platelet lysate at −70° for up to 4 months.

The platelet lysate is thawed rapidly at 37° and immediately centrifuged again at 50,000g in the Beckman JA-25.5 rotor. The supernatant is adsorbed to concanavalin A–Sepharose in batch. Aliquots of fresh PMSF (1 mM) and leupeptin (10^{-5} M) are added to inhibit proteolysis during batch adsorption. The lysate is incubated with ConA–Sepharose for 18 hr at 4°. The ConA–Sepharose resin is packed into a low-pressure column and washed with 20 mM Tris-HCl (pH 7.4), 150 mM NaCl, 1 mM CaCl$_2$, 1 mM MgCl$_2$ containing 1% Triton X-100 (buffer A) until the optical density of the eluate stabilizes. Although Triton X-100 absorbs at 280 nm, we have found monitoring of the wash at this wavelength to be an excellent indicator of the level of nonspecifically bound proteins eluting during the wash. Integrin (both AS-1 and AS-2) is subsequently eluted from ConA–Sepharose in buffer A containing 200 mM methyl-α-D-mannopyranoside. The eluate is examined by Coomassie staining of sodium dodecyl sulfate (SDS)–polyacrylamide gels, and fractions containing integrin $\alpha_{IIb}\beta_3$ are pooled.

The partially purified integrin is recirculated for 18 hr at 4° over an affinity resin containing the peptide KYGRGDSP linked to CNBr–agarose. We have found that a column containing 8 ml of resin linked to approximately 1 mg of peptide per milliliter is sufficient to retain the vast majority of the AS-2 from 100 units of platelets. The flowthrough fraction, containing the nonbinding AS-1, is retained for further purification (see below). To obtain AS-2, the KYGRGDSP affinity resin is washed with 10 column volumes of buffer A, and is then eluted by addition of 0.5 mM GRGDSP–amide in buffer A. Fractions containing AS-2 are pooled and dialyzed exhaustively against buffer A to remove free RGD peptide, and are then stored at −70° in 5% (v/v) glycerol.

The flowthrough from the KYGRGDSP affinity resin is then used as a source for AS-1. This material is recirculated over heparin–agarose for 18 hr at 4° to remove heparin-binding proteins as described.[5] The nonbinding fraction, containing AS-1, is concentrated to approximately 8 ml by placing it into a dialysis bag and overlaying with Aquacide (M_r 500,000 cutoff). The concentrated sample is dialyzed against buffer A, and then passed over a Sephacryl S-300 column equilibrated in the same buffer. Highly purified AS-1 elutes as a single broad peak, and can be detected by Coomassie-stained SDS–polyacrylamide gel electrophoresis (PAGE). Fractions containing AS-1 are pooled and stored with 5% (v/v) glycerol as a stabilizing agent at $-70°$ until use.

Quantifying Free Sulfhydryls in Integrin

The quantification of free sulfhydryls within the integrin redox site is an essential first step. As described previously,[3] the number of unpaired cysteines changes as a consequence of integrin activation, and is likely to be mechanistically linked to activation. We have modified standard carboxymethylation procedures to enable the quantification of free sulfhydryls within integrin. Integrin, at a concentration of greater than 500 μg/ml in buffer A (although 50–100 mM octylgucoside is also acceptable), is first denatured by the addition of 8 volumes of 6 M guanidine-HCl. After a 30-min incubation, EDTA is added to a final concentration of 0.2 M (diluted from a 2 M stock, pH 8.0). We have found this chelation step to be essential for the accurate quantification of free sulfhydryls. After a 15-min incubation at ambient temperature with EDTA, free sulfhydryls in the integrin are alkylated by addition of [14C]iodoacetamide (New England Nuclear) at a 500 : 1 (iodoacetamide : integrin) molar ratio. Alkylation is allowed to proceed for 1 hr at 25°. The alkylated integrin is precipitated by the addition of a 1/5 volume of ice-cold 50% (w/v) trichloroacetic acid (TCA) to bring the final concentration of TCA to 10% (w/v). The sample is maintained on ice for 1 hr, and is then precipitated by centrifugation for 20 min at 14,000 rpm in a microcentrifuge maintained at 4°. It is anticipated that in some cases it will be difficult to achieve integrin concentrations that are sufficient to promote precipitation. Therefore, the addition of carrier protein, to a concentration of 1 mg/ml, to facilitate precipitation would seem prudent. It is important, however, that the carrier be devoid of unpaired cysteines, as are found in bovine serum albumin. We suggest horse myoglobin as a suitable alternative. After centrifugation, the pellet is washed six times with 50 mM Tris-HCl (pH 7.4), 100 mM NaCl. After precipitation, the alkylated integrin can be brought into solution by the addition of 10% (w/v) SDS (in water) and heating to 50°. Frequent agitation over a period of 30 min is normally sufficient to solubilize the precipitated integrin. The sample is subjected to scintillation counting, from which the molar ratio of [14C]iodoacetamide to integrin can be derived. In our experience, it is best to compare an identical sample of integrin that is alkylated with unlabeled iodoacetamide before alkylation with [14C]iodoacetamide to control for nonspecific incorporation of the isotope.

Site-Specific Modification of Integrin Redox Site

Tagging unpaired cysteines with site-specific sulfhydryl modification reagents has also facilitated analysis of the integrin redox site. By using modification reagents linked to biotin, the modification can be readily detected by standard blotting approaches. A number of such reagents are commercially available. In our experience, only reagents greater than 29 Å in length are able to label the redox site on the two β_3-integrins. This limitation may be due to the inherent depth of the site. Two biotinylated modification reagents have yielded good results. These are 1-biotinamido-4-[4'(maleimidomethyl)-cyclohexane-carboximido]butane (biotin–BMCC) and N-[6-biotinamido)hexyl]-3'-(2'-pyridyldithio)propionamide (biotin–HPDP), both purchased from Pierce (Rockford, IL). Biotin–BMCC generates a maleimide link with free sulfhydryl that is insensitive to reduction, whereas biotin–HPDP creates a disulfide link with free sulfhydryls that is readily released by reduction. Although we have found both reagents to reproducibly label $\alpha_{IIb}\beta_3$, our analysis indicates that not all the unpaired cysteine residues within the redox site are modified by these reagents. Rather, alkylation of these cysteine residues of AS-1 with biotin–BMCC blocks the incorporation of between 35 and 60% of the [^{14}C]iodoacetamide incorporated during alkylation. Hence, biotin–BMCC is tagging about one-half of the free sulfhydryls within the redox site. Because biotin–HPDP is sensitive to reduction, we have been unable to measure what percentage of the free cysteines this reagent will modify.

We have found the following method to reproducibly modify the redox site within $\alpha_{IIb}\beta_3$, and $\alpha_v\beta_3$, and consequently expect that the same approach could be applied to examine the redox site in other integrins. Biotin–BMCC is made fresh as 4 mM stock in dimethyl sulfoxide. Biotin–HPDP should be made in dimethylformamide. Either reagent can be added directly to purified integrin (1–10 μg) in buffer A. Modification is allowed to proceed for 1 hr at 25°. We have found the redox site of $\alpha_{IIb}\beta_3$ to label across a concentration range of these reagents, with the labeling saturating at a reagent concentration of approximately 100 μM. The reaction can be quenched by addition of a 100-fold molar excess of reduced glutathione, or L-cysteine. After modification, the integrin can be separated by SDS–PAGE and transferred to polyvinylidene difluoride (PVDF) filters, and the modification assessed by probing the membrane with HRP–avidin (Sigma). It is conceivable that an enzyme-linked immunosorbent assay (ELISA) could also be devised to quantify the extent of modification with biotin–BMCC or biotin–HPDP.

This labeling strategy can also be adapted to examine the redox site within integrin on the surface of cells. For this application, the sulfhydryl modification reagents are added directly to cells in a pellet, making sure that the final concentration of dimethylsulfoxide does not exceed 5% (v/v). Best results are obtained when the modified integrin is subsequently immunoprecipitated from cell lysates, and then analyzed by SDS–PAGE and blotting to PVDF. Integrins are known to have a number of regulatory divalent cation-binding sites, but we have yet to perform rigorous experiments to determine whether the occupation of these sites influences the accessibility of the redox site to site-specific modification reagents. Therefore, to reduce variability in outcome, particular attention should be given to maintaining consistent levels and types of divalent cations.

Identification of Free Sulfhydryls with Mass Spectrometry

All integrin β subunits contain 56 conserved cysteine residues, many of which are located within a 200-residue cysteine-rich domain near the C terminus. The redox site appears to be positioned within this domain, where approximately 40% of all residues are cysteine. Consequently, obtaining definitive information about the position of the unpaired cysteines will be particularly challenging. To identify cysteines within the integrin redox site that are tagged with biotin–BMCC, we have begun to explore the use of precursor ion scanning with a triple quadrupole mass spectrometer.[8,9] Precursor ion scanning allows parent ions with specified properties to be identified on the basis of the mass spectrum of their daughter ions. Hence, by fragmenting the parent ion, and applying mass filters to identify daughters with a fragmentation pattern consistent with the conjugate between biotin–BMCC and cysteine, the parent peptide could presumably be identified. This parent peptide could then be immediately subjected to tandem mass spectrometry (MS/MS) sequencing to pinpoint the position of the tagged cysteine within the whole integrin.

To this point we have focused on establishing a robust method for such analysis, using glutathione (GSH) as a model peptide that contains an unpaired sulfhydryl. Our objectives have been to establish the fragmentation pattern of biotin–BMCC when linked to cysteine, and to test the concept that the conjugate between biotin–BMCC and GSH can be identified from a precursor ion scan of its daughter ions. The conjugate between biotin–BMCC (Fig. 1A) and GSH was analyzed with an ABI 3000 electrospray mass spectrometer (Applied Biosystems, Foster City, CA) in positive ion mode. A 20 μM sample of BMCC–GSH conjugate was infused into the mass spectrometer, using a Harvard syringe pump, at a flow rate of 20 $\mu l/min$. This yielded the Q1 spectrum shown in Fig. 1B. This spectrum reveals the singly charged parent ion ($m/z = 841$) of the conjugate. This conjugate was selected in the first quadrupole and then subjected to MS/MS fragmentation in the second quadrupole. The first quadrupole was set to transmit only the parent ion ($m/z = 841$). Collision with nitrogen gas [collisionally activated dissociation (CAD) gas setting $= 7$) at -80 V in the second quadrupole (RO2

[8] M. Mann and M. Wilm, *Trends Biochem. Sci.* **20,** 219 (1995).
[9] K. Chatman, T. Hollenbeck, L. Hagey, M. Vallee, R. Purdy, F. Weiss, and G. Siuzdak, *Anal. Chem.* **71,** 2358 (1999).

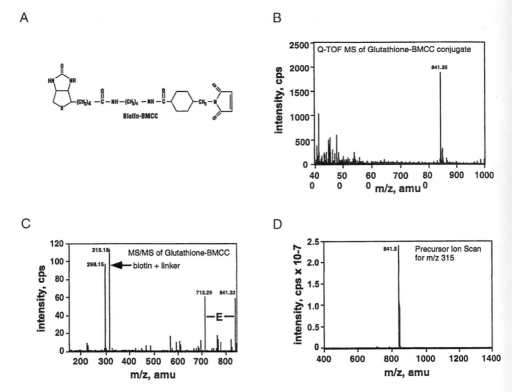

FIG. 1. Identification of biotin–BMCC-conjugated peptides, using precursor ion scanning. (A) Chemical structure of biotin–BMCC. (B) Detection of the conjugate between glutathione and biotin–BMCC (mass = 841.35), using a Q-TOF mass spectrometer. (C) Fragmentation of GSH–biotin–BMCC into daughter ions by MS/MS. (D) Identification of GSH–biotin–BMCC as the parent ion, using a precursor ion scan.

voltage) produced three distinct daughter ions (Fig. 1C). The largest is derived from fragmentation at the peptide bond between the glutamic acid and serine at the N terminus of GSH ($m/z = 712.3$). The loss of glutamic acid is noted as an "E" in Fig. 1C. Two additional fragments are created by fragmentation within the linker region between the biotin and the maleimide groups ($m/z = 315$ and $m/z = 298$). On occasion, we have also observed a fragment with an m/z ratio of 226 (biotin moiety of BMCC). The specific fragmentation and intensity of each daughter ion can be optimized by increasing the RO2 voltage settings to -80, -95, and -105 V for the m/z 315, 298, and 226 ions, respectively. The daughter ions with m/z values of 315 and 298 were then used as specific diagnostic indicators to identify peptides containing a cysteine residue modified by BMCC. This is accomplished by the use of precursor ion scanning (available on mass spectrometers equipped with triple quadrupole mass filters). BMCC–GSH conjugate is detected in the first quadrupole for parent ions, while holding the third quadrupole at a constant m/z value (315, daughter ion) after subjecting all ions to CAD in the second quadrupole. Using this scan mode, the conjugate between GSH– and biotin–BMCC with an m/z value of 841 is readily identified as the parent ion (Fig. 1D). We have also performed experiments to detect the presence of the BMCC–GSH modification in complex peptide mixtures. Using these scanning techniques, at least 100 fmol of BMCC–GSH conjugate parent ion ($m/z = 841$) can be detected in the presence of a 10-pmol mixture of tryptic peptides derived from an in-gel digest, using a nanoelectrospray ionization source (data not shown). Precursor ion scanning for $m/z = 315$ is most specific for the presence of the BMCC–GSH conjugate in complex peptide mixtures. In principle, this approach could be used to identify the position of biotin–BMCC-tagged sulfhydryls in any protein, including integrin.

Acknowledgment

This work was supported by Grants HL-58925, CA-69306, and CA-30199 from the NIH.

[16] Heme Oxygenase 1 in Regulation of Inflammation and Oxidative Damage

By SHAW-FANG YET, LUIS G. MELO, MATTHEW D. LAYNE, and MARK A. PERRELLA

Introduction

Heme oxygenase (HO) enzymes catalyze the initial reaction in heme catabolism.[1] Heme is a diverse metalloporphyrin molecule that is incorporated into important hemoproteins, such as hemoglobin and cytochromes, in the form of iron protoporphyrin IX.[2] Distinct forms of heme oxygenase have been identified[2–4]: (1) HO-1, a 32-kDa protein induced by diverse stimuli, including inflammatory cytokines and factors that promote oxidative stress; (2) HO-2, a 36-kDa protein that is predominantly a constitutive enzyme; and (3) HO-3, a 33-kDa protein highly homologous to HO-2 ($\sim$90%) at the amino acid level.[4] HO-1 and HO-2 are clearly products of distinct genes as they differ in chromosomal localization, gene

[1] R. Tenhunen, H. S. Marver, and R. Schmid, *Proc. Natl. Acad. Sci. U.S.A.* **61,** 748 (1968).

[2] M. D. Maines, *Annu. Rev. Pharmacol. Toxicol.* **37,** 517 (1997).

[3] B. A. Schacter, *Semin. Hematol.* **25,** 349 (1988).

[4] W. K. McCoubrey, Jr., T. J. Huang, and M. D. Maines, *Eur. J. Biochem.* **247,** 725 (1997).

Copyright 2002, Elsevier Science (USA).
All rights reserved.
0076-6879/02 $35.00

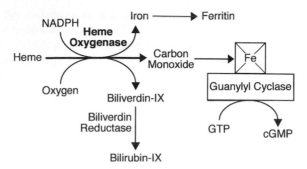

FIG. 1. Catabolism of heme by heme oxygenase. Heme oxygenase degrades heme to generate iron, carbon monoxide, and biliverdin-IX. Biliverdin-IX is subsequently reduced to bilirubin-IX by biliverdin reductase. Iron is sequestered by ferritin. Carbon monoxide can bind to the heme moiety of guanylyl cyclase to increase cellular levels of cGMP.

organization, and structure.[5–7] Although HO-1 and HO-2 are products of distinct genes, these two enzymes catabolize heme in an analogous fashion.[8,9] After transfer of an electron from NADPH to the substrate heme, the HO enzymes utilize molecular oxygen for cleavage of the α-meso carbon bridge of the heme molecule (Fig. 1). In contrast to HO-1 and HO-2, HO-3 has poor heme-degradative activity.[4]

HO-1 and HO-2 are cytoprotective enzymes[10–12] that degrade heme (a potent oxidant) to generate carbon monoxide (CO, a vasodilatory gas that has antiinflammatory properties), bilirubin (an antioxidant derived from biliverdin), and iron (sequestered by ferritin) (Fig. 1). This chapter focuses on HO-1, the inducible isoform that is upregulated by oxidative stress. Because of properties of HO-1 and its products, it is believed that HO-1 may play an important role in protecting cells and tissues in the settings of increased oxidative stress.

We have been interested in the regulation of inflammation and oxidative stress by HO-1. We use several *in vivo* models that induce oxidative stress in animals to study the role of HO-1 in response to pathophysiological stress. HO-1 expression level, inflammatory cell infiltration, and oxidative damage are assessed in tissues of interest.

[5] R. M. Muller, H. Taguchi, and S. Shibahara, *J. Biol. Chem.* **262,** 6795 (1987).

[6] W. K. J. McCoubrey and M. D. Maines, *Gene* **139,** 155 (1994).

[7] R. K. Kutty, G. Kutty, I. R. Rodriguez, G. J. Chader, and B. Wiggert, *Genomics* **20,** 513 (1994).

[8] M. D. Maines, G. M. Trakshel, and R. K. Kuttey, *J. Biol. Chem.* **261,** 411 (1986).

[9] G. M. Trakshel, R. K. Kutty, and M. D. Maines, *J. Biol. Chem.* **261,** 11131 (1986).

[10] S. M. Keyse and R. M. Tyrrell, *J. Biol. Chem.* **262,** 14821 (1987).

[11] K. A. Nath, G. Balla, G. M. Vercellotti, J. Balla, H. S. Jacob, M. D. Levitt, and M. E. Rosenberg, *J. Clin. Invest.* **90,** 267 (1992).

[12] L. E. Otterbein, F. H. Bach, J. Alam, M. Soares, H. Tao Lu, M. Wysk, R. J. Davis, R. A. Flavell, and A. M. Choi, *Nat. Med.* **6,** 422 (2000).

Animal Models

Three mouse models are used as examples to demonstrate the importance of HO-1 in the settings of oxidative stress: (1) endotoxemia, (2) myocardial ischemia and reperfusion, and (3) hypoxia. The models can also be applied to rats with minor adjustments. Animal use conformed to Federal guidelines and institutional policies.

Endotoxemia

Endotoxemia is induced in mice by intraperitoneal injection of *Salmonella typhosa* lipopolysaccharide (LPS). LPS (Sigma, St. Louis, MO) is dissolved in 0.9% (w/v) saline and stored in aliquots at $-20°$. Inject LPS (5 mg/kg body weight) intraperitoneally with a 25-gauge needle.[13] This dose of LPS produces a 15–20% decrease in systolic blood pressure after 4 hr compared with control 0.9% (w/v) saline-injected mice.

Myocardial Ischemia and Reperfusion

The inflammatory response is an important part of reperfusion injury.[14,15] We have observed that hearts from HO-1 null mice exposed to hypoxia showed increased inflammatory cell infiltration.[16] The role of HO-1 in cardiac reperfusion injury is assessed by subjecting mice to a myocardial ischemia and reperfusion model.[17,18] All instruments are sterilized by autoclaving, and surgery is performed with a dissection microscope [Nikon (Thornwood, NY) or Zeiss (Bensheim, Germany)] with fiberoptic illumination. Mice are anesthetized by intraperitoneal injection of sodium pentobarbital (4 mg/ml), 10 μl/g weight of mouse. A rodent ventilator (model 683; Harvard Apparatus, Holliston, MA) is used with 100% oxygen during the surgical procedure. Ventilation is provided by passing a blunt ended 22 gauge catheter (Johnson & Johnson, Arlington, TX) into the trachea via the mouth. The skin on the neck can be opened to guide the placement of the needle. The chest is opened by a vertical incision along the right or left side of the sternum, cutting through the ribs to approximately midsternum. Slight rotation

[13] P. Wiesel, A. P. Patel, N. DiFonzo, P. B. Marria, C. U. Sim, A. Pellacani, K. Maemura, B. W. LeBlanc, K. Marino, C. M. Doerschuk, S.-F. Yet, M.-E. Lee, and M. A. Perrella, *Circulation* **102**, 3015 (2000).

[14] M. L. Entman, K. Youker, T. Shoji, G. Kukielka, S. B. Shappell, A. A. Taylor, and C. W. Smith, *J. Clin. Invest.* **90**, 1335 (1992).

[15] M. L. Entman and C. W. Smith, *Cardiovasc. Res.* **28**, 1301 (1994).

[16] S.-F. Yet, M. A. Perrella, M. D. Layne, C.-M. Hsieh, K. Maemura, L. Kobzik, P. Wiesel, H. Christou, S. Kourembanas, and M.-E. Lee, *J. Clin. Invest.* **103**, R23 (1999).

[17] L. H. Michael, M. L. Entman, C. J. Hartley, K. A. Youker, J. Zhu, S. R. Hall, H. K. Hawkins, K. Berens, and C. M. Ballantyne, *Am. J. Physiol.* **269**, H2147 (1995).

[18] L. H. Michael, C. M. Ballantyne, J. P. Zachariah, K. E. Gould, J. S. Pocius, G. E. Taffet, C. J. Hartley, T. T. Pham, S. L. Daniel, E. Funk, and M. L. Entman, *Am. J. Physiol.* **277**, H660 (1999).

of the animal to the right orients the heart to better expose the left ventricle. The left atrium is slightly retracted, exposing the entire left main coronary artery. A 1-mm section of PE-10 tubing is placed on top of the left anterior descending branch of the left coronary artery 1–3 mm from the tip of the normally positioned left auricle, and a knot is tied at the top of the tubing to occlude the coronary artery with a 7-0 silk suture. After occlusion for 1 hr, reperfusion occurs by cutting the knot at the top of the PE-10 tubing. This allows release of the occlusion and reperfusion of the formerly ischemic bed. The chest wall is then closed by a 5-0 Ticron blue polyester fiber suture with one layer through the chest wall and muscle and a second layer through the skin and subcutaneous material. The animal is removed from the ventilator, the endotracheal tube is withdrawn, and the animal is kept warm by placing it on a 37° warm plate and allowed 100% oxygen via nasal cone. The animal is given butorphanol tartrate (0.1 mg/kg per dose) as an analgesic. After surviving the experimental infarct, the recovery of the animal is monitored for hours to weeks postoperatively, depending on the experimental design.

Hypoxia Exposure

Mice are exposed to normobaric hypoxia (a 10% oxygen environment) in a ventilated Plexiglas box (dimensions, 58 × 40 × 28 cm) for various periods of time to induce pulmonary hypertension.[16,19] The chamber is maintained at 10% oxygen by controlling the in-flow rates of room air and nitrogen from a liquid nitrogen reservoir. The chamber environment is monitored daily with an oxygen analyzer (Fyrite gas analyzer; Bacharach, Pittsburgh, PA). Soda lime and Drierite granules are used to remove CO_2 and excess humidity, respectively. Boric acid is used to minimize ammonia levels within the chamber. Hypoxic exposure is continuous, with less than 10–15 min of interruption every 3–4 days for animal care.

After exposing the animals to the above-described models of endotoxemia, myocardial ischemia and reperfusion injury, and hypoxia we analyze the tissues both by biochemical and histological techniques.

Biochemical Analyses

Preparation of Total Protein from Mouse Tissues

Before harvest, prepare homogenization buffer [25 mM Tris (pH 7.5), 50 mM NaCl, and 10 mM EDTA] with freshly added Complete protease inhibitors (Roche, Indianapolis, IN) and chill on ice. Aliquot 0.5 to 2 ml (0.5 ml/100 mg tissue) of buffer into 5-ml tubes (Sarstedt, Newton, NC) and place the tubes on ice. Mice are killed and tissues are harvested and briefly rinsed in ice-cold phosphate-buffered saline (PBS) to remove excess blood. Grasp tissue with a pair of forceps and hold

[19] H. Christou, A. Yoshida, V. Arthur, T. Morita, and S. Kourembanas, *Am. J. Respir. Cell Mol. Biol.* **18,** 768 (1998).

it just above the tube containing homogenization buffer, mince the tissue with scissors, and let the minced tissue fall (or push it) into the buffer. Homogenize the tissue with a Polytron homogenizer (PT2100 with a 7-mm probe; Brinkmann Instruments, Westbury, NY) (~5–10 sec/burst, three or four bursts) at maximum speed while securing the tube in a beaker filled with ice. After homogenization, transfer the homogenate to a microcentrifuge tube and spin at 4000g at 4° for 15 min. Transfer the supernatant to a new tube and repeat the spinning. Aliquot the supernatant in microcentrifuge tubes and store at −80°.

Western Blot Analysis of Heme Oxygenase 1

Western blot analysis is used to detect HO-1 protein induction in response to LPS-induced endotoxemia, myocardial ischemia and reperfusion, or hypoxic stimulation. Protein concentrations of samples are measured by the bicinchoninic acid (BCA) method according to manufacturer instructions (Pierce, Rockford, IL).

1. Thaw protein extracts on ice.
2. Mix and remove 5 μl in duplicate into 1.5-ml tubes.
3. Add 20 μl of water to each tube.
4. The standard is BSA (25–0.78 μg) at 2-fold dilutions.
   a. The stock provided with the kit is 2 mg/ml.
   b. Remove 25 μl and add to 25 μl of distilled H_2O in duplicate.
   c. Mix and transfer 25 μl to the next tube, repeat and discard 25 μl from the last tube.
   d. Include a blank (no BSA).
5. Mix BCA reagent (0.5 ml/tube will be needed).
   a. Add 1 part of blue solution to 50 parts of clear solution.
   b. Mix and let stand at room temperature for 5 min (this reagent is stable for 1 day).
6. Add 0.5 ml of reagent to each tube and mix.
7. Incubate at 37° for 30 min.
8. Transfer 200 μl to 96-well plate and read on at plate reader at OD_{562}.
9. Plot a standard curve and calculate the concentration of samples.

Aliquots containing 25 μg of tissue protein are fractionated on 10% (w/v) Tricine–sodium dodecyl sulfate (SDS)–polyacrylamide minigels as follows. For one protein minigel, prepare the running gel by mixing the following:

Gel buffer (3 M Tris-HCl, pH 8.45)	1.25 ml
Acrylamide solution [48% (w/v) acrylamide,	
1.5% (w/v) bisacrylamide]	0.775 ml
Glycerol	0.5 ml
Distilled H_2O	1.225 ml

N,N,N',N'-Tetramethylethylenediamine (TEMED)   1.25 μl
Ammonium persulfate (APS; 10%, w/v)   25 μl

After adding TEMED and APS the solution is quickly mixed and approximately 3.2 ml of the final mixture is loaded into a minigel cassette (1.5 mm thick; RPI, Mt. Prospect, IL), followed by gently layering with 95% (v/v) ethanol. After the running gel is polymerized (~20 min) the ethanol is removed, the top of the gel is rinsed with distilled H_2O, and excess water is removed with Whatman (Clifton, NJ) 3MM paper.

Prepare the stacking gel:

Gel buffer (3 *M* Tris-HCl, pH 8.45)	0.375 ml
Acrylamide solution [48% (w/v)	
acrylamide, 1.5% (w/v) bisacrylamide]	0.125 ml
Distilled H_2O	1.0625 ml
TEMED	3.75 μl
APS (10%, w/v)	25 μl

After adding the TEMED and APS, the solution is quickly mixed and approximately 1.2 ml of the mixture is loaded on top of the running gel. A 10-well comb is inserted into the top of the cassette and the stacking gel is allowed to polymerize.

Samples are prepared in 1× sample buffer [5× sample buffer: 10% (w/v) SDS, 10% (v/v) 2-mercaptoethanol, 0.2 *M* Tris (pH 6.8), 50% (v/v) glycerol, 0.5% (w/v) bromphenol blue] and heated at 100° for 5 min and then placed on ice. Rainbow molecular weight markers (Amersham Pharmacia Biotech, Piscataway, NJ) are also used to determine the molecular weights of proteins and to monitor the efficiency of the transfer.

Samples are electrophoresed with cathode (upper chamber) buffer [0.1 *M* Tris (pH 8.25), 0.1 *M* Tricine, and 0.1% (w/v) SDS] and anode (lower chamber) buffer (0.2 *M* Tris, pH 8.9) at 80 V to let samples run through the stacking gel. Increase the voltage to 100 V to run the gel until the dye front reaches the bottom of the gel.

After electrophoresis, the gel cassette is disassembled and the proteins are transferred to Protran nitrocellulose filters (Schleicher & Schuell, Keene, NH), using a Mini Trans-Blot cell (Bio-Rad, Hercules, CA) in transfer buffer [25 m*M* Tris base, 0.2 *M* glycine, and 15% (v/v) methanol] at 20 V, overnight at 4° with gentle stirring.

After transfer, disassemble and mark the molecular weight standards (prestained protein standards) with a laundry pen or nitrocellulose marker. Equilibrate the blots in TBST [25 m*M* Tris (pH 8), 125 m*M* NaCl, and 0.1% (v/v) Tween 20] for 10 min on a shaker at room temperature. The blots are then blocked with 4–10% (w/v) Carnation nonfat dry milk in TBST for 40 min, followed by incubation with polyclonal anti-HO-1 antiserum (SPA-895; StressGen Biotechnologies, Victoria, BC, Canada) diluted 1 : 1000 in TBST plus milk, at room temperature for 2 hr. Blots are then washed three times (15 min each) with 4% (w/v) milk in TBST

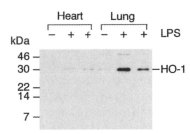

FIG. 2. LPS induces HO-1 protein in heart and lung tissues. Total heart and lung protein was extracted from mice treated with (+) or without (–) LPS for 24 hr. Aliquots (25 μg) were subjected to Western blotting with polyclonal anti-HO-1 antiserum.

before incubation with horseradish peroxidase-conjugated goat anti-rabbit serum (Amersham, Piscataway, NJ), diluted 1 : 4000, for 1 hr at room temperature. The blots are washed with 4% (w/v) milk in TBST two times (15 min each), and then with TBST two times (10 min each). The blots are then processed with an enhanced chemiluminescence reagent (Pierce) and exposed to X-ray film. Include fluorescent tape on the edge of the gel to determine the position of the molecular weight standards. An example of an HO-1 Western blot analysis of protein extracts isolated from LPS-treated mice is shown in Fig. 2.

Heme Oxygenase Enzyme Activity Assay

Two methods are routinely used to measure HO enzyme activity: (1) measurement of bilirubin generation by spectrophotometry[20,21] and (2) determination of CO production by gas chromatography.[22–24] The easier access to a spectrophotometer compared with gas chromatography equipment renders the colorimetric assay the method of choice for many investigators. Neither assay distinguishes the relative contribution of each HO isoform to total enzyme activity.

For the spectrophotometric determination of HO enzyme activity in the tissue and cultured cell extracts, use the following protocol.[25,26]

Preparation of Microsomal Extracts

1. Harvest tissues as described under Preparation of Total Protein from Mouse Tissues (above). Place the tissues (3 ml/g tissue) in ice-cold homogenization buffer

[20] M. D. Maines, N. G. Ibrahim, and A. Kappas, *J. Biol. Chem.* **252,** 5900 (1977).

[21] M. D. Maines and A. Kappas, *J. Biol. Chem.* **253,** 2321 (1978).

[22] H. J. Vreman and D. K. Stevenson, *Anal. Biochem.* **168,** 31 (1988).

[23] M. N. Cook, K. Nakatsu, G. S. Marks, B. E. McLaughlin, H. J. Vreman, D. K. Stevenson, and J. F. Brien, *Can. J. Physiol. Pharmacol.* **73,** 515 (1995).

[24] D. M. Suttner and P. A. Dennery, *FASEB J.* **13,** 1800 (1999).

[25] S.-F. Yet, A. Pellacani, C. Patterson, L. Tan, S. C. Folta, L. Foster, W.-S. Lee, C.-M. Hsieh, and M. A. Perrella, *J. Biol. Chem.* **272,** 4295 (1997).

[26] A. Pellacani, P. Wiesel, A. Sharma, L. C. Foster, G. S. Huggins, S.-F. Yet, and M. A. Perrella, *Circ. Res.* **83,** 396 (1998).

[30 mM Tris-HCl, 0.25 M sucrose, 0.15 M NaCl (pH 7.5)] containing Complete protease inhibitor (Roche) (other protease inhibitor cocktails maybe substituted, from example, Sigma protease inhibitor). For cultured cells, wash the cells with cold PBS twice, and scrape with 1.5–3 ml of homogenization buffer per 150-mm dish. Homogenize the tissues or cells with a Polytron on ice as described under Preparation of Total Protein from Mouse Tissues (above). Filter the homogenate through two layers of cheesecloth to remove large debris if present.

2. Centrifuge the homogenates at 10,000g for 15 min at 4°.

3. Collect the supernatant and centrifuge at 100,000g for 1 hr at 4°.

4. Resuspend the microsomal pellet in 50 mM potassium phosphate buffer (pH 7.4) containing protease inhibitor and sonicate. Determine the protein concentration by the BCA method as described under Western Blot Analysis of Heme Oxygenase 1 (above).

Rat Liver Microsomal Supernatant as Source of Biliverdin Reductase

1. Perfuse rat liver through the portal vein with cold PBS to flush out blood. Take 3 g ($\sim$2 small lobes) of liver and wash twice in ice-cold PBS.

2. Transfer the liver tissue to a small beaker containing 10 ml of ice-cold homogenization buffer (plus protease inhibitors); mince the tissue with scissors.

3. Transfer the minced tissue to a 50-ml tube and homogenize on ice with a Polytron equipped with a 12-mm probe.

4. Filter the homogenate through two layers of cheesecloth to remove large debris.

5. Centrifuge at 10,000g for 15 min at 4°. Filter the supernatant through two layers of cheesecloth. Repeat the spinning.

6. Collect the supernatant and centrifuge at 100,000g for 1 hr at 4°. The microsomal supernatant fraction is used as a source of biliverdin reductase for the HO enzyme assay. Aliquot the supernatant and store at −80° until use.

Heme Oxygenase Activity Assay

1. All reactions should be prepared on ice, using precooled reagents.

2. Prepare 500-μl reactions in duplicate in amber 1.5-ml microcentrifuge tubes as follows. Prepare a blank reaction for each sample by replacing the NADPH-generating system with an equivalent volume of 0.1 M phosphate buffer, pH 7.4.

Microsomal fraction	x μl
Potassium phosphate buffer (pH 7.4), 50 mM	90–x μl
Potassium phosphate buffer (pH 7.4), 0.1 M	177 μl
NADPH-generating system (see *note 1*)	167 μl
(substitute with 167 μl of 0.1 M phosphate buffer, pH 7.4, for blank)	
Hemin solution, 2.5 mM (see *note 2*)	6.7 μl

Add hemin solution last and vortex immediately; return the tubes to ice. *Note 1:* Assemble the NADPH-generating system by mixing equal volumes of each of the following components:

MgCl$_2$	64 mM
NADP (Sigma)	12.5 mM
Glucose 6-phosphate (Sigma)	13.8 mM
Glucose-6-phosphate dehydrogenase (Sigma; diluted in 0.1 M phosphate buffer, pH 7.4)	20 units/ml
Potassium phosphate buffer, pH 7.4	0.1 M

Note 2: Prepare hemin (Sigma) solution in the dark immediately before use: to 1.63 mg of hemin add 0.2 ml of 0.2 N NaOH, vortex to dissolve, and then add 0.8 ml of 0.1 M potassium phosphate buffer, pH 7.4.

3. Incubate the reactions for 10–20 min at 37° in the dark.

4. Terminate the reaction by placing the samples on ice for at least 2 min.

5. The absorbance spectrum of each reaction is scanned between 450 and 550 nm, using a prechilled quartz cuvette with a spectrophotometer (Beckman, Fullerton, CA).

6. Calculate the difference in optical density between 530 nm and the peak (~462–464 nm) for each sample. The amount of bilirubin formed is determined as ΔOD464–530 nm (extinction coefficient, 40 mM^{-1} cm^{-1} for bilirubin).

7. To calculate the total HO enzyme activity, subtract the bilirubin concentration in each sample reaction from the corresponding blank reaction. HO activity is expressed as nanomoles of bilirubin formed per milligram of protein per hour.

Lipid Peroxidation Measurements

A method to assess oxidative damage in tissues is to measure lipid peroxidation products from tissue extracts. Mouse tissues are harvested and washed in ice-cold 20 mM Tris, pH 7.4, blotted dry, and weighed. Homogenize the tissue in 20 mM Tris, pH 7.4 (10%, w/v), using a Polytron as described above. Centrifuge the homogenate at 3000g for 10 min at 4°. Transfer the supernatant to new tubes. The colorimetric lipid peroxidation assay kit (Calbiochem, San Diego, CA) is used to measure malondialdehyde (MDA) and 4-hydroxy-2-nonenal (4-HNE) levels in the supernatant from LPS-treated mouse tissues[13] according to the manufacturer instructions.

1. Prepare diluted reagent 1 (R1): 3 parts R1 to 1 part methanol.

2. To 650 μl of diluted R1 in a 1.5-ml polypropylene microcentrifuge tube add 200 μl of sample (brought to a final volume of 200 μl with 20 mM Tris,

pH 7.4) or standard solution (see below). Vortex for 3–4 sec. For standards, dilute 4-HNE standard (S1) or MDA standard (S2) solutions 100-fold (v/v) in sample buffer to 100 μM. To cover the standard concentration range of 0–20 μM, use 0–200 μl of 100 μM standard solution, brought to a final volume of 200 μl with the same sample buffer (for samples containing low levels of MDA/4-HNE, prepare a standard concentration range of 0–5 μM).

3. To assay MDA plus 4-HNE:
   a. Add 150 μl of reagent 2 (R2).
   b. Mix well and close the caps.
   c. Incubate at 45° for 40 min.
   d. Chill the samples on ice.
   e. Spin the samples at 13,000g for 10 min to pellet cellular debris if samples are cloudy.
   f. Transfer 200-μl aliquots of supernatant to a 96-well plate.
   g. Determine the OD$_{586}$ value.

4. To assay MDA only:
   a. Add 150 μl of 12 N HCl.
   b. Mix well and close the caps.
   c. Incubate at 45° for 60 min.
   d. Chill the samples on ice.
   e. Spin the samples at 13,000g for 10 min to pellet cellular debris if samples are cloudy.
   f. Transfer 200-μl aliquots of supernatant to a 96-well plate.
   g. Determine the OD$_{586}$ value.

5. Plot a standard curve and calculate the concentration of samples.

Histological Analysis

Tissue Preparation for Histology

Tissues from mice subjected to endotoxemia, myocardial ischemia and reperfusion, or hypoxia are harvested for histological analysis to measure HO-1 expression and inflammatory markers. Before harvest, mice are anesthetized and the chest is opened. The heart is perfused with a 22-gauge catheter through the left ventricle, advancing the catheter toward the outflow of the aorta. The right atrium is nicked with scissors to permit outflow of the blood and solutions. The hearts are first perfused with PBS (5–10 ml) to flush out blood, followed by 10% (w/v) neutral buffered formalin (NBF; Sigma). Good perfusion is indicated by a uniform pale appearance of the kidneys and a contraction of the skeletal muscle with the NBF perfusion Remove the heart and other organs and continue to fix in at least 20 volumes of 10% (w/v) NBF at 4° overnight. If samples are not immediately processed as described below, transfer the tissue to PBS and store at 4° to prevent overfixation.

For certain histological stains and immunostaining, hearts are removed after perfusing with PBS only, fixed in methyl Carnoy's (MC) solution [60% (v/v) methanol, 30% (v/v) chloroform, 10% (v/v) acetic acid] at 4° on a shaker for 5 hr, and then changed to 70% (v/v) ethanol and allowed to incubate at 4° overnight. Heart samples are then processed in an automated tissue processor (Hypercenter XP; Shandon, Pittsburgh, PA) according to the following protocol: 50% (v/v) ethanol for 20 min; 70% (v/v) ethanol for 20 min; 80% (v/v) ethanol for 20 min; 95% (v/v) ethanol for 20 min; 95% (v/v) ethanol for 20 min; 100% ethanol for 20 min; 100% ethanol for 30 min; xylene for 30 min; xylene for 30 min; paraffin for 45 min; paraffin for 45 min.

Tissue is then removed from the processor and embedded in paraffin (Paraplast X-tra; Electron Microscopy Sciences, Fort Washington, PA). The specimens are sectioned on a microtome (Leica, Bannockburn, IL) at a thickness of 5 μm, using standard techniques.

Immunohistochemical Detection of Heme Oxygenase 1 Protein and Inflammatory Cell Markers

Immunohistochemistry is performed as follows to detect the expression of HO-1 protein in the heart and other tissues. An example of HO-1 expression in the heart is shown in Fig. 3.

1. Paraffin is removed from the tissue sections, using a hot air dryer (Shandon), until the paraffin around the sections has melted (do not exceed 60°).

2. The sections are deparaffinized and rehydrated with quick rinses through xylene (three times), 100% ethanol (twice), 95% (v/v) ethanol (twice), 70% (v/v) ethanol (once), 50% (v/v) ethanol (once), and distilled H_2O.

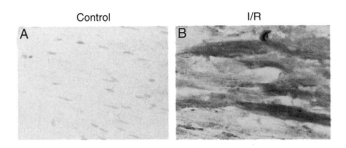

FIG. 3. Immunohistochemical localization of HO-1 expression in the heart after ischemia and reperfusion. Heart sections from (A) a control, nonoperated mouse and (B) a mouse subjected to 1 hr of ischemia and 24 hr of reperfusion (I/R) were stained with a polyclonal anti-HO-1 antibody. Brown staining indicates HO-1 expression. Original magnification: ×400.

3. Slides can be mounted in an IHC Rack&Rinse rack (Shandon) to conserve antibodies and minimize solution volumes. Alternatively, use a PAP pen to outline (Electron Microscopy Sciences) the area surrounding the section and perform immunostaining in a humidified chamber. With either method, it is essential that sections do not dry out during the staining procedure. Incubate the sections with Cadenza buffer (Shandon; 5 min each, two times). *Note:* PBS or Tris-buffered saline may be substituted for Cadenza buffer.

4. Incubate the sections in 3% (v/v) H_2O_2 in Cadenza buffer [900 μl of Cadenza plus 100 μl of 30% (v/v) H_2O_2] at room temperature for 20 min.

5. Wash with Cadenza buffer (twice, 5 min each time).

6. Incubate the sections for 20 min at room temperature with 10% (v/v) normal goat serum (Vector Laboratories, Burlingame, CA) in Cadenza buffer, 200 μl/slide.

7. Incubate the sections with polyclonal anti-HO-1 antiserum (SPA-895; StressGen Biotechnologies) diluted 1 : 400 in Cadenza buffer (200 μl/slide) at room temperature for 1 hr, and then at 4° overnight.

8. Rinse the slides in Cadenza buffer (twice, 5 min each time).

9. Incubate the sections in secondary antibody [biotinylated anti-rabbit IgG(H+L), made in goat; Vector Laboratories] at a 1 : 200 dilution at room temperature for 1 hr.

10. Rinse the slides in Cadenza buffer (twice, 5 min each time).

11. Incubate the sections with Elite Vectastain ABC HP solution (200 μl/slide; Vector Laboratories) at room temperature for 1 hr. This solution should be made up at least 30 min before use and keep at 4°.

12. Rinse the slides as in step 8.

13. Remove from the IHC Rack&Rinse rack a slide to be used for determining the optimal developing time. Develop the slide with a Vector 3,3'-diaminobenzidine (DAB) substrate kit by putting drops of substrate on the slide to cover the sections. Once the sections are covered by substrate, monitor the enzyme reaction closely under a microscope to see the color gradually develop. Stop the development by immersion in water as soon as signal is seen and before any nonspecific (background) staining appears. Develop the rest of the slides, which are still mounted on the IHC Rack&Rinse rack (200 μl of substrate per slide), for the same period of time.

14. Wash the slides in water for 5 min.

15. Counterstain the sections with methyl green [1% (w/v) in water] for 10–15 min for nuclear staining. Wash under running water to rinse out excess staining. Slides are dehydrated through 70% (v/v) ethanol for 30 sec; 95% (v/v) ethanol for 30 sec; 95% (v/v) ethanol for 30 sec; 100% ethanol for 30 sec; 100% ethanol for 2 min; xylene for 2 min; and xylene for 2 min.

16. Mount the slides with coverslips, using Permount/xylene (50 : 50, v/v) and allow to air dry.

WT   TG

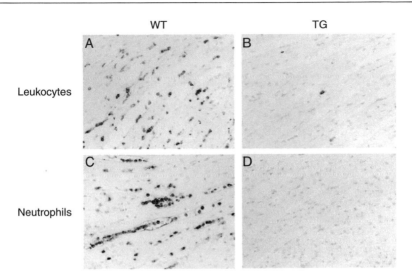

FIG. 4. Inflammatory cell infiltration in mouse hearts after ischemia and reperfusion. Heart sections from (A) wild-type (WT) and (B) cardiac-specific HO-1 transgenic (TG) mice subjected to 1 hr of ischemia and 24 hr of reperfusion were stained with an anti-mouse CD45 (leukocyte common antigen) antibody to detect leukocytes. A neutrophil-specific anti-mouse Ly-6G antibody was used to stain neutrophils in WT (C) and TG (D) heart sections from mice subjected to ischemia and reperfusion. Brown staining indicates positive reaction. Original magnification: ×200.

Immunostaining of Inflammatory Cells

The infiltration of inflammatory cells into tissues of mice treated according to the endotoxemia, myocardial ischemia and reperfusion, and hypoxia models is assessed by immunohistochemistry as described above for HO-1 immunostaining. Examples of inflammatory cell infiltration in the heart after ischemia and reperfusion are shown in Fig. 4. To detect leukocytes, heart sections (we use MC-fixed tissue sections) from wild-type and cardiac-specific HO-1 transgenic mice[27] are stained with an anti-mouse CD45 (leukocyte common antigen, Ly-5) antibody diluted 1 : 1000 (PharMingen, San Diego, CA) (Figs. 4A and B). To further determine the cell type of the inflammatory cells, heart sections (MC-fixed tissue sections) are stained with a neutrophil-specific anti-mouse Ly-6G antibody diluted 1 : 50 (PharMingen) (Figs. 4C and D).

Immunohistochemical Assessment of Oxidative Damage

To assess oxidative damage in tissues from mice treated according to the models of oxidative stress, histological analysis using polyclonal antibody MAL-2 can be

[27] S.-F. Yet, R. Tian, M. D. Layne, Z. Y. Wang, K. Maemura, M. Solovyeva, B. Ith, L. G. Melo, L. Zhang, J. S. Ingwall, V. J. Dzau, M.-E. Lee, and M. A. Perrella, *Circ. Res.* **89,** 168 (2001).

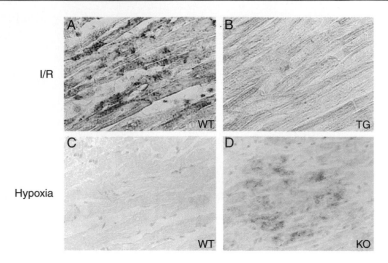

FIG. 5. Oxidative damage in mouse hearts subjected to reperfusion injury or hypoxic stress. Shown is MAL-2 immunostaining (brown) of heart sections from (A) wild-type (WT) or (B) HO-1 transgenic (TG) mice subjected to 1 hr of ischemia and 24 hr of reperfusion. Heart sections from (C) wild-type or (D) HO-1 null mice housed in a hypoxia chamber (10% oxygen) for 7 weeks were stained with MAL-2 antibody. MAL-2-positive cells stained pink. Original magnification: ×400.

performed. The MAL-2 antibody (kindly provided by J. Witztum, Immunology Core of the La Jolla SCOR Program in Molecular Medicine and Atherosclerosis, La Jolla, CA)[28] recognizes oxidation-specific lipid–protein adduct. In tissues being assessed by MAL-2 staining, consider including antioxidant in the perfusion solution as described.[28] Extensive oxidative damage is present after reperfusion in the left ventricles of wild-type heart (Fig. 5A) compared with HO-1 transgenic heart (Fig. 5B). The importance of HO-1 in response to cardiac stress is further demonstrated in hypoxia experiments. In contrast to minimal lipid peroxidation in the right ventricles of wild-type hearts (Fig. 5C), hypoxic exposure causes significant lipid peroxidation in the right ventricles of HO-1 null mouse hearts (Fig. 5D).

Acknowledgments

We thank Stella Kourembanas (Division of Newborn Medicine, Children's Hospital, Boston, MA) for the use of hypoxia chambers and for expertise. This work was supported in part by an American Diabetes Association Research Award (to S.-F. Yet); the National Institutes of Health Grants HL-57977 (to S.-F. Yet), HL-10113 (to M. D. Layne), and HL-60788 and GM-53249 (to M. A. Perrella); and a postdoctoral fellowship from the Medical Research Council of Canada (to L. G. Melo).

[28] M. E. Rosenfeld, W. Palinski, S. Ylä-Herttuala, S. Butler, and J. L. Witzum, *Arteriosclerosis* **10**, 336 (1990).

[17] Redox Properties of Vanillyl-Alcohol Oxidase

By ROBERT H. H. VAN DEN HEUVEL, MARCO W. FRAAIJE, and
WILLEM J. H. VAN BERKEL

Redox Properties of Flavoproteins

Flavins are a ubiquitous class of redox-active coenzymes that are able to catalyze a number of different chemical reactions when bound to apoproteins. They play an important role in (de)hydrogenation and hydroxylation reactions, in oxygen activation, and in one- and two-electron transfer processes from and to redox centers.[1,2] Because of their chemical versatility, flavins are involved in a wide range of biological processes. They have been shown to be involved in programmed cell death by signal transduction[3] and in detoxification of a wide variety of aromatic compounds.[4] They also have a function in regulating biological clocks,[5] in DNA damage repair,[6] and plant phototropism.[7] These unique properties of flavins are always controlled by specific noncovalent or covalent interactions with the apoproteins to which they are bound.

The first flavin-containing protein was isolated in 1933 by Warburg and Christian.[8] Since then, a large number of flavoproteins have been purified and characterized. Most of these proteins contain either a tightly but noncovalently bound flavin mononucleotide (FMN) or flavin adenine dinucleotide (FAD) prosthetic group as redox-active center. However, a significant number of flavoproteins contain a covalently linked flavin.[9] Mammalian succinate dehydrogenase was the first protein recognized to have such a covalent protein–flavin linkage.[10] At present, about 30 covalent flavoproteins have been reported. The precise function of the covalent linkage in most flavoproteins is unknown, but it has been suggested that the anchoring prevents flavin dissociation, increases protein stability, and improves

---

[1] V. Massey, *FASEB J.* **9**, 473 (1995).

[2] V. Massey, *Biochem. Soc. Trans.* **28**, 283 (2000).

[3] S. A. Susin, H. K. Lorenzo, N. Zamzami, I. Marzo, B. E. Snow, G. M. Brothers, J. Mangion, E. Jacotot, P. Costantini, M. Loeffler, N. Larochette, D. R. Goodlett, R. Aebersold, D. P. Siderovski, J. M. Penninger, and G. Kroemer, *Nature* (*London*) **397**, 441 (1999).

[4] B. Entsch and W. J. H. van Berkel, *FASEB J.* **9**, 476 (1995).

[5] A. R. Cashmore, J. A. Jarillo, Y. J. Wu, and D. Liu, *Science* **284**, 760 (1999).

[6] M. S. Jorns, B. Wang, and S. P. Jordan, *Biochemistry* **26**, 6810 (1987).

[7] M. Salomon, J. M. Christie, E. Knieb, U. Lempert, and W. R. Briggs, *Biochemistry* **39**, 9401 (2000).

[8] O. Warburg and W. Christian, *Biochem. Z.* **266**, 377 (1933).

[9] M. Mewies, W. S. McIntire, and N. S. Scrutton, *Protein Sci.* **7**, 7 (1998).

[10] E. B. Kearney and T. P. Singer, *Biochim. Biophys. Acta* **17**, 596 (1955).

Copyright 2002, Elsevier Science (USA).
All rights reserved.
0076-6879/02 $35.00

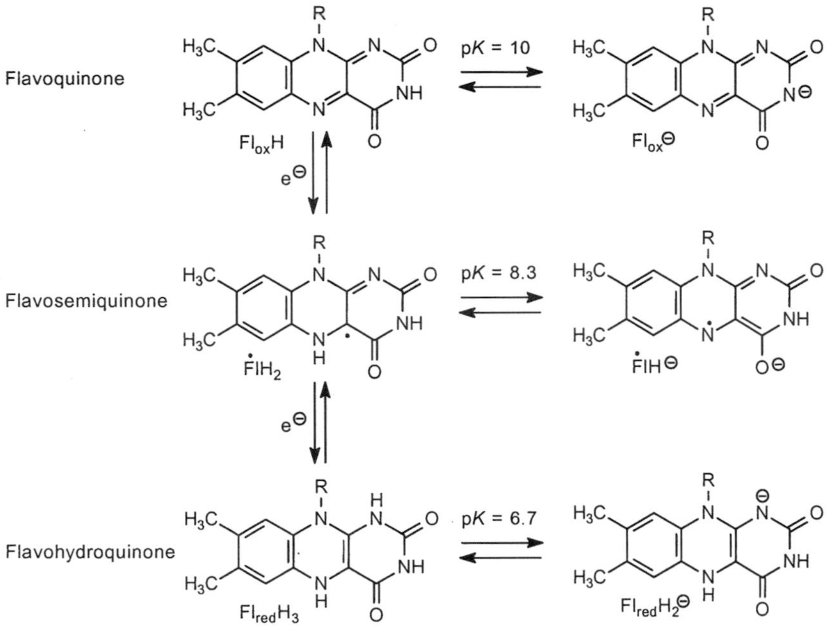

FIG. 1. The structures of biologically relevant flavin species in three different redox states.

resistance against proteolysis. Moreover, we have demonstrated that the covalent bond can raise the oxidative power of the flavin.[11]

Flavins are capable of performing both one- and two-electron transfer processes. The redox potentials for the two-electron reductions of FAD and FMN free in solution at pH 7.0 are -219 mV[12] and -205 mV,[13] respectively. Because a small percentage of the flavin semiquinone (one-electron reduced) is formed in thermodynamic equilibrium with the quinone (oxidized) and hydroquinone (two-electron reduced) (Fig. 1), the overall two-electron reduction can be analyzed for the two single-electron transfer steps, and a redox potential for both one-electron reductions can be determined. The semiquinone species can exist in either the blue neutral form (absorbance maximum in the 500- to 600-nm region) or the red anionic form (absorbance maximum in the 370- to 400-nm region), depending on the pH ($pK = 8.3$) (Fig. 1). When bound to apoprotein the flavin redox characteristics can change dramatically. Some flavoenzymes show essentially no stabilization of semiquinone on reduction, whereas others give nearly 100% stabilization. In the latter case, the protein may stabilize the neutral semiquinone over a wide pH range,

[11] M. W. Fraaije, R. H. H. van den Heuvel, W. J. H. van Berkel, and A. Mattevi, J. Biol. Chem. 274, 35514 (1999).
[12] H. J. Lowe and W. M. Clarke, J. Biol. Chem. 221, 983 (1956).
[13] R. D. Draper and L. L. Ingraham, Arch. Biochem. Biophys. 125, 802 (1968).

indicative of a high pK value, whereas others stabilize only the anionic form, indicative of a low pK. When the semiquinone is stabilized the redox potentials of the two steps in reduction are more separated.[14]

At present, the redox potentials of about 40 flavoenzymes are known, varying from -367 mV for nitroalkane oxidase[15] to $+55$ mV for vanillyl-alcohol oxidase (VAO).[11] For a selected number of flavoproteins the effects of substrate binding and pH have been studied[16] to obtain a better understanding of the catalytic mechanism. Especially when a crystal structure and active site mutant enzymes are available, such studies can reveal important information about the regulation of the flavin redox potential. As an example, we describe the redox properties of VAO, a covalent flavoprotein with an exceptionally high redox potential. However, we first briefly review the most common methods of flavin redox potential determination.

Determination of Flavin Redox Potentials

Electrochemical Methods

Electrochemical methods have proved to be effective approaches for studying the redox properties of flavoenzymes.[17,18] In these methods the oxidation state of the enzyme is changed electrochemically, using an electrode, while measuring the change in some property of the enzyme-bound flavin responsive to its redox state. The redox state of the flavin can be monitored, for example, by absorption spectroscopy, cyclic voltammetry, circular dichroism, or electron spin resonance. However, as the three different redox states of flavoenzymes have distinct spectral properties, absorption spectroscopy is most often used.

The electrochemical technique also allows the determination of the total number of electrons transferred from the electrode to the flavin. The power of this method lies in the fact that it gives quantitative information, which is especially useful when analyzing multi-redox-center enzymes.

Spectrophotometric Method

In the spectrophotometric method the equilibrium of partially reduced enzyme-bound flavin and a redox reference dye is analyzed optically. This is perhaps the simplest method for determining the redox potential of a flavoprotein because

[14] W. M. Clark, *in* "Oxidation–Reduction Potentials of Organic Systems." Williams & Wilkins, Baltimore, MD, 1960.

[15] G. Gadda and P. F. Fitzpatrick, *Biochemistry* **37**, 6154 (1998).

[16] M. T. Stankovich, *in* "Chemistry and Biochemistry of Flavoenzymes" (F. Mueller, ed.), Vol. I, p. 401. CRC Press, Boca Raton, FL, 1991.

[17] W. R. Heineman, *Anal. Chem.* **50**, 390A (1978).

[18] M. T. Stankovich, *Anal. Biochem.* **109**, 295 (1980).

there is no need for any specialized equipment. This optical method only requires a procedure to introduce reducing equivalents into the solution, a system to make a cuvette anaerobic, and a thermostatted scanning spectrophotometer.

In addition to the discussed electrochemical technique, several other procedures are available for introducing reducing equivalents into the reaction system. Chemical reduction by sodium dithionite is widely used for this purpose, as dithionite is a strong reducing agent.[19] Major disadvantages of dithionite, however, are its high reactivity with molecular oxygen, its changes in redox potential with pH, and its instability at low pH values (pH < 6.5). Moreover, the oxidized form of dithionite can form an adduct with several flavoproteins.[20] Photoreduction with 5-deazaflavin is another procedure for introducing reducing equivalents in a system. Catalysis of 5-deazaflavin in photoreduction is due to the formation of the strongly reducing 5-deazaflavosemiquinone species, which is generated by photolysis of the preformed 5-deazaflavin dimer.[21] Photoreduction is technically less complex to use than chemical reduction, because after the initial addition of 5-deazaflavin, reducing equivalents will be introduced continuously after exposing the system to light. A major drawback is that 5-deazaflavin is not commercially available and other photocatalysts do not always reduce the enzyme-bound flavins completely. A third commonly used method for introducing reducing equivalents into a system is the xanthine–xanthine oxidase method.[22] This method is extremely convenient for the determination of redox potentials of many flavoproteins. It uses xanthine oxidase, a low-potential reference dye, and xanthine to introduce reducing equivalents. We discuss this procedure in more detail, as we used it to determine the redox properties of VAO.

Xanthine–Xanthine Oxidase Method

Xanthine oxidase is a flavin-containing enzyme, which oxidizes xanthine to urate with the subsequent reduction of molecular oxygen to hydrogen peroxide. Xanthine oxidase can use, besides molecular oxygen, a number of low-potential dyes as electron acceptors, such as benzyl viologen ($E_m = -359$ mV at pH 7.0) and methyl viologen ($E_m = -449$ mV at pH 7.0). These dyes can react with other electron acceptors and also with most enzyme-bound flavins.

The assay described here is based on the method described by Massey.[22] Equal concentrations of the flavoprotein and the reference dye, typically 8–10 μM, 400 μM xanthine, and 2.5 μM methyl viologen or benzyl viologen, are placed

[19] S. G. Mayhew, *Eur. J. Biochem.* **85,** 535 (1978).

[20] V. Massey, F. Muller, R. Feldberg, M. Schuman, P. A. Sullivan, L. G. Howell, S. G. Mayhew, R. G. Matthews, and G. P. Foust, *J. Biol. Chem.* **244,** 3999 (1969).

[21] V. Massey, M. Stankovich, and P. Hemmerich, *Biochemistry* **17,** 1 (1978).

[22] V. Massey, *in* "Flavins, and Flavoproteins 1990" (B. Curti, S. Ronchi, and G. Zanetti, eds.), p. 59. Walter de Gruyter, Berlin, 1991.

in an anaerobic cuvette (total volume, 800 μl). The use of a Hellma (Plainview, NY) QS-117-104 cuvette sealed with a Subaseal 13 septum gives excellent results. A 10 mM xanthine stock solution is prepared by mixing xanthine with water and adjusting the pH to pH 11 with potassium hydroxide. The mixture of flavoprotein, reference dye, and xanthine is made anaerobic by either continuous flushing with oxygen-free argon gas or repeated cycles of evacuation and flushing with oxygen-free argon. The reduction of the flavoprotein and the reference dye is initiated by the anaerobic addition of xanthine oxidase, using a Hamilton syringe. Anaerobic conditions during the reductive process are maintained by continuously flushing the headspace of the cuvette with oxygen-free argon. To ensure equilibration between the oxidized and reduced species of enzyme and reference dye, the reduction must be sufficiently slow. The rate of reduction can be regulated by the concentration of xanthine oxidase. The final concentration needed depends on the redox potentials of the enzyme-bound flavin and dye, but is generally within 0.5–2.0 μg/ml. During the xanthine oxidase-mediated reduction, typically lasting 1–2 hr for full reduction of both enzyme and dye, spectra are recorded every 30 sec with a scanning diode-array spectrophotometer at 25°. Ideally, the concentrations of the oxidized and reduced forms of the flavoprotein and the reference dye are determined at a wavelength at which the other redox partner has an isosbestic point between the oxidized and reduced forms or has no absorbance at all. It is important to note that the redox potential of the reference dye should be within 30 mV of the flavoprotein to obtain reliable data. A number of useful redox reference dyes are given in Durst et al.[23] and Mayhew.[24] By using this method we could estimate the redox potentials of VAO variants, usually within 5 mV.

The potentials at 50% reduction of the flavoenzyme can be calculated by the Nernst equation[14]:

$$E_h(dye) = E_m(dye) + 2.303(RT/n_{dye}F) \log(dye_{ox}/dye_{red})$$

$$E_h(E) = E_m(E) + 2.303(RT/n_E F) \log(E_{ox}/E_{red})$$

$$E_h(dye) = E_h(E) \qquad \text{(at equilibrium)}$$

where E_h is the observed potential, E_m is the potential when the concentrations of oxidized and reduced forms are equal, R is the gas constant (8.31 J K^{-1} mol^{-1}), T is the temperature in degrees Kelvin, n is the number of electrons needed to convert the oxidized form to the reduced form, and F is the Faraday constant (96,496 J V^{-1} mol^{-1}). Thus, at 25°, 2.303(RT/nF) is equal to 0.059/n V.

[23] R. A. Durst, E. A. Blubaugh, M. L. Fultz, W. A. MacCrehan, and W. T. Yap, *Clin. Chem.* **28**, 1922 (1982).

[24] S. G. Mayhew, in "Flavoprotein Protocols" (S. K. Chapman and G. A. Reid, eds.), Vol. 131, p. 49. Humana Press, Totowa, NJ, 1999.

Redox Properties of Vanillyl-Alcohol Oxidase

Vanillyl-Alcohol Oxidase

Vanillyl-alcohol oxidase (VAO; EC 1.1.3.38) is a homooctameric flavo-enzyme[25] that catalyzes the oxidative demethylation of methyl ethers of *p*-cresol, providing the ascomycete *Penicillium simplicissimum* with a tool for metabolizing these lignin-derived aromatic compounds.[26] Each VAO subunit contains an 8α-N^3-histidyl-FAD as covalently bound prosthetic group.[25] The crystal structure of VAO has revealed that each VAO monomer consists of two domains: the larger domain comprises residues 1–270 and 500–560 and forms the FAD-binding domain, whereas residues 271–499 form the smaller cap domain.[27] The catalytic center of VAO is located at the interface of the two domains in the core of the protein. From sequence alignments it was recognized that VAO is a representative of a widespread family of structurally related oxidoreductases sharing a conserved flavin-binding domain.[28] A similarly folded domain is, for example, also present in the peripheral membrane respiratory flavoenzyme D-lactate dehydrogenase[29] and in UDP-*N*-acetylenolpyruvylglucosamine reductase (MurB), a flavoenzyme involved in peptidoglycan biosynthesis.[30]

Redox catalysis of VAO involves two half-reactions, in which the flavin is reduced by the substrate (reduction) and subsequently the reduced flavin is re-oxidized by molecular oxygen (oxidation). The reductive half-reaction involves the initial transfer of a hydride from the substrate to the flavin, resulting in a binary complex between the two-electron reduced enzyme and the *p*-quinone methide of the substrate. In the oxidative part of the reaction, the reduced flavin is reoxidized by molecular oxygen with the concomitant hydration of the quinone methide intermediate (Fig. 2). For the reaction with the physiological substrate 4-(methoxymethyl)phenol, the reductive half-reaction is rate limiting in overall catalysis.[31]

Redox Properties

The midpoint redox potential of ligand-free VAO was determined by the xanthine–xanthine oxidase method in the presence of the redox reference dye

[25] E. de Jong, W. J. H. van Berkel, R. P. van der Zwan, and J. A. M. de Bont, *Eur. J. Biochem.* **208,** 651 (1992).

[26] M. W. Fraaije, M. Pikkemaat, and W. J. H. van Berkel, *Appl. Environ. Microbiol.* **63,** 435 (1997).

[27] A. Mattevi, M. W. Fraaije, A. Mozzarelli, L. Olivi, A. Coda, and W. J. H. van Berkel, *Struct. Fold. Des.* **5,** 907 (1997).

[28] M. W. Fraaije, W. J. H. van Berkel, J. A. E. Benen, J. Visser, and A. Mattevi, *Trends Biochem. Sci.* **23,** 206 (1998).

[29] O. Dym, E. A. Pratt, C. Ho, and D. Eisenberg, *Proc. Natl. Acad. Sci. U.S.A.* **97,** 9413 (2000).

[30] T. E. Benson, D. J. Filman, C. T. Walsh, and J. M. Hogle, *Nat. Struct. Biol.* **2,** 644 (1995).

[31] M. W. Fraaije and W. J. H. van Berkel, *J. Biol. Chem.* **272,** 18111 (1997).

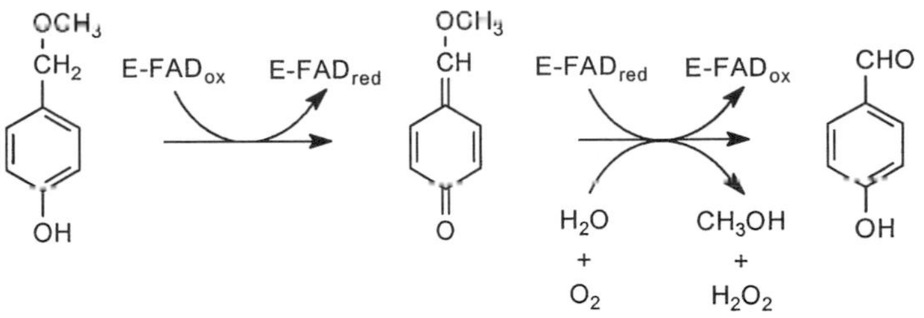

FIG. 2. Oxidative demethylation of 4-(methoxymethyl)phenol by VAO.

thionin ($E_m = +60$ mV) at pH 7.5. Figure 3 clearly shows that the enzyme-bound flavin and the dye reduce simultaneously via a single two-electron process, indicating that the dye and the flavin have a similar redox potential. The observed two-electron process fits well with the catalytic function of VAO, which involves the initial hydride transfer from the substrate to the flavin. When $\log(\text{VAO}_{ox}/\text{VAO}_{red})$ versus $\log(\text{dye}_{ox}/\text{dye}_{red})$ was plotted according to Minnaert[32] a midpoint redox potential of $+55$ mV was estimated (see inset, Fig. 3).[33] This redox potential is exceptionally high compared with other flavin-dependent enzymes. Only thiamine oxidase, also containing an 8α-N^3-histidyl-FAD, has an equally high redox potential ($+55$ mV).[34] The redox potential of VAO complexed with the substrate analog isoeugenol was determined in the presence of methylene blue ($E_m = +11$ mV). The enzyme reduced in a single two-electron reduction process and the redox potential was estimated to be $+15$ mV. Thus, when the active site cavity is saturated with the phenolic ligand isoeugenol the midpoint redox potential decreases by 40 mV.

In our studies we mainly focused on the rationale for the exceptionally high redox potential of VAO. A recurrent feature on analyzing flavin-containing oxidoreductases is the presence of a hydrogen bond donor near the N5 atom of the flavin prosthetic group. In most oxidoreductases this hydrogen bond donor is a backbone or a side-chain nitrogen atom.[35] However, in VAO such a hydrogen bond donor is absent. Instead, a negatively charged residue, Asp-170, is located near the flavin N5 atom (distance, 3.5 Å) (Fig. 4).[27] The side chain of Asp-170 is positioned in such a way that during catalysis it might interact with the reduced FAD cofactor, thereby stabilizing the reduced form. Intriguingly, glycolate oxidase also lacks a hydrogen bond donor to flavin N5 and, like VAO, this enzyme has a relatively high redox potential ($E_m = -21$ mV).[36]

[32] K. Minnaert, *Biochim. Biophys. Acta* **110**, 42 (1965).
[33] R. H. H. van den Heuvel, M. W. Fraaije, A. Mattevi, and W. J. H. van Berkel, *J. Biol. Chem.* **275**, 14799 (2000).
[34] C. Gomez-Moreno, M. Choy, and D. E. Edmondson, *J. Biol. Chem.* **254**, 7630 (1979).
[35] M. W. Fraaije and A. Mattevi, *Trends Biochem. Sci.* **25**, 126 (2000).
[36] C. Pace and M. Stankovich, *Biochemistry* **25**, 2516 (1986).

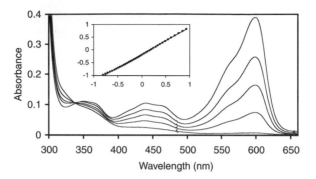

FIG. 3. Determination of the redox potential of wild-type VAO. VAO, 9 μM in potassium phosphate buffer, pH 7.5 at 25°, was reduced in the presence of 9 μM thionin by the xanthine–xanthine oxidase method. The reduction was finished after 100 min. *Inset:* $\log(VAO_{ox}/VAO_{red})$ (measured at 439 nm after correction for thionin) versus $\log(dye_{ox}/dye_{red})$ (measured at 600 nm). [Data from R. H. H. van den Heuvel, M. W. Fraaije, A. Mattevi, and W. J. H. van Berkel, *J. Biol. Chem.* **275,** 14799 (2000).]

The role of Asp-170 in VAO has been studied in detail by site-directed mutagenesis. When Asp-170 is replaced by glutamate or serine the flavin redox potential decreases 50 and 160 mV, respectively.[33] Kinetic characterization of the mutants revealed that Asp-170 is required for rapid substrate-mediated flavin reduction. D170E and D170S are 50-fold and 1000-fold less active, respectively.[33] Moreover, structural analysis of D170S suggested that the mutation does not induce any significant structural perturbations. These results indicate that both the presence

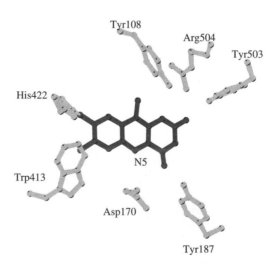

FIG. 4. The active site cavity of VAO. Distance from the flavin N5 atom to Asp-170 is 3.5 Å.

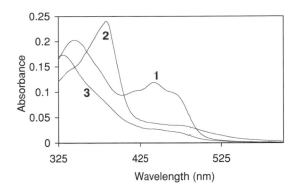

FIG. 5. Reduction of D170S by the xanthine–xanthine oxidase method. Curve 1, quinone form; curve 2, semiquinone anionic form; curve 3, hydroquinone form of D170S. [Data from R. H. H. van den Heuvel, M. W. Fraaije, A. Mattevi, and W. J. H. van Berkel, *J. Biol. Chem.* **275,** 14799 (2000).]

and the orientation of the negative charge near flavin N5 are crucial for the high redox potential of the flavin in VAO, and thus for the oxidative power of the enzyme. The behavior of the D170S mutant in the xanthine oxidase-mediated reduction is different from wild-type VAO, as this mutant highly stabilizes the red flavin semiquinone anion with a typical absorption maximum at 385 nm (Fig. 5). This thermodynamic stabilization of the semiquinone is different from kinetic substrate-mediated reduction experiments, in which no semiquinone formation is observed in the stopped-flow time scale.[33]

Redox potential determinations of flavin derivatives have revealed that modifications at the 8α position of the isoalloxazine ring result in increased redox potentials by 50–60 mV[37,38] and that ionization of the imidazole of 8α-N-imidazolylflavins results in changed redox properties.[39] To investigate the effect of the 8α-(N^3-histidyl)-flavin linkage in VAO, we substituted His-422, the residue to which the flavin is bound to the protein, for alanine.[11] The produced VAO mutant H422A binds the flavin tightly but noncovalently. The H422A mutation does not have any significant effects on the structure of the enzyme, but lowers the turnover rate with the physiological substrate 4-(methoxymethyl)phenol by one order of magnitude. The estimation of the redox potential of H422A revealed a 120-mV lower potential than for wild-type VAO. As Ala-422 is relatively far from the catalytic center of VAO, additional effects on the kinetics of the enzyme are unlikely. Therefore, the lower activity of the enzyme must be fully attributed to the decreased redox potential of the flavin.[11] This is the first report in which it has been

[37] D. E. Edmondson and T. P. Singer, *J. Biol. Chem.* **248,** 8144 (1973).
[38] D. E. Edmondson and R. De Francesco, *in* "Chemistry and Biochemistry of Flavoenzymes" (F. Mueller, ed.), Vol. I, p. 73. CRC Press, Boca Raton, FL, 1991.
[39] G. Williamson and D. E. Edmondson, *Biochemistry* **24,** 7918 (1985).

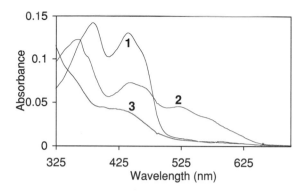

FIG. 6. Reduction of H422A by the xanthine–xanthine oxidase method. Curve 1, quinone form; curve 2, semiquinone neutral form; curve 3, hydroquinone form of H422A. [Data from M. W. Fraaije, R. H. H. van den Heuvel, W. J. H. van Berkel, and A. Mattevi, *J. Biol. Chem.* **274**, 35514 (1999).]

demonstrated that a covalent protein–flavin bond raises the redox potential of the flavin, and, therefore, the oxidative power of the flavoenzyme. Interestingly, the xanthine oxidase-mediated reduction of H422A occurs in two discrete one-electron reduction steps. First, a blue neutral flavin semiquinone species is formed, which is further reduced to the flavin hydroquinone (Fig. 6). Thus, the H422A mutation shifts the pK of the flavin semiquinone to a higher value compared with the D170S mutation, resulting in the neutral form of the flavin semiquinone at pH 7.5.

Conclusions

In the described research we have investigated the redox properties of the flavin prosthetic group in VAO. Site-directed mutagenesis studies have revealed that the exceptional redox properties of VAO are, at least in part, due to two rare structural determinants: the negative charge of Asp-170 near the flavin N5 atom and the covalent histidyl linkage between the protein and the flavin. The negative charge at hydrogen bond distance from the flavin N5 atom (3.5 Å) stabilizes the reduced form of the cofactor, thus facilitating the oxidation of substrates. Except for glycolate oxidase,[36] none of the other flavoenzymes with known three-dimensional structure harbor a hydrogen bond acceptor near flavin N5.[35] In agreement with earlier suggestions,[9] a covalent linkage between the protein and the isoalloxazine ring of the flavin can markedly increase the redox potential of the flavin prosthetic group.[11] This indicates that the histidyl–flavin bond in certain flavoproteins has evolved as a tool for raising the oxidative power of the flavin cofactor.

[18] Anaerobic Oxidations of Myoglobin and Hemoglobin by Spectroelectrochemistry

By Céline H. Taboy, Celia Bonaventura, and Alvin L. Crumbliss

Introduction

A study of the redox properties of the myoglobins (Mbs) and hemoglobins (Hbs) provides insights into heme protein electron transfer processes, including the influence of subunit–subunit interactions and cooperativity. Studying the changes in redox behavior of various Mbs and Hbs has also allowed us and others to obtain important information regarding the involvement of iron in the transport of O_2, NO, and other small molecules.[1–68] In general terms,

[1] C. Bonaventura, S. Tesh, K. M. Faulkner, D. Kraiter, and A. L. Crumbliss, *Biochemistry* **37**, 496 (1998).

[2] E. Antonini, J. Wyman, M. Brunori, J. F. Taylor, A. Rossi-Fanelli, and A. Caputo, *J. Biol. Chem.* **239**, 907 (1964).

[3] K. M. Faulkner, A. L. Crumbliss, and C. Bonaventura, *Inorg. Chim. Acta* **226**, 187 (1994).

[4] E. Antonini and M. Brunori, "Hemoglobin and Myoglobin in Their Reactions with Ligands." Elsevier, North-Holland, Amsterdam, 1971.

[5] E. W. Kristensen, D. H. Igo, R. C. Elder, and W. R. Heineman, *J. Electroanal. Chem.* **309**, 61 (1991).

[6] M. Brunori, G. M. Giacometti, E. Antonini, and J. Wyman, *J. Biol. Chem.* **63**, 139 (1972).

[7] M. Brunori, U. Saggese, G. C. Rotitio, E. Antonini, and J. Wyman, *Biochemistry* **10**, 1604 (1971).

[8] F. Riggs and R. A. Walbach, *J. Gen. Physiol.* **39**, 585 (1956).

[9] M. Bolognesi, S. Onesti, G. Gatti, A. Coda, P. Ascenzi, and M. Brunori, *J. Mol. Biol.* **205**, 529 (1989).

[10] M. Brunori, A. Alfsen, U. Saggese, E. Antonini, and J. Wyman, *J. Biol. Chem.* **243**, 2950 (1968).

[11] M. Peruta, "Science Is Not a Quiet Life." World Scientific Publishing, Singapore, 1997.

[12] R. J. Kassner, *Proc. Natl. Acad. Sci. U.S.A.* **69**, 2263 (1972).

[13] A. G. Mauk and G. R. Moore, *J. Biol. Inorg. Chem.* **2**, 119 (1997).

[14] M. R. Gunner, E. Alexov, E. Torres, and S. Lipovaca, *J. Biol. Inorg. Chem.* **2**, 126 (1997).

[15] E. Lloyd, B. C. King, F. M. Hawkridge, and A. G. Maulk, *Inorg. Chem.* **37**, 2888 (1998).

[16] J. Baldwin and C. Chothia, *J. Mol. Biol.* **129**, 175 (1979).

[17] A. Bellelli and M. Brunori, *Methods Enzymol.* **232**, 56 (1994).

[18] M. Brunori, Á. F. Cutruzzola, C. Savino, C. Travaglini-Allocatelli, B. Vallone, and Q. H. Gibson, *Biophys. J.* **76**, 1259 (1999).

[19] D. H. Doherty, M. P. Doyle, S. R. Curry, R. Vali, T. J. Fattor, J. S. Olson, and D. D. Lemon, *Nat. Biotechnol.* **16**, 672 (1998).

[20] R. F. Eich, T. Li, D. D. Lemon, D. H. Doherty, S. R. Curry, J. F. Aitken, A. J. Mathews, K. A. Johnson, R. D. Smith, G. J. Phillips, Jr., and J. S. Olson, *Biochemistry* **35**, 6976 (1996).

[21] K. Imai, "Allosteric Effects in Hemoglobin." Cambridge University Press, Cambridge, 1982.

[22] J. Monod, J. Wyman, and J. P. Changeux, *J. Mol. Biol.* **12**, 88 (1965).

[23] C. Travaglini-Allocatelli, Á. F. Cutruzzola, A. Brancaccio, B. Vallone, and M. Brunori, *FEBS Lett.* **352**, 63 (1994).

Copyright 2002, Elsevier Science (USA).
All rights reserved.
0076-6879/02 $35.00

important issues such as the number of electrons involved in the redox process, cooperative effects between redox centers and between allosteric sites and redox centers, specific formal or midpoint potentials ($E^{0\prime}$ or $E_{1/2}$, respectively), and the shift in $E_{1/2}$ due to chemical processes coupled to the redox process or to diverse effectors can be probed during a controlled redox

[24] B. Vallone, P. Vecchini, V. Cavalli, and M. Brunori, *FEBS Lett.* **324,** 117 (1993).

[25] Z. Otwinowsky and W. Minor, *Methods Enzymol.* **276,** 307 (1996).

[26] F. A. Rossi and E. Antonini, *Arch. Biochem. Biophys.* **77,** 478 (1958).

[27] W. Zhang, Á. F. Cutruzzola, C. Travaglini-Allocatelli, M. Brunori, and G. La Mar, *Biophys. J.* **73,** 1019 (1997).

[28] J. E. Cradock-Watson, *Nature (London)* **215,** 630 (1967).

[29] B. Giardina, O. Brix, M. E. Clementi, R. Scatena, B. Nicoletti, R. Cicchetti, G. Argentin, and S. G. Condo, *Biochem. J.* **266,** 897 (1990).

[30] F. A. Armstrong, *J. Biol. Inorg. Chem.* **2,** 139 (1997).

[31] L. D. Kwiatkowski, H. L. Hui, A. Wierzba, R. W. Noble, R. Y. Walder, E. S. Peterson, S. G. Sligar, and K. E. Sanders, *Biochemistry* **37,** 4325 (1998).

[32] Y. Wu, E. Y. T. Chien, S. G. Sligar, and G. N. La Mar, *Biochemistry* **37,** 6979 (1998).

[33] J. F. Christian, M. Unno, J. T. Sage, P. M. Champion, E. Chien, and S. G. Sligar, *Biochemistry* **36,** 11198 (1997).

[34] S. J. Smerdon, S. Krzywda, A. M. Brzozowski, G. J. Davies, A. J. Wilkinson, A. Brancaccio, F. Cutruzzola, C. T. Allocatelli, M. Brunori, T. Li, R. E. Brandley, Jr., T. E. Carver, R. F. Eich, E. Singelton, and J. S. Olson, *Biochemistry* **34,** 8715 (1995).

[35] C. Bonaventura, G. Godette, S. Tesh, D. E. Holm, J. Bonaventura, A. L. Crumbliss, L. L. Pearce, and J. Peterson, *J. Biol. Chem.* **274,** 5499 (1999).

[36] C. Bonaventura and J. Bonaventura, *Hemoglobin* **4,** 275 (1980).

[37] C. Poyart, E. Bursaux, A. Arnone, J. Bonaventura, and C. Bonaventura, *J. Biol. Chem.* **255,** 9465 (1980).

[38] E. Bucci, A. Salahuddin, J. Bonaventura, and C. Bonaventura, *J. Biol. Chem.* **253,** 821 (1978).

[39] G. Amiconi, C. Bonaventura, J. Bonaventura, and E. Antonini, *Biochim. Biophys. Acta* **495,** 279 (1977).

[40] F. Bossa, D. Barra, M. Coletta, F. Martini, A. Liverzani, R. Petruzzelli, J. Bonaventura, and M. Brunori, *FEBS Lett.* **64,** 76 (1976).

[41] E. Chiancone, J. E. Norne, S. Forsen, J. Bonaventura, M. Brunori, E. Antonini, and J. Wyman, *Eur. J. Biochem.* **55,** 385 (1975).

[42] C. Bonaventura, J. Bonaventura, G. Amiconi, L. Tentori, M. Brunori, and E. Antonini, *J. Biol. Chem.* **250,** 6273 (1975).

[43] J. Bonaventura, C. Bonaventura, G. Amiconi, L. Tentori, M. Brunori, and E. Antonini, *J. Biol. Chem.* **250,** 6278 (1975).

[44] J. Bonaventura, C. Bonaventura, M. Brunori, B. Giardina, E. Antonini, F. Bossa, and J. Wyman, *J. Mol. Biol.* **82,** 499 (1974).

[45] D. Barra, F. Bossa, J. Bonaventura, and M. Brunori, *FEBS Lett.* **35,** 151 (1973).

[46] M. Brunori, J. Bonaventura, C. Bonaventura, B. Giardina, F. Bossa, and E. Antonini, *Mol. Cell. Biochem.* **1,** 189 (1973).

[47] J. S. Kavanaugh, D. R. Chafin, A. Arnone, A. Mozzarelli, C. Rivetti, G. L. Rossi, L. D. Kwiatkowski, and R. W. Noble, *J. Mol. Biol.* **248,** 136 (1995).

[48] G. E. Borgstahl, P. H. Rogers, and A. Arnone, *J. Mol. Biol.* **236,** 817 (1994).

[49] C. Poyart, E. Bursaux, A. Arnone, J. Bonaventura, and C. Bonaventura, *J. Biol. Chem.* **255,** 9465 (1980).

reaction.[3,7,12,13,32,35,69–83] The spectroelectrochemical technique applied to Hbs and Mbs presented in this chapter offers an opportunity to specifically investigate the cooperativity in electron transfer between the four iron(II/III) sites in the Hbs and to compare the degree of redox cooperativity with that found for O_2 binding. Comparison between oxygenation and anaerobic oxidation can provide important information regarding the mechanism of hemoglobin function as well as enhanced understanding of heme protein electron transfer reactions. The purpose of this chapter is to outline the spectroelectrochemical technique used to study these various issues.

[50] S. O'Donnell, R. Mandaro, T. M. Schuster, and A. Arnone, *J. Biol. Chem.* **254,** 12204 (1979).

[51] A. Arnone and M. F. Perutz, *Nature (London)* **249,** 34 (1974).

[52] A. Arnone, *Annu. Rev. Med.* **25,** 123 (1974).

[53] A. Arnone, *Nature (London)* **237,** 146 (1972).

[54] A. M. Nigen, J. M. Manning, and J. O. Alben, *J. Biol. Chem.* **255,** 5525 (1980).

[55] M. H. Garner, R. A. Bogardt, Jr., and F. R. Gurd, *J. Biol. Chem.* **250,** 4398 (1975).

[56] I. M. Russu and C. Ho, *Biochemistry* **25,** 1706 (1986).

[57] M. Perella, L. Benazzi, M. Ripamonti, and L. Rossi-Bernardi, *Biochemistry* **33,** 10358 (1994).

[58] E. Chiancone, J. E. Norne, S. Forsen, J. Bonaventura, M. Brunori, and E. Antonini, *Biophys. Chem.* **3,** 56 (1975).

[59] S. Bettati, A. Mozzarelli, and M. F. Perutz, *J. Mol. Biol.* **281,** 581 (1998).

[60] M. F. Perutz, D. T. B. Shih, and D. Williamson, *J. Biol. Chem.* **239,** 555 (1994).

[61] M. F. Perutz, G. Fermi, C. Poyard, J. Pagnier, and J. Kister, *J. Mol. Biol.* **233,** 536 (1993).

[62] E. Chiancone, J. E. Norne, S. Forsen, E. Antonini, and J. Wyman, *J. Mol. Biol.* **70,** 675 (1972).

[63] R. Benesch and R. E. Benesch, *Biochem. Biophys. Res. Commun.* **26,** 162 (1967).

[64] C. H. Tsai, T. J. Shen, N. T. Ho, and C. Ho, *Biochemistry* **38,** 8751 (1999).

[65] A. Desbois and R. Banerjee, *J. Mol. Biol.* **92,** 479 (1975).

[66] C. Ho and I. M. Russu, *Biochemistry* **26,** 6299 (1987).

[67] C. Bonaventura and J. Bonaventura, *in* "Biochemical and Clinical Aspects of Hemoglobin Abnormalities" (W. S. Caughey, ed.), p. 647. Academic Press, New York, 1978.

[68] M. L. Meckstroth, B. J. Norris, and W. R. Heineman, *Bioelectrochem. Bioenerget.* **8,** 63 (1981).

[69] K. M. Faulkner, A. L. Crumbliss, and C. Bonaventura, *J. Biol. Chem.* **270,** 13604 (1995).

[70] R. J. Kassner, *J. Am. Chem. Soc.* **95,** 2674 (1973).

[71] G. R. Moore, G. W. Pettigrew, and N. K. Rogers, *Proc. Natl. Acad. Sci. U.S.A.* **83,** 4998 (1986).

[72] A. K. Churg and A. Warshel, *Biochemistry* **25,** 1675 (1986).

[73] T. Pascher, J. P. Chesick, J. R. Wrinkler, and H. B. Gray, *Science* **271,** 1558 (1996).

[74] A. Warshel, A. Papazyan, and I. Muegge, *J. Biol. Inorg. Chem.* **2,** 143 (1997).

[75] E. Lloyd, D. P. Hildebrand, K. M. Tu, and A. G. Mauk, *J. Am. Chem. Soc.* **117,** 6434 (1995).

[76] A. W. Bott, *Curr. Separations* **18,** 47 (1999).

[77] L. D. Dickerson, A. Sauer-Masarwa, N. Herron, C. M. Fendick, and D. H. Busch, *J. Am. Chem. Soc.* **115,** 3623 (1993).

[78] C. Brunel, A. Bondon, and G. Simonneaux, *J. Am. Chem. Soc.* **116,** 11827 (1994).

[79] J.-M. Lopez-Castillo, A. Filali-Mouhim, N. E. Van Binh-Otten, and J.-P. Jay-Gerin, *J. Am. Chem. Soc.* **119,** 1978 (1997).

[80] J. Feitelson and G. McLendon, *Biochemistry* **30,** 5051 (1991).

[81] H. Durliat and M. Comtat, *Anal. Chem.* **54,** 856 (1982).

[82] S. J. Dong, J. J. Niu, and T. M. Cotton, *Methods Enzymol.* **246,** 701 (1995).

[83] M. L. Fultz and R. Durst, *Anal. Chim. Acta* **140,** 1 (1982).

Although there exists a large body of literature that describes the mechanisms associated with oxygen binding to Hbs and Mbs, relatively little information has been reported with respect to the mechanism associated with tuning of the redox potential ($E_{1/2}$) of these proteins.[2,4,10,65] Information regarding the influence of the prosthetic group environment on the overall propensity of the protein to accept or deliver electrons has been shown to lead to a more fundamental understanding of both O_2 binding and anaerobic oxidation.[1,3,35,69,84,85] Although a few results are presented here to illustrate these points, we direct the reader to Taboy et al.[84,85] for detailed discussions.

Spectroelectrochemical Technique and Cell Design

A limited number of techniques are available for investigating the redox properties of proteins. One of these is based on spectroelectrochemistry, in which spectral changes are monitored as a function of an applied potential. In other words, if a spectral difference exists between the oxidized and reduced form of the protein, information regarding its thermodynamic ease of oxidation (i.e., its half-potential, $E_{1/2}$) and the number of electrons being transferred can be obtained. A combination of three components must be specifically chosen for each protein studied: (1) an electrode material and configuration, (2) a mediator or electron shuttle that facilitates electron transfer between the electrode and the protein, and (3) a detection system that can probe the presence of the oxidized and/or reduced state of the protein.[5,81,82,84,86–93]

Electrode

Different electrode materials can be used to perform a spectroelectrochemical experiment. The specific surface is chosen as a function of the applied potential window required to access the oxidation–reduction of the particular protein studied. This information is readily available in electrochemistry monographs such as that by Bard and Faulkner.[94] Different electrode materials provide different

[84] C. H. Taboy, C. Bonaventura, and A. L. Crumbliss, *Bioelectrochem. Bioenerget.* **48,** 79 (1999).

[85] C. H. Taboy, K. M. Faulkner, D. Kraiter, C. Bonaventura, and A. L. Crumbliss, *J. Biol. Chem.* **275,** 39048 (2000).

[86] K. Ashley and S. Pons, *Chem. Rev.* **88,** 673 (1988).

[87] J. D. Brewster and J. L. Anderson, *Anal. Chem.* **54,** 2560 (1982).

[88] J. A. Cooper and R. G. Compton, *Electroanalysis* **10,** 141 (1998).

[89] S. J. Dong, *Denki Kagaru* **59,** 664 (1991).

[90] W. R. Heineman, F. M. Hawkridge, and H. N. Blount, *Electroanalysis* **13,** 1 (1984).

[91] J. J. Niu and S. J. Dong, *Rev. Anal. Chem.* **15,** 1 (1996).

[92] R. S. Robinson, C. W. McCurdy, and R. L. McCreery, *Anal. Chem.* **54,** 2356 (1982).

[93] L. R. Sharpe, W. R. Heineman, and R. C. Elder, *Chem. Rev.* **90,** 705 (1990).

[94] A. J. Bard and L. R. Faulkner, "Electrochemical Methods: Fundamentals and Applications," 2nd Ed. John Wiley & Sons, New York, 2001.

overpotentials for solvent breakdown and therefore determine the effective electrochemical window available for a particular solvent/buffer system. For instance, platinum and gold mesh electrodes have been used successfully to study heme proteins and the transferrins, respectively, in our laboratory. Gold allows for a more negative potential window before breakdown of the buffer/solvent. The electrode material, of course, needs to be configured so that monochromatic light can be used as an oxidation state probe.

Optically Transparent Thin-Layer Cell

The most direct method for probing a redox change in a spectrally active metalloprotein is to record its UV–visible (UV–Vis) spectrum as the prosthetic group is being reduced or oxidized. A set of these spectra is most readily obtained by monitoring the changes in absorbance associated with the metal–ligand electronic properties at various applied potentials. An optically transparent thin-layer electrode (OTTLE) can be used and has the advantages of combining a small sample volume with maxium electrode surface–solution contact. This combination decreases the time required for the system to come to equilibrium at each applied potential. The experiment is essentially based on a bulk electrolysis process in which the bulk solution is delineated in space by the thickness of the OTTLE, typically leading to an ~0.03-cm diffusion layer.

Spectroelectrochemical experiments may be carried out using a two- or three-electrode arrangement (working, reference, and auxiliary) in an OTTLE cell as illustrated in Fig. 1. The cell is constructed from a 1 × 2 cm piece of 52 mesh platinum or gold gauze placed between the inside wall of a 1-cm path length cuvette and a piece of silica or quartz glass held in place by a small Tygon spacer positioned so as not to interfere with the spectral measurement.[3] The OTTLE assembly results in an optical path length of 0.025–0.040 cm, depending on the electrode material and mesh size. A platinum (or gold) wire connects the platinum (or gold) gauze working electrode by insertion through a septum covering the top of the cuvette. A Pasteur pipette salt bridge plugged at the bottom with an agar gel containing 0.2 M KCl is then prepared so as to connect an Ag/AgCl reference microelectrode to the solution containing the working electrode. The salt bridge solution is usually composed of the buffer used for the preparation of the working solution with a minimum of 0.2 M background electrolyte (e.g., KCl). The auxiliary electrode is a 2 × 50 mm platinum (or 1 × 50 mm gold) wire inserted into one of the corners of the cuvette, so as not to obstruct the light path, and separated from the working gauze electrode by the piece of silica (or quartz). The platinum wire is held in place by the Tygon spacer at the bottom of the cell and the septum on top and presses against the silica (or quartz) piece, thereby sandwiching the working electrode tightly between the cell window and the silica (or quartz) piece. The sample solution fills the thin-layer portion of the cell surrounding the working

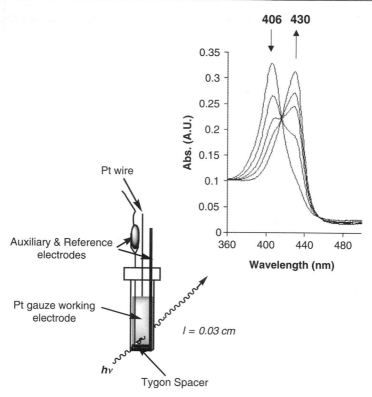

Fig. 1. Schematics of the OTTLE cell and of a representative data set obtained by spectroelectrochemistry during the reduction of Hb A_0, showing an increase in the absorbance at 430 nm in conjunction with a decrease in the absorbance at 406 nm.

electrode by capillary action. The auxiliary, reference, and working electrodes are connected to each other through the working solution added in excess ($\sim$0.3 ml) at the bottom of the cuvette. The OTTLE cell can be purged with N_2 for anaerobic measurement before injecting the protein solution. The use of an OTTLE cell such as described here has the following advantages.

1. Solution volumes are small ($\sim$450 μl), leading to rapid electrolysis and attainment of equilibrium at each applied potential. This is particularly important in protein electrochemistry, where only limited sample quantities are available, and time-dependent protein denaturation is an issue.

2. Anaerobic conditions are more easily achieved.

3. The use of a minigrid in a thin layer of working solution decreases the level of the ohmic drop associated with the electrochemical cell.

Mediator

Protein electrochemistry is made difficult by a number of factors. Most proteins are relatively sluggish compared with small molecules with regard to electron transfer. Moreover, proteins are prone to denaturation on contact with the electrode surface, which can change their redox potential. To increase the electron transfer yield and therefore decrease the time necessary to achieve equilibrium, an electron shuttle is used.[83] This electron shuttle, also called a mediator, consists of an electroactive species that exhibits ideal reversible redox behavior. Its redox potential ($E_{1/2}$) needs to be in the general region of the expected $E_{1/2}$ of the protein being studied.

The electron transfers, both homogeneous (between the protein and the mediator) and heterogeneous (between the mediator and the bare electrode), must be extremely fast in order to allow a rapid turnover of the reduced and oxidized forms of the mediator.[68,83] The main characteristics of a good mediator for this type of study are as follows.

1. Kinetic characteristics include rapid reversible electron transfer with the electrode and with the protein to avoid being the limiting factor during the redox process.

2. Thermodynamic considerations require that a sufficient amount of the oxidized and reduced forms of the mediator be present in solution at all times.

3. The mediator must have a redox potential in the vicinity of the expected redox potential of the protein under investigation,

$$\left| E_m^0 \quad E_p^0 \right| \leq \frac{2RT}{mF}$$

where E_m^0 is the redox potential of the mediator, E_p^0 is the redox potential of the protein, and m is the number of electrons transferred for the mediator.

4. The mediator should be spectroscopically silent in the region of interest for the redox protein, or at least not create interferences with the protein absorbance spectrum in the oxidized or reduced form.

5. The mediator should not interact with the protein in a way that influences the protein redox behavior. (For highly allosteric proteins such as hemoglobin, care must be taken as redox behavior can often be mediator dependent.)

Compilations of mediators for protein electrochemistry are available.[68,83] We have used hexaammineruthenium(II/III) chloride, $Ru(NH_3)_6Cl_3$, as a mediator for heme protein studies.[1,3,35,69,84] This mediator satisfies all the criteria outlined above, and we have demonstrated that it does not influence the redox properties of myoglobin or hemoglobin. Furthermore, its positive charge prevents interaction at the allosteric site in the hemoglobin tetramer. Hemoglobin is known to interact with various anions that can influence both the oxygenation and oxidation of the protein, whereas the protein is rather insensitive to cations.[2,4] The small quantity of chloride ion due to the counterion associated with $Ru(NH_3)_6^{2+/3+}$ was shown not to influence the redox potential of most Hbs (± 2 mV for a 10-fold increase in mediator-to-iron ratio).[3,4,69]

Although we have shown that the presence of the mediator facilitates the protein–electrode electron exchange and allows oxidation–reduction of the Hbs and Mbs without influencing their structure and integrity,[3] the kinetics and mechanism of electron exchange between hemoglobin and $Ru(NH_3)_6^{2+/3+}$ have not been fully described at this point. However, it is clear that because of its large size the ruthenium complex does not enter the heme pocket.

Detection System

Various detection systems can in principle be used to investigate the redox properties of proteins. Although we are describing a UV–Vis spectroscopy-based detection system here, infrared (IR) and emission spectroscopic detection can be developed on the same principle. Extensive reviews have been published on infrared, Raman, transmission, UV–Vis, and EXAFS (extended X-ray absorption fine structure)-based spectroelectrochemistry.[5,81,82,84,86–93] Because both Mbs and Hbs have readily available UV–Vis spectral properties that differ upon oxidation–reduction of the prosthetic groups, we have based our technique on UV–Vis spectroscopic detection.

Specific Examples

We illustrate the principles of anaerobic spectroelectrochemistry using an OTTLE cell by describing a typical experiment used to investigate several myoglobins. In all experiments, the changes in absorbance at one or two wavelengths associated with the oxidized and reduced species resulting from altered electrode potentials are recorded. The absorbance changes indicate changes in the concentrations of the oxidized and reduced species and are used to create a plot according to the Nernst equation [Eq. (1)]:

$$E_{app} = E_{1/2} + \frac{RT}{nF} \ln \frac{[Ox]}{[Red]} = E_{1/2} + \frac{RT}{nF} \ln \frac{[metMb]}{[deoxyMb]} \tag{1}$$

where E_{app} is the applied potential; $E_{1/2}$ is the midpoint potential, at which 50% of the protein is oxidized and 50% is reduced under our specific conditions; R is the gas constant (J K^{-1} mol^{-1}); T is temperature in degrees Kelvin; n indicates the number of electrons involved in the redox process for an ideal system with nernstian behavior; F is the Faraday constant; and [Ox] and [Red] are the concentrations of oxidized and reduced species (M). The details associated with data analysis are described in Data Analysis (below).

Heme Proteins with Nernstian Response

Myoglobins

Myoglobins are a class of metalloproteins that contain a single iron(II) in a heme prosthetic group that can undergo a reversible one-electron transfer. The absence of subunits, such as found in the hemoglobins, precludes a cooperative response to redox, thus providing us with a system that exhibits nernstian behavior.

The spectral properties of horse, sperm whale, and *Aplysia* myoglobins (Mbs) are described briefly here. Sperm whale and horse Mbs have similar amino acid sequences.[4] They have identical absorption maxima in the Soret region with small differences in molar absorptivity. The iron oxidation state of these Mbs can be monitored at 410 nm for oxidized iron(III) Mb and 435 nm for reduced iron(II) Mb (wavelengths of highest extinction coefficients), although other probe wavelengths can be used if the sample is sufficiently concentrated. *Aplysia* Mb, however, has a different sequence of amino acids, particularly in the heme pocket environment,[4,6,7,9] with an absorbance maximum at 435 nm in its iron(II) state and at 399 nm in its iron(III) state. This corresponds to a shift of approximately 10 nm in comparison with the maxima observed in the case of both iron(III) horse and sperm whale Mbs. This slight shift is described by Brunori *et al.* as being associated with a reversible equilibrium between a partially opened globular structure, which exhibits a λ_{max} at 390 nm, and the native structure, which exhibits a λ_{max} at 410 nm.[6]

Methods

Sample and Reagent Preparation. A stock solution of the electrochemical mediator, $Ru(NH_3)_6Cl_3$, is prepared in a 0.05 M 4-morpholinepropanesulfonic acid (MOPS) buffer solution adjusted to pH 7.1 to give a concentration of 4.5 to 5.5 mM. MOPS is selected as the buffer for its noncomplexing nature and stability, as well as the absence of spectral and electrochemical interferences. KNO_3 (0.2 M) is used in this particular set of experiments as the background electrolyte. The presence of a background electrolyte facilitates solution conductivity necessary for an electrochemical experiment. It is important to be systematic in the choice of the electrolyte in order to minimize the different kinds of species in solution. For example, because the bridge solution that connects the reference electrode and the working solution is composed of 0.2 M KCl (as described in Spectroelectrochemical Technique and Cell Design above), a potassium salt is used as the background electrolyte. The presence of K^+ on both sides of the agar gel connection minimizes the errors associated with a junction potential across the reference–working solution interface. Nanopure water is used at all times and all solutions are stored under an N_2 atmosphere at 4°.

For each experiment, a solution containing 0.2 M KNO_3, 1 mM $Ru(NH_3)_6Cl_3$, and 0.05 M MOPS at pH 7.1 in a 5-ml pear-shaped flask is connected to a vacuum line for repeated pump purging with N_2, followed by addition of Mb and additional pump purging with gentle swirling to minimize bubbling. Final concentrations are typically 0.06–0.08 mM in heme.

Spectroelectrochemical Experiment. Spectroelectrochemical experiments for myoglobins described here are carried out in the previously described anaerobic OTTLE cell (Fig. 1). In a typical experiment, about 0.5 ml of the working solution (protein, mediator, buffer with background electrolyte) is injected at the bottom of the OTTLE cell via a gas-tight syringe. The cell is then placed in the temperature-controlled cell holder of a spectrophotometer linked to a potentiostat [e.g., a Cary 2300 UV–Vis–NIR (near infrared) spectrophotometer (Varian, Palo Alto, CA) and a Princeton Applied Research (Oakridge, TN) model 75 potentiostat]. Spectra are collected from 340 to 700 nm, with specific emphasis on the Soret region. Absorbance changes are monitored at 410 nm [absorbance maximum for iron(III) Mb] and 435 nm [absorbance maximum for iron(II) Mb]. The full region between 340 and 700 nm is recorded and the five isosbestic points (420, 462, 522, 606, and 662 nm) are scrutinized to detect any problems associated with the nature or concentration of the protein (e.g., denaturation). The absorbances of the fully oxidized (A_o) and fully reduced (A_r) Mb are obtained by applying a potential of +400 and −250 mV [vs. normal hydrogen electrode (NHE)], respectively, and the absorbance is recorded when the system reaches equilibrium (15 to 45 min may be required to obtain a stable equilibrium absorbance reading, depending on the system). The optical path length can vary from cell to cell, but can be precisely determined for each experiment by using the Soret band absorbance, the known concentration of the working solution, and the extinction coefficients of the protein. The concentration is determined independently, typically after addition of dithionite to a portion of the unused working solution. The extinction coefficients for the reduced species [deoxy-Mb; iron(II) Mb] are as follows: $\varepsilon^{435}_{Aplysia} = 113$ mM^{-1} cm^{-1}, $\varepsilon^{435}_{horse} = 121$ mM^{-1} cm^{-1}, and $\varepsilon^{435}_{sperm\ whale} = 115$ mM^{-1} cm^{-1}.[4,6] The extinction coefficients for the oxidized species [met-Mb; iron(III) Mb] at 435 nm are negligible.

A typical increment of 20 mV is applied to the system, starting at approximately +300 mV (fully oxidized met-Mb) and ending at −120 mV (vs. NHE) (fully reduced deoxy-Mb). Although most experiments are performed by proceeding from fully oxidized to fully reduced Mb, the system is reversible under our experimental conditions and can be performed in either direction. From the recorded spectral changes (Fig. 2) and applied potential, Nernst plots are developed as described below.

Data Analysis

The set of absorbance data obtained for the system at equilibrium at various electrode potentials can be converted to the concentration ratio of oxidized to

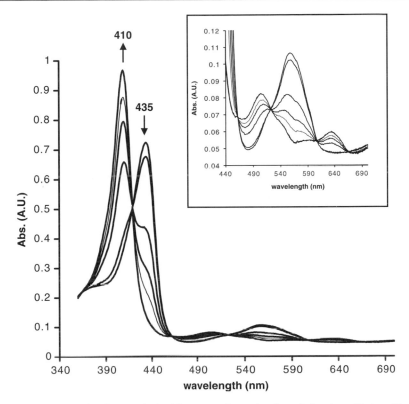

FIG. 2. Representative data set obtained by spectroelectrochemistry during the oxidation of horse Mb, showing a decrease in the absorbance at 435 nm in conjunction with an increase in the absorbance at 410 nm. *Inset:* Enlarged area showing the presence of four of the five isosbestic points at 462, 522, 606, and 662 nm, respectively.

reduced species, using Beer's law [Eq. (2)]:

$$\frac{[\text{Ox}]}{[\text{Red}]} = \frac{A(E) - A_R}{A_O - A(E)} \tag{2}$$

where $A(E)$ is the absorbance of the solution at equilibrium for any applied potential E_{app}, A_O is the absorbance of the fully oxidized protein (at +400 mV vs. NHE), and A_R is the absorbance of the fully reduced protein (at −250 mV vs. NHE).

The [Ox]/[Red] ratio is plotted as a function of the applied potential E_{app}, according to a rearranged form of the Nernst equation [Eq. (1)]. This results in a direct determination of $E_{1/2}$, the potential at which 50% of the protein is oxidized and 50% is reduced, at log[Ox]/[Red] = 0. Representative Nernst plots for horse

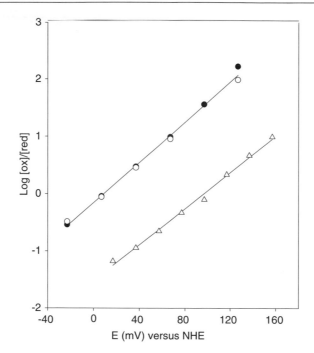

FIG. 3. Representative Nernst plots for horse Mb in 0.05 M MOPS, 1 mM Ru(NH$_3$)$_6$Cl$_3$, 1.0 M KNO$_3$ at pH 7.1, 20°. Data analysis was performed at two different wavelengths, 410 nm (●) and 435 nm (○), and shows the overlap of the Nernst plots in comparison with a representative Nernst plot for *Aplysia* Mb (△) obtained at 435 nm.

and *Aplysia* myoglobins are illustrated in Fig. 3. Although only one wavelength needs to be monitored to obtain a Nernst plot, it is good practice, when possible, to monitor the wavelengths of maximum absorbance of the reduced and oxidized forms of the protein (i.e., absorbances at 410 and 435 nm in the case of the myoglobins). The Nernst plot derived from each monitored absorbance change should overlap to give the same $E_{1/2}$ and n values (Fig. 3). In addition, if a protein has characteristic isosbestic points for its reduced and oxidized forms, denaturation and a change in concentration of the protein can be monitored by evaluating how "clean" these isosbestic points are as the oxidation or reduction experiment is performed (Fig. 2). This is an important spectroscopic tool that we have used systematically to evaluate the quality of our data for both the Mbs and Hbs.

The results obtained for sperm whale, horse, and *Aplysia* Mbs are presented in Figs. 3 and 4. As shown for solutions of a single protein, we obtained linear Nernst plots throughout the applied potential range with an n value of 1, consistent with a well-behaved one-electron transfer process for each of these myoglobins. These

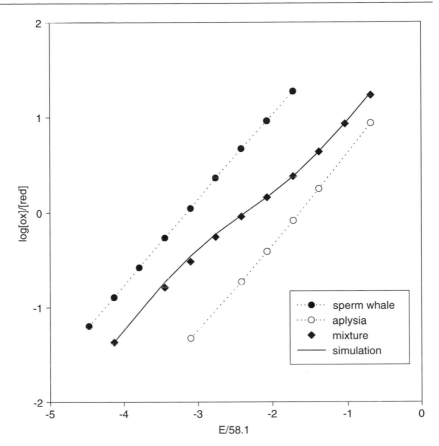

FIG. 4. Nernst plots for sperm whale (●) and *Aplysia* (○) myoglobins, and a 1 : 1 mixture of *Aplysia* : sperm whale (◆) myoglobins. The x axis represents applied potential versus Ag/AgCl according to a rearrangement of Eq. (1), where $E/58.1 = E_{app}F/RT$. Conditions: 1 mM Ru(NH$_3$)$_6$Cl$_3$, 0.06 mM heme, 0.2 M KNO$_3$, 0.05 M MOPS, pH 7.1, and 20°; data collected at $\lambda = 435$ nm. Simulated data (—) calculated according to Eq. (12), where $\varepsilon_{rA} = \varepsilon^{435}_{\text{sperm whale}} = 115$ mM^{-1} cm^{-1}, $\varepsilon_{rB} = \varepsilon^{435}_{Aplysia} = 113$ mM^{-1} cm^{-1}, and $b = 0.3$ mm. [Reprinted from C. H. Taboy, C. Bonaventura, and A. L. Crumbliss, *Bioelectrochem. Bioenerget.* **48,** 79 (1999), with permission from Elsevier Science.]

spectroelectrochemical results are in close agreement with $E_{1/2}$ values previously determined by potentiometric titration in the presence of redox mediators.[6,7,95]

Heme Proteins with Nonnernstian Response

Although the preceding examples represent three well-behaved Nernst plots with a slope corresponding to an n value of 1 for met-Mb reduction to deoxy-Mb,

[95] J. F. Taylor and V. E. Morgan, *J. Biol. Chem.* **144,** 15 (1942).

it is important to be alert for possible deviations from linearity in the Nernst plot. Such deviations can signal a poorly designed or malfunctioning experiment, or may indicate the presence of a more complex system. In the latter case, there are two scenarios that can lead to a nonnernstian response. The first case (scenario 1) involves two (or more) noninteracting redox groups. This results in a deviation from linearity of the slope in the Nernst plot and is sometimes referred to as negative cooperativity. However, this behavior is, in fact, associated with the presence of two noninteracting electroactive species with different $E_{1/2}$ values. This scenario can occur within a unique protein possessing two noninteracting electroactive centers or within a mixture of two or more redox-active proteins, as long as a difference in their respective $E_{1/2}$ values is present. Below, we show that a theoretical model describing this behavior accurately predicts and precisely the experimental data obtained for a mixture of *Aplysia* and sperm whale myoglobins, for which a difference in $E_{1/2}$ of 80 mV has been demonstrated.[84] The second case (scenario 2) involves two (or more) redox centers interacting with one another and is observed experimentally when studying the hemoglobins. Intracellular hemoglobin typically possesses four interdependent heme centers that have the ability to communicate with each other. These heme–heme interactions aid in the efficient uptake and release of dioxygen, a phenomenon known as cooperativity. The presence of a slope corresponding to $n > 1$ (nonnernstian response) for the hemoglobins is linked to the intrinsic electronic properties of the protein and gives information regarding the ability of the four prosthetic groups to influence each other electronically.

These two examples (scenarios 1 and 2) of nonnernstian response are different in nature and cannot be overlooked when analyzing the results from a spectroelectrochemical experiment.

Noninteracting Electroactive Centers: Development of Model

The following model is designed to predict the behavior of a Nernst plot when a mixture of two noninteracting but well-behaved nernstian electroactive species is studied by spectroelectrochemistry (scenario 1 described above). Two cases are developed. In one case the two species have identical spectra, but different midpoint potentials, and in the other case the spectra and midpoint potentials are both different.

For a mixture of two noninteracting species A and B we can represent the oxidized and reduced forms as O_A and O_B, and R_A and R_B, respectively. Beer's law at a fixed applied potential and fixed wavelength may be used to describe the total absorbance at equilibrium, A_e, as shown in Eq. (3):

$$A_e = A_{eA} + A_{eB} = (\varepsilon_{rA}[R_A] + \varepsilon_{oA}[O_A])b + (\varepsilon_{rB}[R_B] + \varepsilon_{oB}[O_B])b \qquad (3)$$

where A_{eA} and A_{eB} are the equilibrium absorbance values for species A and B at each applied potential, respectively; ε_{oA}, ε_{rA}, ε_{oB}, and ε_{rB} are the extinction coefficients of species A and B in their oxidized and reduced forms, respectively; and b is the optical path length through the OTTLE cell. This relationship may then be used to derive an expression for the ratio of the mixture of oxidized to reduced species in terms of absorbance values as shown in Eqs. (4) and (5):

$$[Ox]/[Red] = ([O_A] + [O_B])/([R_A] + [R_B]) \tag{4}$$

$$[Ox]/[Red] = \left(\sum A_e - \sum A_r\right) \big/ \left(\sum A_o - \sum A_e\right) \tag{5}$$

where $\sum A_r$ and $\sum A_o$ are the sums of the absorbances of the fully reduced and fully oxidized species A and B, respectively, and $\sum A_e$ is the sum of the equilibrium absorbances of the oxidized and reduced forms of species A and B at a specific applied potential.

Because the oxidized Mb species have a negligible absorbance at 435 nm, Eq. (5) may be simplified to Eq. (6).

$$[Ox]/[Red] = -1 + \left(\sum A_r \big/ \sum A_e\right) \tag{6}$$

The Nernst equation for species A and B [Eqs. (7) and (8)],

$$[O_A] = [R_A] \exp\{n_A[E - E_{1/2(A)}]F/RT\} \tag{7}$$

$$[O_B] = [R_B] \exp\{n_B[E - E_{1/2(B)}]F/RT\} \tag{8}$$

may be substituted for $[O_A]$ and $[O_B]$ in Eq. (3). Using the mass balance Eqs. (9) and (10), we can relate the applied and midpoint potentials E and $E_{1/2}$ to the absorbance levels via Eq. (11). The symbols c_A and c_B represent the total concentrations of species A and B, respectively.

$$c_A = [R_A] + [O_A] \tag{9}$$

$$c_B = [R_B] + [O_B] \tag{10}$$

$$\sum A_e = [(\varepsilon_{rA} b c_A)/(1 + \exp\{n_A[E - E_{1/2(A)}]F/RT\})] $$
$$+ [(\varepsilon_{rB} b c_B)/(1 + \exp\{n_B[E - E_{1/2(B)}]F/RT\})] \tag{11}$$

Finally, Eq. (12) is derived by substituting Eq. (11) into Eq. (6):

$$[Ox]/[Red] = -1 + \left\{\sum A_r/[(\varepsilon_{rA} b c_A)/(1 + \exp\{n_A[E - E_{1/2(A)}]F/RT\}) \right. $$
$$\left. + (\varepsilon_{rB} b c_B)/(1 + \exp\{n_B[E - E_{1/2(B)}]F/RT\})]\right\} \tag{12}$$

Data simulations derived from the log plots of Eq. (12) are shown as solid lines in Figs. 4, 5 and 6.

Figures 5 and 6 represent simulations of two factors influencing Nernst plots obtained from spectroelectrochemical data. In Fig. 5, the influence of two noninteracting systems with identical spectra, but different midpoint potentials ($\Delta E_{1/2}$), on the Nernst plot is explored. This situation can be related to studying a system possessing two heterogeneous electroactive centers that are spectroscopically

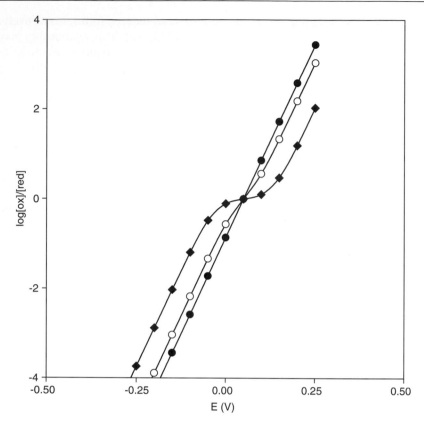

FIG. 5. Simulation showing the influence of $\Delta E_{1/2}$ on a log plot of Eq. (12). For each case, $\varepsilon_{rA} = \varepsilon_{rB}$. ($\bullet$) $E_{1/2(A)} = 50\,\text{mV}$, $E_{1/2(B)} = 50\,\text{mV}$, $\Delta E_{1/2} = 0\,\text{mV}$; ($\bigcirc$) $E_{1/2(A)} = 10\,\text{mV}$, $E_{1/2(B)} = 90\,\text{mV}$, $\Delta E_{1/2} = 80\,\text{mV}$; ($\blacklozenge$) $E_{1/2(A)} = -50\,\text{mV}$, $E_{1/2(B)} = 150\,\text{mV}$, $\Delta E_{1/2} = 200\,\text{mV}$. [Reprinted from C. H. Taboy, C. Bonaventura, and A. L. Crumbliss, *Bioelectrochem. Bioenerget.* **48,** 79 (1999), with permission from Elsevier Science.]

equivalent, but do not influence the electronic properties of one another. As the difference between the midpoint potentials ($\Delta E_{1/2}$) of the two independent species increases, a clear increase in the deviation from linearity around the midpoint potential can be observed, as predicted on theoretical grounds.[5] For this hypothetical system, where $\varepsilon_A = \varepsilon_B$, it is important to note that the minimum slope is still found at the midpoint potential ($E_{1/2}$).

Figure 6 illustrates the additional effect on the Nernst plot for a mixture of two species with homologous spectra that differ in their molar absorptivity coefficients at the probe wavelength (e.g., in our experimental examples at 435 nm, $\varepsilon_A \neq \varepsilon_B$) and with different midpoint potentials; $\Delta E_{1/2} = 100\,\text{mV}$. Two

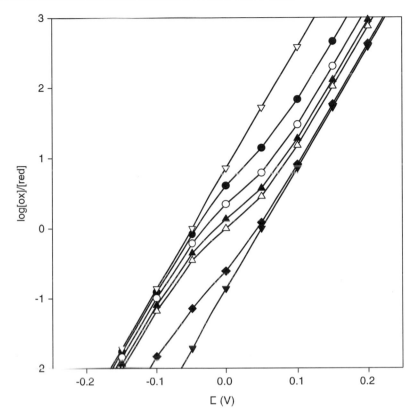

FIG. 6. Simulation showing the influence of the extinction coefficient for species A (ε_{rA}) and B (ε_{rB}) on a log plot of Eq. (12). For all data simulated, $E_{1/2(A)} = +50$ mV, $E_{1/2(B)} = -50$ mV, $\Delta E_{1/2} = 100$ mV. Single species A (▼); single species B (▽). Mixtures (1 : 1) of species A and B with the following ratios of ε_{rA} to ε_{rB}: (●) 1 : 9; (○) 1 : 3; (▲) 2 : 3; (△) 1 : 1; (◆) 9 : 1. [Reprinted from C. H. Taboy, C. Bonaventura, and A. L. Crumbliss, *Bioelectrochem. Bioenerget.* **48**, 79 (1999), with permission from Elsevier Science.]

important points are illustrated in Fig. 6: (1) One of the redox species must have an extinction coefficient at the probe wavelength that is much larger than the other (e.g., $\varepsilon_A/\varepsilon_B = 1 : 9$ or $9 : 1$) in order for the Nernst plot to become tangential to the respective single-component plot at extreme applied potentials; and (2) if there is a difference between the extinction coefficients for the two species at the probe wavelength (ε_A and ε_B), the minimum slope will not be present at $E_{1/2}$, but will shift as a function of the ratio of ε_A to ε_B.

Clearly, the Nernst plots for a mixture of two noninteracting redox systems of different spectra and midpoint potentials will deviate from ideality (linearity).

If the system is functioning properly the shapes of the Nernst plots are diagnostic of the presence of two or more heterogeneous electroactive centers, as illustrated in Figs. 5 and 6.

Noninteracting Electroactive Centers: 1:1 Mixture of Myoglobins

We can experimentally verify the model calculations described above, using a mixture of Mbs. When a 1 : 1 mixture of horse and sperm whale Mbs is studied by spectroelectrochemistry, a well-behaved Nernst plot is obtained with a slope corresponding to $n = 1$ and $E_{1/2} = 17\,\text{mV}$, which is equivalent to the $E_{1/2}$ obtained for the individual species (results not shown). This is the expected result as these two Mbs have identical spectra and $E_{1/2}$ values when isolated from each other and act therefore as a single homogeneous protein, spectrally and electronically "identical" to each other.

Mixtures containing *Aplysia* and sperm whale Mb, however, give a reproducible nonnernstian plot, as anticipated by the model described in the previous section. This is due to a difference in their respective midpoint potentials ($\Delta E_{1/2} = 80\,\text{mV}$). Nernst plots for *Aplysia* and sperm whale Mb alone, and in 1 : 1 mixture, are presented in Fig. 4 along with a simulation for the mixture.

These results demonstrate that a mixture of two noninteracting (and therefore noncooperative) heme proteins with identical $E_{1/2}$ values (horse and sperm whale Mbs) gives a well-behaved linear nernstian response at both 410 and 435 nm (λ_{max} for the oxidized and reduced forms of Mb), with a slope corresponding to $n = 1$ and $E_{1/2}$ overall equal to the $E_{1/2}$ of each independent Mb. By comparison, a mixture of two noninteracting heme proteins with notably different $E_{1/2}$ values gives a nonnernstian response. This nonnernstian behavior, illustrated in Fig. 4, demonstrates the influence on Nernst plots of mixtures containing noninteracting heme sites with significant $E_{1/2}$ differences. Protein heterogeneity will lead to a Nernst plot with minimum slope of less than unity ($n < 1$). This depression in n gives important information regarding the composition of the protein or mixture studied. A representative example illustrating this lack of homogeneity in a structure was presented by Malmström, who rationalized the observed value of $n = 0.5$ reported in early work for cytochrome c oxidase oxidation–reduction by developing a two-site model based on a difference between the $E_{1/2}$ value for each site.[96]

Interacting Electroactive Centers: Hemoglobins

Spectroelectrochemical results obtained from hemoglobin samples differ widely from the nonnernstian behavior observed for myoglobin mixtures, and for other noninteracting electroactive centers. Nernst plots for human Hb A_0 actually show a drastic increase in their slope, corresponding to n values as high as 2.0

[96] B. G. Malmström, *Q. Rev. Biophys.* **6**, 389 (1974).

in some cases. For systems with subunit–subunit interactions the Nernst plot slope (n) no longer strictly corresponds to the number of electrons transferred (as is also the case for noninteracting systems with chain heterogeneity, as discussed above). In this section, we summarize specific representative responses obtained when studying human hemoglobin (Hb A_0). More extensive results on Hb A_0 and other hemoglobins are presented elsewhere.[1,3,35,69,84,85]

The experimental design for evaluation of the redox behavior of the hemoglobins (Hbs) is similar to that described for the Mbs. The extinction coefficient and specific Soret band maximum are shifted relative to Mb, with maxima at 406 nm for iron(III) Hb and at 430 nm for iron(II) Hb and isosbestic points at 415, 455, 524, and 598 nm. An example of the nonnernstian response obtained for Hb A_0 is presented in Fig. 7, along with a well-behaved nernstian response for

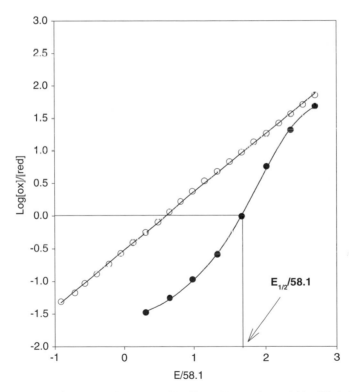

FIG. 7. Nernst plot for sperm whale myoglobin (○) and human hemoglobin (Hb A_0) (●). The x axis represents applied potential versus NHE according to a rearrangement of Eq. (1) where $E/58.1 = E_{app}F/RT$. $E_{1/2}$ is indicated for Hb A_0. Conditions: 0.06–0.08 mM heme, 0.05 M MOPS, 1 mM Ru(NH$_3$)$_6$Cl$_3$, 0.2 M KNO$_3$, pH 7.1, and 20°. [Reprinted from C. H. Taboy, C. Bonaventura, and A. L. Crumbliss, *Bioelectrochem. Bioenerget.* **48**, 79 (1999), with permission from Elsevier Science.]

sperm whale Mb, illustrating the sigmoidal shape of the Nernst plot for Hb A_0. Four parameters are used to describe Hb responses: the midpoint potential ($E_{1/2}$) and the potential at which the maximum slope of the Nernst plot is observed (E_{max}) and, because the slope of the Nernst plot is constantly changing, we also define the n value at the midpoint potential as $n_{1/2}$ and the n value at the maximum slope as n_{max}. These parameters permit comparison of each data set as a function of the specific midpoint potential ($E_{1/2}$) of the system and its level of cooperativity (as described by the slope at the midpoint potential, $n_{1/2}$), and give a general sense of the asymmetry of the curve ($|E_{1/2} - E_{max}|$).

For the hemoglobins, as in the case of any nonnernstian response, the meaning of the midpoint slope ($n_{1/2}$) is not straightforward, and in any case should not be strictly interpreted as the number of electrons involved in the redox process. It can, however, be used to evaluate, within a set of experiments, the effect or influence of the medium composition on site–site interactions. For instance, in the case of hemoglobins, the presence of anions in the medium influences the apparent level of "cooperativity" of the system and gives fundamental clues regarding the mechanism used by the protein to propagate or relay important information between its surface (medium–protein interaction) and its embedded heme groups.[84] As shown in many previous studies, the globin chain exerts a profound influence on the redox potential of the heme site of hemoglobins (Hbs) and myoglobins (Mbs), and protects them from rapid oxidation, which, in turn, allows for the reversibility observed in dioxygen binding to the prosthetic group.[4,11–14,22,24,34,65,70,71,74,77,79,97–103]

Oxygenation and oxidation–reduction studies of iron(II)/iron(III) centers for diverse Hbs have reported interesting interrelationships between these two processes.[2,4,10,69] One of the important observations we have made regarding the parallels between these two processes links the shift from the T [deoxy, iron(II)] to the R (oxygenated) or R-like [met, iron(III)] conformation of the Hb tetramer to the cooperativity observed in both oxygenation and oxidation processes. Numerous studies have shown that structural changes stabilizing either the T or the R conformation of the Hb tetramer are typically reflected by comparable alterations of both oxidation and oxygenation processes.[4,11,69] Preferential binding to the low-affinity

[97] A. Szabo and M. Karplus, *J. Mol. Biol.* **72,** 163 (1972).

[98] K. D. Vandegriff and R. M. Winslow, "Blood Substitutes: Physiological Basis of Efficacy." Birkhauser, Boston, 1995.

[99] J. P. Collman, R. R. Gagne, C. A. Reed, T. R. Halbert, G. Lang, and W. T. Robinson, *J. Am. Chem. Soc.* **97,** 1427 (1975).

[100] X. Hu and T. G. Spiro, *Biochemistry* **36,** 15701 (1997).

[101] H. X. Zhou, *J. Biol. Inorg. Chem.* **2,** 119 (1997).

[102] E. Di Cera, *Chem. Rev.* **98,** 1563 (1998).

[103] F. A. Tezcan, J. R. Winkler, and H. B. Gray, *J. Am. Chem. Soc.* **120,** 13383 (1998).

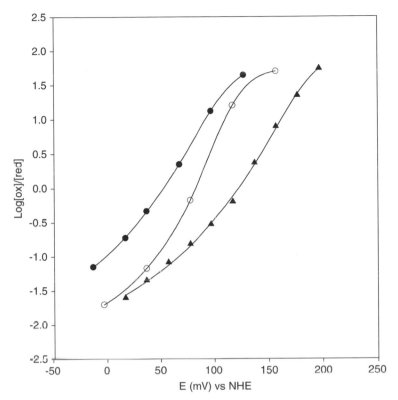

FIG. 8. Nernst plots for Hb A_0 in the absence (○) ($E_{1/2} = 85$ mV, $n_{1/2} = 2.0$) and presence (▲) of 0.2 M KCl ($E_{1/2} = 125$, $n_{1/2} = 1.2$). Conditions: 1 mM Ru(NH$_3$)$_6$Cl$_3$, 0.06 mM heme, in 0.05 M MOPS, pH 7.1, and 20°. Nernst plot for horse Hb (●) ($E_{1/2} = 55$ mV, $n_{1/2} = 1.7$) is also shown. Conditions: 1 mM Ru(NH$_3$)$_6$Cl$_3$, 0.06 mM heme, in 0.05 M HEPES, pH 7.5, and 20°.

(T-state) quaternary conformation has traditionally explained anion effects on Hb oxygenation.[104,105] The presence of heterotropic effectors (e.g., protons, anions, and carbon dioxide) bound at spatially remote sites of the Hb influences the oxygenation process and has been shown to affect the oxidation process in a similar manner.[1,3,17,41,69]

An example of the influence of the medium composition and globin amino acid sequence on heme redox properties is presented in Fig. 8. A typical Nernst

[104] R. E. Benesch, R. Benesch, and C. I. Yu, *Biochemistry* **8**, 2567 (1969).
[105] R. Benesch and R. E. Benesch, *Nature (London)* **221**, 618 (1969).

plot for Hb A_0 (sigmoidal curve) is presented in contrast with a Nernst plot for horse Hb (shifted to more negative potentials with respect to Hb A_0, but keeping its sigmoidal curvature), on the one hand, and for Hb A_0 in the presence of 0.2 M KCl in the medium (flattening of the Nernst plot and $E_{1/2}$ shifted to more positive potentials with respect to Hb A_0), on the other hand. This data set illustrates the sensitivity of the Hbs to the presence of anions in solution (allosteric effectors) as well as the importance of the globin structure around the heme site in fine-tuning the redox response of the protein.

We should note that, in the case of the Hbs, the observed sigmoidal Nernst plot is actually influenced by both scenario 1 and scenario 2 presented above, as the α and β chains of the tetramer are not identical, but cannot be differentiated spectroscopically.[2,4,106] If the four chains had exactly the same midpoint potentials, the Nernst plot would show a pure positive cooperativity. How heterogeneous these two α and β chains are electronically when bound to each other is not known. The extent to which this heterogeneity influences the overall shape of the Nernst plot under various conditions is only speculative. That said, the relative effect of the redox cooperativity typically dominates chain heterogeneity, as illustrated by the observed $n_{1/2} > 1$.

Conclusions

The redox behavior of the Hbs and Mbs and the shape of their Nernst plots are sensitive indicators of globin alterations and/or changes in their environment (e.g., medium composition, presence of allosteric effectors, etc). Studying the effect of these variations on the protein oxidation curve has proved to be a useful tool in understanding the basis of the Hb allosteric mechanism on oxidation, but has also provided insight with respect to the mechanism involved in O_2 binding.[85] The existence of parallels and differences between the oxygenation of the Hbs and their anaerobic oxidation has allowed us to discriminate between the electronic and steric consequences brought about by changes in globin structure.[69] We have published results of spectroelectrochemical assays that probed the influence of broad concentration ranges of allosteric effectors (Cl^-, NO_3^-, PO_4^{3-}, and ClO_4^-) on a number of Hbs that further clarified the impact of these anions on the electronic properties of the active sites of Hbs.[1,85] This, in turn, has elucidated a possible mechanism linked to anion-induced restriction of the conformational fluctuations of the Hbs in controlling oxygen affinity.[1] These studies have been made possible by the use of a combination between an OTTLE and UV–Vis spectroscopic monitoring, as described in this chapter. The technique and methods of data analysis described here have made a significant impact on the study of redox-active metalloproteins.

[106] S. Beychok, I. Tyuma, R. E. Benesch, and R. Benesch, *J. Biol. Chem.* **242,** 2460 (1967).

Acknowledgments

Our work in this area is supported by the NIH (RO1 HL58248) and through the support of a Marine/Freshwater Biomedical Center Grant (ESO-1908) funded by the NIEHS.

[19] Cysteine–Nitric Oxide Interaction and Olfactory Function

By MARIE-CHRISTINE BROILLET

Introduction

The transfer of a nitric oxide group to cysteine sulfhydryls on proteins, known as S-nitrosylation (Fig. 1), is becoming more and more recognized as a ubiquitous regulatory reaction comparable to phosphorylation. It represents a form of redox modulation in diverse tissues including the brain. An increasing number of proteins have been found to undergo S-nitrosylation *in vivo*.[1] These proteins are called S-nitrosothiols. The S-nitrosothiol proteins may play an important role in many processes ranging from signal transduction, DNA repair, host defense, and blood pressure control to ion channel regulation and neurotransmission.

A number of ion channels have been reported to be directly NO sensitive, among them the N-methyl-D-aspartate (NMDA) receptor–channel complex,[2] Ca^{2+}-activated K^+ channels,[3] Na^+ channels in baroreceptors,[4] cardiac Ca^{2+} release channels,[5] L-type Ca^{2+} channels,[6] and our own work on cyclic nucleotide-gated (CNG) channels.[7] However, in only a few cases have the specific cysteine residues been determined.[8,9] Only the CNG channels have been shown to be opened by S-nitrosylation, in other ion channels, it appears rather that the activation and inactivation parameters are being altered.

[1] M.-C. Broillet, *Cell. Mol. Life Sci.* **55,** 1036 (1999).

[2] Y.-B. Choi, L. Tenneti, D. A. Le, J. Ortiz, G. Bai, H.-S. V. Chien, and S. A. Lipton, *Nat. Neurosci.* **3,** 15 (2000).

[3] V. M. Bolotina, S. Najibi, J. Palacino, P. Pagano, and R. A. Cohen, *Nature (London)* **368,** 850 (1994).

[4] Z. Li, M. W. Chapleau, J. N. Bates, K. Bielefeldt, H.-C. Lee, and F. M. Abboud, *Neuron* **20,** 1039 (1998).

[5] L. Xu, J. P. Eu, G. Meissner, and J. S. Stamler, *Science* **279,** 234 (1998).

[6] D. C. Campbell, J. S. Stamler, and H. C. Strauss, *J. Gen. Physiol.* **108,** 277 (1996).

[7] M.-C. Broillet and S. Firestein, *Neuron* **16,** 377 (1996).

[8] J. S. Stamler, L. Jia, J. P. Eu, T. J. McMahon, I. T. Demchenko, J. Bonaventura, K. Gernert, and C. A. Piantadosi, *Science* **276,** 2034 (1997).

[9] M.-C. Broillet, *J. Biol. Chem.* **275,** 15135 (2000).

Copyright 2002, Elsevier Science (USA).
All rights reserved.
0076-6879/02 $35.00

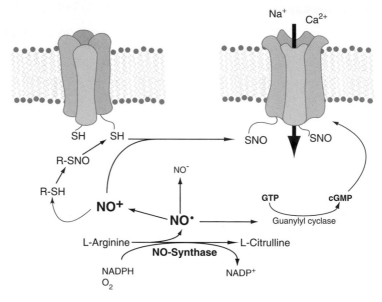

FIG. 1. Biosynthesis pathway for NO and nitrosothiols. Nitric oxide (NO˙) is synthesized from L-arginine by NO synthases (NOS). One NO˙ redox form, the nitrosonium ion (NO⁺), can activate by S-nitrosylation an array of target proteins (such as the cyclic nucleotide-gated channel represented here) either directly or via the formation of intermediate nitrosothiols (R-SNO). The CNG channel can also be activated by cGMP produced after the stimulation of guanylyl cyclase activity by NO˙.

This chapter includes a detailed description of the biochemical, molecular, and electrophysiological methods that we have used to study and identify the mechanism involved in the direct activation of the olfactory CNG channel by cysteine–nitric oxide (NO) interaction.

Cyclic Nucleotide-Gated Channels

CNG channels form a family of nonselective cation channels that are structurally related to voltage-gated channels, but that require the binding of either cAMP or cGMP for activation.[10] Although they have been identified in an assortment of cell types and tissues, they are most prevalent in the peripheral sensory receptor cells of the visual and olfactory systems, where they play a crucial role in transducing sensory information into changes in membrane potential.

[10] W. N. Zagotta and S. A. Siegelbaum, *Annu. Rev. Neurosci.* **19,** 235 (1996).

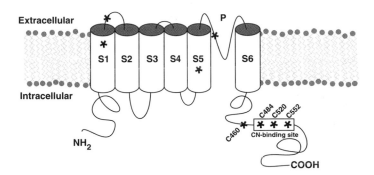

FIG. 2. Hypothetical model of the two-dimensional architecture of a cyclic nucleotide-gated channel subunit. S1–S6 are the putative transmembrane domains, and P is the putative pore region. The cyclic nucleotide (CN)-binding site is defined by homology to the sequences of cAMP- and cGMP-binding proteins. The asterisks (∗) indicate the positions of the cysteine residues in the olfactory CNGA2 subunit. The putative NO target sites are the cysteine(s) located intracellularly (C460, C484, C520, and C552) that have been successively mutated into serine residues. Reproduced with permission from M. C. Broillet, *J. Biol. Chem.* **275,** 15135 (2000).

In olfactory neurons, the channel is probably a tetramer formed by the assembly of three different types of subunits called CNGA2 (principal, or functional subunit) and CNGA4 and CNGB1b (modulatory subunits).[11] This channel can be activated by either cAMP or cGMP, although it is generally believed that under normal physiological conditions it is a rise in intracellular cAMP that is responsible for channel activation.[12] However, because of their sensitivity to cGMP, NO has been proposed as a signaling molecule in the olfactory system[13] and in the visual system.[14] In this model, NO activates a soluble guanylyl cyclase, producing cGMP, which then activates the ion channel (Fig. 1). However, we have shown that NO is able to activate the CNG channels from olfactory neurons in a cGMP-independent manner, providing a possible alternate pathway for channel activation by NO.

In summary, there are eight cysteine residues distributed throughout the rat olfactory CNGA2 channel subunit (Fig. 2). First, biochemical evidences obtained with membrane-impermeant SH-modifying reagents helped us focus on the cysteines located at the intracellular face of the channel. A series of mutant subunit constructs, in which each of these intracellular cysteines was changed to a serine residue, were constructed. We were able to test each of these channels for activation

[11] M. J. Richards and S. E. Gordon, *Biochemistry* **39,** 14003 (2000).

[12] G. M. Shepherd, *Neuron* **13,** 771 (1994).

[13] H. Breer and G. M. Shepherd, *Trends Neurosci.* **16,** 5 (1993).

[14] D. E. Kurenny, L. L. Moroz, R. W. Turner, K. A. Sharkey, and S. Barnes, *Neuron* **13,** 315 (1994).

by cAMP and/or NO after their expression in HEK 293 cells. We found that the cysteine at position 460, within the C-linker region just N-terminal to the CN-binding region, is the critical residue in the reaction that leads to channel activation by NO.[9]

Functional Identification of the S-Nitrosylated Cysteine(s)

Electrophysiological Recordings

Electrophysiological patch–clamp recordings should be performed under the inside-out configuration.[15] Electrodes are fabricated from thin-walled borosilicate glass (World Precision Instruments, Sarasota, FL) and fire polished to tip resistances of 10 to 20 MΩ. The pipette solution is the same as the Ca^{2+}-free solution (see below), so that both faces of the membrane patch are bathed in symmetrical solutions. The Ca^{2+}-free solution contains (in mM) NaCl, 145; EGTA, 0.5; EDTA, 0.5; HEPES, 10; pH 7.6. Single-channel currents are recorded with an Axopatch 1D patch–clamp amplifier (Axon Instruments, Foster City, CA) and stored on digital audio tape (DAS 75; Dagan, Minneapolis, MN). Signals are played back, digitized, and stored on a Power Mac G4 using the software Acquire (Instrutech, Great Neck, NY). The filter bandwidth is 0 to 2000 Hz and the sampling interval is 100 μsec. The signals can be analyzed with the software programs MacTac (Instrutech), pClamp 6 (Axon Instruments), and Igor (WaveMetrics, Lake Oswego, OR). Channel detection is by the 50% threshold protocol. All recordings are performed on isolated rat olfactory neurons or transfected HEK 293 cells (see below). The recordings are made at room temperature.

Olfactory Neurons

Native rat olfactory receptor neurons are isolated from the nasal epithelium of adult rats (Sprague-Dawley) as previously described.[16]

The dissociated cells are then maintained in normal Ringer's solution (NaCl, 135 mM; KCl, 5 mM; $CaCl_2$, 1 mM; $MgCl_2$, 4 mM; HEPES, 10 mM; glucose, 10 mM) and transferred to concanavalin A (1 mg/ml; Sigma, St. Louis, MO)-coated coverslips for recording. During electrophysiological recordings, a Ca^{2+}-free solution is used (NaCl, 145 mM; EGTA, 0.5 mM; EDTA, 0.5 mM; HEPES, 10 mM; pH 7.6).

Drugs

All pharmacological agents are prepared fresh daily and dissolved in the Ca^{2+}-free solution for perfusion via a multibarreled perfusion pipette (Warner

[15] O. P. Hamill, A. Marty, E. Neher, B. Sakmann, and F. J. Sigworth, *Pflugers Arch.* **391,** 85 (1981).
[16] J. W. Lynch and B. Lindemann, *J. Gen. Physiol.* **103,** 87 (1994).

Instrument, Hamden, CT) positioned close to the membrane patch. This rapid perfusion system is computer-controlled by the software Pulse (HEKA Elektronik, Lambrecht, Germany) and has a mean rise time in response to a solution change of ~29 msec.

NO donors or releasing agents are used increasingly in *in vivo* or *in vitro* studies to mimic the effects of ˙NO. A number of them are commercially available [Cayman Chemical (Ann Arbor, MI) and Alexis (San Diego, CA)]. We use the S-nitrosothiol: S-nitroso cysteine (SNC) as NO donor because it is not expensive, easy to prepare, and reliable. Moreover, it probably represents a common *in vivo* donor of NO because the most abundant source of free thiol groups in mammalian tissue is cysteine (in either the free or peptide form). SNC must be prepared on ice from the combination in acid solution of equimolar amounts of L-cysteine hydrochloride and sodium nitrite[17] (Sigma). SNC is formed quantitatively within 1 min. The result of this acid-catalyzed reaction is a red liquid.

The stability of SNC is greatly influenced by ambient conditions including pH, oxygen tension, and contaminant metals. The stock SNC solution remains effective for approximately 2 hr. The mechanism of NO release from SNC is assumed to involve its spontaneous photochemical degradation from two molecules of SNC to one cystine and two NO molecules.

Cyclic adenosine 3' : 5'- monophosphate (cAMP) and the SH-modifying agents iodoacetamide (IAA), *N*-ethylmaleimide (NEM), dithiothreitol (DTT), and 5,5'-dithiobis-(2-nitrobenzoic acid) (DTNB) are obtained from Sigma.

Biochemical Approach

It is necessary to verify first the presence of CNG channels in the patch. Therefore, cAMP (20 or 500 μM) is perfused as a control on every patch. cAMP is then rinsed away and another drug (SNC, IAA, NEM, DTNB, or DTT) can be perfused.

There should be no GTP or ATP in any of the solutions in order to demonstrate that the activation by NO is due to a direct effect on the channel. The experiments can also be performed in the presence of guanylyl cyclase inhibitors.

NO donors such as SNC are then perfused on patches containing CNG channels and the precise electrophysiological characteristics of the activated channels should be measured. In the case of the olfactory CNG channel, the single-channel conductance and the mean amplitude are similar for cAMP or SNC. Amplitude histograms are nearly monotonic in the presence of either cAMP or SNC. Thus, it appears that SNC and cAMP are acting on the same population of channels. Moreover, the NO donors can open the channels in the complete absence of cyclic nucleotides.

[17] S. Z. Lei, Z. H. Pan, S. K. Aggarwal, H. S. V. Chen, J. Hartman, N. J. Sucher, and S. A. Lipton, *Neuron* **8**, 1087 (1992).

When observed on isolated patches of membrane, NO donors such as SNC induce channel activity that is only slowly reversible, with occasional channel openings occurring for up to 30 min after removal of the drugs, suggesting that a persistent modification of the channel has been caused by NO.

To show that the effects observed are dependent on the production of NO, inactive NO donors (i.e., solutions of SNC more than 24 hr old) should be checked on patches containing CNG channels, as well as the effects from the by-product cystine (100 μM) or 100 μM cysteine alone. It should also be possible to mimic the effects of NO donors by SH-modifying agents. The membrane patches can be exposed, for example, to iodoacetamide (IAA). In Fig. 3, application of 2 mM IAA produces immediate channel openings with activity similar to that observed after the addition of SNC. The effect of IAA is not reversible. When SNC is added together with IAA no increase in channel activity is observed, suggesting that these two substances may be activating the CNG channel by the same mechanism. As with SNC, when CNG channels are absent from the patch, as determined by the failure of saturating cAMP to activate a conductance, no effect of IAA can be observed. Experimental results of this type rule out the possibility that IAA, even at the high concentrations used here, activate a nonspecific conductance producing an increase in the patch permeability.

To further investigate the mechanism underlying channel activation by NO several other reagents that modify sulfhydryl groups with known chemistry can be tried.[18-21] On the assumption that the NO group have modified a free SH group, dithiothreitol (DTT) can be applied to determine whether this oxidizing agent could reverse the activation. DTT (2–10 mM) has no effect on the activity of SNC applied directly to a patch containing CNG channels. DTT applied before SNC also has no protective effect. Another sulfhydryl-alkylating agent, N-ethylmaleimide (NEM, 50 μM), also activates the channel irreversibly and, as expected, its effect is not reversed by DTT either. On the other hand, 5,5'-dithiobis-(2-nitrobenzoic acid) (DTNB), a reducing agent that acts by formation of a disulfide bond with a free SH group, also activates the channel; its effect is reversed by application of DTT. These results, taken together, suggest that the activation of the CNG channel by NO groups is by reduction of a free SH group without the subsequent formation of a disulfide bond. Moreover, the results described above, based on the use of sulfhydryl-modifying agents (IAA, NEM, and DTNB), suggest that the

[18] F. H. J. White, in "Reduction and Reoxidation at Disulfide Bonds" (C. H. W. Hirs and S. N. Tinasheff, eds.), p. 481. Academic Press, San Diego, CA, 1972.

[19] G. L. Kenyon and T. W. Bruice, in "Novel Sulfhydryl Reagents" (C. H. W. Hirs and S. N. Timasheff, eds.), p. 407. Academic Press, San Diego, CA, 1977.

[20] P. C. Jocelyn, in "Spectrophotometric Assay of Thiols" (W. B. Jakoby and O. W. Griffith, eds.), p. 44. Academic Press, San Diego, CA, 1987.

[21] P. C. Jocelyn, in "Chemical Reduction of Disulfides" (W. B. Jakoby and O. W. Griffith, eds.), p. 246. Academic Press, San Diego, CA, 1987.

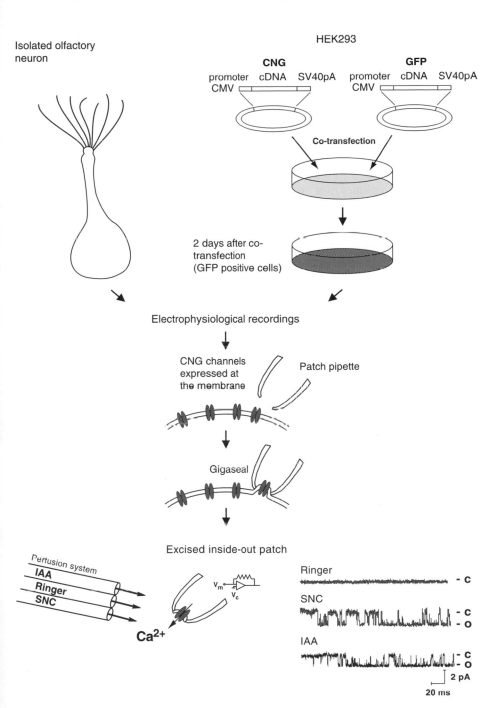

FIG. 3. Experimental steps for establishing the direct activation of CNG channels by NO groups or SH-modifying reagents.

donor cysteine groups must be on the intracellular portion of the channel protein because, unlike NO, these reagents are only poorly membrane permeable,[22] and so must have acted at the intracellular face of the membrane patch. In addition, application of a membrane-impermeant form of NEM, 1 mM dextran–NEM (kind gift of S. J. Kleene, University of Cincinnati; College of Medicine, Cincinnati, OH), can also activate the channel in inside-out patches.

From the known amino acid sequences, the olfactory CNG channel contains eight cysteine residues.[9] Of these, four are both highly conserved and reside intracellularly; three of them are located in the cAMP-binding site of the channel.

Molecular Approach

Site-Directed Mutagenesis

Cysteine-to-serine CNG channel mutants can be generated by substituting the specific cysteine residues with serines by polymerase chain reaction (PCR)-based mutagenesis described by Nelson and Long.[23] *Pfu* polymerase (Stratagene, La Jolla, CA) should be used to reduce the rate of contaminating mutations. All constructs must be verified by sequencing.

Channel Protein Expression

Human embryonic kidney (HEK) 293 cells grown at 37° in minimal essential medium supplemented with 10% (v/v) horse serum and 1% (w/v) gentamicin can be used as an expression system for the wild-type and mutant CNG channels (Fig. 3). A pCIS expression vector (Genentech, South San Francisco, CA) containing either the wild-type rat olfactory CNG channel subunit (CNGA2)[24] or one of the different types of mutant channel is used to perform transient transfections according to a standard calcium phosphate protocol.[25] The cells are cotransfected with a vector carrying the gene for the green fluorescent protein (GFP) (Stratagene) at a 1 : 1 molar ratio. Patch–clamp recordings are made 2–3 days after transfection. GFP is used as an indicator of transfection success,[26,27] efficiency, and probable expression of wild-type or mutant CNG channels. GFP fluorescence can be visualized in living cells, without histological processing, by using a fluorescence

[22] K. Donner, S. Hemila, G. Kalamkarov, A. Koskelainen, I. Pogozheva, and T. Rebrik, *Exp. Eye Res.* **51**, 97 (1990).

[23] R. M. Nelson and L. L. Long, *Anal. Biochem.* **180**, 147 (1989).

[24] T.-Y. Chen, Y.-W. Peng, R. S. Dhallan, B. Ahamed, R. R. Reed, and K.-W. Yau, *Nature (London)* **362**, 764 (1993).

[25] C. M. Gorman, D. R. Gies, and G. McCray, *DNA Protein Eng. Techn.* **2**, 3 (1990).

[26] A. B. Cubitt, R. Heim, S. R. Adams, A. E. Boyd, L. A. Gross, and R. Y. Tsien, *Trends Biochem. Sci.* **20**, 448 (1995).

[27] M.-C. Broillet and S. Firestein, *Neuron* **18**, 951 (1997).

microscope equipped with fluorescein isothiocyanate (FITC) filter sets that span the excitation wavelengths of 450–500 nm. Observation of the degree of fluorescence after cotransfection with a vector carrying the gene for GFP allows us to reliably select for cells with a high probability of either single-channel or macroscopic currents (hundreds of channels) in the membrane patch. Macroscopic currents can also be recorded with an Axopatch 1D patch–clamp amplifier (Axon Instruments) and stored on a Power Mac G4 using the software Pulse (HEKA Elektronik) and analyzed with the software Igor (WaveMetrics).

Northern Blot Analysis

It is necessary to verify by Northern blot analysis that the mutant CNG channels are still expressed in HEK 293 cells. The cells can be harvested 2 days after transfection with wild-type or mutant cDNAs of the CNG channel subunits. The total RNA can be extracted by TRIzol reagent (GIBCO-BRL, Gaithersburg, MD). Fifteen micrograms of total RNA is then size fractionated on a formaldehyde gel and blotted. The blot is hybridized at 42° with a digoxigenin (dig)-11-dUTP-labeled 500-bp cDNA fragment located at the wild-type CNG channel (CNGA2) bp 1251–1751 coding region. The hybridized blot is detected with a digoxigenin nucleic acid detection kit (Boehringer Mannheim, Indianapolis, IN). As a control, nontransfected HEK 293 cells should be used.

Specificity of S-Nitrosylation

It is not uncommon for several cysteine residues on a given protein to be candidates for nitrosylation. In the ryanodine receptor, of a total of 364 cysteines, 84 provide free SH groups, but only 12 are thought to undergo nitrosylation.[5] Although the precise parameters governing accessibility by NO are unknown, the existence of a consensus nitrosylation acid–base motif has been postulated on the basis of large database screenings.[28] The proposed motif is XYCZ, where X can be any of G, S, T, C, Y, N, or Q; Y can be K, R, H, D, or E; and Z can be D or E. The most important element of the sequence is believed to be the aspartate/glutamate residues following the cysteine. In spite of this rather degenerate motif, in the CNG channel only Cys-460, identified by our biochemical and mutation experiments as the NO target site, possesses the required motif (i.e., Q, D, C, E) (Fig. 4).

Because a functional CNG channel is most probably made up of four subunits,[29,30] there are four potential nitrosylation sites per channel. However, factors other than those noted above may also determine the likelihood of NO

[28] J. S. Stamler, *Neuron* **18**, 691 (1997).

[29] D. T. Liu, G. R. Tibbs, and S. A. Siegelbaum, *Neuron* **16**, 983 (1996).

[30] D. T. Liu, G. R. Tibbs, P. Paoletti, and S. A. Siegelbaum, *Neuron* **21**, 235 (1998).

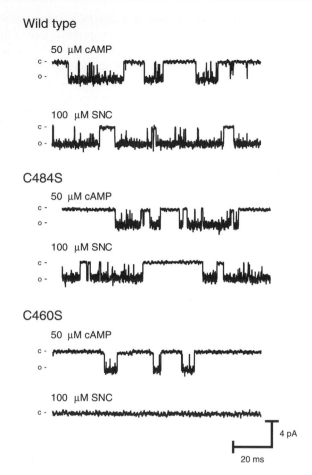

Wild type

50 μM cAMP

100 μM SNC

C484S

50 μM cAMP

100 μM SNC

C460S

50 μM cAMP

100 μM SNC

4 pA

20 ms

FIG. 4. Cys-460 is the nitric oxide target site. A comparison is shown of single-channel recordings from channels composed of either wild-type CNG subunits (CNGA2) or C460S mutant or C484S mutant subunits expressed in HEK 293 cells. For the wild-type and C484S mutant channels, cAMP (50 μM) and SNC (100 μM) elicited high-frequency channel activity. In the case of the C460S mutant channel, no activation could be observed after SNC treatment and a decrease in open probability is observed after cAMP treatment compared with the wild-type channel. All recordings were performed at -60 mV, holding potential. Adapted with permission from M.-C. Broillet, *J. Biol. Chem.* **275,** 15135 (2000).

activity at particular cysteines. Different degrees of accessibility to NO due to protein conformation, different reaction rates with NO at different cysteines due to redox status of the immediate environment, or cysteines in positions that may have no functional consequences on nitrosylation, could also account for the observation that in most proteins relatively few free thiols are in fact involved in nitrosylation-induced activity.[28] In the CNG channel, our concentration–response data indicate a Hill coefficient of less than 2, suggesting that as few as two of the four target cysteines may actually interact with NO. It does not, however, preclude activity at all four sites.

An important role for cysteine residues in CNG channel activation has been proposed by Gordon *et al.*[31] These authors have shown, by work on the rod CNGA1 subunit of the CNG channel, that the N- and C-terminal regions of each subunit interact in the tetrameric channel. This interaction involves particular cysteine residues with the formation of a disulfide bond between Cys-35 (N terminal) and Cys-481 (C terminal). Brown *et al.*[32] have also found that Cys-481 is necessary for the potentiation of the rod CNGA1 channel response to cAMP and cGMP. This cysteine residue, located in the C-terminal region, corresponds exactly to the NO target site in the olfactory CNG channel (Cys-460), confirming that this amino acid, which is highly conserved among the different cloned CNG channels, likely plays a critical role in channel gating and in the potentiation of cyclic nucleotide action.

The fact that Cys-460 is a potent activator of channel gating by NO and also affects channel gating by cAMP indicates the importance of this residue for channel function.

Nitric Oxide and Olfaction

Nitric oxide synthase has been shown to be expressed in olfactory neurons during development and possibly during neuronal regeneration in adult life.[33,34] Developing or regenerating olfactory neurons are dependent on outside stimuli, that is, odors, for their activity, and these odors may not always be present in the environment during critical growth periods. Thus, NO production may be important in maintaining activity in developing neurons independent of the stimulus environment. Because the CNG channels are highly permeable to Ca^{2+} ions[35] their activation could also provide a pathway for increasing intracellular Ca^{2+} concentration, a first step in many cellular processes.

Acknowledgments

I sincerely thank O. Randin for preparing the figures. This work was supported by grants from the Fonds National Suisse de la Recherche (31-51061.97 and 3130-051920.97).

[31] S. E. Gordon, M. D. Varnum, and W. N. Zagotta, *Neuron* **19,** 431 (1997).
[32] L. A. Brown, S. D. Snow, and T. L. Haley, *Biophys. J.* **75,** 825 (1998).
[33] A. J. Roskams, D. S. Bredt, T. M. Dawson, and G. V. Ronnett, *Neuron* **13,** 289 (1994).
[34] H. Zhao, S. Firestein, and C. A. Greer, *NeuroReport* **6,** 149 (1994).
[35] F. Zufall and S. Firestein, *J. Neurophysiol.* **69,** 1758 (1993).

[20] Functional Evaluation of Nonphagocytic NAD(P)H Oxidases

By FRANCIS J. MILLER, JR. and KATHY K. GRIENDLING

In neutrophils, NADPH-dependent superoxide ($O_2^{\cdot-}$) formation is catalyzed by a membrane-bound flavoprotein via the following reaction:

$$NAD(P)H + 2O_2 \rightarrow NAD(P)^+ + H^+ + 2O_2^{\cdot-}$$

It has become apparent that many cell types have an NAD(P)H oxidase activity similar to that of the neutrophil enzyme. Major differences between phagocytic and nonphagocytic oxidase activities include the kinetics of the reaction, the amount and location of the $O_2^{\cdot-}$ produced, the substrate specificity, and the identity of the enzyme subunits (Table I).

Considerations for Measurement of NAD(P)H Oxidase Activity

To accurately assign a role for an NAD(P)H oxidase in the generation of $O_2^{\cdot-}$, oxidase activity must be measured and shown to be inhibited by appropriate pharmacological or molecular agents. This section addresses tissue preparation and special considerations to ensure that the activity is attributable to this enzyme, and the latter half of this chapter deals mainly with accurate measurement of $O_2^{\cdot-}$ production, a prerequisite for assessment of NAD(P)H oxidase activity.

Preparation of Homogenate and Membrane Fractions

The nonphagocytic NAD(P)H oxidase is sensitive to methods of homogenate preparation. The enzyme resides in the membrane fraction, and often the signal-to-noise ratio is significantly enhanced by careful preparation of membrane fractions. Rough handling of cells or tissue can dramatically increase baseline $O_2^{\cdot-}$ production, so mild lysis methods are preferable. Care must be taken to minimize disruption of mitochondria to avoid the confounding contribution of mitochondrial NADH-dependent enzymes.

Cultured Cells. After washing with ice-cold phosphate-buffered saline (PBS), cells are scraped from the plate in 5 ml of this same solution, and centrifuged at 750g at 4° for 10 min.[1] The supernatant is discarded, and the pellet is resuspended in lysis buffer containing protease inhibitors. Both a phosphate buffer [20 mM monobasic potassium phosphate (pH 7.4), aprotinin (10 μg/ml), leupeptin

---

[1] K. K. Griendling, C. A. Minieri, J. D. Ollerenshaw, and R. W. Alexander, *Circ. Res.* **74,** 1141 (1994).

Copyright 2002, Elsevier Science (USA).
All rights reserved.
0076-6879/02 $35.00

TABLE I
CHARACTERISTICS OF PHAGOCYTIC AND NONPHAGOCYTIC NAD(P)H OXIDASES

	Phagocytic oxidase	Nonphagocytic oxidase
Substrate K_m	NADPH (30 μM)	NADPH (50 μM)[a]
	NADH (0.5 mM)	NADH (10 μM)
Substrate supporting maximal activity	NADPH	NADPH
Kinetics of activation	Seconds	Minutes–hours
Capacity	130 nmol of $O_2^{\cdot-}$ per minute per milligram of cell protein	40 nmol of $O_2^{\cdot-}$ per minute per milligram of cell protein
Fate of released $O_2^{\cdot-}$	Extracellular	Intracellular or extracellular
Cellular location	Phagosomes	Plasma membrane/microsomes
Subunit structure	$p22^{phox}$	Variable, often contains one or more
	$gp91^{phox}$	neutrophil subunits and/or a $gp91^{phox}$
	$p47^{phox}$	family member (nox-1, -4, or duox)[b]
	$p67^{phox}$	
	rac-1 or rac-2	

[a] Estimated from nonpurified enzyme preparations.
[b] From J. D. Lambeth, G. Cheng, R. Arnold, and W. A. Edens, *Trends Biochem. Sci.* **25,** 459 (2000).

(0.5 μg/ml), pepstatin (0.7 μg/ml), 0.5 mM phenylmethylsulfonyl fluoride (PMSF)] and a Tris buffer [50 mM Tris (pH 7.4), aprotinin, leupeptin, pepstatin, PMSF] have been used successfully. Sucrose (0.34 M) can be included to stabilize the enzyme. The cell suspension is then homogenized by 100 strokes in a Dounce homogenizer (Wheaton, Millville, NJ) on ice, or sonicated (three times, 10 sec each; power, 4 W), and the homogenate is stored on ice until use (usually less than 2 hr). Protein content is measured by standard procedures, and homogenate samples are used in the assays described below for $O_2^{\cdot-}$.

Tissue. Tissue samples are placed in a chilled, modified Krebs–HEPES buffer (99 mM NaCl, 4.7 mM KCl, 1.9 mM CaCl$_2$, 1.2 mM MgSO$_4$, 1.0 mM K$_2$HPO$_4$, 25 mM NaHCO$_3$, 20 mM Na-HEPES, and 11 mM glucose, pH 7.4).[2] A 10% (w/v) tissue homogenate is prepared in a 50 mM phosphate buffer by homogenization in a glass-to-glass motorized homogenizer. The homogenate is then subjected to low-speed centrifugation (1000g) for 10 min to remove unbroken cells and debris. An aliquot is kept for protein determination, and supernatants (20 μl) are assayed immediately for $O_2^{\cdot-}$ production.

Membrane Preparation. In many cases, it has proved advantageous to remove mitochondrial membranes before measurement of NAD(P)H oxidase activity in order to exclude contributions from NADH-dependent mitochondrial enzymes.

[2] S. Rajagopalan, S. Kurz, T. Münzel, M. Tarpey, B. A. Freeman, K. K. Griendling, and D. G. Harrison, *J. Clin. Invest.* **97,** 1916 (1996).

Although centrifugation through a sucrose gradient to isolate a relatively pure plasma membrane fraction is often used for neutrophils, most investigators simply use sequential high-speed centrifugation in nonphagocytic cells. After lysis, homogenates are centrifuged at 20,000–29,000g for 20 min at 4° to pellet mitochondria, lysosomes, peroxisomes, Golgi membranes, and rough endoplasmic reticulum. The supernatant is then subjected to a 100,000g spin for 60 min at 4° to collect plasma membranes and microsomes, and this pellet is resuspended at a concentration of approximately 1–3 μg of protein per microliter and used to assay NAD(P)H oxidase activity.

Pyridine Nucleotide Dose Response to NAD(P)H

Initial characterization of enzyme activity in a given experimental system should include a substrate dose–response curve, as the concentrations of NADPH or NADH used to assay NAD(P)H oxidase activity are critically important. In general, a range of 1–1000 μM NADPH or NADH elicits a measurable signal. Estimates of K_m values for the vascular smooth muscle enzyme are 50 μM for NADPH and 10 μM for NADH, although the maximal activity supported by NADPH is much higher.[3] Maximal activity is achieved with concentrations of 300 μM of either substrate.

NADPH Consumption

Because NADPH and, to a lesser extent, NADH serve as the substrate for non-phagocytic NAD(P)H oxidases, it is important to verify that increases in enzyme activity are accompanied by a rise in consumption of these pyridine nucleotides. Measurement of NADH and NADPH consumption can be made by a modification of the method described by Brightman and co-workers.[4] Intact cells or tissue are incubated with 250 μM NADH or NADPH in phenol red-free medium (pH 7.4) for the time interval relevant to the biology of a particular sample. The rate of NADH or NADPH consumption is monitored spectrophotometrically by the decrease in absorbance at $\lambda = 340$ nm. The absorption extinction coefficient used to calculate the amount of NADH or NADPH consumed is 6.22 mM^{-1} cm^{-1}. For correlation with measurements of specific oxidase activity, samples must be included in which the NAD(P)H oxidase is maximally inhibited (see below), and then only the portion of the signal that can be inhibited by these strategies is used to calculate the rate of consumption of NADH and NADPH by this enzyme

[3] D. Sorescu, M. J. Somers, B. Lassègue, S. Grant, D. G. Harrison, and K. K. Griendling, *Free Radic. Biol. Med.* **30**, 603 (2001).

[4] A. O. Brightman, J. Wang, R. K. Miu, I. L. Sun, R. Barr, F. L. Crane, and D. J. Morre, *Biochim. Biophys. Acta* **1105**, 109 (1992).

system. Measurements can be expressed as (nano)moles of substrate per minute per 10^6 cells.

Strategies for Specific Inhibition of NAD(P)H Oxidase

Because many intracellular enzymes utilize NADPH or NADH as cofactors or substrates, the greatest difficulty with an assay of the nonphagocytic NAD(P)H oxidase is attributing the measured $O_2^{\cdot-}$ production and/or NAD(P)H consumption to the specific enzyme complex. Therefore, it is imperative that experiments of this type include either a specific inhibitor or molecular manipulation of enzyme expression in order to assign activity to the NAD(P)H oxidase.

Pharmacologically, the choice of inhibitor is extremely limited. The most widely used inhibitor is diphenylene iodonium (DPI), which was originally identified as a specific inhibitor of the neutrophil NADPH oxidase.[5] DPI is normally used at concentrations between 1 and 100 μM. A major drawback, however, is that DPI inhibits other flavin-containing enzymes as well (e.g., cytochromes, nitric oxide synthase), so care must be taken to rule out these other possibilities with inhibitors specific to those enzymes. Apocynin (4-hydroxy-3-methoxyacetophenone) has also been used as a specific oxidase inhibitor at concentrations of 200–500 μM, on the basis of its ability to interfere with translocation of the cytosolic components to the membrane.[6] The drawback of this inhibitor is that in many cases a role for cytosolic components in nonphagocytic oxidases has not been rigorously established. Another possible inhibitor is PR-39, a proline-rich peptide that has been reported to inhibit the neutrophil oxidase by binding to the Src homology domains of p47^{phox}.[7] The specificity of PR-39 has also been questioned because of its tendency to inhibit any SH2 domain-containing protein.

Currently, the best methods of inhibition of the nonphagocytic NADPH oxidase are the use of antisense to one of the subunits, the overexpression of a dominant-negative construct to a specific subunit, or the use of tissues from knockout mice. However, before any of these strategies can be applied, one must clearly identify which oxidase subunits participate in $O_2^{\cdot-}$ generation in the tissue of interest. Marked success has been attained in vascular tissues by using antisense against p22^{phox}. Both creation of cell lines stably expressing p22^{phox} and delivery of antisense phosphorothioate oligonucleotides have been used successfully.[8,9] Antisense

[5] J. Doussiere and P. V. Vignais, *Eur. J. Biochem.* **208**, 61 (1992).

[6] J. Stolk, T. J. Hiltermann, J. H. Dijkman, and A. J. Verhoeven, *Am. J. Respir. Cell Mol. Biol.* **11**, 95 (1994).

[7] J. Shi, C. R. Ross, T. L. Leto, and F. Blecha, *Proc. Natl. Acad. Sci. U.S.A.* **93**, 6014 (1996).

[8] M. Ushio-Fukai, A. M. Zafari, T. Fukui, N. Ishizaka, and K. K. Griendling, *J. Biol. Chem.* **271**, 23317 (1996).

[9] C. Viedt, U. Soto, H. I. Krieger-Brauer, J. Fei, C. Elsing, W. Kubler, and J. Kreuzer, *Arterioscler. Thromb. Vasc. Biol.* **20**, 940 (2000).

oligonucleotides against p47[phox] have also proved effective.[10] Choice of oligonucleotide sequence is dependent on the species used and many other factors including RNA site selection, stability, and modification of bases to improve stability. A discussion of design and delivery techniques for oligonucleotides is beyond the scope of this chapter; however, a review by Stein describes many of the pitfalls of this type of experiment.[11] The dominant-negative strategy has been used successfully so far only for Rac-1, for which structure–function relationships have been well defined.[12]

As more knockout mice are created, many investigators are turning to tissues from these animals to definitively prove or disprove a role for the NAD(P)H oxidase in a particular biological function. For these studies, tissues can be removed from knockout and wild-type animals and used directly for measurement of oxidase activity. Another strategy is to culture cells from these mice so that a more complete characterization of the oxidase can be made. These approaches have been used successfully for tissues/cells from homozygous negative p47[phox] and gp91[phox] mice.[13,14] A caveat to studies of this type is that an independent measure of the potential involvement of the oxidase at the study end point should be evaluated, because knockout animals are notoriously good at compensating, and may in fact induce other subunits of the enzyme to correct for the loss of a critical function. This is of particular concern for homozygous negative gp91phox mice, because of the identification of a family of gp91[phox] homologs with similar function.[15]

Measurement of Superoxide Production

As noted above, assessment of NAD(P)H oxidase activity is primarily directed at detecting $O_2^{\cdot-}$, the product of the oxidase. In general, the approach to demonstrating NAD(P)H oxidase activity is first to demonstrate the generation of $O_2^{\cdot-}$, confirming specificity by showing inhibition of the signal with superoxide dismutase (SOD) or an SOD mimetic. Next, the contribution of the specific NAD(P)H oxidase enzyme to this $O_2^{\cdot-}$ generation is confirmed by one of the strategies described above. The major challenge of measuring nonphagocytic NAD(P)H oxidase activity is that the amount of $O_2^{\cdot-}$ generated is low compared with that generated by the phagocytic NADPH oxidase.

[10] E. A. Bey and M. K. Cathcart, *J. Lipid Res.* **41,** 489 (2000).

[11] C. A. Stein, *Biochim. Biophys. Acta* **1489,** 45 (1999).

[12] M. Sundaresan, Z. X. Yu, V. J. Ferrans, D. J. Sulciner, J. S. Gutkind, K. Irani, P. J. Goldschmidt-Clermont, and T. Finkel, *Biochem. J.* **318,** 379 (1996).

[13] S. L. Archer, H. L. Reeve, E. Michelakis, L. Puttagunta, R. Waite, D. P. Nelson, M. C. Dinauer, and E. K. Weir, *Proc. Natl. Acad. Sci. U.S.A.* **96,** 7944 (1999).

[14] S. H. Jackson, J. I. Gallin, and S. M. Holland, *J. Exp. Med.* **182,** 751 (1995).

[15] J. D. Lambeth, G. Cheng, R. Arnold, and W. A. Edens, *Trends Biochem. Sci.* **25,** 459 (2000).

There are several techniques available for the detection of reactive oxygen species (ROS) in biological systems (see Volumes 105, 186, 233, and 234 of this series and Weber[16]). The selection of a specific technique is dependent on the strengths and weaknesses of the individual assays, the biological sample being studied (intact tissue vs. cellular homogenate), and the amount of $O_2^{\cdot-}$ generated. Depending on the technique used, NAD(P)H oxidase activity can be demonstrated in intact tissue, isolated cells, or tissue homogenate.

Unlike the phagocyte NADPH oxidase, in some nonphagocytic cells the addition of the cytosolic fraction to the membrane fraction does not augment $O_2^{\cdot-}$ generation.[1] Although this may reflect differences between the two oxidases, cytosolic SOD can make detection of $O_2^{\cdot-}$ difficult. Inhibition of SOD by pretreatment with diethyl dithiocarbamate (10 mM) can help to unmask $O_2^{\cdot-}$ generation in nonphagocytic cells.

NAD(P)H oxidase activity is demonstrated under basal conditions and after addition of NADH or NADPH. Basal production of $O_2^{\cdot-}$ is measured in intact cells or tissues before inhibition of NAD(P)H oxidase function. In contrast, NADH- or NADPH-stimulated $O_2^{\cdot-}$ is measured in homogenized fractions. Because NADH and NADPH do not cross membranes, their addition to intact cell preparations is not a sufficient means for detection of oxidase activity.

Cytochrome c Reduction

Reduction of cytochrome c is a well-accepted technique for measurement of $O_2^{\cdot-}$; unfortunately, this assay is relatively insensitive for the detection of NAD(P)H oxidase activity in nonphagocytic cells, where enzymatic activity is low. Another potential disadvantage is that cytochrome c is unable to detect intracellular $O_2^{\cdot-}$. This assay is used to detect extracellular $O_2^{\cdot-}$ in cultured cells or $O_2^{\cdot-}$ production in cell homogenate. Superoxide is measured as the SOD-inhibitable reduction of cytochrome c, determined in a spectrophotometer by the increase in absorbance at 550 nm.[17]

However, biologic tissues contain reductases and oxidases that interfere with measurement of $O_2^{\cdot-}$ by cytochrome c. Membrane reductases artifactually increase A_{550}, whereas oxidation of reduced cytochrome c by oxidases will underestimate $O_2^{\cdot-}$ levels. In contrast, acetylated cytochrome c is much less susceptible to tissue oxidases and reductases and should be employed for the detection of $O_2^{\cdot-}$ in biologic membranes.[18]

Intact Cells. To measure extracellular production of $O_2^{\cdot-}$ by cells, paired samples of cultured cells are washed and allowed to equilibrate in phenol red-free medium or Hanks' balanced salt solution (HBSS) at 37°. Acetylated cytochrome c

[16] G. F. Weber, *J. Clin. Chem. Clin. Biochem.* **28,** 569 (1990).

[17] J. M. McCord and I. Fridovich, *J. Biol. Chem.* **244,** 6049 (1969).

[18] A. Azzi, C. Montecucco, and C. Richter, *Biochem. Biophys. Res. Commun.* **65,** 597 (1975).

(0.1 mM) is added and absorbance is measured at 550 nm. Ideally, the assay is performed in a split beam spectrophotometer containing paired samples, one of which is pretreated with SOD (100 U/ml). If production of $O_2^{\cdot-}$ is low, then the samples can be incubated with cytochrome c at 37° for 15–30 min before measuring A_{550}.

Membrane Fractions. Adapted from assays of phagocyte NADPH oxidase activity,[19] the membrane fraction is resuspended in buffer [8 mM piperazine-N,N'-bis(2-ethanesulfonic acid) (PIPES), 100 mM KCl, 3 mM NaCl, 3.5 mM MgCl$_2$, and 1.25 mM EGTA] containing proteolytic inhibitors. In a multiwell plate, flavin adenine dinucleotide (FAD, 0.01 mM), acetylated cytochrome c (0.1 mM), membrane sample (0.1 mg), and GTPγS (0.01 mM) are added. SOD (100 U/ml) is included in half the samples. After 2 min at room temperature, sodium dodecyl sulfate (SDS, 0.1 mM) is added. After 3 min, NAD(P)H (0.2 mM) is added and A_{550} is measured every 15 sec for 10 min.

Data are plotted kinetically as the maximum velocity (V_{max}) of the reaction, and expressed as the change in A_{550} per minute. The change in A_{550} (ΔA_{550}) is determined by the difference in cytochrome c reduction between mixtures containing tissue sample with and without SOD. An extinction coefficient (dependent on the light path length) is used to convert V_{max} results to nanomoles of $O_2^{\cdot-}$ per minute per cell number or protein concentration. The millimolar extinction coefficient is 21.1 for a 1-cm light path length.

Chemiluminescence

Because of its sensitivity, chemiluminescence is a commonly used method for the detection of $O_2^{\cdot-}$ in tissue. Several substances are available that interact with $O_2^{\cdot-}$ and emit a photon that can be detected in a luminometer or spectrophotometer. Many of these probes are cell permeable and therefore able to detect both intracellular and extracellular $O_2^{\cdot-}$. Of the available probes, lucigenin (bis-N-methylacridinium nitrate) has been the most widely used to measure $O_2^{\cdot-}$. However, lucigenin is susceptible to redox cycling with cellular reductases, resulting in artifactual generation of $O_2^{\cdot-}$.[20] Autoxidation is more likely to occur with lucigenin concentrations greater than 5 μM and, therefore, higher concentrations should not be used. Although concentrations of lucigenin less than 5 μM also have the potential to redox cycle, the significance in altering $O_2^{\cdot-}$ measurements in biologic samples is not clear. Coelenterazine [2-(4-hydroxybenzyl)-6-(4-hydroxyphenyl)-8-benzyl-3,7-dihydroimidazo[1,2-α]pyrazin-3-one] and methyl-cypridina-luciferin analog, that is, MCLA [2-methyl-6-(p-methoxyphenyl)-3,7-dihydroimidazo[1,2-α]pyrazin-3-one], are two compounds gaining popularity in chemiluminescence

[19] F. R. DeLeo, J. Renee, S. McCormick, M. Nakamura, M. Apicella, J. P. Weiss, and W. M. Nauseef, *J. Clin. Invest.* **101,** 455 (1998).

[20] S. I. Liochev and I. Fridovich, *Arch. Biochem. Biophys.* **337,** 115 (1997).

detection of $O_2^{\cdot-}$ in tissues.[21,22] In addition to $O_2^{\cdot-}$, coelenterazine also reacts with $ONOO^-$. The specificity of MCLA for $O_2^{\cdot-}$ has not been evaluated with the same degree of scrutiny as for lucigenin and coelenterazine. The use of another chemiluminescence agent, luminol (5-amino-2,3-dihydro-1,4-phthalazinedione), is problematic not only because its activation is peroxidase dependent, but also because it does not react with $O_2^{\cdot-}$, but instead with its product H_2O_2. It is important to note that the chemiluminescence signal from these agents is sensitive to changes in pH.[23]

An advantage of chemiluminescence measurements is the ability to detect $O_2^{\cdot-}$ in multiple preparations, including intact tissue, cultured cells, and tissue homogenates. Samples are added to polypropylene tubes containing balanced salt solution (137 mM NaCl, 2.7 mM KCl, 4.3 mM Na_2HPO_4, 1.5 mM KH_2PO_4) and chemiluminescence probe (5 μM lucigenin, 10 μM MCLA or 10 μM coelenterazine). Photon emission is recorded with a luminometer or a scintillation counter (out-of-coincidence mode). Some luminometers accept multiwell plates or 35-mm culture dishes, allowing measurements of $O_2^{\cdot-}$ in attached cultured cells. After 5 min of dark adaptation, basal chemiluminescence in tissue or intact cells is measured continuously for 10 min. In samples of tissue homogenate, NAD(P)H is added and luminescence is measured for 10 min. Samples can be pretreated with the cell-permeable $O_2^{\cdot-}$ scavenger Tiron (4,5-dihydroxy-1,3-benzenedisulfonic acid, 10 mM) to confirm that the chemiluminescence signal reflects $O_2^{\cdot-}$, or with any of the NAD(P)H oxidase inhibition methods described above. Data are expressed as relative light units (RLU) per unit of time, normalized to surface area, tissue weight, or protein, after subtraction of background chemiluminescence signal (cell-free preparation).

Fluorescent Dyes

Although several fluorescent probes are available that react with ROS, the most applicable in determining nonphagocytic NAD(P)H oxidase activity are dihydroethidium (DHE) and 2′,7′-dichlorofluorescein diacetate (DCFH-DA).[24,25] DHE is cell permeable until reacting with $O_2^{\cdot-}$ to form ethidium. Ethidium intercalates with DNA, primarily providing a nuclear fluorescence (excitation at 520 nm, emission at 610 nm). DHE is specific for $O_2^{\cdot-}$, but in some

[21] M. M. Tarpey, C. R. White, E. Suarez, G. Richardson, R. Radi, and B. A. Freeman, *Circ. Res.* **84,** 1203 (1999).

[22] L. Pronai, H. Nakazawa, K. Ichimori, Y. Saigusa, T. Ohkubo, K. Hiramatsu, S. Arimori, and J. Feher, *Inflammation* **16,** 437 (1992).

[23] M. M. Oosthuizen, M. E. Engelbrecht, H. Lambrechts, D. Greyling, and R. D. Levy, *J. Biolumin. Chemilumin.* **12,** 277 (1997).

[24] V. P. Bindokas, J. Jordan, C. C. Lee, and R. J. Miller, *J. Neurosci.* **16,** 1324 (1996).

[25] W. O. Carter, P. K. Narayanan, and J. P. Robinson, *J. Leukoc. Biol.* **55,** 253 (1994).

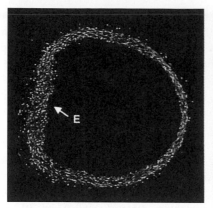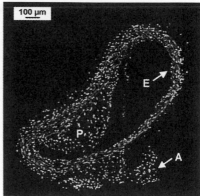

FIG. 1. *In situ* detection of $O_2^{\cdot-}$. Confocal fluorescence photomicrographs of cross-sections of aortas from a normal mouse (*left*) and from a hypercholesterolemic mouse (*right*) after DHE staining. Samples were processed as described in text. Fluorescence reflects cellular $O_2^{\cdot-}$ levels. A, Adventitia; E, endothelium; P, atherosclerotic plaque.

situations has the ability to react with $OH^{\cdot}$ and H_2O_2. In contrast, DCFH-DA is a more general indicator of ROS, reacting with H_2O_2, $OH^{\cdot}$, $ONOO^-$, lipid hydroperoxides, nitric oxide, and, to a much lesser extent, $O_2^{\cdot-}$. DCFH-DA is cleaved in the cell by esterases to DCFH, which reduces its cellular permeability. DCFH reacts with radicals to form fluorescent 2′,7′-dichlorofluorescein (DCF; excitation at 488 nm, emission at 525 nm). Because H_2O_2 is a secondary product of $O_2^{\cdot-}$, DCF fluorescence has been used to implicate $O_2^{\cdot-}$ production.

After reaction with ROS, these fluorescent compounds are generally not cell permeable, making them ideal for detection of intracellular $O_2^{\cdot-}$ (Fig. 1). Other strengths include the ability to localize cellular production of ROS within the tissue, and the sensitivity to detect ROS within a single cell. Unlike other available assays of $O_2^{\cdot-}$, which provide a composite average of $O_2^{\cdot-}$ levels, DHE fluorescence allows for distinction of subpopulations of cells with varying levels of $O_2^{\cdot-}$ production. Also, by taking advantage of differences in the emission wavelengths, detection of $O_2^{\cdot-}$ by DHE can be colocalized with other fluorescent markers.

As with any assay for detection of ROS, the user must be attentive to potential limitations of fluorescent probes. Because ethidium binds to DNA, conditions that alter DNA content (e.g., apoptosis) will affect the detection of $O_2^{\cdot-}$ by DHE. Autofluorescence can be high in many tissues, but can be digitally subtracted from the fluorescent image by image analysis software [e.g., Adobe (San Jose, CA) Photoshop]. To reduce photoreduction, working conditions should minimize exposure of fluorescent probes to light, in particular while viewing the specimens by microscopy. It has been reported that DHE may catalyze the dismutation of $O_2^{\cdot-}$, thus underestimating $O_2^{\cdot-}$ levels.[26]

It has been shown that during the formation of DCF, an intermediate radical can generate $O_2^{\cdot-}$.[27] Dismutation of this $O_2^{\cdot-}$ to H_2O_2 will result in self-amplification of DCF fluorescence. Another potential weakness of DCFH is that formation of DCF from DCFH requires peroxidase activity. Therefore, conditions that change cellular peroxidase levels may alter DCF fluorescence independent of ROS levels.[28,29] For these reasons, DCFH-DA should be used with caution, and DCF fluorescence interpreted in the context of other assays for ROS.

Dihydroethidium Fluorescence

DHE stock (10 mM) is prepared by adding 315 μl of dimethyl sulfoxide (DMSO) to 1 mg of DHE (Molecular Probes, Eugene, OR) in a light-protected container under nitrogen or argon to avoid oxidation. The solution is vortexed for 30 sec to ensure that the DHE is dissolved, and working aliquots are stored at $-20°$.

Tissue Samples. Tissues are harvested, rinsed in cold HBSS, and snap frozen in O.C.T. compound (Miles, Elkhart, IL).[30] Tissues that will be compared directly with each other are processed in parallel. Sections (30 μM thick to preserve the integrity of cells) are placed on glass slides and covered with DHE (2 μM, diluted in HBSS). Slides are incubated in a light-protected humidified chamber for 30 min at 37°. DHE is rinsed and sections are coverslipped. Images are obtained with a laser scanning confocal fluorescence microscope (excitation at 488 nm and detection of fluorescence at 585 nm, using a long-pass filter). Some sections are incubated in parallel with a cell-permeable SOD mimetic or NAD(P)H oxidase inhibitor before incubation with DHE. For a given magnification, confocal settings (laser power, iris, gain, and black level) are optimized while imaging the control sample, such that ethidium fluorescence is only just detectable. All other tissues are imaged with the same confocal settings. Relative fluorescence intensity indicates $O_2^{\cdot-}$ levels.

Cultured Cells. Cells are grown under the desired experimental conditions and generally studied at ~70% confluence after overnight serum deprivation. Cells are washed in phenol red-free culture medium or HBSS, and DHE (2 μM) is added. If studying NAD(P)H oxidase activity in response to an agonist, the agonist can be added several hours before DHE (prolonged stimulation) or after the cells are loaded with DHE (acute stimulation). After 30 min at 37°, cells are rinsed.

[26] L. Benov, L. Sztejnberg, and I. Fridovich, *Free Radic. Biol. Med.* **25,** 826 (1998).

[27] C. Rota, C. F. Chignell, and R. P. Mason, *Free Radic. Biol. Med.* **27,** 873 (1999).

[28] C. Rota, Y. C. Fann, and R. P. Mason, *J. Biol. Chem.* **274,** 28161 (1999).

[29] H. Zhu, G. L. Bannenberg, P. Moldeus, and H. G. Shertzer, *Arch. Toxicol.* **68,** 582 (1994).

[30] F. J. Miller, Jr., D. D. Gutterman, C. D. Rios, D. D. Heistad, and B. L. Davidson, *Circ. Res.* **82,** 1298 (1998).

At this point, ethidium fluorescence within the cells can be imaged with a fluorescence microscope or by flow cytometry after preparing a cell suspension. If using fluorescence microscopy, it is convenient to grow the cells on a chamber slide. After incubation in DHE, the cells are rinsed and then coverslipped. To measure ethidium fluorescence by flow cytometry, the cells are detached from the dish gently with trypsin, added to serum to inhibit trypsin activity, centrifuged at 400g for 5 min at 4°, and then washed and resuspended in cold PBS or HBSS. Alternatively, the cell suspension can be prepared before incubation with DHE. In this case, placing the cells on ice stops the reaction of DHE with $O_2^{\cdot-}$. In tissues with low $O_2^{\cdot-}$ levels, higher concentrations of DHE (10 μM) or longer incubation times (60 min) will increase sensitivity for $O_2^{\cdot-}$.

Fluorescent dyes and nitroblue tetrazolium (see below) can complement other methods of $O_2^{\cdot-}$ measurement by demonstrating *in situ* localization of the radical (Fig. 1). Although image analysis software (NIH Image) allows measurement of fluorescence intensity, caution should be exercised in quantifying ROS levels by fluorescent probes in intact tissue. There are limitations in the linearity of fluorescence with $O_2^{\cdot-}$, especially with saturation of the fluorescence signal, and with standardization of fluorescence.

Nitroblue Tetrazolium

Nitroblue tetrazolium (NBT) undergoes reduction by $O_2^{\cdot-}$ to form diformazan, a dark blue insoluble precipitate.[31] Similar to DHE, NBT detects intracellular $O_2^{\cdot-}$; however, it is less sensitive and specific for $O_2^{\cdot-}$ than is DHE. NBT is susceptible to reduction by several tissue reductases,[18] and therefore specificity for $O_2^{\cdot-}$ should be confirmed by inhibition of NBT staining by SOD. Like lucigenin, NBT also has been shown to artifactually generate $O_2^{\cdot-}$ by autoxidation.[32] For these reasons, detection of $O_2^{\cdot-}$ in biologic samples should not rely exclusively on NBT reduction.

Tissue samples are placed in phenol red-free medium or HBSS with or without addition of inhibitors (SOD mimetic, DPI, etc.) at 37° for 30 min. NBT (0.1 mM) is added and the incubation is continued for 1–3 hr, depending on the amount of $O_2^{\cdot-}$ production. An equal volume of 0.5 N HCl is added to stop the reaction and the tissues are rinsed with cold PBS.

Formazan Extraction. Levels of $O_2^{\cdot-}$ generated by the tissue are quantified by measuring the absorbance of blue formazan. Formazan is extracted by crushing the tissue in liquid nitrogen and dissolving it in 100% pyridine at 80° for 30 min. Samples are centrifuged at 20,000g for 20 min at 4° and absorbance of supernatants is

[31] C. Auclair and E. Voison, *in* "Handbook of Methods for Oxygen Radical Research," (R. A. Greenwald, ed.), p. 123. CRC Press, Boca Raton, FL, 1985.

[32] J. Vasquez-Vivar, P. Martasek, N. Hogg, H. Karoui, B. S. Masters, K. A. Pritchard, Jr., and B. Kalyanaraman, *Methods Enzymol.* **301**, 169 (1999).

determined at 540 nm. Results are reported as picomoles of $O_2{}^{\cdot-}$ per minute per milligram wet weight of tissue, calculated as $A_{540} \times$ volume/(time $\times$ weight $\times E \times L$), where $E = 0.72 \, \mathrm{m}M^{-1} \, \mathrm{mm}^{-1}$ and L is the length of the light path.

In Situ Localization. After staining in NBT, tissues can be sectioned (8 μm thick) and visualized by light microscopy to determine the cellular location of $O_2{}^{\cdot-}$ in the tissue. Samples are fixed in 4% (v/v) formalin and embedded in paraffin. The sectioning and subsequent preparation of sections are complicated by solubilization and leaching of blue formazan from the tissue. This can be avoided by imaging the tissue block while epiluminated on the microscope stage, or by rinsing sections in Clear-Rite 3 solution (Stephens Scientific, Riverdale, NJ) after deparaffinization.

Electron Spin Resonance Measurement of Superoxide Production

Electron spin resonance (ESR) is currently the "gold standard" for measuring $O_2{}^{\cdot-}$ production. It is well suited to measurement of NAD(P)H oxidase activity, because it is easily adaptable to cell homogenates, and the concentrations of pyridine nucleotides in the assay buffer can be varied, permitting evaluation of NAD(P)H-dependent $O_2{}^{\cdot-}$ production. Different spin traps have been used for $O_2{}^{\cdot-}$, including DEPMPO (5-diethoxyphosphoryl-5-methyl-1-pyrroline-*N*-oxide), DMPO (5,5'-dimethyl-1-pyrroline-*N*-oxide), and CPH (1-hydroxy-3-carboxypyrrolidine), but published studies focusing on measurement of oxidase activity have mainly utilized DEPMPO.[3,33] The advantage of DEPMPO for low-output systems is the relatively long half-life of the DEPMPO–OOH adduct, so that cumulative measures of $O_2{}^{\cdot-}$ can be made. For a more complete discussion of spin trapping of $O_2{}^{\cdot-}$, see Rosen and Rauckman.[34]

ESR can be used in cell homogenates or membrane preparations, with the latter giving better signals. Homogenate or membrane proteins are prepared as described above, except that the phosphate-based lysis buffer is first treated for 2 hr with Chelex (5 g/100 ml; Bio-Rad, Hercules, CA) and filtered. Twenty micrograms of protein is incubated with 25 mM DEPMPO, 1–1000 μM NADPH or NADH, and 0.2 mM diethylenetriaminepentaacetic acid in Chelex-treated PBS (pH 7.4) at 37° for 30 min. Samples can be assayed immediately or snap frozen in liquid nitrogen until use.

ESR measurements are performed at room temperature with a Bruker (Billerica, MA) EMX spectrometer with the following instrument settings: modulation amplitude, 1 G; time constant, 160 msec; modulation frequency, 100 kHz; microwave power, 20 mW; microwave frequency, 9.78 GHz. The scanning time is 84 sec. Frozen samples should be thawed at room temperature immediately before measurement, and scanned five times. Comparative quantification can

[33] M. J. Somers, J. S. Burchfield, and D. G. Harrison, *Antioxid. Redox Signal.* **2,** 779 (2000).
[34] G. M. Rosen and E. J. Rauckman, *Methods Enzymol.* **105,** 198 (1984).

be performed by addition of all eight peaks of the spectrum and subtraction of the value obtained from buffer blank spectrum (without cell membrane), or by calculating the integral of the area under the curve for each spectra. Because the DEPMPO–OOH adduct is rapidly converted to the hydroxyl adduct DEPMPO–OH, parallel samples should be measured in the presence of SOD to identify the spectrum as deriving originally from $O_2^{\cdot-}$.

Aconitase

The citric acid cycle enzyme aconitase belongs to the family of $O_2^{\cdot-}$-sensitive, [4Fe–4S]-containing dehydratases. Both the mitochondrial and cytosolic forms of aconitase are inactivated by $O_2^{\cdot-}$, so that activity can be used to monitor $O_2^{\cdot-}$ production. The drawback of this assay is that it measures intracellular $O_2^{\cdot-}$ production from all sources, not only from NAD(P)H oxidases, and it is not easily adaptable to the measurement of NADPH-dependent $O_2^{\cdot-}$ production, except with the use of antisense or knockout strategies (see above).

Cell extracts are prepared according to Gardner et al.[35] by sonication in hypotonic Tris buffer [50 mM Tris-HCl (pH 7.4), 0.6 mM MnCl$_2$, 20 μM ($\pm$)fluorocitrate]. The lysate is then immediately stored at $-80°$. Fluorocitrate and MnCl$_2$ are included to limit the inactivation of aconitase by $O_2^{\cdot-}$ during extract preparation and storage, and lysates can be stored for up to 2 weeks. Cysteine (1 mM) can be substituted for fluorocitrate.[36]

Aconitase activity is measured spectrophotometrically by monitoring the increase in absorbance at 240 nm due to the formation of *cis*-aconitate from isocitrate. For this assay, lysates are incubated in 50 mM Tris-HCl (pH 7.4), 20 mM isocitrate, and 0.5 mM MnCl$_2$. After addition of the extract, A_{240} is determined for 1–2 min ($\varepsilon_{240} = 3.6 \, \text{m}M^{-1} \, \text{cm}^{-1}$).

Aconitase activity can also be assessed by monitoring the formation of NADPH at 340 nm. Here, citrate is used as the substrate for aconitase, and its product, isocitrate, is converted to α-ketoglutarate by the NADP$^+$-dependent enzyme isocitrate dehydrogenase. Lysates are clarified by centrifugation for 20 sec at 14,000g, and 10- to 100-μg aliquots of the supernatant are incubated in a reaction mixture containing 50 mM Tris-HCl (pH 7.4), 5 mM sodium citrate, 0.6 mM MnCl$_2$, 0.2 mM NADP$^+$, and isocitrate dehydrogenase (1 U/ml). Absorbance at 340 nm ($\varepsilon_{340} = 6.22 \, \text{m}M^{-1} \, \text{cm}^{-1}$) is monitored at 25°, and rates are measured in the latter half of a 60-min assay. One milliunit of aconitase activity is defined as the amount catalyzing the formation of 1 nmol of isocitrate per minute. To obtain a quantitative measure of $O_2^{\cdot-}$, one must also determine the rate constant of aconitase reactivation according to Hausladen and Fridovich[36] and then

[35] P. R. Gardner, D. D. Nguyen, and C. W. White, *Proc. Natl. Acad. Sci. U.S.A.* **91,** 12248 (1994).
[36] A. Hausladen and I. Fridovich, *Methods Enzymol.* **269,** 37 (1996).

$[O_2^{\cdot-}] = [\text{aconitase}_{\text{inactive}}]k_{\text{reactivation}}/[\text{aconitase}_{\text{active}}]k_{\text{inactivation}}$, although many investigators simply report the relative ability of two samples to inactivate aconitase to obtain a comparative measure of $O_2^{\cdot-}$ levels.

Acknowledgments

This work was supported by NIH Grants HL38206 (K.K.G.), HL58000 (K.K.G.), HL58863 (K.K.G.), HL62984 (F.M.), and HL03669 (F.M.). We are grateful to Dr. Dan Sorescu and Marjorie Akers for critical reading of the manuscript, and to Carolyn Morris for excellent secretarial assistance.

[21] Purification and Assessment of Proteins Associated with Nitric Oxide Synthase

By Charles J. Lowenstein

Introduction

Calmodulin was the first polypeptide shown to interact with nitric oxide synthases (NOS).[1] A variety of polypeptides have subsequently been identified that interact with NOS (Table I[1–18]). Some of them, such as calmodulin, interact with all three of the NOS isoforms: neuronal NOS (nNOS or NOS1), inducible NOS (iNOS or NOS2), and endothelial NOS (eNOS or NOS3). Other proteins interact with

[1] D. S. Bredt and S. H. Snyder, *Proc. Natl. Acad. Sci. U.S.A.* **87,** 682 (1990).

[2] H. J. Cho, Q. W. Xie, J. Calaycay, R. A. Mumford, K. M. Swiderek, T. D. Lee, and C. Nathan, *J. Exp. Med.* **176,** 599 (1992).

[3] R. Dusse and A. Mulsch, *FEBS Lett.* **265,** 133 (1990).

[4] V. J. Venema, H. Ju, R. Zou, and R. C. Venema, *J. Biol. Chem.* **272,** 28187 (1997).

[5] E. A. Ratovitski, M. R. Alam, R. A. Quick, A. McMillan, C. Bao, C. Kozlovsky, T. A. Hand, R. C. Johnson, R. E. Mains, B. A. Eipper, and C. J. Lowenstein, *J. Biol. Chem.* **274,** 993 (1999).

[6] G. Garcia-Cardena, R. Fan, D. F. Stern, J. Liu, and W. C. Sessa, *J. Biol. Chem.* **271,** 27237 (1996).

[7] J. B. Michel, O. Feron, D. Sacks, and T. Michel, *J. Biol. Chem.* **272,** 15583 (1997).

[8] V. J. Venema, R. Zou, H. Ju, M. B. Marrero, and R. C. Venema, *Biochem. Biophys. Res. Commun.* **236,** 155 (1997).

[9] D. S. Chao, J. R. Gorospe, J. E. Brenman, J. A. Rafael, M. F. Peters, S. C. Froehner, E. P. Hoffman, J. S. Chamberlain, and D. S. Bredt, *J. Exp. Med.* **184,** 609 (1996).

[10] E. A. Ratovitski, C. Bao, R. A. Quick, A. McMillan, C. Kozlovsky, and C. J. Lowenstein, *J. Biol. Chem.* **274,** 30250 (1999).

[11] O. Feron, C. Dessy, D. J. Opel, M. A. Arstall, R. A. Kelly, and T. Michel, *J. Biol. Chem.* **273,** 30249 (1998).

[12] J. E. Brenman, D. S. Chao, S. H. Gee, A. W. McGee, S. E. Craven, D. R. Santillano, Z. Wu, F. Huang, H. Xia, M. F. Peters, S. C. Froehner, and D. S. Bredt, *Cell* **84,** 757 (1996).

[13] M. Bucci, F. Roviezzo, C. Cicala, W. C. Sessa, and G. Cirino, *Br. J. Pharmacol.* **131,** 13 (2000).

Copyright 2002, Elsevier Science (USA).
All rights reserved.
0076-6879/02 $35.00

TABLE I
PROTEINS THAT INTERACT WITH NITRIC OXIDE SYNTHASE

NOS1		NOS2		NOS3	
Protein	Ref.	Protein	Ref.	Protein	Ref.
Calmodulin	1	Calmodulin	2	Calmodulin	3
Caveolin 3	4	Kalirin	5	Caveolin 1	6–8
Syntrophin	9	NAP-110kDa	10	Caveolin 3	11
PSD-93	12			Heat shock protein 90	13
PSD-95	12			Bradykinin receptor	14,15
PIN	16			Dynamin 2	17
CAPON	18				
Bradykinin receptor	14				

Abbreviations: PSD, Postsynaptic density protein; NAP, neutrophil-activating protein.

only one NOS isoform. For example, syntrophin interacts only with NOS1 and not NOS2 or NOS3; 110-kDa NOS-associated protein (NAP-110) interacts only with NOS2; and 90-kDa heat shock protein (Hsp90) interacts only with NOS3.[10,19,20] The interactions of polypeptides with NOS have been shown to have a variety of functions. Some interactions serve to localize NOS: syntrophin localizes NOS1 to neuronal synapses, and caveolin-1 localizes eNOS to caveolae of endothelial cells.[6,7,12,21] Some interactions regulate NOS activity: the protein inhibitor of NOS1 (PIN) inactivates NOS1, NAP inactivates NOS2, and caveolin not only localizes eNOS but also inactivates it.[7,8,10,16,22–24]

Although a variety of polypeptides inhibit NOS, they have been identified by similar techniques. This chapter describes a set of methods used to identify a novel

[14] R. Golser, A. C. Gorren, A. Leber, P. Andrew, H. J. Habisch, E. R. Werner, K. Schmidt, R. C. Venema, and B. Mayer, *J. Biol. Chem.* **275,** 5291 (2000).

[15] H. Ju, V. J. Venema, M. B. Marrero, and R. C. Venema, *J. Biol. Chem.* **273,** 24025 (1998).

[16] S. R. Jaffrey and S. H. Snyder, *Science* **274,** 774 (1996).

[17] S. Cao, J. Yao, T. J. McCabe, Q. Yao, Z. S. Katusic, W. C. Sessa, and V. Shah, *J. Biol. Chem.* **276,** 14249 (2001).

[18] S. R. Jaffrey, A. M. Snowman, M. J. Eliasson, N. A. Cohen, and S. H. Snyder, *Neuron* **20,** 115 (1998).

[19] D. S. Bredt, *Proc. Soc. Exp. Biol. Med.* **211,** 41 (1996).

[20] G. Garcia-Cardena, R. Fan, V. Shah, R. Sorrentino, G. Cirino, A. Papapetropoulos, and W. C. Sessa, *Nature (London)* **392,** 821 (1998).

[21] H. Ju, R. Zou, V. J. Venema, and R. C. Venema, *J. Biol. Chem.* **272,** 18522 (1997).

[22] O. Feron, J. B. Michel, K. Sase, and T. Michel, *Biochemistry* **37,** 193 (1998).

[23] J. P. Gratton, J. Fontana, D. S. O'Connor, G. Garcia-Cardena, T. J. McCabe, and W. C. Sessa, *J. Biol. Chem.* **275,** 22268 (2000).

[24] G. Garcia-Cardena, P. Martasek, B. S. Masters, P. M. Skidd, J. Couet, S. Li, M. P. Lisanti, and W. C. Sessa, *J. Biol. Chem.* **272,** 25437 (1997).

polypeptide that interacts with NOS2, and inhibits its activity by regulating NOS2 homodimerization. However, these methods have been used by others to identify proteins that interact with the other isoforms of NOS as well.

Strategies for Experimental Design

Techniques to identify protein–protein interactions involve methods that are either relatively specific or relatively sensitive. Coimmunoprecipitation of interacting proteins from mammalian cells is more specific, but may fail to identify proteins that interact with the target at low affinity (false negatives). The yeast two-hybrid system is more sensitive, but may include proteins that do not actually interact with the target (false positives). We have employed the yeast two-hybrid system to screen several cDNA libraries for polypeptides that interact with NOS2. To eliminate false positives obtained by screening in yeast, we then rescreened potentially positive clones expressed in mammalian cells by coimmunoprecipitation.

Yeast Two Hybrid Library Screening for Nitric Oxide Synthase-Associated Proteins

The yeast two-hybrid system is a method of genetic screening that relies on the interaction of two fusion polypeptides to form a GAL4 transcription factor from its separate modules.[25] Yeast are first transformed with a bait plasmid encoding a polypeptide consisting of NOS2 fused to the GAL4 DNA-binding domain (GAL4-BD). These yeast are then transformed a second time with a plasmid library consisting of random cDNA fragments fused to the coding sequence of the GAL4 activation domain (GAL4-AD). If the fusion polypeptide consisting of GAL4-AD–cDNA interacts with GAL4-BD–NOS2, the GAL4-AD polypeptide is physically linked to the GAL4-BD polypeptide, and can activate a reporter gene. Yeast expressing reporter genes are expanded, and the cDNA obtained from the cDNA library plasmid is isolated.

We employ the yeast two-hybrid system to identify novel polypeptides that interact with NOS2.[5,10] We first construct a cDNA library from mouse bone marrow-derived primary macrophages. Mouse bone marrow-derived primary macrophages are isolated from female C57BL/6 mice (F_2 wild type; Jackson Laboratories, Bar Harbor, ME) and cultured in Dulbecco's modified Eagle's medium (DMEM) supplemented with 10% (v/v) low-lipopolysaccharide (LPS) fetal bovine serum (FBS), 15% (v/v) filter-sterilized L-cell conditioned medium, and 1% (w/v) penicillin–streptomycin/1% (w/v) glutamine at 37° under 8% CO_2. Cells are then stimulated with interferon γ (IFN-γ; 20 units/ml) and LPS (1 μg/ml) for 16 hr. RNA is prepared from stimulated macrophages. (*Note:* We have found that the most critical step in library construction is the careful preparation of RNA.)

[25] S. Fields and O. Song, *Nature (London)* **340,** 245 (1989).

cDNA prepared with Moloney murine leukemia virus (Mo-MuLV) reverse transcriptase and oligo(dT)–primer is inserted into the HybriZAP-Gal4-AD λ vector (Stratagene, La Jolla, CA). The resultant pGal4-AD–cDNA library plasmids encode fusion proteins consisting of the GAL4 activation domain and random macrophage polypeptides. The titer of the primary unamplified two-hybrid cDNA library is estimated as 1.6×10^6 plaque-forming units (PFU). The two-hybrid cDNA library is first excised from λ phage in the presence of ExAssist filamentous phage (Stratagene). The library is then amplified by introduction into XLOLR *Escherichia coli* cells to a final titer of 3.3×10^9 colony-forming units (CFU). This produces a library of plasmids containing cDNA fragments upstream of GAL4-AD. The plasmids encode a fusion polypeptide consisting of a library of polypeptides fused to GAL4-AD.

One critical step in designing interaction experiments is to choose the appropriate target. Using the entire protein as a bait may not be the optimal strategy. For example, a screen of proteins interacting with the entire NOS2 polypeptide as bait ultimately yielded only calmodulin, which has a high affinity for NOS2 (our unpublished data, 2000). However, a screen for proteins interacting with selected domains of NOS2 yields other polypeptides but not calmodulin, as described below.

We have constructed a bait plasmid by ligating fragments of NOS2 3′ to the GAL4-BD. Murine NOS2 cDNA sequences (1–549 or 1–210 bp) are prepared by the polymerase chain reaction (PCR) and inserted 3′ to the cDNA for the GAL4-binding domain of plasmid pGal4-BD (Stratagene). The resultant pGal4-BD–NOS2 bait plasmids encode a fusion protein composed of the GAL4-binding domain and an amino-terminal fragment of NOS2 (amino acid residues 1–183 or 1–70). This produces a plasmid encoding a fusion polypeptide consisting of NOS2 fragments fused to GAL4-BD.

We have used the two sets of plasmids to screen for polypeptides that interact with the amino-terminal domain of NOS2. YRG-2 yeast cells are cotransformed with pGal4-BD–NOS2 and the HybriZAP-Gal4-AD cDNA library, and grown on a selective medium lacking tryptophan, leucine, histidine, or combinations thereof. Colonies that contain cDNA encoding target library proteins interacting with the bait fusion protein are identified by transcription of the *HIS3* and *lacZ* genes. In our experiment, 1.4×10^6 yeast transformants were placed under selection. Approximately 50 cotransformed yeast clones were able to synthesize histidine. These 50 were screened for the ability to make β-galactosidase (β-Gal), revealing 18 clones that could synthesize both histidine and β-Gal. Plasmids from these 18 β-Gal-positive yeast colonies were isolated and retransformed into competent *E. coli* DH5α cells.

Analysis of these plasmids revealed that four clones contained overlapping cDNA sequences encoding a 110-kDa protein, which we called 110-kDa NOS-associated protein (NAP-110).

In Vitro Analysis of Interacting Domains

In vitro analysis of protein–protein interactions can be used to confirm interactions detected in yeast.[26] *In vitro* analysis can also be used to determine the domains of proteins responsible for interacting with other proteins.

Expression of polypeptide fragments in bacteria is technically challenging when the goal is expression of large polypeptides (greater than approximately 50 kDa). One particular challenge is to induce correct folding of the expressed polypeptide. To produce protein that is folded correctly, we express proteins in bacteria at lower temperatures and for longer times than is usually reported.[27–29]

We examine the ability of iNOS to interact *in vitro* with kalirin.[5] Bacterial expression vectors are constructed that express a hexahistidine [(His)$_6$]–iNOS fusion protein or a glutathione-*S*-transferase (GST)–kalirin (amino acid residues 447–1124) fusion protein. We have used several expression vectors, including the pET system (Novagen, Madison, WI) and pSG04.[29] JM109 bacteria are transformed and plated on Luria–Bertani (LB)–ampicillin plates, fresh colonies are isolated and grown in LB medium overnight, and then inoculated into 1 liter of LB–ampicillin. The bacteria are grown until the optical density (OD) at 600 nm > 0.6, and then induced with isopropyl-β-D-thiogalactopyranoside (IPTG) and grown at 16° for 72 hr. The bacteria are then harvested by centrifugation and sonicated, and then the supernatant is cleared by ultracentrifugation. The proteins are then purified by metal chelate column chromatography. Purified proteins are then mixed, and agarose–glutathione pulldown assays are used to demonstrate interactions. For example, we have used this method to show that (His)$_6$–NOS2 interacts with a GST–kalirin fusion polypeptide (but does not interact with GST alone).[5]

In Vivo Confirmation of Nitric Oxide Synthase-Associated Proteins

Sensitive screening methods such as the yeast two-hybrid system may uncover potential interactors that are actually false positives. Therefore all positive results obtained by the yeast two-hybrid system must be screened by other methods to assess the *in vivo* validity of the interaction. We screen all putative interactors for interactions with NOS2 in mammalian cells, using coimmunoprecipitation techniques.[5,10] RAW 264.7 macrophages are stimulated with LPS and IFN to induce NOS2. Cells are suspended in lysis buffer for 30 min on ice. Supernatants are recovered by centrifugation at 15,000g for 15 min at 4°, and 500 μl

[26] J. Estojak, R. Brent, and E. A. Golemis, *Mol. Cell. Biol.* **15**, 5820 (1995).
[27] E. A. Richard, S. Ghosh, J. M. Lowenstein, and J. E. Lisman, *Proc. Natl. Acad. Sci. U.S.A.* **94**, 14095 (1997).
[28] S. Ghosh, T. Pawelczyk, and J. M. Lowenstein, *Protein Expr. Purif.* **9**, 262 (1997).
[29] S. Ghosh and J. M. Lowenstein, *Gene* **176**, 249 (1996).

of supernatant is incubated with 10 μl of normal rabbit serum for 30 min, and then incubated with 50 μl of protein A–Sepharose 4B for 30 min. After centrifugation, 500 μl of supernatant is incubated for 3 hr with primary antibodies, and then with 40 μl of a 50% (v/v) suspension of protein A–Sepharose 4B (or goat anti-mouse agarose) for 4 hr at 4°, and washed three times with 1 ml of cold 20 mM Tris-HCl (pH 7.4), 125 mM NaCl, 1 mM Na$_3$VO$_4$, 50 mM NaF, 1 mM EDTA, 0.2% (v/v) Triton X-100, 0.2 mM phenylmethylsulfonyl fluoride (PMSF). Samples are boiled with sodium dodecyl sulfate (SDS) and 2-mercaptoethanol, fractionated by SDS–polyacrylamide gel electrophoresis (PAGE), and transferred onto polyvinylidene difluoride (PVDF) membranes. Membranes are incubated for 1–2 hr at room temperature with antibodies to iNOS2, washed, incubated for 1–2 hr with goat anti-mouse or goat anti-rabbit antibody coupled to horseradish peroxidase, and visualized with an enhanced chemiluminescence (ECL) reagent (Amersham, Arlington Heights, IL).

Coimmunoprecipitation experiments reveal that NOS2 physically associates with NAP-110 in mammalian cells. Negative controls include noninduced RAW cells. Additional negative controls include peritoneal macrophages isolated from NOS2 null mice. These experiments confirm the data derived from experiments in yeast: NAP-110 interacts with NOS2.

Functional Relevance of Nitric Oxide Synthase-Associated Proteins

To determine whether NOS2-associated proteins regulate NOS2 activity, we cotransfect HeLa cells with expression plasmids for NOS2 and expression plasmids for interacting proteins.[5,10] Accordingly, we next explored the effect of NAP-110 interactions with NOS2.

HeLa cells are transiently transfected with pCI-mu-NOS2 or pcDNA3.1-NAP110 or both plasmids. Cells are maintained in minimal essential medium (MEM) supplemented with 10% FBS (Gemini Biotech, The Woodlands, TX), penicillin, and streptomycin. Briefly, 2.5×10^6 cells are plated in 100-mm tissue culture dishes 1 day before transfection. The cells are washed twice with Opti-MEM (Life Technologies, Rockville, MD) and incubated for 4 hr with 1 ml of Opti-MEM containing 5 μg of DNA and 1 μl of LipofectAMINE. The DNA–LipofectAMINE solution is then aspirated, 3 ml of Opti-MEM is added, and the cells are incubated overnight at 37° and then harvested. Cell lysates are analyzed for NOS activity, using the arginine-to-citrulline conversion assay previously described by us.[30]

Cotransfection experiments reveal that NAP inhibits NOS2 activity (Fig. 1). We next explored whether NAP inhibits NOS2 by regulating NOS2 homodimerization. Cell lysates are prepared from cells transfected with NOS2 expression vectors alone, or from cells cotransfected with expression vectors for NOS2 and expression

[30] C. J. Lowenstein and S. H. Snyder, *Methods Enzymol.* **233**, 264 (1994).

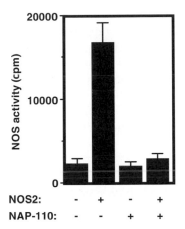

FIG. 1. NOS-interacting protein inhibits NOS activity. HeLa cells were transfected with empty vector alone, vector expressing NOS2 alone, vector expressing NAP-110 alone, or vector expressing NOS2 plus vector expressing NAP-110. Lysates of these cells were assayed individually for NOS2 activity.

vectors for NAP-110. Lysates are fractionated on a gel-filtration column in the following manner. Lysates of cells are loaded onto a Superdex 200 chromatography column, equilibrated and eluted with 20 mM Tris-HCl (pH 7.4), 125 mM NaCl, 1 mM Na$_3$VO$_4$, 50 mM NaF, 1 mM EDTA, 0.2% (v/v) Triton X-100, 0.2 mM PMSF buffer, and 1-ml fractions are collected. Molecular weight markers are analyzed separately to determine approximate molecular weights of substances in the eluted fractions. Aliquots are then electrophoresed on a

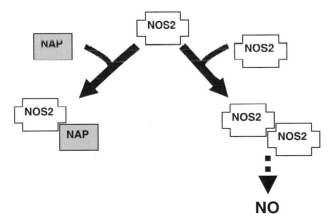

FIG. 2. NAP interacts with NOS2 monomers and prevents NOS2 homodimerization and NO synthesis.

denaturing gel and then immunoblotted, using antibody to NOS2. NOS2 is normally found in cells as a monomer (approximate M_r 140,000) and as a homodimer (approximate M_r 280,000). However, coexpression of NAP-110 alters the relative mobility of NOS2: no NOS2 monomers are detected, and a novel NAP–NOS2 complex is detected—and the proportion of NOS2 homodimers are greatly decreased.

These experiments demonstrate that NAP-110 interacts with NOS2: NAP-110 inhibits NOS2 by inhibiting formation of the active NOS2 homodimer (Fig. 2).

Conclusions

The use of the yeast two-hybrid system has facilitated the identification of polypeptides that potentially interact with the NOS isoforms. Because the yeast two-hybrid system can produce false positives, demonstration of an interaction in mammalian cells is necessary. The functional consequences of any protein interacting with NOS can be explored by assessing NOS localization, NOS activity, and NOS dimerization in the absence and presence of the interacting protein. Identification of the biological effects of NOS interactors may reveal novel physiological mechanisms of regulating NOS.

[22] Detection of Redox Sensor of Ryanodine Receptor Complexes

By WEI FENG and ISAAC N. PESSAH

Introduction

A prominent property of the three genetic isoforms of ryanodine receptors (RyRs) is their exquisite sensitivity to functional modification by sulfhydryl-modifying reagents.[1] Sensitivity to sulfhydryl-modifying agents appears to be a general property of Ca^{2+} channels targeted to sarcoplasmic and endoplasmic reticulum (SR/ER), including all three genetic isoforms of RyR and the inositol 1,4,5-trisphosphate receptor (IP_3R). A large number of chemically dissimilar sulfhydryl-oxidizing, -reducing, and -arylating reagents have been used to activate and inhibit RyR channel activity. The net influence of sulfhydryl modification is highly dependent on the concentration of the reagent utilized, the length of time the reaction is permitted to proceed, and the nature of the chemical reaction the

[1] I. N. Pessah and W. Feng, *Antioxidants Redox Signal.* **2**, 17 (2000).

Copyright 2002, Elsevier Science (USA).
All rights reserved.
0076-6879/02 $35.00

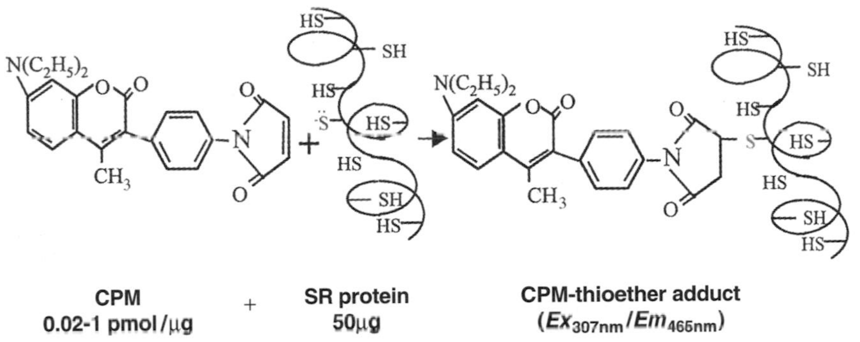

CPM   +   SR protein   CPM-thioether adduct
0.02-1 pmol/μg   50μg   (Ex_{307nm}/Em_{465nm})

FIG. 1. CPM forms a highly fluorescent thioether adduct with protein thiols by Michael addition. For clarity a single adduct is shown. However, by severely limiting the mole ratio of CPM to protein in the reaction (i.e., 0.02–1 pmol/μg), CPM adducts will form rapidly with the most reactive class of thiol, if present (shown as –S– in the example).

reagent undertakes with sulfhydryl groups. What has proved most challenging is to gain understanding of how specific sulfhydryl moieties ascribe specific aspects of channel function, and to ascribe their physiologic or pathophysiologic importance. In this respect, one methodological strategy has aimed to define whether RyR complexes possess a class of cysteine moieties that can be distinguished on the basis of their exceptional chemical reactivity[2] and to understand their contribution to channel function. The following sections present detailed biochemical and functional methods that have been employed to detect the presence of hyperreactive sulfhydryl moieties within RyR complexes and to define their role in redox sensing.

Measurement and Localization of Hyperreactive Sulfhydryls Associated with Junctional Sarcoplasmic Reticulum

CPM [7-diethylamino-3-(4'-maleimidylphenyl)-4-methylcoumarin; Molecular Probes, Eugene, OR] is a membrane-permeable coumarin maleimide that readily undergoes Michael addition with protein sulfhydryl residues.[3] CPM has low intrinsic fluorescence until it forms a thioether adduct on protein cysteines (Fig. 1). Although CPM lacks ligand specificity for a particular binding domain on the RyR complex, its chemical properties impart high specificity toward cysteines within both hydrophilic and hydrophobic regions of structure. Therefore CPM, when utilized with isolated SR/ER membrane vesicles, will readily form fluorescent CPM–thioether adducts with cysteines residing within both accessible (e.g., cytoplasmic) and inaccessible (e.g., lumenal) domains of intrinsic membrane proteins.

[2] G. Liu, J. J. Abramson, A. C. Zable, and I. N. Pessah, *Mol. Pharmacol.* **45,** 189 (1994).
[3] T. O. Sipple, *J. Histochem. Cytochem.* **29,** 314 (1981).

ST   LSR   JSR

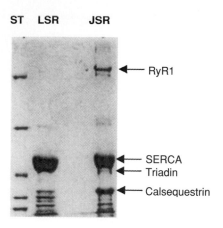

FIG. 2. SDS–PAGE pattern of light and junctional SR fractions isolated by sucrose density gradient centrifugation, using the method of A. Saito, S. Seiler, A. Chu, and S. Fleischer, *J. Cell Biol.* **99,** 875 (1984).

However, we have found that when [CPM] is limiting (0.02–1 pmol/μg SR) the kinetics of thioether adduct formation differ significantly with protein composition of the SR (light SR vs. junctional SR; see below) and the composition of the assay medium (presence of RyR activators or inhibitors).

Isolation of Junctional Sarcoplasmic Reticulum Membranes: Importance of Redox-Buffered Solutions

The most commonly used experimental preparation consists of membrane vesicle fractions isolated from either fast twitch skeletal muscle of New Zealand White rabbit or cardiac muscle from the right ventricle obtained from a variety of species (rat, mouse, rabbit, sheep, etc.). For example, skeletal membranes derived primarily from junctional regions of SR (JSR) and enriched in ryanodine receptor type 1 (RyR1) and its associated proteins are routinely isolated by sucrose density gradient centrifugation within the 38–45% interface as described in detail by Saito *et al.*[4] and Chu *et al.*[5] A light SR (LSR) membrane fraction enriched in SR/ER Ca^{2+} ATPase (SERCA) pump but lacking RyR1 complex can also be isolated from the 32–34% sucrose interface from the same five-step gradient. Sodium dodecyl sulfate–polyacrylamide gel electrophoresis (SDS–PAGE) analysis reveals a clear distinction between the protein composition of the two fractions (Fig. 2). Junctional SR enriched in RyR2 complex and longitudinal SR deficient in RyR2 complex can be isolated from heart muscle in a similar manner.[6]

---

[4] A. Saito, S. Seiler, A. Chu, and S. Fleischer, *J. Cell Biol.* **99,** 875 (1984).

[5] A. Chu, M. C. Dixon, A. Saito, S. Seiler, and S. Fleischer, *Methods Enzymol.* **157,** 36 (1988).

[6] M. Inui, S. Wang, A. Saito, and S. Fleischer, *Methods Enzymol.* **157,** 100 (1988).

We have included reduced and oxidized glutathione (GSH and GSSG, respectively) in the initial skeletal muscle homogenization solution to buffer its redox potential.[7] GSH (3.5 mM) and GSSG (59 μM) are added to homogenization solution consisting of 0.3 M sucrose, 5 mM imidazole-HCl (pH 7.4), 100 μM phenylmethylsulfonyl fluoride (PMSF), and leupeptin (10 μg/ml) to give a redox potential (E_h) of -0.22 V (see below for calculation of redox potential). This redox potential mimicks the typical cytoplasmic redox potential *in vivo* and is likely to prevent overoxidation of the RyR complex during the early stages of membrane isolation. The redox buffer is included only in the initial homogenization and first 11,000g pellet wash; it is omitted in subsequent steps. The inclusion of a redox buffer in the initial homogenization solution yields JSR preparations that possess a high fraction of reconstituted channels exhibiting a low open probability (low-P_o) gating mode (>40%) that can be tightly regulated by transmembrane redox potential. By comparison, conventional preparations lacking redox buffer possess a high fraction of reconstituted channels exhibiting a high-P_o gating mode whose behavior appears to stem from overoxidization of protein thiols in the isolation protocol.[7–10] This observation is consistent with a report that a physiological concentration of GSH partially protects the RyR1 complex from oxidation in room air. Under ambient oxygen tension (PO_2 ~150 mmHg) RyR1 loses about six free thiols per subunit compared with when it is under physiological PO_2 (~10 mmHg, normally found in muscle).[10] Therefore the high-P_o gating mode more frequently observed with conventional JSR preparations may actually represent channels in an overoxidized state.

Kinetic Measurement of Hyperreactive Sulfhydryls of
   Junctional Sarcoplasmic Reticulum, Using
   7-Diethylamino-3-(4′-maleimidylphenyl)-4-methylcoumarin

CPM is dissolved in dry dimethyl sulfoxide (DMSO) at a final stock concentration of 100 μM, and 50-μl aliquots are divided into opaque plastic microcentrifuge tubes and stored for up to 1 year at $-20°$. When needed, an aliquot of stock is diluted serially 100-fold with DMSO just before use. All stocks are protected from light by wrapping the container with aluminum foil. The kinetics of CPM–thioether adduct formation with JSR or LSR are determined with a fluorimeter equipped with a 3-ml cuvette holder, which can provide precise control of temperature and stirring of the contents [e.g., SLM (Urbana, IL) model 8000 or Hitachi (Tokyo, Japan) model F-2000]. Excitation and emission are set at 397 and 465 nm (width of slit, 4 nm), respectively. It is advantageous to interface the

[7] W. Feng, G. Liu, P. D. Allen, and I. N. Pessah, *J. Biol. Chem.* **275,** 35902 (2000).

[8] J. J. Marengo, C. Hidalgo, and R. Bull, *Biophys. J.* **74,** 1263 (1998).

[9] T. Murayama, T. Oba, E. Katayama, H. Oyamada, K. Oguchi, M. Kobayashi, K. Otsuka, and Y. Ogawa, *J. Biol. Chem.* **274,** 17297 (1999).

[10] J. P. Eu, J. Sun, L. Xu, J. S. Stamler, and G. Meissner, *Cell* **102,** 499 (2000).

instrument with a computer possessing data acquisition and analysis software. In our laboratory, the Hitachi F-2000 captures fluorescence data at 1 Hz. The data files are exported for nonlinear regression analysis by Origin 6.0 software (Microcal Software, Northampton, MA).

To quantify the kinetics of CPM–thioether adduct formation, SR membrane vesicles (50 μg of protein per milliliter) are incubated in a solution consisting of 100 mM KCl and 20 mM 3-(N-morpholino)propanesulfonic acid (MOPS), pH 7.0, at 37°. Modulators of RyR channel activity are added to the assay as small aliquots from a $\geq$100× stock, using a Hamilton syringe, and permitted to equilibrate for 1 min. CPM is quickly added by Hamilton syringe through the opening in the cover to the spectrofluorometer directly into the center of the stirring sample to give a final concentration of 1–50 nM. Under these conditions (0.02–1 pmol of CPM per microgram of SR protein), the concentration of free SR sulfhydryls greatly exceeds that of CPM, and permits direct analysis of highly reactive cysteine residues under the conditions that enhance or inhibit RyR channel activity.[2,10,11] Figure 3 shows that the modulation of the rate of thioether adduct formation with JSR is highly dependent on whether a channel inhibitor (2 mM Mg^{2+}) or a channel activator (50 μM Ca^{2+}) is included in the assay buffer. By contrast, LSR deficient in RyR complex is insensitive to buffer conditions and exhibits only slow kinetics for forming CPM–thioether adducts regardless of whether channel-activating or -inhibiting conditions are present (Fig. 3, bottom). SDS–PAGE analysis of CPM-labeled skeletal JSR quenched within 1 min of initiation of the reaction reveals that RyR1 and triadin form CPM–thioether adducts primarily during the rapid phase (when channel conditions favor the closed state). When channel activation is favored, the slow phase of adduct formation is primarily on the abundant SERCA pump (Fig. 3, top, inset).

The time course of the increase in fluorescence intensity (F_t) obtained under conditions promoting channel closure (millimolar Mg^{2+} or Ca^{2+} buffered to <100 nM by EGTA^2) or channel activation (in the presence of micromolar Ca^{2+} or redox-active quinone[12]) is fitted with single or multiexponentials, respectively, leading to the corresponding time constants (k) from which apparent half-times ($t_{1/2}$) are calculated (Fig. 3, top). The rate constant (k) was considered to be proportional to the number of free sulfhydryl groups available for CPM conjugation (i.e., $k = k_m[\text{SH}]_t$).[2] A unique feature of the hyperreactive thiol moieties is that Ca^{2+} and ryanodine, known to activate and inhibit channel activity depending on concentration, allosterically influence the rate of adduct formation in a biphasic manner.[2] Studies reveal the existence of a transmembrane redox sensor within the RyR1 channel complex that confers tight regulation of channel activity in response to changes in transmembrane redox potential produced by cytoplasmic and lumenal

[11] G. Liu and I. N. Pessah, *J. Biol. Chem.* **269**, 33028 (1994).
[12] W. Feng, G. Liu, R. Xia, J. J. Abramson, and I. N. Pessah, *Mol. Pharmacol.* **55**, 821 (1999).

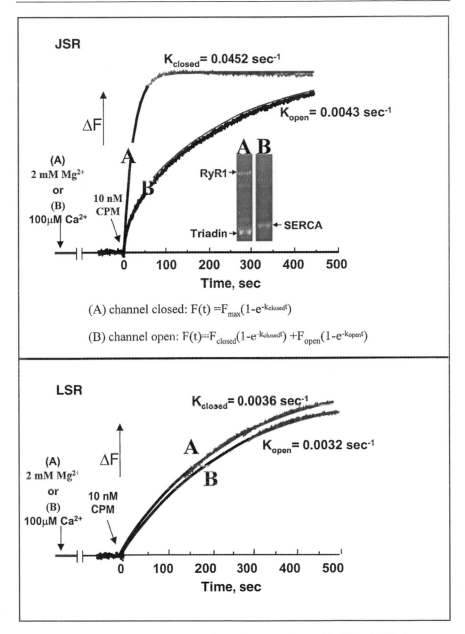

FIG. 3. Kinetic measurement of forming CPM–thioether adducts with JSR and LSR. Data traces show that JSR rapidly forms an adduct with CPM in the presence of physiologic channel inhibitor Mg^{2+} (*top*) whereas LSR exhibits only slow adduct formation regardless of which modulator is present (*bottom*). *Inset, top:* SDS–PAGE separation of JSR labeled for 1 min with 10 nM CPM in the presence of either (A) 2 mM Mg^{2+} or (B) 100 μM Ca^{2+}.

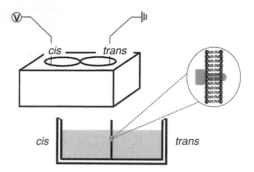

FIG. 4. Schematic representation of the bilayer chambers with incorporated single channel in planar phospholipid membrane.

glutathione.[7] The hyperreactive cysteine moieties constitute the essential component of the redox sensor, which conveys information about localized changes in redox potential produced by physiologic (e.g., glutathione) and pathophysiologic (e.g., quinones) channel modulators of the Ca^{2+} release process.[10,12,13]

Measurement of Transmembrane Redox Sensor of Channels Reconstituted in Bilipid Layer Membrane

The bilayer lipid membrane (BLM) preparation affords precise control of the redox state on both the cytoplasmic and lumenal faces of the reconstituted channel. Adjustment of the GSH : GSSG ratio to form various redox potentials on either side of the membrane has turned out to be a powerful approach for detecting the RyR redox sensor. Our modified method for preparing junctional SR (i.e., including GSH/GSSH in the homogenization solution to buffer redox potential to ~0.22 V) protects against overoxidation of critical sulfhydryls, and thus provides a higher percentage of "intact" channels that monitor and functionally follow transmembrane redox fluctuation.

Incorporation of Ryanodine Receptor 1 in Planar Lipid Bilayer

Reconstitution of RyR1 into BLM and recording of the channel activity are performed at room temperature (24°). The typical BLM setup is composed of a chamber and a removable cup and can be purchased commercially (Warner, Hamden, CT) (Fig. 4). The bilayer cup possesses a small circular hole (200–300 μm) and can be snugly inserted into one of the machined holes of the chamber, thereby forming two compartments whose only electrical connection is the hole in the

[13] I. N. Pessah, C. Beltzner, S. W. Burchiel, G. Srihar, T. Penning, and W. Feng (2001). *Mol. Pharmacol.* **59,** 507 (2001).

cup. Typically Cs^+ in the form of CsCl is used as charge carrier, and the current is measured as a holding potential is applied to the *cis* (cytoplasmic) chamber of the reconstituted channel whereas the *trans* (lumenal) chamber is the virtual ground. In the standard commercially available BLM setup (e.g., Warner), the *cis* and *trans* chambers each have a 0.7-ml capacity. Specifically, two standard stock solutions are prepared and stored at 4° for up to 4 weeks:

cis solution: 500 mM CsCl, 100 μM CaCl$_2$, 20 mM HEPES; adjusted to pH 7.4

trans solution: 50 mM CsCl, 100 μM CaCl$_2$, 20 mM HEPES; adjusted to pH 7.4

Once the *cis* and *trans* solutions are in place (10 : 1 Cs^+ gradient), the BLM is made with 5 : 2 phosphatidylethanolamine : phosphatidylcholine (50 mg/ml in decane; Northern Lipids, Vancouver, BC, Canada). It is formed by adsorbing a small amount of the lipid solution on either a small sable brush trimmed to two or three bristles or a Pasteur pipette that has been fired to shape a small ball at its end. The brush or pipette is gently brought across the hole within the bilayer cup. Lipid solutions are made fresh daily and used once. To reconstitute a channel, the *cis* side is held to a positive potential relative to the *trans* (ground) side, which is favorable for JSR vesicle fusion with the BLM. On fusion of JSR with BLM, the *cis*-to-*trans* Cs^+ gradient is reversed to 1 : 10, by perfusing the *cis* chamber with 15 volumes of 50 mM CsCl, 20 mM HEPES, pH 7.4 (no added CaCl$_2$), and the *trans* chamber with 500 mM CsCl, 20 mM HEPES, pH 7.4. A negative holding potential is applied to the *cis* side and single-channel activity is measured with a bilayer amplifier [e.g., Dagan (Minneapolis, MN) 3910 Expander]. The data are filtered at 1 kHz before being acquired at 10 kHz by a Digidata 1200 (Axon Instruments, Union City, CA). The data are analyzed by pCLAMP 6 (Axon Instruments) without additional filtering.

Preparation of Reduced and Oxidized Glutathione Stock Solutions

GSH is dissolved in degassed HEPES (20 mM) buffer, and the solution is adjusted to pH 7.0. Aliquots ($\sim$0.2 ml) are portioned into vials briefly gassed with argon and sealed. Vials can be stored at $-20°$ for up to 60 days without measurable oxidation of GSH. Each vial is used once. GSSG solutions are made and stored in a similar manner except that degassing and argon are omitted. When used as transmembrane glutathione redox buffer in bilayer lipid membrane (BLM) experiments, GSH and GSSG are sampled from both sides of the BLM chamber at the end of channel recordings to verify that the initial redox potential has not changed during the course of the experiments. Redox buffers have been found to be stable for at least 1 hr. GSH and GSSG contents are determined according to the method of Senft et al.,[14] using the fluorescence indicator o-phthalaldehyde (OPA).

Experimental Design to Obtain Redox Potential

The glutathione redox potential is calculated by the Nernst equation from the GSH : GSSG ratios and the total glutathione concentration according to

$$E_h = E_0 + 2.303 \times RT/Fn \times \log \frac{[GSSG]}{[GS^-]^2}$$

where E_h is the redox potential referred to the normal hydrogen electrode, V; E_0 is the standard potential of glutathione, -0.24 V; R is the gas constant, 8.31 J/deg · mol; T is the absolute temperature (K); n is the number of electrons transferred, $n = 2$ for SH–SS exchanges; F is the Faraday constant, 96,406 J/V; and [GSSG] and [GSH] are the molar concentrations of oxidized and reduced glutathione, respectively.

At the typical temperature for bilayer measurements (22°),

$$E_h = -0.24 + 29.28 \times \log \frac{[GSSG]}{[GS^-]^2}$$
$$[GS^-] = \left([GSSG]/10^{(E_h+0.24)/29.28}\right)^{1/2}$$
$$[GSSG] = [GS^-]^2 \times 10^{(E_h+0.24)/29.28}$$

Instillation of Glutathione Redox Gradient across Bilipid Layer Membrane

Single-channel studies have shown that glutathione used as an individual reducing (GSH) or oxidizing (GSSG) agent can inhibit or stimulate RyR gating activity, respectively.[15] However, this approach does not address the physiological importance of glutathione in the regulation of RyR channel function. In mammalian cells, the redox buffer system is composed of the glutathione couple, GSH and GSSG. Furthermore, the microsomal membrane within which RyR and IP$_3$R reside is normally subject to a large transmembrane redox potential difference with a high ratio of GSH to GSSG in the cytosol, $\geq 30:1$, and a much lower and stabilized ratio of 3 : 1 to 1 : 1 maintained within lumen.

These physiological conditions can be easily mimicked in the BLM preparation. A precise combination of GSH and GSSG can be instilled on each side of the BLM to precisely set the transmembrane glutathione redox potential. Furthermore, the redox potential in *trans* can be maintained at a physiologic -180 mV whereas the *cis* potential can be varied by perfusion of solutions possessing defined ratios of GSH to GSSG. Using this approach the typical reconstituted RyR1 channel in

[14] A. P. Senft, T. P. Dalton, and H. G. Shertzer, *Anal. Biochem.* **280**, 80 (2000).
[15] A. C. Zable, T. G. Favero, and J. J. Abramson, *J. Biol. Chem.* **272**, 7069 (1997).

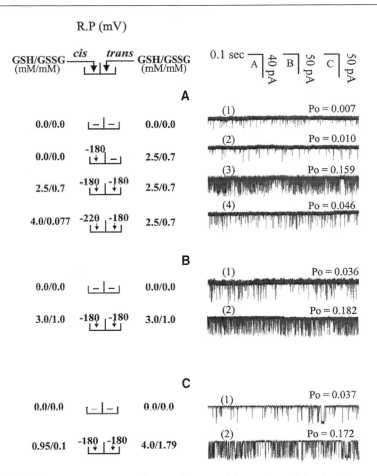

FIG. 5. RyR1 responses to transmembrane redox potential regardless of the glutathione concentration applied in the buffer. Free Ca^{2+} concentrations present in *cis* buffer are 7, 50, and 7 μM for channels in (A), (B), and (C), respectively. Holding potentials of -30, -40, and -40 mV were applied to channels shown in (A), (B), and (C), respectively. Current fluctuation is downward. R.P. (mV), Redox potential (in millivolts).

BLM exhibits a tight response to changes in transmembrane redox potential independent of absolute concentrations of GSH and GSSG. It is important to note that control of reconstituted RyR1 channels has been observed to be lost with high (non-physiologic) concentrations of glutathione (>5 mM GSH and GSSG). Figure 5 demonstrates the modulation of RyR1 channel activity by experimentally setting transmembrane redox potential. In Fig. 5A, the current through the RyR1 channel is recorded in the presence of 7 μM free Ca^{2+} at a holding potential of -40 mV. One second of representative current traces from a total of 2–5 min of recording is displayed. The channel in the absence of defined redox potential (without glutathione redox buffer on either side of the channel) displays a low open probability, P_o (trace 1). GSH (2.5 mM) and GSSG (0.7 mM) introduced into the *cis* solution give a redox potential of -180 mV (calculated according to the Nernst equation), and has a negligible influence on channel activity (trace 2). Subsequent introduction of the same GSH : GSSG ratio (approximate redox potential of -180 mV) into the *trans* chamber significantly increases channel P_o (trace 3). Perfusion of the *cis* side with a reduced redox potential of -220 mV (4.0 mM GSH and 77 μM GSSG) results in a dramatic decrease in P_o (trace 4). Figure 5B and C show that the RyR channel is activated on instillation of a symmetric -180 mV transmembrane redox potential, and is independent of the individual GSH and GSSG concentration used in the *cis* and *trans* chambers to obtain -180 mV (compare trace 1 with trace 2 in Fig. 5B and C).

Measurement of Redox Sensing of Ryanodine Receptor Channel in High Open Probability Gating Mode

Studies have revealed that RyR1 channels in a high-P_o gating mode reflect an overoxidized state of the protein. Even though glutathione is included in the initial homogenization buffer to avoid this potential problem, critical free sulfhydryls are likely to be oxidized in the room air during subsequent handling of JSR preparations, and frequently channels exhibiting a high-P_o gating mode are observed. With many but not all channels, instillation of transmembrane redox potential resets the redox sensor of high-P_o type channels. Figure 6 shows examples of two channels, one that recovers redox sensing and another that does not. Figure 6A shows a typical high-P_o channel in the presence of 50 μM *cis* Ca^{2+} but with undefined transmembrane redox potential (P_o of 0.637; trace 1). Instillation of a relatively reduced redox potential gradient *cis* to *trans* (-220 to -180 mV, respectively) immediately results in a low-P_o gating mode (P_o of 0.031; trace 2, Fig. 6A). A subsequent change in the *cis/trans* transmembrane redox potential to $-180/-180$ mV clearly demonstrated recovery of the transmembrane redox sensor (P_o increases to 0.125; trace 3, Fig. 6A).

By contrast, Fig. 6B shows another channel exhibiting the high-P_o gating mode ($P_o = 0.320$ with 40 μM *cis* Ca^{2+}; trace 1). Introducing a *cis/trans* redox potential of -220 mV/-180 mV does not affect channel gating activity ($P_o = 0.32$; trace 2, Fig. 6B). The lack of responsiveness to transmembrane redox potential exhibited by this type of channel may reflect the fact that the critical (hyperreactive) cysteines that are an essential component of the redox sensor may be damaged by overoxidation to disulfides. Addition of a low concentration of redox-active 1,4-naphthoquinone (NQ) has been shown to produce time-dependent biphasic actions on RyR gating activities. NQ activates RyR channels rapidly (a reversible effect) followed in time by complete inactivation (an irreversible effect). It is

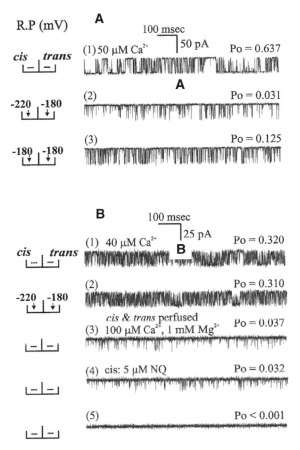

FIG. 6. Responses of the high P_o gating mode channel to transmembrane redox signaling. In (A), after instillation of transmembrane redox potential, the *trans* side is held constant at -180 mV potential by 4.0 m*M* GSH and 1.792 m*M* GSSG. On the *cis* side, -220 mV is formed by 4.0 m*M* GSH and 0.077 m*M* GSSG (trace 2), and -180 mV is formed by 4.0 m*M* GSH and 1.792 m*M* GSSG. In (B), -220 mV on the *cis* side is formed by 4.0 m*M* GSH and 0.077 m*M* GSSG, and -180 mV on the *trans* side is formed by 4.0 m*M* GSH and 1.792 m*M* GSSG. The voltage holding potential is held at -30 and -40 mV for (A) and (B), respectively.

probable that NQ initially enhances channel activity through a redox-sensing mechanism whereas the subsequent effects are the result of protein arylation.[12] The unique response of channels exhibiting the high-P_o gating mode to NQ can be tested by perfusing both sides of the BLM with buffer lacking GSH and GSSG (100 μ*M* Ca^{2+}, 1 m*M* Mg^{2+} in *cis*, $P_o = 0.037$; trace 3, Fig. 6B). Addition of 5 μ*M* NQ reveals that high-P_o channels that fail to respond to glutathione transmembrane redox potential also fail to exhibit an initial activation phase with

NQ (P_o = 0.032; trace 4, Fig. 6B). These channels, however, are eventually inhibited irreversibly ($P_o < 0.001$; trace 5, Fig. 6B). Taken together these results suggest that redox-sensing cysteine moieties within the RyR1 complex are easily overoxidized (probably to disulfides) with an attendant loss of the redox sensor.

Association of Hyperreactive Sulfhydryls with Ryanodine Receptor Transmembrane Redox Sensor

As shown above, conditions promoting the closed state of the channel permit the rapid formation of thioether adducts between CPM and RyR thiols. Whether these hyperreactive thiols are an essential component of the redox sensor can be tested. A single channel is reconstitued in BLM and tested for redox sensing as described above. Nanomolar CPM is then added to the *cis* chamber under low-P_o conditions to promote CPM arylation of hyperreactive thiols. Although CPM does not appear to alter channel P_o and gating kinetics, it does disable the transmembrane

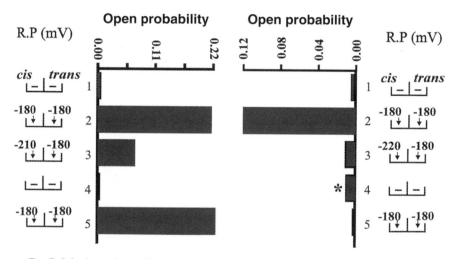

FIG. 7. Selective arylation of hyperreactive thiols by CPM disables the transmembrane redox sensor. Recording of channel activity starts from undefined transmembrane redox potential in the presence of 7 μM Ca^{2+} (bar 1, both panels). In left panel [GSH]/[GSSG] = 3 mM/1 mM (giving −180 mV redox potential) is symmetrically applied into both *cis* and *trans* (bar 2); subsequent addition of GSH 9.72 mM introduces −210 mV into *cis* (bar 3); extensive perfusion of both *cis* and *trans* is performed to remove transmembrane redox potential (bar 4) followed by reestablished with [GSH]/[GSSG] = 3 mM/1 mM to both *cis* and *trans* (bar 5). In right panel [GSH]/[GSSG] of 0.95 mM/0.10 mM, and 4.0 mM/1.79 mM are introduced into the cytoplasmic chamber for −180 mV in *cis* and *trans*, respectively (bar 2). Further reduction of *cis* to −220 mV is made by addition of [GSH]/[GSSG] to 4.55 mM/0.1 mM (bar 3). After perfusion of both chambers to removed GSH and GSSG, CPM (20 nM) is introduced into *cis* for 2 min before being terminated by perfusion (bar 4, marked with an asterick). Reinstillation of [GSH]/[GSSG] is made with 0.95 mM/0.10 mM and 4.0 mM/1.79 mM in *cis* and *trans*, respectively (symmetrical −180 mV; bar 5).

redox sensor.[7] Figure 7 shows P_o graphs of two separate channels before and after labeling with CPM. Both channels respond strongly to symmetric -180 mV transmembrane redox potential and are negatively regulated by a $-220/-180$ mV cis/trans gradient [compare Fig. 7, bars 1–3 (left) and P_o graphs (right)]. After exposing the channels to 20 nM CPM for 2 min [Fig. 7, bar 4 (marked with an asterisk), CPM removed from the bath by extensive perfusion of the chamber with 25 volumes of buffer], neither channel responds to a $-180/-180$ mV symmetric cis/trans redox gradient (compare bars 4 and 5 in Fig. 7).

Acknowledgments

This work was supported by NIH Grants 2RO1AR43140 and 1POAR17605.

[23] Redox Control of 20S Proteasome

By Bertrand Friguet, Anne-Laure Bulteau, Mariangela Conconi, and Isabelle Petropoulos

Introduction

Oxidative modifications of proteins have been implicated in age- and disease-related impairment of cellular functions and are known to affect protein turnover.[1–3] Because the 20S proteasome has been shown to be the major actor in the degradation of oxidized protein[4] and consequently to be important in the regulation of the steady state level of altered proteins in the cell, the fate of proteasome subjected to oxidative processes has deserved specific attention.[5–9] On oxidative stress, an increase in intracellular proteolysis of oxidized protein is well documented in different cell systems although no upregulation of proteasome subunits synthesis has

[1] B. S. Berlett and E. R. Stadtman, *J. Biol. Chem.* **272,** 20313 (1997).

[2] T. Grune, T. Reinheckel, and K. J. Davies, *FASEB J.* **11,** 526 (1997).

[3] B. Friguet, A. L. Bulteau, N. Chondrogianni, M. Conconi, and I. Petropoulos, *Ann. N.Y. Acad. Sci.* **908,** 143 (2000).

[4] T. Grune, T. Reinheckel, M. Joshi, and K. J. Davies, *J. Biol. Chem.* **270,** 2344 (1995).

[5] M. Conconi, L. I. Szweda, R. L. Levine, E. R. Stadtman, and B. Friguet, *Arch. Biochem. Biophys.* **331,** 232 (1996).

[6] P. R. Strack, L. Waxman, and J. M. Fagan, *Biochemistry* **35,** 7142 (1996).

[7] M. Conconi and B. Friguet, *Mol. Biol. Rep.* **24,** 45 (1997).

[8] M. Conconi, I. Petropoulos, I. Emod, E. Turlin, F. Biville, and B. Friguet, *Biochem. J.* **333,** 407 (1998).

[9] T. Reinheckel, N. Sitte, O. Ullrich, U. Kuckelkorn, K. J. Davies, and T. Grune, *Biochem. J.* **335,** 637 (1998).

Copyright 2002, Elsevier Science (USA).
All rights reserved.
0076-6879/02 $35.00

been reported.[2] This increase in proteasome-dependent proteolysis is likely the result of an increased proteolytic susceptibility of oxidized proteins, although a transient activation of the proteasomal system cannot be excluded.[6,8] The proteasomal system is made of a catalytic core, the 20S proteasome (EC 3.4.25.1), and several regulatory components that affect its specificity and activity.[10,11] The 20S proteasome is a high molecular weight (700,000) multicatalytic proteinase found both in the cytosol and nucleus of eukaryotic cells. The eukaryotic proteasome is made up of 28 (2 × 14) different subunits arranged as 4 stacked rings, each containing 7 subunits. The α-type subunits form the outer rings and the β-type subunits form the inner rings; three of the β-type subunits have an N-terminal threonine that is critical for proteolytic activity.[12,13] Indeed, the hydroxyl group acts as a nucleophile in the active site and defines a new type of protease as compared with serine and cysteine proteases. The 20S proteasome is characterized by three main proteolytic activities with distinct specificities against short synthetic peptides: the trypsin-like activity (which cleaves after basic residues like arginine or lysine) expressed by the Z and MECL1 subunits, the chymotrypsin-like activity (which cleaves after large hydrophobic residues such as tyrosine or phenylalanine) expressed by the X, LMP2, and LMP7 subunits, and the peptidylglutamyl peptide-hydrolyzing activity (which cleaves after acid residues such as glutamic acid) expressed by the Y subunits.[14] Exposure to metal-catalyzed oxidation of purified 20S proteasome *in vitro* or of rat hepatoma cells results in the inactivation of certain peptidase activities.[5,8] However, the alterations in peptidase activities observed *in vitro* depend on whether the purified proteasome is in its latent or active form before treatment with the reactive oxygen species-generating system.[8] Appropriate protocols aimed at isolating the 20S proteasome from different sources, as well as methods for promoting the conversion from latent to active proteasome, are described. Different ways of monitoring oxidation-mediated functional and structural changes of the 20S proteasome, either purified or within tissue and cell homogenates, are then presented.

20S Proteasome Isolation

When assayed in crude homogenates, it is not always simple to distinguish between the proteasome peptidase activities and other intracellular protease activities. Even by using specific proteasome inhibitors that permit selectively measurement of its peptidase activities, these activities may still be affected by the presence of endogenous inhibitors, activators, and competing denatured protein substrates. In addition, the analysis of structural modifications of the 20S

[10] O. Coux, K. Tanaka, and A. L. Goldberg, *Annu. Rev. Biochem.* **65,** 801 (1996).

[11] G. N. DeMartino and C. A. Slaughter, *J. Biol. Chem.* **274,** 22123 (1999).

[12] J. Lowe, D. Stock, B. Jap, P. Zwickl, W. Baumeister, and R. Huber, *Science* **268,** 533 (1995).

[13] M. Groll, L. Ditzel, J. Lowe, D. Stock, M. Bochtler, H. D. Bartunik, and R. Huber, *Nature (London)* **386,** 463 (1997).

[14] M. Orlowski and S. Wilk, *Arch. Biochem. Biophys.* **383,** 1 (2000).

proteasome can be conducted only after its isolation. Depending on the source and the amount of starting biological material, different purification protocols must be set up in order to obtain both good yields and purification factors. Accordingly, ammonium sulfate precipitation followed by conventional ion-exchange and gel-filtration chromatographic methods is recommended with such organs as human placenta or rat liver. Ultracentrifugation followed by ion-exchange or affinity chromatography is advisable when dealing with smaller organs or tissues such as rat heart, human epidermis biopsies, or cell cultures (e.g., fibroblasts or keratinocytes). Addition of EDTA to the homogenization buffer and working at a low temperature (4°) are required to prevent inactivation and oxidative modification of the proteasome during the sample preparation procedure.

Purification by Precipitation and Conventional Chromatographic Methods

Organs or tissues that have been kept frozen at −70° are thawed and homogenized with either a Potter–Elvehjem glass homogenizer with a Teflon pestle (for rat liver) or a Waring Blender (for human placenta) in 20 mM HEPES, pH 7.8, supplemented with 0.1 mM EDTA and 1 mM 2-mercaptoethanol. The homogenate is centrifuged at 15,000g for 1 hr at 4° and the supernatant is subjected to a first precipitation with 35% (w/v) saturated ammonium sulfate followed by a second precipitation of the resulting supernatant with 60% (w/v) saturated ammonium sulfate. The pellet obtained after centrifugation at 10,000g for 30 min at 4° is then resuspended in 10 mM Tris-HCl, pH 7.2, supplemented with 100 mM KCl, 0.1 mM EDTA, and 1 mM 2-mercaptoethanol, and dialyzed overnight at 4° against the same buffer. A first ion exchange is performed on a DEAE-5PW column (Toso-Haas, Stuttgart, Germany) in 10 mM Tris-HCl, pH 7.2, with a linear gradient of KCl from 0.1 to 0.5 M on a Beckman Gold liquid chromatograph (Beckman-Coulter, Fullerton, CA). For the different chromatographic steps, the fractions are assayed for 20S proteasome activity with the synthetic substrate LLVY-MCA (see below). The pooled fractions containing the 20S proteasome are then subjected to a second ion exchange on a Mono Q HR 5/5 column (Amersham Pharmacia Biotech, Uppsala, Sweden) in 20 mM Tris-HCl, pH 7.2, with a linear gradient of KCl from 0.1 to 0.5 M. Finally, the pooled fraction containing the 20S proteasome is chromatographed on a Superose 6 HR column (Amersham Pharmacia Biotech) in 50 mM potassium phosphate (pH 7), 100 mM KCl. Purified 20S proteasome is then dialyzed against a buffer such as 20 mM HEPES, pH 7.8, and stored at −70°. For long storage periods, the addition of glycerol up to 20% (v/v) is recommended.

Purification by Ultracentrifugation and Ion-Exchange or Affinity Chromatography

The handling of smaller amounts of starting biological material prevents the use of the conventional purification procedure described above. Therefore, a two-step procedure consisting of ultracentrifugation followed by chromatography, either ion-exchange chromatography on a Mono Q HR 5/5 column (Amersham Pharmacia Biotech) or affinity chromatography with human proteasome monoclonal antibody (MCP 21) coupled to a HiTrap NHS-activated column according to the manufacturer instructions (Amersham Pharmacia Biotech), has proved to be useful for purifying 20S proteasome from rat heart and human epidermis, respectively. After homogenization of the tissue and/or lysis of the cells, the homogenate is first centrifuged at 15,000g for 1 hr at 4° to remove cellular debris. The supernatant is then centrifuged at 100,000g for 16 hr at 4° to pellet the 20S proteasome. The pellet is dissolved in 20 mM Tris-HCl, pH 7.2, supplemented with 100 mM KCl, 0.1 mM EDTA, and 1 mM 2-mercaptoethanol before being loaded on a Mono Q HR 5/5 column (Amersham Pharmacia Biotech) as decribed above. Alternatively, the pellet is dissolved in 25 mM Tris-HCl, pH 7.5, before being loaded on the MCP 21 affinity column. After washing with 25 mM Tris-HCl, pH 7.5, the 20S proteasome is eluted from the affinity column with 2 M NaCl in 25 mM Tris-HCl, pH 8, and then dialyzed against 25 mM Tris-HCl, pH 7.5. The original affinity purification of the human 20S proteasome using MCP 21 has been described by Hendil and Uerkvitz,[15] and MCP 21 is now available as purified antibody from Affiniti (Exeter, UK) or as a hybridoma cell line from either the European collection of cell cultures (ECACC, Salisbury, UK) or the American Type Culture Collection (ATCC, Manassas, VA).

Conversion of 20S Proteasome from Latent to Active Form

After purification, the proteasome is generally obtained in a latent form that can be further activated *in vitro* by various treatments such as incubation with poly-L-lysine or fatty acids, heating, freezing and thawing, long storage in the absence of glycerol, addition of a low concentration of sodium dodecyl sulfate (SDS), or dialysis against water.[16–19] Depending on the treatment, 20S proteasome activation is characterized by increased proteolytic activity against both synthetic peptide substrates and protein substrates (e.g., oxidized protein and casein). This activation presumably results from conformational rearrangements of the proteasome complex.[17,18,20,21] Activation of the 20S proteasome is also achieved by physiological activators of the proteasome, such as PA 28 (or 11S regulator) and PA 700 (or 19S regulator).[10,11,22] The precise way these effectors regulate proteasome

[15] K. B. Hendil and W. Uerkvitz, *J. Biochem. Biophys. Methods* **22**, 159 (1991).
[16] M. J. McGuire, M. L. McCullough, D. E. Croall, and G. N. DeMartino, *Biochim. Biophys. Acta* **995**, 181 (1989).
[17] P. E. Falkenburg and P. M. Kloetzel, *J. Biol. Chem.* **264**, 6660 (1989).
[18] Y. Saitoh, H. Yokosawa, and S. Ishii, *Biochem. Biophys. Res. Commun.* **162**, 334 (1989).
[19] T. Tokumoto and K. Ishikawa, *Biochem. Biophys. Res. Commun.* **192**, 1106 (1993).
[20] H. Djaballah, A. J. Rowe, S. E. Harding, and A. J. Rivett, *Biochem. J.* **292**, 857 (1993).
[21] M. E. Figueiredo-Pereira, W. E. Chen, H. M. Yuan, and S. Wilk, *Arch. Biochem. Biophys.* **317**, 69 (1995).
[22] D. Voges, P. Zwickl, and W. Baumeister, *Annu. Rev. Biochem.* **68**, 1015 (1999).

activity is still unclear, but it has been suggested that binding of these activators promotes a different conformation of the proteasome. Addition of SDS at concentrations varying from 0.01 to 0.05% (w/v) has shown that full activation of the three proteasome peptidase activities can be achieved at about 0.03% (w/v) for 20S proteasome purified from rat liver or human placenta. On overnight dialysis of 20S proteasome from rat liver or human placenta against water, the proteasome peptidase activities become activated, as does the oxidized glutamine synthetase or casein degradation activity. In the following sections dealing with the effects of different oxidative treatment on 20S proteasome peptidase activity, the proteasome active form that has been used is that obtained after dialysis against water because, in this case, the proteasome preparation is devoid of any additives, such as SDS, that may interfere with the treatment.

20S Proteasome Functional Changes on Oxidative Treatments

Proteolytic Assays Using Peptide and Protein Substrates

Proteasome peptidase activities can be monitored by using fluorogenic peptide substrates (available from Sigma, St. Louis, MO): Ala-Ala-Phe-amidomethylcoumarin (AAF-MCA) or succinyl-Leu-Leu-Val-Tyr-amidomethylcoumarin (Suc-LLVY-MCA) at 20 μM for the chymotrypsin-like activity, N-tert-butyloxycarbonyl-Leu-Ser-Thr-Arg-amidomethylcoumarin (Boc LSTR-MCA) at 20 μM for the trypsin-like activity, and N-benzyloxycarbonyl-Leu-Leu-Glu-β-naphthylamide (Cbz-LLE-NA) at 100 μM for the peptidylglutamyl-peptide hydrolase activity. In cell or tissue homogenates, the use of the AAF-MCA fluorogenic peptide should be avoided because it has an unprotected N terminus and is also a good substrate for aminopeptidases. Incubation is achieved with 1 to 5 μg of 20S proteasome in 200 μl of 0.1 M HEPES, pH 7.8, or 25 mM Tris-HCl, pH 7.5, at 37° for 20 min, at which point the reaction is stopped with 300 μl of acid (30 mM sodium acetate, 70 mM acetic acid, 100 mM sodium chloroacetate, pH 4.3) or ethanol. After addition of 2 ml of distilled water, the fluorescence is monitored with a spectrofluorimeter at excitation and emission wavelengths of 350/440 and 333/410 nm for aminomethylcoumarin (MCA) and β-naphthylamine (NA) products, respectively. A calibration curve is obtained by using different concentrations, from 0.1 to 1 μM, of the product, MCA or NA. For a large number of samples (up to 96) the assay can be conducted with a temperature-controlled microplate fluorimetric reader (e.g., FLUOstar Galaxy; B & L Systems, Maarssen, The Netherlands), which has the additional advantage of allowing the kinetics with which the fluorescent products appear to be monitored.

Oxidized proteins and casein are considered good protein substrates for the 20S proteasome. Proteolysis of these substrates is monitored by measuring the resulting small peptides as a function of incubation time of the protein with the proteasome. As opposed to the protein substrate and the proteasome, such small peptides are soluble in 10% (w/v) trichloroacetic acid (TCA), and several methods can be used to quantify the concentration of the released peptides. A substrate of choice is radiolabeled [^{14}C]casein (DuPont-NEN, Zaventem, Belgium) or oxidized proteins (glutamine synthetase).[23] Otherwise, the TCA supernatant is first neutralized with an equal volume of 2 M sodium borate, pH 10. When the substrate is casein–fluorescein isothiocyanate (Sigma), the concentration of small peptides is monitored directly with a spectrofluorimeter (excitation at 495 nm, emission at 515 nm). Alternatively, when protein substrates are not labeled by radioactivity or fluorescence, the concentration of small peptides may also be determined after reaction of their free NH$_2$ groups with fluorescamine (0.06-mg/ml final concentration; stock solution, 0.3 mg/ml in acetone) by spectrofluorimetry (excitation at 375 nm, emission at 475 nm).

Oxidation-Mediated Changes in Activity of Purified 20S Proteasome

20S proteasome from rat liver (at 0.1 mg/ml) has been exposed to metal-catalyzed oxidation on incubation in 0.1 M HEPES, pH 7.8, in the presence of 0.1 mM FeCl$_3$ and 25 mM ascorbate. At the indicated times, 50-μl aliquots are diluted in 200 μl of 0.1 M HEPES, pH 7.8, containing the fluorogenic peptide of interest, with the intention of monitoring each specific peptidase activity. The results shown in Fig. 1A indicate that all three peptidase activities are increased when the 20S proteasome is in its latent form before oxidation, whereas the peptidase activities are decreased when carried by the 20S proteasome active form (Fig. 1B). The trypsin-like and peptidylglutamyl-peptide hydrolase activities are much more sensitive to oxidation than the chymotrypsin-like activity, which is only slightly inactivated. It is interesting to note that the trypsin-like activity is known to be the most sensitive to thiol-blocking agents (e.g., N-ethylmaleimide),[24] which indicates that there is an important thiol for the proteolytic activity that may be a good target for reactive oxygen species. Of additional interest is the fact that the chaperone proteins Hsp90 (heat shock protein 90) and α-crystallin (a member of the Hsp27 family) were found to specifically protect in $vitro$ the trypsin-like activity and also a nonconventional N-benzyloxycarbonyl-Leu-Leu-Leu-amidomethylcoumarin (Cbz-LLL-MCA) hydrolyzing activity of the 20S proteasome active form. Indeed, a 4-fold molar excess of either Hsp90 or α-crystallin was enough to prevent the oxidative inactivation of the 20S proteasome, whereas no protection was observed with thyroglobulin and glucose-6-phosphate dehydrogenase, used as controls.[8] This finding suggests that these chaperone proteins can bind to the proteasome and may serve as specific protectors against inactivation by reactive oxygen species.

[23] B. Friguet, E. R. Stadtman, and L. I. Szweda, $J.$ $Biol.$ $Chem.$ **269,** 21639 (1994).
[24] P. J. Savory and A. J. Rivett, $Biochem.$ $J.$ **289,** 45 (1993).

A
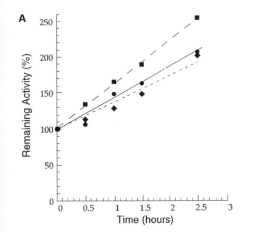

B
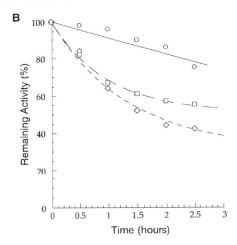

FIG. 1. Peptidase activities after metal-catalyzed oxidation of 20S proteasome in latent and active forms. Latent and active forms of proteasome at 0.1 mg/ml in 0.1 M HEPES, pH 7.8, were incubated with 0.1 mM FeCl$_3$ and 25 mM ascorbate at 37°. At the indicated times peptidase activities were assayed as described. For each peptidase, 100% activity was defined as the peptidase activity at oxidation time 0. (A) Activities of the latent form: chymotrypsin-like activity (●); trypsin-like activity (■); peptidylglutamyl-peptide hydrolase activity (◆). (B) Activities of the active form: chymotrypsin-like activity (○); trypsin-like activity (□); peptidylglutamyl-peptide hydrolase activity (◇). [Reproduced with permission from M. Conconi, I. Petropoulos, I. Emod, E. Turlin, F. Biville, and B. Friguet, *Biochem. J.* **333,** 407 (1998), copyright © 1998 the Biochemical Society.]

In addition to the production of hydroxyl radicals by metal-catalyzed oxidation, the UV-A-mediated production of a variety of reactive oxygen species such as hydrogen peroxide and singlet oxygen is able to induce functional changes in the 20S proteasome. Indeed, on irradiation by UV-A (10 J/cm^2) of the 20S proteasome (at 0.2 mg/ml in 25 mM Tris-HCl, pH 7.5) in its latent form, the trypsin-like activity undergoes a 2-fold increase whereas the chymotrypsin-like activity is only slightly increased and the peptidylglutamyl-peptide hydrolase activity remains unaffected. The same UV-A irradiation on the proteasome in its active form results in only a 30% decrease of the peptidylglutamyl-peptide hydrolase activity. As observed for metal-catalyzed oxidation, the peptidylglutamyl-peptide hydrolase activity is also a target for UV-A-mediated oxidative damage. Protein modifications may also originate from reactions with small aldehydes such as malonaldehyde or 4-hydroxy-2-nonenal (HNE), which are the main products of lipid peroxidation. 4-Hydroxy-2-nonenal readily reacts at physiological pH with cystein, lysine, and histidine residues to form Michael adducts.[25–27] Not surprisingly, HNE was found

[25] H. Esterbauer, R. J. Schaur, and H. Zollner, *Free Radic. Biol. Med.* **11,** 81 (1991).
[26] K. Uchida and E. R. Stadtman, *Proc. Natl. Acad. Sci. U.S.A.* **89,** 5611 (1992).
[27] L. I. Szweda, K. Uchida, L. Tsai, and E. R. Stadtman, *J. Biol. Chem.* **268,** 3342 (1993).

to inactivate the trypsin-like activity of the 20S proteasome active form,[7] which corroborates the fact that this activity is the most sensitive to thiol-blocking agents.[24]

20S Proteasome Activity in Cells Exposed to Metal-Catalyzed Oxidation

Subconfluent FAO rat hepatoma cells are exposed to metal-catalyzed oxidation on addition of 25 mM ascorbate and 0.1 mM FeCl$_3$ in 0.5 mM ADP to the culture medium. After various durations of treatment, cells are harvested and disrupted by five sonications (5 sec each) in 0.25 M Tris-HCl, pH 7.8. After centrifugation at 20,000g for 20 min, the supernatant containing total soluble cellular protein is collected and the trypsin-like and peptidylglutamyl-peptide hydrolase activities are measured with 50 μg of total proteins in 0.1 M HEPES, pH 7.8, and the appropriate flurogenic peptide at the usual concentration in a final volume of 200 μl. 20S proteasome proteolytic activity is determined as the difference between the total activity and the remaining activity of the crude extract in the presence of 20 μM proteasome inhibitor MG-132 [N-carbobenzoxy-L-leucyl-L-leucyl-leucinal; available from Calbiochem (Meudon, France) or Affiniti (Mamhead, Exeter, UK)]. The results shown in Fig. 2A indicate that both trypsin-like and peptidylglutamyl-peptide

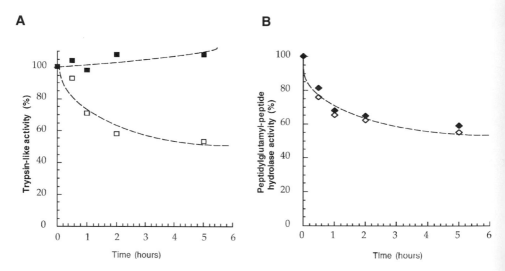

Fig. 2. Peptidase activities of proteasome after metal-catalyzed oxidation in FAO cells pretreated or not pretreated with FeCl$_3$. FAO cells were subjected to pretreatment with 0.1 mM FeCl$_3$ for 16 hr. At the indicated times, 0.1 mM FeCl$_3$ and 0.5 mM ascorbate were added to the cells. Crude extracts were prepared and peptidase activities of the proteasome were assayed as described. For each peptidase activity, 100% was defined as the value obtained at treatment time 0. (A) Trypsin-like activity; (B) peptidylglutamyl-peptide hydrolase activity. Open symbols, naive cells; solid symbols, FeCl$_3$-preincubated cells. [Reproduced with permission from M. Conconi, I. Petropoulos, I. Emod, E. Turlin, F. Biville, and B. Friguet, *Biochem. J.* **333**, 407 (1998), copyright © 1998 the Biochemical Society.]

hydrolase activities of the proteasome are inactivated on exposure of FAO cells to metal-catalyzed oxidation. Induction of Hsp90 by challenging the cells with 0.1 mM FeCl$_3$ before exposure to metal-catalyzed oxidation results in the protection of the 20S proteasome trypsin-like and Cbz-LLL-MCA-hydrolyzing activities (Fig. 2B).[8]

20S Proteasome Structural Changes on Oxidative Treatments

A method of choice for investigating structural changes of the 20S proteasome is to analyze the two-dimensional (2D) gel electrophoresis pattern of subunits before and after oxidative treatment. The Multiphor system from Amersham Pharmacia Biotech has been used with Immobilines Drystrips (pH 3–10; length, 13 cm) for the first dimension. Purified proteasome (15 μg) is diluted in sample buffer [9 M urea, 2% (w/v) 3-[(3-cholamidopropyl)-dimethyl-ammonio]-1-propanesulfonate (CHAPS), 2% (v/v) Pharmalytes (pH 3–10), 20 mM dithiothreitol, and bromphenol blue]. The Drystrips are rehydrated in this solution in a reswelling tray (Amersham Pharmacia Biotech) overnight at room temperature and then focused for 50,000 V · hr for 23 hr. After focusing, the Immobilines Drystrips are equilibrated for 10 min in equilibration buffer [50 mM Tris-HCl (pH 6.8), 6 M urea, 30% (v/v) glycerol, 1% (w/v) SDS] supplemented with 1% (w/v) dithiothreitol and for 10 min in equilibration buffer containing 2.5% (w/v) iodoacetamide and 0.01% (w/v) bromphenol blue. The second dimension is performed by the Laemmli method of SDS–polyacrylamide gel electrophoresis (PAGE)[28] on a 12% (w/v) polyacrylamide gel, using the Protean II system (Bio-Rad, Hercules, CA). Proteins are stained with silver nitrate[29] and the gel is digitized on a JX 330 scanner (Sharp, Hamburg, Germany). Protein spot detection and quantification are performed with Imagemaster 2D Elite software (Amersham Pharmacia Biotech).

Mobility shift and/or differences of staining intensity were observed for certain subunits of 20S proteasome that were exposed to different oxidative treatments (metal-catalyzed oxidation and UV-A irradiation), reflecting the occurrence of structural modifications for these subunits. Specific differences of staining of four subunits were also observed when we compared proteasome isolated from epidermis of young and old donors, suggesting an oxidative modification of these particular subunits during aging.[30]

The appearance of oxidative modifications on proteasome subunits can be checked by detecting the presence of carbonyl groups,[31,32] using the Oxyblot

[28] U. K. Laemmli, *Nature* (*London*) **227**, 680 (1970).
[29] C. R. Merril, D. Goldman, S. Sedman, and H. Ebert, *Science* **211**, 1437 (1981).
[30] A. Bulteau, I. Petropoulos, and B. Friguet, *Exp. Gerontol.* **35**, 767 (2000).
[31] R. L. Levine, D. Garland, C. N. Oliver, A. Amici, I. Climent, A. G. Lenz, B. W. Ahn, S. Shaltiel, and E. R. Stadtman, *Methods Enzymol.* **186**, 464 (1990).
[32] R. L. Levine, J. A. Williams, E. R. Stadtman, and E. Shacter, *Methods Enzymol.* **233**, 346 (1994).

technique after 1D (Quantum-Appligene, Illkirch, France) or 2D gel[33] electrophoresis of purified 20S proteasome. Alternatively, the presence of specific modifications can be detected by Western blotting after 1D or 2D gel electrophoresis of purified 20S proteasome, using an antibody specific for the modification of interest. As an example, the proteasome isolated from kidney of mice treated with a renal carcinogen (ferric nitrilotriacetate) was shown to be modified by HNE adduct in Western blot experiments, using a monoclonal antibody specifically raised against this modification.[34] A polyclonal antibody against HNE adducts can also be used[35] and is available from Calbiochem. Identification of modified 20S proteasome subunits can be achieved by 2D Western blot with monoclonal antibodies specific for the different subunits and available from Affiniti.

Conclusion

Growing evidence argues for the 20S proteasome being a target for oxidative modification on different *in vitro* and physiological oxidative stresses. Alteration of 20S proteasome peptidase activities as observed on aging and during several types of oxidative stress is likely to impair proteasome function and to have important physiological consequences. Indeed, the proteasomal system is pivotal not only for oxidized protein degradation and general protein turnover but also for specific cellular processes including activation of transcription factors such as NF-κB, antigen processing, progression of the cell cycle, and apoptosis through the activation of caspases. Activation of the proteasomal system by specific stimuli, including those affecting the redox status, remains a yet to be solved but interesting possibility.[6,8] Such activation processes are likely to involve transient interactions with other components of the proteasomal system (e.g., PA 28 and PA 700) that may not be observed when analyzing only the 20S proteasome catalytic core. However, studying the pattern of 20S proteasome modifications and their consequences on 20S proteolytic activity has already yielded valuable information about the structure–function relationship of this key enzyme for intracellular proteolysis.

Acknowledgments

Our laboratory is supported by funds from the MENRT (Institut Universitaire de France and Université Denis Diderot-Paris 7) and by a European Union QLRT "Protage" Grant (QLK6-CT1999-02193).

[33] J. M. Talent, Y. Kong, and R. W. Gracy, *Anal. Biochem.* **263**, 31 (1998).
[34] K. Okada, C. Wangpoengtrakul, T. Osawa, S. Toyokuni, K. Tanaka, and K. Uchida, *J. Biol. Chem.* **274**, 23787 (1999).
[35] L. I. Szweda, P. A. Szweda, and A. Holian, *Methods Enzymol.* **319**, 562 (2000).

[24] Measuring Reactive Oxygen Species Inhibition of Endothelin-Converting Enzyme

By CHARLES J. LOWENSTEIN

Introduction

Oxygen and nitrogen radicals serve important physiological roles in the cardiovascular system.[1-3] For example, hydrogen peroxide mediates platelet-derived growth factor signal transducion.[4] Superoxide mediates $p21^{ras}$ regulation of the cell cycle.[5] Reactive nitrogen intermediates such as nitric oxide (NO) regulate vasodilation, platelet adhesion and aggregation, and apoptosis.

However, high levels of reactive oxygen species (ROS) can be cytotoxic. Superoxide generated by the NADPH oxidase in neutrophils can damage endothelial cells.[1,3,6] Ischemia and reoxygenation can lead to the production of high levels of ROS inside endothelial and smooth muscle cells, which can also damage vascular cells. Superoxide may also play a role in the pathophysiology of hypertension: angiotensin II activates endothelial cell production of superoxide, which decreases the bioavailability of NO, leading to vasoconstriction.[7,8]

We hypothesized that endogenous counterregulatory mechanisms might defend the vasculature from the harmful effects of ROS. We focused our attention on the effect of radicals on the endothelin-converting enzyme (ECE). ECE is a membrane-bound, zinc metalloproteinase that converts the precursor big endothelin 1 into the vasoactive polypeptide endothelin 1 (ET-1).[9] We hypothesized that ROS would inhibit ECE conversion of big ET-1 to ET-1. Accordingly, we examined the effect of ROS on ECE purified from cultured endothelial cells.

[1] P. J. Goldschmidt-Clermont and L. Moldovan, *Gene Expr.* **7,** 255 (1999).

[2] T. Finkel, *Curr. Opin. Cell Biol.* **10,** 248 (1998).

[3] D. G. Harrison, *J. Clin. Invest.* **100,** 2153 (1997).

[4] M. Sundaresan, Z. X. Yu, V. J. Ferrans, K. Irani, and T. Finkel, *Science* **270,** 296 (1995).

[5] K. Irani, Y. Xia, J. L. Zweier, S. J. Sollott, C. J. Der, E. R. Fearon, M. Sundaresan, T. Finkel, and P. J. Goldschmidt-Clermont, *Science* **275,** 1649 (1997).

[6] K. K. Griendling and D. G. Harrison, *Circ. Res.* **85,** 562 (1999).

[7] K. K. Griendling, C. A. Minieri, J. D. Ollerenshaw, and R. W. Alexander, *Circ. Res.* **74,** 1141 (1994).

[8] S. Rajagopalan, S. Kurz, T. Munzel, M. Tarpey, B. A. Freeman, K. K. Griendling, and D. G. Harrison, *J. Clin. Invest.* **97,** 1916 (1996).

[9] T. Miyauchi and T. Masaki, *Annu. Rev. Physiol.* **61,** 391 (1999).

Copyright 2002, Elsevier Science (USA).
All rights reserved.
0076-6879/02 $35.00

Culture of Endothelial Cells

To prepare protein extracts containing ECE, we first grow bovine aortic endothelial cells (BAECs).[10] We purify endothelial cells from freshly harvested bovine aortas and also purchase cells from commercial sources (Clonetics, Walkersville, MD). BAECs are grown on gelatin-coated plates, and fed every 2 days with RPMI 1640 medium supplemented with 15% (v/v) calf serum (CS), penicillin (100 U/ml), and streptomycin (100 μg/ml). Cells are fed every 2 days until they are 90% confluent, rinsed with Hanks' balanced salt solution (HBSS) for 10 min, trypsinized, and then split at a ratio of 1 : 4.

Purification of Membranes Containing Endothelin-Converting Enzyme

ECE is a type II membrane glycoprotein. We purify endothelial cell membrane proteins containing ECE in order to assess the effects of radicals on ECE activity. Confluent monolayers of BAECs are grown in 10-cm dishes. BAECs are homogenized in 1 ml of homogenization buffer [20 mM Tris-HCl (pH 7.5), 5 mM MgCl$_2$, 0.1 mM phenylmethylsulfonyl fluoride (PMSF), 20 μM leupeptin, 20 μM aprotinin]. The homogenates are sonicated and then centrifuged at 100,000g for 45 min. The membranes are washed three times in homogenization buffer. The protein concentration in each supernatant is determined according to the bicinchoninic acid (BCA) protein assay (Pierce, Rockford, IL).

Measurements of Endothelin-Converting Enzyme Activity

We wanted to measure the effects of ROS on endothelin signal transduction. Endothelin 1 (ET-1) is synthesized from the large precursor prepro-ET-1. A signal peptidase cleaves prepro-ET-1 into proET-1, which in turn is cleaved by a furin-like enzyme into big ET-1. Endothelin-converting enzyme (ECE) then cleaves big ET-1 into ET-1, a 21-amino acid residue polypeptide. All the precursors of ET-1 are inactive; only ET-1 is biologically active. We therefore chose to measure the conversion of big ET-1 into ET-1, which is mediated by ECE.

We use an enzyme-linked immunosorbent assay (ELISA) to measure ECE activity. We add big ET-1 to BAEC membrane proteins containing ECE, and measure ET-1 production by ELISA in the following manner. Membrane proteins are extracted from confluent layers of BAECs as described above. Bovine big ET-1 (100 ng) (Sigma, St. Louis, MO) is then added and the mixture is incubated for 4 hr at 37° in 250 μl of a reaction mixture containing 50 mM Tris-HCl buffer, pH 7. The reaction is stopped by adding 600 μl of ice-cold ethanol. After centrifugation at 10,000g for 10 min at 4°, the resulting supernatant is lyophilized. The dried

[10] P. A. Marsden, T. A. Brock, and B. J. Ballermann, *Am. J. Physiol.* **258**, F1295 (1990).

residues are reconstituted with assay buffer, and the production of ET-1 in each sample is measured by an ELISA for ET-1, using a 96-well microtiter plate reader (Amersham, Buckinghamshire, UK). To generate a standard curve for ET-1, serial dilutions of ET-1 (Amersham) ranging from 1 to 16 fmol (2.49–39.9 pg) are added to assay buffer and placed in the ELISA wells. A cubic-spline curve is fitted to the standards and unknown values are interpolated from the standard curves automatically.

As a control, we measure ECE activity in BAECs in the absence and presence of the fungal metabolite phosphoramidon, which is an ECE inhibitor.

Addition of Radicals to Purified Endothelin-Converting Enzyme

To study the effects of ROS on ECE, we pretreat purified BAEC membrane proteins with various radical donors. We use a mixture of xanthine and xanthine oxidase to generate superoxide; a mixture of glucose and glucose oxidase to generate hydrogen peroxide; and S-nitrosopenicillamine as an NO donor. For example, to treat ECE with superoxide, purified BAEC membranes (30 μg) are pretreated with buffer or with mixtures of xanthine (0–0.1 mM) and xanthine oxidase (0–1 mU/ml) at 37° for 1 hr. To treat ECE with hydrogen peroxide, we either directly add hydrogen peroxide (Sigma), or add 0.1 mM glucose and glucose oxidase (1 mU/ml) to 30 μg of BAEC membrane proteins and incubate the mixture at 37° for 1 hr.

Addition of superoxide decreases ECE activity (Fig. 1). Other ROS or reactive nitrogen intermediates (RNIs) do not inhibit ECE, suggesting this is an effect

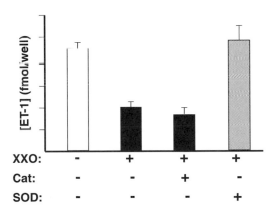

XXO:	-	+	+	+
Cat:	-	-	+	-
SOD:	-	-	-	+

Fig. 1. Superoxide generated by xanthine and xanthine oxidase inhibits ECE activity. BAEC membrane proteins containing ECE were incubated with big ET-1, and the production of ET-1 was measured by ELISA. Reaction mixtures were incubated with buffer alone, or with 0.1 mM xanthine and xanthine oxidase (10 mU/ml) (XXO), or with XXO and catalase (Cat, 80 U/ml), or with XXO and SOD (100 U/ml).

specific to superoxide. To confirm the specificity of the effects of superoxide, some reactions are also incubated with superoxide dismutase (SOD, 100 U/ml) or catalase (80 U/ml). SOD but not catalase blocks the inhibition of ECE by superoxide donors, confirming that superoxide inhibits ECE (Fig. 1).

It is important to show that ROS do not affect the substrate and products of the assay. For example, if ROS modified the product ET-1 and decreased the sensitivity of the ET-1 ELISA, a false-positive result would be obtained. We therefore have measured a fixed amount of ET-1 that has been treated or not treated with ROS, to confirm that the ET-1 ELISA is not affected by ROS. We also have treated a fixed amount of the substrate big ET-1 with buffer or with ROS, and measured its conversion by BAEC membrane proteins, in order to confirm that ROS do not modify the substrate of ECE. These two important controls have increased our confidence that any effects of ROS on ECE conversion of big ET-1 into ET-1 are indeed true-positive results.

Radicals and Dimerization of Endothelin-Converting Enzyme

We next explored various mechanisms by which superoxide might inhibit ECE. Radicals can disrupt disulfide bridges, altering the tertiary structure of proteins. We therefore sought to examine the effect of superoxide on the homodimerization of ECE. We have treated ECE-containing protein extracts with buffer alone or superoxide as described above, fractionated the proteins on nondenaturing gels, and then determined the relative mobility by immunoblotting with an antibody to ECE. Radical treatment of ECE does not alter its mobility.

Radicals and Zinc: Indirect Measurements of Zinc in Endothelin-Converting Enzyme

ECE is a zinc metalloproteinase, and shares with other zinc metalloproteinases a zinc-binding HEXXH motif. Zinc is required for ECE activity. We hypothesized that superoxide might eject Zinc from ECE.

Therefore, we have indirectly assessed the effect of superoxide on the zinc content of ECE. We first confirmed that ECE requires a divalent cation for activity. We performed the ECE assay as described above in the presence or absence of EDTA ($0-100~\mu M$). We found that $1~\mu M$ EDTA inhibits ECE activity by more than 60%. To show that zinc is essential for ECE activity, we performed the ECE assay in the presence of $1~\mu M$ EDTA and $10~\mu M$ ZnCl$_2$. The addition of zinc restores ECE activity. To determine whether superoxide ejects zinc from ECE, we treated ECE-containing BAEC extracts with superoxide as described above for 1 hr, and then added increasing amounts of zinc ($1-10~\mu M$) for 10 min. Zinc restores ECE activity after inhibition with superoxide.

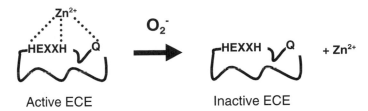

Active ECE   Inactive ECE

FIG. 2. Superoxide ejects zinc ion from ECE. ECE interacts with zinc ion through two histidine residues in an HEXXH motif and through a third zinc ligand, glutamate. Superoxide may eject zinc ion by reducing it.

These experiments suggest that superoxide inhibits ECE by ejecting zinc from the enzyme (Fig. 2). However, these experiments do not directly assess the zinc content of ECE.

Radicals and Zinc: Direct Measurements of Zinc in Endothelin-Converting Enzyme

Direct measurements of the content of zinc in a protein can be challenging. We use instrumental neutron activation analysis (INAA) to quantify zinc in ECE. A full discussion of INAA is outside the scope of this chapter. In brief, INAA involves neutron irradiation of a sample, and the measurement of the decay of various activation products.

We prepare BAEC membrane proteins as described above, and ECE is then immunoprecipitated with antibody to ECE. Purified ECE is treated with superoxide generators, or not so treated. We place our ECE preparations into small acid-washed polyethylene vials, using an additional 0.2 ml of buffer solution to ensure complete transfer. (*Note:* It is essential to use double-distilled, deionized water in buffer solutions, to avoid the possible contamination of the samples with zinc from tap water.)

The samples and standards are irradiated together for 12 hr at a neutron flux of 8×10^{12} n/cm^2 sec; transferred to irradiated vials, using two additions of 0.5 ml of unirradiated buffer; and then allowed to decay for approximately 4 weeks. The presence of the activation product ^{65}Zn (half-life, 243.8 days; γ energy, 1115.5 keV) is measured in the samples and standards by using four high-purity germanium (HPGe) detectors coupled to a dedicated personal computer (all hardware and software are from Canberra Industries, Meriden, CT). The zinc levels in the proteins are calculated on the basis of the ^{65}Zn activities in the protein samples and the reference materials, the reference sample masses, and the certified zinc reference concentrations.

We found that treatment of ECE with 0.1 mM xanthine and xanthine oxidase (1.0 mU/ml) for 1 hr removes all zinc from ECE (Fig. 2). Thus INAA can be used to assess the content of zinc in proteins such as metalloproteinases.

Conclusion

ROS regulate a variety of vascular signaling pathways. Using an ELISA, we have shown that superoxide regulates ECE activity. Indirect measurements and direct measurements show that superoxide inhibits ECE by ejecting zinc from the enzyme. Several pathways in the cardiovascular system involve the activity of converting enzymes, including the renin–angiotensin–aldosterone pathway. Techniques that measure protease activity, and measure the effect of ROS on proteases, may reveal novel mechanisms by which radicals regulate cardiovascular signaling.

[25] Redox Sensor Function of Metallothioneins

By JAMES P. FABISIAK, GREGORY G. BORISENKO, SHANG-XI LIU, VLADIMIR A. TYURIN, BRUCE R. PITT, and VALERIAN E. KAGAN

Introduction

Metallothioneins (MTs) are low molecular weight (approximately 6000) cysteine-rich (30%) metal-binding proteins.[1] Originally discovered and isolated as a cadmium-binding protein from horse kidney, it is now apparent that the multi-isoform family of MT proteins can serve to protect cells and animals from the toxic effects of metals,[2] reactive electrophiles,[3] and reactive oxygen/nitrogen species (ROS/RNS).[4,5]

The 20 cysteines contained in MTs are highly conserved across species; are arranged into two distinct thiolate clusters, termed α and β; and are fundamental to the metal-binding function through the formation of sulfur–metal bonds.[6] For metal binding to occur, the cysteine SH groups must be in their reduced state. Copper binding to MT has been shown to be directly proportional to the content of reduced SH groups[7] and oxidative or nitrosative stress can, in fact, promote copper

[1] J. H. R. Kagi and A. Schaffer, *Biochemistry* **27**, 8509 (1988).
[2] M. P. Waalkes and P. L. Goering, *Chem. Res. Toxicol.* **3**, 281 (1990).
[3] J. S. Lazo and B. R. Pitt, *Annu. Rev. Pharmacol. Toxicol.* **35**, 635 (1995).
[4] B. R. Pitt, M. Schwarz, E. S. Woo, E. Yee, K. Wasserloos, S. Tran, W. Weng, R. J. Mannix, S. A. Watkins, Y. Y. Tyurina, V. Tyurin, V. E. Kagan, and J. S. Lazo, *Am. J. Physiol.* (*Lung Cell. Mol. Physiol.*) **273**, L856 (1997).
[5] M. A. Schwarz, J. S. Lazo, J. C. Yalowich, I. Reynolds, V. E. Kagan, V. Tyurin, Y.-M. Kim, S. Watkins, and B. R. Pitt, *J. Biol. Chem.* **269**, 15238 (1994).
[6] J. D. Otvos and I. M. Armitage, *Proc. Natl. Acad. Sci. U.S.A.* **77**, 7094 (1985).
[7] J. P. Fabisiak, V. A. Tyurin, Y. Y. Tyurina, G. G. Borisenko, A. Korotaeva, B. R. Pitt, J. S. Lazo, and V. E. Kagan, *Arch. Biochem. Biophys.* **363**, 171 (1999).

Copyright 2002, Elsevier Science (USA).
All rights reserved.
0076-6879/02 $35.00

release and potentiate copper-induced cytotoxicity.[8,9] NO has been demonstrated to release intracellular cadmium bound to MT and potentiate the toxicity of this metal as well.[10]

It has been speculated, as well, that MT can play an active role in maintaining metal ion homeostasis through delivery or sequestration of important endogenous metal ions. For example, Maret and Vallee[11] and Jacob and co-workers[12] have suggested that MT can deliver zinc to a variety of enzymes that require this metal for biological activity. Similarly, we have shown that MT can deliver copper ion to zinc-replete apo-superoxide dismutase (apo-SOD) in a nitric oxide-dependent fashion.[13] Alternatively, apometallothionein can serve to remove zinc from specific sites on proteins, such as caspase and glyceraldehyde-3-phosphate dehydrogenase and, thus, activate their function by removal of the zinc-dependent inhibition.[14] All these studies have invoked a critical role for redox regulation in determining metal binding and delivery by MT. Remarkably, MTs are not prone to the traditional posttranslational modifications that typically modulate protein function. No evidence of MT phosphorylation, palmitoylation, farnesylation, or controlled proteolysis has been presented. This leaves the thiolate clusters and their redox conversions as a likely mode for regulation of MT function. Thus, knowledge of redox conversion of MT cysteines is central to understanding the regulatory roles and pathways of this important metal-binding protein. The purpose of this chapter is to describe contemporary approaches to assess the redox status of MT cysteines and their relationship to MT metal sequestration and release.

Cell-Free Model System Experiments with Metallothionein *in Vitro*

MTs can be isolated from animal sources in relatively large amounts. For example, MTs can be obtained highly purified from bovine liver, as well as from the livers of mice and rabbits pretreated with MT inducers such as zinc.[15–17] The

[8] J. P. Fabisiak, L. L. Pearce, G. G. Borisenko, Y. Y. Tyurina, V. A. Tyurin, J. Razzack, J. S. Lazo, B. R. Pitt, and V. E. Kagan, *Antioxidants Redox Signal.* **1,** 349 (1999).

[9] S.-X. Liu, K. Kawai, V. A. Tyurin, Y. Y. Tyurina, G. G. Borisenko, J. P. Fabisiak, P. J. Quinn, B. R. Pitt, and V. E. Kagan, *Biochem. J.* **354,** 397 (2001).

[10] R. R. Misra, J. F. Hochadel, G. T. Smith, J. C. Cook, M. P. Walkes, and D. A. Wink, *Chem. Res. Toxicol.* **9,** 326 (1996).

[11] W. Maret and B. L. Vallee, *Proc. Natl. Acad. Sci. U.S.A.* **31,** 3478 (1998).

[12] C. Jacob, W. Maret, and B. L. Vallee, *Proc. Natl. Acad. Sci. U.S.A.* **95,** 3489 (1998).

[13] S.-X. Liu, J. P. Fabisiak, V. A. Tyurin, G. G. Borisenko, B. R. Pitt, J. S. Lazo, and V. E. Kagan, *Chem. Res. Toxicol.* **13,** 922 (2000).

[14] W. Maret, C. Jacob, B. L. Vallee, and E. H. Fischer, *Proc. Natl. Acad. Sci. U.S.A.* **96,** 1936 (1999).

[15] P. Chen, A. Munoz, D. Nettesheim, and C. F. Shaw, Jr., *Biochem. J.* **317,** 395 (1996).

[16] D. T. Minkel, K. Poulsen, S. Wielgus, C. F. Shaw III, and D. H. Petering, *Biochem. J.* **191,** 475 (1980).

[17] M. Vasak, *Methods Enzymol.* **205,** 41 (1991).

availability of large amounts of pure protein permits the application of conventional analytical techniques to study metal binding and cysteine redox status in cell-free model systems. In fact, such preparations of purified MT are available commercially. We have found, however, that the redox state of many of these available MTs needs to be carefully characterized before use. It is commonly observed that the number of reduced cysteines is significantly lower than that theoretically predicted by the amino acid sequence (20 mol of SH groups per mole of MT). Therefore, an obligatory treatment of MTs with a reducing agent such as dithiothreitol (DTT) is required in order to reconstitute the full complement of cysteines. It is necessary to be cautious and verify that the number of cysteines after this treatment is indeed close to the theoretical maximum for the amount of protein present. Moreover, systematic monitoring of MT SH groups needs to be conducted during its storage. We have, indeed, observed that some MTs stored for prolonged periods of time fail to reconstitute the full complement of cysteines after DTT treatment, suggesting their oxidation beyond simple disulfides. A detailed protocol for the preparation of reduced MT and characterization of thiol group status is given below.

Protocol 1: Preparation of Freshly Reduced Metallothionein and Measurement of SH Group Content

1. Five milligrams of rabbit Zn-MTII (Sigma, St. Louis, MO) is dissolved in 4.165 ml of 10 mM Tris-HCl, pH 8, to yield an approximately 200 μM MT stock solution and frozen in 200-μl aliquots at $-80°$. Tris buffer and all other solutions are prepared with Chelex-treated water to remove any adventitious metals. Solutions are also purged for 5 min with N_2 to remove dissolved oxygen. These precautions minimize the spontaneous oxidation of MT during storage.

2. To prepare freshly reduced MT, 10 μl of 100 mM DTT prepared in water is added to 200 μl of MT stock and incubated at room temperature for 30 min.

3. The resultant mixture is split into two equal portions and applied to the tops of Microcon YM-3 centrifuge filters (3000 MW cutoff; Millipore, Bedford, MA) and centrifuged at 1300g for 30 min at $4°$. The filtrate is discarded and another 100 μl of fresh 10 mM Tris, pH 8.0, is added to the top of the filter. This is repeated for a total of three centrifugations and the final retentates are combined, transferred to a clean microcentrifuge tube, and brought to 200 μl with 10 mM Tris buffer.

4. The exact protein concentration of MT should be determined by preparing a 100-fold dilution of MT in 0.1 M HCl and measuring the absorbance at 220 nm ($\varepsilon_{220} = 48,200\ M^{-1}\ cm^{-1}$). Recovery of MT is, on the basis of original weight, approximately 30–50%.

5. To determine reduced thiol content the MT solution is adjusted to a MT concentration of 10 μM. Five microliters of this solution is mixed with 45 μl of 2 mM dithiodipyridine (DTDP) made by dilution of freshly prepared 100 mM DTDP stock in 0.1 M acetic acid. After incubation at room temperature for 30 min the

TABLE I

EFFECTS OF HYDROGEN PEROXIDE ON MODULATION OF COPPER-DEPENDENT
REDOX CYCLING BY METALLOTHIONEIN[a]

Assay	$\pm H_2O_2$	Copper : MT ratio
Cis-PnA oxidation	No H_2O_2	12 : 1
Luminol oxidation	H_2O_2	5 : 1
Ascorbate oxidation	No H_2O_2	12 : 1
Hydroxyl radical formation (DMPO–OH adduct)	H_2O_2	8 : 1
Formation of Cu^{2+} in Cu-MTs by EPR	No H_2O_2	10 : 1
	H_2O_2	4 : 1
HL-60 cytotoxicity	No H_2O_2	10 : 1
	H_2O_2	4 : 4

Abbreviations: PnA = parinaric acid; DMPO = 5,5-dimethyl-1-pyrroline-N-oxide.
[a] From J. P. Fabisiak, V. A. Tyurin, Y. Y. Tyurina, G. G. Borisenko, A. Korotaeva,
B. R. Pitt, J. S. Lazo, and V. E. Kagan, *Arch. Biochem. Biophys.* **363**, 171 (1999); and
J. P. Fabisiak, L. L. Pearce, G. G. Borisenko, Y. Y. Tyurina, V. A. Tyurin, J. Razzack,
J. S. Lazo, B. R. Pitt, and V. E. Kagan, *Antioxidants Redox Signal.* **1**, 349 (1999).

absorbance at 343 nm is read with a SpectroMate UV–visible (UV–Vis) fiberoptic microspectrophotometer (World Precision Instruments, Sarasota, FL) equipped with a 2-μl micropipetter cuvette. The number of reduced SH groups per molecule of MT can be calculated on the basis of the extinction coefficient of $\varepsilon_{343} = 7600 \ M^{-1} \ cm^{-1}$ and the actual protein content determined as described in step 4. Aliquots of freshly reduced MT can be stored frozen at $-80°$ for several weeks without appreciable loss of DTDP reactivity. The effects of prolonged storage or varying conditions on MT protein content and SH integrity should be assessed.

When these rules are obeyed quantitative studies relating SH status and metal-binding capacity can be performed.[15,18] As an example, freshly reduced MT can be shown to bind 12 mol of copper (its maximum capacity) with little or no accompanying redox-cycling activity. Conversely, oxidizing conditions that are associated with SH modification of 50% of the MT cysteines also decrease safe sequestration of redox-active copper such that copper-dependent redox reactions become apparent at copper : MT ratios as low as 6 : 1. Table I essentially summarizes our results, in which the abilities of fully reduced and H_2O_2-oxidized MT to mitigate copper-dependent redox cycling are compared over a range of copper : MT ratios.[7,8] The various end points to measure copper-dependent redox cycling include direct measurement of radical formation by electron paramagnetic resonance (EPR), specific oxidation of reporter molecules, and direct cytotoxicity. All these assays consistently reveal that oxidation of MT cysteines by H_2O_2 results in a

[18] P. Chen, P. Onana, C. F. Shaw III, and D. H. Petering, *Biochem. J.* **317**, 389 (1996).

decrement in the molar ratio of copper : MT at which copper-dependent redox cycling can first be observed. This implies an inverse correlation between the degree of MT SH modification and the ability of MT to bind copper in a redox-inert manner.

SH groups are also physiologically important sites for modification by NO; however, the potential of MT to regenerate reduced SH groups after nitrosative modification has received little attention. We, therefore, investigated the ability of the endogenous reductant dihydrolipoic acid (DHLA) to regenerate MT thiols after exposure to NO. Freshly reduced native MT (0.03 mM) is exposed to the NO donor, 8 mM NOC-15 (Alexis, San Diego, CA), in 10 mM Tris, pH 7.4, for 15 min at 37°. NOC-15 is removed and MT is recovered in the retentate after centrifugation four times through Microcon YM-3 filters. To test the effect of DHLA, NO-exposed MT is then incubated with 0.6 mM DHLA (Sigma) in buffer containing 100 mM desferrioxamine and 100 mM diethylenetriaminepentaacetic acid to chelate any adventitious iron or copper, respectively, for 30 min at 37°. DHLA-regenerated MT is recovered by filter ultracentrifugation as described above. Exposure of purified MT containing copper (5 mol of Cu : 1 mol of MT) to the NO donor, NOC-15, enhances redox-cycling activity of copper as shown by increases in its ability to catalyze the one-electron oxidation of ascorbate to form ascorbate radical, which is detectable by EPR spectroscopy (Fig. 1A). This effect can be reversed by the physiologically relevant reductant, DHLA. Figure 1A further shows that changes in copper-dependent redox cycling are paralleled by concomitant reductions and restoration of MT SH group content (Fig. 1B) after NO exposure and DHLA treatment, respectively. Essentially identical results are obtained if DHLA itself is replaced with an enzymatic recycling system consisting of lipoamide dehydrogenase (0.1 U/ml), lipoic acid (1.2 mM), and NADH (4 mM) (data not shown). Because the nitrosation and regeneration steps are carried out separately these effects represent a direct effect of DHLA on the modified protein and not the ability to quench RNS/ROS. Thus, DHLA could serve to couple MT to an efficient protein repair system *in vivo,* and because lipoic acid can readily be taken up by cells it is feasible to use it as a pharmacologic agent to augment protein repair, as it would be converted to bioactive DHLA by intracellular metabolism.

In this simple cell-free system the physiologically relevant functions of Cu-MT can also be examined. It is plausible that Cu-MTs can act as sources of catalytically active copper for copper-dependent enzymes such as Cu,ZnSOD. In particular, treatment of Cu$_5$-MT with NO donor NOC-15, in the presence of apo/ZnSOD, results in reconstitution of SOD activity as shown in Fig. 2. SOD activity is assessed by the ability to inhibit superoxide-driven reduction of nitroblue tetrazolium measured spectrophotometrically, as well as in native sodium dodecyl sulfate (SDS)–polyacrylamide activity gels. This reconstitution is accompanied by modification of approximately 70% of the MT cysteines. To determine whether copper transfer occurs in a redox-inert or redox-active manner similar incubations

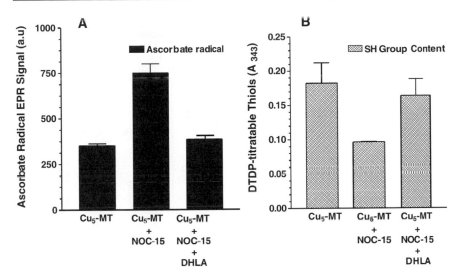

FIG. 1. DHLA regeneration of reduced sulfhydryls in MT exposed to NO and restoration of copper-binding capacity. Native, NO-exposed, and DHLA-regenerated MT were exposed to $CuSO_4$ (5 : 1 molar ratio of copper : MT) and sodium ascorbate (500 μM). EPR spectra for ascorbate radical were obtained by repeated scanning, using 335.4 mT, center field; 0.3 mT, sweep width; 0.05 mT, field modulation; 10 mW, microwave power; 0.3 sec, time constant; 20 sec, time scan. (A) Amplitude of the ascorbate radical signal obtained 8 min after incubation of the various Cu–MTs plus ascorbate. (B) MT thiol content in the retentate measured by optical absorbance at 343 nm after reaction with DTDP. Data in both (A) and (B) represent means ± SD obtained from at least three observations.

are performed in the presence of ascorbate, whose one-electron oxidation by copper can be monitored quantitatively by EPR measurement of ascorbate radical. Importantly, negligible redox-cycling activity of copper is observed during copper transfer to the apoenzyme (Fig. 2B). A detailed protocol describing the *in vitro* assessment of intramolecular transfer of copper from MT to apo/ZnSOD is provided below.

Protocol 2: Intramolecular Transfer of Copper between Cu-MT and Apo/ZnSOD

1. Apo/ZnSOD is prepared by dissolving Cu,ZnSOD prepared from bovine erythrocytes (Sigma) to a final concentration of 0.1 mM in 0.1 M potassium phosphate buffer (pH 7.4).[19] The copper chelator diethyl dithiocarbamate is added to achieve a chelator : protein ratio of 10 : 1 and incubated for 3 hr at 37°. The yellow insoluble diethyl dithiocarbamate–copper complex is removed by centrifugation at 39,000g for 30 min at 4°. The resultant supernatant is then exhaustively dialyzed against phosphate buffer and stored at 4°.

[19] D. Cocco, L. Calabrese, A. Rigo, F. Marmocchi, and G. Rotilio, *Biochem. J.* **199,** 675 (1981).

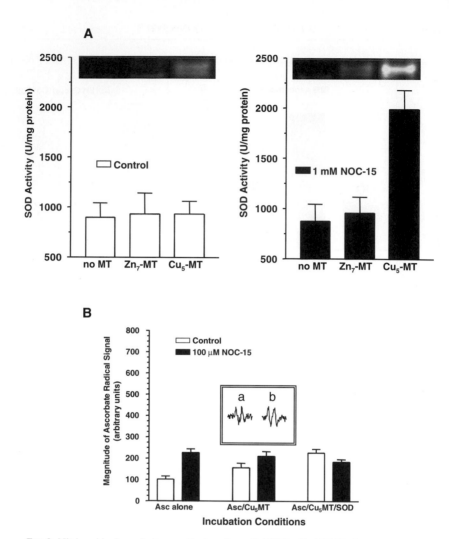

FIG. 2. Nitric oxide-dependent reconstitution of apo-ZnSOD by Cu-MT. The intramolecular transfer of copper from Cu_5-MT to apo-ZnSOD in the presence and absence of 100 μM NOC-15 was performed as outlined in protocol 2. (A) Resultant SOD activity measured on the basis of the inhibition of superoxide-driven nitroblue tetrazolium reduction. *Insets:* Corresponding SOD activity gels developed after native polyacrylamide gel electrophoresis was applied to the samples. (B) Similar incubation with Cu_5-MT and apo-ZnSOD (both 1 μM) with and without 100 μM NOC-15 but conducted in the presence of 40 μM ascorbate (Asc) as a reporter molecule of copper-dependent redox cycling. Redox-active copper catalyzes the one-electron oxidation of ascorbate, which was measured by electron paramagnetic resonance (EPR) of ascorbate radical. Data represent the magnitude of the EPR ascorbate radical signal recorded over 8 min. *Inset:* Magnitude of the ascorbate radical signals obtained with Cu_5-MT alone in the absence (a) and presence (b) of the NO donor; the magnitudes were essentially indistinguishable from background signal obtained in the absence of any MT. Note that the transfer of copper from MT to apo-ZnSOD was not accompanied by any enhanced redox activity of copper.

2. Cu MT containing various molar ratios of copper to MT ranging from Cu_1-MT to Cu_{10}-MT can be prepared by incubating Zn-MT (freshly reduced as described in protocol 1) in the presence of varying concentrations of $CuSO_4$ in 50 mM phosphate buffer. For example, to prepare Cu_5-MT 40 μM Zn-MT is incubated with 200 μM $CuSO_4$ in 50 mM phosphate buffer containing 800 μM sodium ascorbate for 20 min at room temperature. Ascorbate is included to maintain copper ions in the cuprous (Cu^+) state required for binding to MT. The much higher affinity of MT for copper relative to zinc assures the rapid equilibration and stoichiometric addition of copper to MT at subsaturating copper concentrations.

3. Coincubations of Cu-MT and apo/ZnSOD are performed by combining Cu-MT with apo/ZnSOD (both at 10 μM) in 0.1 ml of 50 mM phosphate buffer and incubating for 2 hr at room temperature. Oxidants, NO generators, or other thiol-active agents can be added to parallel tubes to assess the requirement of SH modification for copper transfer. EDTA (0.1 mM, final concentration) is added at the end of incubation to prevent further reconstitution. After incubation the activity of SOD can be directly measured by the inhibition of superoxide-driven nitroblue tetrazolium reduction[20] or by SOD activity gels developed after native polyacrylamide gel electrophoresis.[21]

Experiments with Intact Cells. The modification of MT SH groups has been implicated as a mechanism to account for the reported antioxidant activity of MT, as well as modulation of metal binding and release. However, describing the quantitative relationships between MT oxidation/nitrosation and MT function within cells is often compromised by lack of sensitivity and the problems inherent in complex mixtures of intracellular protein that contain relatively small amounts of MT. Therefore, chromatographic separation of MTs in conjunction with sensitive biochemical analysis of cysteine redox status is required to achieve fully quantitative results under these conditions. Protocol 3, given below, describes our approach to assess the redox status of MT in intact cells.

Protocol 3: NO-Dependent Modification of Metallothionein Thiols in Intact HL-60 Cells

1. MT can be induced in actively growing HL-60 cells by treatment with 150 μM $ZnCl_2$ in RPMI 1640 containing 12% (v/v) fetal bovine serum (FBS) at 37° under a 5% CO_2 atmosphere for 48 hr. $ZnCl_2$ is removed by centrifugation and the recovered cells can then be returned to culture for exposure to SH-modifying conditions, such as the NO donor NOC-15 (1 mM) for 1 hr.

2. Cells (1×10^8 per treatment) are then recovered by centrifugation (400g, 10 min, 4°), washed twice with cold phosphate-buffered saline (PBS), and

[20] J. F. Ewing and D. R. Janero, *Anal. Biochem.* **232,** 243 (1995).

[21] C. Beauchamp and I. Fridovich, *Anal. Biochem.* **44,** 276 (1971).

subjected to sonication for 1 min in 1 ml of N_2-saturated 50 mM phosphate buffer, pH 7.8, containing 2 mM EDTA, using a 4710 series ultrasonic homogenizer (Cole-Palmer Instrument, Chicago, IL).

3. The resulting homogenate is cleared by centrifugation at 40,000g for 20 min at 4°. The supernatant is then applied to a Sephadex G-75 column (1 × 40 cm) and eluted with N_2-saturated 10 mM Tris-HCl, 2 mM EDTA, pH 7.8. The column is run at 4° at a flow rate of 0.5 ml/min and fractions (0.5 ml) are collected.

4. The SH group content of each fraction can be determined by adding a 20-μl aliquot of each fraction to 5 μl of 10 mM dithiodipyridine in 0.5 M sodium acetate, pH 4.0. After incubation at room temperature for 30 min the absorbance of each sample is read at 343 nm, using a SpectroMate UV–Vis fiberoptic micro-spectrophotometer (World Precision Instruments) equipped with a 15-μl capillary microcuvette with internal reflecting surface and a resulting optical band path of 10 cm.

5. In addition, the presence of MT protein in the fractions can be assessed by an immuno-dot-blot procedure.[22] Briefly, a 20-μl aliquot of each fraction is mixed with 20 μl of 3% (v/v) glutaraldehyde and applied to a nitrocellulose membrane, using a Bio-Dot microfiltration apparatus (Bio-Rad, Hercules, CA). The membrane is blocked with 5% (w/v) fat-free milk in TBST [50 mM Tris-HCl (pH 7.5), 200 mM NaCl, 0.05% (v/v) Tween 20] for 1 hr and then incubated with E9 mouse monoclonal anti-MT antibody (Dako, Carpinteria, CA)(1 : 500 in TBST). After washing (six times with TBST), horseradish peroxidase-conjugated polyclonal goat anti-mouse IgG (PharMingen, San Diego, CA) is added for 1 hr (1 : 5000 in TBST). The signal is developed by enhanced chemiluminescence, using a SuperSignal West Pico chemiluminescence kit (Pierce Chemical, Rockford, IL) and exposure to X-ray film.

The first hurdle to overcome is the relative lack of sensitive biochemical methods to measure small changes in the relatively low amounts of MT available in cells. Various fluorigenic thiol-specific probes such as ThioGlo (Covalent Associates, Woburn, MA) are available but we have found their specific application to MT to be problematic. Such probes appear to grossly underestimate the amount of available thiols, presumably because of difficulty in removing bound metal ions that may competitively inhibit ThioGlo interaction with SH groups. In addition, nearly all the MT cysteines within the two thiolate clusters represent contiguous (Cys-Cys) or vicinal (Cys-X-Cys) pairs. Therefore, reaction between ThioGlo and one cysteine may limit the accessibility of SH-active reagent to the neighboring cysteine. Alternatively, a significant self-quenching of fluorescence may also limit

[22] C. A. Mizzen, N. J. Cartel, W. H. Yu, P. E. Fraser, and D. R. McLaughlin, *J. Biochem. Biophys. Methods* **32,** 77 (1996).

fluorescence yield when multiple reporter molecules exist in close proximity to each other within the thiolate clusters of MT.

We, therefore, have chosen to apply traditional spectrophotometric titration with the thiol reagent 2,2′-dithiodipyridine (DTDP) combined with contemporary fiberoptic-based sensitive spectrophotometry. We utilize a SpectroMate UV–Vis fiberoptic microspectrophotometer (World Precision Instruments) supplied with a 15-μl capillary microcuvette with internal reflecting surface and a resulting optical band path of 10 cm. This provides for highly sensitive measurement of weakly absorbing samples in microvolumes.

The second difficulty is to specifically measure MT SH groups in the presence of much higher amounts of glutathione (GSH) and high molecular weight (HMW) protein thiols. For this, we have taken advantage of the small size of MT and applied gel-filtration chromatography to resolve an MT-enriched fraction separate from other proteins as well as low molecular weight thiols. Figure 3 shows typical elution profiles of DTDP-reactive thiols contained in cell lysates derived from intact control HL-60 cells (solid stars) and cells treated with the MT inducer zinc for 24 hr (solid circles). Note the prominent appearance of an intermediate peak of SH groups between HMW proteins and glutathione observed after zinc pretreatment, which most likely represents MT cysteines. It is important, however, to firmly establish the presence of MT protein by an independent measure. To this end immunological identification of MT protein, which is likely to be independent of its redox status, may be performed. Both monoclonal and polyclonal MT-specific antibodies have been used.[23,24] As shown in Fig. 3B, treatment of HL-60 cells with zinc causes a manyfold increase in the amount of immunoreactive MT (detected with E9 monoclonal antibody; Dako) within fractions corresponding to the intermediate peak of thiols. No immunoreactivity is detected in untreated control cells. In addition, MT immunoreactivity is never observed in the HMW or GSH fractions. This confirms that MT is confined to these intermediate fractions. This alone, however, does not exclude the presence of other SH-containing proteins similar in molecular mass to MT within this fraction. This issue can be resolved, at least in part, by quantifying the total protein content in the fraction and comparing its molar content of SH groups with the predicted 20 : 1 ratio in intact MTs.

After characterizing this peak as representing MT thiols, assessment of redox status after oxidative/nitrosative challenge can be performed. Figure 3A demonstrates that the amount of cysteines within the MT peak is decreased by about 70% after exposure of live zinc-pretreated cells to a donor of nitric oxide, NOC-15. Essentially no change is observed in other protein thiols, and the decrement in GSH is less than that observed in MT, suggesting that MT cysteines

[23] B. Jansani and M. E. Elmes, *Methods Enzymol.* **205,** 95 (1991).
[24] J. S. Garvey, *Methods Enzymol.* **205,** 141 (1991).

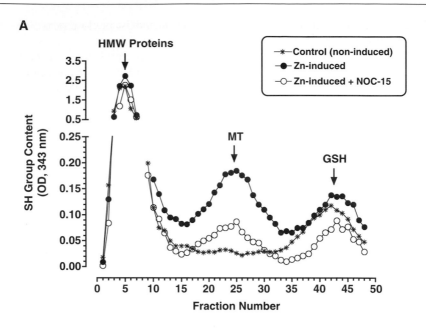

FIG. 3. NO-induced modification of MT thiols in HL-60 cells. (A) Typical column elution profiles of reduced thiols from cell homogenates obtained from HL-60 cells under control conditions (solid stars) and after MT induction by 48-hr exposure to 150 μM ZnCl$_2$ (solid circles). Note the prominent appearance of an intermediate peak of reduced SH groups between high molecular weight (HMW) proteins and glutathione (GSH) in zinc-induced cells. In addition, it appears the treatment of zinc-induced cells with 1 mM NOC-15 for 1 hr (open circles) produced a substantial reduction in the SH group content of the MT peak. To specifically identify the middle peak as MT, the various column fractions were analyzed for MT immunoreactivity. (B) Only the fractions corresponding to the middle peak obtained from zinc-induced cells contain significant MT-immunoreactive protein. Furthermore, in contrast to the effect of thiol content, exposure of zinc-induced cells to NO did not alter the amount of immunoreactive MT contained within the middle peak.

represent hypersensitive thiols that may be preferentially modified during oxidative/nitrosative challenge. No change in immunoreactive protein is observed after NO treatment (Fig. 3B), indicating that the loss of SH groups, indeed, reflects the degree of cysteine modification within MT. Further details regarding domain specificity for these effects could be obtained by proteolytic mapping of modified MT.

Although the approaches outlined above facilitate identification of modifications of MT from intact cells, it is oftentimes desirable to obtain such information from live cells in a more dynamic fashion. To this end, we followed the example of the Ca^{2+} indicator, cameleon-1,[25] and constructed a chimera in which a yellow–green fluorescent protein (GFP) variant, that is, enhanced yellow fluorescent protein (EYFP), and a cyan GFP variant (ECFP) were fused to the C and N termini, respectively, of human MTIIa (see Fig. 4). We used liposome-mediated transfer of this expression vector into primary cultures of sheep pulmonary artery endothelial cells and imaged these cells on a Nikon (Garden City, NY) inverted microscope with a Photometrics (Roper Scientific, Trenton, NJ) cooled charge-coupled device camera controlled by ISEE software (Inovision, Raleigh, NC).[26] The cyan and yellow GFPs acted as donor and acceptor, respectively, for fluorescence resonance energy transfer (FRET) and, hence, revealed changes in intramolecular proximity and relative orientation of the fluorophores. By constantly monitoring the emissions ratio of the acceptor (535 nm) to donor (480 nm) molecule, conformational changes in MT can be inferred, including a relative decrease in FRET when MT is modified in such a way as to lose metal. Examples of this include exposure of the purified protein for several hours to the metal chelator EDTA (and NaCl) or the exposure of intact cells to medium saturated with gaseous NO.

Although these approaches permit quantitative assessment of the amount of cysteine modification, they give no information regarding the chemical nature or identity of the resulting products. Generally speaking, a fraction of these modified cysteines may harbor covalently adducted NO to form S-nitrosothiols (MT-SNO). In addition, other oxidation species may be formed such as disulfides or other higher S-oxides (sulfenic, sulfinic, and sulfonic acids).[27,28] The latter pathway may be prevalent in highly oxygenated tissues such as lung. Direct analysis of S-NO within MT can be accomplished by a variety of techniques, one of which is described elsewhere in this series.[29] This technique is based on UV-induced

[25] A. Miyawaki, J. Llopos, R. Heim, J. M. McCaffrey, J. A. Adams, M. Ikura, and R. Y. Tsien, *Nature* (*London*) **388**, 882 (1997).

[26] L. L. Pearce, R. E. Gandley, W. Han, K. Wasserloos, M. Stitt, A. J. Kanai, M. K. McLaughlin, B. R. Pitt, and E. S. Levitan, *Proc. Natl. Acad. Sci. U.S.A.* **97**, 477 (2000).

[27] C. T. Aravindakumar, J. Ceulemans, and M. De Ley, *Biochem. J.* **344**, 253 (1999).

[28] A. R. Quesada, R. W. Byrnes, S. O. Krezoski, and D. H. Petering, *Arch. Biochem. Biophys.* **334**, 241 (1996).

[29] V. A. Tyurin, Y. Y. Tyurina, S.-X. Liu, H. Bayir, C. A. Hubel, and V. E. Kagan, *Methods Enzymol.* **352**, 347 (2002).

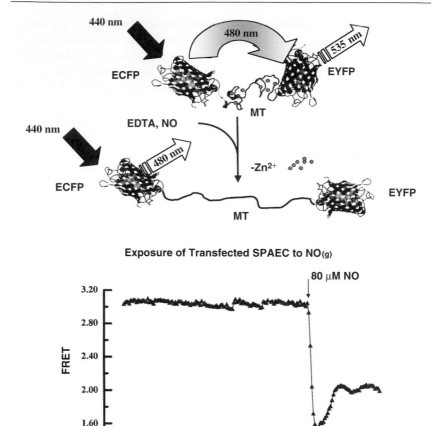

FIG. 4. Modulation by nitric oxide of fluorescence resonance energy transfer (FRET) in an MT–green fluorescent protein chimera. *Top:* Schematic of MT chimera, in which cDNA for hMTIIa is flanked by enhanced cyan fluorescent protein (ECFP) and enhanced yellow fluorescent protein (EYFP) on the N and C termini. Under control conditions, MT is associated with seven zinc molecules and is in a relative folded conformation. *Bottom:* After exposure to EDTA (and NaCl) or medium that was gassed with NO to reach a final concentration of 80 μM, MT is predicted to lose zinc molecules and unfold. The conformational change is monitored by exciting the cells with a source at 440 nm and continuously monitoring the emission ratio of acceptor (535 nm) to donor (480 nm). SPAEC, sheep pulmonary artery endothelial cells. [The authors acknowledge Dr. Linda L. Pearce (University of Pittsburgh) for providing Fig. 4.]

liberation of free NO from nitrosothiol and its subsequent specific detection with the fluorogenic reagent 4,5-diaminofluorescein to produce a highly fluorescent triazole derivative.[30] Interestingly enough, this approach reveals that only a small fraction (between 2 and 3%) of the total modified thiols in MT could be accounted for by S-NO. This suggests that the remainder of these modified MT cysteines underwent further oxidation after the initial nitrosylation event. A simple estimation of the disulfide portion among the oxidized cysteines can be performed with standard disulfide reducing reagents, such as DTT. Under the specific conditions outlined in Fig. 3 exposure of HL-60 cells to NO resulted in approximately 30% of DTT-recoverable cysteines in MT. The remaining 70% of the modified SH groups are presumably represented by higher S-oxides that were insensitive to DTT reduction. This is not surprising, given that the NO exposure was performed in aerobically incubated cell suspensions. It remains to be clarified whether this relatively high level of nonregenerable cysteines in MT is typical *in vivo*.

Summary

In summary, the redox conversions of MT cysteines are likely to be the principal mechanisms for regulation of metal binding and release by this protein. Oxidative and/or nitrosative challenges can serve to promote metal ion release from MT to render their delivery to specific target proteins. It is tempting to consider the potential roles of MTs as redox sensors because of their high sensitivity to cysteine modification, as well as their potential to amplify signals by releasing multiple metal ions. In other words, MTs may act early in a biological signaling cascade that triggers metal-dependent biochemical and cellular responses. Alternatively, uncontrolled release of metals by excessive oxidative stress may contribute to metal toxicity. Because oxidative and nitrosative signaling is ubiquitous within cells, the physiological function of MT demands that efficient recycling of modified cysteines be operative. Little is known regarding the potential mechanisms for the regeneration of MT after oxidative/nitrosative modification, but they may involve endogenous dithiols, such as thioredoxin, and pharmacologically relevant dithiols, such as dihydrolipoate.

Acknowledgments

This work was funded in part by an EPA STAR Grant (R827151) and by the NIH (HL-32154).

---

[30] H. Kojima, N. Nakatsubo, K. Kikuchi, S. Kawahara, Y. Kirino, H. Nagoshi, Y. Hirata, and T. Nagano, *Anal. Chem.* **70,** 2446 (1998).

[26] SIR2 Family of NAD$^+$-Dependent Protein Deacetylases

By Jeffrey S. Smith, Jose Avalos, Ivana Celic, Shabazz Muhammad, Cynthia Wolberger, and Jef D. Boeke

Introduction

Nicotinamide adenine dinucleotide (NAD$^+$) is a key cofactor in numerous metabolic pathways, including glycolysis and the citric acid cycle. The reduction of NAD$^+$ in these reactions produces NADH, which subsequently acts as the key electron carrier in the oxidative phosphorylation process of ATP generation. Oxidative phosphorylation oxidizes NADH to regenerate NAD$^+$. The NAD$^+$: NADH and NADP$^+$: NADPH concentration ratios largely define the redox state of a cell. Changes in the redox state can dramatically affect multiple cellular processes including stress response, apoptosis, and aging. Some of these responses are due to the production of reactive oxygen species (ROS), but others are directly related to NAD$^+$ itself.

The link between NAD$^+$ and chromatin structure actually goes back to the 1970s, when it was demonstrated that a large amount of NAD$^+$ synthesis occurs in the nucleus of vertebrate cells.[1] Furthermore, one of the enzymes involved in NAD$^+$ synthesis was shown to be associated with chromatin.[2,3] It turns out there are at least two chromatin-related enzymatic activities that utilize NAD$^+$ as substrates in the nucleus. The first is poly(ADP-ribose) polymerase (PARP, EC 2.4.2.30), which hydrolyzes NAD$^+$ and transfers the ADP-ribose moiety to itself and probably some other proteins. This activity is triggered by double-stranded DNA breaks that are bound by PARP.[4] In DNA-damaged cells PARP activity can quickly deplete the cellular NAD$^+$ pool, leading to necrotic cell death. The second chromatin-related enzymatic activity, identified more recently, is the NAD$^+$-dependent deacetylase activity of Sir2p.[5–7] The silent information regulator 2 (Sir2) family of proteins is highly conserved, from bacteria to humans, and plays critical roles in

---

[1] M. Rechsteiner and V. Catanzarite, *J. Cell Physiol.* **84,** 409 (1974).

[2] G. Magni, N. Raffaeli, M. Emanuelli, A. Amici, P. Natalini, and S. Raggieri, *Methods Enzymol.* **280,** 248 (1997).

[3] G. Hogeboom and W. Schneider, *J. Biol. Chem.* **197,** 611 (1952).

[4] N. A. Berger, *Radiat. Res.* **101,** 4 (1985).

[5] J. Landry, A. Sutton, S. T. Tafrov, R. C. Heller, J. Stebbins, L. Pillus, and R. Sternglanz, *Proc. Natl. Acad. Sci. U.S.A.* **97,** 5807 (2000).

[6] S.-J. Lin, P.-A. Defossez, and L. Guarente, *Science* **289,** 2126 (2000).

[7] J. S. Smith, C. B. Brachmann, I. Celic, M. A. Kenna, S. Muhammad, V. J. Starai, J. L. Avalos, J. C. Escalante-Semerena, C. Grubmeyer, C. Wolberger, and J. D. Boeke, *Proc. Natl. Acad. Sci. U.S.A.* **97,** 6658 (2000).

Copyright 2002, Elsevier Science (USA).
All rights reserved.
0076-6879/02 $35.00

transcriptional silencing and cellular life span regulation.[6,8,9] In this chapter we describe SIR2 function and the relevant experimental techniques used in its study.

Sir2p as General Regulator of Transcriptional Silencing in Yeast

SIR2 is the founding member of the SIR2 family of proteins. It was originally identified from the budding yeast, *Saccharomyces cerevisiae*, as a gene required for transcriptional silencing of the *HML* and *HMR* silent mating type loci.[10,11] Mutations in *SIR2* result in derepression of α and **a** mating-type information genes located in *HML* and *HMR*, respectively. The result is expression of α and **a** mating-type information in the same haploid cell, which results in formation of the **a**1/α2 repressor complex, repression of haploid-specific genes, and a nonmating phenotype. There are three other *SIR* genes (*SIR1*, *SIR3*, and *SIR4*) that are required for silencing at *HMR* and *HML*. *SIR2*, along with *SIR3* and *SIR4*, are required for transcriptional silencing at yeast telomeres.[12] In this case, mutations in *SIR2* derepress silenced reporter genes that are integrated adjacent to a telomere. More recently, silencing in the ribosomal DNA locus was shown to require *SIR2*, but not *SIR1*, *SIR3*, or *SIR4*. In this case, *sir2* mutations cause a loss of recombinational suppression between rDNA repeats,[13] and also derepression of polymerase II (Pol II)-transcribed reporter genes that are artificially integrated in the rDNA.[14,15] Silencing in the rDNA is significant because it demonstrates that Sir2p can function in silencing independently of the other Sir proteins, and therefore must have an important activity, which turns out to be NAD$^+$-dependent histone deacetylase activity.

Identification and Comparisons of SIR2 Gene Family

Additional *SIR2* family members have been identified through a combination of degenerate polymerase chain reaction (PCR), low-stringency hybridization, and database searches. In yeast alone, four additional homologs were identified: *HST1*, *HST2*, *HST3*, and *HST4*.[8,9] All Sir2-like proteins share a common core domain of approximately 250 amino acids that contains several highly conserved blocks of amino acids. The two motifs that are most diagnostic of the Sir2p family are GAGISTS(L/A)GIPDFR and YTQNID. See Fig. 1A[15a] for a sequence alignment of the Sir2 core domain from various eukaryotic species. For clarity, an alignment

[8] C. B. Brachmann, J. M. Sherman, S. E. Devine, E. E. Cameron, L. Pillus, and J. D. Boeke, *Genes Dev.* **9**, 2888 (1995).

[9] M. K. Derbyshire, K. G. Weinstock, and J. N. Strathern, *Yeast* **12**, 631 (1996).

[10] J. Rine and I. Herskowitz, *Genetics* **116**, 9 (1987).

[11] J. M. Ivy, A. J. Klar, and J. B. Hicks, *Mol. Cell. Biol.* **6**, 688 (1986).

[12] O. M. Aparicio, B. L. Billington, and D. E. Gottschling, *Cell* **66**, 1279 (1991).

[13] S. Gottlieb and R. E. Esposito, *Cell* **56**, 771 (1989).

[14] M. Bryk, M. Banerjee, M. Murphy, K. E. Knudsen, D. J. Garfinkel, and M. J. Curcio, *Genes Dev.* **11**, 255 (1997).

[15] J. S. Smith and J. D. Boeke, *Genes Dev.* **11**, 241 (1997).

[15a] R. A. Frye, *Biochem. Biophys. Res. Commun.* **273**, 793 (2000).

A

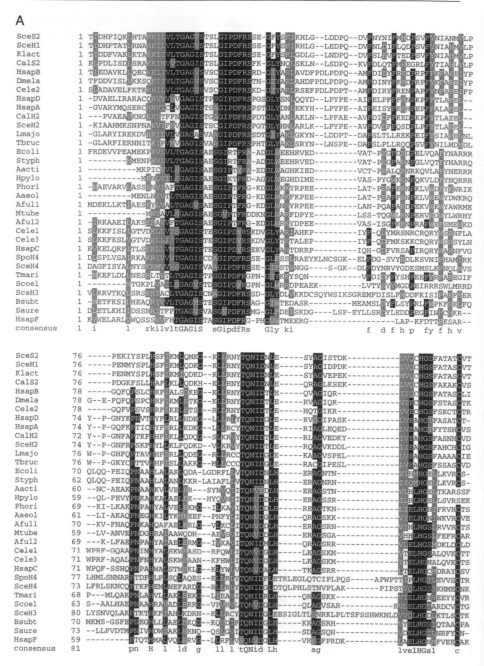

FIG. 1.

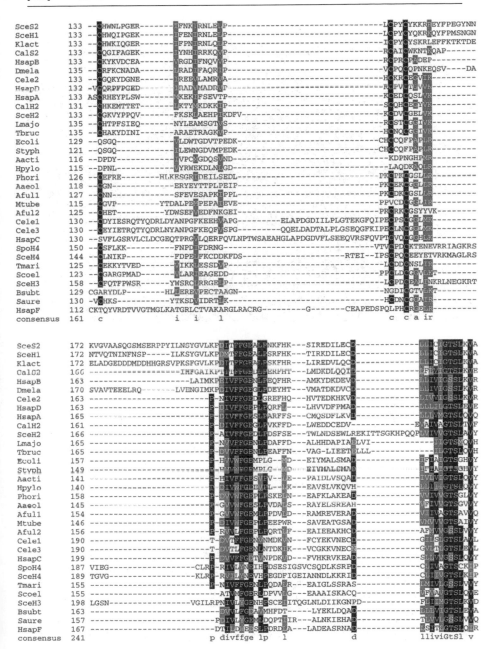

FIG. 1.

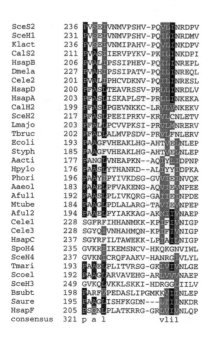

FIG. 1. (A) Sequence alignment of a wide variety of Sir2 proteins from eukaryotic and prokaryotic sources. Alignments are courtesy of the CLUSTAL W server at the BCM search launcher (Baylor College of Medicine) and are shaded with Boxshade, using default parameters. Abbreviations are as follows: eukaryotes include SceS2, *S. cerevisiae* Sir2; SceH1-4, *S. cerevisiae* Hst1-4; Klact, *Kluyveromyces lactis* Sir2; CalS2, *Candida albicans* Sir2; Cele3, *Caenorhabiditis elegans* Sir2; Dmela, *Drosophila melanogaster;* HsapA-F, human Sir2A–F; CalH2, *Candida albicans* Hst2; Lmajo, *Leishmania major;* Tbruc, *Trypanosoma brucei.* Prokaryotes and archaea include Ecoli, *Escherichia coli;* Styph, *Salmonella typhimurium;* Aacti, *Actinobacillus actinomycetecomitans;* Hpylo, *Helicobacter pylori;* Phori, *Pyrococcus horikoshii;* Aaeol, *Aquifex aeolius;* Aful1 and Aful2, *Archaeoglolobus fulgidus* Sir2-Af1,2; Mtube, *Mycobacterium tuberculosis;* SpoH4, *Schizosaccharomyces pombe* Hst4; Tmari, *Thermotoga maritima,* Scoel, *Streptomyces coelicolor;* Bsubt, *Bacillus subtilis;* Saure, *Staphylococcus aureus.* The human Sir2 proteins have also been referred to with an alternative nomenclature by R. A. Frye, *Biochem. Biophys. Res. Commun.* **273**, 793 (2000). Sir2A = SIRT2; Sir2B = SIRT1; Sir2C = SIRT4; Sir2D = SIRT3; Sir2F = SIRT6. (B) Sequence alignment of a wide variety of Sir2 proteins from archaea and prokaryotes only.

B

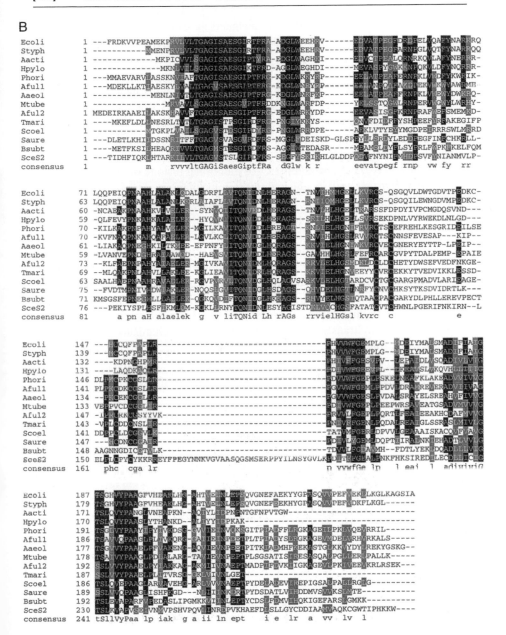

FIG. 1. (*continued*)

of just the bacterial Sir2p species is included as well. Note that the bacterial species have much shorter N and C termini and are more conserved as a group. Interestingly, the region surrounding the GAGISTS motif does have some weak similarity to a region of bacterial 6-phosphogluconate dehydrogenases, an NAD^+-binding protein.[16] This enzyme uses $NADP^+$ or NAD^+ as a cofactor in its activity and one-half of its binding site is similar to Sir2p. Indeed, in the crystal structure of an archaeal Sir2 protein, these motifs contain NAD^+-binding residues.[17] Sir2p has been reported to have a weak mono(ADP)-ribosyltransferase activity in addition to its deacetylase activity.[16] Beyond these small features, the Sir2 protein family does not have any other characteristics of traditional histone deacetylases, which is actually consistent with its unusual reaction mechanism.

Several publications have examined the reaction mechanism and discovered that NAD^+ is not simply a cofactor, but is actually turned over.[18–20] For every acetyl group removed from a lysine residue, one molecule of NAD^+ is hydrolyzed to form one molecule of nicotinamide and one molecule of a novel compound called 1'-O-acetyl-ADP-ribose.[19,20] This compound is formed by transfer of the acetyl group from the acetyllysine-containing peptide substrate to the ADP-ribose moiety of NAD^+. Several different chemical mechanisms have been proposed for this reaction, including both covalent enzyme–ADP-ribose intermediates as in the case of ADP-ribosyltransferases and more complex mechanisms that do not require such intermediates. The stereochemistry of the acetyl-ADP-ribose should definitively settle the matter. The existence of this novel compound raises a whole new series of questions, as this molecule could play a signaling role in the cell.[20]

Sir2p as NAD^+-Dependent Protein/Histone Deacetylase

Even though Sir2p does not have any homology to other known histone deacetylases, its high degree of evolutionary conservation suggested that it might still be an enzyme. Speculation that it might be a deacetylase originated from the Broach laboratory, which demonstrated that *SIR2* overexpression resulted in bulk histone hypoacetylation,[21] which was consistent with Sir2p being a deacetylase. Multiple laboratories unsuccessfully attempted to show that Sir2p was a conventional histone deacetylase. The key breakthrough came when a protein involved in vitamin B_{12} (cobalamin) synthesis in *Salmonella typhimurium* (CobB) was found to be a SIR2 family member.[22] This was important because CobB can carry out a phosphoribosyltransferase-like reaction utilizing an NAD^+ intermediate (NaMN)

[16] J. C. Tanny, G. J. Dowd, J. Huang, H. Hilz, and D. Moazed, *Cell* **99,** 735 (1999).

[17] J. Min, J. Landry, R. Sternglanz, and R. M. Xu, *Cell* **105,** 269 (2001).

[18] J. Landry, J. T. Slama, and R. Sternglanz, *Biochem. Biophys. Res. Commun.* **278,** 685 (2000).

[19] J. C. Tanny and D. Moazed, *Proc. Natl. Acad. Sci. U.S.A.* **98,** 415 (2001).

[20] K. G. Tanner, J. Landry, R. Sternglanz, and J. M. Denu, *Proc. Natl. Acad. Sci. U.S.A.* **97,** 14178 (2000).

[21] M. Braunstein, A. B. Rose, S. G. Holmes, C. D. Allis, and J. R. Broach, *Genes Dev.* **7,** 592 (1993).

[22] A. W. Tsang and J. C. Escalante-Semerena, *J. Biol. Chem.* **273,** 31788 (1998).

as a substrate. It was subsequently demonstrated that CobB and a human Sir2 family member could transfer a radiolabel from NAD^+ to an exogenous protein (bovine serum albumin, BSA) *in vitro*.[23] This was interpreted to mean that Sir2p had protein-ADP-ribosyltransferase activity. Yeast Sir2p was then reported to have a weak mono(ADP)-ribosyltransferase activity on histones and itself *in vitro*.[16]

NAD^+ is the substrate for ADP-ribosyltransferase activity of Sir2p and therefore is included in the *in vitro* reactions. In testing whether acetylated histones are substrates for ADP-ribosylation, the Guarente and Sternglanz laboratories found that purified Sir2p also had an NAD^+-dependent histone deacetylase activity with specificity for Lys-16 of histone H4 and Lys-9 and Lys-14 of histone H3.[5,24] Other Sir2 family members, including the bacterial CobB protein and an Archea bacterial homolog, also harbored NAD^+-dependent deacetylase activity.[7] This activity can easily be detected *in vitro* with whole cell yeast extracts and hyperacetylated chicken histones, which allows direct testing of mutant cells for the presence of activity.[7] Interestingly, the Hst2 protein contributes most of the NAD^+-dependent deacetylase activity from whole cell extracts.[7] This may reflect the fact that Hst2p is localized in the cytoplasm and is simply extracted more efficiently.[25] A study that directly compares the activity of multiple purified Sir2 family members has not been carried out to date.

Another significant question remaining concerns the *in vivo* targets for the Sir2 family members. At this time it is still not clear whether Sir2p deacetylates histones H3 and H4 *in vivo*, even though some *in vitro* targets (histones H3 and H4) have been identified. Bacteria do not have histones, and therefore the targets for their Sir2-like proteins will definitely not be histones. Furthermore, the CobB protein already has a known function in cobalamin synthesis that, interestingly, is more similar to ADP-ribosyltransferase activity. Interestingly, yeast SIR2 and a human SIR2 (SIR2A) can partially complement a *cobB* mutant *Salmonella* cell line, suggesting that whatever activity CobB protein has, the yeast and human proteins can also do it (C. B. Brachmann, E. Caputo, J. S. Smith, I. Celic, V. J. Starai, J. C. Escalante-Semerena, and J. D. Boeke, unpublished data, 2000). The deacetylation activity of Sir2p has been definitively shown to be important for silencing, but the putative ADP-ribosyltransferase activity was not required. This does not rule out the possibility that ADP-ribosylation could occur *in vivo* and be involved in a different process.

Three-Dimensional Structure of Archaeal Sir2p

The structure of an Sir2 protein from the archaebacterium *Archaeoglobus fulgidus* has been determined.[17] This species contains two Sir2-like genes; the

[23] R. A. Frye, *Biochem. Biophys. Res. Commun.* **260,** 273 (1999).

[24] S.-i. Imai, C. M. Armstrong, M. Kaeberlein, and L. Guarente, *Nature (London)* **403,** 795 (2000).

[25] S. Perrod, M. M. Cockell, T. Laroche, H. Renauld, A. L. Ducrest, C. Bonnard, and S. M. Gasser, *EMBO J.* **20,** 197 (2001).

encoded proteins share 64% sequence identity, suggesting that the two proteins have diverged, possibly to recognize different substrates. Indeed, one of the two proteins, Sir-Af2, can recognize and deacetylate histones,[7] whereas the other cannot.[17]

The structure of Sir2p, like that of many other nucleotide-binding enzymes, is based on a Rossman fold (Fig. 2). The NAD^+ sits on the surface of the Rossman fold but in an unusual conformation relative to that of other NAD^+-binding enzymes. The nicotinamide moiety is inferred to inhabit a sizable cavity in the surface, within which it is presumed that catalysis occurs. The roof of the cavity is formed by a flexible loop that assumes different conformations in two different crystal forms

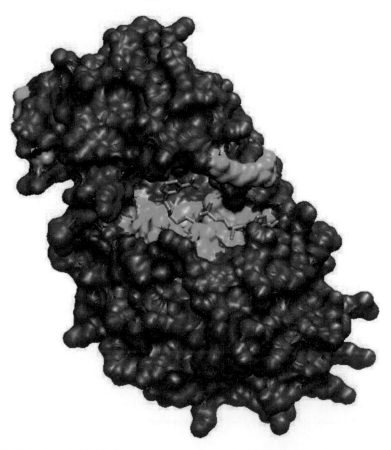

FIG. 2. Surface of Sir2-Af1 from crystal structure, showing invariant residues in green, chemically conserved residues in blue, and NAD^+ in yellow. The Rossman fold is the lower half of the structure; the deep cleft on the left side is a proposed acetyllysine side chain-binding site.

studied. Protruding from the Rossman fold domain are two inserted sequences: a zinc-binding domain and a helical domain that connects the loop to the Rossman fold.

Regulation of Silencing by Cellular NAD$^+$ Concentration

The finding that Sir2p is an NAD$^+$-dependent histone deacetylase raises the possibility that chromatin structure associated with silencing is linked to cellular metabolism. Indeed, mutations in an NAD$^+$ salvage pathway gene, *NPT1*, have been shown to compromise all three forms of silencing in yeast.[7] Deletion of *npt1* also causes a 3- to 4-fold decrease in the intracellular NAD$^+$ concentration.[7] Because Sir2p depends on NAD$^+$ for its silencing function, the reduction in cellular NAD$^+$ concentration likely represses the deacetylation activity of Sir2p. Another metabolism/silencing/chromatin link is that the extension of yeast life span by caloric restriction depends on *SIR2* and *NPT1*.[6] On the basis of these findings it has been proposed that Sir2p is a molecular sensor of changes in the cellular redox/metabolic state that translates this information to alterations in chromatin structure.[6,26] In this chapter we describe methods associated with the analysis of SIR2 function in silencing and the deacetylation reactions.

Methods

A variety of genetic and biochemical methods useful for the study of Sir2 and silencing is included below.

Silencing Assays

Transcriptional silencing can be assayed by several different methods including RNA blot analysis, efficiency of plating (spot assays), and colony color assays. The assays are based on repression of a *URA3* or *MET15* reporter gene that is integrated into the ribosomal DNA tandem array. Examples of the two different assays showing the effect of deleting *SIR2* are shown in Fig. 3.

Spot Assays

Grow the appropriate reporter strains as 1- to 2-cm^2 patches on rich yeast extract-peptone-dextrose (YPD) agar plates overnight at 30°. Use selective medium when the strains harbor a selectable plasmid. All media contain glucose as the carbon source. Using a wooden applicator stick, scrape cells from the agar surface and resuspend them in 1 ml of sterile water in a 1.5-ml microcentrifuge tube. Vortex the cell mixture and measure the absorbance of the mixture at 600 nm

---

[26] L. Guarente, *Genes Dev.* **14,** 1021 (2000).

A

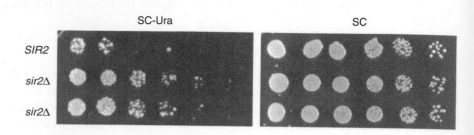

B

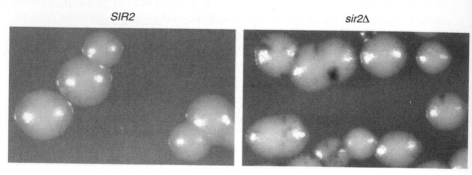

FIG. 3. The requirement of *SIR2* for rDNA silencing. (A) Spot assay. Reporter strains containing *mURA3* in the rDNA were spotted onto either SC-Ura or nonselective SC medium and allowed to grow for 3 days. Loss of silencing in the *sir2Δ* strains is indicated by increased growth on the SC-Ura plates. (B) Colony color assay. Wild-type or *sir2Δ* strains containing *MET15* in the rDNA were streaked onto rich medium containing lead nitrate. A lighter colony color and increased numbers of dark brown sectors indicate loss of silencing in the *sir2Δ* strain.

in a spectrophotometer. If testing multiple strains on the same plate, normalize each culture to an A_{600} of 1.0 in a new 1.5-ml microcentrifuge tube. Starting with an A_{600} of 1.0, serially dilute the cells at a 1 : 5 ratio (40 μl into 160 μl) with water. The serial dilutions are conveniently carried out in a sterile 96-well microtiter dish, using an 8-channel multipipette. Using the multipipette, spot 5 μl of each dilution onto a nonselective control plate such as YPD or synthetic complete (SC) medium, and onto SC-Ura medium for the analysis of *URA3* silencing. The number of colonies growing on the selective medium compared with the nonselective SC medium is directly correlated with the expression level of *URA3* in the rDNA. Colonies are grown at 30° for 2 days, and then photographed (Fig. 3A). The same procedure can be used to measure the silencing of telomeric or *HM*-specific reporter genes.

Colony Color Assays

In colony color assays, a white colony color indicates no silencing of the reporter gene, and a dark colony color indicates silencing. For telomeres and the *HM* loci, the *ADE2* gene is used as the reporter. Cells mutated for *ade2* form red colonies when grown on medium containing a limiting amount of adenine (64 μM). Therefore, when a telomeric *ADE2* gene is expressed in an *ade2* strain background, the resulting cells will grow white. Silencing of telomeric *ADE2* results in a red color. Two or 3 days of growth at 30° followed by 4–5 days at 4° results in optimal color formation. In a wild-type strain, the normal phenotype is a red-and-white variegated colony. A *sir2* mutation turns on telomeric *ADE2* and turns all the cells white. For the rDNA, a different colony color reporter gene is used, called *MET15*. Cells that are *MET15*⁺ have a white colony color when grown on rich medium containing lead nitrate (0.1%, w/v). When *MET15* is missing or silenced in the rDNA, the colony color turns to brown or tan, respectively. Colonies are grown for 5 days at 30° for optimal colony color development (Fig. 3B).

Cloning of SIR2 Homologs

We developed a set of degenerate PCR primers based on the conserved motifs in Sir2 proteins. These were originally used to clone *HST3* and *HST4* from *S. cerevisiae,* and subsequently to clone homologs from *Schizosaccharomyces pombe* and other organisms. The primer JB 710 [5′-GGNRTNCCNGAYTTY(A/C) G-3′], which recognizes the sequence encoding the peptide G(I/V)PDFR, is used in combination with either JB 708 [5′-RTC(A/G/T)ATRTTYTGNGTRTA-3′], which recognizes the sequence encoding the peptide YTQNID, or JB 709 [5′-RTC(A/G/T) ATRTTYTGNGT(A/G/T)AT-3′], which recognizes the sequence encoding the peptide ITQNID (see Fig. 1 for peptide sequences).

These oligonucleotides are were used in a touchdown PCR protocol using AmpliTaq polymerase (PerkinElmer, Norwalk, CT) under standard PCR conditions and the following cycling parameters:

(1 min at 94°, 30 sec at 54°, 30 sec at 72°), 2 cycles
(1 min at 94°, 30 sec at 50°, 30 sec at 72°), 2 cycles
(1 min at 94°, 30 sec at 46°, 30 sec at 72°), 2 cycles
(1 min at 94°, 30 sec at 42°, 30 sec at 72°), 2 cycles
(1 min at 94°, 30 sec at 38°, 30 sec at 72°), 2 cycles
(1 min at 94°, 30 sec at 48°, 30 sec at 72°), 25 cycles
5 min at 72°

PCR products are then cloned by standard methods.

TABLE I
Sir2p HOMOLOGS EXPRESSED SUCCESSFULLY

Protein	Organism	MW ($\times 10^3$)	Vector/tag	*Escherichia coli* strain
Sir2p	*Saccharomyces cerevisiae*	63	pGEX-4T-1/GST	BL21
Hst3p	*Saccharomyces cerevisiae*	51	pGEX-4T-3/GST	BL21
Sir2Ap	*Homo sapiens*	41	pET30c/His$_6$	BL21(DE3)
			pGEX-4T-3/GST	BL21
Sir2-Af2	*Archaeoglobus fulgidus*	28.5	pET11a/none	BL21(DE3)
Sir2-Tm	*Thermotoga maritima*	27.5	pET11a/none	BL21(DE3) codon[+]

Protein Expression

Several Sir2p homologs from various organisms have been expressed in a variety of vectors. Table I shows the Sir2p homologs that have been successfully expressed at high levels (at least 10 mg/liter).

Basic Protocol for Expression and Purification of Sir2-Af2

1. Inoculate 1 liter of glucose-enriched M9ZB medium (see Buffers and Solutions for Purifications, below, for all medium and buffer recipes) with 10 ml of overnight growth of *Escherichia coli:* BL21(DE3) (Stratagene, La Jolla, CA) in a selective medium containing carbenicillin at 0.1 mg/ml.

2. Grow the cells at 37° in carbenicillin (0.1 mg/ml). When the OD$_{660}$ = 0.6–0.8, induce the cells with 1 mM final concentration of isopropyl-β-D-thiogalacto-pyranoside (IPTG). Induce for 4 hr at 37°.

3. Spin the cells at 5000 rpm for 30 min at 4° in an RC-3B Sorvall (Newtown, CT) centrifuge and discard the supernatant carefully.

4. Resuspend the pellet in 50 ml of lysis buffer Af and run the cells through a microfluidizer or French press. Spin the lysate at 13,000 rpm for 30 min at 4° in a Sorvall GSA or similar rotor and carefully collect the supernatant.

5. Incubate the supernatant at 85° for 30 min, and then cool the lysate in ice water for 10 min and spin again at 13,000 rpm for 30 min at 4° in the same rotor. Collect the supernatant carefully and dialyze in buffer A.

6. Run the solution in buffer A over an NAD$^+$ affinity column packed with Cibacron Blue beads (Pharmacia Biotech, Piscataway, NJ) and wash the column with buffer A-NaCl. Finally, elute the protein with buffer B. Dialyze the eluate in buffer C.

7. Run the solution in buffer C over a MonoQ 10/10 anion-exchange column (Pharmacia) and elute the protein in a salt gradient with buffer C-NaCl. Dialyze the protein in buffer D.

8. Run the protein in buffer D over a Superdex 75 sizing column and collect fractions of pure protein (see Fig. 4A). This protocol yields 10–20 mg/liter of culture medium.

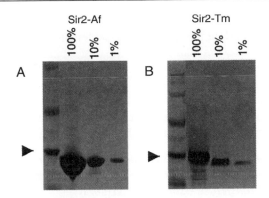

FIG. 4. (A) SDS–polyacrylamide gel of Sir2-Af2 with 10-fold dilutions showing more than 99% purity. (B) SDS–polyacrylamide gel of TmSir2p with 10-fold dilutions showing approximately 99% purity. Arrowheads indicate position of 30 kDa marker.

Basic Protocol for Expression and Purification of Sir2-Tm

1. Inoculate 1 liter of glucose-enriched M9ZB medium with 10 ml of overnight growth of BL21 codon$^+$ (DE3) (Stratagene) in a selective medium of carbenicillin (0.1 mg/ml).

2. Grow the cells at 37° in carbenicillin (0.1 mg/ml). When the OD$_{660}$ − 0.5–0.6, induce the cells with a 1 mM final concentration of IPTG. Induce for 4–6 hr at 37°.

3. Spin the cells at 5000 rpm for 30 min at 4° in an RC-3B Sorvall centrifuge and discard the supernatant carefully.

4. Resuspend the pellet in 50 ml of lysis buffer Tm and run the cells through a microfluidizer or French press. Spin the lysate at 13,000 rpm for 30 min at 4° in a Sorvall GSA or similar rotor and carefully discard the supernatant.

5. Wash the pellet by resuspending completely in 100 ml of lysis buffer Tm and spin at 12,000g for 30 min at 4°.

6. Resuspend the washed pellet in 100 ml of lysis buffer Tm, spin at 760g for 10 min at 4°, and collect inclusion bodies in the supernatant. Spin the supernatant at 12,000g for 30 min at 4° to collect inclusion bodies in the pellet.

7. Wash the inclusion bodies five times by resuspending them in 100 ml of lysis buffer Tm, and spinning at 12,000g for 30 min at 4°. After a final wash resuspend them again and spin at 760g for 10 min at 4°, discard the pellet, and spin the supernatant at 12,000g for 30 min at 4° to collect inclusion bodies in the pellet.

8. After the final spin, solubilize the inclusion bodies in 5 ml of buffer E by rocking them in a tube overnight at 4°. Determine the protein concentration and dilute in buffer E to a final concentration of 1 mg/ml or less. Renature the protein by dialyzing into 0 M urea, keeping everything else in the buffer E constant.

9. Dialyze the protein into buffer C and run over a Q-Sepharose Fast Flow ion-exchange column (Pharmacia). Elute the protein in a salt gradient with buffer C-NaCl.

10. *Optional:* Dialyze protein fractions in buffer F and run through a MonoQ 10/10 anion exchange column (Pharmacia) and elute the protein in a salt gradient with buffer C-NaCl.

11. Dialyze the protein in buffer D, run it through a Superdex-75 sizing column, and collect fractions of pure protein (see Fig. 4B).

Basic Protocol for Purification of Saccharomyces cerevisiae Sir2p
　　and Homo sapiens Sir2A Expressed as N-Terminal Fusions to
　　Glutathione S-Transferase Using pGEX Vectors

1. Inoculate 50 ml of Luria–Bertani (LB) medium supplemented with ampicillin (100 μg/ml) with DH5α containing Sir2 expression vector and grow overnight at 37°.

2. Inoculate 500 ml of LB medium supplemented with ampicillin (100μg/ml) with 5 ml of overnight culture and grow at 30° until the culture reaches an OD_{600} of about 0.7.

3. At that point induce with 100 μM IPTG and transfer the culture to a 20° incubator. Grow overnight.

4. In the morning pellet the cells in a GS3 Sorvall rotor (4000 rpm, 10 min, 4°.)

5. Wash the cells once with cold phosphate-buffered saline (PBS).

6. Resuspend the cell pellet in PBS supplemented with 0.5 mM dithiothreitol (DTT), 0.5 mM phenylmethylsulfonyl fluoride (PMSF), and protease inhibitors.

7. Disrupt the cells by French press.

8. To the cell suspension add NaCl to increase the final salt concentration to 350 mM. Sir2 from *S. cerevisiae* has a tendency to precipitate during subsequent steps of purification, so a final concentration of 350 mM NaCl is maintained in the buffers used during the purification.

9. Spin the cell suspension in an ultracentrifuge equipped with a in Ti70 rotor (Beckman, Fullerton, CA) at 40,000 rpm for 1 hr at 4°.

10. Fill a disposable column (Bio-Rad, Hercules, CA) with an appropriate volume of glutathione–Sepharose 4B beads (Pharmacia).

11. Equilibrate the column with 10 volumes of PBS.

12. Apply the supernatant.

13. Wash the column with 20 volumes of 50 mM Tris-HCl (pH 8.0), 350 mM NaCl.

14. Elute the protein from the column with 10 volumes of 50 mM Tris-HCl, 350 mM NaCl, 20 mM reduced glutathione (Pharmacia).

15. Dialyze the eluant against 2 liters of storage buffer [25 mM HEPES-NaOH (pH 8.0), 200 mM NaCl, 20% (v/v) glycerol], concentrate, and store in small aliquots.

Buffers and Solutions for Purifications

Glucose-enriched M9ZB medium
    M9 minimal salt 1× solution
    NaCl: final concentration, 5 g/liter
    Tryptone or Casamino Acids (Difco Laboratories, Detroit, MI): 10 g/liter
    Glucose (added after autoclaving): 0.4% (w/v)
    MgSO$_4$ (added after autoclaving): 2 mM

Lysis buffer Af
    HEPES (pH 7.0), 20 mM
    EDTA, 2 mM
    DTT, 1 mM
    Pefabloc, 0.1 mg/ml
    Aprotinin, 5 μg/ml
    Leupeptin, 2 μg/ml

Buffer A
    Bis-Tris (pH 6.0), 20 mM
    DTT, 1 mM
    ZnCl$_2$ 25 μM

Buffer A-NaCl
    Bis-Tris (pH 6.0), 20 mM
    DTT, 1 mM
    ZnCl$_2$, 25 μM
    NaCl, 1 M

Buffer B
    Bis-Tris propane (pH 8.5), 20 mM
    DTT, 1 mM
    ZnCl$_2$, 25 μM

Buffer C
    Tris (pH 8), 40 mM
    DTT, 1 mM
    ZnCl$_2$, 25 μM

Buffer C-NaCl
    Tris (pH 8), 40 mM
    NaCl, 1 M
    DTT, 1 mM
    ZnCl$_2$, 25 μM

Buffer D
    HEPES (pH 8), 20 mM
    NaCl, 150 mM
    DTT, 1 mM
    ZnCl$_2$, 25 μM

Lysis buffer Tm: Keep at 4°
    Tris (pH 7.5), 50 mM
    EDTA, 1 mM
    NaCl, 100 mM

Buffer E
    Tris (pH 8), 50 mM
    Urea, 4 M
    NaCl, 100 mM
    DTT, 1 mM
    ZnCl$_2$, 25 μM

Buffer F
    Tris (pH 8), 40 mM
    NaCl, 100 mM
    DTT, 1 mM
    ZnCl$_2$, 25 μM

Purification of Acetylated Chicken Histones as Histone Deacetylase Substrate

Chicken histones used for labeling with [^{3}H]acetic acid are isolated from the blood of white Leghorn chickens injected with 1% (w/v) phenylhydrazine.

Chickens were fasted for 48 hr, followed by daily injections of 1% (w/v) phenyl-hydrazine in 10 mM sodium phosphate buffer, pH 7.2, for 7 days. Birds were killed and blood was collected with heparinized syringes.

1. Spin 10 ml of chicken blood at 1000g for 5 min at room temperature.
2. Resuspend the cells in 50 ml of Swim's S-77 medium (Sigma, St. Louis, MO).
3. Spin the cells (1000g, 5 min, room temperature) and wash them with 50 ml of the same medium one more time.
4. Resuspend the cells in 100 ml of Swim's S-77 medium supplemented with 10% (v/v) newborn calf fetal serum and 10 mM sodium butyrate.
5. Add 10 mCi of [^{3}H]acetic acid (9.4 Ci/mmol; ICN, Costa Mesa, CA) and incubate at 37° for 30 min.
6. Spin the cells at 1000g for 5 min at room temperature and resuspend them in 100 ml of Swim's S-77 medium supplemented as described above and incubate for 1 hr at 37°.
7. Cool the cells on ice and spin them at 3000g for 5 min at 4°.
8. Resuspend the cells in 20 ml of NIB buffer (see Media Required for Purification of Acetylated Chicken Histones, below, for buffer and solution recipes) and incubate for 30 min on ice.
9. Spin at 3000g for 5 min at 4° and wash twice in 20 ml of NIB buffer.
10. After a final spin (3000g, 5 min, 4°), resuspend the pellet in 20 ml of Tris–NaCl solution.
11. Add 2.3 ml of 4 M sulfuric acid and incubate on ice for 1 hr.
12. Spin at 3000g for 5 min at 4°.
13. Precipitate with 10 volumes of ice-cold acetone–5 M HCl (99 : 1, v/v).
14. Spin at 10,000g for 10 min at 4°.
15. Lyophilize the pellet and resuspend in 5 ml of 20 mM Tris-HCl (pH 6.0), 10 mM sodium butyrate.
16. Clarify the slurry by centrifugation at 8000g for 10 min at 4°, aliquot the supernatant, and freeze at −80°.

This labeling procedure yields histones with specific activity between 700 and 1300 cpm/μg.

Media Required for Purification of Acetylated Chicken Histones

NIB buffer:
20 mM piperazine-N,N'-bis(2-ethanesulfonic acid) (PIPES, pH 6.8), 0.25 M sucrose, 60 mM KCl, 5 mM MgCl$_2$, 1 mM CaCl$_2$, 10 mM sodium butyrate, 1 mM PMSF, 0.5% (v/v) Triton X-100
Tris–NaCl solution:
100 mM Tris-HCl (pH 8.0), 0.4 M NaCl, 1 mM PMSF

Histone Deacetylation Assays

The assays are done in a final volume of 100 μl. Each reaction contains the following:

Tris-HCl (pH 8.0), 50 mM
NaCl, 50 mM
β-NAD, 200 μM
^{3}H-labeled chicken erythrocyte histones, 13,000 cpm/μg
1–5 μg Sir2,

1. The reaction mixture is incubated at 30° for *S. cerevisiae* Sir2, at 37° for human Sir2A, and at 55° for *A. fulgidus* Sir2.
2. The reaction is stopped by adding 36 μl of 1 M HCl, 0.4 M acetic acid and extracted with 0.8 ml of ethyl acetate with 10 min of incubation on ice.
3. Organic phase (600 μl) is added to 3 ml of scintillation fluid and the samples are counted in a scintillation counter.

NAD$^+$ Measurements from Intact Yeast Cells

1. Inoculate 50-ml cultures in appropriate growth medium, and shake overnight at 30°.
2. Measure the A_{600} of the overnight culture and dilute to an A_{600} of 0.2 in a final volume of 500 ml in a 1-liter flask. Continue shaking the culture at 30° and allow it to grow to an A_{600} of approximately 1.0. The protocol can be scaled down.
3. Transfer an equal number of cells for each culture into a 500-ml centrifuge bottle (on ice). Example: If a 450-ml culture is at an A_{600} of 1.0, take 400 ml of a culture that is at an A_{600} of 1.125. Balance the bottles with fresh medium.
4. Pellet the cells in the GS3 Sorvall rotor at 2500 rpm for 10 min at 4°. Discard the supernatant down the sink.
5. Resuspend each of the cell pellets with 40 ml of ice-cold water by vortexing and pour the mixtures into 50-ml conical (blue cap/Falcon) tubes (again on ice).
6. Pellet the cells in an Eppendorf swinging-bucket centrifuge at 2500 rpm for 5 min at 4°.
7. Discard the supernatant down the sink and resuspend the cells in 5 ml of 1 M formic acid saturated with butanol (on ice) (see Buffers Needed, below, for recipe). Mix by swirling, not by vortexing. Incubate on ice for 30 min, swirling at three equally spaced intervals to prevent the cells from settling to the bottom of the tubes.
8. Transfer the mixtures by pouring into 30-ml Oakridge centrifuge tubes (on ice).
9. Add a 1/4 volume (1.25 ml) of ice-cold 100% trichloroacetic acid (TCA) and mix by swirling. Incubate on ice for 15 min.

10. Pellet the cells in the SS-34 rotor at 6000 rpm (4000g) for 20 min at 4°.

11. Transfer the supernatants to 50-ml Falcon tubes, using a 10-ml glass pipette. Try to avoid taking any cell debris. (*Note*: The NAD$^+$ is in this fraction.) Store on ice. In the case of the mock culture, avoid taking the lower liquid phase.

12. Wash the pellets with 2.5 ml of ice-cold 20% (w/v) TCA. (Vortex to resuspend the pellets.)

13. Pellet the cells again in the SS-34 rotor at 6000 rpm for 20 min, 4°.

14. Carefully remove the supernatant with a 5-ml glass pipette and combine with the first TCA supernatant in the 50-ml Falcon tube.

15. Spin the tube containing the supernatants in an Eppendorf tabletop centrifuge at 3000 rpm for 5 min at 4°. This will pellet any remaining cell debris.

16. Set up alcohol dehydrogenase (ADH) reactions:

Water	338 μl
ADH buffer (2×)	500 μl
Ethanol, 100%	2 μl
Alcohol dehydrogenase (15-mg/ml stock in 50 mM Tris, pH 7.5)	10 μl
Acid extract from cells	150 μl
	1000 μl

Mix each tube by briefly vortexing and incubate in a 30° water bath for 20 min.

17. Using a single quartz cuvette, blank the spectrophotometer at 340 nm with the mock extract reaction without ADH added, and then read the A_{340} for each sample.

Buffers Needed

TCA, 100%
TCA, 20% (w/v)
Formic acid (1 M) saturated with butanol (250-ml stock): Combine 212 ml of water, 13 ml of formic acid [88% (w/v) stock], and 25 ml of 1-butanol; mix and allow the phases to separate, and then bring the volume up to 250 ml with water and top off with 1-butanol. Store at 4°.
ADH buffer (2×)
    Tris (pH 9.7), 0.6 M
    Lysine-HCl, 0.4 M

[27] Defining Redox State of X-Ray Crystal Structures by Single-Crystal Ultraviolet–Visible Microspectrophotometry

By Carrie M. Wilmot, Tove Sjögren, Gunilla H. Carlsson, Gunnar I. Berglund, and Janos Hajdu

Introduction

Exciting results have been emerging from the field of single-crystal X-ray crystallography, giving unprecedented detail of freeze-trapped reaction intermediates from important classes of macromolecules that contain chromophores[1–10] (for reviews see Refs. 11–13). The reason for the current excitement is that these structures have been coupled with single-crystal UV–visible microspectrophotometry. This has defined the distinct catalytic intermediates present in the crystal structures, allowing the correlation of electronic transitions with the observed structural transitions. Of particular note is that many of these structures have been generated "on the fly" during kinetic turnover in the crystal. Most enzymatic reactions proceed through distinct catalytic intermediates that, under favorable conditions, may accumulate transiently in the crystal during turnover. In some cases, the physical constraints of the contacts within crystals may also lead to a significant slowing of

[1] P. A. Williams, V. Fülöp, E. F. Garman, N. F. W. Saunders, S. J. Ferguson, and J. Hajdu, *Nature* (*London*) **389**, 406 (1997).

[2] K. Edman, P. Nollert, A. Royant, H. Belrhali, E. Pebay-Peyroula, J. Hajdu, R. Neutze, and E. M. Landau, *Nature* (*London*) **401**, 822 (1999).

[3] H. Lücke, B. Schobert, H. T. Richter, J. P. Cartailler, and J. K. Lanyi, *Science* **286**, 255 (1999).

[4] A. Royant, K. Edman, T. Ursby, E. Pebay-Peyroula, and R. Neutze, *Nature* (*London*) **406**, 645 (2000).

[5] H. J. Sass, G. Büldt, R. Gessenich, D. Hehn, D. Neff, R. Schlesinger, J. Berendzen, and P. Ormos, *Nature* (*London*) **406**, 649 (2000).

[6] C. M. Wilmot, J. Hajdu, M. J. McPherson, P. F. Knowles, and S. E. V. Phillips, *Science* **286**, 1724 (2000).

[7] N. I. Burzlaff, P. J. Rutledge, I. J. Clifton, C. M. H. Hensgens, M. Pickford, R. M. Adlington, P. L. Roach, and J. E. Baldwin, *Nature* (*London*) **401**, 721 (1999).

[8] I. Schlichting, J. Berendzen, K. Chu, A. M. Stock, A. S. Maves, D. E. Benson, R. M. Sweet, D. Ringe, G. A. Petsko, and S. G. Sligar, *Science* **287**, 1615 (2000).

[9] K. Chu, J. Vojtechovsky, B. H. McMahon, R. M. Sweet, J. Berendzen, and I. Schlichting, *Nature* (*London*) **403**, 921 (2000).

[10] A. Ostermann, R. Waschipky, F. G. Parak, and G. U. Nienhaus, *Nature* (*London*) **404**, 205 (2000).

[11] J. Hajdu, R. Neutze, T. Sjögren, K. Edman, A. Szöke, R. C. Wilmouth, and C. M. Wilmot, *Nat. Struct. Biol.* **7**, 1006 (2000).

[12] I. Schlichting, *Acc. Chem. Res.* **33**, 532 (2000).

[13] G. A. Petsko and D. Ringe, *Curr. Opin. Chem. Biol.* **4**, 89 (2000).

METHODS IN ENZYMOLOGY, VOL. 353

Copyright 2002, Elsevier Science (USA).
All rights reserved.
0076-6879/02 $35.00

the reaction at certain points along the pathway where conformational changes are required. This can lead to a transient build-up of spectrally distinct intermediates in the crystal that can be trapped by flash freezing in liquid nitrogen, allowing a complete single-crystal data set to be collected to the highest possible resolution at a later time. Similar build-up of intermediates may be achieved by altering the pH, temperature, or the solvent environment around the protein in the crystal, or by producing engineered variants that build up an intermediate of interest.

This chapter focuses on the technical considerations required to carry out UV–visible microspectroscopy of single crystals.

Historical Problems

The large solvent channels found in macromolecular crystals have been allowing crystallographers to soak drugs/products/substrate analogs into enzyme crystals for decades. These single-crystal X-ray structures of enzyme complexes have aided our understanding of function. Although these can give insight to the chemical groups involved and suggest mechanisms, the structures represent beginning or end points of the reaction. Critics also suggested that the constraints of crystal contacts could lead to kinetically artificial structures, not only via kinetic dead-ends, but by favoring a minor reaction pathway over the major pathway observed in solution. Particularly in redox enzymes the reactions often lead to forms that are X-radiation sensitive. In this case the oxidation state of redox centers may be affected by X-rays during data collection. Figure 1 shows spectral changes in a crystal of the compound III form of horseradish peroxidase during the collection of 90° of X-ray data. These results demonstrate a general problem with redox proteins; without spectral controls it is practically impossible to obtain reliable structures for oxidized catalytic intermediates, because electrons liberated during X-ray exposure alter the oxidation state of the redox center in the active site. Absorption spectroscopic studies of single crystals of numerous heme enzymes show that, contrary to general belief, the redox state of the heme iron is affected by electrons liberated in the sample even at the shortest X-ray wavelengths.[14,15] In addition, the medium surrounding the crystal can influence X-ray sensitivity. The original structure of an "oxidized" ribonucleotide reductase had metal ligation identical to that of the reduced enzyme. Glycerol, a good electron hole stabilizer, was present in the crystal mother liquor, and had facilitated reduction of the iron in the X-ray beam. Eventually the structure of the oxidized enzyme was solved by excluding glycerol from the crystal-stabilizing solution.[16] As radiation damage is a major concern in the structure determination of redox intermediates of proteins, it is useful to understand the basis for this.

---

[14] G. I. Berglund, G. H. Carlsson, A. T. Smith, H. Szöke, A. Henriksen, and J. Hajdu, *Nature (London)*, in press (2002).

[15] B. Ziaja, D. van der Spoel, A. Szöke, and J. Hajdu, *Phys. Rev.* **B64**, 214104 (2001).

[16] M. Eriksson, A. Jordan, and H. Eklund, *Biochemistry* **37**, 13359 (1998).

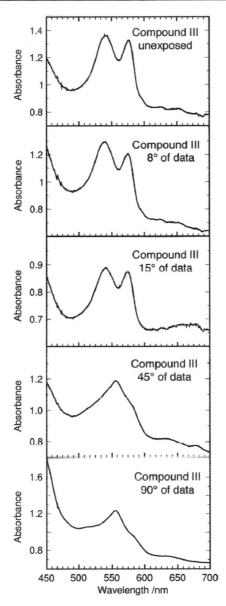

FIG. 1. X-ray-induced reduction of the compound III form of horseradish peroxidase during data collection at cryogenic temperatures. Absorbance changes in a single crystal are shown during data collection at 0.98 Å wavelength [beam line 14 of the European Synchrotron Radiation Facility (ESRF), Grenoble, France].

Primary Interactions in Radiation Damage

X-rays interact with matter through one of the following primary channels.

Photoelectric Effect and Its Satellite Processes

At 1 Å wavelength, about 90% of the interacting photons will deposit energy into a biological sample, causing damage mainly through the photoelectric effect. The photoelectric effect causes an X-ray photon to be absorbed and a tightly bound electron to be ejected from an inner shell of the atom. The Auger effect is the emission of a second electron after a high-energy photon has expelled a photoelectron. The departing photoelectron leaves a hole in a low-lying orbital, and an upper shell electron falls into it. This electron may emit an X-ray photon to produce X-ray fluorescence (dominant process with heavy elements) or may give up its energy to another electron, which is then ejected from the ion as an Auger electron (dominant process with light elements). The probability of fluorescence emission or Auger emission depends on the binding energy of the electron, and through this, on the Z number of the element. In biologically relevant light elements, the predominant relaxation process (>99%) is through the Auger effect. Thus, most photoelectric events ultimately remove two electrons from carbon, nitrogen, oxygen, and sulfur. The secondary Auger electrons have about two orders of magnitude lower energies than the primary photoelectrons.

During photoemission, the inner shell electron leaves so abruptly that often the outer shell electrons have no time to relax. The result of this is that the outer shell electrons find themselves in a state that is not an Eigenstate of the atom in its surroundings. Quantum mechanics describes such a state as a superposition of proper Eigenstates, which include states in which one or more of the electrons are unbound. Such a collective effect is called *shakeup*. The release of the unbound electron "competes" with Auger electrons. After the Auger process is completed, the electric field in the inner shells of the atom returns to its original value. We estimate that shakeup excitations may contribute about 20–30% more electrons than the number released in the Auger process.[17]

Compton (or Inelastic) Scattering

Compton (or inelastic) scattering is the predominant source of energy deposition with hydrogen, and represents about 3% of all interactions between X-rays and biological material at 1 Å wavelength.

Elastic Scattering

Elastic scattering is a relatively rare event, and at 1 Å wavelength, only about 10% of X-ray photons that actually interact with a biological sample are scattered elastically and carry useful structural information.

[17] P. Persson, S. Lunell, A. Szöke, B. Ziaja, and J. Hajdu, *Protein Sci.* **10,** 2480 (2001).

Secondary Cascade Processes in Radiation Damage

During conventional X-ray data collection on a macroscopic protein crystal, both the photoelectrons and the secondary Auger electrons become thermalized and trapped in the sample. In small samples, most of the photo- and Auger electrons may escape, and, therefore, electron impact cascades play a smaller role.[18] In macroscopic samples, the process of thermalization involves inelastic electron–electron interactions and, to a lesser degree, electron–nuclear interactions. Thermalization is a longer process than primary photoemission or secondary Auger emission, and produces a large number of "cascade electrons." The total number of cascade electrons is roughly proportional to the energy of the first impact electron entering the cascade. A photoelectron of 12 keV energy liberates about 1000 cascade electrons, whereas an Auger electron of 0.25 keV energy (a typical value for carbon) produces about 20 cascade electrons as it comes to thermal equilibrium.[15] Subsequent chemical damage takes place on a much longer time scale than that produced by cascade electrons.

The primary, secondary, and tertiary electron emission processes liberate a total of about 10,000 electrons for each elastically scattered photon in macroscopic samples exposed to X-rays of 1 Å wavelength. Most of these are cascade electrons. Once thermalized, electrons can be recaptured by solvent or protein molecules, where they tend to settle in low-lying orbitals on electron-deficient atoms/ions. The result is a redistribution of these electrons within the sample, and the potential for changing the oxidation state of the redox center. Luckily, many redox enzymes have spectroscopically distinct chromophores, and therefore the availability of tools to allow the spectroscopic probing of crystals to prove their oxidation state before and after X-ray data collection was the initial driving force to expand spectroscopy to the crystalline state.

Turning Problems into Advantages

It soon became apparent that single-crystal microspectrophotometry could deliver much more than static spectra of a crystal before and after data collection. The crystal acts as a porous cage holding the enzyme molecules in place, and allows a multitude of different solutions to be passed over the crystal. The slowing down of stages of the complex reaction pathway in the enzyme crystal means that enzyme kinetics in the crystal can usually be monitored at leisure by spectroscopic techniques. Although the rates of formation of intermediate states may be drastically altered in the crystal, an analysis of spectroscopically distinct species and their order of appearance in the crystal allows the determination of whether the major kinetic reaction pathway observed in solution is that followed in the crystal. Structural build-up of intermediates that may be transient in solution, but

[18] R. Neutze, R. Wouts, D. van der Spoel, E. Weckert, and J. Hajdu, *Nature (London)* **406,** 752 (2000).

have life-times of minutes in the crystal, allows trapping by flash freezing in liquid nitrogen, giving a unique opportunity to observe kinetic species by single-crystal X-ray crystallography.

Tracking the redox state of crystals during data collection by mounting the microspectrophotometer on the goniometer of the X-ray camera has also given some unique advantages. Work has shown that redox transitions initiated by X-rays can be substantial and may happen early during data collection. As a consequence, only a small amount of data (not more than a few degrees) can be collected from a crystal before the redox state is significantly affected.[11,14] Studies of reaction intermediates of the oxidase reaction in crystals of cytochrome cd_1 nitrite reductase showed that about $10°$ of data could be obtained from a crystal without changing the redox state. The structure of an early intermediate in oxygen reduction by cytochrome cd_1 nitrite reductase could thus be obtained by merging partial data sets from 11 crystals.[19]

Single-Crystal Ultraviolet–Visible Microspectrophotometers

The general components are similar to those of any solution UV–visible spectrophotometer: a light source, optics, and a detector. In the case of single-crystal work, the sample mounting is achieved via a goniometer equivalent to that used in X-ray cameras (Fig. 2).[20] As such, samples can be mounted either in loops or quartz capillaries in an analogous fashion to the techniques used for X-ray data collection. A cryostream can be positioned so that measurements can be taken at liquid nitrogen or liquid helium temperatures.

The light source should be sufficiently intense to penetrate through the crystal, and the optics should focus the beam into a spot that is smaller than the crystal for effective measurements. Because of the demands of crystal alignment; the bringing in of additional components, such as a cryostream; and the ability to mount the optics on an X-ray camera, the sample is not enclosed, and so appropriate precautions need to be taken if using a UV light source.

Portability of the single-crystal microspectrophotometer can be a distinct advantage, allowing transport to other X-ray data collection facilities, such as synchrotrons.

Mounting Crystals for Taking Single Spectra

In general, large crystals containing the greatest concentration of chromophore will give the cleanest spectral signal. However, for crystals with intense spectral bands (such as the Soret band of heme-containing proteins) a small crystal may

---

[19] T. Sjögren and J. Hajdu, *J. Biol. Chem.* **276,** 13072 (2001).
[20] A. T. Hadfield and J. Hajdu, *J. Appl. Crystallogr.* **26,** 839 (1993).

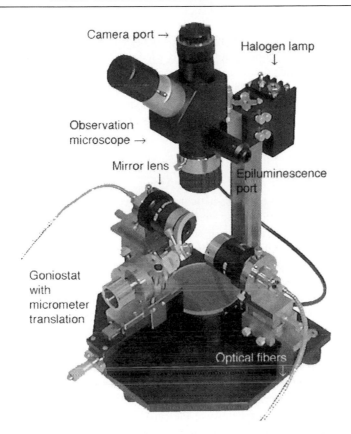

FIG. 2. UV–visible single-crystal microspectrophotometer (http://www.4dx.se).

need to be selected to prevent complete absorption of the light source by the crystal. Although the instrument light source is intense and focused, a good "rule of thumb" for obtaining clear spectra in the visible spectral region is that a crystal that appears black under a microscope is probably too thick. The crystal selected should transmit light so that the color of the chromophore is clear to the observer. If all crystals are too large, a razor blade or microsurgical knife can be used to cut off a piece of crystal.

For room temperature single-crystal spectra, crystals can be mounted in quartz X-ray capillaries (not glass, which would absorb the UV radiation) in the same way as for an X-ray experiment. Before mounting the capillary on the goniometer head, the outside of the capillary should be gently wiped down with distilled water to make sure there are no contaminants on the outside. Care should be taken to align the crystal as much as possible along the capillary axis to aid in crystal alignment on the spectrophotometer, thus reducing prism effects.

For low-temperature single-crystal spectra, the standard mounting loops, cryo-protectants, and mounting pins used in X-ray crystallographic work can be used. As a standard X-ray goniometer head is used on the microspectrophotometer, transfer of the crystals onto the instrument is exactly the same as for transferring frozen crystals on and off X-ray data collection equipment. A fine paint brush, first cooled for several seconds in the cryocooler gas stream, can be used to gently push bent loops containing crystals so that they lie along the rotation axis for easier crystal alignment.

Crystal Alignment

Crystal alignment is the most critical and time-consuming part of single-crystal spectroscopy, and should not be underestimated. It is not uncommon for crystal alignments to take half an hour or so. This is due to three particular difficulties encountered by the single-crystal spectroscopist. The first problem is that crystals are anisotropic (except for the cubic system) and spectral features of a crystalline sample vary with the direction of the incident light. The second difficulty arises from crystal prism effects, and the third from inaccuracies in path length between the reference and experimental spectra in absorbance measurements.

Chromophore Nonrandomness in Crystals

The solution spectrum represents an average of spectra recorded from all possible orientations of the studied sample. The ordered nature of the crystal means that the intensity of the spectrum is not identical when measured in different directions. This can pose problems especially if the chromophore is a large π-system such as a heme group and if several chromophores are present. In this case anisotropic spectra will be obtained from orientating the crystal in different directions. A simple solution is to use a microcrystalline slurry in a flow cell to establish conditions for trapping of intermediates (see below). A more elegant method is to use polarized absorption spectroscopy. Absorption of plane-polarized light with the electric vector parallel to a principal optical direction in a crystal obeys the Beer–Lambert law, and the solution spectrum can be reproduced by averaging spectra taken along the main optical directions in a crystal.[21] However, if the electrons are isotropically distributed at the redox center, or the combined orientations of the chromophores in the crystal are random enough, then the resulting spectrum will be relatively isotropic, and the equivalent solution spectra can be used for comparison. Even in anisotropic crystal spectra, the observed peaks can be assigned to the equivalent peaks in the solution spectra and monitored for changes. In this case the orientation of the crystal to the beam must be maintained for valid comparison. Having

[21] A. Merli, D. E. Brodersen, B. Morini, Z. Chen, R. C. E. Durley, F. S. Mathews, V. L. Davidson, and G. L. Rossi, *J. Biol. Chem.* **271,** 9177 (1996).

comparison spectra are particularly useful in aiding in the minimization of crystal prism effects and inaccurate path lengths.

Crystal Prism Effects

An additional factor that may modulate the spectrum recorded from a crystal (and even from a cubic crystal) is the prism effect. This is a general feature of all spectral measurements when light enters a dispersive material at an angle. In single-crystal microspectroscopy, this manifests itself as small spectral changes while the crystal is rotated in the light beam, and is due to the dispersive nature of the little crystal prism in the optical path. Crystals with parallel faces offer easily reproducible measurements when the light beam falls normal to one of the crystal faces. In the general case, when the faces are not parallel, accurate measurements can still be made by performing relative measurements at a fixed orientation, for example, as a function of time, temperature, or pH. By comparison with the solution or anisotropic spectra, the best crystal orientation match can be found.

The user should always remember that the electronic state of the chromophore in the crystal may not be what is expected, particularly after data collection, when oxidation/reduction in the X-ray beam is not uncommon. Therefore, the user should keep an open mind with regard to which spectra should match the crystal. In addition, there is also the possibility of a genuine change in the environment of the chromophore on crystallization that may modulate the spectrum of a crystal compared with solution. Finally, the crystalline state often alters the equilibrium and rate constants between reaction intermediates, particularly if a conformational change is involved that may be restricted by crystal contacts. This can potentially reveal an intermediate that is so transient in solution that it has never been observed spectrally, or lead to species that are either off the reaction pathway or belong to a minor reaction solution pathway that becomes the major pathway in the crystal.

Reducing Inaccuracies in Path Length

Absorbance spectra represent the data most commonly measured from crystals. This involves removing any absorbing signals generated by the crystal mother liquor or cryoprotectant from those of the chromophores in the crystal itself. Unlike solution work, there is no defined path length created by a cell, so the user must try to match the path length in the mother liquor to the path length in the crystal. Mismatched path lengths lead to either residual signals from absorbers in the crystal mother liquor appearing in the final absorbance spectrum, or loss of signal in coincident areas of absorbance between the mother liquor and the crystal chromophores, depending on whether the path length is longer for the reference or the experimental spectra, respectively.

To help identify a good path length match between reference and crystal spectra, the characteristics of the reference spectrum should be examined. If there is a strong peak, then different references should be taken until the final absorbance spectrum shows no anomalies from the general form of the comparison spectrum around the wavelength of the peak. In contrast, the crystal solution may have a broad absorbance. When the absorbance increases/decreases across the spectral range, the mismatched path lengths will lead to a skewing of the spectrum, giving higher absorbance than expected at one end and less absorbance than expected at the other. Similar effects can also be observed when the beam is not perpendicular to the sample. Again, different references should be taken until the absorbance spectrum exhibits the correct overall slope of the comparison spectrum. If the absorbance is relatively even across the wavelength range of interest, then the overall form of the final corrected absorbance spectrum will not be affected, but the relative position of the entire spectrum in relation to zero absorbance will change. In this case air can act as a simple and effective reference. Even absorbance across the spectral range from the surrounding medium is not a problem in terms of being convinced that the electronic species of interest is present in the crystal, but demonstrates that single-crystal UV–visible spectroscopy cannot be used to assess the amount of chromophore quantitatively, as absolute peak height will always contain an undetermined error associated with inaccurate path lengths. As this is a constant error, relative spectral changes between species during the reaction will be more accurate.

For room temperature spectra in capillaries, a wet mounting can be helpful. In this case translating along the rotation axis to the solution just beyond the crystal should give a reasonable match. To aid in crystal alignment, a scale that allows the user to return to a previously noted position along the crystal length and rotation is useful. Putting a small drop of mother liquor about the size of the crystal elsewhere in the capillary can also be used to obtain a reference. This also has the advantage that working from the thin edge of the drop to the middle can vary the path length to help match reference and experimental spectra.

For crystals at low temperature, an empty loop of the same size as that containing the crystal, but with cryoprotectant alone, tends to give a relatively good reference match. If the crystal-containing loop is much larger than the crystal, and can be translated so that the cryoprotectant alone is in the incident light of the spectrophotometer, then a reference can be attempted from this area. Experience has shown that this area tends to be much thinner than the crystal (the crystal often can be seen to bulge out of the cryoprotectant if the loop is viewed edge on), and in these cases the references obtained rarely match that of the experimental spectrum.

Mounting Crystals for Kinetic Experiments

A variety of custom-made flow cells have been used, but the simplest is an open, solution-filled quartz capillary used as the cell. Capillaries should be carefully selected, as variation between them is significant. The capillary needs to taper to a diameter less than that of the crystal, so that the crystal will wedge in the capillary. This provides a conveniently solid mounting framework for good-quality rigid crystals. A capillary needs to be selected that tapers slowly, so that the cross-section where the crystal wedges will remain fairly constant. This will allow a good match between the path length for the reference and crystal in absorbance spectra. A diamond cutter can be used to cut the end off the capillary to give an even end. It can then be filled with mother liquor, being careful to avoid air bubbles. The whole capillary can be dropped into a tube containing mother liquor for storage. If the crystal is either negatively or positively buoyant in the mother liquor, gravity can be used to allow the crystal to float into position. Otherwise the capillary can be lifted out into the air, and allowed to drip, using the solution flow to pull the crystal into position. The capillary should be topped up regularly until the crystal is in position, and then transferred back to the tube of mother liquor.

If the crystal is fragile or the mother liquor is especially viscous, it may not be easily wedged into a tapered capillary tube. In this case the crystal can be embedded in a small Sephadex column (Fig. 3a). Platelike crystals may also need to be similarly mounted in a larger bore capillary. This prevents them from bending with the curve of the capillary wall, which can cause problems related to prism effects. A thin wire can be used to push a couple of threads of cotton wool into the capillary, and another wire from the bottom can be used to pack the cotton wool loosely in the capillary (packing too tightly can block solution flow). A few granules of Sephadex G-25 Superfine are then added to 0.5 ml of mother liquor in an Eppendorf tube until the solution is cloudy when shaken, but fluid. The capillary with the cotton wool bung is then filled with mother liquor. One microliter of the Sephadex solution is added, and the capillary is returned to the tube of mother liquor and allowed to settle. The Sephadex should settle into a layer on top of the cotton wool, and should not leak through the plug. If the layer is too thin more Sephadex can be added. The crystal is then maneuvered onto the Sephadex, either by gravity or solution flow, while trying to align the crystal along the capillary wall to aid alignment on the instrument. More Sephadex is added to the capillary to embed the crystal, and this is allowed to settle in the tube of mother liquor, so that a layer of Sephadex sits above the crystal. The capillary is removed from the mother liquor and allowed to stand for about 10 min to let fluid flow through and compact the column. Rolling the capillary occasionally between two fingers helps the Sephadex to settle evenly. The capillary should be topped up frequently with mother liquor. Finally, a small plug of cotton wool should be added to the top of the column.

Experience shows that Sephadex granules in the beam path do not interfere with spectral measurements in the near UV–visible spectral range.[22] The

[22] V. Fülöp, P. Phizackerley, M. Soltis, I. J. Clifton, S. Wakatsuki, J. Erman, J. Hajdu, and S. L. Edwards, *Structure* **2,** 201 (1994).

a

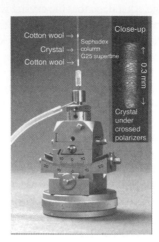

b

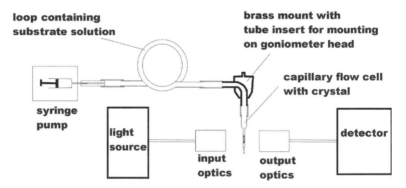

FIG. 3. (a) Simple flow cell for single-crystal kinetics. (b) Schematic of kinetics setup for a single crystal microspectrophotometer.

single-crystal microspectrophotometer should be set up so that the quartz capillary is vertical, with the solution dripping out of the bottom into a container (Fig. 3b). If the capillary is horizontal, the dripping of the solution from the end during kinetics runs will lead to oscillation of the capillary and distortion of the spectra. Mother liquor-filled tubing is used to connect the capillary via the goniometer head to a syringe pump. Tubing lengths should be kept to a minimum to reduce potential oscillation of the solution in the tubing. Making sure that there are positive menisci when connecting tubing will help ensure that no air bubbles, which must be eliminated, are formed. Depending on the resistance from the tubing and viscosity of the solution, a syringe pump speed of 1–5 ml/hr generally gives a good flow rate. Before measurements the outside of the capillary should be gently cleaned with distilled water to remove any salt crystals that may have dried on the outside of the capillary during the set-up.

Solution should be kept flowing through the capillary to prevent salt formation at the open end, while aligning the crystal to give the starting spectrum. At this point a loop of substrate solution (250 μl) can be introduced into the tubing, between the goniometer head and the syringe pump. If tolerated by the crystal, the substrate concentration should be high (100 mM) so that penetration of the substrate into the crystal is rate limited by diffusion alone. In this case introduce an air bubble at the back of the loop so that the end of the substrate/mother liquor interface can be tracked, and the flow of substrate over the crystal is maintained. Spectra are then collected in kinetics mode to record subsequent spectra at required time intervals. When the substrate reaches the crystal, spectral changes should begin to occur. When the steady state, or a kinetic end point, is reached, the spectra will become constant. Finally the substrate loop can be removed, and the substrate may be washed off the crystal with mother liquor to check that the crystal returns to its native spectrum, and has not been irreversibly trapped in an off-reaction pathway.

To track spectroscopic changes on the addition of the gaseous substrate oxygen to cytochrome P450$_{cam}$, Schlichting et al.[8] used a microcrystalline slurry in a flow cell. This removes all the problems associated with crystal alignment, such as crystal prism and chromophore orientation effects due to the random orientations of all the microcrystals, and inaccuracies in path length between reference and experimental spectra, providing the microcrystals are tightly packed into the cell.

Based on time points of maximal build-up of spectrally distinct intermediates from the kinetics, crystals can be exposed to substrate for the appropriate time period, and then flash frozen in liquid nitrogen in loops for X-ray data collection. In this case the reaction will be run in a substrate/cryoprotectant solution, and therefore the kinetics should be repeated with this solution to check that the timings observed are the same.

Copper-Containing Amine Oxidase: A Practical Example

Copper-containing amine oxidases (CuAOs) are ubiquitous and convert primary amines to aldehydes in a variety of biological roles. As well as the copper ion, these redox enzymes contain an organic cofactor derived from a constitutive tyrosine, 2,4,5-trihydroxyphenylalanine quinone (TPQ). During catalysis to generate product aldehyde, TPQ is reduced to the aminoquinol form. During the oxidative half of the reaction (Fig. 4), the aminoquinol (species 1) is reoxidized back to TPQ (species 5), via a semiquinone intermediate (species 2 and 3) and a postulated iminoquinone intermediate (species 4).

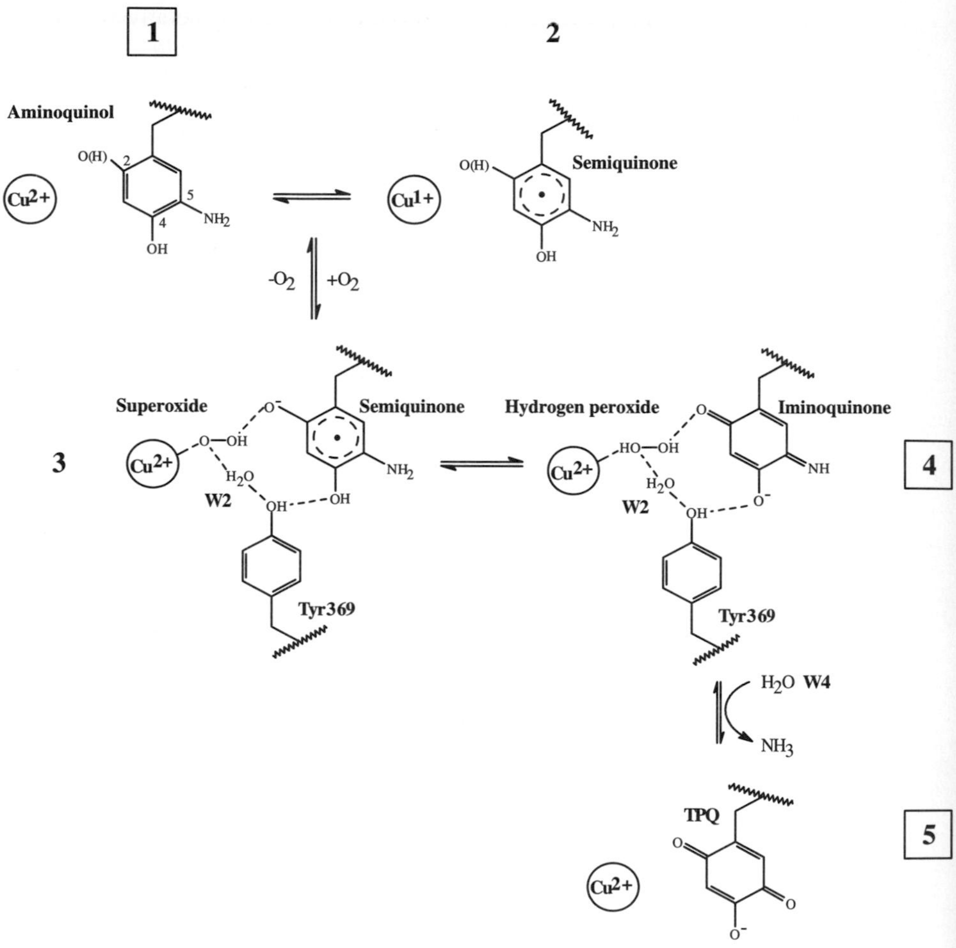

FIG. 4. Catalytic intermediates in the oxidative half-reaction of *E. coli* amine oxidase. X-ray single-crystal structures [C. M. Wilmot, J. Hajdu, M. J. McPherson, P. F. Knowles, and S. E. V. Phillips, *Science* **286,** 1724 (1999); M. R. Parsons, M. A. Convery, C. M. Wilmot, K. D. S. Kapil, V. Blakeley, A. S. Corner, S. E. V. Phillips, M. J. McPherson, and P. F. Knowles, *Structure* **3,** 1171 (1995); J. M. Murray, C. G. Saysell, C. M. Wilmot, W. S. Tambyrajah, J. Jaeger, P. F. Knowles, S. E. V. Phillips, and M. J. McPherson, *Biochemistry* **38,** 8217 (1999)] (indicated by boxes) are available for species 1 (PDB code 1D6U), 4 (PDB code 1D6Z), and 5 (PDB code 1DYU).

Crystals of *Escherischia coli* CuAO are grown under conditions that match the pH optimum for enzymatic activity (pH 7.2).[23] The crystals are pale pink in color (peak at 480 nm), which is that expected for the TPQ chromophore, and

[23] M. R. Parsons, M. A. Convery, C. M. Wilmot, K. D. S. Kapil, V. Blakeley, A. S. Corner, S. E. V. Phillips, M. J. McPherson, and P. F. Knowles, *Structure* **3,** 1171 (1995).

are rod-shaped (0.15 × 0.1 × 0.6 mm), making them ideal for crystal alignment on a single-crystal UV–visible microspectrophotometer to remove crystal prism effects. The crystal-stabilizing mother liquor contains 1.4 M sodium citrate and is relatively viscous, so the crystal is embedded in a Sephadex column to prevent it moving during kinetics.

The catalytic intermediate species for CuAOs have distinct spectroscopic features in the UV–visible range, based on protein and model studies. Using crystals of $E.\ coli$ CuAO, the kinetics of the reaction with the substrate β-phenylethylamine were monitored by single-crystal UV–visible microspectrophotometry under aerobic conditions (Fig. 5). In solution, $E.\ coli$ CuAO turns over β-phenylethylamine at a rate of 118 molecules per second,[24] but in the crystal the steady state is not reached for 8 min. A semiquinone species builds up in the crystal (twin peaks at 435 and 460 nm) before reaching a featureless steady state spectrum. The steady state could consist of a mixture of species, but in this case the spectrum was due to a single rate-determining species in the crystal, that of the iminoquinone, which has never been observed in solution, probably because it is transient. In the single-crystal X-ray structure of this rate-determining freeze-trapped intermediate, dioxygen can be seen bound to the enzyme, probably as the product, hydrogen peroxide.[6] Proton transfer pathways are evident from the structure (species 4; Fig. 4): one by direct transfer from the O2 position of the quinone cofactor to dioxygen, and the other via the hydroxyl of a conserved tyrosine (Tyr-369 in $E.\ coli$ CuAO) and a water (W2) from the O4 position of cofactor to dioxygen.

Unexpectedly, the product phenylacetaldehyde is observed at the back of the active site (Fig. 6). The exit channel for the product is formed at the interface of two protein domains. One of these domains is involved in a crystal contact, and this restricts the movement of one domain in relation to the other, thereby limiting "breathing" of the exit channel and making it more difficult for the product to exit the protein. The presence of the product explains why this species is rate determining in the crystal. The phenylacetaldehyde is keeping the catalytic base, Asp-383, protonated, and thereby preventing activation of a water (W4) that is present in the crystal structure for nucleophilic attack to release ammonia and regenerate fully oxidized TPQ (Fig. 6).

The semiquinone intermediate observed during turnover in the crystals, which could potentially be either species 2 or 3 in Fig. 4, can also be freeze trapped in liquid nitrogen. However, this species is sensitive to X-radiation at a wavelength of 0.9 Å, reducing back to the aminoquinol form of the cofactor during data collection. This can be demonstrated by taking single-crystal spectra both before and after X-ray data collection, and shows the importance of always measuring a crystal spectrum after exposure to the X-ray beam.

[24] J. M. Murray, C. G. Saysell, C. M. Wilmot, W. S. Tambyrajah, J. Jaeger, P. F. Knowles, S. E. V. Phillips, and M. J. McPherson, *Biochemistry* **38**, 8217 (1999).

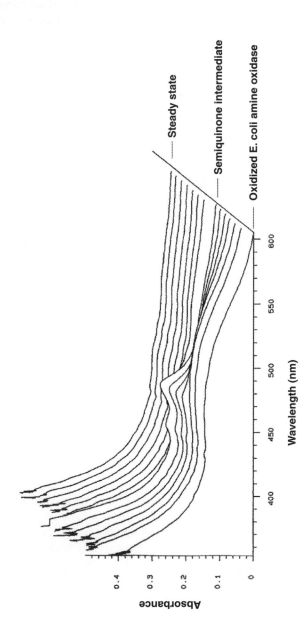

FIG. 5. Spectral changes spaced 1 min apart during enzyme catalysis in crystals of *E. coli* amine oxidase. Spectra were collected every 20 sec, and every third spectrum is shown.

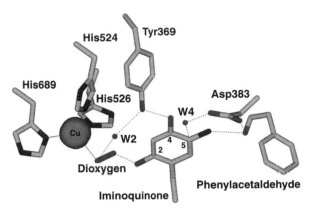

FIG. 6. The rate-determining intermediate in crystals of *E. coli* amine oxidase (PDB code 1D6Z) [C. M. Wilmot, J. Hajdu, M. J. McPherson, P. F. Knowles, and S. E. V. Phillips, *Science* **286**, 1724 (1999)] corresponding to species 4 in Fig. 4. Dashed lines indicate key hydrogen bonds and van der Waals interactions. (Reprinted with permission, copyright 1999 American Association for the Advancement of Science.)

Other Hardware for Single-Crystal Optical Microspectrophotometry

Single-crystal microspectrophotometers have been adapted so that the optics can be mounted directly onto the goniometers of X-ray cameras. This allows the crystal to be monitored during data collection for any redox changes in the beam.[14,19]

A flow system that keeps a moist atmosphere around a naked crystal bathed in cryoprotectant in a loop has been developed. A cryostream can be set up with the system, with the cold stream blocked from the crystal. Substrate can be added directly to the loop, allowing kinetics to be monitored directly, and the cold stream can be unblocked to freeze the crystal instantaneously at the point of maximal build-up of the desired intermediate, and enable X-ray data collection. In addition, this system allows the atmosphere around the crystal to be controlled, for example, at a defined oxygen concentration.

By linking the single-crystal microspectrophotometer stage (Fig. 2) with an infrared (IR) source and detector using appropriate optics, single-crystal Fourier transform infrared (FTIR) spectroscopy can be performed.[25] The identity of small ligands observed in protein crystal structures can then be determined by difference IR.[26,27] In principle, there is no reason why the microspectrophotometers cannot

[25] J. M. Hadden, D. Chapman, and D. C. Lee, *Biochim. Biophys. Acta* **1248**, 115 (1995).
[26] J. C. Fontecilla-Camps, M. Frey, E. Garcin, C. Hatchikian, Y. Montet, C. Piras, X. Vernede, and A. Volbeda, *Biochimie* **79**, 661 (1997).
[27] H. Khachfe, M. Mylrajan, and J. T. Sage, *Cell. Mol. Biol.* **44**, 39 (1998).

also be adapted for fluorescence measurements, while single-crystal micro-Raman spectroscopy was demonstrated in the early 1990s.[28]

This is an exciting time in redox biochemistry: flash freezing of crystals, coupled with supporting spectroscopies, is allowing "snapshots" from the catalytic cycles of enzymes, obtained by single-crystal X-ray crystallography, to be accurately placed along the reaction pathway, giving structural insight into the controlling factors of biological catalysis.

[28] G. Smulevich and T. G. Spiro, *Methods Enzymol.* **226,** 397 (1993).

Section II

Nucleic Acids and Genes

[28] Model System for Developing Gene Therapy Approaches for Myocardial Ischemia–Reperfusion Injury

By JUSAN YANG, TERESA C. RITCHIE,* and JOHN F. ENGELHARDT

Introduction

Ischemic heart diseases arise from a number of clinical conditions, including transient coronary artery occlusion, thrombotic stroke, cardiac surgery, and heart transplantation. Cardiac muscle damage from these episodes is attributable to a period of ischemia followed by reperfusion with oxygenated blood.[1] Reactive oxygen species (ROS) formed during reperfusion may cause direct tissue damage and/or may act as second messengers in redox-sensitive signal transduction pathways. Direct tissue damage caused by ROS is thought to be an important component of immediate necrosis in the ischemic zone, whereas ROS-activated signal transduction pathways are thought to initiate a cascade of events leading to apoptosis of cardiac myocytes.[2–4] Although endogenous free radical-scavenging enzymes are capable of clearing damaging ROS generated during normal cellular metabolism, abnormal ischemia–reperfusion episodes can lead to an overwhelming production of ROS that exceeds cellular metabolic resources to clear these toxic compounds. This is reflected by a variety of pathologic features such as myocardial stunning, reperfusion arrhythmias, impaired reflow, and calcium overload.[1,5]

Potentially, therapeutic interventions to block ROS-sensitive signal transduction pathways can prevent a major component of ischemia–reperfusion (IR) damage. Current treatments for myocardial IR injury include the administration of exogenous antioxidant reagents, such as nonenzymatic (e.g., vitamins A, E, and C) and enzymatic ROS scavengers [e.g., superoxide dismutase (SOD) and catalase]. Although protective effects have been observed with these pharmaceutical reagents, the results have been highly variable. For example, both positive and negative effects were observed with enzymatic ROS scavengers, most likely due to limited uptake of exogenous enzymes at the relevant cellular targets or

---

* Deceased.
[1] S. R. Maxwell and G. Y. Lip, *Int. J. Cardiol.* **58,** 95 (1997).
[2] N. Maulik, T. Yoshida, and D. K. Das, *Free Radic. Biol. Med.* **24,** 869 (1998).
[3] N. Maulik, H. Sasaki, and N. Galang, *Ann. N.Y. Acad. Sci.* **874,** 401 (1999).
[4] R. A. Gottlieb and R. L. Engler, *Ann. N.Y. Acad. Sci.* **874,** 412 (1999).
[5] A. Ar'Rajab, I. Dawison, and R. Fabia, *New Horiz.* **4,** 224 (1996).

Copyright 2002, Elsevier Science (USA).
All rights reserved.
0076-6879/02 $35.00

dose-dependent toxicity.[5–11] An alternative method for altering the cellular redox environment and protecting the myocardium from IR damage is the direct delivery of transgenes expressing redox-scavenging enzymes.[12] The development of gene therapy techniques to deliver such genes to heart muscle will help refine therapeutic approaches by allowing for direct modulation of pathophysiologically relevant ROS pathways at appropriate subcellular sites.

Unique features of the heart render it an organ presenting both opportunity and challenge for the application of gene therapy strategies. Historically, gene transfer to cardiomyocytes has been largely unsuccessful, because systemic administration of viral vectors into the circulation primarily results in their delivery to the liver and lung.[13] This chapter discusses technical aspects of our efforts to determine the most effective mode of delivery of viral vectors to cardiomyocytes and the application of these techniques in gene therapy for ischemia–reperfusion injury in the heart. Methods are also presented for validating the rat IR model, and for assessing the efficacy of transgene expression in reducing the size of an IR-induced infarct.

Choice of Gene Transfer Vector and Mode of Delivery

In the development of strategies for gene delivery to cardiac muscle, several issues must be considered. First is the choice of an appropriate viral vector system for this organ. Potential barriers for applying gene therapy techniques include the efficiency of gene transfer, the duration of transgene expression, and organ inflammation that may follow infection with viral gene therapy vectors. Another issue is the mode of delivery necessary to achieve widespread transgene expression in heart muscle, but that is preferably restricted to this organ. In preliminary stages of the development of this model system, two recombinant viral vectors expressing green fluorescent protein (GFP) reporter genes, adenovirus and adeno-associated virus (AAV), were evaluated. Both vectors have been widely used in gene transfer applications and possess different advantages for specific applications. Adenovirus exhibits a wide tropism for many cells and tissues, and achieves high-level, although short-term, transgene expression. A disadvantage with first-generation E1-deleted adenoviral vectors includes cellular inflammatory responses

[6] J. T. Flaherty and M. L. Weisfeldt, *Free Radic. Biol. Med.* **5,** 409 (1988).

[7] C. Abadie, A. Ben Baouali, V. Maupoil, and L. Rochette, *Free Radic. Biol. Med.* **15,** 209 (1993).

[8] H. H. Klein, S. Pich, S. Lindert-Heimberg, K. Nebendahl, and P. Niedmann, *J. Mol. Cell. Cardiol.* **25,** 103 (1993).

[9] S. T. Sinatra and J. DeMarco, *Conn. Med.* **59,** 579 (1995).

[10] G. J. Gross, N. E. Farber, H. F. Hardman, and D. C. Warltier, *Am. J. Physiol.* **250,** H372 (1986).

[11] J. M. Downey, B. Omar, H. Ooiwa, and J. McCord, *Free Radic. Res. Commun.* **12–13,** 703 (1991).

[12] H. L. Zhu, A. S. Stewart, M. D. Taylor, C. Vijayasarathy, T. J. Gardner, and H. L. Sweeney, *Mol. Ther.* **2,** 470 (2000).

[13] M. Y. Alexander, K. A. Webster, P. H. McDonald, and H. M. Prentice, *Clin. Exp. Pharmacol. Physiol.* **26,** 661 (1999).

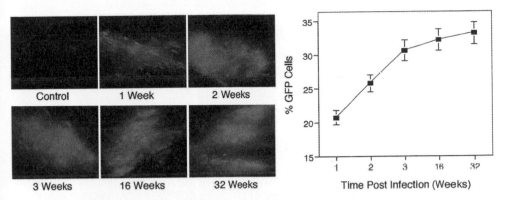

FIG. 1. Recombinant AAV gene transfer after direct myocardial injection. Photomicrographs (*left*) show GFP expression in control uninfected and rAV.EGFP-infected myocardium at 1 to 32 weeks postinfection (as marked). *Right:* Morphometry was used to quantify the extent of gene transfer. Results depict the mean (±SEM) for four animals.

to residual viral proteins expressed in the target cells. AAV, on the other hand, is a nonpathogenic parvovirus that elicits no immune reactions and mediates long-term, stable transgene expression, particularly in skeletal muscle.[14] However, the transduction efficiency with AAV is variable among tissue types and can be limited by the anatomical organization of the tissue target and/or the abundance of viral receptors. Furthermore, unlike adenoviral vectors, which provide immediate transient high levels of expression, AAV vectors have a substantial lag time for maximal transgene expression. Both adenoviral and AAV transduction of heart muscle were evaluated via direct cardiac muscle injection, subpericardial infusion, and coronary artery infusion.

GFP reporter gene expression in heart muscle was achieved with both recombinant AAV (rAV.EGFP) and first-generation recombinant adenoviral (rAd.GFP, E1/E3 deleted) vectors. With AAV, transgene expression in rat heart increased over time. GFP expression can be detected 1 week postinfection with AAV, peaks by 3–5 weeks postinfection, and reaches 30% of the cardiomyocytes in the area of the intramyocardial injection (Fig. 1). Hearts infected with the rAV.EGFP vector exhibited no evidence of any inflammatory response in regions associated with transgene expression at any of the time points evaluated (data not shown). A major disadvantage with AAV-mediated gene delivery for IR injury in the heart was the limited local expression at the site of injection. Furthermore, efforts to increase distribution of the virus by either subpericardial or coronary artery infusion were ineffective (data not shown). In contrast to AAV, recombinant adenovirus mediated a greater level of transgene expression, which was evenly distributed after coronary

[14] J. Yang, W. Zhou, Y. Zhang, T. Zidon, T. Ritchie, and J. F. Engelhardt, *J. Virol.* **73**, 9468 (1999).

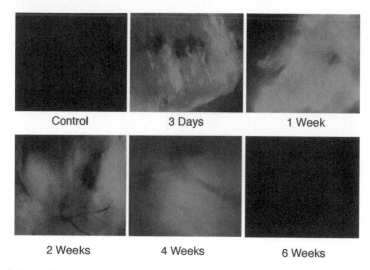

Fig. 2. Recombinant adenoviral infection of the heart by coronary artery infusion. Photomicrographs show GFP fluorescent images in control uninfected and rAd.GFP-infected myocardium 3 days to 6 weeks after infection by coronary artery infusion (as marked).

artery infusion (Fig. 2). When comparing the three modes of delivery (direct injection, subpericardial injection, and coronary artery infusion) the highest efficiency of delivery was achieved with coronary artery infusion. With coronary artery infusion of recombinant adenovirus (rAd), GFP reporter expression was predominantly confined to the left ventricular wall, which is the area of ischemia produced by transient coronary artery occlusion in our rat IR model. Thus, transgene expression with rAd includes the majority of the risk area. rAd-mediated transgene expression was achieved as early as 3 days, and approximately 50–80% of the left ventricle was transduced by 1 week after infection. Maximal expression was seen at 1 week with a significant decline by 2–4 weeks. Complete loss of transgene expression was noted by 6 weeks. In contrast to rAAV, large T cell-mediated inflammatory responses were noted with this E1-deleted adenovirus vector by 2 weeks postinfection. These cellular inflammatory responses are likely responsible for the decline in transgene expression that occurs 2 weeks after infection.

On the basis of these preliminary experiments, the model system described in this chapter utilizes coronary artery infusion of adenovirus 3 days before the initiation of experimental ischemia–reperfusion to avoid complications with cellular immune reactions. Ultimately, the development of gene therapy approaches for the heart will require the use of "gutted" rAd, in which all the viral genes except the long terminal repeats (LTRs) and ψ signal packaging sequence are deleted. This approach will overcome most of the disadvantages associated with the cellular immune responses to this vector.

Rat Model of Transient Ischemia–Reperfusion in Heart

The rat model system for ischemia–reperfusion in the heart involves the surgical implantation of "occluder" and "releaser" sutures around the left main coronary artery, using a modified procedure by Himori and Matsuura.[15] The sutures exit the body wall and are left accessible after the surgery (Figs. 3 and 4). Infection with viral vectors can be performed during the surgical procedure before placement of the sutures. After 3 days of postoperative recovery, IR is performed by manipulation of the sutures in anesthetized or conscious animals. In this model, 1 hr of ischemia is followed by 5 hr of reperfusion, using alterations in the electrocardiogram (EKG) to confirm ischemia and reperfusion, and the serum level of creatine kinase (CK) to assess IR damage. As shown in Fig. 5, left coronary artery occlusion produces an ST-T wave depression or elevation of more than 0.1 mV 1 min after the initiation of ischemia. This change is maintained during the entire ischemic period and is used to confirm the success of the occlusion technique. The normalization of the S-T segment is observed 1–5 min after release of the occlusion and confirms the success of reperfusion. The serum level of CK (Fig. 6) increases approximately 6- to 7-fold after ischemia–reperfusion injury. Serum CK levels before IR were similar to those of control animals that had not undergone surgery 3 days previously, suggesting that this experimental model does not invoke detectable levels of damage that might otherwise complicate experiments.

On the basis of these results, the protocol described for the IR model system utilizes 1 hr of ischemia and 5 hr of reperfusion. The short, 3-day interval between viral infection and analysis also assures that no cellular inflammatory responses to adenovirus have occurred during the experimental window.

Experimental Methods

Surgical Methods

All animal surgery for the development of this model is performed under a protocol approved by the University of Iowa Animal Care and Use Committee, and conforms to the National Institutes of Health (NIH, Bethesda, MD) guidelines. All surgical procedures described should be performed by aseptic technique with sterilized instruments, materials, and buffers.

MATERIALS

Ketamine (Fort Dodge Animal Health, Fort Dodge, IA)
Xylazine (Phoenix Pharmaceutical, St. Joseph, MO)
Phosphate-buffered saline (PBS), 1×
Normal saline
Small animal surgical board

[15] N. Himori and A. Matsuura, *Am. J. Physiol.* **256,** H1719 (1989).

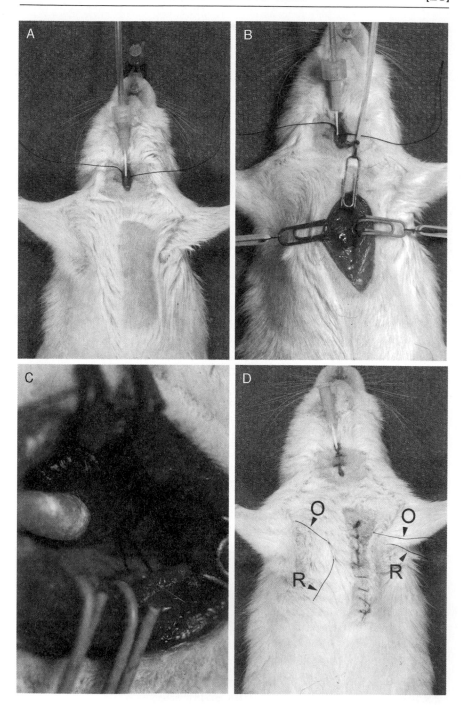

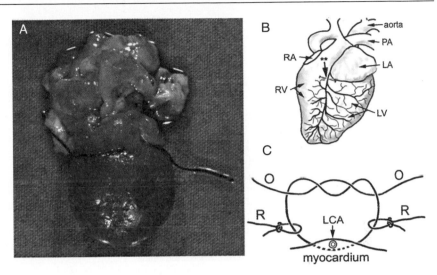

FIG. 4. Placement of the occluder and releaser sutures. The placement of the occluder suture is shown in a dissected rat heart (A) and in a schematic illustration of the heart (B). The left atrium is used as a landmark, because the left main coronary artery runs along its medial side. The occluder suture is paced just at the lower edge of the atrium about 2–3 mm lateral on either side of its medial margin [large arrow in (B), marked by asterisks (**)], and at a depth of about 2 mm. Placement of the occluder suture around the left main coronary artery (LCA) is marked by a dashed line in (C). The drawing in (C) illustrates the knots tied in the occluder and releaser sutures during surgery. The occluder suture runs just under the LCA (marked by an arrow), as indicated by a dashed line. LA, Left atrium; LV, left ventricle; PA, pulmonary artery; RA, right atrium; RV, right ventricle; O, occluder suture; R, releaser suture.

Intravenous catheter (21 gauge; Terumo Medical, Tokyo, Japan)
Animal respirator (CIV-101; Columbus Instruments, Columbus, OH)
High-temperature cautery (model CRS-232; Roboz Surgical, Rockville, MD)
4-0 coated Vicryl braided suture with cutting PS-2 needle (Ethicon, Cincinnati, OH)
5-0 silk black braided suture with taper C-1 needle (Ethicon)
2-0 and 7-0 black braided silk suture (Ethicon)

FIG. 3. Surgical preparation of the rat ischemia–reperfusion model. (A) Male Wistar rats are anesthetized to a surgical plane and intubated with a catheter tied tightly in place with a suture. The rat is maintained on artificial respiration for the duration of the surgery. (B and C) A left thoracotomy is performed and the chest cavity is opened and gently spread with three small retractors. Sterilized bent paper clips hooked to rubber bands and anchored to the surgery board with pins can serve as effective retractors. After the pericardium is pierced to expose the heart, adenoviral infection is accomplished via coronary infusion, as described in text. (C) The occluder and releaser sutures are then implanted around the left coronary artery and tested. The ends of these sutures are inserted through the body wall and skin before closing all incisions, but are then reinserted under the skin, leaving only a small loop exposed on the exterior. (D) An illustration of the freed ends of the occluder (O) and releaser (R) sutures, which are accessible for experimental induction of ischemia and reperfusion 3 days after the surgical preparation.

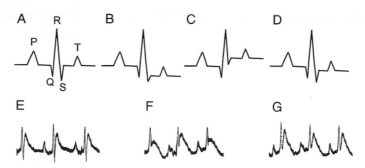

FIG. 5. EKG analysis of coronary artery-induced ischemia and reperfusion. Schematic illustration of a normal EKG trace (A) and the diagnostic EKG changes characteristic of ischemia (B and C) and reperfusion (D). The two theoretical possibilities in (B) and (C) may occur depending on placement of the EKG leads and the channel analyzed. (E–G) EKG recordings performed on an anesthetized rat before ischemia (E), 10 min after the initiation of ischemia (F), and 2 min after the start of reperfusion (G). The recordings in (E) and (F) are from lead II in the same animal.

9-0 black monofilament nylon with CS-175-8 needle (Ethicon)

Surgical instruments: Fine surgical scissors, fine forceps (curved or strait), mosquito hemostatic forceps (straight or curved), disposable scalpels (Feather Safety Razor, Osaka, Japan, Medical Division), small animal retractors, and modified surgical clamps (fine curved surgical forceps with the tips bent to 90° and covered with PE20 tubing)

Hamilton microinjection syringe with a 33-gauge needle

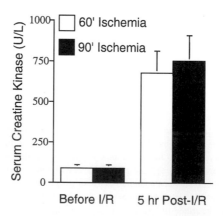

FIG. 6. Serum CK levels before and after ischemia–reperfusion. After performing either 60 or 90 min of ischemia followed by 5 hr of reperfusion, the serum creatine kinase level increases approximately 6- to 7-fold. These findings confirm that the model system is capable of producing significant IR damage to the myocardium, which results in the release of creatine kinase into the serum.

Surgical Preparation
See Fig. 3.

1. A male Wistar rat weighing between 250 and 300 g is anesthetized with 100 μl/100 bgw (body gram weight) of a mixture of ketamine (60 mg/ml) and xylazine (8 mg/ml) in PBS administered by intraperitoneal injection. When a surgical plane of anesthesia is reached, the rat is positioned on its back on an animal surgical board.

2. Intubation and respiration: After the neck is shaved and disinfected with 75% ethanol, a midline 1-cm incision is made in the subhyoid region. The soft tissue and muscles are blunt-dissected to expose the trachea, taking care to avoid damaging the thymus and thyroid glands overlying the trachea, as well as the vagus nerves laterally. A 1-mm incision is made in the trachea, using fine surgical scissors. A 21-gauge intravenous catheter (1.5 cm in length) is inserted into the trachea and is tied tightly in place with 2-0 silk. This catheter is then attached to an animal respirator (CIV-101; Columbus Instruments). The rat is maintained on artificial respiration with ambient air at a tidal volume of 6 ml and a rate of 60 strokes/min.

3. Left thoracotomy: After the left chest is shaved and disinfected with 75% ethanol, a 2-cm longitudinal incision is made with a disposable scalpel in the left side of the chest parallel and 0.5 cm lateral to the sternum. After sharp dissection of the muscles, the left chest cavity is opened by cutting the fourth and the fifth ribs and gently spreading the tissue with three small retractors. The beating heart and the left lung moving with the respirator can be clearly seen. The lung is fragile, and care should be taken to avoid damage. Stop all bleeding with a cotton tip applicator wetted with normal saline or by cauterizing if necessary.

Coronary Artery Infusion of Adenovirus

4. The pericardium is opened carefully to avoid tearing vessels, using two fine curved forceps. The pulmonary artery is occluded by pressing with one prong of the modified surgical clamp. After a few seconds to allow the evacuation of blood from the left ventricle, both prongs of the clamp are closed to clamp both the pulmonary artery and the aorta briefly ($\sim$10 sec), while 100 μl of adenovirus (2×10^{11} particles) is injected into the chamber of the left ventricle. This procedure generates the coronary artery infusion necessary to deliver virus to the left ventricular wall. The clamps are then promptly removed.

Note. Other methods of direct coronary artery infusion, such as without clamping the aorta and pulmonary artery, were evaluated. However, the distribution of virus in the heart was far less effective and most of the virus exited the heart into the general circulation.

Placement of Silk Occluder and Releaser Sutures
See Fig. 4.

5. After thoracotomy and incision of the pericardium as described, the beating heart is gently reflected to the animal's right side, using a cotton tip applicator. This allows visualization of the left coronary vein, which lies between the pulmonary artery and the root of the left atrial appendage. The left main coronary artery lies just beneath this vein.

6. A tapered c-1 needle threaded with 5-0 silk suture is used to pierce the epicardium (0.5-mm depth) just distal to the left atrium and 2–3 mm on either side of the left coronary artery (and vein). A loose, overhand knot is made in the silk to form the occluder suture around the artery, leaving a 10-cm length of suture on each end.

7. Two 7-0 silk sutures (releasers), 10 cm in length, are tied on either side of the silk occluder, as illustrated in Fig. 4.

8. Before the incision is closed, the occluder and releaser knots are tested by a brief (~5 sec) ischemic event. When the occluder is tightened, the ischemic area on surface of the left ventricle lightens in color. After tightening of the releaser sutures, normal color is recovered.

9. Each of the four silk suture ends is externalized through the chest wall and skin to allow for experimentally induced occlusion 3 days later. The ends of the four silk leads are then reinserted under the skin, leaving only a small loop exposed on the exterior. This will prevent the animal from biting them during the postoperative period.

10. Saline-wetted cotton tip applicators are used to gently place the heart and tissues in their normal positions in the chest cavity and to stop any bleeding as well. A chest tube (PE20 tubing) is then introduced into the midthoracic cavity. Approximation of the ribs is accomplished with three individual 4-0 coated Vicryl sutures. The deep and superficial muscle layers, and then the skin, are then closed with 4-0 coated Vicryl sutures.

11. After closing all the incisions, a 5-ml syringe with a 21-gauge needle is inserted into the chest tube to allow removal of the air and fluid from the cavity and to reinflate the lungs. The chest is evacuated until it is well sealed and no additional air or fluid can be removed. The chest tube is slowly removed from the chest cavity while gently pulling back on the syringe.

Note. This procedure is optional but may benefit survival and increase the number of animals available for ischemia–reperfusion experiments.

12. The tracheal catheter is left in place, but is detached from the respirator. The animal is stimulated to breath on it own by rhythmically pressing both sides

of the chest three to five times until spontaneous breathing is recovered. After spontaneous breathing is stable for 5–10 min, the catheter is removed from the trachea.

13. The incision in the trachea is closed with 9-0 coated Vicryl suture. After repositioning the muscles and glands, the skin above the tracheal incision is then closed with 4-0 coated Vicryl suture.

14. Buprenorphine (0.05–0.1 mg/kg, subcutaneous or intravenous) can be administered for postoperative analgesia immediately after surgery, on the morning after surgery and at other times as necessary. Rats are housed appropriately for 3 days after surgery, at which time ischemia–reperfusion is performed.

Note. Prophylactic antibiotic treatment may be used, but we have not found this necessary if aseptic technique is scrupulously adhered to in all surgical procedures.

Coronary Artery Ischemia–Reperfusion

MATERIALS

Electrocardiograph recorder
Ketamine (Fort Dodge Animal Health)
Xylazine (Phoenix Pharmaceutical)
Heparin (1000 U/ml), 50 μl
Forceps
Metoprolol tartrate (5 mg/5 ml; Novartis Pharma, Basel, Switzerland)
Lidocaine-HCl (5 mg/ml; Abbott Laboratories, North Chicago, IL)

PROCEDURE

1. Three days after infusion of adenovirus and placement of the occluder and releaser sutures, the rat is reanesthetized with ketamine–xylazine mixture as described, and 50 μl of heparin (1000 U/ml) is administered by tail vein injection.

Note. If required, the ischemia–reperfusion protocol can be performed in conscious animals. This procedure requires immobilizing the animals in a restrainer, and is not described here.

2. EKG measurements are made with an EKG recorder. The leads are modified for rats, using pins inserted into the inner side of the four legs. The analysis uses standard and augmented limb leads and includes I, II, III, aVR, aVL, and aVF. EKGs are assessed before and during ischemia and after reperfusion for each animal (Fig. 5).

Note. The reader is referred to *The Rat Electrocardiogram in Pharmacology and Toxicology*[16] and *Electrocardiographic Analysis*[17] for details on methodology for EKG recording and analysis.

3. The ischemia–reperfusion protocol is initiated. During the surgical implantation of the occluder and releaser sutures, the ends of the sutures were reintroduced under the skin, leaving only a small externalized loop. Before the initiation of ischemia–reperfusion, the four silk leads are externalized by gently pulling on the loops. The occluder leaders are then pulled to tighten the knot, inducing the coronary artery ligature.

4. The effectiveness of the left coronary ligature is evaluated and monitored by standard EKG indices of myocardial ischemia (Fig. 5). Within 1 min after induction of ischemia, the ST-T wave will be elevated or depressed by at least 0.1 mV (depending on the lead being evaluated). On pulling the sutures, the orientation of the heart may be changed, and the shape of the QRS wave will be correspondingly altered. These changes are maintained throughout the 1-hr duration of ischemia.

Note. In some animals, tightening of the coronary ligature will produce rapid premature ventricular beating or ventricular vibration, which decreases survival. To prevent this complication, 300–500 μl of metoprolol tartrate (5 mg/5 ml) and 300–500 μl of lidocaine-HCl (5 mg/ml) can be injected via the tail vein before the induction of ischemia.

5. After 1 hr of ischemia, the releaser silks are pulled to loosen the occluder ligature, resulting in coronary artery reperfusion. Reperfusion is maintained for the 5-hr duration of the experiment. EKG analysis is also used to monitor the reperfusion (Fig. 5). Typically, the reperfusion EKG has distinctive features, and thus is used to confirm the success of reperfusion. One characteristic is reperfusion arrhythmia, which consists of premature ventricular beating that sometimes occurs in the first minute of reperfusion. After 1–5 min of reperfusion, the ST-T wave will return to approximately normal.

Methods for Analysis

Measurement of Serum Creatine Kinase
See Fig. 6.

---

[16] R. Budden, D. K. Detweiler, and G. Zbinden, "The Rat Electrocardiogram in Pharmacology and Toxicology." Pergamon Press, Oxford, 1981.
[17] R. H. Bayley, "Electrocardiographic Analysis." Hoeber, New York, 1958.

Materials

Disposable 1-ml syringes
Needles (25 gauge)
Microcentrifuge tubes (1.5 ml)
Microcentrifuge
Deionized H_2O
Normal saline
Plates (96 well)
Creatine kinase assay kit (Sigma, St. Louis, MO)
Microplate, visible wavelength spectrophotomoter (Molecular Devices, Menlo Park, CA)

Procedure

1. Blood (300 μl) is collected via the tail vein before ischemia, at various time points during reperfusion, and just before killing the animals with an overdose of anesthetic after 5 hr of reperfusion. The serum is collected by centrifugation 2 min at 10,000 rpm, and then is stored in a $-20°$ freezer until analysis.

2. CK assay procedures are slightly modified from the original manufacturer protocol to accommodate the small volumes of serum. Reagents A and B are dissolved with 10 and 4 ml of deionized water, respectively, and then mixed at a 10 : 1 (A : B) ratio.

3. Serum (10 μl) is added to each well of a 96-well plate, followed by the addition of A : B mixture (250 μl/well).

4. The plate is placed in the prewarmed $37°$ sample chamber of the microplate reader (Molecular Devices). After 3 min of incubation and mixing, optical density (OD) readings at 340-nm wavelength are taken at exactly 1, 2, and 3 min, blanked against normal saline.

5. The serum CK level is calculated by changes in OD over time, as described in the protocol supplied with the kit. The following formula is used:

$$\text{CK activity (U/liter)} = \frac{(\Delta A/\text{min}) \times TV \times 1000}{6.22 \times LP \times SV}$$

where TV is the total volume (ml), SV is the sample volume (ml), and LP is the light path (1 cm).

Green Fluorescent Protein Reporter Gene Expression Assay

Materials

Fluorescence microscope, equipped with a fluorescein isothiocyanate (FITC) filter

Ketamine (Fort Dodge Animal Health)

Xylazine (Phoenix Pharmaceutical)

Surgical instruments: Scalpel, forceps, scissors, and fine razor blades

PBS, 1×

Neutralized formalin, 10% (v/v)

Graded [10, 20, and 30% (w/v)] sucrose solutions in PBS

Plastic embedding molds

Tissue Tek optimal cutting temperature (O.C.T.) medium (Sakura Finetek, U.S.A., Torrance, CA)

Cryostat

Probe On Plus microscope slides (Fisher Scientific, Pittsburgh, PA)

Cover glass (No. 1; Surgipath Medical Industries, Richmond, IL)

Aqueous mounting medium (Citifluor; UKC Chemical Laboratory, Canterbury, UK)

PROCEDURE

1. Rats are killed by an overdose of ketamine–xylazine mixture, and the heart is perfused and washed with 1× PBS two or three times.

2. After the removal of all traces of blood, the expression of GFP can be visualized *en bloc* in freshly excised heart tissue under a fluorescence microscope.

3. The GFP-positive area of the heart is dissected and fixed with 10% (v/v) neutralized formalin for 2 hr, followed by cryoprotection in graded sucrose solutions [10, 20, and 30% (w/v)] at 4° for 4–6 hr each, or until the tissue sinks.

4. The tissue is embedded in appropriately sized embedding molds, using O.C.T. medium. The tissue blocks are quick-frozen in a shallow isopentane–dry ice bath, and stored at −80°.

5. Sections (6 μm) are cut with a cryostat, and the slides are coverslipped with Citifluor as mounting medium. The sections are evaluated for GFP expression with a fluorescence microscope.

Note. In the present protocol, only recombinant adenovirus harboring the GFP transgene has been discussed. This approach to reduce IR-mediated damage can be applied with therapeutic transgene products such as MnSOD and Cu,ZnSOD encoded in identical recombinant vector backbones. In these instances, localization of the transgene product can be performed by immunocytochemistry or immunofluorescence methodology compatible with the antibodies for a given transgene product.

Determination of Infarct Size by 2,3,5-Triphenyl-2H-Tetrazolium Chloride Staining

See Fig. 7.

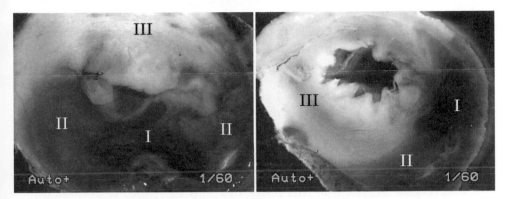

FIG. 7. Visualization of ischemic myocardial tissue with Evan's blue and TTC staining. After IR injury, cardiac tissue can be divided into three zones based on the color of Evan's blue and TTC staining. Zone I, nonischemic tissue (blue regions) stained by Evan's blue. This tissue represents normal myocardium not included in the area-at-risk (AAR) (i.e., the ischemic zone); zone II, AAR within the ischemic zone that stains red with TTC is viable and nonnecrotic; zone III, AAR within the ischemic zone that does not stain with TTC (white regions) is necrotic, nonviable tissue. The total AAR (i.e., the total ischemic area) includes both TTC-positive and -negative regions, which do not stain with Evan's blue. Two examples of cardiac slices after 60 min of ischemia and 5 hr of reperfusion are shown.

MATERIALS

    Evan's blue dye (Sigma-Aldrich, Milwaukee, WI)
    2,3,5-Triphenyl-2H-tetrazolium chloride (TTC; Sigma-Aldrich)
    Water bath, 37°
    Phosphate buffer (20 mM, pH 7.5)
    Neutralized formalin, 10% (v/v)
    Vibratome

PROCEDURE

1. At the termination of the IR experiment, the left coronary artery is reoccluded with the attached sutures (see above), and 200 μl of 2% (w/v) Evan's blue dye in 1× PBS is infused into the left ventricle. This dye will stain normal, nonischemic cardiac tissue dark blue, while the ischemic zone will not be stained.

2. The heart is removed and perfused with 1× PBS at room temperature. After draining the heart on dry gauge, the atrial appendages and great vessels are trimmed off. The heart tissue is then directly frozen *en bloc* in liquid nitrogen and stored at −80° for up to 3–6 months.

3. The heart tissue is warmed in a −20° freezer for 2 hr, and then the heart is sectioned transversely in 1-mm-thick slices, using a Vibratome. The slices are

warmed to 0–4° by placement in a plastic petri dish on ice water, and then are rinsed briefly in ice-cold PBS to remove traces of blood from the cut surfaces.

4. The heart slices are incubated with a solution consisting of 1% (w/v) TTC in phosphate buffer (20 mM), pH 7.5, for 30 min at 37° in a water bath. TTC imparts a red stain to viable ischemic tissue.

5. After 24 hr of fixation and enhancement of the contrast in 10% (v/v) neutralized formalin, the slices are rinsed with PBS and photographed. The nonrisk area will appear blue from the Evan's blue dye, the area-at-risk (AAR) will be stained red with TTC or will be unstained (white), and the infarction area will appear white. Viable tissue within the AAR of the ischemic zone will be red.

6. Digital analysis with NIH Image 1.62 software is used to calculate the percentage of infarction size and AAR in the heart.

Note. As an alternative method for determining the infarct size, NBT staining alone can be used. The heart slices are incubated with a solution consisting of Sorensen's phosphate buffer (0.1 M), pH 7.4, and 2,2′-di-p-nitrophenyl-5, 5′-diphenyl-3,3′-(3,3′-dimethoxy-4,4′-biphenylene) ditetrazolium chloride (NBT, 0.5 mg/ml; Sigma) for 30 min at 37° in a water bath. Normal myocardium stains dark blue within 15 min, while the infracted area remains unstained or is stained only faintly. Although this method will give a direct indication of the infarct size, this procedure is not compatible with Evan's blue demarcation of the ischemic zone. Because variability in the vascular branching between animals can lead to different sizes of ischemic zones, it is preferable to indicate the extent of infarct size as the percentage of the AAR in the ischemic zone.

Acknowledgments

This work was supported by an American Heart Association postdoctoral fellowship (#0020354Z; J.Y.), by NIH SCOR P50 HL60316 (G.H.), and by the Center for Gene Therapy funded by the NIH and the CFF (P30 DK54759; J.F.E.).

[29] Cloning and Characterization of Soluble Decoy Receptors

By JOHANNA LAUKKANEN and SEPPO YLÄ-HERTTUALA

Introduction

Cytokines, growth factors, and hormones exert their action via membrane-bound receptors or, sometimes, via soluble receptors.[1] These soluble receptors are formed either by shedding of the membrane-bound receptor or by alternative splicing. Soluble receptors can act as carrier proteins for their ligands and protect ligands against proteolytic degradation.[2] Soluble receptors can also act as suppliers of cytokines and growth factors, and they can act as antagonists [interleukin 4 (IL-4) and tumor necrosis factor (TNF)][3,4] or agonists to the membrane-bound receptors. Increases in soluble receptor levels have been detected in some disease states.

One extensively studied *agonistic* soluble receptor is soluble IL-6α receptor (sIL-6R).[5,6] IL-6 receptor consists of two different proteins: IL-6α and homodimeric gp130. gp130 is the transmembrane signaling protein and IL-6α is the cytokine-binding protein. IL-6α also exists in a soluble form. It is formed both by shedding and by alternative splicing and it can mediate the effect of IL-6 on cells that express only gp130. This process is termed transsignaling. In mice chimeric IL-6/IL-6R complex has induced liver regeneration after severe hepatocellular injury when neither IL-6 nor IL-6R was able to do so alone.[6]

One example of *antagonistic* soluble receptors is the truncated form of the receptor for advanced glycation end products (RAGE).[7,8] It has been shown that administration of soluble RAGE protein inhibits accelerated diabetic atherosclerosis[7] and suppresses tumor growth and metastasis in mice.[8] We have used a similar

[1] S. Rose-John and P. C. Heinrich, *Biochem. J.* **300,** 281 (1994).

[2] R. Fernandez-Botran and E. S. Vitetta, *J. Exp. Med.* **174,** 673 (1991).

[3] C. R. Maliszewski, T. A. Sato, T. Vanden Bos, S. Waugh, S. K. Dower, J. Slack, M. P. Beckmann, and K. H. Grabstein, *J. Immunol.* **144,** 3028 (1990).

[4] L. Ozmen, G. Gribaudo, M. Fountoulakis, R. Gentz, S. Landolfo, and G. Garotta, *J. Immunol.* **150,** 2698 (1993).

[5] V. Modur, Y. Li, G. A. Zimmerman, S. M. Prescott, and T. M. McIntyre, *J. Clin. Invest.* **100,** 2752 (1997).

[6] E. Galun, E. Zeira, O. Pappo, M. Peters, and S. Rose-John, *FASEB J.* **14,** 1979 (2000).

[7] L. Park, K. G. Raman, K. J. Lee, Y. Lu, L. J. J. Ferran, W. S. Chow, D. Stern, and A. M. Schmidt, *Nat. Med.* **4,** 1025 (1998).

[8] A. Taguchi, D. C. Blood, G. del Toro, A. Canet, D. C. Lee, W. Qu, N. Tanji, Y. Lu, E. Lalla, C. Fu, M. A. Hofmann, T. Kislinger, M. Ingram, A. Lu, H. Tanaka, O. Hori, S. Ogawa, D. M. Stern, and A. M. Schmidt, *Nature (London)* **405,** 354 (2000).

Copyright 2002, Elsevier Science (USA).
All rights reserved.
0076-6879/02 $35.00

approach with human macrophage scavenger receptor type A.[9] Macrophage scavenger receptor (MSR) type AI, AII, and AIII are trimeric membrane glycoproteins involved in the deposition of lipids in the arterial wall during atherosclerosis.[10,11] MSR expression is upregulated in atherosclerotic lesions, where it takes up modified low-density lipoprotein (LDL), which leads to foam cell formation and the progression of atherosclerosis.[12,13] MSR is also an adhesion molecule. *In vitro* studies have shown that cation-independent adhesion of macrophages to cell culture plastic is mediated by MSR,[14] as is macrophage adhesion to apoptotic thymocytes[15,16] and glycosylated collagen.[17]

We have constructed a chimeric fusion protein that consists of the bovine growth hormone signal sequence and the human MSR AI extracellular domains.[18,19] This soluble "decoy" MSR (sMSR) was cloned into an adenoviral vector, and sMSR recombinant adenoviruses were produced. It was found that sMSR inhibits the degradation of modified LDL by 70–80% and inhibits foam cell formation *in vitro*. Tail vein injection of 1×10^9 PFU of sMSR adenoviruses reduced atherosclerotic lesion formation in hypercholesterolemic LDL receptor knockout mice, which is an example of the therapeutic use of antagonistic soluble receptors. A similar approach would be useful for blocking the action of several growth factors and cytokines. In this chapter we present general guidelines for the cloning, production, and characterization of soluble decoy receptors, using sMSR and adenoviral gene transfer as an example.

Materials

Selected Reagents and Suppliers

5-Bromo-4-chloro-3-indolyl phosphate and nitroblue tetrazolium (BCIP–NBT; Roche, Indianapolis, IN)

[9] T. Kodama, M. Freeman, L. Rohrer, J. Zabrecky, P. Matsudaira, and M. Krieger, *Nature (London)* **343,** 531 (1990).
[10] M. Krieger, *Curr. Opin. Lipidol.* **8,** 275 (1997).
[11] D. R. Greaves, P. J. Gough, and S. Gordon, *Curr. Opin. Lipidol.* **9,** 425 (1998).
[12] S. Ylä-Herttuala, M. E. Rosenfeld, S. Parthasarathy, E. Sigal, T. Särkioja, J. L. Witztum, and D. Steinberg, *J. Clin. Invest.* **87,** 1146 (1991).
[13] T. P. Hiltunen, J. S. Luoma, T. Nikkari, and S. Ylä-Herttuala, *Circulation* **97,** 1079 (1998).
[14] I. Fraser, D. Hughes, and S. Gordon, *Nature (London)* **364,** 343 (1993).
[15] N. Platt, H. Suzuki, Y. Kurihara, T. Kodama, and S. Gordon, *Proc. Natl. Acad. Sci. U.S.A.* **93,** 12456 (1996).
[16] V. Terpstra, N. Kondratenko, and D. Steinberg, *Proc. Natl. Acad. Sci. U.S.A.* **94,** 8127 (1997).
[17] J. el Khoury, C. A. Thomas, J. D. Loike, S. E. Hickman, L. Cao, and S. C. Silverstein, *J. Biol. Chem.* **269,** 10197 (1994).
[18] J. Laukkanen, P. Lehtolainen, P. J. Gough, D. R. Greaves, S. Gordon, and S. Ylä-Herttuala, *Circulation* **101,** 1091 (2000).
[19] J. Laukkanen, P. Leppänen, P. J. Gough, D. R. Greaves, S. Gordon, and S. Ylä-Herttuala, *Circ. Res.* submitted (2002).

Bovine serum albumin (BSA; Sigma, St. Louis, MO)

Dulbecco's modified Eagle's medium (DMEM; GIBCO-BRL, Gaithersburg, MD)

DyNAzyme II DNA polymerase (Finnzymes, Espoo, Finland)

Goat anti-scavenger receptor AI antibody (Chemicon International, Temecula, CA)

Horseradish peroxidase (HRP)-conjugated donkey anti-goat antibody (Chemicon International)

M2 Mouse anti-Flag monoclonal antibody (Sigma)

M2 Rabbit anti-Flag polyclonal antibody (Sigma)

Minimal essential medium (MEM; GIBCO-BRL)

Moloney murine leukemia virus (Mo-MuLV) reverse transcriptase (MBI Fermentas, Hanover, MD)

Opti-MEM (GIBCO-BRL)

Poly(G) (Sigma)

RAW 264.7 monocyte-macrophages (ATCC TIB-71)

Ribonuclease inhibitor (RNase A inhibitor; MBI Fermentas)

RQ1 DNase (Promega, Madison, WI)

Streptavidin–alkaline phosphatase (AP) (Roche)

3,3'5,5'-Tetramethylbenzidine (Sigma)

Stock Solutions

Agar, 2.5% (w/v)

$AgNO_3$, 10% (w/v)

$CaCl_2$, 1 M

Dextrin, 1% (w/v) in deionized water

Glycerol

HEPES (pH 7.4), 1 M

NaCl, 5 M

$NaHCO_3$ (pH 9.5), 1 M

NaN_3, 1% (w/v)

Oil red O, 1% (w/v) in 2-propanol

Sodium dodecyl sulfate (SDS), 10% (w/v)

Tris-HCl (pH 7.4), 1 M

Trichloroacetic acid (TCA), 100%

Working Solutions

HEPES (5 mM) and 5 mM HEPES–20% (v/v) glycerol

Buffer A: 20 mM Tris-HCl (pH 8), 150 mM NaCl, 1 mM $CaCl_2$

Buffer B: 20 mM Tris-HCl (pH 8), 150 mM NaCl, 1 mM $CaCl_2$, BSA (2 mg/ml), 0.01% (w/v) NaN_3

Western blot sample buffer: 62.5 mM Tris-HCl (pH 6.8), 10% (v/v) glycerol, 2% (w/v) SDS, 5% (v/v) 2-mercaptoethanol, 0.05% (w/v) bromphenol blue (in nonreducing buffer, 2-mercaptoethanol is replaced by deionized water)

TCA, 50% (w/v)

AgNO$_3$, 10% (w/v)

Oil red O: 60 ml of Oil red O and 40 ml of dextrin are combined and filtered through filter paper

Agar: 1× minimal essential medium (MEM), 0.5% (w/v) agar, 20 mM MgCl$_2$, 1% (v/v) fetal bovine serum (FBS), penicillin–streptomycin (5 U/ml and 5 μg/ml, respectively). Agar is melted and kept liquid in a 60° water bath and added to 37° medium containing reagents.

Tris-buffered saline (TBS): 0.05 M Tris base, 0.15 M NaCl, pH 7.4

General Procedures

In Vitro

Cloning of Soluble Receptor

sMSR, consisting of the bovine growth hormone signal sequence and the extracellular parts of the human macrophage scavenger receptor,[9] is constructed and an eight-amino acid Flag peptide is added to enable easy detection of the transgene. sMSR is constructed in three steps.[18] First, a 109-bp fragment of the bovine growth hormone signal sequence is created with a *Hin*dIII site at the 5' end, and an *Eco*47III site encoding the last two amino acids of the signal peptide and an *Xba*I site at the 3' end. The fragment is cloned into similarly cut pRc/CMV plasmid. Second, an oligonucleotide encoding the Flag epitope (DYKDDDDK), with *Eco*47III at the 5' end, a *Not*I site following the Flag epitope, and an *Xba*I site at the 3' end, is synthesized. Fragments from steps 1 and 2 are ligated. Third, an oligonucleotide containing a *Not*I restriction site in frame with the coding sequence of the first eight amino acids of the human MSR extracellular region[9] and a 3' oligonucleotide specific for the C terminus of the type AI MSR with an *Xba*I restriction site after the stop codon is used in a polymerase chain reaction (PCR) to generate the extracellular domains of the receptor. The fragment is ligated into the similarly cut vector from step 2. Fourth, the receptor is cloned into the pcDNA3 vector with *Hin*dIII and *Xba*I.

A replication-deficient E1/E3-deleted adenovirus is generated, using a pAdvBgl plasmid.[20] The receptor, along with the cytomegalovirus (CMV) promoter and bovine growth hormone poly(A), is excised from the pcDNA3 vector with *Nru*I and *Pvu*II, blunted, and subcloned into the 3' end of the first map unit of the adenovirus genome (Fig. 1). The correct structure of the construct is verified by sequencing.

[20] E. Barr, J. Carroll, A. M. Kalynych, S. K. Tripathy, K. Kozarsky, J. M. Wilson, and J. M. Leiden, *Gene Ther.* **1**, 51 (1994).

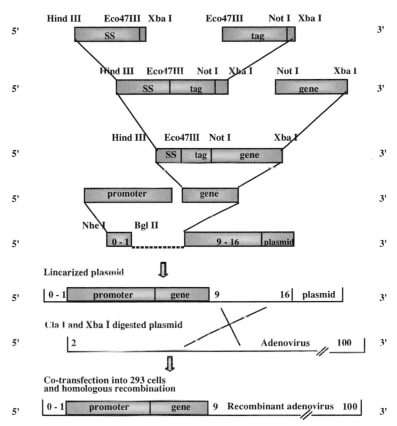

FIG. 1. Construction of sMSR plasmid and adenovirus. Bovine growth hormone signal sequence (SS), FLAG epitope (tag), and extracellular domains of the human macrophage scavenger receptor cDNA (gene) are ligated. Promoter and sMSR are subcloned in an adenoviral plasmid and sMSR recombinant adenoviruses are produced. The adenovirus genome is divided into 100 map units, with each map unit corresponding to 360 bases. Numbers shown in the plasmid and adenovirus constructs correspond to these map units (diagrams are not to scale; see text for details).

Generation of Adenoviruses

Ten micrograms of adenovirus plasmid psMSRA-1-flag containing sMSR and Flag tag is linearized with *Nhe*I restriction enzyme, and 0.5 μg of sub360 adenovirus genomic DNA is digested with *Cla*I and *Xba*I restriction enzymes. Plasmids are cotransfected into 293 cells by CaPO$_4$ precipitation.[21] After the appearance of the full cytopathic effect cells are harvested and suspended in 3 ml of DMEM–2% (v/v) FBS containing antibiotics. Virus is released from the cells by freezing and thawing in liquid nitrogen and a 37° water bath three times. Cells are centrifuged at

[21] F. Ausubel, R. Brent, R. Kingston, D. Moore, J. Seidman, J. Smith, and K. Struhl (eds.), "Current Protocols in Molecular Biology," Parts I and II. John Wiley & Sons, New York, 1995.

$200g$ for 10 min at room temperature and the supernatant is saved. A dilution series ranging from 10^{-4} to 10^{-10} is created. One milliliter of the dilution is added to the confluent 6-cm plate of 293 cells. Cells are incubated with the virus for 1 hr, washed once with medium, and covered with 5 ml of 0.5% (w/v) agar. Single plaques are picked 3 days later and diluted in 1 ml of DMEM–2% (v/v) FBS containing antibiotics. Virus from single plaques is expanded: 80% confluent 293 cells growing in 10-cm plates are transduced with the virus for 1 hr. After the appearance of the full cytopathic effect the virus is harvested as indicated above. The plaque purification step is repeated once.

Genomic PCR is used to confirm the presence of the transgene, with primers 5′-TAC AAG GAC GAC GAT GAC-3′ (upper primer) and 5′-GTA AAC ACG CTC CTC TAA-3′ (lower primer) annealing to the Flag epitope and sMSR, respectively (50-μl reaction with 1 mM MgCl$_2$ and 1.4 U of DNA polymerase per reaction: hot start, 94° for 5 min; 30 cycles of 94° for 30 sec, 57° for 30 sec, 72° for 90 sec; final extension at 72° for 10 min). PCR is also used to detect the presence of possible wild-type virus.[22] Virus from one expanded plaque is used for a large-scale preparation of recombinant adenovirus in 15-cm plates as indicated above. After the appearance of the cytopathic effect, cells are harvested, centrifuged, and diluted in 50 ml of DMEM–2% (v/v) FBS and the virus is released by freezing and thawing. Cells are removed by centrifugation at $2000g$ for 20 min at room temperture. Virus is purified by two-gradient centrifugation. The first centrifugation is at $40,000g$ for 2 hr at 20°. The lowest band is harvested and diluted with an equal volume of 5 mM HEPES, pH 7.8. The second centrifugation is at $111,000g$ overnight (18 hr or more) at 20°. The lowest band is harvested, diluted with HEPES, and dialyzed against two changes of HEPES and one change of HEPES containing 20% (v/v) glycerol overnight at 4°. Viruses are analyzed for toxicity and for the absence of wild-type virus, using a cytopathic effect assay.[23] A purified virus preparation is also tested for the absence of microbiological contaminants, mycoplasma, and lipopolysaccharide according to standard methods.[24]

Protein Characterization

RAW 264.7 macrophages or RAASMC[25] cells (6×10^6) are seeded into tissue culture plates and transduced with recombinant adenoviruses expressing sMSR at a multiplicity of infection of 100–1000 for 1 hr in serum-free medium. After

[22] W. W. Zhang, P. E. Koch, and J. A. Roth, *Biotechniques* **18,** 444 (1995).

[23] M. S. Horwitz, *in* "Fields Virology" (B. N. Fields and D. M. Knipe, eds.), p. 2149. Raven Press, New York, 1996.

[24] M. Laitinen, K. Mäkinen, H. Manninen, P. Matsi, M. Kossila, R. S. Agrawal, T. Pakkanen, J. S. Luoma, H. Viita, J. Hartikainen, E. Alhava, M. Laakso, and S. Ylä-Herttuala, *Hum. Gene Ther.* **9,** 1481 (1998).

[25] S. Ylä-Herttuala, J. Luoma, H. Viita, T. Hiltunen, T. Sisto, and T. Nikkari, *J. Clin. Invest.* **95,** 2692 (1995).

the transduction, 10% (v/v) serum is added and the cells are incubated for 16 hr. Medium is replaced with five milliliters of Opti-MEM containing 10 or 0.5% (v/v) lipoprotein-deficient serum (LPDS) and collected at 12-hr intervals. The presence of correctly sized protein in the medium is analyzed by Western blot and the functionality of the protein is characterized by ligand-binding assay and by degradation assay.[26]

Ligand-Binding Assay. A 750-μl volume of conditioned medium with 0.5% (v/v) LPDS is mixed with 250 μl of buffer B and 25 μl of poly(G)-conjugated resin in buffer A and incubated overnight at 4°. The mixture is centrifuged and the supernatant, containing only trace amounts of sMSR (sample 1), is saved and mixed with Western blot sample buffer. The resin pellet, containing bound sMSR, is washed two times with buffer A and mixed with Western blot sample buffer. sMSR is released from the resin by boiling for 4 min. The supernatant is separated by brief centrifugation and stored on ice (sample 2).

Western Blot. Conditioned medium containing 0.5% (v/v) LPDS is subjected to standard SDS–polyacrylamide gel electrophoresis (PAGE) with 4% (w/v) stacking gel and 8% (w/v) separating gel under reducing or nonreducing conditions along with samples 1 and 2 from the ligand-binding assay. Expressed protein is detected with mouse anti-Flag monoclonal antibody M2 (1 μg/ml), biotinylated horse anti-mouse IgG secondary antibody, and streptavidin–alkaline phosphatase (0.3 U/ml), using BCIP–NBT as substrate.

Degradation Assay. The degradation assay measures the amount of lipoprotein taken up and degraded by the cells. After precipitation of free ^{125}I and nondegraded ^{125}I-labeled acetylated LDL the trichloroactic acid-soluble noniodide radioactivity in the medium is measured.

RAW 264 macrophages are grown in 24-well plates to 80% confluency. Cells are incubated with ^{125}I-labeled acetylated LDL[26] (10 μg/ml) in conditioned medium containing sMSR (see Protein Characterization, above) or control medium in a total volume of 500 μl for 9 hr.

MEDIUM. 400-μl volume of medium is transferred to a new tube. First, 50 μl of 3% (w/v) BSA is added, mixed, and incubated for 5 min. Then, 100 μl of ice-cold 50% (w/v) TCA is added and mixed. Finally, 100 μl of 10% (w/v) AgNO$_3$ is added, mixed, and incubated for 5 min. The mixture is centrifuged at 2500g for 30 min. A 325-μl volume of the supernatant is analyzed with a γ counter.

SPECIFIC ACTIVITY. Specific activity (cpm/ng of protein) is determined by measuring radioactivity from a portion of ^{125}I-labeled acetylated LDL of known concentration.

CELLS. Cells are washed three times with 500 μl of 0.2% (w/v) BSA and two times with PBS. A 200-μl volume of 0.2 N NaOH is added and cells are incubated

[26] U. P. Steinbrecher, S. Parthasarathy, D. S. Leake, J. L. Witztum, and D. Steinberg, *Proc. Natl. Acad. Sci. U.S.A.* **81**, 3883 (1984).

overnight at $37°$. A 200-μl volume of distilled water is added and the radioactivity in 100 μl of the sample is measured with a γ counter.

The results are given as nanograms of degraded protein per microgram of cell protein.

$$\left(\frac{cpm}{\text{Total cell protein } (\mu g) \times \text{specific activity of } ^{125}\text{I-labeled acetylated LDL (cpm/ng)}} \right)$$

$$\times\, 2.5 = ng \text{ of degraded } ^{125}\text{I-labeled acetylated LDL}/\mu g \text{ of cell protein}$$

In Vivo

Gene Transfer

The *in vivo* functionality of the sMSR is tested in mice. Mice are transduced via the tail vein with 1×10^9 PFU of sMSR adenoviruses or control *lacZ* adenoviruses diluted in physiological saline in a total volume of 200 μl. Tail vein injection directs transgene expression mainly to liver. Three days later the mice are killed and analyzed for the expression of sMSR.

Polymerase Chain Reaction

Total RNA and DNA are isolated from liver and other organs of interest. RNA samples (10 μg) are treated with RQ1 DNase in RQ1 buffer containing DNase enzyme (1 U/μg of RNA) and RNase A ribonuclease inhibitor (1 U/μl) for 30 min at $37°$ and the reaction is stopped with EGTA and heating for 10 min at $65°$. Half the DNase-treated RNA is used for cDNA synthesis. RNA and 0.4 μg of random primers are incubated for 5 min at $70°$ and cooled on ice. Buffer, 1 mM dNTPs, and RNase inhibitor (1 U/μl) are added and incubated for 5 min at $25°$. Forty units of Moloney murine leukemia virus reverse transcriptase is added and incubated for 10 min at $25°$ and then for 60 min at $37°$. The reaction is stopped by heating at $70°$ for 5 min. cDNA (2 μl) is used for PCR as described in Generation of Adenoviruses (above). A nested PCR can be done if the expression level in tissues is low. Reaction product (2 μl) from the first PCR is taken for the second PCR with primers 5'-GGA CGA CGA TGA CAA GGC GG-3' (upper primer) and 5'-TTG CAT TCC CAT GTC CCT GG-3' (lower primer), both annealing to the sMSR (50-μl volume, containing 4 mM MgCl$_2$ and 1.4 U of DNA polymerase per reaction: $95°$ for 3 min; 30 cycles of $95°$ for 45 sec, $72°$ for 30 sec, $74°$ for 1 min; and final extension at $72°$ for 10 min).

Histology

Mice transduced with sMSR or control *lacZ* adenoviruses[24] are killed and perfused with PBS and then with 4% (w/v) paraformaldehyde in phosphate buffer.[25] Organs are embedded in paraffin and 6- to 10-μm sections are cut. M2 antibody against the Flag epitope encoded by the transgene is used to detect protein in paraffin sections, using standard immunohistochemical methods.

Enzyme-Linked Immunosorbent Assay

Enzyme-linked immunosorbent assay (ELISA) plates are coated overnight at 4° with 1 μg of polyclonal rabbit M2 antibody diluted in 0.1 M NaHCO$_3$ and washed with PBS–0.05% (v/v) Tween 20 three times before blocking with 1% (v/v) human serum albumin (HSA)–PBS for 1 hr at room temperature. After washing the plates three times with PBS–Tween 20, conditioned medium or mouse plasma samples diluted in 0.2% (v/v) HSA–0.05% (v/v) Tween 20 are incubated on the plates for 2 hr at room temperature. Plates are washed and incubated with polyclonal goat anti-scavenger receptor antibody at a 1 : 2000 dilution for 1 hr. After washing the plates are incubated with HRP-conjugated anti-goat antibody at a 1 : 4000 dilution. After washing the plates are incubated with peroxidase substrate (3,3′,5,5′-tetramethylbenzidine as chromogen) 100 μl per well, for 30 min in the dark. Color development is stopped with 0.5 M H$_2$SO$_4$, 100 μl per well. Absorbance is measured at 450 nm (Multiscan).

Conclusions

Recombinant soluble receptors offer the chance to study the function of transmembrane and soluble receptors *in vitro* and *in vivo* and also to use them for therapeutic purposes. We have described the construction of sMSR. Other recombinant soluble receptor proteins can also be used, as has been reported for RAGE.[7,8] Careful design of the recombinant soluble receptor is important and may require the excision of the membrane-spanning region of the transmembrane receptor. Therefore it is crucial to know detailed properties of the receptor in order to preserve domains important for ligand binding. Depending on the gene/protein of interest different approaches can be taken to design the construct, for example, the molecule can be secreted as a nonactive precursor protein and certain proteases can then activate it. A Flag or Myc epitope can be cloned in the construct for distinction from native transmembrane receptors. However, these epitopes may be immunogenic or alter intracellular processing of the soluble receptor in *in vivo* experiments.

Secreted receptors are especially useful for gene therapy purposes.[18,19] Even though it is difficult to achieve an adequate concentration of the transgene in blood, and to obtain sufficiently long expression of the transgene in the target tissue, secreted molecules have the advantage that all cells need not be transfected in order to obtain an *in vivo* effect.[27] It is anticipated that secreted receptors could prove useful in disease treatment, because even a relatively small amount of protein delivered systemically or locally could have a therapeutic effect.

Acknowledgments

This study was supported by grants from the Sigfrid Juselius Foundation, the Finnish Foundation for Cardiovascular Research, and the Finnish Academy, and by Kuopio University Hospital/EVO Grant 5130.

[27] S. Ylä-Herttuala and J. F. Martin, *Lancet* **355,** 213 (2000).

[30] Using Genetically Engineered Mice to Study Myocardial Ischemia–Reperfusion Injury

By Dipak K. Das, Wolfgang Dillmann, Ye-Shih Ho, Kurt M. Lin, and Bernd R. Gloss

Introduction

Constitutive cellular protection against acute stress can be provided by various intracellular antioxidants such as glutathione, α-tocopherol, ascorbic acid, and β-carotene as well as antioxidant enzymes that include superoxide dismutase (SOD), catalase, and glutathione peroxidase. These cellular compounds reduce or eliminate the oxidative stress by directly quenching or inhibiting the formation of reactive oxygen species before they can damage vital cellular components. Several heat shock proteins (HSPs), including HSP 32 (also known as heme oxygenase), HSP 70, and αB-crystallin, can directly or indirectly inhibit the development of oxidative stress and also be considered members of the defense system protecting against acute stress.

Mammalian tissues, especially mammalian hearts, are likely to undergo ischemia–reperfusion injury when subjected to ischemic insult.[1] Reperfusion of ischemic myocardium is associated with a reduction of antioxidants in concert with the production of a large amount of reactive oxygen species, thus subjecting the heart to oxidative stress. Laboratory studies demonstrated that oxygen free radical scavengers and/or antioxidants can reduce or ameliorate ischemia–reperfusion injury, at least in part, by removing oxidative stress in the myocardium.[2] Heme oxygenase, HSP 70, as well as αB-crystallin were also found to reduce the amount of oxidative stress in the ischemic myocardium.[3]

Evidence exists in the literature supporting a role of reactive oxygen species in the pathophysiology of ischemic heart disease. Most of the evidence is based on the fact that free radical scavengers and/or antioxidants reduce the severity of ischemic injury. However, the pathogenesis of ischemic injury is multifactorial, and many factors, in addition to the oxygen free radicals, for example, lipids, calcium, and many other mediators, play a crucial role in the development of ischemic heart disease leading to congestive heart failure.[4] To derive more conclusive evidence of the role of free radicals in ischemic heart disease and to demonstrate whether the

[1] D. K. Das, "Pathophysiology of Reperfusion Injury." CRC Press, Boca Raton, FL, 1992.

[2] D. K. Das and N. Maulik, Methods Enzymol. 233, 601 (1994).

[3] D. K. Das and N. Maulik, in "Heat Shock Proteins and the Cardiovascular System" (A. A. Knowlton, ed.), p. 159. Kluwer Academic, Norwell, MA, 1997.

[4] D. Bagchi, G. J. Wetscher, M. Bagchi, P. R. Hinder, G. Perdikis, S. J. Stohs, R. A. Hinder, and D. K. Das, Chem. Biol. Interact. 104, 65 (1997).

Copyright 2002, Elsevier Science (USA).
All rights reserved.
0076-6879/02 $35.00

constitutive cellular protection against ischemic stress is indeed provided by one or more of these antioxidants, use of genetically engineered animals is the most desirable method. Thus, it can be be assumed that transgenic animals bearing extra copies of one or more antioxidant genes will be resistant to ischemia–reperfusion injury whereas knockout animals bearing no copies of an antioxidant gene will be susceptible to ischemic injury. The following sections briefly describe the methods of preparing transgenic mice bearing extra copies or no copies (gene knockout) of the genes of a few important members of the antioxidant defense system against ischemic injury.

GSHPx-1 Transgenic and Knockout Mice

Glutathione (GSH) is one of the essential components of myocardial defense against ischemic stress. The presence of GSH ensures the conversion of toxic lipid peroxides into nontoxic products, utilizing the necessary reducing equivalents from the reduced GSH, which becomes oxidized through the action of glutathione peroxidase (GSHPx). GSHPx can detoxify H_2O_2, which may be produced in the ischemic myocardium from O_2^-. To define the precise role of this antioxidant enzyme in myocardial protection, transgenic mice overexpressing GSHPx-1 and knockout mice devoid of any copy of the GSHPx-1 gene are produced.

Generation of GSHPx-1 Transgenic Mice

Mouse genomic clones for cellular GSHPx-1 are initially isolated from a bacteriophage FIX II genomic library, prepared with DNA from a 129/SvJ mouse (Stratagene, La Jolla, CA) by hybridization screening with a corresponding rat cDNA clone. A 5.3-kb SacI genomic fragment is found to contain the entire mouse GSHPx-1 gene, with the sequence virtually identical to that published previously[5] except for a few base substitutions in the regions of introns. This piece of DNA[6] (Fig. 1), including approximately 2.0 kb of 5' flanking sequence, is microinjected into fertilized eggs derived from B6C3F1 × B6C3F1 mice. Two lines of transgenic mice carrying extra copies of the mouse GSHPx-1 gene are generated. The transgenic line Tg[MGP]-41 is used for ischemia–reperfusion study. To create the GSHPx-1 gene knockout mice, the fragment of DNA is used in the construction of targeting vector (Fig. 1). The coding sequence is disrupted by insertion of a neomycin resistance gene cassette (neo) derived from pMCIpolA into the EcoRI site located in exon 2. A herpesvirus thymidine kinase gene expression cassette is placed 3' to the targeting sequence in order to perform negative selection with

[5] L. J. Chambers, J. Frampton, P. Goldfarb, N. Affara, W. McBain, and P. R. Harrison, EMBO J. 5, 1221 (1986).
[6] T. Yoshida, M. Watanabe, D. T. Engelman, R. M. Engelman, J. A. Schley, N. Maulik, Y.-S. Ho, T. D. Oberley, and D. K. Das, J. Mol. Cell. Cardiol. 28, 1759 (1996).

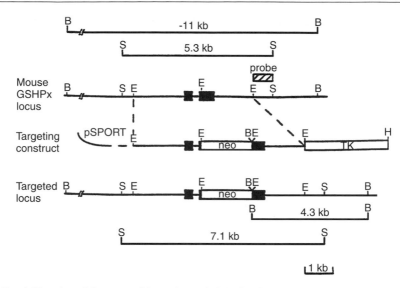

Fig. 1. Targeting of the mouse *GSHPx-1* gene in R1 ES cells. Genomic structure and partial restriction map of the mouse *GSHPx-1* locus (*top*), the targeting vector (*middle*), and the targeted locus (*bottom*) are shown. Solid boxes represent exons. Probe used for DNA blot analysis is shown as a hatched box on top of the restriction map of the *GSHPx-1* locus. B, *Bam*HI; S, *Sac*I; E, *Eco*RI; H, *Hin*dIII; neo, neomycin resistance cassette; TK, herpesvirus thymidine kinase gene under the control of the mouse promoter of the phosphoglycerate kinase 1 (PGK-1) gene. The sizes of *Sac*I and *Bam*HI restriction fragments of normal and targeted loci hybridized with the probe are shown at the top and bottom, respectively.

ganciclovir (Syntex, Palo Alto, CA). The targeting vector linearized with *Hin*dIII is transfected into R1 embryonic stem (ES) cells and selected with both G418 and ganciclovir. Resistant colonies are screened by DNA blot analysis, using a probe 3′ external to the targeting sequence. In our work, 30% of the colonies screened were found to contain the desired mutated structure in one of the two *GSHPx-1* alleles. We isolated approximately 100 homologous recombinant clones from about 300 colonies screened. Three clones were microinjected into C57B1 blastocysts, and embryos were reimplanted into the uterine horns of foster mothers. Twenty-three chimeric mice were generated. Four male chimeric mice with more than 95% agouti coat color chimerism were bred with C57B1 female mice. One hundred percent of their F_1 offspring carried the agouti coat color, suggesting that the reproductive organs of these chimeric mice are basically developed from the microinjected ES cells. Germ line transmission of the targeted *GSHPx-1* allele was evident by DNA blot analysis.

Characterization of GSHPx-1 Transgenic Mice

RNA blot analysis shows a dramatic increase (600%) in GSHPx-1 mRNA in the hearts of transgenic mice compared with nontransgenic littermates. An increase in GSHPx-1 mRNA is also found in eye, brain, lung, muscle, spleen, and tongue of transgenic mice. The specific activity of *GSHPx-1* in the hearts of transgenic mice is found to increase by approximately 500% relative to controls.[6] Compared with normal heart homogenate, which has a GSHPx-1 activity of 0.037 ± 0.003 μM NADPH/min per milligram of protein, the transgenic hearts show an activity of 0.188 ± 0.014 μM NADPH/min per milligram of protein. Increased GSHPx-1 activity is also noticed in brain, lung, and muscle of the transgenic mice. Over-expression of cardiac GSHPx-1 has no effects on the levels of other antioxidant enzymes, including MnSOD, Cu,ZnSOD, GSH reductase, catalase, and glucose-6-phosphate dehydrogenase, in the hearts of the transgenic mice.

The specific types of heart cells overexpressing GSHPx-1 are further revealed by immunoperoxidase staining with anti-rat GSHPx-1 antibodies. Increased immunostaining is found in the endothelial and smooth muscle cells of arteries in the hearts of transgenic mice compared with nontransgenic littermates. Overex-pression of GSHPx-1 is not found in the cardiac myocytes of the transgenic mice at the light microscopic level.

For *GSHPx-1* gene knockout mice, the $3'$ external probe is hybridized with a 5.3-kb *Sac*I fragment and an approximately 11-kb *Bam*HI genomic fragment of normal mouse DNA. Insertion of the *neo* selective marker results in hybridizing *Sac*I and *Bam*HI fragments with sizes of 7.1 and 4.3 kb, respectively. Germ line transmission of the targeted *GSHPx-1* allele is similar. RNA blot analysis shows an approximately 50% reduction of the GSHPx-1 mRNA in heart, brain, liver, and lung of *GSHPx-1* heterozygous knockout mice compared with normal littermates. The total GSHPx activities in brain, heart, kidney, liver, and lung of heterozygous *GSHPx-1* knockout mice are approximately 50 to 60% of those in corresponding tissues of normal littermates.

Sod1 Gene Knockout Mice

In biological tissue, superoxide anions are readily dismutated by superoxide dismutase (SOD) into hydrogen peroxide, which can then be converted into hydroxyl radical by a Fenton-type reaction. SOD contains a transition metal at the active site center and, depending on the transition metal, three types of SOD are known to exist: manganese SOD (MnSOD), copper/zinc SOD (Cu,ZnSOD), and iron SOD (FeSOD). The best characterized SOD is MnSOD, which is located primarily in the soluble matrix of the mitochondria. In contrast, Cu,ZnSOD is located exclusively in the cytosol. Whereas the cardioprotective role of MnSOD is well established,[7] little is known about the role of Cu,ZnSOD in myocardial pro-tection. Several studies have documented that Cu,ZnSOD plays a crucial role in cellular protection associated with ischemia and reperfusion.[8] To confirm the role

---

[7] D. K. Das, R. M. Engelman, and Y. Kimura, *Cardiovasc. Res.* **27,** 578 (1993).
[8] T. Yoshida, N. Maulik, R. M. Engelman, Y.-S. Ho, and D. K. Das, *Circ. Res.* **86,** 264 (2000).

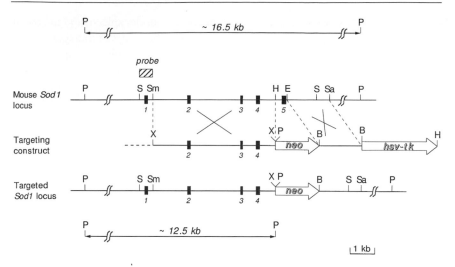

FIG. 2. Targeted disruption of the mouse *Sod1* gene. Schematic diagram showing the genomic structure and partial restriction map of the mouse *Sod1* locus (*top*), the targeting vector (*middle*), and the predicted structure of the targeted locus (*bottom*). Numbered solid boxes represent exons. The hatched box represents the 5' external sequence used as a hybridization probe. B, *Bam*HI; E, *Eco*RI; H, *Hin*dIII; P, *Pst*I; S, *Sac*I; Sa, *Sal*I; Sm, *Sma*I. *neo,* Neomycin resistance gene cassette; *tk,* herpesvirus thymidine kinase gene cassette. The predicted sizes of hybridizing *Pst*I genomic fragments of the wild-type allele and the targeted allele are indicated at the top and bottom, respectively.

of Cu,ZnSOD (SODI) in ischemic heart disease, we have created SOD1 knockout mice by disrupting the *Sod1* gene.

Targeted Disruption of Mouse Sod1 Gene

Mouse *Sod1* genomic clones are isolated from a 129/SvJ genomic library (Stratagene) by screening with a rat *Sod1* cDNA probe.[9] An approximately 7.2-kb *Sac*I genomic fragment, which contains the entire mouse *Sod1* gene, is isolated from clone 30 and used in the construction of the gene-targeting vector. To inactivate the mouse *Sod1* gene, the *Sma*I and *Hin*dIII restriction sites flanking the *Sma*I–*Hin*dIII fragment, which contains sequences from intron 1 to intron 4, are converted into *Xho*I sites by linker ligation, and then the fragment is inserted into the *Xho*I site of plasmid vector pPNT (Fig. 2). Similarly, linker ligation is also used to clone the *Eco*RI–*Sal*I fragment containing the 3' flanking sequence of the gene into the *Bam*HI site of the pPNT vector.

The *Sod1* targeting vector, in which exon 5 is deleted, is linearized by *Hin*dIII digestion and transfected into R1 ES cells.[9] Clones resistant to G418 and

---

[9] P. Wang, H. Chen, H. Qin, S. Sankarapandi, M. W. Becher, P. C. Wong, and J. L. Zweier, *Proc. Natl. Acad. Sci. U.S.A.* **95,** 4556 (1998).

ganciclovir (Syntex) are screened by Southern blot analysis, using a probe 5' external to the genomic sequence present in the targeting vector (Fig. 2). Targeted clones are microinjected into C57BL/6 blastocysts according to standard procedure. Chimeric mice with near 100% chimerism are generated with *Sod1* knockout clone 5, and show 100% transmission of the 129/SvJ chromosomes.

Generation and Characterization of Sod1 Knockout Mice

As shown in Fig. 2, exon 5 of the mouse *Sod1* gene (which encodes the C terminus of the protein, i.e., amino acid residues 120 to 154, which constitute both the structure and function of the active site channel and some of the flanking intron sequences) is replaced with the neomycin resistance cassette (*neo*). This also creates a new *Pst*I restriction site, resulting in a shorter *Pst*I genomic fragment from the targeted allele (12.5 kb) than from the wild-type allele (16.5 kb). Mice heterozygous ($Sod1^{+/-}$) for the targeted allele are interbred to generate homozygous knockout ($Sod1^{-/-}$) mice. Male and female $Sod1^{-/-}$ mice grow normally and are apparently healthy under routine animal husbandry.

Inactivation of the functional mouse *Sod1* gene in mouse hearts by gene targeting is initially determined by RNA blot analysis. An approximate 50% reduction of Cu,ZnSOD mRNA is found in the hearts of $Sod1^{+/-}$ mice compared with wild-type ($Sod1^{+/+}$) mice. Furthermore, no Cu,ZnSOD mRNA can be detected in $Sod1^{-/-}$ heart, indicating that the truncated Cu,ZnSOD or Cu,ZnSOD–*neo* fusion mRNA is degraded rapidly in the heart. Reduction of heart Cu,ZnSOD enzyme activity in $Sod1^{+/-}$ and $Sod1^{-/-}$ mice is also confirmed by SOD activity staining on a native polyacrylamide gel. Cu,ZnSOD activities in the hearts of $Sod1^{+/+}$, $Sod1^{+/-}$, and $Sod1^{-/-}$ mice are proportional to the mRNA levels in these mice.[9] A decrease in Cu,ZnSOD activity apparently has no effect on the activity of other heart antioxidant enzymes such as MnSOD (Fig. 2, middle), catalase, glutathione peroxidase, and enzymes involved in the recycling of oxidized glutathione including glutathione reductase and glucose-6-phosphate dehydrogenase. Male and female $Sod1^{+/+}$ and $Sod1^{-/-}$ mice at 10 to 12 weeks of age are used for myocardial ischemia–reperfusion study.

Heme Oxygenase 1 Gene Knockout Mice

Heme oxygenase (HO) catalyzes the reaction for heme metabolism, yielding equimolar quantities of CO, iron, and biliverdin. The HO system consists of two isoforms: oxidative stress-inducible HO-1 (Hmox-1), also known as HSP 32, and the constitutive isozyme HO-2. Besides oxidants, a variety of other environmental stresses including heat stress, hypoxia, metals, endotoxin, and certain hormones can induce Hmox-1.[10] Evidence suggests that Hmox-1 induction plays a role in

---

[10] G. F. Vile and R. M. Tyrrell, *J. Biol. Chem.* **268,** 14678 (1993).

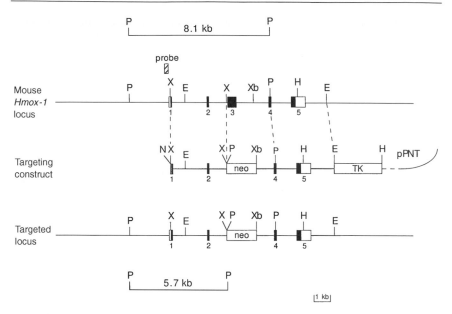

FIG. 3. Targeted disruption of the mouse heme oxygenase 1 (*Hmox-1*) gene. Genomic structure and partial restriction map of the mouse *Hmox-1* locus (*top*), the targeting vector (*middle*), and the predicted structure of the targeted locus (*bottom*) are shown. Solid and open boxes represent the coding and noncoding regions of the exons, respectively. The hatched box indicates the DNA fragment used as the hybridization probe. E, *Eco*RI; H, *Hind*III; N, *Not*I; P, *Pst*I; X, *Xho*I; Xb, *Xba*I. The expected sizes of the *Pst*I genomic fragments from the wild-type and targeted alleles hybridized with the probe are shown at the top and bottom, respectively.

cellular protection against injury caused by the reactive oxygen species. Reactive oxygen species produced in the ischemic–reperfused myocardium have been shown to induce *Hmox-1* gene expression.[11]

    The physiological significance of *Hmox-1* gene induction during myocardial ischemia remains unknown. It seems reasonable to speculate that *Hmox-1* gene induction during ischemia–reperfusion is the heart's own stress signal for survival against oxidative stress. Although a role for Hmox-1 protein in myocardial protection has been speculated, a definitive cardioprotective role for Hmox-1 is missing. To fill this gap, we have developed *Hmox-1*$^{+/-}$ mice by targeted disruption of the mouse *Hmox-1* gene. The isolated hearts with one functional copy of the *Hmox-1* gene and matched wild-type mice are subjected to ischemia–reperfusion.

Targeted Disruption of Mouse Heme Oxygenase 1 Gene

    As shown in Fig. 3, the *Xho*I fragment of the mouse *Hmox-1* gene, spanning from part of exon 1 to intron 2, is cloned into the *Xho*I site of the targeting vector

[11] N. Maulik, H. S. Sharma, and D. K. Das, *J. Mol. Cell. Cardiol.* **28,** 1261 (1996).

pPNT. The 3′ XbaI–EcoRI fragment of the gene containing sequence from intron 3 to the 3′ flanking region of the gene is subcloned into the corresponding XbaI and EcoRI sites in pPNT.[12] The targeting vector is linearized by NotI digestion and then transfected into R1 embryonic stem cells by electroporation. In our work, 5 homologous recombinant ES clones were identified from among 224 clones by DNA blot analysis, using the 5′ external probe encompassing sequences from nucleotides −133 to +73. Male chimeric mice, generated by microinjecting C57BL/6 blastocytes with three independent clones, are used in breeding with C57BL/6 female mice. Germ line transmission of 129/SvJ chromosomes is achieved by chimeric mice derived from each of the homologous recombinant ES clones. Line 183 Hmox-1$^{+/-}$ mice are used in this study.[12]

Protein Blot Analysis of Hmox-1 in Hearts

Hearts are homogenized in 50 mM phosphate buffer (pH 7.8) containing 0.1% (v/v) Triton X-100, using a Polytron (Brinkmann, Westbury, NY) homogenizer. The homogenates are centrifuged at 20,000g, and stored at −70°. Protein content is determined with an assay kit (Pierce, Rockford, IL). For protein blot analysis, total cellular protein is separated on a sodium dodecyl sulfate (SDS)– 12% (w/v) polyacrylamide gel and electrophoretically transferred onto nitrocellulose paper. The protein blot filter paper is incubated in 1× Hanks' balanced salt solution (HBBS) plus a 1 : 2000 dilution of a monoclonal anti-human Hmox-1 antibody (StressGen, Victoria, BC, Canada) for 2 hr. The filter paper is washed with HBSS containing 1% (w/v) nonfat milk, incubated with a 1 : 3000 dilution of horseradish peroxidase-conjugated goat anti-mouse IgG (Bio-Rad, Hercules, CA) in HBSS plus 5% (w/v) nonfat milk for 1 hr, washed again with HBSS, and then subjected to autoradiography in the presence of an enhanced chemiluminescent substrate (Pierce). Western blot data are quantified by densitometric scanning.

Generation and Characterization of Hmox-1$^{+/-}$ Mice

As shown in Fig. 3 (top), the mouse *Hmox-1* gene is inactivated by replacing exon 2 with a neomycin resistance (*neo*) cassette. Insertion of the *neo* gene creates a new *Pst*I restriction site in the *Hmox-1* gene, resulting in a shorter *Pst*I genomic fragment from the targeted allele than from the wild-type allele (Fig. 3, middle). Heterozygous *Hmox-1$^{+/-}$* mice are apparently healthy and show no pathologic or phenotypic differences from the wild-type littermates. However, no homozygous knockout mice can be detected in the progeny from a heterozygote intercross, indicating that embryonic lethality occurs in the homozygous mice. A 40% reduction of Hmox-1 was evident in the hearts of heterozygous mice compared with

[12] T. Yoshida, N. Maulik, Y.-S. Ho, J. Alam, and D. K. Das, *Circulation* **103**, 1695 (2001).

wild-type mice (Fig. 3, bottom). We reasoned that a decrease in expression of Hmox-1 protein due to the targeted mutation in *Hmox-1$^{+/-}$* mice might render the mice more susceptible to increased oxidative stress compared with control mice.

Heat Shock Protein 70 and αB-Crystallin Transgenes

Several reports in the literature indicate a role for HSP 70 in mediating cardioprotective effects against myocardial ischemia–reperfusion injury.[13,14] The hearts of the animals subjected to heat shock are associated with resistance against ischemic injury. In concert, several HSPs are induced after the heat shock, including HSP 27, HSP 60, HSP 70, and HSP 90. Among the HSPs, HSP 70 is well characterized and has not only been found to play a significant role in ischemia–reperfusion injury, but is also implicated in the reduction of oxidant injury.

The potential cardioprotective role of the small heat shock protein family began to receive attention with the observation that phosphorylation of MAPKAP (mitogen-activated protein kinase-activated protein) kinase 2, leading to the phosphorylation of HSP 27, plays a crucial role in myocardial adaptation to ischemic stress.[15] This observation is further strengthened by the fact that HSP 27 is involved in cytoskeletal stabilization and that cytoskeletal injury plays a crucial role in the pathogenesis of ischemia–reperfusion injury.[16] Consistent with these findings is a report demonstrating that enhanced expression of either HSP 27 or αB-crystallin in cardiomyocytes results in decreased cytosolic enzyme release after simulated ischemia.[17] αB-Crystallin is a member of the small HSP family, and shares many properties in common with HSP 27. Studies have demonstrated that αB-crystallin is the most abundantly expressed stress protein in the heart, and its production is induced by agents that promote the disassembly of microtubules.[18] It has also been demonstrated that αB-crystallin helps to preserve microtubular integrity in the face of simulated ischemia in cardiomyocytes.[19] To assess whether the cardioprotective effects of enhanced expression of αB-crystallin against cellular ischemic injury as suggested by *in vitro* studies are sufficient to preserve myocardial contractile

[13] R. Mestril, S. H. Chi, M. R. Sayen, K. O'Rielly, and W. H. Dillmann, *J. Clin. Invest.* **93,** 759 (1994).

[14] X. Liu, R. M. Engelman, I. I. Moraru, J. A. Rousou, J. E. Flack, D. W. Deaton, N. Maulik, and D. K. Das, *Circulation* **86** (Suppl. 2), 358 (1992).

[15] N. Maulik, T. Yoshida, Y. L. Zu, M. Sato, A. Banerjee, and D. K. Das, *Am. J. Physiol.* **275,** H1857 (1998).

[16] J. L. Martin, R. Mestril, R. Hilal-Daudem, L. L. Brunton, and W. H. Dillmann, *Circulation* **96,** 4343 (1997).

[17] W. F. Bluhm, J. L. Martin, R. Mestril, and W. H. Dillmann, *Am. J. Physiol.* **27,** H2243 (1998).

[18] F. A. van de Klundert, M. L. Gijsen, P. R. van den Ijssel, L. H. Snoeckx, and W. W. de Jong, *Eur. J. Cell. Biol.* **75,** 38 (1998).

[19] N. Golenhofen, W. Ness, R. Koob, P. Htun, W. Schaper, and D. Drenckhahn, *Am. J. Physiol.* **274,** H1457 (1998).

function and reduce infarct size when challenged by an ischemic insult, transgenic mice overexpressing αB-crystallin were generated and characterized.

Generation of Heat Shock Protein 70 and αB-Crystallin Transgenic Mice

Transgene expression in the mouse heart has been reported with a variety of enhancer-promoter constructs upstream of a cDNA to be overexpressed. Naturally these enhancers and promoters were derived from genes that are functionally expressed in the heart such as the genes encoding myosins and troponins. For example, the mouse myosin heavy chain α promoter has been extensively characterized,[20] and the mouse myosin light chain 2v promoter has been studied.[21] In most cases these promoters are subject to substantial regulation by pathophysiological changes in heart function or are specific for certain compartments of the heart. But it is often the intent to explore the effect of overexpressing a cDNA in the whole heart during a pathophysiological state, and therefore it is desirable to construct a transgene whose expression is largely independent of changes in heart function. The human cytomegalovirus (CMV) enhancer coupled to the chicken β-actin promoter is a strong driver in all compartments of the heart and is largely independent in its expression during pathophysiological changes in heart function. This hybrid promoter, which we chose to use for overexpression of the αB-crystallin and inducible HSP 70 genes, was originally described by Niwa et al.[22] and is composed of a human CMV enhancer linked to a chicken β-actin promoter containing an intron, followed by a rabbit β-globin 3' flanking sequence downstream of the cDNA insertion point (Fig. 4). The cDNA for the rat αB-crystallin gene has been described[23] and was available to us in a plasmid from which we could derive the coding region by digestion with XbaI and NotI. To this fragment are ligated EcoRI–XbaI and NotI–EcoRI converters (sequences 5'-AATTCGATCTCGAT-3' and 5'-GGCCGCATATTATG-3', respectively, in the upper coding strand) and both are annealed to complementary oligonucleotides to generate a 5' EcoRI site and 3' XbaI site or a 5' NotI site and 3' EcoRI site, respectively. The αB-crystallin cDNA fragment modified in this way can be ligated to the EcoRI-digested plasmid pCAGGS, which contains the human cytomegalovirus enhancer–chicken β-actin promoter upstream of the insertion point of the cDNA (Fig. 4, left). The transgene construct is then digested with SalI and BglII to yield a linear fragment that contains the enhancer–promoter followed by an intron upstream of the αB-crystallin cDNA containing its own polyadenylation signal.

[20] H. Rindt, A. Subramaniam, and J. Robbins, Transgenic Res. **4,** 397 (1995).

[21] S. A. Henderson, M. Spencer, A. Sen, C. Kumar, M. A. Siddiqui, and K. R. Chien, J. Biol. Chem. **264,** 18142 (1989).

[22] H. Niwa, K. Yamamura, and J. Miyazaki, Gene **108,** 193 (1991).

[23] A. Iwaki, T. Iwaki, J. E. Goldman, and R. K. Liem, J. Biol. Chem. **265,** 22197 (1990).

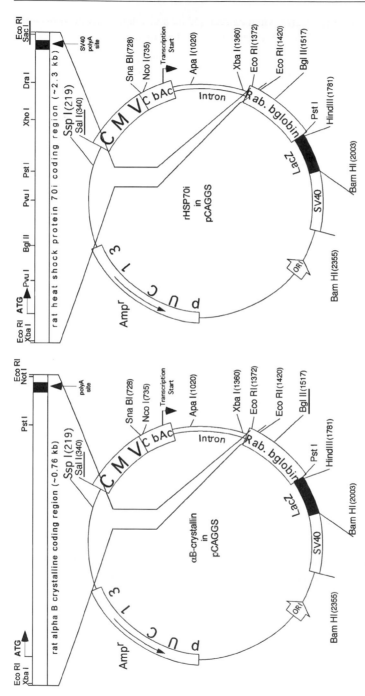

FIG. 4. *Left:* Map of the plasmid construct from which the rat αB-crystallin transgene was derived. It consists of the CAGGS enhancer–promoter (human cytomegalovirus enhancer (CMV), chicken β-actin promoter (CbAc) and αB-crystallin cDNA followed by a rabbit β-globin (Rab. bglobin) polyadenylation signal. The unique *Sal* I and *Bgl* II sites with which the plasmid was cut to purify the transgene fragment for oocyte injection are underlined. *Right:* Map of the plasmid construct from which the inducible rat heat shock protein70 transgene (rHSP70i) was derived. It consists of the CAGGS enhancer-promoter and the rHSP70i cDNA followed by a SV40 polyadenylation signal. The unique *Sal* I and *Sac* I sites with which the plasmid was cut to purify the transgene fragment for oocyte injection are underlined.

The cDNA for the inducible rat heat shock protein 70 gene has been cloned in our laboratory[24] and is cleaved from a plasmid by digestion with XbaI and SacI. To this fragment EcoRI–XbaI and SacI–EcoRI converters are ligated, as described above. The EcoRI–XbaI converter has the sequence 5'-AATTCGATCTCGAT-3' on the upper coding strand and is also annealed to a complementary oligonucleotide so that a 5' EcoRI site and a 3' XbaI site are generated. The SacI–EcoRI converter has the sequence 5'-CGATCTGCATTGAG-3' on the upper coding strand and is also annealed to a complementary oligonucleotide so that a 5' SacI site and a 3' EcoRI site are generated. The rHSP 70i cDNA fragment modified in this way can be ligated to the EcoRI-digested plasmid pCAGGS as described above and the resulting transgene construct is shown in Fig. 4 (right). This transgene construct is then digested with SalI and SacI to yield a linear fragment that contains the enhancer–promoter followed by an intron upstream of the rHSP 70i cDNA. On the 3' side of the cDNA a simian virus 40 (SV40) polyadenylation signal is present to confer proper termination and stabilization of the expressed transgene mRNA.

Both transgene fragments are purified from the rest of the vector backbone in a 0.8% (w/v) low melting point agarose gel from which they are subsequently isolated by passing the melted and diluted agarose through an Elutip (Schleicher & Schuell, Keene, NH) column. After dialysis against injection buffer [7.5 mM Tris (pH 7.4), 0.15 mM EDTA], 1–2 μl of the DNA fragment is injected into the pronuclei of fertilized eggs at a concentration of 2 μg/ml. In the case of αB-crystallin we have derived two transgenic lines, αB1 and αB2, of which the αB1 line has the highest transgene expression levels and is used primarily for the reported studies. In the case of rHSP 70i we have several founder lines that overexpress the inducible HSP 70 mRNA and protein at fairly equal levels. Representative Southern, Northern, and Western blots are discussed below.

Characterization of αB-Crystallin and Heat Shock Protein 70 Transgenic Mice

αB-Crystallin transgenic mice are identified by Southern blot analysis of mouse tail DNA with a transgene-specific probe[25] (Fig. 5A). As shown in Fig. 5B, in the hearts of heterozygous αB1 mice, αB-crystallin mRNA is 3.1-fold elevated above the level of that in control mice. There is a 6.9-fold increase at the protein level (Fig. 5C). Endogenous αB-crystallin expression in the rodent heart is relatively high. In mouse heart extract, a basal level of αB-crystallin constitutes close to 0.1' of total protein. The level of transgene expression in the heart is high. In addition, significant transgene expression occurs in skeletal muscle and the increase is even more dramatic than that observed in the heart. A 4.8-fold increase is seen at the mRNA level and a 7.9-fold increase is seen at the protein level. In other organs such

[24] R. Mestril, S. H. Chi, M. R. Sayen, and W. H. Dillmann, *Biochem. J.* **298**, 561 (1994).
[25] P. S. Ray, J. L. Martin, E. A. Swanson, H. Otani, W. H. Dillmann, and D. K. Das, *FASEB J.* **15**, 393 (2001).

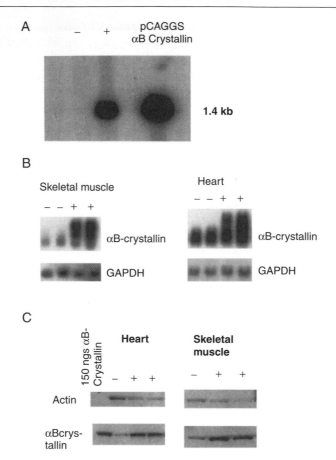

FIG. 5. (A) Southern blot of mouse tail DNA. Mouse tail DNA was digested with *Apa*I and *Pst*I and hybridized with the transgene-specific probe that corresponds to the first chicken β-actin intron generated by *Apa*I-*Xba*I digestion of the transgene. The (−) lane represents DNA from transgene-negative littermates and the (+) lane contains DNA from a αB-crystallin transgene-positive mouse. The right lane contains an *Apa*I–*Pst*I digest of the plasmid containing the αB-crystallin transgene. (B) Northern blot analysis. Total RNA was isolated from mouse heart ventricles and skeletal muscle, and hybridized to ^{32}P-labeled probes corresponding to αB-crystallin. GAPDH was used to demonstrate loading levels. In transgene-positive animals, significant increases in the amount of αB-crystallin occurred in both skeletal muscle and heart. (C) Western blot analysis. A total of 100 μg of protein from homogenates of mouse ventricle and skeletal muscle was probed with antibodies against actin and αB-crystallin. In lane 1, 150 ng of recombinant αB-crystallin was loaded as a positive control. A significant increase in the amount of αB-crystallin was found in the heart and skeletal muscle of transgene-positive mice.

as the liver, kidney, and spleen, no αB-crystallin-derived transgene expression can be identified, with the exception of brain, where significant increases in transgene expression also occur.

HSP 70i transgene-positive and -negative littermates are screened by Southern analysis of genomic DNA obtained from tail clips, and their hearts are further analyzed by Northern and Western blotting. Proteins harvested from transgenic mouse heart is probed in a Western blot with a polyclonal antibody that recognizes both constitutive and inducible forms of HSP 70 and with a monoclonal antibody (C92F3A-5, StressGen) that recognizes only the inducible form of HSP 70. As shown in Fig. 6A, the hearts of transgene-positive mice have appreciable HSP 70i immunoreactivity, and the amount of constitutive HSP 70 (HSP 70c) protein does not appear to be altered by the expression of transgene. Figure 6B demonstrates the Northern blot of cardiac and skeletal muscle RNA from a transgene-positive mouse, a transgene-negative mouse, and a transgene-negative mouse subjected to whole body heat stress (8 hr after 42°, 15 min).[26] The chimeric transgene (containing the rHSP 70i B form) is transcribed into an mRNA of a unique size due to the addition of sequences derived from the chicken β-actin promoter upstream of the translation start site and from SV40 after the translation stop site. The resulting transcript has a size of 2.6 kb and migrates between the mRNAs for the two endogenous mouse HSP 70i transcripts with sizes of 2.7 kb (mHSP 70i A) and 2.5 kb (mHSP 70i B). The novel chimeric rHSP 70i RNA is the transcript responsible for the excess HSP 70i immunoreactivity seen in Fig. 6A. Similar to αB-crystallin transgenic mice, the expression of rHSP 70i transgene is predominantly at heart and skeletal muscle, and some regions of brain.

Using Transgenic Mice to Study Myocardial Ischemia–Reperfusion Injury

The most common method for studying ischemia–reperfusion injury, using transgenic mice, involves an isolated perfused mouse heart preparation. Isolated hearts can be studied in the nonworking Langendorff mode or in working mode. The working mouse heart preparation is difficult and requires sophisticated techniques and skills. Whereas working mouse heart preparations are used in only a handful of laboratories, nonworking mouse hearts are extensively used.

Isolated Nonworking Mouse Heart Preparation and Measurement of Contractile Function

Mice are anesthetized and the hearts are excised after thoracotomy. The aorta is cannulated, and the heart is perfused with Krebs–Henseleit bicarbonate (KHB)

---

[26] R. Hotchkiss, I. Nunnally, S. Lindquist, J. Taulien, G. Perdrizet, and G. Karl, *Am. J. Physiol.* **265**, R1447 (1993).

A

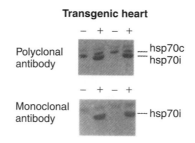

Transgenic heart

− + − +

Polyclonal
antibody
— hsp70c
— hsp70i

− + − +

Monoclonal
antibody
— hsp70i

B

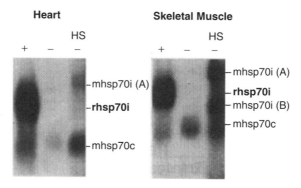

Heart   **Skeletal Muscle**

HS   HS

+   −   −   +   −   −

−mhsp70i (A)   −mhsp70i (A)

−**rhsp70i**   −**rhsp70i**
                                    −mhsp70i (B)

−mhsp70c   −mhsp70c

FIG. 6. (A) Western blot analysis. Proteins from hearts of transgene-positive and transgene-negative mice were probed with polyclonal and monoclonal antibodies. The polyclonal antibody recognizes constitutive (HSP 70c) and inducible (HSP 70i) forms of the 70-kDa heat shock protein. The monoclonal antibody recognizes only HSP 70i. A strong HSP 70i signal is seen in the hearts of transgene-positive mice but not in transgene-negative mice. (B) Northern blot analysis. RNA isolated from heart and skeletal muscle of transgene-positive (lane 1) and transgene-negative (lane 2) mice was analyzed by hybridization of ^{32}P-labeled rHSP 70i cDNA. High levels of rHSP 70i were seen in heart and skeletal muscle of transgene-positive animals but not in transgene-negative animals. In lane 3, samples from transgene-negative mice subjected to whole body heat stress (8 hr after 42° for 15 min) were loaded for comparison. Heat stress has a greater effect in inducing mHSP 70i in skeletal muscle than in heart. Both forms (A and B) of mHSP 70i are indicated and the size of the chimeric transgene is different from that of the endogenous mouse HSP 70i transcript.

buffer [composed of (in mM): 118 NaCl, 24 NaHCO$_3$, 4.7 KCl, 1.2 KH$_2$PO$_4$, 1.2 MgSO$_4$, 1.7 CaCl$_2$, 10 glucose, and gassed with 95% O$_2$:5% CO$_2$, pH 7.4 at 37°] by the retrograde Langendorff method at a constant perfusion pressure of 80 cmH$_2$O[6,12] (Fig. 7A). A small incision is made at the main trunk of the pulmonary artery to drain coronary effluent. The effluent is collected to measure the

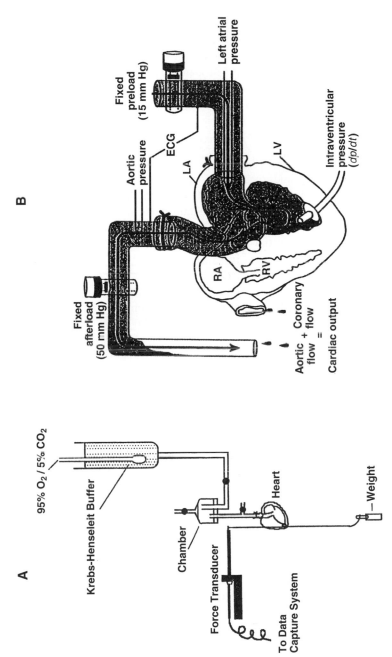

FIG. 7. (A) Langendorff heart preparation. (B) Working heart preparation.

release of creatine kinase (CK) and lactic acid dehydrogenase (LDH) (to assess cell necrosis) and malonaldehyde (MDA), a presumptive marker for oxidative stress. A 4-0 silk suture on a round-bodied needle is passed through the apex of the heart and attached to the apex, which in turn is attached to a force transducer. The heart rate (HR), force developed by the heart (DF), and first derivative of developed force (dF/dt) are recorded.[16] The hearts of all mice are generally subjected to 30 min of ischemia followed by 2 to 3 hr of reperfusion.

Isolated Working Mouse Heart Preparation and Measurement of Contractile Function

Wild-type or transgenic mice are first anesthetized with sodium pentobarbital (80 mg/kg body weight, intraperitoneal injection; Abbott Laboratories, North Chicago, IL) and anticoagulated with sodium heparin (500 U/kg body weight, intraperitoneal injection; Elkins-Sinn, Cherry Hill, NJ). After ensuring an adequate depth of anesthesia, the thoracotomy is performed and the heart is excised and immediately immersed in ice-cold perfusion buffer. The aortic arch is quickly isolated and incised. The aorta is cannulated and retrograde perfusion in the Langendorff mode through the aortic cannula is initiated at a perfusion pressure of 60 mmHg. The perfusion buffer used in this study consists of a modified Krebs–Henseleit bicarbonate (KHB) buffer [composed of (in mM): 118 NaCl, 4.7 KCl, 1.2 MgSO$_4$, 25 NaHCO$_3$, 10 glucose, and 1.7 CaCl$_2$, gassed with 95% O$_2$:5% CO$_2$, filtered through a 5-μm filter to remove any particulate contaminants, pH 7.4] maintained at a constant temperature of 37° and gassed continuously for the entire duration of the experiment.

Next, the pulmonary venous opening is located. The beveled sharp end of a PE-50 catheter is inserted through this opening and passed through the left atrium and mitral valve and pushed out the apex so that its fluted end remains in the left ventricular lumen (Fig. 7B). This catheter is connected to a pressure transducer to monitor left ventricular pressure. The pulmonary venous opening is then cannulated with a short piece of PE-50 tubing and is connected to the left atrial inflow line. Sidearms of the aortic and left atrial cannulas are connected to pressure transducers to permit continuous monitoring of the respective chamber pressures. At this time, perfusion is switched to the working heart mode by stopping retrograde perfusion through the aortic cannula, opening the aortic outflow line and initiating antegrade perfusion through the left atrial inflow line, in that order. A bubble trap/pressure chamber containing an air cushion of 1.5 ml is located in the aortic output line to allow for adequate elastic recoil. The method followed is essentially the same as that described previously[27] except for a slight modification,

---

[27] N. S. Gauthier, G. P. Matherne, R. R. Morrison, and J. P. Headrick, *J. Mol. Cell. Cardiol.* **30,** 453 (1998).

in that for our model we employ a fixed preload of 15 mmHg and a fixed afterload of 50 mmHg maintained by hydrostatic columns.[25] Left ventricular, aortic, and left atrial pressures (LVP, AOP, and LAP, respectively) are monitored, analyzed, and recorded in real time with the Digimed data acquisition and analysis system (Micromed, Louisville, KY). Heart rate (HR), left ventricular developed pressure (DP), defined as the difference between the maximum and minimum left ventricular pressures, and maximum positive $dLVP/dt$ are all derived or calculated from the continuously obtained left ventricular pressure signal.

The preparation is allowed to attain steady state values of functional parameters. After a 15-min stabilization period, baseline functional data are recorded and perfusate samples are collected. The hearts are then subjected to 30 min of global ischemia by arresting buffer flow through the left atrial cannula. Reperfusion for the first 10 min is commenced first in the retrograde Langendorff mode to allow for postischemic stabilization. Thereafter perfusion is switched to the antegrade working heart mode in the same way as described above to allow for monitoring of functional parameters which are recorded at various time points during reperfusion. Perfusate samples are also collected at these time points and, in addition, at 1, 3, 7, and 10 min of reperfusion in order to allow for the assessment of oxidative stress in the immediate early reperfusion period, during which it is typically the greatest.

Measurement of Myocardial Infarction by Estimating Necrosis and Apoptosis

Necrosis and apoptosis have been found to contribute independently to the development of myocardial infarction. At the end of the 2- to 3-hr reperfusion period, hearts are removed from the apparatus and the atrial tissue is dissected away. The ventricles are either fixed in 10% buffered formalin or are immersed in 1% triphenyl tetrazolium chloride (TTC) solution in phosphate buffer (88 mM Na_2HPO_4, 1.8 mM NaH_2PO_4) at 37° for 10 min. Hearts to be used for infarct size calculations ($n = 3$) are taken on termination of the experiment and immersed in 1% triphenyl tetrazolium solution in phosphate buffer for 10 min at 37° and then stored at $-70°$ for later processing. Frozen hearts (including only ventricular tissue) are sliced transversely in a plane perpendicular to the apicobasal axis into approximately 0.5-mm-thick sections, blotted dry, placed between microscope slides, and scanned on a Hewlett-Packard (Palo Alto, CA) Scanjet 5p single-pass flat-bed scanner. Using NIH Image 1.6.1 image-processing software, each digitized image is subjected to equivalent degrees of background subtraction, brightness and contrast enhancement for improved clarity and distinctness. Risk as well as infarct zones of each slice is traced and the respective areas are calculated in terms of pixels. The weight of each slice is then recorded to facilitate the expression of total and infarct masses of each slice in grams in order to remove the introduction of any errors due to nonuniformity of heart slice thickness. The individual risk masses and infarct masses of each slice are summed to obtain the risk and infarct masses for the whole heart. Infarct size is expressed as a percentage of the area at risk for any one heart.

To determine apoptotic cardiomyocytes, hearts kept in formalin are later embedded in paraffin according to standard procedures, and 3-μm-thick transverse ventricular sections are obtained to perform TUNEL [TdT (terminal deoxynucleotidyltransferase)-medicated dUTP nick and labeling] assays for the detection of apoptosis. Immunohistochemical detection of apoptotic cells is carried out using TUNEL in which residues of digoxigenin-labeled dUTP are catalytically incorporated into the DNA by terminal deoxynucleotidyltransferase, an enzyme that catalyzes a template-independent addition of nucleotide triphosphate to the $3'$-OH ends of double- or single-stranded DNA. The incorporated nucleotide is incubated with a sheep polyclonal anti-digoxigenin antibody followed by a fluorescein isothiocyanate (FITC)-conjugated rabbit anti-sheep IgG as a secondary antibody as described by the manufacturer (Apop Tag Plus; Oncor, Gaithersburg, MD). The sections ($n = 3$) are washed in PBS three times, blocked with normal rabbit serum, and incubated with mouse monoclonal antibody recognizing cardiac myosin heavy chain (Biogenesis, Poole, UK) followed by staining with tetramethylrhodamine isothiocyanate (TRITC)-conjugated rabbit anti-mouse IgG ($200 : 1$ dilution; Dako Japan, Tokyo, Japan). The fluorescence staining is viewed with a confocal laser microscope (Olympus, Tokyo, Japan).

Assessment of Cellular Injury

Cellular injury in the myocardium exposed to ischemia–reperfusion is determined by estimating the amount of creatine kinase (CK) and lactic acid dehydrogenase (LDH) release from the coronary effluent. Estimation of CK and LDH is carried out with commercially obtained specific assay kits according to the manufacturer instructions.

Measurement of Malondialdehyde for Assessment of Oxidative Stress

Malondialdehyde (MDA) is assayed as described previously[28] to monitor the development of oxidative stress during ischemia–reperfusion. Coronary perfusates are collected at the time baseline functional parameters are recorded and thereafter at time points of 1, 3, 5, 7, 15, 30, 60, 90, 120, 150, and 180 min into reperfusion for the measurement of MDA. The MDA in the collected coronary perfusate samples is derivatized with 2,4-dinitrophenylhydrazine (DNPH). One milliliter of perfusate is added to 0.1 ml of DNPH reagent (310 mg of DNPH in 100 ml of 2 N HCl, 1.56 mmol of DNPH) in a 20-ml Teflon-lined screw-capped test tube, contents are vortexed, and 10 ml of pentane is added before intermittent rocking for 30 min.

[28] G. A. Cordis, N. Maulik, and D. K. Das, *J. Mol. Cell. Cardiol.* **27**, 1645 (1995).

The aqueous phase is extracted three times with pentane, blown down with N_2, and reconstituted in 200 μl of acetonitrile. Aliquots of 25 μl in acetonitrile are injected onto a Beckman (Fullerton, CA) Ultrasphere C_{18} (3 mm) column in a Waters (Milford, MA) HPLC. The products are eluted isocratically with a mobile phase containing acetonitrile–H_2O–CH_3COOH (34 : 66 : 0.1, v/v/v) and detected at three different wavelengths: 307, 325, and 356 nm. The peak for malondialdehyde is identified by cochromatography with a DNPH derivative of the authentic standard, peak addition, comparison of the UV patterns of absorption at the three wavelengths, and by GC-MS. The amount of MDA is quantitated by performing peak area analysis with Maxima software program (Waters) and is expressed as picomoles per milliliter.

[31] Transgenic Model for the Study of Oxidative Damage in Huntington's Disease

By José Segovia

Introduction

Huntington's disease (HD) is a hereditary neurodegenerative disorder characterized by motor, psychiatric, and cognitive symptoms. To reproduce some of the biochemical, morphological, and behavioral alterations of HD several acute animal models have been developed. Intrastriatal injection of glutamate analogs, such as kainic acid,[1] ibotenic acid,[2] and quinolic acid,[3] or of a mitochondrial inhibitor, 3-nitropropionic acid,[4] forms the basis of some of the best studied models used in rodents.

More recently, however, a new system for modeling HD has emerged. The disease is caused by an abnormal expansion of the CAG repeats that encode a polyglutamine tract in a novel protein called huntingtin (htt).[5] Because the genetic defect responsible for the onset of the disease has been unambiguously identified, the concept of generating animal models of the human disease by manipulating

[1] J. T. Coyle and R. Schwarcz, *Nature (London)* **263**, 244 (1976).

[2] R. Schwarcz, T. Hokfelt, K. Fuxe, G. Jonsson, M. Goldstein, and L. Terenius, *Exp. Brain Res.* **37**, 199 (1979).

[3] M. F. Beal, N. W. Kowall, D. W. Ellison, M. F. Mazurek, R. Schwarcz, and J. B. Martin, *Nature (London)* **321**, 168 (1986).

[4] E. Brouillet, B. G. Jenkins, B. T. Hyman, R. J. Ferrante, N. W. Kowall, R. Srivastava, D. S. Roy, B. R. Rosen, and M. F. Beal, *J. Neurochem.* **60**, 356 (1993).

[5] Huntington's Disease Collaborative Research Group, *Cell* **72**, 971 (1993).

Copyright 2002, Elsevier Science (USA).
All rights reserved.
0076-6879/02 $35.00

the expression of the affected gene has been widely used. Early embryonic death of knockout mice for the murine homolog of htt has demonstrated the fundamental role of the protein during development.[6–8] However, the function of the protein in the CNS could not be determined from these experiments. On the other hand, precise knowledge of the genetic defect present in HD has also allowed the generation of transgenic animals, particularly mice, that express the HD mutation. Several of the transgenic mouse lines present biochemical and morphological alterations in the striatum, and display progressive neurological phenotypes.[9–12] The mechanism(s) by which the expanded polyglutamine tract causes cell death, particularly of the medium spiny γ-aminobutyric acid (GABA)-containing neurons in the caudate putamen of patients, remains unclear. However, it has been proposed that defects in the energy metabolism of the affected cells, which may cause the formation of free radicals, are important components in the etiology of the disease.[13,14]

Oxidative damage is one of the major consequences of defects in energy metabolism, and it is present in HD models induced by the injection of excitotoxins and mitochondrial inhibitors.[15] Moreover, we have observed a correlation between the onset of the neurological phenotype and striatal free radical-induced damage in one line of transgenic mice, R6/1,[16] expressing a human mutated *htt* exon 1 with 116 CAG repeats.[9] Other studies have shown that mice transgenic for the HD mutation present mitochondrial defects and increased NO production.[17,18] All these results suggest that mice transgenic for HD are an important tool for the study of the cellular mechanisms underlying the onset of HD, and particularly

[6] M. P. Duyao, A. B. Auerbach, A. Ryan, F. Persichetti, G. T. Barnes, S. M. McNeil, P. Ge, J.-P. Vonsattel, J. F. Gusella, A. L. Joyner, and M. E. MacDonald, *Science* **269,** 407 (1995).
[7] J. Nasir, S. B. Floresco, J. R. O'Kuskey, V. M. Diewert, J. M. Richman, J. Zeisler, A. Borowski, J. D. Marth, A. G. Philips, and M. R. Hayden, *Cell* **81,** 811 (1995).
[8] S. Zeitlin, J.-P. Liu, V. E. Papaionnou, and A. Efstradiatis, *Nat. Genet.* **11,** 155 (1995).
[9] L. Mangiarini, K. Sathasivam, M. Seller, B. Cozens, A. Harper, C. Hetherington, M. Lawton, Y. Trottier, H. Lehrach, S. W. Davies, and G. P. Bates, *Cell* **87,** 493 (1996).
[10] P. H. Reddy, M. Williams, V. Charles, L. Garrett, L. Pike-Buchanan, W. O. Whetsell, G. Miller, and D. A. Tagle, *Nat. Genet.* **20,** 198 (1998).
[11] J. G. Hodgson, N. Agopyan, C. A. Gutekunst, B. R. Leavitt, F. LePiane, R. Singaraja, D. J. Smith, N. Bissada, K. McCutcheon, J. Nasir, L. Jamot, X. J. Li, M. E. Stevens, E. Rosemond, J. C. Roder, A. G. Philips, E. M. Rubin, S. M. Hersch, and M. R. Hayden, *Neuron* **23,** 181 (1999).
[12] A. Yamamoto, J. J. Lucas, and R. Hen, *Cell* **101,** 57 (2000).
[13] M. F. Beal, *Biochim. Biophys. Acta* **1366,** 211 (1998).
[14] A. H. V. Schapira, *Biochim. Biophys. Acta* **1410,** 159 (1999).
[15] A. Petersén, K. Mani, and P. Brundin, *Exp. Neurol.* **157,** 1 (1999).
[16] F. Pérez-Severiano, C. Ríos, and J. Segovia, *Brain Res.* **862,** 234 (2000).
[17] S. J. Tabrizi, J. Workman, P. E. Hart, L. Mangiarini, A. Mahal, G. Bates, J. M. Cooper, and A. H. V. Schapira, *Ann. Neurol.* **47,** 80 (2000).
[18] M. Chen, V. O. Ona, M. Li, R. J. Ferrante, K. B. Fink, S. Zhu, J. Bian, L. Guo, S. M. Hersch, W. Hobbs, J.-P. Vonsattel, J.-H. J. Cho, and R. M. Friedlander, *Nat. Med.* **6,** 797 (2000).

in the evaluation of the participation of oxidative damage. This chapter discusses some techniques we have used to determine the oxidative status of striata of mice transgenic for HD.

Experimental Procedure

Animal Handling and Genotype

We employ R6/1 males of the CBA × C57BL/6 strain, which carry a human mutated exon 1 with approximately 116 CAG repeats.[9] R6/1 male mice and nontransgenic CBA female mice are purchased from Jackson Laboratory (Bar Harbor, ME), and a colony has been established in our vivarium. The macro-environmental conditions of the room are maintained by a heating, ventilation, and air-conditioning (HVAC) system controlled by Excell 5000 system software (Honeywell). Relative humidity is kept at $50 \pm 10\%$, and 10–15 changes of air volume are performed per hour. The HVAC system is equipped with 35% efficiency prefilters, and 95% efficiency HEPA filters. Mice are housed in cages isolated with a microbarrier system with electrostatic filter (MBS 7105 and MBS 10196; Allentown Caging Equipment, Allentown, NJ); this equipment is autoclavable. Mice are kept under controlled temperatures ($20 \pm 2°$) with a regulated 12-hr light–dark cycle and with *ad libitum* access to food and water. Food pellets (lab rodent breeder diet 5013; Purina Mills, St. Louis, MO) and bedding (shredded aspen; Northeastern Products, Warrensburg, NY) are autoclaved, and water is ozone purified and autoclaved. Mice are changed twice a week and are under permanent veterinary observation. All handling of animals is performed under a hood.[19] All animal procedures have been approved by the Institutional Review Committee, and are in accordance with the National Institutes of Health (NIH, Bethesda, MD) *Guide for the Care and Use of Laboratory Animals.*

Transgenic R6/1 mice are hemizygous for the HD mutation, and carry a single copy of the transgene. Male R6/1 transgenic mice are crossed to noncarrying CBA females and the genotype is determined by using the polymerase chain reaction (PCR). Transgenic mice, and nontransgenic littermates, are identified and used for the following experiments. All mice that develop the neurological phenotype have been genotyped as transgenics. To obtain DNA, samples from ear tissue are obtained from mice 4 to 5 weeks old. Each sample is approximately 3 mm in diameter, and is added to 20 μl of lysis buffer. The buffer consists of 50 mM Tris-HCl–20 mM NaCl–0.3% (w/v) sodium dodecyl sulfate (SDS), and 2 μl of proteinase K is added (from a 10-mg/ml stock solution) to obtain a 22-μl volume. Samples are incubated at 55° for 15 min, and then vigorously shaken and centrifuged at 7780g for 30 sec. This last procedure is repeated twice. Distilled sterile

---

[19] J. H. Fernández, J. Segovia, M. Flores, Y. Heuze, and F. Pérez, *Anim. Exp.* **5,** 17 (2000).

water (28 μl) is added to obtain a 50-μl final volume. Samples are finally boiled for 7 min, cooled, and stored at $-20°$ until PCR assays are performed.[20] Isolated DNA is checked by running samples in a 1% agarose gel and staining with ethidium bromide.

To identify the presence of the transgene, PCR assays are performed for each mouse. We have already outlined the assay,[16] which is based on the assay described by Mangiarini *et al.* in 1996,[9] and on a personal communication from P. Schweitzer (Jackson Laboratory). To prepare 10 ml of 10× reaction buffer, the components are as follows: 670 mM Tris-base, 166 mM NH_4SO_4, 20 mM $MgCl_2$, bovine serum albumin (BSA, 1.7 mg/ml), and 10 mM 2-mercaptoethanol, all dissolved in 9 ml of Tris–EDTA (TE), pH 8.8. The final volume of 10× reaction buffer is brought up to 10 ml, and sterile filtered. For each PCR, 5 μl of the buffer is mixed with 5 μl of dimethyl sulfoxide (DMSO), 1 μl (50–100 pmol) of each of the primers, and a 0.5 mM concentration of each dNTP (1 μl of a 25 mM stock). We have also tested the results of adding more $MgCl_2$ to the reaction mix—from 2.0 to 4 mM—and concluded that 2.0 mM is the optimal concentration. A 4-μl sample of DNA previously obtained from a mouse is added to the reaction mix. One unit of *Taq* polymerase (diluted to 1 U/5 μl of sterile water) is used, and a final volume of 50 μl per reaction is obtained by the addition of sterile water. The use of appropriate controls, as in any PCR assay, is critical. For positive controls we utilize as DNA templates the pGemHDEL plasmid, which contains 4 kb of human genomic DNA including the first exon of the *htt* gene (a gift from A. J. Tobin and G. Lawless, University of California, Los Angeles, CA), and DNA from a patient (obtained from E. Alonso, Instituto Nacional de Neurología y Neurocirugía, Mexico City, Mexico). As negative controls, we routinely substitute DNA for water, and also use DNA from nontransgenic mice, preferably from another strain. Primer sequences are GCAGCAGCAGCAGCAACAGCCGCCACCGCC and CGGCT-GAGGCAGCAGCGGCTGT. Samples are overlaid with mineral oil and placed in a Stratagene (La Jolla, CA) RoboCycler 40. Samples are denatured at 94° for 90 sec, and then 35 cycles of the following protocol are run: 30 sec of denaturation (94°), 30 sec of annealing (65°), and 30 sec of extension (72°). A final extension step of 10 min is performed. Eighteen to 20 μl of each sample is run on a 3% (w/v) agarose gel, and the PCR products are stained with ethidium bromide and observed with UV light (Fig. 1).

Behavioral Analyses and Phenotype

Four lines of transgenic mice carrying the huntingtin exon 1 with expanded CAG repeats were originally reported, and three of the lines, R6/1, R6/2, and R6/5, show a progressive neurological phenotype.[9] However, R6/2 mice present

---

[20] R. López-Revilla, L. Chávez-Dueñas, and Y. Azamar, *Focus* **21,** 14 (1999).

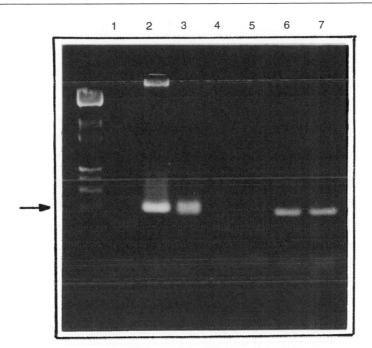

FIG. 1. Transgene detection by PCR. Products correspond to the following templates: lane 1, negative control, no DNA added; lane 2, pGemHDEL plasmid; lane 3, DNA from a patient; lanes 4 and 5, DNA from nontransgenic mice; lanes 6 and 7, DNA from transgenic mice. Arrow indicates 100 bp.

the phenotype earlier, and their motor activity has been characterized in greater detail than that of the other lines.[9,21,22] The progressive neurological phenotype of the R6/2 mice includes weight loss, resting tremor, abrupt movements, seizures, progressive hypoactivity, and the presence of dyskinesia of the limbs, when held by the tail, that develops to clasping feet.[9,21,22] We use the R6/1 mice, because we consider that the longer time that elapses to the manifestation of the phenotype may more closely resemble HD, whereas the R6/2 line may be more similar to juvenile HD, in which it is common to observe seizures.[15] Because the behavior of the R6/1 mice has not been studied so thoroughly as that of the R6/2 line, and we are interested in determining the relationship between oxidative damage and the expression of the neurological phenotype, we have examined the development of motor behavior of R6/1 mice starting at 11 weeks of age (a time at which R6/1

[21] R. J. Carter, L. A. Lione, L. Humby, L. Mangiarini, A. Mahal, G. P. Bates, S. B. Dunnett, and J. A. Morton, *J. Neurosci.* **19,** 3248 (1999).
[22] L. A. Lione, R. J. Carter, M. J. Hunt, G. P. Bates, J. A. Morton, and S. B. Dunnett, *J. Neurosci.* **19,** 10428 (1999).

mice are undistinguishable from wild-type mice by simple observation), up to 35 weeks of age.

To determine the neurological phenotype, both transgenic and nontransgenic littermates are tested for motor activity and for feet clasping. Mice at various ages, ranging from 11 to 35 weeks of age, are tested for motor behavior by an automated system (electronic motility meter 40Fc; Motron Products, Stockholm, Sweden). Mice are placed in a clear plastic box, and both horizontal and vertical movements (rearing up onto their back legs) are automatically recorded for a 10-min session for each animal. Each mouse is subjected to only one motor activity test. Feet clasping is determined by holding the mice by the tail. If a mouse does not clasp its feet within a maximum period of 2 min, it is considered as not presenting the behavior. Mice are tested only once for feet clasping. We also record mouse body weight every week. These simple observations allow us to relate the onset of the neurological phenotype with striatal oxidative damage.

The difference in body weight between R6/1 and wild-type mice becomes evident by 24 weeks of age, and by 35 weeks the weight of transgenic mice is 63% of that of age-matched nontransgenic controls. Decreased horizontal motor activity becomes manifest by 19 weeks of age in transgenic mice. At that age R6/1 mice show a 45% reduction in horizontal motor activity when compared with nontransgenic littermates. By 35 weeks of age, motor behavior of transgenic mice is only 18% compared with age-matched controls. With respect to vertical motor behavior, transgenic mice show a 43% reduction, compared with controls at 24 weeks of age, and by 35 weeks they had only 21% of the activity of wild-type mice. Nontransgenic control mice never present feet clasping behavior, whereas R6/1 mice show the behavior starting at 19 weeks of age. Transgenic mice also present resting tremor.[16] It is worth pointing out that all mice that presented the phenotype have been identified as transgenics; behavioral evaluations are relevant to validate the use of this model.

Oxidative Damage

Lipid Peroxidation

We determine lipid peroxidation (LP) as a measure of oxidative cell damage induced by the formation of free radicals. In pathological conditions, free radicals cause different cellular toxic effects. One of the most relevant is LP, the most commonly used index to measure the biological effect of free radicals.[23] LP is oxidative damage of the polyunsaturated membrane lipids, which are susceptible to oxidation by molecular oxygen through a free radical chain process.[24] The final result of LP is a change in the lipid composition of the cell membrane that

[23] J. M. Gutteridge and B. Halliwell, *Trends Biochem. Sci.* **15**, 129 (1990).
[24] B. Halliwell and J. M. Gutteridge, *Mol. Aspects Med.* **8**, 89 (1985).

induces alterations of its physicochemical properties, including increased membrane rigidity, which may eventually lead to cell death due to energy and structural changes.[25,26] At different ages both R6/1 mice and nontransgenic littermates are killed by decapitation. Brains are obtained and the striata and other regions (cortices and cerebella) are rapidly dissected out on an ice-cold surface. The formation of lipid-soluble fluorescence is determined on the basis of the method of Triggs and Willmore,[27] with modifications.[16] Tissue (i.e., one striatum) is homogenized in 3 ml of distilled water, and 1-ml aliquots of the homogenates are added to 4 ml of a chloroform–methanol [high-performance liquid chromatography (HPLC) grade] mixture (2 : 1, v/v). After stirring, mixtures are ice-cooled for 30 min, and the top phase is gently removed by aspiration; the phase must be completely removed so that it will not interfere with the fluorescence determination. Fluorescence of the chloroform phase is measured at 350-nm excitation and 430-nm emission wavelengths. To calibrate the sensitivity of the assay, the spectrophotometer is adjusted to 140 fluorescence units with a 0.001-mg/ml concentration of a quinine standard prepared in 0.05 M H_2SO_4. Protein content is determined by the bicinchoninic acid (BCA) assay (Pierce, Rockford, IL). Results are expressed as relative fluorescence units per milligram of protein.

Nitric Oxide Synthase Activity

Nitric oxide is synthesized by the oxidation of a terminal guanidino nitrogen atom of L-arginine by the enzyme nitric oxide synthase (NOS).[28] Two major forms of NOS, a constitutive and an inducible form, have been described.[29-34] The constitutive form requires calcium, calmodulin, NADPH, and tetrahydrobiopterin (BH_4) for full activity,[30-35] whereas the inducible form (iNOS) is calcium and calmodulin independent.[36-38] The method described to measure NOS activity is

[25] P. H. Chan, M. Yurko, and R. A. Fishman, *J. Neurochem.* **38**, 531 (1982).

[26] K. Kogure, B. D. Watson, R. Busto, and K. Abe, *Neurochem. Res.* **7**, 1405 (1982).

[27] W. P. Triggs and L. J. Willmore, *J. Neurochem.* **42**, 976 (1984).

[28] R. M. J. Palmer, D. S. Ashton, and S. Moncada, *Nature (London)* **333**, 664 (1988).

[29] D. S. Bredt and S. H. Snyder, *Proc. Natl. Acad. Sci. U.S.A.* **87**, 682 (1990).

[30] B. Mayer, M. John, and E. Bohme, *FEBS Lett.* **277**, 215 (1990).

[31] R. Busse and A. Mulsch, *FEBS Lett.* **265**, 133 (1990).

[32] U. Forstermann, J. S. Pollock, H. H. H. W. Schmidt, M. Heller, and F. Murad, *Proc. Natl. Acad. Sci. U.S.A.* **88**, 1788 (1991).

[33] R. G. Knowles, M. Merrett, M. Salter, and S. Moncada, *Biochem. J.* **270**, 833 (1990).

[34] S. Hauschldt, A. Lackhoff, A. Mulsch, J. Kohler, W. Bessler, and R. Busse, *Biochem. J.* **270**, 354 (1990).

[35] J. Giovannelli, K. L. Campos, and S. Kaufman, *Proc. Natl. Acad. Sci. U.S.A.* **88**, 7091 (1991).

[36] S. S. Gross, E. A. Jaffe, R. Levi, and G. Kilbourn, *Biochem. Biophys. Res. Commun.* **178**, 823 (1991).

[37] R. Busse and A. Mulsch, *FEBS Lett.* **275**, 87 (1990).

[38] D. J. Stuehr, H. J. Cho, N. S. Kwon, M. F. Weise, and C. F. Nathan, *Proc. Natl. Acad. Sci. U.S.A.* **88**, 7773 (1991).

based on the stoichiometric conversion of L-arginine to L-citrulline, a stable by-product of NO production from L-arginine.[29,39]

For each assay we use one striatum. Tissue from R6/1 mice and from nontransgenic littermates is dissected out. Tissue is homogenized on ice in 250 μl of a buffer containing a cocktail of protease inhibitors. A 50 mM Tris-HCl–0.1 mM EDTA–0.1 mM EGTA–0.1% (v/v) 2-mercaptoethanol solution (pH 7.5) is prepared and kept in the refrigerator. Before homogenizing, the protease inhibitors are mixed to the following final concentrations: 100 μM leupeptin, 1 mM phenylmethylsulfonyl fluoride (PMSF) from a 100 mM stock dissolved in ethanol (ethanol may cloud the homogenizing buffer, which must be vortexed frequently), aprotinin (2 μg/ml), soybean trypsin inhibitor (SBTI; 10 μg/ml), and 0.1% (v/v) Tergitol type NP-40 (Sigma, St. Louis, MO). Protein content is measured by the BCA assay. After the tissue has been homogenized in the presence of the protease inhibitor cocktail, volumes containing 500 μg of protein are taken for each reaction, and the following reagents are added to the mix (all final concentrations): 1 mM NADPH (100 mM stock is 100 mg/1.2 ml of homogenizing buffer without protease inhibitors; prepare fresh); 100 nM calmodulin, from a 10 μM stock in buffer without protease inhibitors kept at $-20°$; 30 μM tetrahydrobiopterin (stock solution is 100 mM in sterile water; prepare fresh); and 2.5 mM CaCl$_2$ from a 25 mM solution in Tris buffer. Reaction mix also contains 10 μM L-arginine-HCl, from a 10 mM stock (2.1 mg/ml in water; prepare fresh), and 0.2 μCi of L-[^{3}H]arginine (approximately 66 Ci/mmol, 1 μCi/μl; Amersham, Arlington Heights, IL) is added per 100-μl reaction (labeled arginine is diluted 1 : 5 in the original Tris buffer, and the necessary amount is added to the master mix). Master mix must be prepared according to the number of samples that will be assayed. Total arginine is 10 μM unlabeled plus 3 pmol labeled for each reaction. Final reaction volumes are adjusted to 100 μl with homogenizing buffer. Starting with a 250-μl volume should be sufficient to run a duplicate of each reaction both in the presence and in the absence of calcium. This final mixture is incubated for 30 min at 37°. Reactions are stopped by adding 1 ml of ice-cold stop buffer (2 mM EGTA, 2 mM EDTA, 20 mM HEPES, pH 5.5). The reaction mixture, now 1.1 ml, is applied to a 1-ml column of Dowex-50W resin that had been previously equilibrated with stop buffer. The cation-exchange resin retains labeled arginine and allows L-[^{3}H]citrulline to flow through the column. Wash the incubation tube with 1 ml of distilled water, vortex, and pass it over the column to wash through the sample. Samples are measured by liquid scintillation spectroscopy [5 ml of Aquasol-2 (New England Nuclear, Boston, MA) per approximately 2 ml of eluate]. Check, by duplicate, total counts of tube containing 0.2 μCi [^{3}H]arginine of master mix. To determine background radioactivity, tubes are prepared without tissue, and the procedure described is followed. Background should not exceed about 4% of total counts added. Subtract background

[39] A. Rengasamy and R. A. Johns, *Neuroprotocols Companion Methods Neurosci.* **1,** 159 (1992).

counts and calculate nanomoles of citrulline generated, knowing the specific activity of the labeled L-arginine added (cpm/nmol). We have expressed results as nanomoles of L-[^{3}H]citrulline per milligram of protein per 30 min. To test for calcium dependency of NOS activity, and thus differentiate NOS and iNOS activities, enzyme activity is measured, both in the presence (2.5 mM) and in the absence of CaCl$_2$, with 1.0 mM EGTA (from a 10 mM stock prepared in water; a drop or two of 1 N NaOH may be needed to dissolve) to chelate residual calcium in the incubation mixture. To prepare the ion-exchange column 100–200 g of Dowex-50W (50X8-200; Sigma) resin is placed in distilled water. The mixture is swirled into a slurry. After the gel settles, the water is removed and the resin is washed twice with 50–100 ml of 1 N NaOH to convert the acid form of the resin to a salt. NaOH is removed, and the resin is washed with abundant water until the supernatant reaches a pH lower than pH 8.0 (we have observed best and most consistent results when the pH is between pH 7.5 and 7.9). A few minutes before stopping the incubation of the tissue 1 ml of preequilibrated resin is placed in filtration columns (S/P screening columns P 5194; Scientific Products, McGraw Park, IL). Alternatively, the preequilibrated resin can be stored in stop buffer at 4°.

Concluding Remarks

Mice transgenic for the HD mutation represent interesting systems for studying the cellular and molecular mechanisms that underlie the onset of the human disease. Moreover, they may provide relevant models to test therapeutic treatments, including treatments aimed at protecting cells from damage induced by free radicals. This chapter has described the manner in which this laboratory handles R6/1 transgenic mice, and a reliable PCR method to genotype them. Simple behavioral analyses and observations were also discussed that indicate the progression of the neurological phenotype and that can be related to the degree of oxidative damage sustained by the striata of transgenic mice. Finally, two biochemical assays were described that provide information regarding the oxidative status of the brains of transgenic mice. These assays indicate the degree of oxidative damage (LP and NOS activity). These data provide an overview of the relationships between the expression of the HD mutation, the neurological phenotype, and the degree of oxidative damage sustained by the brain.

Acknowledgment

This work was supported by a CONACYT grant (33042-N).

[32] Heme Oxygenase 1 Transgenic Mice as a Model to Study Neuroprotection

By Mahin D. Maines

Introduction

The heme oxygenase (HO) system is the most effective biological mechanism for degradation of heme of hemoglobin, myoglobin, and cytochromes (reviewed in Maines[1]). To date, three isozymes, HO-1, HO-2, and HO-3, which are also known as heat shock protein 32 (HSP32) cognates, have been described.[2,3] HO-1 and HO-2 are the catalytically active forms and have been fully characterized. HO-3 was more recently described and has marginal activity.[3] HO-1 and HO-2 represent different gene products and, except for the heme-binding domain, share little similarity in primary structure, gene organization, or regulation.[1,4] HO-1 is the heat shock/stress-responsive form, whereas HO-2 is primarily responsive to glucocorticoids.[5]

Traditionally the HO system was considered only in the context of heme catalysis. This view has changed because of findings by our laboratory and others that the products of HO activity—biliverdin and CO, as well as the released iron—are all biologically active molecules. Biliverdin and its reduction product, bilirubin, possess potent antioxidant properties.[6] Also, bilirubin and biliverdin are inhibitors of protein phosphorylation and modulators of immune effector functions and the inflammatory response (reviewed in Willis[7]). Reversible phosphorylation/dephosphorylation of proteins is a key mechanism for the regulation of neuronal (as well as nonneuronal) functions.[8] CO is suspected to function as a signaling molecule and gaseous modulator for the second messenger cGMP production in neurons (reviewed in Maines[5] and Snyder *et al.*[9]). cGMP is essential for vasodilation responses and also affects ion channels, the activity of phosphodiesterases, and protein phosphorylation by cGMP-dependent kinases. Iron released in the

---

[1] M. D. Maines, "Heme Oxygenase: Clinical Applications and Functions." CRC Press, Boca Raton, FL, 1992.

[2] M. D. Maines, G. M. Trakshel, and R. K. Kutty, *J. Biol. Chem.* **261,** 411 (1986).

[3] W. J. McCoubrey, T. J. Huang, and M. D. Maines, *Eur. J. Biochem.* **247,** 725 (1997).

[4] M. D. Maines, *FASEB J.* **2,** 2557 (1988).

[5] M. D. Maines, *Annu. Rev. Pharmacol. Toxicol.* **37,** 517 (1997).

[6] P. A. Dennery, A. F. McDonagh, D. R. Spitz, and D. K. Stevenson, *Free Radic. Biol. Med.* **19,** 395 (1995).

[7] D. Willis, *in* "Inducible Enzymes in the Inflammatory Response" (D. A. Wiloughby and A. Tomlinson, eds.), p. 55. Birkhäuser Verlag, Basel, 1999.

[8] S. I. Walaas and P. Greengard, *Pharmacol. Rev.* **43,** 299 (1991).

[9] S. H. Snyder, S. R. Jaffrey, and R. Zakhary, *Brain Res. Brain Res. Rev.* **26,** 167 (1998).

Copyright 2002, Elsevier Science (USA).
All rights reserved.
0076-6879/02 $35.00

course of heme oxidation is sequestered by transferrin or ferritin, or is reutilized.[10] When free, the iron is a potent catalyst for oxygen free radical formation.

The results of gene targeting and gene transfer studies are for the most part supportive of the role of the HO system in cellular defense mechanisms.[11-14] However, because of differences in gene regulation and the primary structure of HO-1 and HO-2, the enzymes may have differential function in defense mechanisms[15]; HO-2 is a hemoprotein and is among a select group of proteins that have a motif known as the "heme regulatory motif" and has been suggested to function as a "sink" for nitric oxide.[16,17]

To test the neuroprotective role of HO-1 we have developed[18] mice that express HO-1 under the control of enolase [HO-1 transgenic (Tg) mice]. These mice have higher neuronal levels of cGMP and Bcl-2; and, when subjected to middle cerebral artery occlusion (MCAo), show both a reduced volume of edema and stroke as well as an inhibition of nuclear localization of the oncogene $p53$.[12] Further studies have shown that granule neurons isolated from HO-1 Tg mice are more resistant to oxidative stress caused by glutamate or H_2O_2 exposure.[19] The oxygen radicals activate signaling pathways that result in activation of many genes, including those of oncogenes that are effectors of cell apoptosis. The p53 protein is among the set of regulatory molecules whose expression is influenced by the redox state of the cell. p53-associated apoptosis is suspected to be a common mechanism of cell death in neurodegenerative diseases.[20] Furthermore, protection by HO-1 has been proposed against global and transient ischemia and against delayed cerebral vasospasm.[21]

On the basis of findings with ischemia–reperfusion kidney injury,[22] it is reasonable to suspect that protection afforded by upregulation of HO-1 involves, in part, degradation of hemoglobin heme and increased ability to generate bilirubin.

[10] P. Ponka, C. Beaumont, and D. R. Richardson, *Semin. Hematol.* **35,** 35 (1998).

[11] K. D. Poss and S. Tonegawa, *Proc. Natl. Acad. Sci. U.S.A.* **94,** 10925 (1997).

[12] N. Panahian, M. Yoshiura, and M. D. Maines, *J. Neurochem.* **72,** 1187 (1999).

[13] S. Dore, K. Sampei, S. Goto, N. J. Alkayed, D. Guastella, S. Blackshaw, M. Gallagher, R. J. Traystman, P. D. Hurn, R. C. Koehler, and S. H. Snyder, *Mol. Med.* **5,** 656 (1999).

[14] L. E. Otterbein, J. K. Kolls, L. L. Mantell, J. L. Cook, J. Alam, and A. M. K. Choi, *J. Clin. Invest.* **103,** 1047 (1999).

[15] N. Panahian and M. D. Maines, *J. Neurochem.* **76,** 539 (2000).

[16] W. J. McCoubrey, T. J. Huang, and M. D. Maines, *J. Biol. Chem.* **272,** 1568 (1997).

[17] Y. Ding, W. K. McCoubrey, and M. D. Maines, *Eur. J. Biochem.* **264,** 854 (1999).

[18] M. D. Maines, B. Polevoda, T. Coban, K. Johnson, S. Stoliar, T. J. Huang, N. Panahian, D. Cory Slechta, and W. J. McCoubrey, *J. Neurochem.* **70,** 2057 (1998).

[19] K. Chen, K. Gunter, and M. D. Maines, *J. Neurochem.* **75,** 304 (1999).

[20] S. M. deLaMonte, Y. K. Sohn, N. Ganju, and J. R. Wands, *Lab. Invest.* **22,** 401 (1998).

[21] H. Suzuki, K. Kanamaru, H. Tsunoda, H. Inada, M. Kuroki, H. Sun, S. Waga, and T. Tanaka, *J. Clin. Invest.* **104,** 59 (1999).

[22] M. D. Maines, R. D. Mayer, J. F. Ewing, and W. J. McCoubrey, *J. Pharmacol. Exp. Ther.* **264,** 457 (1993).

It is notable that under MCAo ischemic conditions, cellular ability to generate bilirubin, as indicated by an increase in biliverdin reductase protein levels, is increased in the most vulnerable parts of the brain.[23] In addition, CO may afford neuroprotection by inhibiting platelet aggregation,[24] and by stimulating generation of cGMP and vasorelaxation.[25]

The focus of this chapter is on methods that this laboratory has successfully used to detect neuroprotection against oxidative stress. Although the point of reference is HO-1 Tg mice, the described procedures can be applied in other settings where the beneficial or detrimental effects of altered gene expression—any gene— are examined subsequent to exposure to a neurotoxin—any neurotoxin. The methods are divided into two parts: Part 1 pertains to experiments with intact brain and Part 2 pertains to studies with isolated granule neurons.

Methods

Studies with Intact Mouse

Immunocytochemical and Histochemical Protocols

1. It is essential to perfuse brain completely for use in immunocytochemical (ICC) experiments. For this, mice are perfused through the left atrium with heparinized saline administered through a 26-gauge angiocatheter positioned into the apex of the left ventricle.

2. This is followed by reperfusion with 40 ml of a chilled solution of 4% (w/v) paraformaldehyde (PFA) in 0.1 M phosphate-buffered saline (PBS, pH 7.4).

3. The brain is postfixed in 4% (w/v) PFA for 4 hr at 4–6°.

4. This is followed by treatment with a cryoprotection solution, consisting of 30% (v/v) ethylene glycol and 20% (w/v) sucrose in 0.1 M PBS (pH 7.4) at 4° for 2–3 days.

5. The frozen specimens are cut serially into 35-μm-thick longitudinal or horizontal sections, using a cryostat sliding microtome (e.g., Microm HM400R; Carl Zeiss, Thornwood, NY). The brain tissue of Tg and non-Tg mice should be processed at the same time under identical conditions and using the same reagents and solutions.

HO-1 is detected with anti-HO-1 monoclonal or polyclonal antibodies (StressGen, Victoria, BC, Canada) as follows.

6. After 60 min of blocking in a solution of 5% (v/v) normal horse serum in 0.1 M Tris-buffered saline (TBS), followed by a wash in 0.25% (v/v) Triton X-100

[23] N. Panahian, T. Huang, and M. Maines, *Brain Res.* **850**, 1 (1999).

[24] B. Brune and V. Ullrich, *Mol. Pharmacol.* **32**, 497 (1987).

[25] W. Durante, N. Christodoulides, K. Cheng, K. J. Peyton, R. K. Sunahara, and A. I. Schafer, *Am. J. Physiol.* **273**, H317 (1997).

solution in TBS, all specimens are transferred [in Costar (Cambridge, MA) net wells] into the primary HO-1 antibody diluted 1 : 1000 in carrier solution [0.1 M TBS containing 0.5% (v/v) horse serum, 0.25% (v/v) Triton X-100], and incubated for 48 hr at 4–6°.

7. The specimens are then rinsed five times (10 min each) with 0.1 M TBS containing 0.25% (v/v) Triton X-100 and placed into a biotinylated secondary antibody reagent according to the manufacturer recommendations (e.g., Vectastain Elite, mouse IgG kit; Vector Laboratories, Burlingame, CA). For peroxidase reactions, the typical incubation is 3 hr at room temperature. After five 10-min washes in 0.1 M TBS, specimens are incubated at room temperature for 90 min with the avidin biotin reagent prepared in 0.1 M PBS (ABC solution; Vector Laboratories).

8. After consecutive 10-min rinses in TBS and Tris-HCl, the sections are placed into a filtered solution of 0.04% (v/v) 3',3'-diaminobenzidine (DAB) and 0.06% (v/v) H_2O_2 in 0.1 M Tris buffer for 4–5 min. A stereomicroscope (e.g., MZ-8; Leica, Bensheim, Germany) can be used to detect the reaction product during development.

9. Sections are rinsed in phosphate buffer, dehydrated serially in 95% and 100% alcohol, incubated in histological grade xylene, mounted on Superfrost Plus slides, and coverslipped with Permount (Fisher Scientific, Pittsburgh, PA).

As exemplified below, oncogene products that are linked to cell death can be visualized[15] with monoclonal antibodies. Sections are incubated overnight at 4–6° with the desired antibodies at the dilution indicated below. After several consecutive washes with TBS containing 0.3% (v/v) Triton X-100, the sections are processed for peroxidase detection. Anti-mouse p53 monoclonal IgG$_1$ (sc-100) is obtained from Santa Cruz Biotechnologies (Santa Cruz, CA) and used at a 1 : 500 dilution. Anti-TRAIL (sc-1891) and anti-FADD (sc-6036), affinity-purified goat polyclonal antibodies of mouse origin, can be obtained from the same supplier. The TRAIL antibody is raised against a peptide corresponding to amino acids 273–291 mapping at the carboxy terminus of a TNF-related apoptosis-inducing ligand, whereas anti-FADD is raised against a peptide corresponding to amino acids 187–205 mapping at the carboxy terminus of the Fas-associated death domain-containing protein.

Double immunolabeling using HO-1 monoclonal antibody with either p53, FADD, or TRAIL can be performed[15] using the Vector Laboratories peroxidase detection system. As recommended by the supplier (Vector Laboratories), DAB is used as a primary chromogen, whereas Vector SG is used as the secondary chromogen.

Fe^{3+} is detected by Perl's reaction followed by DAB enhancement.[26] Perl's reaction is based on the formation of ferric ferrocyanide (Prussian Blue). Ferric ion released from iron-containing compounds by HCl reacts with potassium

[26] L. M. Sayre, G. Perry, and M. A. Smith, *Methods Enzymol.* **309,** 133 (1999).

ferrocyanide. The reaction is further enhanced by oxidation with DAB. Staining for Fe^{3+} is incompatible with tissue processing for ICC procedures.[15] Adjacent sections are stained to detect oxygen radical-mediated lipid peroxidation.[12] Free-floating brain sections are incubated for 45 min in the dark at room temperature in Schiff's reagent (paraosaniline base, thionyl chloride) prepared as described by Pearse.[27] Staining for iron and lipid peroxidation are not compatible with tissue processing for ICC.

Fixatives Used for Immunocytochemistry and Histochemistry

Choice of fixative is an empirical process. Fixation is generally done overnight, although for low-affinity antibodies that are sensitive to the fixative, times as short as 2 hr may be employed. They are listed below and are assessed for a given application. Vector Laboratories (Burlingame, CA) is a good source for many fixatives as well as histochemical stains.

Equal parts solution: A mixture of equal parts doubly distilled H_2O, glacial acetic acid, acetone, and ethanol. Used as a fixative for hematoxylin and eosin staining as well as for thionin, nuclear and fast red, and methyl green or other nuclear histostains. It should never be used for ICC

Formalin (10%, w/v): This is the easiest and most common fixative for tissue. It can be purchased inexpensively as a 10% (w/v) solution, powder, or 37% (w/v) stock. It is stable and can be stored at room temperature almost indefinitely. Its main drawback is that some antibodies, particularly monoclonals, work poorly in formalin-fixed tissue. It is important to note that often tissue sent to a pathology department for paraffin embedding will be treated with 10% (w/v) formalin during the automated processing and, therefore, may not be suitable for ICC with some antibodies

Methacarm: Developed specifically for iron staining[26] but works for other histochemistry and works best on free-floating paraffin sections. It is a $60:30:10$ (v/v/v) mixture of methanol–chloroform–acetic acid

Michel's fixative: Commercial fixative used for histochemical stains for metals (e.g., iron, calcium, and zinc)

Paraformaldehyde (4%, w/v): This is the first choice of fixative for ICC and also works for histochemistry. It is prepared by heating 100 ml of 0.1 M phosphate buffer to 60°, adjusting to pH 7.5 with 2 N NaOH, and adding 4 g of paraformaldehyde. Stir until dissolved (about 20 min) while continuing to heat, but do not allow the temperature to rise above 65–70°. Place on ice for 2–4 hr, filter immediately with a Nalgene filter before use, and use within 12–16 hr. Addition of glutaraldehyde to 0.1% (w/v) can enhance

---

[27] A. G. E. Pearse, "Histochemistry: Theoretical and Applied," 4th Ed., Vol. 1, p. 655. Churchill Livingstone, New York, 1980.

the quality of the fixation. Some antibodies are unable to penetrate tissue fixed with 4% (w/v) paraformaldehyde. In those cases, a 2% (w/v) solution is supplemented with parabenzaquinone

Paraformaldehyde (2%, w/v)–parabenzoquinone (0.2%, w/v): Paraformaldehyde (4%, w/v) is prepared as described above. A 0.4% (w/v) parabenzoquinone solution is prepared in water and both solutions are chilled for 2–3 hr. Equal volumes are mixed and filtered immediately before use. The combination makes the tissue almost rubbery and facilitates sectioning

Zamboni's fixative (PAF): The commercial solution is 2% (w/v) paraformaldehyde containing acetic acid and picric acid. It is particularly useful for preserving fine structure such as neuronal processes. However, it is used primarily as a secondary fixative because of high background when used as the primary agent. Works well for ICC and histochemistry

Coverslipping Agents

Sections must be xylene processed before coverslipping. The agent is spread across the bottom of the slide and the coverslip is applied by slowly rolling from bottom to top to spread the agent. The commonly used agents are listed below.

Aquapolymount: A water-soluble mount for cases in which alcohol and xylene cannot be used on slides (some chromogens are sensitive to alcohol). Works well for fluorescence, but because of viscosity bubbles can be a problem

Cytoseal: This is the most commonly used agent and easy to apply. Because of problems with optical quality, it does not work well for high-magnification (e.g., $\geq \times 400$) photomicroscopy

Permount: This has better optical properties than Cytoseal. Excellent for high-magnification microscopy especially when mixed with 2 parts xylene

Vectashield: This is not truly a mounting solution as it does not solidify. Useful for fluorescence as it helps prolong the half-life of the fluorescent compound.

Hematoxylin and Eosin Staining

General tissue/cell morphology can be visualized by hematoxylin and eosin (H&E) staining. It can be performed on formalin- or paraformaldehyde-fixed sections, but not on Zamboni-fixed tissue. It works best on fresh frozen, cryostat-cut, or paraffin-embedded sections. All solutions mentioned below are available from Sigma (St. Louis, MO).

1. Fix sections for 2 min in equal parts solution. Omit this fixation if the sections have previously been stained for ICC.

2. Rinse with copious amounts of tap water.

3. Stain for 3 min in Harris hematoxylin solution.

4. Rinse for 2 min with water.

5. Destain with 1% (v/v) acid alcohol by immersing the sections once in the solution.

6. Stop the destaining by immersing the sections in water. Rinse 2–3 min.

7. Fix the membranes with 0.05% (w/v) lithium carbonate for 1 min.

8. Wash with water.

9. Stain with eosin solution for 45 sec to 1 min.

10. Wash twice (11 or 12 dips per wash) with 95% (v/v) ethanol and then with 100% ethanol.

11. Place the sections in xylene for a minimum of 5 min (paraffin) or 15 min (fresh frozen) or, if convenient, overnight; slides can be stored for several days and up to 1 month before coverslipping.

Analysis of Apoptosis

There are a large number of assays that can be used to detect damaged cells and apoptosis.[28,29] Here, only detection of DNA damage is described. The formation of high molecular weight (>50 kb) DNA fragments and intranucleosomal fragmentation or DNA laddering (200-bp DNA fragments) are described. DNA from necrotic cells has a random and general cleavage and appears smeared when electrophoresed. DNA is obtained from flash-frozen tissues and quantitated spectrophotometrically by measuring absorption at 260 nm. For the detection of ladder formation, 15-μg samples are electrophoresed at 45 V for 15 min and then at 65 V for an additional 2 hr. After ethidium bromide staining, gels are photographed. High molecular weight fragmentation is detected by heating 1 μg of DNA to 65° for 10 min before agarose gel electrophoresis in Tris–acetate–EDTA (0.5%, v/v). Pulse-field agarose gel electrophoresis[30] is carried out at 4° for 96 hr at 20 V with the buffer circulating. For sizing, high molecular weight DNA markers (Invitrogen, Carlsbad, CA) are used.

Morphological visualization of DNA single-strand breaks for detection of early events in apoptosis in the brain can be carried out by *in situ* nick translation.[31] For this, frozen brain sections are cut with a cryostat into 10-μm sections. They are fixed for 5 min in phosphate-buffered saline–4% (w/v) paraformaldehyde and dehydrated with 70 and 100% ethanol. Thereafter, sections are treated with a

[28] J. Chen, K. Jin, M. Chen, W. Pei, K. Kawaguchi, D. A. Greenberg, and R. P. Simon, *J. Neurochem.* **69,** 232 (1997).

[29] B. Zhivotovsky, A. Samali, and S. Orrenius, in "Current Protocols in Toxicology" (M. D. Maines, L. Costa, D. J. Reed, S. Sassa, and J. G. Sipes, eds.), p. 2.1.1. John Wiley & Sons, New York, 1999.

[30] R. Arand and E. M. Southern, in "Pulse Field Gel Electrophoresis of Nucleic Acids: A Practical Approach" (D. Rickwood and B. D. Haines, eds.), 2nd Ed., p. 101. IRL Press, Oxford, 1990.

[31] B. Gavrieli, Y. Sherman, and S. BenSasson, *J. Cell Biol.* **119,** 493 (1992).

reaction mixture containing 2.5–5 μCi of α-^{32}P-labeled dCTP, incubated for 1–2 hr at room temperature in the presence or absence of DNA polymerase I, washed 3 or 4 times for 10 min each with buffered saline, dehydrated with 70 and 100% ethanol, and exposed to X-ray film. All manipulations (*in situ* nick translation, exposure to X-ray film, etc.) should be performed under the same conditions for a given experiment. Autoradiograms can be evaluated by image analysis and by measuring the intensity of the labeling. No radioactivity should be detected in sections incubated without DNA polymerase.

Studies with Isolated Neurons

Preparation of Neuronal Cell Cultures

The method of cell preparation is based on that described by Miller *et al.*[32]

1. Seven-day-old mice are killed by decapitation.
2. Dissect out the cerebellum and place it in 1 ml of digestion buffer in a microcentrifuge tube. The digestion buffer consists of papain (1 mg/ml), DNase I (12.5 μg/ml), bovine serum albumin (BSA, 0.2 mg/ml), L-cysteine HCl (0.2 mg/ml), and glucose (5 mg/ml).
3. Mince the tissue into small pieces with fine scissors.
4. Incubate in a 37° water bath for 20 min.
5. Transfer to a 15-ml conical centrifuge tube, add 4 ml of culture medium, and mix with a pipette. The medium consists of Eagle's basal medium, 10% (v/v) fetal bovine serum, 25 mM KCl, 1:100 penicillin–streptomycin (Pen/Strep) solution.
6. Centrifuge in Beckman J2-M1 at 1000 rpm for 10 min at 4°.
7. Decant the supernatant and resuspend the cells in 2.5 ml of culture medium.
8. Determine the cell concentration with a hemocytometer.
9. Seed poly-L-lysine (10 μg/ml)-coated culture plates with 3 × 10^5 cells/cm^2.
10. Incubate at 37° in a 5% CO_2 atmosphere overnight.
11. Add cytosine arabinofuranoside (AraC; 1 M in water) to a concentration of 10 μM to inhibit growth of nonneuronal cells.
12. Incubate the neurons for 7 days before experimentation. Cells can be used only on day 7 or 8 and after that must be discarded.

Glutamate Treatment of Neuronal Cells

There are a number of ways that oxidative stress can be induced. The following is that caused by glutamate exposure.[33]

[32] T. M. Miller, K. L. Moulder, C. M. Knudson, D. J. Creedon, M. Deshmukh, S. J. Korsmeyer, and E. M. Johnson, *J. Cell Biol.* **139**, 205 (1997).
[33] M. Ankarcrona, J. M. Dypbukt, E. Bonfoco, B. Zhivotsky, S. Orrenius, S. A. Liptos, and P. Nicotera, *Neuron* **15**, 961 (1995).

1. Remove and save the culture medium from cells grown for 7 days in 48-well plates.

2. Wash the cells with Locke's solution [5 mM HEPES (pH 7.4) containing 134 mM NaCl, 25 mM KCl, 4 mM NaHCO$_3$, 2.3 mM CaCl$_2$, and 5 mM glucose].

3. Treat cells at 37° with 150 μl of 30 μM or 3mM glutamate in Locke's solution for 30 min.

4. Remove glutamate and return to culture medium. Do not use fresh medium as fresh serum is toxic for the cells at this stage.

5. Incubate at 37° for 3 hr before measuring viability, necrosis or nuclear condensation.

Neuronal Viability Assessed by MTT Assay

The MTT assay measures cell viability as assessed by the mitochondrial ability to metabolize MTT.[34]

1. Dissolve MTT (3-4,5-dimethylthiazol-2,5-diphenyltetrazolium bromide) in serum-free medium at 0.25 mg/ml.

2. Remove the medium from cells and add 150 μl of MTT per well.

3. Incubate at 37° for 30 min in a 5% CO$_2$ atmosphere, and then add 150 μl of solubilization solution [20% (w/v) sodium dodecyl sulfate (SDS) in 50% (v/v) dimethyl formamide, adjusted to pH 4.8] to each well.

4. Incubate at 37° for 4–5 hr (CO$_2$ is not required at this step).

5. Measure the absorbance at 570 nm. Viable cells metabolize MTT, producing a purple color. There is a direct correlation between the absorbtion value and cell viability.

Neuronal Necrosis

1. Add a 1/100th volume of a solution of 1-mg/ml fluorescein diacetate and 1-mg/ml propidium iodide to the culture medium (final concentration is 10 μg/ml for each dye).

2. Incubate at 37° for 5 min.

3. Examine by fluorescence microscopy, using a UV lamp and fluorescein isothiocyanate (FITC) filter. Viable cells are green whereas necrotic cells are red.

Analysis of Chromatin Condensation

1. Remove the medium from cells and permeabilize for 15 min with 80% (v/v) methanol.

2. Wash briefly with phosphate-buffered saline (PBS).

3. Stain for 5 min with propidium iodide (10 μg/ml in PBS).

4. Rinse with PBS.

---

[34] T. Mosmann, *J. Immunol. Methods* **65,** 55 (1983).

5. Examine by fluorescence microscopy, using a UV lamp and FITC filter. DNA in the nuclei of all cells will be red; in normal cells, the distribution will be patchy, reflecting the diffuse nature of chromatin, whereas the chromatin of apoptotic cells will appear to be highly condensed.

Measurement of Reactive Oxygen Species

The assay can be used to detect reactive oxygen species (ROS) formation in response to any number of agents[35]; herein it is described with respect to glutamate-induced ROS formation.

1. Remove medium and wash briefly with Locke's solution.
2. Add 200 μl of Locke's solution containing 3 mM glutamate and dichlorofluorescein diacetate (DCF, 1 μg/ml) to each well.
3. Incubate at 37° and read the fluorescence of the solution at wavelengths of 485 nm (excitation) and 535 nm (emission) at 30-min intervals for a total of 90 min after addition of the glutamate solution. An instrument such as a Wallac (Turku, Finland) 1420 multilable counter can be used.
4. Plot fluorescence versus time. Fluorescence is proportional to ROS, so a flatter slope indicates slower generation of ROS. It should be noted that fluorescence units are relative and therefore only samples in the same plate can be compared.

Measurement of Intracellular Free Calcium

1. Sterilize 25-mm glass cloverslips overnight under a UV lamp in a tissue culture hood.
2. Place a coverslip at the bottom of each well of a six-well tissue culture dish.
3. Precoat the coverslips overnight with 250 μl of poly-L-lysine (10 μg/ml in water).
4. In accordance with the protocol for culturing neurons, prepare the cells and seed 3×10^5 cells/cm^2 in a 250-μl volume onto each coverslip.
5. Incubate at 37° for 30 min, and then add 2 ml of culture medium without antibiotics.
6. Add AraC and grow the cells for 7 days as for standard neuron culture.
7. Add 1 mM fura-2 acetomethyl ester (Molecular Probes, Eugene, OR) to a final concentration of 5 μM (10 μl for 2 ml of culture medium).
8. Incubate at 37° for 1 hr.
9. Wash with Locke's solution and then transfer the coverslip to a microscope holder. We have used a Nikon fluorescence photometric system with an inverted Fluor ×40, 1.3 numerical aperture objective.

[35] D. J. Kane, T. A. Sarafian, R. Anton, H. Hahn, E. B. Gralla, J. S. Valentine, T. Ord, and D. E. Bredesen, *Science* **262**, 1274 (1993).

10. Add sufficient (0.5 ml) 30 μM glutamate in Locke's solution to cover the surface of the coverslip.

11. Collect data for up to 10 min. The excitation wavelength is set at 340 nm and the emission wavelength is set at 375 nm. Data are collected at 510 nm.

Immunocytochemistry of Neuronal Cells

1. Cells are grown on coverslips as for calcium measurements or on 18-mm coverslips in 12-well plates.

2. Wash the cells with PBS.

3. Fix the cells with 4% (w/v) paraformaldehyde on ice for 10 min.

4. Wash three times with PBS.

5. Block with 5% (v/v) horse serum in PBS containing 0.25% (v/v) Triton X-100 (PBS-T) for 1–2 hr at room temperature.

6. Remove the blocking solution and add primary antibody diluted in PBS-T containing 3% (v/v) horse serum. Dilutions used are as follows: HO-1 monoclonal, 1 : 250; monoclonal neuronal microfilament, 1 : 200; and HO-1 polyclonal, 1 : 5000.

7. Incubate at 4° overnight.

8. Wash three times with PBS-T.

9. Add 980 μl of PBS-T, 15 μl of horse serum, and 5 μl of secondary antibody Vector ABC [horseradish peroxidase (HRP)-conjugated anti-mouse IgG for monoclonals or antirabbit for rabbit polyclonal primary antibody] to each well.

10. Incubate at room temperature for 1 hr.

11. During the 1-hr incubation, prepare A/B solution. Mix 1 drop each of solutions A and B into 2.5 ml of PBS (each well will receive 0.6 ml) and allow the solution to remain at room temperature for 30 min before use.

12. Wash the wells four times with PBS-T and then add 0.6 ml of A/B solution to each well.

13. Incubate at room temperature for 30 min.

14. Wash three times with PBS-T.

15. Prepare chromogen [VIP substrate for peroxidase (SK4600; Vector Laboratories)] by adding 3 drops each of reagents 1, 2, and 3 to 5 ml of PBS. Mix and add 3 drops of H_2O_2 and vortex briefly.

16. Add 1 ml of chromogen to each well and watch the color development under a microscope.

17. When color development is sufficient, stop the reaction by replacing the staining solution with water or PB.

18. Remove the coverslips and dehydrate by dipping then six times in 95% (v/v) ethanol and then six times in 100% ethanol.

19. Dip the coverslips six times in xylene and mount, inverted, onto a slide with Permount.

Analysis of Protein Expression

For Western analysis, the contents of all six wells of a six-well plate are pooled.

1. Wash the cells in PBS.
2. Scrape the cells from the wells, using a rubber policeman, and place the cell in a microcentrifuge tube.
3. Centrifuge at 1000 rpm for 5 min at 4°.
4. Resuspend in a minimal volume (50 μl if possible) of lysis buffer.
5. Sonicate twice, 10–15 sec each.
6. Centrifuge at 10,000 rpm for 10 min at 4°.
7. Remove the supernatant and store at −80°.
8. Use approximately 30 μl for a Western blot. Use the ECL system (Amersham, Arlington Heights, IL) for detection of transferred proteins. Follow the manufacturer instructions.

RNA Analysis

1. Add 1 ml of TRIzol reagent to 100 mg of cells. This amount of cells is obtained from a 25-cm^2 flask (use 0.5 ml for a six-well plate).
2. Homogenize the tissue with a Dounce homogenizer and transfer to a microcentrifuge tube.
3. Add 0.2 ml of chloroform, mix, and hold at room temperature for 3 min.
4. Centrifuge at maximum speed in the microcentrifuge for 15 min at 4°.
5. Add an equal volume of isopropanol to the supernatant in a fresh tube.
6. Mix and place on ice for 10 min.
7. Centrifuge at maximum speed for 10 min at 4°.
8. Add 0.5 ml of ice-cold 70% (v/v) ethanol to the pellet.
9. Vortex vigorously and repeat the centrifugation.
10. Air dry the pellet for 10–20 min.
11. Dissolve the pellet in water (typically 200 μl for 100 mg of tissue).
12. Quantitate by UV spectroscopy.

REVERSE TRANSCRIPTASE-POLYMERASE CHAIN REACTION

1. In a 0.5-ml microcentrifuge tube mix 2 μg of total RNA and 1 μl of oligo(dT) primer [Invitrogen (San Diego, CA) cDNA cycle kit] with sufficient water to make 10 μl.
2. Place the tube at 70° for 10 min.
3. Add 4 μl of 5× SuperScript buffer, 1 μl of 0.1 M dithiothreitol (DTT), 4 μl of water, 0.5 μl of 100 mM dNTPs, and 0.5 μl of SuperScript reverse transcriptase to the tube. (For multiple samples, it is easiest to prepare a master mix by adding 10 μl to the tube containing RNA.)

4. Incubate the tube at 42° for 50 min and then heat denature it for 2 min at 95° and place on ice.

5. PCR is carried out with 1 μl of the cDNA in a 25-μl reaction volume containing 0.5 μl (50 pmol) of each primer and 0.25 μl of *Taq* polymerase (2.5 U/ml).

6. The number of cycles varies with tissue and primers; usually 20–30 cycles are used. Products are analyzed on 1.2–1.6% (w/v) agarose gels in Tris–acetate buffer with EDTA (TAE).

RNA PROBE LABELING. The RNA probe labeling protocol can be used for any cloned cDNA along with appropriate vectors and restriction enzymes.

1. Set up four restriction digests of 20 μl, each containing 2–3 μg of plasmid DNA. Use appropriate restriction enzyme and buffer. Incubate at 37° overnight.

2. Gel purify the DNA fragment on a 2% (w/v) agarose gel in TAE.

3. Phenol extract the pooled products and ethanol precipitate.

4. Resuspend the pellet in 20 μl of water; 2–4 μl is used for a standard labeling reaction.

5. Using the Stratagene (La Jolla, CA) RNA transcription kit, mix the following: 1 μl each 10 mM ATP, GTP, and CTP, 1 μl of 0.75 M DTT, 5 μl of [α-^{32}P]UTP (3000 Ci/mmol), 2 μl of T7 RNA polymerase, 2–4 μl of template (add this last), and water to a final volume of 25 μl.

6. Incubate at 37° for 1 hr and then heat denature the polymerase at 95° for 2 min.

7. Place on ice for 5 min, and then add 1 μl of DNaseI (10 mg/ml).

8. Incubate at 37° for 15 min and then heat denature at 85° for 4 min before loading on a gel. Loading dye is not required because of the glycerol present in the reaction buffer. A complete reaction mixture can be loaded in a single lane. The gel is 8% (w/v) acrylamide [19 : 1 (w/w)] plus 7 M urea in 1× TBE and is poured with 1.5-mm spacers. Load bromphenol blue (BPB) dye in an adjacent lane to allow monitoring of the gel progress.

9. Run the gel at 175 V until the tracking dye is approximately two-thirds down the gel.

10. Cut the gel to a convenient size, wrap it in plastic wrap, and perform a rapid autoradiogram exposure.

11. Using a razor blade, cut out the labeled probe band, which is the fainter band about one-third of the way down the gel, and place it in 1.5-ml microcentrifuge tube. The unincorporated label is the intense band near the bottom of the gel.

12. Using a rotating mixer, elute the probe overnight at room temperature into 250 μl of 0.5 mM ammonium acetate containing 1 mM EDTA and 0.1% (w/v) sodium dodecyl sulfate (SDS).

13. Transfer 200 μl of the supernatant to a fresh tube, add 100 μl of 5 M ammonium acetate and 0.6 ml of ice-cold ethanol, and place the tube on dry ice for 10–15 min, until frozen.

14. Centrifuge for 15 min at 4° at 2000 rpm, rinse the pellet with cold 70% (v/v) ethanol, centrifuge for 5 min at 2000 rpm, and air dry the pellet.

15. Resuspend the labeled RNA in 50 μl of water and count 1 μl in scintillation fluid. Adjust to approximately 5000 cpm/μl and store in aliquots at −80° for up to 1 week.

Detecting Transgenic Genotype

Preparation of Mouse Tail DNA

1. Add 600 μl of digestion buffer [10 mM Tris HCl (pH 7.5), 400 mM NaCl, 0.1 mM EDTA, and 0.6% (w/v) SDS] and 35 μl of proteinase K (10 mg/ml) to a microcentrifuge tube containing a tail clipping (about 1 cm). DNA recovery from adult mice is not as efficient as from newborn mice. Three-week-old mice are best for obtaining samples.

2. Place at 50–65° overnight until all tillus is gone (hair will remain).

3. Add 200 μl of 5 M NaCl; vortex and centrifuge at maximum speed for 5 min.

4. Transfer the supernatant to a fresh tube and add an equal volume of ice-cold ethanol.

5. Freeze on dry ice and then centrifuge at maximum speed for 20–25 min at 4°.

6. Rinse the pellet with 70% (v/v) ethanol and centrifuge for 5 min.

7. Discard the ethanol and air dry the pellet briefly.

8. Resuspend in 500 μl of water or TE [10 mM Tris-HCl (pH 7.9), 1 mM EDTA] and quantitate by UV spectroscopy. Because the DNA is of high molecular weight, resuspension usually requires an elevated temperature (37°) and several hours for completion.

Screening Transgenics by Polymerase Chain Reaction

The protocol is designed to differentiate HO-2 Tg mice, using DNA prepared from tail snips, and to determine whether the animals are homozygous or heterozygous. HO-1 is used as a copy number control and primers are selected such that the HO-2 primers would be separated by an intron in the genomic DNA and thus the endogenous copy either does not generate a product or produces one significantly larger than the transgenic cDNA copy. HO-1 primers come from the same exon, so they produce the same size product as the cDNA would. A similar approach can be used for any transgene as long as the information about the organization of the gene and sequence is known.

1. Set up a 100-μl PCR, using standard reaction conditions and including 1 μl each of the HO-1 and HO-2 primers and 0.5 μg of DNA. Include a control reaction with no template DNA. The primers used will produce a 451-bp HO-1 fragment and a 324-bp HO-2 product. Sequences of the primers are as follows: HO-1 forward, 5'-AATGTTGAGCAGGAAGGCGGT; reverse, 5'-CTGGAAGAGGAGATAGA-GCG; HO-2 forward, 5'-AGGCCAGGTACTGAAGAAGG; reverse, 5'-TGAGC-AGCATAAAAGGGG. The primers are based on published data.[36,37]

2. Split the reaction into four tubes of 25 μl each and overlay with mineral oil.

3. Thermocycling parameters are as follows: an initial denaturation at 94° for 1 min, and then cycles of 94° for 35 sec, 57° for 35 sec, and 72° for 35 sec.

4. Aliquots are removed to ice and stopped after 18, 20, and 24 cycles.

5. Products are separated on a 2% (w/v) agarose gel in 1× TBE and are quantitated by laser densitometry of negatives from the stained gel.

Summary

Bile pigments and CO are formed in the course of heme degradation by the isozymes and are biologically active moieties. In the course of heme degradation the chelated iron is also released. Heme and iron are prooxidants, whereas bile pigments are antioxidants. In addition, CO functions as a signal molecule and HO-2 may serve as an intracellular "sink" for NO. In the balance, the published data suggest that the HO system functions in cellular defense mechanisms. The methods described in this chapter can be used to assess the tissue/cell toxicity of chemicals in general, and as pertains to the defense activity of HO-1, specifically.

Acknowledgments

This research was supported by Grant NIH ES04391. I am grateful to Suzanne Bono for preparation of this manuscript.

[36] W. K. McCoubrey, Jr. and M. D. Maines, *Gene* **139,** 155 (1994).
[37] R. M. Müller, H. Taguchi, and S. Shibahara, *J. Biol. Chem.* **262,** 6795 (1987).

[33] Copper/Zinc Superoxide Dismutase Transgenic Brain in Neonatal Hypoxia–Ischemia

By R. ANN SHELDON, LYNN ALMLI, and DONNA M. FERRIERO

Introduction

Although significant progress has been made in characterizing the mechanisms of cellular injury in models of adult stroke, the neonatal brain is less well understood. Often, discoveries in the adult brain are either not duplicated in the developing brain or the results are quite different. We found this to be the case for copper/zinc superoxide dismutase (SOD1). Whereas research on adult animals has shown reduced cerebral ishemic injury in SOD1-overexpressing mice,[1–3] we found the opposite in a model of hypoxia–ischemia in neonatal animals of the same transgenic strain.[4]

Hypoxia–Ischemia as Model of Neonatal Stroke

The Vannucci model is a standard technique for neonatal stroke.[5,6] It consists of a permanent, unilateral ligation of the common carotid artery combined with a period of hypoxia. The combination of ischemia and hypoxia produces an area of infarct ipsilateral to the ligation. Neither ligation nor hypoxia alone produces cell death discernable by standard histological techniques. Postnatal day 7 (P7) animals are used, as the brain development at this age is considered to be comparable to a term newborn human.

Surgery

P7 mouse pups are removed from the dam as a litter and kept on a 37° heating pad. Pups are individually removed from the group and anesthetized with a continuous flow of 2.5% halothane in 40% oxygen, balance nitrogen. When the animal does not respond to a tail or foot pinch, an incision is made in the midline

[1] H. Kinouchi, C. J. Epstein, T. Mizui, E. Carlson, S. F. Chen, and P. H. Chan, *Proc. Natl. Acad. Sci. U.S.A.* **88,** 11158 (1991).

[2] G. Yang, P. H. Chan, J. Chen, E. Carlson, S. F. Chen, P. Weinstein, C. J. Epstein, and H. Kamii, *Stroke* **25,** 165 (1994).

[3] P. H. Chan, M. Kawase, K. Murakami, S. F. Chen, Y. Li, B. Calagui, L. Reola, E. Carlson, and C. J. Epstein, *J. Neurosci.* **18,** 8292 (1998).

[4] J. S. Ditelberg, R. A. Sheldon, C. J. Epstein, and D. M. Ferriero, *Pediatr. Res.* **39,** 204 (1996).

[5] J. E. D. Rice, R. C. Vannucci, and J. B. Brierley, *Ann. Neurol.* **9,** 131 (1981).

[6] R. Vannucci, J. Connor, D. Mauger, C. Palmer, M. Smith, J. Towfighi, and S. Vannucci, *J. Neurosci. Res.* **55,** 158 (1999).

Copyright 2002, Elsevier Science (USA).
All rights reserved.
0076-6879/02 $35.00

of the neck, no longer than 3–4 mm. The right common carotid artery is then exposed with the use of two curved forceps and permanently ligated with a bipolar coagulator (Codman & Schurtleff, Randolph, MA). The incision is closed and the pup is returned to the heating pad. The ligation procedure should last no longer than 5 min. When all pups are ligated, they are weighed and returned to the dam, where they remain for 1.5 hr for recovery and feeding.

Hypoxia

Pups are again removed from the dam and placed in airtight containers partially submerged in a 37° water bath. Humidifed 8% oxygen, balance nitrogen flows through the containers after passing a flowmeter (Manostat; Barnant, Barrington, IL), which maintains a constant rate of flow. We have determined that the ultimate degree of injury is based not only on the duration of hypoxia, but also on the strain of mouse being used. Hence, the duration of hypoxia used is dependent on the strain.[7]

Superoxide Dismutase-Overexpressing Mouse in Hypoxia–Ischemia

The first application of the Vannucci procedure in mice was in our study of the effect of SOD overexpression in transgenic animals, developed by Epstein *et al.*[8] We modified the original rat model of hypoxia–ischemia for use in mice, which allowed us to investigate the effect of SOD in this model by utilizing transgenic mice that overexpress SOD.[4] Anesthesia and the ligation procedure are the same as for rats, only scaled down for pups about one-quarter the size of rats. The incision is not sutured in mice, but rather pinched together, as the dam will chew out the sutures. A small drop of methyl methacrylate adhesive may also be used to close the incision. We determined that the ideal duration of hypoxia for SOD-overexpressing pups is 90 min. This produces a severe degree of injury. We have subsequently found that 45 min of hypoxia produces a moderate degree of injury, while minimizing mortality, in the background strain CD1.

Adults of the same SOD-overexpressing mouse strain were found to have reduced injury in a model of focal cerebral ischemic injury, leading to the reasonable conclusion that, in these animals, SOD is protective, and that the superoxide radicals generated by ischemia are, at least in part, responsible for pathogenesis.[1] With neonatal hypoxia–ischemia (HI), however, we found that the SOD animals were not only not protected, they were more frequently and more severely injured than nontransgenic littermates.[4] Assuming that superoxide radicals were indeed

---

[7] R. A. Sheldon, C. Sedik, and D. M. Ferriero, *Brain Res.* **810**, 114 (1998).

[8] C. J. Epstein, K. B. Avraham, M. Lovett, S. Smith, O. Elroy-Stein, G. Rotman, C. Bry, and Y. Groner, *Proc. Natl. Acad. Sci. U.S.A.* **84**, 8044 (1987).

responsible for initiating pathogenesis, and that the superoxide dismutase reaction generates hydrogen peroxide (H_2O_2), we would expect an accumulation of H_2O_2 after HI. Under normal conditions, endogenous glutathione peroxidase (GPx) and catalase convert H_2O_2 to oxygen and water. We looked at how these enzymes were affected after HI in the SOD1 transgenic pups, as well as levels of H_2O_2.[9] Activity of both catalase and glutathione peroxidase was the same in the SOD overexpressors and wild-type P7 mice. In response to HI, GPx activity decreased significantly by 24 hr posthypoxia in both the transgenic and wild-type pups. In addition, an increase in H_2O_2 accumulation was seen at 24 hr in the transgenic pups only. Catalase was unchanged in both transgenic and wild-type mice. We also examined the role of iron in oxidative stress due to HI. The iron chelator deferoxamine (DFO; Sigma, St. Louis, MO) has been shown to be neuroprotective in the rat model of HI.[10] Using the same experimental paradigm (DFO at 100 mg/kg 10 min after HI and again at 24 hr) in SOD overexpressors, we showed that wild-type mice were protected by iron chelation, but that there was no significant protection in the SOD transgenic mice.[11]

Histological Methods

Seven days after hypoxia–ischemia, mice are injected with a lethal dose of pentobarbital (50 mg/kg) and perfused transcardially with cold 4% (w/v) paraformaldehyde in 0.1 M phosphate buffer (pH 7.4). Brains are then removed and postfixed in the same fixative for 4 hr. The forebrain is sectioned on a Vibratome (Ted Pella, Redding, CA), and alternate 50-μm sections are collected for cresyl violet and enhanced Perl's iron stain.[12,14] Analysis of brains with both stains increases the ability to visualize dead and dying cells microscopically. For quantification of the degree of injury, we have developed a scoring system for mice, which reflects the greater degree of injury seen in the hippocampus. Eight regions are given a score from 0 to 3: 0, no injury; 1, mild with focal areas of cell loss; 2, moderate, such as a majority of cells dead in the pyramidal cell layer of the hippocampus, or columnar areas of cell loss in the cortex; and 3, cystic infarction. The regions scored are the anterior, middle, and posterior cortex; the striatum/caudate putamen

[9] H. J. Fullerton, J. S. Ditelberg, S. F. Chen, D. P. Sarco, P. H. Chan, C. J. Epstein, and D. M. Ferriero, *Ann. Neurol.* **44,** 357 (1998).

[10] C. Palmer, R. L. Roberts, and C. Bero, *Stroke* **25,** 1039 (1994).

[11] D. Sarco, J. Becker, C. Palmer, R. A. Sheldon, and D. M. Ferriero, *Neurosci. Lett.* **282,** 113 (2000).

[12] C. Palmer, S. L. Menzies, R. L. Roberts, G. Pavlick, and J. R. Connor, *J. Neurosci. Res.* **56,** 60 (1999).

[13] J. R. Connor, G. Pavlick, D. Karli, S. I. Menzies, and C. Palmer, *J. Comp. Neurol.* **355,** 111 (1995).

[14] J. Nguyen-Legros, J. Bizot, M. Bolesse, and J. P. Pulicani, *Histochemistry* **66,** 239 (1980).

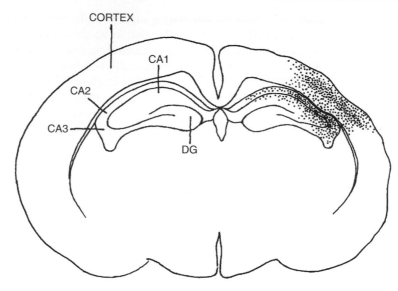

FIG. 1. Schematic representation of a coronal section of a mouse brain, showing areas of injury (dots). CA1, CA2, and CA3 are regions of hippocampus that are commonly injured in this model. DG, Dentate gyrus.

(as a whole); and CA1, CA2, CA3, and dentate gyrus of the hippocampus. Taken together, a score of 0–24 is possible (Figs. 1 and 2).

Cresyl Violet

Cresyl violet is a standard histological stain for neurons. Sections are first mounted onto gelatin-coated slides and dried overnight. Neonatal brains do not need to be delipidized, and after a rinse in H_2O slides are immersed in stain for 3–5 min. Three components are made and then mixed: 0.3 g of cresyl echt violet (Roboz) in 50 ml of H_2O, 3.48 ml of glacial acetic acid in 300 ml of H_2O, and 5.44 g of sodium acetate in 200 ml of H_2O. Slides are then rinsed twice in H_2O, differentiated in 70% (v/v) ethanol with a few drops of acetic acid, followed by dehydration in graded ethanols and two changes of xylene, and coverslipped with Depex (Biomedical Specialties, Santa Monica, CA).

Perl's Iron Stain

Perl's iron stain is enhanced with diaminobenzidine (DAB).[13] Free-floating sections are incubated for 30 min in 2% (w/v) potassium ferrocyanide (Sigma) mixed 1 : 1 with 2% (v/v) hydrochloric acid, rinsed three times in H_2O, and

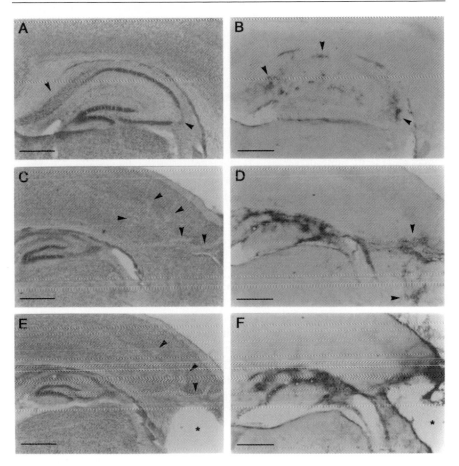

FIG. 2. Three different representative brains from SOD1-overexpressing mice showing potential injury outcome. Cresyl violet stain (A) and iron stain (B) of a brain with small focal areas of cell loss (arrowheads). Injury score, 4. Cresyl violet stain (C) and iron stain (D) of a brain with greater cell loss in the pyramidal cell layer and general shrinkage of the hippocampus and prominent injury to the cortex (arrowheads). Injury score, 16. Cresyl violet stain (E) and iron stain (F) of a severely injured brain. Note the large area of cystic infarction to the cortex (asterisk). Injury score, 23. There is relative sparing of the dentate gyrus in all three brains. Scale bars: 0.25 mm.

then reacted with DAB [20 mg/10 ml of phosphate buffer, 13.3 μl of 30% (v/v) H_2O_2]. When DAB is visibly deposited in injured areas of tissue, the reaction is stopped by rinsing the sections three times in H_2O. Sections are then mounted onto gelatin-coated slides, allowed to dry, dehydrated in graded ethanols, and cover-slipped.

Enzymatic Assays

Right and left cortices and hippocampi are dissected and flash frozen on dry ice.[15]

Glutathione Peroxidase

GPx is assayed according to Roveri et al.,[16] with minor modifications. Samples are homogenized in phosphate buffer (0.1 M potassium phosphate, 2 mM sodium azide, pH 7.0), and centrifuged at 5000g for 5 min. GPx activity is determined kinetically in duplicate samples by monitoring the decrease in absorbance of NADPH at 340 nm.[15] GPx activity is expressed as units per milligram of protein. One unit is defined as 1 nmol of NADPH consumed per minute.

Catalase

Catalase is assayed by previously described methods, again with minor modifications.[17–19] Brain samples are homogenized in 0.2 ml of cold phosphate buffer–0.1 mM EDTA–0.1% (v/v) Triton X-100 (pH 7.8, 4°), and purified by addition to glass wool columns resting in microcentrifuge tubes and centrifugation at 5000g for 5 min at 4°. Samples of cortex are diluted 1 : 10 with buffer, and hippocampus samples are diluted 1 : 5 with buffer. Catalase activity is determined kinetically in duplicate samples by monitoring the decrease in absorbance of a known amount of H_2O_2 at 240 nm.[19] Catalase activity is expressed as units per milligram of protein. One unit is defined as 1 μmol of H_2O_2 reduced per minute. Values are normalized to an internal control consisting of a homogenate of several cortices or hippocampi from naive animals.

Hydrogen Peroxide

H_2O_2 is measured indirectly by treatment of the animals with aminotriazole (Sigma), which selectively and irreversibly inhibits catalase that is bound to H_2O_2, and is therefore directly proportional to H_2O_2 concentration at the time of inhibition.[20,21] We first determined that there was no difference in brain aminotriazole levels between transgenic and wild-type mice 2 hr after treatment. Aminotriazole levels were determined by diazorization of the aminotriazole with sodium

[15] L. Flohe and W. A. Günzler, *Methods Enzymol.* **105,** 114 (1984).

[16] A. Roveri, M. Maiorino, and F. Ursini, *Methods Enzymol.* **233,** 202 (1994).

[17] S. Przedborski, V. Jackson-Lewis, V. Kostic, E. Carlson, C. J. Epstein, and J. L. Cadet, *J. Neurochem.* **58,** 1760 (1992).

[18] P. M. Sinet, R. E. Heikkila, and G. Cohen, *J.Neurochem.* **34,** 1421 (1980).

[19] H. Aebi, *Methods Enzymol.* **105,** 121 (1984).

[20] E. Margoliash, A. Novogrodsky, and A. Schejter, *Biochem. J.* **74,** 339 (1960).

[21] P. Nicholls, *Biochim. Biophys. Acta* **59,** 414 (1962).

nitrate, followed by coupling to a chromotropic acid (Sigma) to form a visible derivative that is determined by measuring absorbance at 595 nm.[18,22] We then performed a time course of catalase activity on the basis of this information and determined that there was a 50% inhibition of catalase 2 hr after injection. Thus, mice are injected intraperitoneally with aminotriazole (200 mg/kg in saline) or saline 2 hr before they are killed and brains are dissected for assay. Catalase is then assayed as described above; greater inhibition of the catalase enzyme indicates a higher H_2O_2 concentration. Results are expressed as percent inhibition of catalase activity by aminotriazole.

Total protein in homogenates is determined for all assays with bicinchoninic acid (BCA) reagents by the BCA method (Pierce, Rockford, IL).

Oxidative Stress on Neurons *in Vitro*

We used primary neuronal cell cultures to further study the mechanisms of H_2O_2 toxicity. In cortical cultures of pure wild-type neurons, we found that immature neurons (6 days *in vitro*) are selectively vulnerable to H_2O_2 exposure and mature neurons (20 days *in vitro*) are relatively resistant.[23] Cultures of hippocampal cells, which show greater injury in HI, showed the same vulnerability as immature cells, using similar methods.[24,25] We also examined the impact of free metal ions in this system, as they are known to catalyze hydroxyl radical formation from H_2O_2 via the Fenton reaction, which is a one-electron nonenzymatic transfer reaction in which transition metals generate hydroxyl radicals from H_2O_2. When cultures are pretreated with the iron chelator deferoxamine (DFO), H_2O_2-mediated hippocampal cell death is attenuated, but treatment of cells at the same time as H_2O_2 exposure is ineffective. Consequently, the broad-spectrum metal chelator N,N,N',N'-tetrakis(2-pyridylmethyl)ethylenediamine (TPEN) is used in a similar fashion to determine whether other metals are involved. TPEN added at the same time as H_2O_2 consistently reduces cell death by more than 50%.

Cell Culture Methods

The preparation of primary neuronal cultures is based on previously described methods,[26] with some modifications. Hippocampal cells are isolated from fetal (embryonic day 16, E16) CD1 mice. Briefly, hippocampi are removed from adjacent cortices and meninges and enzymatically dissociated with trypsin (2 mg/ml; Sigma) in Hanks' buffered salt solution (HBSS) for 10 min at 37°. Dissociation is

[22] F. O. Green and R. N. Feinstein, *Anal. Chem.* **29**, 1658 (1957).

[23] R. E. Mischel, Y. S. Kim, R. A. Sheldon, and D. M. Ferriero, *Neurosci. Lett.* **231**, 17 (1997).

[24] L. M. Almli, S. E. G. Hamrick, A. Koshy, M. Tauber, and D. M. Ferriero, *Dev. Br. Res.* **132**, 121 (2001).

[25] L. M. Almli, P. H. Donohoe, M. Taeuber, and D. M. Ferriero, *Soc. Neurosci. Abstr.* **25**, 1522 (1999).

[26] D. Choi, M. Maulucci-Gedde, and A. Kriegstein, *J. Neurosci.* **7**, 357 (1987).

stopped by the addition of horse serum (10%, v/v) and the cell suspension is centrifuged at low speed (190g). The cells are resuspended in astrocyte conditioned medium (ACM) supplemented with 10% (v/v) horse serum. ACM is prepared from a custom base of minimum essential medium (MEM) with Earle's salts, without L-glutamine, which is supplemented with glucose (15 mM), glutamine (2 mM), and 10% (v/v) fetal bovine serum immediately before plating on a confluent layer of astrocytes in 75-ml flasks, 24 hr before use. After gentle trituration, the cells are diluted in ACM and plated at a density of 1.4×10^6 cells/ml onto 96-well plates (Falcon; Becton Dickinson Labware, Lincoln Park, NJ) that have been coated with poly-D-lysine (5 mg/100 ml of pyrogen-free water; Sigma). Thirty minutes after plating, half of the medium is removed and replaced with fresh ACM without horse serum, reducing the concentration of horse serum to 5% (v/v). Twenty-four hours later (1 day *in vitro*), astrocyte growth is inhibited by the addition of 10 mM cytosine arabinoside, (Ara-C; Sigma). At 2 days *in vitro,* medium is replaced such that the concentration of Ara-C is reduced to 3 mM. This procedure ensures an astrocyte population of less than 5%, which is confirmed in each experiment by immunocytochemistry for astrocytes, using glial fibrillary acidic protein (GFAP) antibody (ICN, Costa Mesa, CA) and neuron-specific enolase (NSE) antibody (Dako, Carpinteria, CA) according to standard methods. The astrocyte cultures are prepared from cortices isolated from postnatal day 1 (P1) mice and processed in a manner similar to the above-described procedure, except that they are plated onto untreated 75-ml flasks without the addition of Ara-C, and allowed to grow to confluency.

Discussion

The neonatal brain in the SOD1 transgenic mouse accumulates H_2O_2 in response to HI, contributing to cell death, as H_2O_2 is known to be toxic to neurons as well as other cell types.[27] Although some adult stroke models have shown protection with increased SOD1, we have found that the immature brain is particularly vulnerable to oxidative stress produced by excess SOD1. Similarly, we have seen that immature neurons (day 6 *in vitro*) are more susceptible to injury and death from H_2O_2 exposure than mature neurons (20 days *in vitro*).[23] The mechanism of H_2O_2 toxicity is unclear, but one likely pathway is the conversion of H_2O_2 to the highly toxic hydroxyl radical in the presence of Fe^{2+} via the Fenton reaction.[28] The immature brain has relatively high levels of iron and it has been shown that there is an accumulation of iron in response to HI by as early as 4 hr.[12,29] We

[27] T. Rando and C. Epstein, *Ann. Neurol.* **46,** 135 (1999).

[28] J. A. Imlay, S. M. Chin, and S. Linn, *Science* **240,** 640 (1988).

[29] A. J. Roskams and J. R. Connor, *J. Neurochem.* **63,** 709 (1994).

have shown that the iron chelator deferoxamine protects wild-type brains from HI, but not SOD overexpressors.[11] Furthermore, our *in vitro* data with metal chelators indicate that the presence of iron and other free metal ions, such as zinc, and the subsequent generation of hydroxyl radicals contribute to neuronal death in culture.

The endogenous antioxidant enzymes glutathione peroxidase and catalase are low in neonates compared with mature animals, which may explain the increased accumulation of H_2O_2 in SOD1 overexpressors in the setting of HI, and subsequently the greater degree of injury seen in these animals. The fact that catalase is unchanged, and GPx decreases, after HI, further indicates that these enzymes are incapable of ameliorating the deleterious effects of an overproduction of H_2O_2.

SOD1 neurons in culture, taken from the same strain of mice that we have used, have been shown to be more vulnerable than wild-type neurons to the superoxide-generating compounds menadione and paraquat.[30] Like us, Ying *et al.* hypothesized that H_2O_2 production is the likely mechanism of this toxicity.

It has been shown clinically that SOD levels increase after stroke in adults, indicating that superoxide radicals are formed and a compensatory effort is made by SOD production.[31] However, SOD activity in the sera of stroke patients has been shown to be reduced and is inversely correlated with infarct size, indicating a need for replacement of antioxidants in these patients.[32] Our data indicate that babies may be even more susceptible to ischemic injury, and that injury could be exacerbated by SOD. Also, we need to understand the pathogenesis of hypoxic–ischemic injury in order to find therapies that can be administered after the initial insult, but before irreparable cellular injury has occurred. Although treatments that are administered after HI may shed light on the mechanisms of injury, they have limited clinical relevance. Targeting the downstream reactive oxygen species is a part of this goal of finding appropriate therapies.

[30] W. Ying, C. M. Anderson, Y. Chen, B. A. Stein, C. S. Fahlman, J.-C. Copin, P. H. Chan, and R. A. Swanson, *J. Cereb. Blood Flow Metab.* **20,** 359 (2000).

[31] N. Gruener, B. Gross, O. Gozlan, and M. Barak, *Life Sci.* **54,** 711 (1994).

[32] M. Spranger, S. Krempien, S. Schwab, S. Donneberg, and W. Hacke, *Stroke* **28,** 2425 (1997).

[34] Manganese Superoxide Dismutase Transgenic Mice: Characteristics and Implications

By CHING K. CHOW, HSIU-CHUAN YEN, WISSAM IBRAHIM, and DARET K. ST. CLAIR

Introduction

Aerobic organisms are protected from oxidative stress by a variety of interacting antioxidant systems under normal conditions. Among the antioxidant systems, superoxide dismutase (SOD) is considered the first line of defense against oxidative stress.[1,2] There are two types of intracellular SOD: the manganese-containing SOD (MnSOD) and the copper–zinc-containing SOD (Cu,ZnSOD), and one extracellular SOD (EC-SOD), mainly the copper–zinc form, in mammalian organs. By rapidly removing superoxide, SOD reduces the tissue concentration of superoxide and prevents the production of reactive hydroxyl radical and peroxynitrite. MnSOD is a critical antioxidant enzyme in aerobic organisms because the superoxide is mainly generated on the matrix side of the inner mitochondrial membrane where MnSOD is located.[3] The essentiality of MnSOD is evidenced by the findings that deficiency in superoxide dismutases, but not catalase, shortens the life span of yeast cells[4]; that MnSOD [but not Cu,ZnSOD, catalase, or glutathione (GSH) peroxidase] knockout mice die within the first few weeks of life[5,6]; that overexpression of Cu,ZnSOD does not prevent neonatal lethality in mutant mice that lack MnSOD[7]; and many other reports.

As a large variety of antioxidant systems are involved in protecting aerobic organs against oxidative stress, the contribution of an individual antioxidant system to the overall antioxidant defense in intact cell/animals is difficult to assess. The application of genetic manipulation to create cells/animals that are specifically enriched with or lack one or more of the antioxidant enzymes represents a physiological avenue to understand their specific roles *in situ*. A number of genetically altered mice with increased or decreased expression of important antioxidant

[1] B. Halliwell, *Free Radic. Res.* **31,** 261 (1999).
[2] Y. S. Ho, J. L. Magnenat, M. Gargano, and J. Cao, *Environ. Health Perspect.* **106,** 1219S (1998).
[3] R. Balzan, D. R. Agius, and W. H. Bannister, *Biochem. Biophys. Res. Commun.* **256,** 63 (1999).
[4] J. Wawryn, A. Krzepilko, A. Myszka, and T. Bilinski, *Acta Biochim. Pol.* **46,** 249 (1999).
[5] R. M. Libovitz, H. Zhang, H. Voguel, J. Cartwright, Jr., L. Dionne, N. Lu, S. Huang, and M. M. Martzuk, *Proc. Natl. Acad. Sci. U.S.A.* **93,** 9782 (1996).
[6] Y. Li, T. T. Huang, E. J. Carlson, S. Melov, P. C. Ursell, J. L. Olsion, L. J. Olson, L. J. Noble, M. P. Yoshimura, C. Berger, P. H. Chan, D. C. Wallace, and C. J. Epstein, *Nat. Genet.* **11,** 376 (1995).
[7] J. C. Copin, Y. Gasche, and P. H. Chan, *Free Radic. Biol. Med.* **28,** 1571 (2000).

Copyright 2002, Elsevier Science (USA).
All rights reserved.
0076-6879/02 $35.00

enzymes, MnSOD,[5,6,8-11] Cu,ZnSOD,[12-14] EC-SOD,[15-17] catalase,[18,19] and/or GSH peroxidase,[20,21] have been produced. The use of these transgenic mice and mutant to study free radical-induced oxidative damage has provided a better understanding of the role of individual antioxidant enzymes in antioxidant defense. However, over- or underexpression of a single antioxidant enzyme may shift the balance of cellular redox status and/or result in compensatory changes in other antioxidant systems. We describe the procedures employed to identify and characterize MnSOD transgenic mice used in our laboratories.

Identification and Characterization of MnSOD in Transgenic Mice

The MnSOD transgenic mice used in our laboratories were prepared from the F_1 progeny of C57BL/6 × C3H hybrids (B6C3) at the transgenic animal facility of the University of Kentucky (Lexington, KY), using standard procedures.[22] Human MnSOD cDNA under the transcriptional control of the human β-actin promoter as used for the generation of the human MnSOD transgene. The DNA was introduced into pronuclei of fertilized mouse eggs by microinjection. Mice with stable integrated human MnSOD transgenes identified by Southern analysis were selected as founders, and were propagated as heterozygous transgenic mice. Six to 8 weeks after birth, the line of transgenic mice expressing human MnSOD was bred with nontransgenic mice to produce transgenic and nontransgenic offspring. Two

[8] J. R. Wispe, B. B. Warner, J. C. Clark, C. R. Dey, J. Newman, S. W. Glasser, J. D. Crapo, L. Y. Chang, and J. A. Whitsett, *J. Biol. Chem.* **267**, 23937 (1992).

[9] H.-C. Yen, T. D. Oberley, S. Vichitbandha, Y.-S. Ho, and D. K. St. Clair, *J. Clin. Invest.* **98**, 1253 (1996).

[10] Y.-S. Ho, R. Vincent, M. S. Dey, J. W. Slot, and J. D. Crapo, *Am. J. Respir. Cell Mol. Biol.* **18**, 538 (1998).

[11] H. Van Remmen, C. Salvador, H. Yang, T. T. Huang, C. J. Epstein, and A. Richardson, *Arch. Biochem. Biophys.* **363**, 91 (1999).

[12] M. Peled-Kamar, J. Lotem, I. Wirguin, L. Weiner, A. Hermalin, and Y. Groner, *Proc. Natl. Acad. Sci. U.S.A.* **94**, 3883 (1997).

[13] J. L. Cadet, S. F. Ali, R. B. Rothman, and C. L. Epstein, *Mol. Neurobiol.* **11**, 155 (1995).

[14] C. W. White, K. B. Avraham, P. F. Shanley, and Y. Groner, *J. Clin. Invest.* **87**, 2162 (1991).

[15] M. L. Sentman, L. M. Jonsson, and S. L. Marklund, *Free Radic. Biol. Med.* **27**, 790 (1999).

[16] J. D. Oury, Y. S. Ho, C. A. Piantadosi, and J. D. Crapo, *Proc. Natl. Acad. Sci. U.S.A.* **89**, 9715 (1992).

[17] H. Sheng, M. Kudo, G. B. Mackensen, R. D. Pearlstein, J. D. Crapo, and D. S. Warner, *Exp. Neurol.* **163**, 392 (2000).

[18] V. Nilakantan, Y. Li, B. T. Spear, and H. P. Glauert, *Ann. N. Y. Acad. Sci.* **804**, 542 (1996).

[19] Y. J. Kang, Y. Chen, and P. N. Epstein, *J. Biol. Chem.* **271**, 12610 (1996).

[20] J. F. Bilodeau and M. E. Mirault, *Int. J. Cancer.* **80**, 863 (1999).

[21] M. Weisbrot-Lefkowitz, K. Reuhl, B. Perry, P. H. Chan, M. Inouye, and O. Mirochnitchenko, *Brain Res. Mol. Brain Res.* **53**, 333 (1998).

[22] B. Hogan, R. Beddington, R. Costantini, and E. Lacy, "A Laboratory Manual," 2nd Ed. p. 1. Cold Spring Harbor Laboratory Press, Cold Spring Harbor, NY, 1986.

weeks after birth, the presence of introduced human MnSOD gene in the pups was identified by Southern blot analysis, and expression of steady state mRNA from the human MnSOD gene was confirmed by Northern analysis. Translation of human MnSOD mRNA into protein was identified by Western blotting, and immunogold staining was used to determine the intracellular location of the MnSOD protein in the transgenic mice. Human MnSOD activity was verified by native activity gels, which permit the separation of MnSOD isozymes. The activities of SOD, glutathione (GSH) peroxidase, and catalase, as well as the levels of low molecular weight antioxidants, GSH, ascorbic acid, and vitamin E (principally α-tocopherol), in tissues were assessed to determine whether altered expression of MnSOD results in compensatory or adaptive changes in other important antioxidant systems.

Identification of Human MnSOD Transgene

By using a probe that yields distinct bands for both the foreign DNA and an endogenous gene, Southern blot analysis allows for specific detection of the human MnSOD transgene. After isolation from approximately 0.5-cm mouse tail biopsies, genomic DNA is digested in 0.5 ml of 1 M Tris-HCl (pH 8.5), 500 mM EDTA, 200 mM NaCl, 0.2% (w/v) sodium dodecyl sulfate (SDS), and proteinase K (100 μg/ml) at 55° overnight, and is centrifuged at 13,000g for 10 min at room temperature. The supernatant is then mixed with 0.5 ml of isopropanol and centrifuged at 13,000g for 5 min. The precipitated DNA is then washed twice with 0.5 ml of 95% (v/v) ethanol, and allowed to dry at room temperature. The resulting genomic DNA is redissolved in 100 μl of 10 mM Tris-HCl and 1 mM EDTA, pH 7.4, and stored at −20°.

Genomic DNA (6–8 μl) isolated from mouse tail is digested with restriction enzyme *Pst*I at 37° overnight. The digested DNA is then electrophoresed on a 0.8% (w/v) agarose gel. The gels are then denatured, neutralized, and transferred to a nitrocellulose membrane (Nytran paper; Schleicher & Schuell, Keene, NH). The membrane is air dried, baked at 80° for 2 hr in a vacuum oven, and prehybridized for 5–10 hr at 42° in hybridization buffer [5× saline–sodium phosphate–EDTA buffer (SSPE), 50% (v/v) formamide, 5× Denhardt's solution, 0.1% (w/v) SDS, and denatured salmon sperm DNA (0.1 mg/ml)]. After hybridization with the [^{32}P]dCTP-labeled human MnSOD cDNA probe for 24–48 hr at 42° in hybridization buffer, the membrane is washed twice with membrane wash solution [5× saline–sodium citrate buffer (SSC), 0.1% (w/v) SDS, and 0.05% (w/v) sodium pyrophosphate] for 15 min at room temperature and once with 0.1× SSC buffer, 0.1% (w/v) SDS, and 0.05% (w/v) sodium pyrophosphate at 65°. The membrane is then air dried and exposed to Kodak (Rochester, NY) XAR film at −80°. Figure 1 shows a Southern analysis of the human MnSOD gene isolated from tails of our transgenic founder mice. The copy number of the human MnSOD gene in each founder mouse varies significantly.

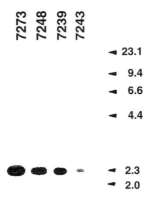

FIG. 1. Southern analysis of human MnSOD gene in transgenic mice. Five micrograms of genomic DNA from the tail of each transgenic mouse was digested with restriction enzyme *Pst*I. After digestion, the DNA was separated on a 0.8% (w/v) agarose gel, transferred to nitrocellulose, and probed with ^{32}P-labeled human MnSOD cDNA.

Alternatively, the presence of the human MnSOD transgene in the mouse genome can be detected by polymerase chain reaction (PCR). In this case the same genomic DNA used for Southern analysis can be used in the PCR. To assure that a negative result is not due to failure in the amplification reactions, the mouse MnSOD gene is also amplified from the same DNA sample. PCR is carried out in a 50-μl reaction mixture containing 20 mM Tris-HCl (pH 8.8), 2 mM MgSO$_4$, 10 mM KCl, 10 mM (NH$_4$)$_2$SO$_4$, 0.1% (v/v) Triton X-100, nuclease-free bovine serum albumin (BSA, 1 mg/ml), 10 mM each of dATP, dCTP, dGTP, and dTTP, 0.2–0.5 μl of *Taq* DNA polymerase, and 5 μl of genomic DNA. The oligonucleotide sequences of the primer sets are as follows:

Human MnSOD (positions 93–280):
5′-AGC ATG TTG AGC CGG GCA GT-3′
5′-AGG TTG TTC ACG TAG GCC GC-3′
Mouse MnSOD (positions 8–280):
5′-AAC GGC CGT GTT CTG AGG AG-3′
5′-AGG TGG TTC ACG TAG GCC GC-3′

Each PCR product of MnSOD is then subjected to thermal cycling. Human MnSOD PCR product is initially denatured at 94° for 2 min, followed by 29 cycles (30 sec of denaturation at 94°, 15 sec of annealing at 65°, 20 sec of extension at 72°), and finished with a final extension at 72° for 7 min. Mouse MnSOD PCR product is initially denatured at 94° for 2 min, followed by 34 cycles (3 sec of denaturation at 94°, 30 sec of annealing at 60°, 40 sec of extension at 72°), and finished with a final extension at 72° for 7 min. The PCR products are then analyzed on a 1% (w/v) agarose gel in Tris–acetate buffer with ethidium bromide staining. The products of human MnSOD PCR and mouse MnSOD PCR are 187 and 272 bp, respectively.

Southern blot analysis screening of MnSOD transgenic mice is preferred over PCR analysis because the presence of bands of the predicted size(s) on a Southern blot clearly indicates a positive transgenic mouse, and under appropriate conditions less than one copy per cell of gene can be detected, whereas PCR can easily give a false-negative result. Also, contamination of a mouse DNA sample with plasmid DNA is likely to produce bands that differ in size from the expected bands on a Southern blot, while contaminating plasmid could cause a false positive on PCR. In addition, a Southern blot provides immediate information about the structure and integrity of the inserted DNA sequences. However, Southern blot analysis is subject to errors due to such factors as nonuniform transfer of DNA to the filter and incomplete digestion with the restriction enzyme. The choice of restriction enzyme and hybridization probe for the analysis of the integrated DNA is important.

Northern Analysis of MnSOD mRNA

The expression of steady state mRNA from the human MnSOD gene in various issues of transgenic mice can be demonstrated by Northern analysis. The levels of human MnSOD mRNA in tissue are detected by Northern analysis after isolation of total RNA by the guanidine isothiocyanate method.[9] Northern analysis of MnSOD mRNA is straightforward and hybridization conditions used in Southern analysis can be applied here. The most critical factor in the success of mRNA analysis is the quality of RNA used. Although many commercial kits for RNA isolation can be used, guanidine isothiocyanate with cesium chloride centrifugation as described by Chirgwin et al.[23] works best for tissues with a high level of ribonuclease. Figure 2 shows Northern analysis of MnSOD mRNA isolated from the skin of transgenic and control mice.

Western Blot and Activity Gel for MnSOD

Western blot allows for direct detection of MnSOD via its recognition by specific antibody raised against the purified protein. Standard conditions for electrophoresis and transfer of proteins are used in our study. However, it is important to note that the specificity of the antibody is important for this assay. The antibody raised against human MnSOD used in our work was kindly provided by L. Oberley at the University of Iowa (Iowa City, IA). Although the antibody also recognizes mouse MnSOD, it consistently provides a single shape band of human MnSOD protein on a blot. To visualize the activity of the MnSOD protein on the

[23] J. M. Chirgwin, A. E. Przybyla, R. J. MacDonald, and W. J. Rutter, Biochemistry 18, 939 (1979).

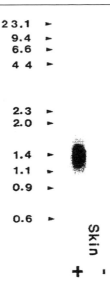

2 3.1 ►
9.4 ►
6.6 ►
4 4 ►

2.3 ►
2.0 ►

1.4 ►
1.1 ►
0.9 ►

0.6 ►

Skin

+ ─

FIG. 2. Northern analysis of MnSOD mRNA in transgenic mice. Poly(A)$^+$ RNA from the skin of control (−) and transgenic (+) mice was extracted on a 1.1% (v/v) formaldehyde–agarose gel, transferred to nitrocellulose, and probed with ^{32}P-labeled human MnSOD cDNA.

gels, an SOD activity gel assay is performed according to the method described by Beauchamp and Fridovich.[24] Tissue homogenate is prepared in 50 mM potassium phosphate buffer, pH 7.8. Two hundred micrograms of protein per lane is electrophoresed through a nondissociating riboflavin–polyacrylamide gel consisting of a 5% (w/v) stacking gel, pH 6.8, and a 10% (w/v) running gel, pH 8.8, at 4°. To visualize SOD activity, gels are first incubated in 2.43 mM nitroblue tetrazolium in deionized water for 15 min and then in 0.028 mM riboflavin and 280 mM N,N,N',N'-tetramethyl ethylenediamine in 50 mM potassium phosphate buffer, pH 7.8, for 15 min in the dark. Gels are then washed with deionized water and illuminated under fluorescent light until clear zones of SOD activity are distinctly evident. The activity gel assay, which facilitates the identification of SOD isoenzymes, demonstrates the presence of human MnSOD activity, and is the only method that permits distinction between MnSOD activity derived from the transgene and endogenous mouse MnSOD.

Immunogold Staining of MnSOD

Tissues are cut into 1-mm^3 blocks, fixed in Carson's modified Millonig fixative [4% (v/v) formaldehyde in 0.16 M sodium phosphate buffer, pH 7.2], and processed

[24] C. Beauchamp and I. Fridovich, *Anal. Biochem.* **44**, 276 (1971).

for immunogold electron microscope localization of MnSOD in the mitochondria as previously described.[9] With the exception of the antibody used, the procedures are identical to those described in detail for the detection of 4-hydroxy-2-nonenal by T. D. Oberley *et al.*[24a]

Assessment of Antioxidant Status

In addition to MnSOD, the activities of Cu,ZnSOD, GSH peroxidases, and catalase, and levels of important low molecular weight antioxidants, i.e., ascorbic acid, GSH, and vitamin E (α-tocopherol), are measured to determine whether altered expression of MnSOD results in compensatory or adaptive changes in antioxidant status.

Immediately after sacrifice, tissues are removed, trimmed, and homogenized with 1.15% (w/v) KCl in 0.05 M phosphate buffer, pH 7.4. Portions of the homogenate, after centrifugation at 400g for 5 min to remove cell debris, are used to measure SOD activity by monitoring the inhibition of nitroblue tetrazonium–bathocuproline sulfonate reduction, according to the method of Spitz and Oberley.[25] The activity of MnSOD is assayed in the presence of 5 mM NaCN, which inhibits Cu,ZnSOD, at 560 nm in 0.05 M potassium phosphate buffer, pH 7.8. NaCN is incubated with the reaction mixture for 30 min before adding xanthine oxidase. In the absence of NaCN, total SOD activity is measured, and the activity of Cu,ZnSOD is calculated. The activity of SOD is expressed as units per minute per milligram protein, and 1 unit is defined as the amount of enzyme required to inhibit nitroblue tetrazonium reduction at 50% maximum activity. Catalase activity in 9000g supernatant is measured spectrophotometrically at 240 nm, using hydrogen peroxide as substrate in 0.05 M potassium phosphate buffer, pH 7.0.[26] The activity of catalase is expressed as micromoles of hydrogen peroxide reduced per minute per milligram protein. The activities of GSH peroxidases in 9000g supernatant are measured by monitoring NADPH oxidation at 340 nm, using hydrogen peroxide or cumene hydroperoxide as substrate in 0.05 M Tris-HCl, pH 7.6.[27] The use of cumene hydroperoxide as substrate allows for measuring total GSH peroxidase, whereas Se-GSH peroxidase activity is measured when hydrogen peroxide is used as substrate. The activity of the enzyme is expressed as nanomoles of NADPH oxidized per minute per milligram protein. The activities of MnSOD, Cu,ZnSOD, catalase, and Se-GSH peroxidase in the heart and skeletal muscle of MnSOD transgenic and control mice are shown in Fig. 3.

Measurements of GSH peroxidases and catalase are relatively simple and straightforward, and are not normally subject to methodological errors. On the

---

[24a] T. D. Oberley, S. Toyokuni, and L. I. Szweda, *Free Rad. Biol. Med.* **27,** 695 (1999).

[25] D. R. Spitz and L. W. Oberley, *Anal. Biochem.* **195,** 133 (1989).

[26] R. F. Beers and J. W. Sizer, *J. Biol. Chem.* **195,** 133 (1952).

[27] R. A. Lawrence and R. F. Burk, *Biochem. Biophys. Res. Commun.* **71,** 952 (1976).

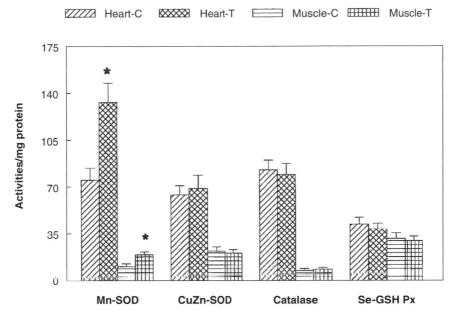

FIG. 3. Activities of antioxidant enzymes in the heart and skeletal muscle of 12-week-old male MnSOD transgenic (T) and control (C) mice. The activity of SOD is expressed as units per minute per milligram protein, that of catalase as micromoles per minute per milligram protein, and that of Se-GSH peroxidase (Px) as nanomoles per minute per milligram protein. The data are shown as means ± standard deviation ($n = 6$). An asterisk denotes a significant difference ($p < 0.01$) between the mean values of transgenic and control mice.

other hand, methodological inaccuracies in SOD measurements can easily result if optimal conditions for inhibiting Cu,ZnSOD activity by NaCN as well as the proper amount of enzyme preparation to be used are not established on a tissue-by-tissue basis before measurement.[25]

Freshly prepared tissue homogenate is used to measure low molecular weight antioxidants ascorbic acid, GSH, and vitamin E (α-tocopherol). After acid precipitation of protein, the level of ascorbic acid in the supernatant is measured after reaction with 2,4-dinitrophenylhydrazine at 515 nm,[28] and that of GSH is measured after reaction with 5,5'-dithiobis(2-nitrobenzoic acid) at 412 nm.[29] After extraction of lipids with hexane, the levels of α-tocopherol are measured by high-performance liquid chromatography (HPLC), using a C18 reversed phase column with fluorescence detection.[30] The spectrophotometric measurement of GSH, ascorbic acid, and α-tocopherol is not as specific as the HPLC procedure. However, it allows

[28] S. T. Omaye, J. D. Turnbull, and H. E. Sauberich, *Methods Enzymol.* **62,** 3 (1979).

[29] J. Sedlack and R. H. Lindsay, *Anal. Biochem.* **25,** 192 (1968).

[30] L. J. Hatam and H. J. Kayden, *J. Lipid Res.* **20,** 639 (1979).

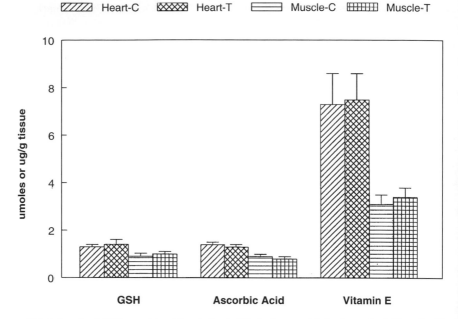

FIG. 4. Levels of GSH, ascorbic acid, and vitamin E in the heart and skeletal muscle of 12-week-old male MnSOD transgenic (T) and control (C) mice. The levels of GSH and ascorbic acid are expressed as micromoles per gram tissue, and that of vitamin E (α-tocopherol) as micrograms per gram tissue. The data are shown as means $\pm$ standard deviation ($n = 6$).

for speedy completion of these assays, which is essential to minimize the oxidation of these compounds during sample processing. Measurement of α-tocopherol by HPLC with fluorescence detection is simple and specific, and is much more sensitive than UV detection. The levels of GSH, ascorbic acid, and vitamin E (α-tocopherol) in the heart and skeletal muscle of MnSOD transgenic and control mice are shown in Fig. 4.

Implications

Free radical-induced oxidative damage may occur when antioxidant potential is decreased and/or when oxidative stress is increased.[31] As a critical antioxidant enzyme in aerobic organisms, increased expression of MnSOD gene is expected to provide increased protection against oxidative stress. Indeed, intratracheal injection of adenovirus containing the human MnSOD gene protects athymic nude mice from irradiation-induced organizing alveolitis,[32] and transgenic mice

---

[31] C. K. Chow, *Am. J. Clin. Nutr.* **32,** 1066 (1979).

[32] M. W. Epperly, J. A. Bray, S. Krager, L. M. Berry, W. Gooding, J. F. Engelhardt, R. Zwacka, E. L. Travis, and J. S. Greenberger, *Int. J. Radiat. Oncol. Biol. Phys.* **43,** 169 (1999).

expressing MnSOD in pulmonary epithelial cells are more resistant to oxygen toxicity.[8] Also, transgenic mice expressing human MnSOD in the heart are protected against doxorubicin (Adriamycin)-induced cardiac toxicity[9] and myocardial ischemia–reperfusion injury.[33] In addition, expression of human MnSOD gene prevents neural apoptosis and reduces ischemic brain injury (suppression of peroxynitrite production, lipid peroxidation, and mitochondrial dysfunction)[34] and attenuates 1-methyl-4-phenyl-1,2,5,6-tetrahydropyridine (MPTP) toxicity[35] in MnSOD transgenic mice. On the other hand, MnSOD knockout mice on a CD1 (outbred) genetic background die within the first 10 days of life with a complex phenotype that includes dilated cardiomyopathy, accumulation of lipid in liver and skeletal muscle, metabolic acidosis and ketosis, and a severe reduction in succinate dehydrogenase (complex II) and aconitase activities in the heart and, to a lesser extent, in other organs.[6,36] Also, heterozygous MnSOD knockout mice with a 50% decrease in MnSOD activity, and no change in GSH peroxidase or Cu,ZnSOD activities, survive hyperoxia as well as their normal littermates, and do not develop any ultrastructural abnormality in the myocardium after 100% oxygen exposure for 90 hr.[36,37] However, these heterozygous MnSOD knockout mice have increased hepatic mitochondrial oxidative damage and altered mitochondrial function. These findings indicate that MnSOD is required to maintain the integrity of mitochondrial enzymes susceptible to direct inactivation by superoxide, and that transgenic mice expressing the human MnSOD gene are protected under oxidative stress-mediated pathogenic conditions.

Induction of antioxidant enzymes, including MnSOD, in mammalian cells and tissues is generally accompanied by an increased tolerance to environmental agents that cause oxidative stress.[38,39] However, as hydrogen peroxide is the dismutation product of SOD, it has been hypothesized that increased expression of MnSOD may enhance mitochondrial production of hydrogen peroxide and increase oxidative damage. This concern is promoted by the findings that overexpression of Cu,ZnSOD is associated with the rapid aging feature of the brains of patients with Down's syndrome,[40,41] and that the pathogenic mechanisms in the motor

[33] Z. Chen, B. Siu, Y. S. Ho, R. Vincent, C. C. Chua, R. C. Hamdy, and B. H. Chua, *J. Mol. Cell Cardiol.* **30**, 2281 (1998).

[34] J. N. Keller, M. S. Kindy, F. W. Holtsberg, D. K. St. Clair, H. C. Yen, A. Germeyer, S. M. Steiner, A. J. Bruce-Keller, J. B. Hutchins, and M. P. Mattson, *J. Neurosci.* **18**, 687 (1998).

[35] P. Klivenyi, D. K. St. Clair, M. Wermer, H. C. Yen, T. Oberley, L. Yang, and M. Flint-Beal, *Neurobiol. Dis.* **5**, 253 (1998).

[36] M. D. Williams, H. Van Remmen, C. C. Conrad, T. T. Huang, C. J. Epstein, and A. Richardson, *J. Biol. Chem.* **273**, 28510 (1998).

[37] M. F. Tsan, J. E. White, B. Caska, C. J. Epstein, and C. Y. Lee, *Am. J. Cell Mol. Biol.* **19**, 114 (1998).

[38] C. K. Chow, *Nature* (*London*) **260**, 721 (1976).

[39] L. B. Clerch and D. Massaro, *J. Clin. Invest.* **91**, 499 (1993).

[40] O. Elroy-Stein and Y. Groner, *Cell* **52**, 259 (1988).

[41] I. Ceballen-Picot, A. Nicole, P. Briand, G. Grimber, A. Delacourte, A. Defossez, F. Javoy-Agid, M. Lafon, J. L. Blouin, and P. M. Sinet, *Brain Res.* **552**, 198 (1991).

neuron disease familial amyotrophic lateral sclerosis are associated with mutation of Cu,ZnSOD.[42,43] An enhancement of free radical formation due to a decrease in K_m for hydrogen peroxide is linked to a gain-of-function of a familial amyotrophic lateral sclerosis-associated Cu,ZnSOD mutant. The reason for the contrasting effects of expression between the MnSOD gene and the Cu,ZnSOD gene is not clear. However, it appears that where oxygen radical is generated and the enzyme present may be critical in determining whether the expression of SOD is beneficial or detrimental.[3]

In the transgenic mice employed in our studies,[9,34,35] heart and skeletal muscle exhibited the most significant amount of human MnSOD expression of all organs. Also, lung and brain had appreciable amounts of expression, whereas the expression in kidney and liver was limited.[44] The activity of Cu,ZnSOD was significantly higher in kidney, but in other tissues analyzed, whereas catalase activity was significantly lower in brain and liver of transgenic mice than in those of their nontransgenic littermates. The activities of Se-GSH peroxidase and non-Se-GSH peroxidase were not significantly different among transgenic and nontransgenic mice in all tissues analyzed. Also, increased expression of human MnSOD genes did not significantly alter the levels of low molecular weight antioxidants GSH, ascorbic acid, and α-tocopherol in all tissues measured.

In agreement with the findings that transgenic mice expressing MnSOD are resistant to free radical-induced tissue injury,[9,34,35] transgenic mice had significantly lower levels of lipid peroxidation product, mainly malondialdehyde, in the heart and muscle than did their nontransgenic littermates.[44] The levels of conjugated dienes and protein carbonyls were not significantly different in tissues that overexpress the human MnSOD gene (heart, muscle, brain, and lung) or in nonoverexpressed tissues (liver and kidney) between transgenic and nontransgenic mice. These findings suggest that the activities of GSH peroxidase and catalase present in various tissues of MnSOD transgenic mice are capable of handling the hydrogen peroxide generated and that small changes in the activities of Cu,ZnSOD and catalase in various tissues of MnSOD transgenic mice do not adversely affect their ability to handle hydrogen peroxide generated or influence the oxidative stress status. It also suggests that expression of the human MnSOD gene provides protection against peroxidative damage to membrane lipids, which may contribute to the increased resistance of MnSOD transgenic mice to the toxic effects of environmental agents. Thus, MnSOD transgenic mice, along with other transgenic mice expressing higher or lower levels of expression of antioxidant enzyme(s), should

[42] M. E. Gurney, F. B. Cutting, P. Zhai, P. K. Andrus, and E. D. Hall, *Pathol. Biol.* **44,** 51 (1996).

[43] M. B. Yim, J. H. Kang, H. S. Yim, H. S. Kwak, P. B. Chock, and E. R. Stadtman, *Proc. Natl. Acad. Sci. U.S.A.* **93,** 5709 (1996).

[44] W. Ibrahim, U.-S. Lee, H.-C. Yen, D. K. St. Clair, and C. K. Chow, *Free Radic. Biol. Med.* **28,** 397 (2000).

be a useful tool for unraveling the identity of reactive oxygen species that cause or promote the pathogenesis of various degenerative disorders and for defining the role of each antioxidant enzyme in cellular defense against free radical-mediated tissue injury.

Acknowledgments

Supported in part by NIH Grant CA80152.

[35] Tissue-Specific Knockout Model for Study of Mitochondrial DNA Mutation Disorders

By Aleksandra Trifunovic and Nils-Göran Larsson

Introduction

Respiratory chain dysfunction is increasingly recognized as an important cause of organ failure in human pathology.[1,2] The biogenesis of the respiratory chain is unique in its bipartite dependence on both nuclear and mitochondrial DNA (mtDNA)-encoded genes (Fig. 1). The mtDNA encodes only 13 of the $\sim$100 respiratory chain subunits; however, the mtDNA-encoded subunits are key components absolutely required for a functional respiratory chain.[1] A large number of genetic syndromes with respiratory chain dysfunction due to mutations of nuclear- or mtDNA-encoded genes have been described.[1] Abundant circumstantial evidence also associates mitochondrial dysfunction with common diseases, such as heart failure, diabetes mellitus, and neurodegeneration, and the naturally occurring process of aging.[2] Mitochondria are not only cellular energy factories but also generate most of the cellular reactive oxygen species (ROS) and perform key regulatory steps in apoptosis signaling. The molecular connection between respiratory chain dysfunction, ROS production, and apoptosis induction is unclear at present and in-depth understanding of these processes will require studies of model organisms, preferably transgenic mice.

Several multisystem disorders of humans are caused by mutations of mtDNA that interfere with the abundance or function of one or several transfer RNA (tRNA) molecules and thus impair mitochondrial translation (Fig. 1).[2] Large-scale deletions of mtDNA (ΔmtDNA) result in the lack of several tRNAs, stalled translation, and severe respiratory chain deficiency.[2] The phenomenon of heteroplasmy,

---

[1] N. G. Larsson and D. A. Clayton, *Annu. Rev. Genet.* **29**, 151 (1995).
[2] N. G. Larsson and R. Luft, *FEBS Lett.* **455**, 199 (1999).

Copyright 2002, Elsevier Science (USA).
All rights reserved.
0076-6879/02 $35.00

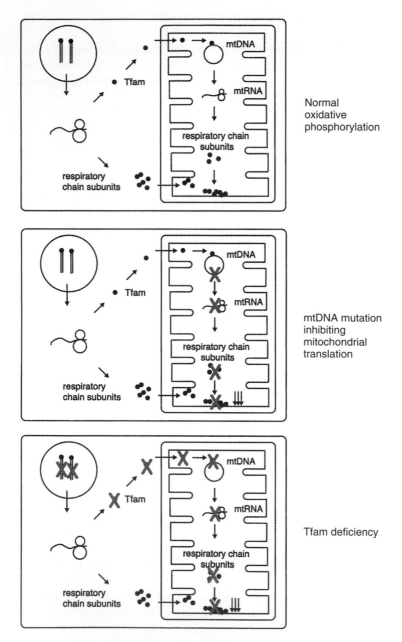

Normal
oxidative
phosphorylation

mtDNA mutation
inhibiting
mitochondrial
translation

Tfam deficiency

FIG. 1. The biogenesis of the respiratory chain. The biogenesis of the respiratory chain is dependent on subunits encoded by both nuclear and mtDNA genes (*top*). Pathogenic mutations of mtDNA often affect transfer RNA genes and impair mitochondrial translation. This results in impaired synthesis of all mtDNA-encoded respiratory chain subunits and a severe respiratory chain dysfunction (*middle*). Homozygous knockout of *Tfam* results in loss of mtDNA and loss of mitochondrial transcripts. This will abolish mitochondrial protein synthesis and result in a severe respiratory chain dysfunction (*bottom*).

whereby normal and mutated mtDNAs coexist within the same cell, creates a mosaic pattern of respiratory chain deficiency due to different levels of mutated mtDNA in different cells of the affected organs. There is evidence suggesting that the distribution of mutated mtDNA is an important determinant of the organ-specific manifestations and that accumulation of mutated mtDNA with time may explain the progressive deterioration of respiratory function in postmitotic organs, for example, brain and heart, of affected patients.

Maintenance and expression of mtDNA are completely dependent on nuclear genes and it is therefore possible to produce a global reduction of mtDNA expression, similar to the reduction observed in patients with mtDNA mutations, by disruption of nuclear genes. We have demonstrated that important pathophysiology associated with mtDNA mutations indeed can be reproduced by disrupting the nuclear *Tfam* gene, which encodes a transcriptional activator that is imported to mitochondria (Fig. 1).[3] The Tfam protein specifically binds mtDNA promoters and activates transcription. Tfam has the ability to bend and unwind DNA and may activate transcription by facilitating binding of mitochondrial RNA polymerase and other factors to the mtDNA promoters. Mitochondrial transcription is not only necessary for gene expression but also for mtDNA replication by providing the RNA primers necessary for initiation of mtDNA replication by mitochondrial DNA polymerase. Transcription is thus a prerequisite for mtDNA replication.

Tfam is absolutely required for mtDNA maintenance *in vivo*, and homozygous germ line *Tfam* knockouts lack mtDNA and die during embryogenesis.[3] Characterization of tissue-specific *Tfam* knockouts has demonstrated that Tfam protein depletion leads to a downregulation of mtDNA copy number, reduced levels of mitochondrial transcripts, and severe respiratory chain deficiency (Fig. 1).[4–6] Interestingly, the phenotype of tissue-specific *Tfam* knockouts faithfully reproduces pathology found in humans with ΔmtDNA disorders, for example, dilated cardiomyopathy with atrioventricular conduction blocks and mitochondrial diabetes.[4–6] It is thus likely that impaired mtDNA expression is a key pathogenesis feature of ΔmtDNA disorders and that the distribution of ΔmtDNA and, as a consequence, the distribution of the respiratory chain deficiency is the main determinant of the phenotype. Work from our laboratory suggests that reduced mtDNA expression increases ROS production and induces apoptosis, but the involved molecular pathways remain to be elucidated.[7]

[3] N. G. Larsson, J. Wang, H. Wilhelmsson, A. Oldfors, P. Rustin, M. Lewandoski, G. S. Barsh, and D. A. Clayton, *Nat. Genet.* **18**, 231 (1998).

[4] J. Wang, H. Wilhelmsson, C. Graff, H. Li, A. Oldfors, P. Rustin, J. C. Brüning, C. R. Kahn, D. A. Clayton, G. S. Barsh, P. Thoren, and N. G. Larsson, *Nat. Genet.* **21**, 133 (1999).

[5] H. Li, J. Wang, H. Wilhelmsson, A. Hansson, P. Thoren, J. Duffy, P. Rustin, and N. G. Larsson, *Proc. Natl. Acad. Sci. U.S.A.* **97**, 3467 (2000).

[6] J. P. Silva, M. Kohler, C. Graff, A. Oldfors, M. A. Magnuson, P. O. Berggren, and N. G. Larsson, *Nat. Genet.* **26**, 336 (2000).

[7] J. Wang, J. Silva, C. M. Gustafsson, P. Rustin, and N. G. Larsson, *Proc. Natl. Acad. Sci. U.S.A.* **98**, 4038 (2001).

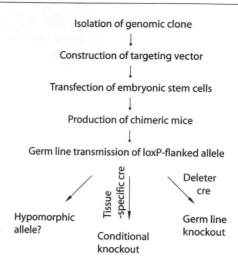

FIG. 2. Strategy for generation of conditional knockout animals. When making a conditional knock-out mouse strain it is possible to obtain a series of different alleles. The first possibility is that the insertion of *loxP* sites may result in a hypomorphic allele with reduced expression of the targeted gene. It is usually preferable to avoid creating a hypomorphic allele because this may produce a phenotype in the homozygous form and thus prevent a clear interpretation of the observed phenotype in the tissue-specific knockouts. Mating of mice with the *loxP*-flanked allele to mice with ubiquitous expression of *cre* (deleter-*cre* mice) will result in a germ line deletion. Tissue-specific knockout mice are obtained by mating mice with the *loxP*-flanked allele with mice with tissue-specific expression of Cre recombinase.

Strategy for Tissue-Specific Knockout of *Tfam*

The *cre–loxP* recombination system of bacteriophage P1 has been exploited to establish a conditional knockout system in the mouse that allows tissue-specific disruption of genes.[8] This simple system is based on the insertion of short DNA sequences, *loxP* sequences, flanking the exons to be removed. The insertion of *loxP* sequences should ideally not affect the function of the gene. The two *loxP* sequences will recombine in the presence of Cre recombinase, thus deleting the intervening sequence.

The strategy we use for conditional knockout of *Tfam* is outlined in Figs. 2 and 3. There are two main considerations when constructing the targeting vector. The first is to ensure that the insertion of *loxP* sites and the neomycin resistance gene does not result in a hypomorphic allele, that is, a *Tfam^{loxP}* allele that has reduced Tfam protein expression. The second consideration is to ensure that *cre*-mediated disruption of the *Tfam^{loxP}* allele results in a null allele without residual Tfam protein expression.

[8] H. Gu, J. D. Marth, P. C. Orban, H. Mossmann, and K. Rajewsky, *Science* **265,** 103 (1994).

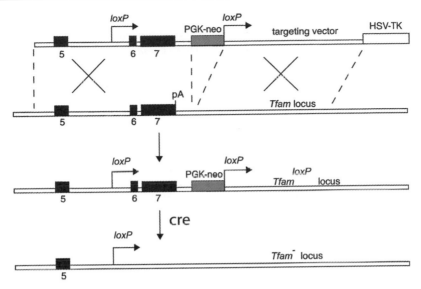

FIG. 3. The strategy used for introduction of *loxP* sites by gene targeting of the *Tfam* locus. The targeting vector is composed of homologous sequences corresponding to the three 3′ exons of *Tfam*, a positive selection marker (PGK-*neo*) and a negative selection marker (HSV-*TK*). Homologous recombination events will result in creation of the *Tfam*^{loxP} locus, where the last two exons of *Tfam* are flanked by *loxP* sites. The knockout allele *Tfam*⁻ is created by *cre* expression, which deletes the two exons flanked by *loxP* sites.

A convenient strategy for disruption of *Tfam* is to insert *loxP* sites flanking the two most 3′ exons, that is, exons 6 and 7, because the corresponding part of the protein is absolutely required for binding of Tfam to the mitochondrial promoters whereby mitochondrial transcription is activated. The neomycin resistance gene with expression driven by the phosphoglycerate kinase promoter (*PGK–neo*) was inserted downstream of the polyadenylation site of the *Tfam* gene. The location of the neomycin resistance gene is of critical importance. It has been observed that insertion of this gene in an intron may result in a hypomorphic allele, possibly because of the presence of cryptic splice sites.[9] It is therefore advisable to remove the neomycin resistance gene if it has been inserted in an intron. This can be achieved by inserting three *loxP* sites and removing *neo* by partial *cre*-mediated recombination in embryonic stem (ES) cells, thus producing an allele where the two remaining *loxP* sites flank exons to be deleted at a later stage.[8] An alternative approach is to utilize the *flp–frt* system and flank the gene with *frt* sites, thus allowing *flp*-mediated germ line excision of the neomycin resistance gene.[9]

[9] E. N. Meyers, M. Lewandoski, and G. R. Martin, *Nat. Genet.* **18,** 136 (1998).

The *cre*-mediated disruption of the $Tfam^{loxP}$ allele (Fig. 3) will remove exons 6 and 7 of *Tfam* as well as *PGK–neo*. We have published data demonstrating that animals homozygous for the $Tfam^{loxP}$ allele ($Tfam^{loxP}/Tfam^{loxP}$) have normal Tfam expression, normal mtDNA copy number, normal mtRNA levels, and normal respiratory chain function.[4] This demonstrates that the insertion of *loxP* sites and *PGK–neo* does not result in a hypomorphic *Tfam* allele. This is a prerequisite for maintaining $Tfam^{loxP}$ animals as homozygous stocks and for setting up matings to produce tissue-specific knockouts.

Construction of Targeting Vector for Conditional Gene Inactivation

Every targeting (or replacement) vector must have some fundamental properties that enable recombination with a specific chromosomal locus. The basic components of such a vector are sequences that are homologous with the chromosomal locus to be targeted and a plasmid vector backbone.

The homologous sequence in the vector should be derived from genomic libraries of the same mouse strain as that of the ES cells, usually the 129 mouse strain, because subtle differences in sequence between different mouse strains can impair homologous recombination.[10] The ideal length of homologous sequences in the targeting vector is 5–10 kb. There is a general rule stating that the greater the length of homology the higher the targeting frequency; however, large inserted sequences will make it difficult to handle the targeting vector and also limit the choice of unique restriction sites.[11] There must be a unique restriction enzyme site located outside the region of homology to allow linearization of the targeting vector before transfection of ES cells. The positive and negative selection markers are also important components of the targeting vector. The purpose of the positive selection marker, usually *neo,* is to isolate stably transfected ES cell clones. The negative selection marker, herpes simplex virus thymidine kinase (HSV-*TK*), is used to enrich for homologous recombination events over random integration. ES cell clones with random integration of the targeting vector express HSV-*TK* and will be killed by ganciclovir, whereas HSV-*TK* is lost in clones that have undergone homologous recombination.

It is also crucial to design a screening strategy to identify ES cell clones produced by homologous recombination. We think that Southern blot analysis is more robust for this purpose than polymerase chain reaction (PCR) analysis. There are two requirements that must be fulfilled to allow screening by Southern blot analysis. First, it is necessary to identify a probe outside the region of homology

[10] L. M. Silver, "Mouse Genetics: Concepts and Application." Oxford University Press, New York, 1995.
[11] M. R. Capecchi, *Science* **244,** 1288 (1989).

included in the targeting vector. Second, the targeting vector should preferably introduce a novel restriction enzyme site.

Screening of Lambda FIX II Mouse Genomic Library

The first step in making the targeting vector is to identify mouse genomic clones containing the locus to be targeted. Most ES cell lines are from the 129 mouse and we therefore used a 129 mouse genomic library for cloning (Lambda FIX II library; Stratagene, La Jolla, CA).

Approximately 1×10^5 phages are mixed with 1 ml of a culture (OD 0.5) of XL1-Blue MRA(P2) bacteria and 8 ml of melted 0.7% (w/v) agarose. This mixture is plated on one 20×20 cm LB (Luria broth) plate and incubated at $37°$ overnight. Eight plates are prepared for one screening. Plate lifts are performed with Hybond N membranes (Amersham Life Science, Buckinghamshire, UK) before the plaques reach confluency. The membranes are denatured for 1 min on trays containing a 3MM paper (Whatman, Clifton, NJ) soaked in 1.5 M NaCl; 0.5 M NaOH. Filters are then neutralized for 1 min on 3MM papers soaked in 1.5 M NaCl–0.5 M Tris-HCl, pH 8.0, and washed for at least 5 min on 3MM papers soaked in 0.2 M Tris-HCl (pH 7.5)–2× saline–sodium citrate (SSC). Finally, the membranes are dried in air for 1–2 hr and baked in a vacuum oven at $80°$ for 2 hr.

Southern blot analyses are carried out by standard protocols and hybridized with a radiolabeled *Tfam* cDNA probe.[12] Positive plaques are identified by autoradiography and films are aligned with plates on a light board. Positive plaques are removed from the plates by using cut 1-ml pipette tips. Each plaque is transferred to 0.5 ml of SM buffer [50 mM Tris-HCl (pH 7.5), 100 mM NaCl, 10 mM MgSO$_4$, and 0.1% (w/v) gelatin] and incubated at room temperature for >1 hr to allow phages to diffuse from the agar to the SM buffer. For the second screening, 10 μl of phage solution is diluted in 1 ml of SM buffer and 3 μl of this diluted phage stock is mixed with 200 μl of bacterial culture (OD 0.5) and plated on 10 × 10 cm plates. Positive clones are isolated and the phage solution is diluted for the third round of screening, performed like the first two screenings except that round 10-cm bacterial plates are used.

Phage DNA Isolation

Obtaining large quantities of high-quality phage DNA is necessary to subclone a gene of interest and thus make a targeting vector. A single phage plaque is placed in a 15-ml tube (Falcon; Becton Dickinson Labware, Lincoln Park, NJ) containing 300 μl of adsorption buffer (10 mM MgCl$_2$, 10 mM CaCl$_2$) and 200 μl

---

[12] J. Sambrook, E. F. Fritsch, and T. Maniatis (eds.), "Molecular Cloning: A Laboratory Manual." Cold Spring Harbor Laboratory Press, Cold Spring Harbor, NY, 1989.

of an overnight XL1-Blue MRA(P2) bacterial culture in LB with 0.4% (w/v) maltose. The infection of bacteria is achieved by incubation at 37° for 20 min. The whole mixture is then used to inoculate 100 ml of LB with 10 mM MgSO$_4$ and 0.1% (w/v) glucose. The mixture is vigorously shaken (250 rpm) for 5–6 hr until lysis is complete. Initially bacterial growth is observed and lysis starts to appear after 2–4 hr. Lysis is indicated by the appearance of bacterial debris, which agglutinates because of the hydrophobic interaction of the lipid portion of the membranes. Finally, 1 ml of chloroform is added when massive lysis is present, and the culture is incubated for an additional 10 min. Chloroform disrupts the remaining intact bacterial cells. Released bacterial DNA and RNA tend to form a compact net that captures phage particles and the culture is therefore treated with 10 μl of a 10-mg/ml DNase I solution and 100 μl of a 50-mg/ml RNase A solution for 30 min. NaCl is added to a final concentration of 1 M and the bacterial debris is precipitated for 1 hr at 4°. After centrifugation at 10,000g for 10 min at 4° the supernatant containing phage particles is collected and immediately transferred to new 200-ml tubes. Phages are precipitated after addition of 10 g of polyethylene glycol (PEG) 8000. It is difficult to dissolve PEG 8000, and the solution may need to be stirred for up to 1 hr. The phage particles are collected after centrifugation at 10,000g for 10 min at 4°. The pellet containing the phage particles is dissolved in 2 ml of SM buffer and the chloroform is extracted. The aqueous phase is transferred to Beckman (Fullerton, CA) SW-40.1 ultracentrifugation tubes and the phages are collected by centrifugation at 25,000 rpm for 2 hr at 4°. The pellet containing the purified phage particles is resuspended in 1.5 ml of SM buffer by overnight shaking. Phage DNA is isolated next morning by formamide extraction followed by phenol–chloroform and chloroform extraction and isopropanol precipitation.

Subcloning of Gene of Interest

After identification of several phage clones containing the gene of interest, it is necessary to map the content of the inserts. Mapping is facilitated by subcloning the phage clone inserts into a plasmid vector, for example, pBluescript. It is not always straightforward to perform this subcloning. We have found it important to have a high concentration of the isolated phage insert when ligating it to linearized plasmid. Possible problems at this step are sequences that are toxic to bacteria and repeats that may create deletions.

The subcloned genomic fragment is then mapped by cutting it with various restriction enzymes, preferably enzymes that cleave relatively infrequently (usually enzymes with hexanucleotide target sequences). This task is made easier if an overlapping set of clones is available, because regions of overlap highlight the relative positions of specific restriction fragments. It is important to find at least some restriction enzymes with unique recognition sites. Restriction mapping combined with probing with known gene sequences will give a rather precise map. There is often a need for partial sequencing of the insert. The restriction map will indicate which fragment(s) can be used to construct the targeting vector. Suitable DNA fragments are isolated by preparative electrophoresis and subcloned into plasmid vectors.

As mentioned previously, the *loxP* sites should be introduced within introns surrounding the exons to be removed by later Cre recombinase excision. It is important that *cre–loxP*-mediated deletion of the gene produce a null allele. There are several ways to accomplish this. Removal of 5′ exons may result in loss of targeting sequences and no mitochondrial import. Ideally, the removal of exons should result in a frameshift, preferably resulting in the generation of a new stop codon close to the recombination site. An alternative approach could involve deletion of 3′ exons and loss of the polyadenylation site. This latter approach was used to generate the *Tfam* knockout allele and subsequent studies showed that loss of the polyadenylation site will result in a null allele, probably due to dramatically reduced stability of the truncated, nonpolyadenylated *Tfam* transcript.

We use the double selection method to enrich for transfected ES cell clones that have undergone homologous recombination. The positive selection marker should be placed within an intron or 3′ to the polyadenylation signal and it should be flanked by *loxP* sites to ensure that it is removed in the knockout allele. The bacterial *neo* gene is by far the most commonly used positive selection marker. The negative selection marker, often the HSV-*TK* gene, should be placed at the end of the targeting vector. The negative selection marker is lost if the vector is integrated by homologous recombination, whereas randomly integrated vectors will retain the negative selection marker. Addition of the base analog ganciclovir to the ES cell selection medium will kill cells harboring the *TK* gene, whereas clones that have undergone homologous recombination will be insensitive to this treatment.

Once vector is introduced to the ES cells, it is important to be able to score for homologous recombination events. Commonly used techniques for screening for desired recombination event are PCR and Southern blot analysis. PCR analysis is quick, requires less material and can be performed on many samples at the same time. On the other hand, even if PCR analysis is used for the screening, final results require us of Southern blot analysis because PCR does not provide a complete picture of the recombination event. The main advantage of Southern blot analysis is its ability to distinguish different recombination events, including those that result in deletions, insertions, and rearrangements. Screening with Southern blot analysis requires the design of a unique probe that can easily distinguish wild-type from recombined allele. The probe can be classified as internal or external depending on its ability to recognize a sequence that is included or not included in the targeting vector. Internal probes should always be used in combination with external probes because internal probes can give false-positive results by detecting randomly integrated targeting vectors. In the ideal situation one should use an external probe combined with restriction digestion with an enzyme that does not cut within sequences included in the targeting vector. In this case, desired recombination events will be readily detected as a fragment of larger size than the wild-type allele. However, it may be difficult to find restriction enzyme sites that do not cut within the homologous region contained in the targeting vector. In this case, it is advantageous to use a restriction enzyme site introduced by the targeting vector.

Transfection of Embryonic Stem Cells and Production of Gene-Targeted Mice

Many universities have core facilities for ES cell culture and production of transgenic mice. Culture of ES cells is a straightforward procedure and can be performed in any laboratory with tissue culture facilities. We describe these procedures below, but there are also several other comprehensive descriptions of transfection and selection of ES cell clones.[13,14] The production of chimeric mice through blastocyst injection or ES cell embryo aggregations is usually performed by core facilities, and these procedures are not described in this chapter.

We use an early passage of the R1 ES cell line, which is suitable both for ES cell embryo aggregation and for blastocyst injection. Briefly, the ES cell medium is composed of DME-H21, glucose (4.5 g/liter; GIBCO-BRL, Gaithersburg, MD), 20% (v/v) fetal calf serum (HyClone, Logan, UT), 1 mM nonessential amino acids (GIBCO-BRL), 1 mM sodium pyruvate (GIBCO-BRL), 2 mM L-glutamine (GIBCO-BRL), penicillin (50 μg/ml; GIBCO-BRL), streptomycin (50 μg/ml; GIBCO-BRL), 100 μM 2-mercaptoethanol (Sigma, St. Louis, MO), and leukocyte inhibitory factor [LIF (2000 U/ml) and ESGRO; Chemicon International, Temecula, CA). The ES cell medium is additionally supplemented with G418 (0.2 mg/ml; Clontech, Palo Alto, CA) and 2 μM ganciclovir (Cytovene; Syntex, Palo Alto, CA) for selection of transfected ES cells. ES cells are grown on gelatinized plastic plates on a layer of irradiated mouse embryonic fibroblasts.

The targeting vector DNA is linearized and purified by phenol, phenol–chloroform, and chloroform extraction. The DNA is ethanol precipitated and recovered by high-speed centrifugation for 5 min, and the DNA pellet is rinsed with a large volume of 70% (v/v) ethanol. Finally, the DNA is resuspended in sterile Dulbecco's modified Eagle's medium (DMEM) without serum in a laminar air flow hood.

For transfection, a 10-cm plate of confluent R1 ES cells is trypsinized, the cells are washed once in ES cell medium, and then resuspended in about 600 μl of ice-cold ES cell medium. Clumps of cells are disaggregated by gentle pipetting. The ES cell suspension (600 μl) is mixed with 100 μl of the DNA solution, containing 25 μg of linearized vector DNA, and kept on ice until it is transferred

[13] P. M. Hassarmam and M. L. DePamphilis (eds.), *Methods Enzymol.* **225** (1993).

[14] A. L. Joyner, "Gene Targeting." Oxford University Press, New York (2000).

to a precooled electroporation cuvette (0.4 mm; Bio-Rad, Hercules, CA). The cells are electroporated with a Bio-Rad Gene Pulser II at 500-μF capacitance and 250 V with an observed time constant of 6–9 ms. After electroporation cells are allowed to recover for 5 min at room temperature and then diluted in ES cell medium and subsequently transferred to six-well plates with irradiated feeder cells. The cell content of one electroporation is distributed among all the wells of a six-well plate. We start selection 24 hr after transfection to allow the cells to recover from the electroporation procedure. Positive selection is performed by adding G418 (0.2 mg/ml) and negative selection is performed by adding fresh 2 μM ganciclovir. The selection medium is changed every 1–2 days. Massive cell death is noted after 2–4 days and drug-resistant colonies are visible after 7–9 days.

Visible ES cell colonies (diameter, ~1 mm) are marked with a circle on the bottom of the plate. ES cell-containing plates are washed twice with phosphate-buffered saline (PBS) and 5 ml of PBS is added. Clones are picked with a micropipette tip, transfered to 96-well plates with 10 μl of 0.02% (w/v) trypsin, incubated for 10–20 min, resuspended by pipetting, and finally transferred to 96-well plates with feeder cells and 200 μl of fresh ES cell medium. From this stage onward the ES cells are grown without selection. The medium is changed every day and cells growing within the well should reach confluency within the next 2–4 days. Confluent colonies are trypsinized and transferred to 5 wells of a 96-well plate. The cells are expanded on plates without feeder cells for DNA analysis or on plates with feeders for ES cell freezing.

Matings for Germ Line and Organ-Specific Gene Disruption

Once germ line transmission of the *loxP*-flanked allele is obtained, the animals should be bred to homozygosity for the *loxP*-flanked allele. Homozygous animals (*loxP/loxP*) should be checked carefully for any phenotype. Gene expression should be monitored by Northern and Western blot analyses of tissues from *loxP/loxP* animals and wild-type animals. The absence of any obvious phenotype and normal expression of the *loxP*-flanked allele strongly suggest that the insertion of *loxP* sites has not created a hypomorphic allele.

Germ line knockouts are obtained by mating animals that are heterozygous for the *loxP*-flanked allele (+/*loxP*) with animals with ubiquitous expression of Cre recombinase, for example, β-*actin cre* transgenic animals. This cross will generate heterozygous knockouts (+/–). The *cre* transgene can be removed by mating these animals with wild-type animals and identifying offspring that are heterozygous knockouts but lack the *cre* transgene. *cre–loxP*-mediated recombination is frequently partial and germ line transmission ensures that the heterozygous knockout is present in every cell.

Tissue-specific knockouts are generated by mating *loxP/loxP* animals with heterozygous *cre*-transgenic animals. From this cross, mice that are heterozygous for the *loxP*-flanked allele and heterozygous for the *cre* transgene (*+/loxP; +/cre*) are recovered and mated with *loxP/loxP* animals. This latter cross will produce 25% knockouts (*loxP/loxP; +/cre*).

It should be noted that *cre–loxP* recombination frequently is partial and it is therefore necessary to estimate the recombination efficiency. This is not always an easy task because the knockout usually is present in only one of several cell types in the tissue. This can be exemplified by heart knockouts produced by *cre* expression from the α-myosin heavy chain promoter. This will result in knockout of *loxP*-flanked alleles in cardiomyocytes but not in other cell types of the heart, for example, endothelial cells and fibroblasts. Mating of *cre*-expressing mice with reporter mice with *cre*-mediated activation of *lacZ* expression will give valuable information about tissue specificity and recombination efficiency. It is also necessary to use Northern blots, Western blots, and perhaps also *in situ* hybridization techniques to determine the knockout efficiency.

Future Perspectives

There are now many published examples demonstrating that conditional knockout technology is a powerful and reliable tool in mouse genetics. The selection of the promoter used for expression of *cre* recombinase determines the cell-type specificity of the knockout. Furthermore, it is also possible to obtain temporal control of the knockout by choosing promoters activated at different developmental stages.[5] Systems for inducible expression of *cre* recombinase are undergoing rapid development and it is likely that future systems will allow full spatial and temporal control of the knockout.[15] The *cre–loxP* recombination system can also be used to induce chromosomal translocations in ES cells by insertion of *loxP* sites on different chromosomes.[16]

The *flp–frt* recombination system of yeast can also be used for recombination in mice, and it has been utilized mainly in germ line deletion of inserted selection markers due to low recombination efficiency.[9] However, there are now improved versions of the Frt enzyme with high catalytic activity at 37°, suggesting that the *flp–frt* recombination system may be as efficient as the *cre–loxP* recombination system in mice.[17]

The conditional knockout technique mediated by *cre–loxP*-mediated recombination is now an established method in mice. There is a wide variety of available

[15] J. Brocard, X. Warot, O. Wendling, N. Messaddeq, J. L. Vonesch, P. Chambon, and D. Metzger, *Proc. Natl. Acad. Sci. U.S.A.* **94,** 14559 (1997).

[16] R. Ramirez-Solis, P. Liu, and A. Bradley, *Nature (London)* **378,** 720 (1995).

[17] C. I. Rodriguez, F. Buchholz, J. Galloway, R. Sequerra, J. Kasper, R. Ayala, A. F. Stewart, and S. M. Dymecki, *Nat. Genet.* **25,** 139 (2000).

mouse strains with tissue-specific *cre* expression and some of these are listed at *http://www.mshri.on.ca/nagy/*. Future development in this area will generate a wide variety of mouse strains with spatially and temporally controlled *cre* expression. This will give us unlimited possibilities of studying the function of genes at different developmental stages and of creating knockouts in old animals to model the pathology of human age-associated diseases.

[36] Antisense Oligodeoxyribonucleotides: A Better Way to Inhibit Monocyte Superoxide Anion Production?

By Erik A. Bey and Martha K. Cathcart

Introduction

Reduced nicotinamide-adenine dinucleotide phosphate (NADPH) oxidases are a group of plasma membrane-associated enzymes, among which the phagocytic leukocyte NADPH oxidase has been most often studied.[1-3] During phagocytosis or exposure to other activating agonists, the NADPH oxidase of neutrophils and other professional phagocytic cells (including monocytes) catalyzes the production of superoxide anion ($O_2^{\cdot-}$) by the one-electron reduction of molecular oxygen. The enzyme complex uses NADPH, provided by the pentose phosphate pathway, as the electron donor.[1,2,4,5] $O_2^{\cdot-}$, generated by the NADPH oxidase of monocytes and neutrophils, is readily converted to other potent reactive oxidants, such as hydrogen peroxide (H_2O_2), hydroxyl radical ($OH^\cdot$), hypochlorite (OCl^-), and singlet oxygen ($^1O_2$).[1,6,7] These more potent oxidants are directly responsible for killing bacterial and fungal pathogens. The production of $O_2^{\cdot-}$ and the more potent oxidants by these cells leads to an abrupt rise in oxygen consumption. For this reason the NADPH oxidase is often referred to as the respiratory burst oxidase.

The leukocyte NADPH oxidase complex consists of a membrane-associated *b*-type cytochrome b_{558} that copurifies with a GTP-binding protein, RAP1A[8];

[1] B. M. Babior, *Blood* **93**, 1464 (1999).

[2] R. A. Clark, *J. Infect. Dis.* **179** (Suppl. 2), S309 (1999).

[3] T. Leto, *in* "Inflammation: Basic Principles and Clinical Correlates" (J. I. Gallin and R. Snyderman, eds.), p. 769. Lippincott Williams & Wilkins, Philadelphia, PA, 1999.

[4] S. J. Chanock, R. el Benna, R. M. Smith, and B. M. Babior, *J. Biol. Chem.* **269**, 24519 (1994).

[5] F. Rossi, *Biochim. Biophys. Acta* **853**, 65 (1986).

[6] J. M. Robinson and J. A. Badwey, *Histochem. Cell Biol.* **103**, 163 (1995).

[7] B. M. Babior, *J. Clin. Invest.* **73**, 599 (1984).

[8] M. T. Quinn, C. A. Parkos, L. Walker, S. H. Orkin, M. C. Dinauer, and A. J. Jesaitis, *Nature (London)* **342**, 198 (1989).

Copyright 2002, Elsevier Science (USA).
All rights reserved.
0076-6879/02 $35.00

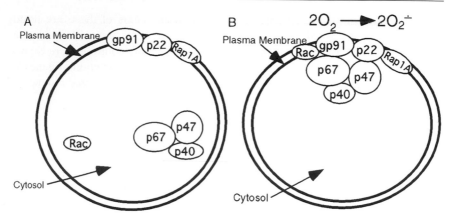

Fig. 1. (A) Components of the phagocytic NADPH oxidase lie unassembled in unactivated mono-cytes. The NADPH oxidase components are both membrane associated and cytosolic in unactivated monocytes. (B) NADPH oxidase components assemble on monocyte activation. After activation by appropriate stimuli, cytosolic components of the phagocytic NADPH oxidase translocate and assemble with membrane-associated NADPH oxidase components and form an active complex resulting in $O_2^{\dot{-}}$ production.

a cytosolic complex consisting of three components, $p47^{phox}$, $p67^{phox}$, and $p40^{phox}$; and a small cytosolic GTP-binding protein, Rac 1/2.[4] The components of the leukocyte NADPH oxidase are unassembled in unactivated cells (Fig. 1A). In response to appropriate stimuli, the cytosolic components translocate and assem-ble at the membrane. The proper assembly of the NADPH oxidase components activates the oxidase and $O_2^{\dot{-}}$ is produced (Fig. 1B).

Our laboratory previously reported that $O_2^{\dot{-}}$ production was required for human monocyte-mediated low-density lipoprotein (LDL) oxidation[9] because monocyte-mediated LDL oxidation was inhibited by superoxide dismutase (SOD), a scav-enger of $O_2^{\dot{-}}$. However, to definitively show that $O_2^{\dot{-}}$ was indeed required for monocyte-mediated LDL oxidation, we thought that it was necessary to perform additional experiments that did not rely on the specificity of action of SOD. Phar-macologic agents have been reported to inhibit NADPH oxidase activity, such as diphenylene iodonium (DPI) and phenylbutazone.[10–12] Studies in our laboratory have shown that both of these agents dose dependently inhibit NADPH oxidase ac-tivity while also causing dose-dependent toxicity to the cells.[13] As an alternative to

[9] M. K. Cathcart, A. K. McNally, D. W. Morel, and G. M. Chisolm III, *J. Immunol.* **142,** 1963 (1989).
[10] A. R. Cross and O. T. Jones, *Biochem. J.* **237,** 111 (1986).
[11] J. T. Hancock and O. T. Jones, *Biochem. J.* **242,** 103 (1987).
[12] D. G. Hafeman and Z. J. Lucas, *J. Immunol.* **123,** 55 (1979).
[13] V. F. Nivar, Ph.D. Thesis. Department of Regulatory Biology, Cleveland State University, Cleveland, OH, 1993.

TABLE I

ANTISENSE OLIGONUCLEOTIDE SELECTION

1. Analyze the predicted secondary structure of the mRNA and select potential target sequences in areas lacking secondary structure (i.e., areas predicted to be single stranded) (Mulfold)
2. Screen these areas of the mRNA for sequences unique to this message (BLAST)
3. Select 20-mer ODNs and eliminate sequences predicted to stably pair or fold (Mfold, energy > −5.0)
4. Exclude 20-mers with high GC : AT ratios, quad-G sequences, etc.
5. Synthesize antisense ODNs and control ODNs with phosphorothioate-modified bases and purify by HPLC (usual controls are sense, scrambled and/or mismatched ODNs)
6. Assess efficacy by Western blots and activity assays

using these rather nonspecific pharmacologic inhibitors, we chose to use antisense oligodeoxyribonucleotides (ODNs) to block the production of an essential protein component of the NADPH oxidase and thereby inhibit NADPH oxidase activity. This essentially creates an *in vitro* "knockout" of expression of this protein. The protein we chose to target was $p47^{phox}$ because it is required for NADPH oxidase activity *in vivo* as determined in humans with mutations in the gene encoding this protein.[14–16] After creating and characterizing the *in vitro* "knockout," we then determined the effect of this knockout on O_2^{-} production as well as on LDL oxidation.

In this chapter we describe protocols for selecting antisense ODNs that can modulate NADPH oxidase activity. We also describe the requisite characteristics of antisense ODNs and how to characterize their effectiveness, using a cytochrome *c* reduction assay to measure O_2^{-} production and Western blot analysis to measure effects on $p47^{phox}$ protein expression. Finally, we describe how we have employed antisense ODNs to identify other upstream regulatory components, namely the protein kinase C α isoform (PKCα) and cytosolic phospholipase A_2 (cPLA$_2$), that modulate NADPH oxidase activity in human monocytes.

Protocols

Selection of Suitable Antisense Sequence

Choosing the sequence of the target mRNA for antisense design should not be a random process. We have devised a series of analyses and prerequisites to be met before selecting the sequence of our antisense ODN reagents (Table I). We routinely begin our antisense ODN design by first examining the predicted secondary structure of the target mRNA. A convenient folding program, predicting

[14] B. D. Volpp, W. M. Nauseef, and R. A. Clark, *Science* **242,** 1295 (1988).
[15] A. W. Segal, P. G. Heyworth, S. Cockcroft, and M. M. Barrowman, *Nature (London)* **316,** 547 (1985).
[16] S. Dusi and F. Rossi, *Biochem. J.* **296,** 367 (1993).

the secondary structure of RNA, has been devised and can be accessed via the Internet (http://bioinfo.math.rpi.edu/~zukerm/).[17-19] The program, termed Mfold, provides the predicted secondary structure of any mRNA on the basis of minimizing the computed free energy of the structure. We have always employed this approach because it made intuitive sense that an area of the mRNA that was predicted to be single stranded would prove to be an ideal and available sequence for interaction with antisense ODNs. However, this approach of targeting areas of the mRNA predicted to exist in single-stranded conformation has been shown to be predictive of antisense ODN efficacy.[20]

An especially advantageous feature of the Mfold analysis of predicted secondary structure is the option to view the frequency of single strandedness of a particular base or area of the mRNA. This feature allows the user to compare, among the numerous predicted secondary structures, the frequency with which a particular base is unpaired, thereby circumventing the need to perform de-tailed comparisons among the several predicted structures, and instead rely on this cross-structure analysis. When using this feature it is desirable to select potential target sites from sequence stretches that are entirely or mostly single stranded.

Once a single-stranded region(s) has been selected as a potential target site for antisense ODN binding, the target sequence must be analyzed further to ensure that it is unique in the database. Requirements are that it does not stably fold or pair with itself, and the overall base content and sequence must be screened to eliminate problem sequences. To determine whether the sequence is unique, the selected sequence is analyzed by BLAST (Basic Local Alignment Search Tool), a search tool available through the National Center for Biotechnology Informa-tion (http://www.ncbi.nlm.nih.gov/BLAST/). BLAST explores all the available sequence databases, using a set of similarity search programs. Once the sequence(s) has been shown to be unique, the selected sequences (usually 20-mers) are screened to ensure that they do not fold or pair with one another in stable arrangements. These latter properties can be assessed by again using the Mfold folding program. Finally, base composition and sequence are evaluated. Sequences containing runs of guanosines can form "G-quartets" that interact with proteins with greater affinity than unstructured nucleotides.[21] These sequences should be avoided. We have always limited the GC content of our selected ODNs as much as possible and prefer to have approximately 50% or less GC base content and limited strings of GC base pairs. The idea behind this is that the higher affinity GC base pairing may allow for stable pairing with incomplete sequence identity.

[17] J. A. Jaeger, D. H. Turner, and M. Zuker, *Proc. Natl. Acad. Sci. U.S.A.* **86,** 7706 (1989).

[18] J. A. Jaeger, D. H. Turner, and M. Zuker, *Methods Enzymol.* **183,** 281 (1990).

[19] D. H. Mathews, J. Sabina, M. Zuker, and D. H. Turner, *J. Mol. Biol.* **288,** 911 (1999).

[20] V. Patzel, U. Steidl, R. Kronenwett, R. Haas, and G. Sczakiel, *Nucleic Acids Res.* **27,** 4328 (1999).

[21] J. R. Wyatt, T. A. Vickers, J. L. Roberson, R. W. Buckheit, T. Klimkait, E. DeBaets, P. W. Davis, B. Rayner, J. L. Imbach, and D. J. Ecker, *Proc. Natl. Acad. Sci. U.S.A.* **91,** 1356 (1994).

Another important issue in antisense ODN treatment is the selection of appropriate control sequences. Similar base content in jumbled order is often used as a control. Sense sequences can be used as well. When possible, another good control is the antisense ODN to a related protein, for example, another isoform of the enzyme, for example, an antisense ODN to PKCβ as a control for an antisense to PKCα.[22] Usually it is desirable to use two different types of control ODN or to assess the effects of the antisense ODN on another, unrelated protein.[23] An advantage of using a sense ODN as a control is that the folding, pairing, and homology analyses, performed to evaluate the antisense ODN sequence, are shared with the antisense ODN, thus obviating the need to analyze the control sequences. In contrast, scrambled or mismatched ODNs must be analyzed as described above.

Modification of Antisense Oligodeoxyribonucleotides

Antisense ODNs are susceptible to enzymatic degradation and, therefore, for experimental purposes are routinely chemically modified to prevent degradation by ubiquitous nucleases during delivery to cells.[24–26] Before synthesizing the antisense it is necessary to select which modification is most appropriate for the studies proposed. Phosphorothioate modification is one of the most commonly used methods[27 31] and one that we have effectively used with human monocytes in our *in vitro* studies for more than 15 years. This method of modification replaces the nonbridging oxygen atoms within the DNA backbone of the oligonucleotide molecule with sulfur. We originally chose this modification because it was easily attainable and did not cause nonspecific effects on monocyte viability, protein expression, or a variety of monocyte enzymatic activities. It has worked so well in human monocytes that we have continued using this modification for most of our studies even though modifications that provide greater ODN stability have been developed.

Phosphorothioate modifications of nucleotides in ODNs can be incorporated throughout the 20-mer or can be incorporated only at the ends of the ODN. Our

[22] Q. Li, V. Subbulakshmi, A. P. Fields, N. R. Murray, and M. K. Cathcart, *J. Biol. Chem.* **274,** 3764 (1999).

[23] E. A. Bey and M. K. Cathcart, *J. Lipid Res.* **41,** 489 (2000).

[24] K. J. Myers and N. M. Dean, *Trends Pharmacol. Sci.* **21,** 19 (2000).

[25] S. T. Crooke and C. F. Bennett, *Annu. Rev. Pharmacol. Toxicol.* **36,** 107 (1996).

[26] C. A. Stein and Y. C. Cheng, *Science* **261,** 1004 (1993).

[27] S. Agrawal and E. R. Kandimalla, *Mol. Med. Today* **6,** 72 (2000).

[28] L. V. Varga, S. Toth, I. Novak, and A. Falus, *Immunol. Lett.* **69,** 217 (1999).

[29] E. G. Marcusson, B. Bhat, M. Manoharan, C. F. Bennett, and N. M. Dean, *Nucleic Acids Res.* **26,** 2016 (1998).

[30] R. W. Wagner, *Nat. Med.* **1,** 1116 (1995).

[31] C. Wahlestedt, *Trends Pharmacol. Sci.* **15,** 42 (1994).

laboratory has shown that end-modified ODNs are as effective as totally modified ODNs (H. Xu, A. R. Heath, and M. K. Cathcart, in preparation, 2002).

Despite substantial advances in ODN synthetic chemistry, final ODN synthetic products are usually only 75% full-length sequences.[5] It is critical to perform additional purification to limit contamination by incomplete sequences that are more promiscuous in their recognition of target sequences. This service is often provided by the company that synthesizes the ODN. We prefer reversed-phase high-performance liquid chromatography (HPLC) because of its efficient and thorough elimination of incomplete sequences from the ODN preparations of 20-mers in the quantities that we usually order (1–10 μmol). The ODN purification is critical for antisense ODN efficacy.

Delivery of Oligodeoxyribonucleotides to Cells

An antisense ODN can inhibit the expression of a specific mRNA only if it successfully reaches its target. The chance that an antisense will interact with its target is substantially increased by uptake enhancers.[24,27] A variety of these agents/methods have been shown to be effective. The list includes, among others, cationic lipids,[32,33] liposomes,[34,35] peptides,[36,37] dendrimers,[38,39] polycations,[40,41] conjugation with cholesterol,[42,43] aggregation with cell surface ligands,[44,45] and electroporation.[46,47]

[32] C. F. Bennett, M. Y. Chiang, H. Chan, J. E. Shoemaker, and C. K. Mirabelli, *Mol. Pharmacol.* **41,** 1023 (1992).

[33] J. G. Lewis, K. Y. Lin, A. Kothavale, W. M. Flanagan, M. D. Matteucci, R. B. DePrince, R. A. Mook, R. W. Hendren, and R. W. Wagner, *Proc. Natl. Acad. Sci. U.S.A.* **93,** 3176 (1996).

[34] J. Y. Legendre and F. C. Szoka, *Pharm. Res.* **9,** 1235 (1992).

[35] A. Bochot, P. Couvreur, and E. Fattal, *Prog. Retin. Eye Res.* **19,** 131 (2000).

[36] J. P. Bongartz, A. M. Aubertin, P. G. Milhaud, and B. Lebleu, *Nucleic Acids Res.* **22,** 4681 (1994).

[37] T. B. Wyman, F. Nicol, O. Zelphati, P. V. Scaria, C. Plank, and F. C. Szoka, *Biochemistry* **36,** 3008 (1997).

[38] A. Bielinska, J. F. Kukowska-Latallo, J. Johnson, D. A. Tomalia, and J. R. Baker, *Nucleic Acids Res.* **24,** 2176 (1996).

[39] R. Delong, K. Stephenson, T. Loftus, M. Fisher, S. Alahari, A. Nolting, and R. L. Juliano, *J. Pharm. Sci.* **86,** 762 (1997).

[40] J. P. Leonetti, G. Degols, and B. Lebleu, *Bioconjug. Chem.* **1,** 149 (1990).

[41] O. Boussif, F. Lezoualc'h, M. A. Zanta, M. D. Mergny, D. Scherman, B. Demeneix, and J. P. Behr, *Proc. Natl. Acad. Sci. U.S.A.* **92,** 7297 (1995).

[42] A. M. Krieg, J. Tonkinson, S. Matson, Q. Zhao, M. Saxon, L. M. Zhang, U. Bhanja, L. Yakubov, and C. A. Stein, *Proc. Natl. Acad. Sci. U.S.A.* **90,** 1048 (1993).

[43] S. K. Alahari, N. M. Dean, M. H. Fisher, R. Delong, M. Manoharan, K. L. Tivel, and R. L. Juliano, *Mol. Pharmacol.* **50,** 808 (1996).

[44] G. Citro, D. Perrotti, C. Cucco, I. D'Agnano, A. Sacchi, G. Zupi, and B. Calabretta, *Proc. Natl. Acad. Sci. U.S.A.* **89,** 7031 (1992).

[45] G. Y. Wu and C. H. Wu, *J. Biol. Chem.* **267,** 12436 (1992).

[46] P. H. Watson, R. T. Pon, and R. P. Shiu, *Exp. Cell Res.* **202,** 391 (1992).

[47] R. Bergan, Y. Connell, B. Fahmy, and L. Neckers, *Nucleic Acids Res.* **21,** 3567 (1993).

The mechanism by which oligonucleotides are internalized by cells *in vivo* and *in vitro* is not clearly understood. It is hypothesized that this process may involve internalization by endocytotic vesicles.[29,48] Thus, one of the more effective ways to deliver antisense ODNs is by mixing the ODNs with cationic lipids or liposomes, masking the negative charge before interaction with cells. The use of cationic lipids and liposomes, by increasing effective delivery of oligonucleotides into the cell, also decreases the effective dose of antisense ODNs required to inhibit a specific target.[29,49,50] Our experience with human monocytes is that they effectively incorporate 20-mer phosphorothioate ODNs without cationic lipids but that cationic lipids can lower effective doses by at least 100-fold.[22,23,51–53]

An innate advantage of using monocytes as the cell target for antisense ODN regulation of protein expression is their characteristic rapid pinocytotic rate. Monocytes constantly sample their extracellular environment, using pinocytosis and endocytosis, as part of their duty in providing immune surveillance and antigen detection and presentation. This is one reason why monocytes may be particularly conducive to regulation by antisense ODNs and why ODNs are effective with monocytes even without delivery agents. Further, circulating monocytes are essentially in G_0 and are not proliferating. Therefore, they have only basal gene expression and many genes are unexpressed until the monocyte becomes activated by external stimuli. If the mRNA target is from one of these latent genes, then antisense ODN inhibition will have a more immediate impact. Also, by not proliferating, the delivered antisense is not diluted on cell division. Thus, the monocyte appears to be an ideal target cell for antisense ODN intervention. This is particularly fortuitous because these cells are essentially nontransfectable with full-length cDNAs. Although this latter problem limits experimental manipulation, the use of antisense ODNs affords new options for regulating gene expression.

Monocyte Isolation

Whole blood (240 ml) is collected from donors into heparinized syringes. The mononuclear cell layer is isolated by centrifugation of diluted blood over a Ficoll-Paque density solution.[54] The isolation of monocytes from mononuclear cells is performed by a modification[54,55] of the method of Kumagai *et al.*[56] The

[48] O. Zelphati and F. C. Szoka, Jr., *Pharm. Res.* **13,** 1367 (1996).
[49] K. Lappalainen, A. Urtti, E. Soderling, I. Jaaskelainen, K. Syrjanen, and S. Syrjanen, *Biochim. Biophys. Acta* **1196,** 201 (1994).
[50] D. Wielbo, N. Shi, and C. Sernia, *Biochem. Biophys. Res. Commun.* **232,** 794 (1997).
[51] B. Roy and M. K. Cathcart, *J. Biol. Chem.* **273,** 32023 (1998).
[52] Q. Li and M. K. Cathcart, *J. Biol. Chem.* **269,** 17508 (1994).
[53] Q. Li and M. K. Cathcart, *J. Biol. Chem.* **272,** 2404 (1997).
[54] M. K. Cathcart, D. W. Morel, and G. M. Chisolm III, *J. Leukoc. Biol.* **38,** 341 (1985).
[55] A. K. McNally, G. M. Chisolm III, D. W. Morel, and M. K. Cathcart, *J. Immunol.* **145,** 254 (1990).
[56] K. Kumagai, K. Itoh, S. Hinuma, and M. Tada, *J. Immunol. Methods* **29,** 17 (1979).

mononuclear cell layer is washed three times with phosphate-buffered saline (PBS). Next, contaminating platelets are removed by two centrifugations through 4 ml of bovine calf serum (BCS; HyClone, Logan, UT) after overlaying the serum with the mononuclear cells as previously described.[54] Monocytes are isolated from the platelet-free mononuclear cells by adherence to 75-cm^2 polystyrene tissue culture flasks (Costar, Cambridge, MA), precoated with 100% BCS and containing Dulbecco's modified Eagle's medium (DMEM) and 10% (v/v) BCS (BCS–DMEM). The flasks are incubated for 2 hr at 37° in 10% CO_2. The nonadherent cells are removed by washing the flasks three times with BCS–DMEM, pelleted by centrifugation at 200g, resuspended in BCS–DMEM, and plated in a separate, serum-coated flask. The flasks are allowed to incubate at 37° in 10% CO_2 for 24 hr. The nonadherent cells are removed and discarded, the flasks are washed three times with BCS–DMEM, and adherent cells are collected from both flasks by removing the medium and placing the cells in the presence of 0.4 mM EDTA. The flasks are incubated for 15 min at 37° in 10% CO_2. The flasks are then shaken and pipetting is performed to release the cells. EDTA is removed by washing the monocytes three times with BCS–DMEM. The cells are resuspended in BCS–DMEM and counted. Although these "monocytes" are cultured for 1–4 days during the course of the following experiments, we refer to them as monocytes to distinguish them from monocyte-derived macrophages, which are typically cultured for 7–10 days before use and are more fully differentiated.

Superoxide Anion Assay

The method used to detect $O_2^{\bar{\cdot}}$ produced by human monocytes is a modification of an assay previously reported by Johnston.[57] The assay measures SOD-inhibitable cytochrome c reduction. Human monocytes are plated in flat-bottomed 96-well tissue culture plates (100 μl, 1×10^6/ml) in BCS–DMEM and allowed to adhere for at least 2 hr. After antisense ODN treatment (see legends to Figs. 2 and 3 for details of dose and length of incubation), the medium in the wells is changed to RPMI without phenol red (BioWhittaker, Walkersville, MD) and monocytes are incubated with opsonized zymosan (ZOP, 2 mg/ml) to activate the monocytes, cytochrome c (160 U/ml; Sigma, St. Louis, MO), ±SOD (300 U/ml; Sigma) for 1 hr at 37°. The absorbance is determined in a Molecular Devices (Menlo Park, CA) Thermomax microplate reader at 550 nm. The amount of $O_2^{\bar{\cdot}}$ produced by monocytes is determined by measuring the SOD-inhibitable reduction of cytochrome c, using the extinction coefficient of 158.73 and expressed as nanomoles per milliliter.[23,58]

---

[57] R. B. Johnston, *in* "Methods for Studying Mononuclear Phagocytes" (D. O. Adams, P. J. Edelson, and H. Koren, eds.), p. 489. Academic Press, New York, 1981.

[58] E. Pick and D. Mizel, *J. Immunol. Methods* **46,** 211 (1981).

Western Blots

Human monocytes are plated in six-well tissue culture plates (Costar) at a density of 2.5×10^6/ml in BCS–DMEM and treated with a $1–10 \ \mu M$ concentration of antisense or sense ODNs for up to 3 days (as indicated). After the incubation period, cells are lysed with 200 μl of hypotonic lysis buffer [50 mM Tris-HCl (pH 7.5), 5 mM MgSO$_4$, 0.5 mM EGTA, 0.1% (v/v) 2-mercaptoethanol, 1 mM phenylmethylsulfonyl fluoride (PMSF), leupeptin (20 μg/ml), and 0.5% (v/v) Nonidet P-40]. The cells are vortexed for 10–15 sec and cellular debri and nuclei are removed by centrifugation at 200g for 10 min. The supernatants are sonicated for 3–5 sec and then centrifuged at 10,000g at 4° for 10 min. Supernatants from these tubes are separated from pellets and 25–100 μg of lysate protein is prepared for sodium dodecyl sulfate–10% (w/v) polyacrylamide gel electrophoresis (10% SDS–PAGE). Proteins from SDS–polyacrylamide gels are transferred to a polyvinylidene difluoride (PVDF) membrane (0.2 μm; Bio-Rad, Hercules, CA) by the semidry method. Nonspecific binding sites on the membrane are blocked with 5% (w/v) milk in Tris-buffered saline [20 mM Tris-HCl (pH 7.4), 150 mM NaCl, 1% (v/v) Nonidet P-40 (NP-40)] at 4° for 24 hr. Human p47^{phox} is detected with goat anti-human p47^{phox} polyclonal antibody (diluted 1 : 1000; generously provided by H. Malech and T. Leto, NIAID, NIH, Bethesda, MD) followed by horseradish peroxidase-conjugated rabbit anti-goat IgG (diluted 1 : 1000; Pierce Biochemicals, Rockford, IL). The protein is detected by developing the membrane with enhanced chemiluminescence (Amersham, Arlington Heights, IL). As a control, Extracellular signal-regulated kinase 1/2 (ERK1/2), a constitutively synthesized protein, is detected with rabbit anti-human Erk1/2 kinase (diluted 1 : 1000; Upstate Biotechnology, Waltham, MA) followed by horseradish peroxidase-conjugated goat anti-rabbit IgG (diluted 1 : 1000; Pierce Biochemicals). The PVDF membrane is developed as described above.

Results

After exposing cells to antisense ODNs, the extent of target knockout must be evaluated. The "gold standard" for evaluating the effect of antisense is to measure, by Western blot analysis, the expression of the protein encoded by the targeted mRNA. The Western blot is usually evaluated first; however, if a microassay for protein function exists, such a functional assay can precede the Western blot as a screen for appropriate antisense ODN treatment conditions. This approach was used in establishing the optimal culture conditions for exposure of monocytes to antisense to p47^{phox}. After finding the optimal conditions for causing maximal inhibition of O$_2^{\underline{\ \cdot\ }}$ production, we then confirmed that the target protein was inhibited in cells treated similarly.

Direct Assessment of Effects of Antisense Oligodeoxyribonucleotides
on Monocyte NADPH Oxidase Activity

To determine the effective dose of p47^{phox} antisense ODNs required to down-regulate NADPH oxidase activity, we used an SOD-inhibitable cytochrome c reduction assay to measure $O_2^{\cdot-}$ production in human monocytes treated with varying doses (1–10 μM) and for various lengths of time (1–3 days) with antisense and sense ODNs. We determined that optimal inhibition was obtained with 5–10 μM antisense ODNs when ODNs were delivered without delivery agents. We also determined that the optimal length of incubation was 72 hr with feedings of ODNs at time 0 and at 48 hr. Our results in Fig. 2 show that 5 μM p47^{phox} antisense ODNs significantly inhibited $O_2^{\cdot-}$ production in human monocytes whereas the sense ODNs had no effect. This result is representative of a minimum of three similar experiments in which the effective dose range was 5–10 μM.

After determining the effective dose and length of incubation required to downregulate NADPH oxidase activity by p47^{phox} antisense ODNs in human monocytes, the effect of antisense ODNs on p47^{phox} protein expression in human monocytes was evaluated by Western blot analysis (Fig. 3A). The results shown

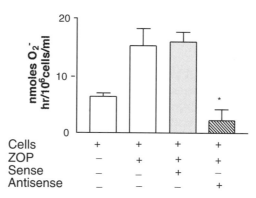

FIG. 2. p47^{phox} antisense ODN treatment inhibits $O_2^{\cdot-}$ production by activated human monocytes. Human monocytes were plated in 96-well tissue culture plates at a concentration of 1.0×10^5 cells/0.1 ml per well. Monocytes were treated with antisense or sense ODNs (5 μM) for 3 days (with refeeding after 2 days). After incubation, the cells were washed with RPMI without phenol red and $O_2^{\cdot-}$ production was determined as described in Protocols. The data represent means ±standard deviation of triplicate determinations. The results are from a representative experiment of three performed. Cells, human monocytes; ZOP, activator; antisense, p47^{phox} antisense ODNs; sense, p47^{phox} sense ODNs. These data were analyzed using the unpaired, one-tailed Student t test. The asterisk indicates that $O_2^{\cdot-}$ values from cells treated with ZOP and antisense ODNs were significantly different from those obtained with cells treated with ZOP alone ($p = 0.01$) or as compared with those obtained with cells treated with ZOP plus sense ODNs ($p = 0.003$). [Adapted from E. Bey and M. K. Cathcart, *J. Lipid Res.* **41,** 489 (2000), with permission from *Journal of Lipid Research.*]

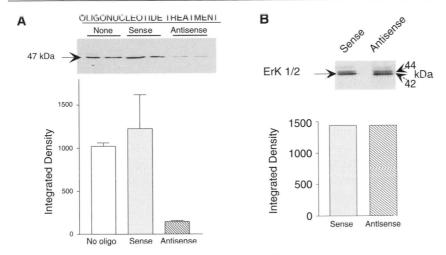

FIG. 3. (A) p47^{phox} antisense ODN treatment inhibits p47^{phox} protein expression in human monocytes. Human monocytes (2.5×10^6/ml) were treated with 5 μM antisense or sense ODNs for 72 hr with a refeeding at 48 hr. Cells were lysed, run on SDS–10% (w/v) polyacrylamide gels, and transferred to a PVDF membrane as described in Protocols. p47^{phox} was detected with human p47^{phox} polyclonal antibody (diluted 1 : 1000) followed by incubation with horseradish peroxidase-conjugated rabbit anti-goat IgG (diluted 1 : 1000). The blot was developed by ECL. Samples were run in duplicate as indicated. The left arrow indicates the migration of p47^{phox} (47 kDa) based on the migration of molecular weight markers that were in adjacent lanes. The samples were loaded in duplicate and the bars represent the average density of two bands per group. The bar graph depicts integrated densities of p47^{phox} bands in the blot as determined by analysis of lightly exposed film by the software program NIH Image. Error bars represent the data range of duplicates. (B) In a separate experiment monocytes were treated as described above with antisense or sense ODNs to p47^{phox} (5 μM) and then Western analysis of p47^{phox} was performed. Again, inhibition of p47^{phox} was observed. The blot was then stripped and reprobed with antibody to ERK1/2 kinase as an unrelated, constitutively produced protein. This blot was similarly developed. Bar graphs represent the integrated density of the relative OD curves derived from NIH Image analysis of lightly exposed films of the blots developed by ECL. [Adapted from E. Bey and M. K. Cathcart, *J. Lipid Res.* **41**, 489 (2000), with permission from *Journal of Lipid Research.*]

here indicate that 5 μM p47^{phox} antisense ODNs significantly inhibited p47^{phox} protein expression, whereas sense ODNs had no effect. The inhibition shown here is approximately 80% as determined by NIH Image densitometry analysis. These results are representative of three experiments in which inhibition by p47^{phox} antisense ODNs ranged from 45 to 80%. Figure 3B shows a separate experiment in which a p47^{phox} antisense/sense Western blot was stripped and reprobed with ERK1/2. This result indicates that p47^{phox} antisense, although effective in inhibiting expression of p47^{phox} protein, did not inhibit an unrelated, constitutively expressed protein. Together, these results suggest that inhibition of p47^{phox}

protein expression correlates with a downregulation of NADPH oxidase activity and that $p47^{phox}$ protein expression in human monocytes is required for O_2^{-} production.

As just demonstrated, the specificity of antisense ODN inhibition can be assessed by evaluating levels of related or unrelated cell proteins. It is essential to remember that antisense ODN effectiveness, in inhibiting the expression of a protein, depends not only on blocking new mRNA translation but on the decay of existing protein. Information regarding the half-life of the protein of interest will aid study design for the duration of antisense ODN treatment. This issue was important in these studies with $p47^{phox}$ because we discovered that this protein has a long half-life. Thus it was necessary to treat for 3 days, with two feedings, in order to observe substantial inhibition of NADPH oxidase activity and related inhibition of $p47^{phox}$ expression.

Indirect Assessment of NADPH Oxidase Function Using Antisense ODN as a Tool

Inhibition of O_2^{-} Production Using Antisense ODN to PKCα. In other studies, we have discovered that NADPH oxidase can be regulated tightly by upstream signaling pathways required for achieving NADPH oxidase activity. Among these are studies with PKC.[22,52] Selective pharmacologic inhibitors of PKC were shown to substantially inhibit the production of O_2^{-} by activated human monocytes. To confirm that this was indeed due to inhibition of PKC, we designed an antisense ODN to a conserved region of PKC isoforms belonging to the conventional PKC (cPKC) group of PKC enzymes. Previous studies indicating requisite roles for calcium influx and release from intracellular stores influenced our choice to target the calcium-dependent group of PKC isoenzymes.[59] This group is composed of PKCα, PKCβI, PKCβII, and PKCγ, although this latter isoform is not expressed in monocytes. The cPKC antisense ODNs blocked O_2^{-} production, on monocyte activation, whereas sense ODNs were without effect. These results confirmed the data obtained with pharmacologic inhibitors. To determine which of the isoenzymes in this cPKC group were required for O_2^{-} production through NADPH oxidase activity, we designed an antisense ODN specific for PKCα and another antisense ODN that was specific for both PKCβI and βII. Only the antisense for PKCα inhibited O_2^{-} production, thus supporting the conclusion that PKCα is the only cPKC isoenzyme required for NADPH oxidase activity.[22]

Inhibition of O_2^{-} Production Using Antisense ODN to cPLA$_2$. Additional studies have revealed a requisite role for another upstream enzyme regulating NADPH oxidase activity, namely cPLA$_2$.[53] Studies performed in our laboratory, using a variety of pharmacologic inhibitors as well as antisense ODNs, have implicated an essential role for this enzyme in allowing the activation of NADPH

---

[59] Q. Li, A. Tallant, and M. K. Cathcart, *J. Clin. Invest.* **91**, 1499 (1993).

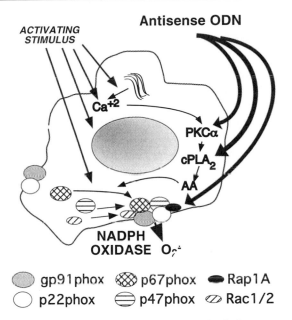

FIG. 4. Antisense ODN inhibition of NADPH oxidase activity in human monocytes. Antisense ODNs can be used to inhibit NADPH oxidase activity either by directly inhibiting the expression of components of the NADPH oxidase (e.g., p47^{phox}) or by inhibiting expression of enzymes shown to provide essential upstream regulatory signals.

oxidase. Thus, specific inhibition of cPLA$_2$ by antisense ODN allowed us to prove that it regulates the activity of this enzyme complex and the related O$_2^-$ formation. An interesting aspect of this finding is that the inhibition of expression of cPLA$_2$ by antisense ODNs can be completely abolished by addition of arachidonic acid (AA) at predicted physiologic levels. It seems that the requisite role for cPLA$_2$ activity is entirely due to the requirement for the generation of AA.[53]

Furthermore, studies by our laboratory have indicated that the requirement for PKCα, mentioned above, is solely for phosphorylating and activating cPLA$_2$ and generating the essential AA.[60] Thus, AA can also completely restore the activity of NADPH oxidase in monocytes treated with antisense ODNs to PKCα.[22,60] The identification of these pathways as requisite regulators of NADPH oxidase activity provides alternative ways to control the activity of this enzyme complex and to regulate O$_2^-$ production. The various antisense ODNs used to modulate NADPH oxidase activity in human monocytes by our laboratory are depicted in Fig. 4. Thus, direct as well as indirect targets for antisense ODN inhibition can efficiently regulate the activity of this important enzyme complex.

[60] Q. Li, V. Subbulakshmi, and M. K. Cathcart, in revision (2002).

Concluding Remarks

We have devised alternative approaches for inhibiting NADPH oxidase activity in human monocytes, using antisense ODNs. Although several pharmacologic inhibitors have been suggested to inhibit the activity of this enzyme complex, antisense ODN-mediated inhibition offers greater specificity and related lower toxicity. We recommend that none of these approaches, outlined in this chapter, be used in isolation. The use of multiple complementary approaches to modulate NADPH oxidase activity, or for that matter any enzyme activity, affords the investigator confidence that the experimental observations are sound.

Acknowledgments

This work was supported by NIH Grants HL-51068 and HL-61971 (M.K.C.). Additional support was provided by Minority Research Supplement HL-51068 (E.A.B.).

[37] Transgenic Shuttle Vector Assays for Determining Genetic Differences in Oxidative B Cell Mutagenesis in Vivo

By Klaus Felix, Lynne D. Rockwood, and Siegfried Janz

Because tumorigenesis is an *in vivo* phenomenon that can be evaluated optimally only in *in vivo* studies, it has been postulated that attempts to link tumor development with mutagenesis may be most informative when mutagenesis can also be assessed *in vivo*. We have chosen two transgenic shuttle vector assays, λLIZ (Big Blue) and pUR288 (placZ), to explore in mice the role of oxidative B cell mutagenesis in the development of the malignant plasma cell tumor, plasmacytoma (PCT). Here we provide a brief introduction to the putative role of oxidative stress during mouse plasmacytomagenesis (PCTG) and describe the utility of the λLIZ and pUR288 assays for determining oxidative B cell mutagenesis *in vivo*. In addition, we introduce a genetic system that employs congenic mouse strains to associate the susceptibility to PCT development with the genetic control of oxidative mutagenesis in the B cell compartment.

BALB/c Plasmacytomas

Our interest in oxidative mutagenesis in B lymphocytes stems from our efforts to learn about the pathogenesis of inflammation-induced PCTs in mice. PCTs are immunoglobulin-producing neoplasms of terminally differentiated

Copyright 2002, Elsevier Science (USA).
All rights reserved.
0076-6879/02 $35.00

B lymphocytes, that is, plasma cells. The tumors can be induced in genetically susceptible, inbred BALB/c (B/c) mice by intraperitoneal injections of a variety of proinflammatory agents, including the C_{19} isoalkane pristane (2,6,10,14-tetramethylpentadecane). Treatment with pristane provokes the formation of the pristane granuloma, a chronic inflammatory tissue in the peritoneal cavity. The granuloma is mainly composed of macrophages and further characterized by extensive infiltrates of neutrophils. The ability of macrophages and neutrophils to secrete copious amounts of reactive oxygen intermediates (ROIs) has long raised suspicions that a pathogenetic link exists between oxidative mutagenesis and PCT development. However, the exact nature of this link has been difficult to elucidate because PCTG is a multifactorial process (reviewed in Potter and Wiener[1]) that has many potential connections to oxidative stress. The following features of PCTG are relevant for contemplating links to inflammation and oxidative mutagenesis: (1) PCTG takes place in the pristane granuloma. Consequently, PCTs do not develop in normal B/c mice not treated with pristane and, thus, devoid of granuloma; (2) PCTG is dependent on the chronic inflammatory process in the granulomatous tissue. Treatment of mice with the antiinflammatory agent, indomethacin, abrogates PCT formation in pristane-primed mice without grossly changing granuloma histology. Furthermore, three injections of pristane spaced 2 months apart are considerably more effective in PCT induction than single injections, because three waves of inflammation are generated over a prolonged period of time instead of one wave; (3) PCTG is a multigenic trait that is under tight genetic control. Among a large panel of common inbred strains of mice injected with pristane, B/c was found to be uniquely susceptible to tumor induction, whereas DBA/2N (D2) and many other strains were found to be solidly resistant. PCT-susceptible B/c mice and PCT-resistant D2 mice have since been used as benchmark strains to identify genes that confer susceptibility and resistance to PCT development (R/S genes). R/S genes have been mapped to several regions of the genome, including three loci to the distal part of chromosome 4[2]; (4) PCT susceptibility is incompletely penetrant. At best, only 40–60% of conventionally maintained pristane-treated B/c mice develop PCTs. Certain changes in the environment, for example, in the diet or antigenic exposure of mice, can drastically reduce the tumor incidence; (5) PCTG is a prolonged process that takes on average 220 days to complete; and (6) the histogenesis of PCT development begins with the formation of small clusters of B lymphocytes, lymphoplasmacytoid cells, and plasma cells approximately 30 days after the first injection of pristane. It proceeds with the appearance of plasmacytic foci (compact aggregates of at least 50 plasma cells in one tissue site ~75 days after pristane administration), and, in some cases, megafoci (aggregates of several hundred to a thousand atypical plasma cells ~150 days after pristane administration)

---

[1] M. Potter and F. Wiener, *Carcinogenesis* **13,** 1681 (1992).
[2] S. Zhang, E. S. Ramsay, and B. A. Mock, *Proc. Natl. Acad. Sci. U.S.A.* **95,** 2429 (1998).

before culminating in incipient PCTs at 150 to 300 days after pristane administration. In the course of this prolonged oncogenic process, B cells are thought to undergo malignant transformation in close proximity to, or possibly by direct cell-to-cell contact with, macrophages and neutrophils, two significant sources of ROIs *in situ*. We hypothesized that B cells may become targets of oxidative DNA damage and mutagenesis if the exposure to ROIs overwhelmed their antioxidative defense and DNA repair capacity. If so, the pristane granuloma may function as a catalyst of PCTG by providing an environment of oxidative mutagenesis. We further postulated that D2-typical R genes and B/c-typical S genes may contribute to the phenotypes of PCT resistance and susceptibility by minimizing and increasing, respectively, the levels of oxidative mutagenesis in the B cell compartment.

λLIZ and placZ Congenics

The λLIZ (Big Blue) and pUR288 (placZ) mutagenesis assays are based on shuttle vectors (λLIZ and pUR288, respectively) that have been inserted as inheritable transgenes into the germ line of C57BL/6 (B6) mice. λLIZ has been integrated as a 40-copy concatemer in the central portion of chromosome 4 near *Tyrp1* (38 cM). The B6–λLIZ mouse was developed by J. Short and colleagues at Stratagene (La Jolla, CA).[3] pUR288 (line 60) is somewhat unusual with regard to transgene integration, as two 10-copy concatemers have been incorporated on chromosomes 3 and 4. The pUR288 mouse was developed by J. Vijg and associates.[4,5] We transferred the shuttle vectors from their original B6 background onto the two benchmark strains for genetic studies of PCTG: PCT-susceptible B/c mice and PCT-resistant D2 mice. Strain C.D2-Idh1-Pep3, a B/c mouse congenic for a 41.2-cM-long D2-derived portion of chromosome 1 spanning *Idh1* and *Pep3*, was chosen as an additional recipient because it demonstrated the unusual phenotype of hypersusceptibility to PCT development (~75% tumors 300 days after the first injection of pristane).[6] To generate the envisioned congenics, a facilitated backcross protocol was employed that combined detection of the λLIZ and pUR288 transgenes with selection for recipient-type paternal chromosomes by means of simple sequence length polymorphic markers detectable by polymerase chain reaction (PCR). Of importance, transgene-positive offspring considered for further breeding were routinely tested for rescue efficiency of the shuttle vector. This ensured that only animals in which the mutagenesis assay worked properly were chosen for the next-generation backcross. We highly recommend this control

[3] S. W. Kohler, G. S. Provost, A. Fieck, P. L. Kretz, W. O. Bullock, D. L. Putman, J. A. Sorge, and J. M. Short, *Environ. Mol. Mutagen.* **18,** 316 (1991).

[4] J. A. Gossen, W. J. de Leeuw, A. C. Molijn, and J. Vijg, *Biotechniques* **14,** 624 (1993).

[5] M. E. Boerrigter, M. E. Dolle, H. J. Martus, J. A. Gossen, and J. Vijg, *Nature* (*London*) **377,** 657 (1995).

[6] M. Potter, E. B. Mushinski, J. S. Wax, J. Hartley, and B. A. Mock, *Cancer Res.* **54,** 969 (1994).

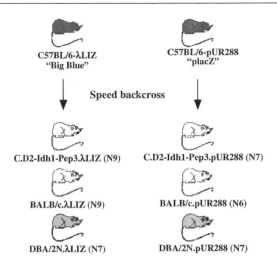

FIG. 1. Development of a genetic system that is conducive for correlating the genetics of oxidative B cell mutagenesis with the susceptibility to PCT development. The system is based on PCT-hypersusceptible C.D2-Idh1-Pep3, PCT-susceptible B/c, and PCT-resistant D2 mice congenic for shuttle vector λLIZ or pUR288. The congenic mice were generated by facilitated backcrossing of λLIZ or pUR288 from B6 donor mice to the indicated recipients. All congenics are homozygous, λLIZ^{+/+} or placZ^{+/+}. However, the pUR288 congenics contain only 20 copies of the transgene instead of the 40 copies in B6–pUR288 parental mice. This loss was caused by limiting the backcross of pUR288 to the 10-copy transgene on chromosome 3. The second 10-copy transgene on chromosome 4 was eliminated during breeding because this chromosome contains as many as three R and S genes for PCTG. The presence of these genes in the neighborhood of pUR288 would have complicated the backcross in a major way and possibly interfered with the desired R/S phenotype of the congenics.

to anyone who contemplates a similar backcross, because we did observe several mice with severe reductions in rescue efficiency. The reasons for that remained unclear; however, deletions in the transgenic concatemer during pairing of homologous chromosomes in meiosis are a possible candidate. Summarizing 3 years of breeding, genotyping, and testing for shuttle vector rescue efficiency, Fig. 1 illustrates the congenic strains currently available.

λLIZ (Big Blue) Assay

Shuttle Vector and Principle of Assay

The principal utility of the λLIZ shuttle vector (Fig. 2A) lies in its ability to transfer a suitable target and reporter gene for mutagenesis (*lacI* and *αlacZ*, respectively) from the mouse genome to an *Escherichia coli* detection system.[7]

---

[7] G. S. Provost, P. L. Kretz, R. T. Hamner, C. D. Matthews, B. J. Rogers, K. S. Lundberg, M. J. Dycaico, and J. M. Short, *Mutat. Res.* **288**, 133 (1993).

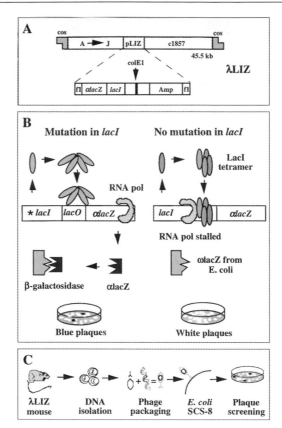

FIG. 2. *In vivo* mutagenesis assay, using λLIZ. (A) Phage λ-derived shuttle vector λLIZ. The partial f1 origins in the vector allow the excision of the target region, pLIZ, as a single-stranded phagemid in *E. coli* coinfected with the helper phage, M13. The colE1 origin of replication allows in a second step the recovery of pLIZ in a suitable *E. coli* host as a convenient double-stranded plasmid. This is a great help for mutational analysis at the DNA sequence level. pLIZ includes in addition to the target gene of mutagenesis, *lacI*, and the reporter gene of mutagenesis, α*lacZ*, plasmid maintenance sequences and the ampicillin resistance gene (Amp). (B) Principle of the assay. α*lacZ* is repressed in the absence of a mutation in *lacI* (scheme to the right) but is derepressed in the presence of a mutation in *lacI* (indicated by an asterisk in the scheme to the left). See text for details. (C) Overall performance of the λLIZ assay. It entails (1) the preparation of high molecular weight genomic DNA from tissue samples or purified cell populations, (2) the recovery of the λLIZ transgenes from the genomic DNA as infectious phages [accomplished by a phage-packaging reaction that utilizes the paired *cos* sites flanking the prophage; see (A)], (3) the infection of restriction-deficient *recA lacZΔM15 E. coli* K12 SCS-8 with the recovered phage, (4) the plating of the infected bacteria in top agarose containing X-Gal (a substrate for β-Gal), and (5) the enumeration of β-Gal-expressing blue mutant plaques after an 18-hr incubation at 37° in relation to the total number of plaques. The efficiency of the phage-packaging reaction ranges in our hands from 2×10^4 to 6×10^4 plaques per reaction.

λLIZ permits the expression, via a derepression mechanism, of the β-galactosidase (β-Gal) reporter gene, $\alpha lacZ$, whenever an inactivating mutation in *lacI* occurs. *lacI* encodes the physiological $lacZ/\alpha lacZ$ repressor, the LacI protein, which binds in its unmutated tetrameric form to the operator, *lacO*, and blocks or represses, thereby, the transcription of $\alpha lacZ$. This is illustrated with the stalled RNA polymerase complex in Fig. 2B (right). A mutation in *lacI*—which either abrogates coding potential altogether (by creating a stop codon or frameshift), or results in the synthesis of a weak dysfunctional protein that is unable to undergo proper tetramer formation and binding to *lacO* (via base substitution)—allows the RNA polymerase to read through and produce the α portion of the *lacZ* transcript. $\alpha lacZ$ is said to be derepressed in this situation (Fig. 2B, left), and the encoded α portion of LacZ can be combined in α-complementing *E. coli* with the ω portion of LacZ (encoded by the bacterial genome) to form a catalytically active β-Gal protein. The overall performance of the λLIZ assay is illustrated in Fig. 2C. Mutant frequencies in gene *lacI* are calculated as the ratio of mutant to total plaques. To identify the nature of individual mutations in the *lacI* target gene, mutant plaques can be cored, verified by replating, and analyzed at the molecular level, which is most commonly done by DNA sequencing. To avoid the loss of *lacI* mutants expressing the phenotype of a light blue color, the inclusion of the CM0 and CM1 faint color mutants is recommended as an internal sensitivity standard of the assay.[8] The λLIZ system is commercially available as the Big Blue *in vivo* mutagenesis assay (Stratagene), which is supported by a range of related products and a useful website (www.stratagene.com/cellbio/toxicology/big_blue_intro.htm) that includes an extensive bibliography. In addition, J. de Boer (Center for Environmental Health of the University of Victoria, BC, Canada) has designed a beautiful Web page (http://esg-www.mit.edu:8001/bio/pge/mutants.html) on the λLIZ assay, which contains useful databases, downloadable software, detailed experimental protocols, and much other beneficial information.

Oxidative Mutagenesis in Lymphoid Tissues and Purified B Cells

The applicability of the λLIZ assay to assessments of oxidative mutagenesis in whole tissue samples or purified cell fractions is illustrated by two pieces of data. The genotypic differences in mutant rates that were observed in spleens of B/c.λLIZ and D2.λLIZ mice after intraperitoneal injection of pristane, or after treatment with the inhibitor of γ-glutamylcysteine synthase, buthionine sulfoximine (BSO), are depicted in Fig. 3. Pristane is thought to cause oxidative stress by attracting inflammatory cells (macrophages and neutrophils) that can mutagenize B cells in the peritoneal cavity before they migrate back to the spleen. BSO is known to increase endogenous oxidative stress by depleting cytosolic

[8] B. J. Rogers, G. S. Provost, R. R. Young, D. L. Putman, and J. M. Short, *Mutat. Res.* **327,** 57 (1995).

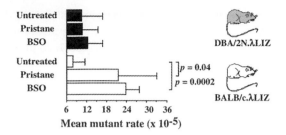

FIG. 3. Mean mutant frequencies (± standard deviations) in the spleens of PCT-resistant D2.λLIZ mice and PCT-susceptible B/c.λLIZ mice. Experimental groups consisted of four to six mice. A total of 3.11×10^5 plaques was screened on average in each group. The number of blue mutant plaques enumerated in these groups ranged from 25 to 63.

glutathione (GSH) stores. PCT-susceptible B/c mice displayed significant elevations in splenic mutant rates after treatment with pristane or BSO, whereas PCT-resistant D2 mice did not.[9] This result is consistent with the view that the genetic susceptibility to PCT development may be associated with elevated levels of oxidative mutagenesis in lymphoid tissues and B cells.[10] The hypothesis that B/c B cells may be particularly susceptible to oxidative mutagenesis was further supported by the result of the experiment summarized in Table I. The λLIZ assay was used to determine the mutant rates in B/c.λLIZ spleen cells after separation by magnetic cell sorting (MACS; Miltenyi Biotec, Bergisch Gladbach, Germany) into B220$^+$ B cells and B220$^-$ non-B cells. Comparison with the mutant rates in whole spleens obtained from age-matched B/c.λLIZ mice was also included. Treatment with pristane resulted in an increase in mutant levels in whole spleens by 64.8% (from 5.88×10^{-5} to 9.69×10^{-5}). The increase was significant ($p = 0.0187$) in the two-tailed paired t test. Furthermore, although no differences in mutant rates between whole spleens and fractionated B and non-B cells were found in normal B/c mice (Table I, rows 1–3), a significant reduction and elevation in mutant rates was found in non-B cells and B cells, respectively, when fractionated cells obtained from pristane-primed animals were compared with whole spleens of pristane-primed animals (Table I, rows 4–6). These results were interpreted to mean that B/c B cells may indeed experience dysproportionately high amounts of oxidative mutagenesis under conditions of pristane-induced inflammation. D2 B cells may be different in this regard, but this has not yet been demonstrated.

Many groups, including our own, have observed that time-consuming procedures for B cell isolation can result in an unwanted loss of the main cellular antioxidant, GSH, and a drop in the biological activity of many other enzymes

[9] K. Felix, K. Kelliher, G. W. Bornkamm, and S. Janz, *Cancer Res.* **58,** 1616 (1998).
[10] K. Felix, K. A. Kelliher, G. W. Bornkamm, and S. Janz, *Cancer Res.* **59,** 3621 (1999).

TABLE I

lacI MUTANT RATES IN WHOLE SPLEENS AND SPLENIC B AND NON-B CELLS

Tissue[a]	n[b]	Treatment[c]	Mutants[d]	PFU[e]	MF[f]	SD[g]	p[h]
Spleen	3	—	64	1,130,600	5.88	0.582	—
B220⁻	3	—	14	234,340	5.98	2.86	NS
B220⁺	3	—	18	233,400	7.66	1.09	NS
Spleen	5	Pristane	104	1,120,400	9.69	1.56	—
B220⁻	5	Pristane	25	388,000	6.82	1.72	<0.004
B220⁺	5	Pristane	45	336,900	14.0	2.18	<0.002

[a] Spleens were obtained from three normal untreated BALB/c.λLIZ (N9F1) mice or five pristane-treated BALB/c.λLIZ (N9F1) mice 67 days after the first injection of pristane. Splenic cells from two additional groups of mice, three untreated mice and five pristane-treated mice, were used to fractionate B220⁺ B cells from B220⁻ non-B cells, using MACS.
[b] Number of mice.
[c] Mice were left untreated or injected with Pristane on days 1 and 60.
[d] Number of blue plaques (plaque-forming units, PFU) with mutations in gene lacI.
[e] Number of total plaques.
[f] Mean mutant frequency determined as the ratio of mutant to total plaques, as multiples of 10^{-5}.
[g] Standard deviation of the mean mutant frequency, as multiples of 10^{-5}.
[h] Significance values were determined with the two-tailed paired t test. Mutant rates in fractionated cells were compared with mutant rates in whole spleens (NS, not significant).

involved in antioxidative defense. To minimize this problem, mice should be stressed as little as possible before sacrifice, which should be performed by cervical dislocation rather than asphyxiation in CO_2. For the fractionation of mouse B cells, we prefer MACS over fluorescence-activated cell sorting (FACS), despite the somewhat lower purity obtainable by MACS. The main benefit of MACS is the speed and ease of the procedure. Among the various protocols for B cell purification by MACS, we favor the positive enrichment afforded by the B220 (CD45R) surface receptor, which is expressed throughout the entire B lineage except plasma cells. Our protocol begins with a splenic perfusion technique,[11] which yielded in six independent experiments an average of 1.58 (±0.236) × 10^8 spleen cells. It continues with MACS VS⁺ separation columns, which permitted us to recover from these single-cell suspensions 5.27 (±0.796) × 10^7 B220⁺ B cells (33.4% recovery) and 5.45 (±0.649) × 10^7 B220⁻ non-B cells (34.5% recovery). One-third (32.1%) of the cells was typically lost. We refer the reader to the Web page of Miltenyi Biotec (www.miltenyibiotec.com) for detailed technical instructions on the fractionation of various cell types by MACS.

[11] K. Felix, S. Lin, G. W. Bornkamm, and S. Janz, Eur. J. Immunol. 27, 2160 (1997).

pUR288 (placZ) Assay

Shuttle Vector and Principle of Assay

In the plasmid-based shuttle vector pUR288, *lacZ* functions as both target and reporter gene of mutagenesis. The 5346-bp vector contains in addition to *lacZ*-coding sequences the 35-bp binding site for the LacI repressor, *lacO*, the binding site for CRP (cAMP receptor protein, which facilitates transcription of *lacZ* by stimulating the formation of an active promoter complex), an origin of replication (*ori*), and an ampicillin resistance gene (Fig. 4A). The principle of the pUR288

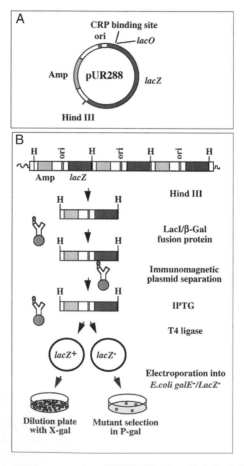

FIG. 4. *In vivo* mutagenesis assay, using pUR288. (A) Plasmid-derived shuttle vector pUR288. CRP, cAMP receptor protein; *ori*, origin of replication; *lacZ*, gene encoding β-Gal; *Hind*III, restriction site used to release the plasmid from genomic DNA. (B) Principle of the assay: LacZ⁻ mutants are positively selected on P-Gal plates. See text for additional details.

mutagenesis assay is illustrated in Fig. 4B. Briefly, shuttle vectors are excised from the transgenic concatemer by restriction with *Hind*III. Linearized plasmids are then separated from bulk genomic DNA with the help of magnetic beads that are coated with a LacI fusion protein that can bind to the *lacO* sequence of the plasmid.[4] Elution from the beads is achieved by adding isopropyl-β-D-thiogalactopyranoside (IPTG), an inactivator of LacI that induces an affinity decreasing conformational change in the protein for *lacO*. Plasmids are circularized at the cohesive *Hind*III sites by ligation with T4 ligase and then electroporated into *E. coli* that is (1) deficient in β-Gal (*lacZ*$^-$), (2) galactose intolerant due to the absence of galactose epimerase (*galE*$^-$), and (3) restriction negative to prevent the degradation of incoming methylated plasmid DNA. The *galE*$^-$ mutation is key[12] because it allows for the positive selection of *lacZ*$^-$ mutants in the presence of the lactose analog P-Gal (phenyl-β-D galactoside), a substrate for β-Gal. LacZ$^-$ mutants are unable to cleave P-Gal; in contrast, wild-type LacZ$^+$ cells are able to cleave it, and thereby release galactose. Galactose is converted to UDP-galactoside, which cannot be further metabolized on the *galE*$^-$ background; instead, it is accumulated intracellularly to toxic and eventually lytic concentrations. Thus, whereas LacZ$^+$ cells are prevented from growth on P-Gal-supplemented agar plates, LacZ$^-$ cells are not. To determine the rescue efficiency of plasmids from genomic DNA, a small aliquot (usually 2 μl) of a 2-ml suspension of pUR288-transfected *E. coli* is plated on a titer plate that has been supplemented with 5-bromo-4-chloro-3-indolyl-β-D-galactopyranoside (X-Gal). X-Gal is cleaved by β-Gal, which produces a blue halo around growing LacZ$^+$ colonies. Note that LacZ$^-$ mutants will be missed on the dilution plate (no blue halo), but this is negligible because the ratio between LacZ$^+$ to LacZ$^-$ colonies is on the order of 10^4 : 1. Note further that the cleavage of X-Gal releases galactose, which is converted to the same toxic UDP-galactoside derived from P-Gal; however, the amounts of galactose liberated from X-Gal are small and therefore compatible with the growth of LacZ$^+$ colonies [the molar concentration of X-Gal (183 μM; 75 μg/ml) is 64 times lower than that of P-Gal (11.7 mM, 3 mg/ml)]. The remaining part of the suspension (1998 μl) is plated on a single P-Gal plate to select for mutants. Mutants grow as small, red, formazan-stained colonies in the presence of a tetrazolium salt that should be added for improved visibility of the sometimes tiny colonies. The mutant frequency is calculated as the ratio of mutants to nonmutants; that is, the number of colonies on the P-Gal selection plate to the number of colonies ($\times$1000) on the X-Gal titer plate. To characterize the mutational spectrum, mutant colonies can be picked and used directly as templates in long-range PCR amplifications of the entire *lacZ* gene. The PCR fragments are useful for restriction analysis and DNA sequencing. Plasmid DNA minipreparations are equally useful if it is decided not to use PCR for template preparation. Excellent reviews of the pUR288 assay including detailed protocols

---

[12] J. A. Gossen, A. C. Molijn, G. R. Douglas, and J. Vijg, *Nucleic Acids Res.* **20**, 3254 (1992).

and sections on troubleshooting are available.[13–15] More specialized articles on the detection of so-called color mutants,[16] the nature of background mutations,[17] and sources of assay variability[18] have also been published. In addition, a commercial kit-based version of the assay—together with technical support, a step-by-step protocol, and accessory services, such as mutational analysis by DNA sequencing and two-dimensional electrophoresis—is being offered by Leven (www.leveninc.com). Our overall experience with the assay is positive, but attention must be paid to preparing DNA of high quality [even small amounts of impurities (proteins, salts, organic solvents) can sometimes decrease the rescue efficiency of the shuttle vector] and avoiding contamination with unrelated plasmids that are $lacZ^-$, $lacO^+$, and ampicillin resistant. Plasmids of this nature (e.g., derivatives of pBluescript) are in widespread use in many laboratories, and they can easily show up as false "pUR288 mutants" after finding their way into reagents or laboratory space where the mutagenesis assay is performed.

Phagocyte-Mediated Oxidative Mutagenesis in B Lymphoblasts

The utility of the pUR288 assay for evaluating the impact of defined genetic mutations on the levels of oxidative mutagenesis in B cells is illustrated in Fig. 5. Pristane-elicited peritoneal exudate cells (PECs; mainly macrophages and neutrophils) were utilized as effectors of mutagenesis in coincubation experiments with lipopolysaccharide (LPS)-stimulated proliferating splenic B lymphoblasts, the targets of mutagenesis. The experiment was designed as the *in vitro* equivalent of the pristane granuloma to test the hypothesis that the close proximity to inflammatory phagocytes may create a mutagenic environment for B cells (Fig. 5A). A benefit of the chosen experimental design was that only mutations in B cells were enumerated. B cells harbored the pUR288 shuttle vector but PECs did not. Two major variables were included in the experiment. First, PECs were obtained from B6 knockout mice that lacked the $p47^{phox}$ protein subunit of the NADPH oxidase complex and were thus unable to undergo an oxidative burst,[19] or were obtained from normal NADPH-proficient B/c mice. Second, lymphoblasts prepared from normal B/c.pUR288 mice were compared with lymphoblasts from B/c.pUR288 mice that carried the *xid* mutation. *xid* is a naturally occurring null mutation in

[13] J. A. Gossen, W. J. de Leeuw, and J. Vijg, *Mutat Res.* **307**, 451 (1994).

[14] J. Vijg, M. E. Dolle, H. J. Martus, and M. E. Boerrigter, *Mech. Ageing Dev.* **99**, 257 (1997).

[15] J. Vijg and G. R. Douglas, *in* "Technologies for Detection of DNA Damage and Mutations" (G. P. Pfeiffer, ed.), p. 391. Plenum, New York, 1996.

[16] M. E. Boerrigter, *Environ. Mol. Mutagen.* **32**, 148 (1998).

[17] M. E. Dolle, H. J. Martus, M. Novak, N. J. van Orsouw, and J. Vijg, *Mutagenesis* **14**, 287 (1999).

[18] M. E. Boerrigter and J. Vijg, *Environ. Mol. Mutagen.* **29**, 221 (1997).

[19] S. H. Jackson, J. I. Gallin, and S. M. Holland, *J. Exp. Med.* **182**, 751 (1995).

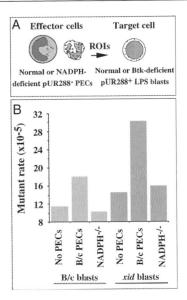

FIG. 5. Phagocyte-mediated mutagenesis in neighboring B cells. (A) Scheme of the coincubation of pUR288⁻ effectors of mutagenesis, peritoneal exudate cells (PECs), with pUR288⁺ target cells of mutagenesis, splenic B lymphoblasts. (B) Mutant frequencies were determined in normal B/c lymphoblasts (triplet of light gray columns to the left) and Btk-deficient B/c lymphoblasts (triplet of dark gray columns to the right) in the absence of PECs (left columns), the presence of normal B/c PECs (middle columns), and the presence of B6 $p47^{phox-/-}$ PECs deficient in NADPH oxidase (right columns). A total of $4–25 \times 10^5$ colonies was screened in the different experimental groups.

Bruton's tyrosine kinase, an important signaling molecule in B cells that has been shown to play a major role in B/c PCTG.[20] Figure 5B shows that normal B/c PECs induced elevations in mutant rates in both normal and *xid* B cells. In contrast, NADPH-deficient B6 PECs ($p47^{phox-/-}$) were unable to mutagenize neighboring B cells. Furthermore, although the overall response pattern was highly similar in normal and *xid* B cells, the latter seemed to be more susceptible to phagocyte-induced mutagenesis. Taken together, the results demonstrate that the pUR288 assay is useful for both assessing cell-mediated oxidative mutagenesis in B cells and studying the effect of genetic mutations on this phenotype.

Comparison of λLIZ and pUR288

In vivo mutagenesis systems are powerful experimental tools that have revolutionized mutagenicity testing. We believe they are worth the initial investment

---

[20] M. Potter, J. S. Wax, C. T. Hansen, and J. J. Kenny, *Int. Immunol.* **11,** 1059 (1999).

of time and resources required to set them up in the laboratory. On the basis of our own experience, we can recommend both the phage λ-based λLIZ assay and the plasmid-based pUR288 assay. Both assays are thought to reflect accurately the mutation rate in the overall genome.[21] To appreciate this point, it is important to realize that transgenic shuttle vectors reside as irrelevant, heavily methylated, nontranscribed passengers in the genome. Mutations in the target genes of mutagenesis, *lacI* (λLIZ) or *lacZ* (pUR288), are therefore neutral, neither beneficial nor damaging to the cell in which they occur. It follows that both assays offer appropriate indicator systems for evaluating general mutagenesis *in vivo* in the absence of any selective pressure of the mutated gene. A major additional argument for the usefulness of transgenic shuttle vectors is the remarkable similarity of mutational patterns in reporter genes to mutation profiles observed in endogenous genes, such as *Hprt*[22] and Ha-*ras* and *Apc*.[23] When it comes to choosing between λLIZ and pUR288 for a particular mutagenicity project, it is necessary to consider some important differences between both assays (Fig. 6). The phage assay detects point mutations, frameshifts, and small deletions (<20 bp) with great efficiency, but it fails to detect larger deletions, the hallmark mutations of oxidative mutagens. However, with the pUR288 system, large deletions of up to 3 kbp, the size of *lacZ*, can be readily detected,[24] as well as recombinations of the reporter gene with mouse genomic DNA.[5] There are also important practical ramifications that distinguish both assays. The λLIZ assay requires at least 20 μg of high-quality genomic DNA, the acquisition of which may be difficult when working with rare cell types such as certain subpopulations of B cells. The pUR288 assay gets by with less DNA; 5 μg, corresponding to ~2 × 10^6 cells, may be enough to obtain reliable data. The λLIZ assay in its original version, which does not utilize a positive selection scheme for mutants, is more labor intensive and time consuming than the plasmid assay, which employs the clever *galE*⁻ method for mutant selection. A positive selection method for the λLIZ cII gene has been developed[25] and validated,[26,27] but we have not yet implemented this version of the assay in our laboratory. The clear strength of the λLIZ assay is the ease with which mutational

[21] L. Cosentino and J. A. Heddle, *Mutagenesis* **14**, 113 (1999).

[22] R. A. Mittelstaedt, M. G. Manjanatha, S. D. Shelton, L. E. Lyn-Cook, J. B. Chen, A. Aidoo, D. A. Casciano, and R. H. Heflich, *Environ. Mol. Mutagen.* **31**, 149 (1998).

[23] H. Okonogi, T. Ushijima, X. B. Zhang, J. A. Heddle, T. Suzuki, T. Sofuni, J. S. Felton, J. D. Tucker, T. Sugimura, and M. Nagao, *Carcinogenesis* **18**, 745 (1997).

[24] S. Nakamura, H. Ikehata, J. Komura, Y. Hosoi, H. Inoue, Y. Gondo, K. Yamamoto, Y. Ichimasa, and T. Ono, *Int. J. Radiat. Biol.* **76**, 431 (2001).

[25] J. L. Jakubczak, G. Merlino, J. E. French, W. J. Muller, B. Paul, S. Adhya, and S. Garges, *Proc. Natl. Acad. Sci. U.S.A.* **93**, 9073 (1996).

[26] S. E. Andrew, L. Hsiao, K. Milhausen, and F. R. Jirik, *Mutat. Res.* **427**, 89 (1999).

[27] D. M. Zimmer, P. R. Harbach, W. B. Mattes, and C. S. Aaron, *Environ. Mol. Mutagen.* **33**, 249 (1999).

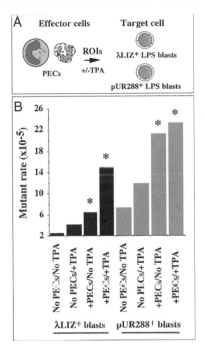

FIG. 6. PEC-mediated mutagenesis in coincubated B cells. (A) Shuttle vector-free PECs, the effectors of mutagenesis, were cocultured with LPS-stimulated splenic B lymphoblasts (LPS blasts), the targets of mutagenesis. LPS blasts contained either λLIZ or pUR288. (B) Mutant frequencies in LPS blasts were measured with the help of the λLIZ or pUR288 assay, either in the absence of PECs (±TPA) or the presence of PECs (±TPA). Significant increases in mutant frequencies (with respect to background levels found in the absence of PECs and TPA) are indicated by asterisks. Note that mutant frequencies were higher in all four samples in the pUR288 assay. This was caused, at least in part, by the ability of the pUR288 assay to detect large deletions and recombinations, two classes of mutations that are missed in the λLIZ assay. Plaques (6–17.6 × 10³) were screened in the λLIZ assay and 0.95–2.0 × 10⁵ colonies were screened in the pUR288 assay in the different experimental groups.

spectra can be analyzed at the DNA sequence level. The approximately 1-kb-long target gene, *lacI*, is not only three times smaller than the target gene of the pUR288 assay, *lacZ*, but most of the mutations are restricted to the 250-bp region of *lacI* that encodes the DNA-binding domain of the LacI repressor. This further reduces the sequencing effort. In addition, the database of known mutations in *lacI* is huge, greatly facilitating interpretations of observed mutation profiles with well-established standards. Thus, the pros and cons of λLIZ and pUR288 should be carefully weighed in order to choose the appropriate assay for an envisioned mutagenesis experiment.

Acknowledgments

We acknowledge Dr. Scott Provost (Stratagene) for teaching K.F. the λLIZ assay and providing λLIZ transgenic B6 mice for backcrossing. We are grateful to Drs. Martijn Dollé and Jan Vijg (Cancer Therapy and Research Center, University of Texas Health Science Center at San Antonio) for training L.R. in the pUR288 assay and providing the pUR288 transgenic B6 mice for backcrossing. We are equally grateful to Dr. Michael Boerrigter (Leven) for sharing his expertise on the pUR288 assay and many fruitful scientific discussions. We are indebted to Drs. Sharon Jackson and Steven Holland (NIAID, NIH) for the kind gift of p47^{phox} knockout mice. We thank our long-term collaborators, Drs. Georg-Wilhelm Bornkamm (Institute for Tumor Genetics and Clinical Molecular Biology, GSF, Munich) and Michael Potter (Laboratory of Genetics, National Cancer Institute, NIH) for supporting these studies.

[38] Redox Control of Cell Cycle-Coupled Topoisomerase IIα Gene Expression

By PRABHAT C. GOSWAMI, RYUJI HIGASHIKUBO, and DOUGLAS R. SPITZ

Introduction

Mammalian topoisomerase IIα (Topo II) is a multifunctional protein involved in many cellular processes including replication, repair, transcription, recombination, chromosome condensation and segregation, and the G_2 cell cycle checkpoint pathway.[1-3] Topo II gene expression during the cell cycle is regulated mainly via posttranscriptional mechanisms of changes in mRNA stability.[4] Topo II mRNA and protein levels increase in late S phase, peak in G_2/M, and rapidly decrease after cell division.[4] Several cancer therapeutic agents including ionizing radiation are known to generate reactive oxygen species and affect Topo II gene expression.[5-8] Because a growing body of literature suggests the importance of oxygen radicals as possible physiological regulators of cell proliferation and expression of Topo II is proliferation dependent, development of methods to assay redox regulation of Topo II gene expression may provide a mechanistic understanding of how the

[1] J. C. Wang, *Annu. Rev. Biochem.* **65,** 635 (1996).

[2] W. C. Earnshaw and M. M. S. Heck, *J. Cell Biol.* **100,** 1716 (1985).

[3] C. S. Downes, D. J. Clarke, A. M. Mullinger, J. F. Gimenez-Abian, A. M. Creighton, and R. T. Johnson, *Nature (London)* **372,** 467 (1994).

[4] P. C. Goswami, J. L. Roti Roti, and C. R. Hunt, *Mol. Cell. Biol.* **16,** 1500 (1996).

[5] P. C. Goswami, M. Hill, R. Higashikubo, W. D. Wright, and J. L. Roti Roti, *Radiat. Res.* **132,** 162 (1992).

[6] S. M. DeToledo, E. I. Azzam, M. K. Gasmann, and R. E. J. Mitchel, *Int. J. Radiat. Biol.* **67,** 135 (1995).

[7] T. A. Jarvinen, J. Kononen, M. Pelto-Huikko, and J. Isola, *Am. J. Pathol.* **148,** 2073 (1996).

[8] D. J. Grdina, J. S. Murley, and J. C. Roberts, *Cell. Prolif.* **31,** 217 (1998).

Copyright 2002, Elsevier Science (USA).
All rights reserved.
0076-6879/02 $35.00

Intracellular redox state influences Topo II gene expression both under normal growth conditions and in response to stress.[9–12]

The purpose of this chapter is to provide a useful description of methods for studying cell cycle-coupled Topo II gene expression and the possible regulatory role of the intracellular redox state in these processes.

Cell Synchronization and Topoisomerase IIα mRNA Stability Assay

HeLa cells (human cervical cancer cell line) are grown in monolayer cultures at 37° and 5% CO_2 in Ham's F10 medium supplemented with 10% (v/v) calf serum (GIBCO, Grand Island, NY), penicillin (100 U/ml), and streptomycin (100 μg/ml). Monolayer cells are grown to 60–70% confluence and synchronous cell populations representing the mitotic phase are obtained by selective detachment of mitotic cells according to previously published procedures (Goswami et al.[4] and Terasima and Tolmach[13]). To minimize perturbation in cell growth, all procedures are performed in a 37° warm room. Approximately 1.5% of mitotic cells are obtained by the shake-off method from an exponentially growing cell culture. Viability, as measured by a colony-forming assay, is more than 95% and the mitotic index is more than 97% as determined by microscopic examination.[4] Synchronized cells (3 5 × 10^5) are plated in 10 ml of prewarmed and CO_2-equilibrated medium in a 100-mm tissue culture dish. Progression through the cell cycle is monitored by fluorescence-activated cell sorting (FACS) analysis. Monolayer cells are pulse labeled with 10 μM bromodeoxyuridine (BrdU) for 30 min, trypsinized, and fixed in 1 ml of ice-cold 70% (v/v) ethanol. Indirect immunostaining of BrdU-labeled cells and subsequent FACS analysis are performed according to a previously published procedure.[14] Briefly, ethanol-fixed cells are washed with phosphate-buffered saline (PBS) and treated with pepsin (0.4 mg/ml in 0.1 N HCl) for 30 min at room temperature. Nuclei are isolated by centrifuging the samples at 530g for 5 min in a Beckman (Fullerton, CA) centrifuge set at 4°. The pellet is washed once with PBS and incubated with anti-BrdU antibody (5 μl of the antibody in a 50-μl sample volume) for 1 hr at room temperature. Antibodies, both primary and secondary, are purchased from BD Immunocytometry Systems (San Jose, CA). At the end of the incubation, samples are diluted with 1 ml of PBS and centrifuged at 1600 rpm for 5 min at 4°. Nuclei are then incubated with fluorescein isothiocyanate (FITC)-conjugated goat

[9] T. Finkel, Curr. Opin. Cell Biol. **10**, 248 (1998).

[10] R. H. Burdon, Free Radic. Biol. Med. **18**, 775 (1995).

[11] P. C. Goswami, J. Sheren, L. D. Albee, A. Parsian, J. E. Sim, L. A. Ridnour, R. Higashikubo, D. Gius, C. R. Hunt, and D. R. Spitz, J. Biol. Chem. **275**, 38384 (2000).

[12] G. Pani, R. Colavitti, B. Bedogni, R. Anzevino, S. Borrello, and T. Galeotti, J. Biol. Chem. **275**, 38891 (2000).

[13] T. Terasima and L. J. Tolmach, Exp. Cell Res. **30**, 344 (1963).

[14] P. C. Goswami, W. He, R. Higashikubo, and J. L. Roti Roti, Exp. Cell Res. **214**, 198 (1994).

anti-mouse IgG for 1 hr and digested with RNase A (0.1 mg/ml) for 30 min at room temperature. The nuclei are counterstained with propidium iodide (PI, 20 μg/ml) for 1 hr and stained cells are analyzed on a FACS 440 (BD Immunocytometry Systems) flow cytometer. The FACS 440 flow cytometer is equipped with a coherent I90-5 UV laser operating at 300 mW and at 488-nm excitation wavelengths. Red fluorescence from PI is detected through a 640-nm long-pass filter, and green fluorescence from FITC is detected through a 525-nm band-pass filter. Data from a minimum of 10,000 nuclei are acquired in list mode and processed by Cytomation software. More than 90% of the cells are in G_1 phase 1 hr after replating (Fig. 1A) and approximately 30% of the cells enter S phase at 12 hr (Fig. 1B). Four hours later, approximately 70% of the cells enter the G_2/M phase (Fig. 1C) and by 20 hr 25% of the cells enter the G_1 phase of the next generation (Fig. 1D). These results demonstrate a synchronous progression of cells during the time period of the experiment and show that the mitotic shake-off method is a suitable technique for obtaining high degree of cell synchrony.

Cells in duplicate dishes are treated with actinomycin D (10 μg/ml) 1 hr (G_1 phase) and 14 hr (S phase) after replating. Cells are cultured in the presence of actinomycin D for an additional 4 hr and scraped into 1 ml of TRI-reagent (MRC, Cincinnati, OH). Total cellular RNA is then isolated and analyzed by Northern blotting (Fig. 1E) according to standard procedures. Radiolabeled Topo II human cDNA probe is prepared by random prime labeling and radioactive bands are visualized by exposing the blot to a PhosphorImager screen (Molecular Dynamics STORM 840 PhosphorImager; Amersham Pharmacia Biotech, Piscataway, NJ). The results presented in Fig. 1E show that Topo II mRNA levels are low in G_1 phase and increase more than 16-fold in late S phase (compare lanes 1 and 3). Inhibition of new transcription in G_1 phase rapidly turns over Topo II mRNA levels (compare lanes 1 and 2 in Fig. 1E). These results are in sharp contrast to those seen in late S phase (14 hr postmitosis). In late S phase, there is essentially no turnover of Topo II mRNA in the absence of new transcription (compare lanes 3 and 4 in Fig. 1E). These results demonstrate that mRNA stability plays a significant role in regulating Topo II mRNA levels during the cell cycle. These results also show the applicability of the mitotic shake-off cell synchronization method in studying cell cycle-coupled gene expression.

Reporter Transfection Assay to Determine Possible Role of Topoisomerase IIα 3′-Untranslated Region in mRNA Levels during Cell Cycle

The plasmid pNASSβ (Clontech, Palo Alto, CA) is used to generate reporter constructs. The pNASSβ vector lacks eukaryotic promoter and enhancer sequences and carries the *Escherichia coli* β-galactosidase gene as the reporter. A 1.6-kb fragment containing 650 nucleotides (nt) of the human Topo II promoter, exon 1 (21 nt),

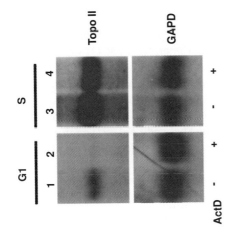

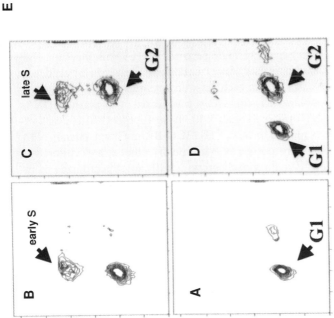

FIG. 1. Cell synchronization and Topo II mRNA stability assay. HeLa cells synchronized by mitotic shake-off were replated and pulse labeled with BrdU at representative times during the cell cycle. (A–D) Contour plots of cell cycle phase distribution assayed by FACS analysis at 1 hr (A), 12 hr (B), 16 hr (C), and 20 hr (D) after mitotic shake-off. Arrows represent cells in G_1, S, and G_2 phases. (E) Northern blot analysis of Topo II mRNA in early G_1, and late S, and G_2 phases. Synchronized HeLa cells either in G_1 phase (1 hr) or S phase (14 hr after plating) were treated with actinomycin D to block new transcription. Total cellular RNA was isolated from cells immediately after addition of the drug and after culture for 4 hr. The blots were rehybridized with human GAPD cDNA to quantitate the amount of RNA loaded per time point. The blots were exposed to a PhosphorImager screen and the abundance of Topo II mRNA was determined relative to that present initially in the untreated cells.

the following intron (860 nt), and exon 2 (150 nt) is amplified by polymerase chain reaction (PCR) from human genomic DNA (Clontech), using primer pairs $5'_{-562}$GGGGCGGGGTTGAGGCAGATGCCAGAATCT$_{-532}$ $3'$ and $5'_{+150}$AGGT GTCTGGGCGGAGCAAAATATGTTCC$_{+121}$ $3'$.[15] The samples are denatured initially at 94° for 2 min and amplification is performed with a DNA thermal cycler (GeneAmp PCR system 9600; PerkinElmer, Norwalk, CT) at 94° for 30 sec, annealing at 60° for 30 sec, and extension at 72° for 30 sec for 25 cycles. The final cycle is followed by a 10-min extension step at 72°. The PCR-amplified fragment is cloned into TA-cloning vector (Invitrogen, Carlsbad, CA) and inserted clones are selected on the basis of blue/white color. The Topo II promoter-containing insert in the TA plasmid DNA is excised with *Spe*I and *Xho*I restriction enzymes and directionally cloned into *Sma*I- and *Xho*I-digested pNASSβ plasmid DNA. The resulting reporter construct (HβgalSV40), when expressed, transcribes a Topo II promoter-driven β-galactosidase reporter mRNA with the simian virus 40 (SV40) $3'$ untranslated region (UTR) at its $3'$ end (approximately 3 kb in size). In the second reporter construct (HβgalTopo), the SV40 $3'$ UTR in HβgalSV40 plasmid DNA is removed by *Sal*I and *Bam*HI restriction enzyme digestion and replaced with the human Topo II $3'$ UTR. The Topo II $3'$ UTR is excised by *Xho*I and *Bam*HI restriction enzyme digestion of plasmid DNA containing the entire human Topo II cDNA.[16] The resulting reporter HβgalTopo construct, when expressed, represents an approximately 4-kb β-galactosidase reporter mRNA with the Topo II $3'$ UTR at its $3'$ end.[11] Orientation and sequence of all inserts are verified by dideoxy sequencing of both strands of DNA in each plasmid construct.

Mouse NIH 3T3 fibroblast cells are stably transfected with plasmid DNAs pSVneo (Clontech), HβgalTopo, or HβgalSV40, using LipofectAMINE according to the manufacturer-supplied protocol (GIBCO-BRL, Grand Island, NY). Geneticin-resistant colonies are pooled or individually cloned, and cultured in G418-containing medium. Asynchronously growing cells are synchronized by incubating monolayer cultures for 30 hr in medium containing 0.2% (v/v) serum and stimulated to reenter the cell cycle in medium containing 10% (v/v) serum. Cells are harvested at various times for analysis of cell cycle position and reporter mRNA levels. For analysis of cell cycle position, cells are harvested by trypsinization and fixed in 70% (v/v) ethanol. The cell pellet is digested with RNase A (1 mg/ml) and stained with PI (10 μg/ml) for FACS analysis. Figure 2A shows representative histograms of cell cycle positions after reentry into the cell cycle. The majority of the cells are in G_1 phase (approximately 85%) at the time of serum stimulation (0 hr) and 6 hr after the stimulation. At 20 hr after serum stimulation 58% of the cells enter S phase and approximately 20–30% of the cells enter G_2/M phase 24 hr

[15] D. Hochhauser, C. A. Stanway, A. L. Harris, and I. D. Hickson, *J. Biol. Chem.* **267,** 18961 (1992).
[16] M. Tsai-Pflugfelder, L. F. Liu, A. A. Liu, K. M. Tewey, J. Whang-Peng, T. Knutsen, K. Huebner, C. M. Croce, and J. C. Wang, *Proc. Natl. Acad. Sci. U.S.A.* **85,** 7177 (1988).

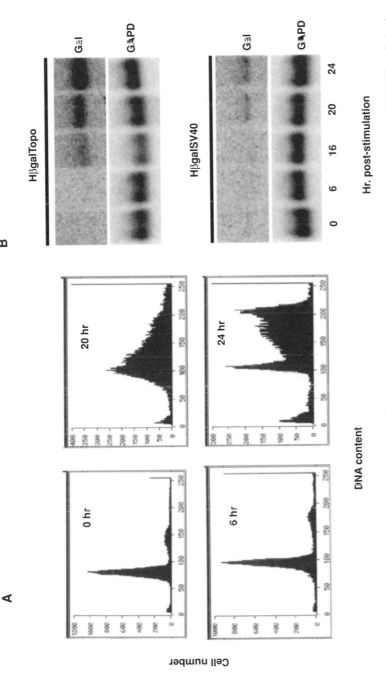

FIG. 2. Reporter transfection assay to determine the possible role of the Topo II 3' UTR in mRNA levels during the cell cycle. (A) FACS analysis of cell cycle position in stably transfected 3T3 cells containing the HβgalTopo reporter construct. Asynchronously growing monolayer cells were incubated for 30 hr with medium containing 0.2% (v/v) serum and serum stimulated to reenter the cell cycle. Cells were trypsinized at the time of serum stimulation (0 hr) and 6, 20, and 24 hr after stimulation and fixed in ethanol. Ethanol-fixed cells were treated with RNase A, stained with propidium iodide and analyzed for DNA content by FACS analysis. (B) Representative Northern blot analysis of HβgalTopo and HβgalSV40 reporter mRNA levels during the cell cycle. Cells in duplicate dishes were harvested at the indicated times for RNA isolation and hybridized with a radiolabeled β-galactosidase probe. Blots were rehybridized with a radiolabeled mouse GAPD cDNA probe.

after serum stimulation. These results are comparable to results obtained with un-transfected control cells (data not shown) and demonstrate that transfection of the reporter construct does not perturb cell cycle progression.

When the total cellular RNA from the HβgalTopo-transfected cell populations is analyzed by Northern blotting, an mRNA band of approximately 4 kb, representing the HβgalTopo reporter construct, is detected with a radiolabeled β-galactosidase probe (Fig. 2B). HβgalTopo reporter mRNA levels are low in G_1 phase (0–6 hr) and increase 12- to 16-fold by late S to G_2 phases (20–24 hr). Results with the HβgalSV40 reporter construct (20–24 hr), which contains the Topo II promoter but not the 3' UTR, indicate a small increase in cell cycle-regulated β-galactosidase reporter mRNA levels. This can be attributed to the 2-fold increase in Topo II transcription that has been reported previously to occur during late S phase by us and other investigators.[4,15,17] Because the two reporter constructs differ only in the 3' UTR sequence, the above-described results show a regulatory role for the Topo II 3' UTR in cell cycle-coupled expression of Topo II. These results demonstrate the feasibility of the reporter assay, which could be used to determine the possible role of 3' UTRs in the regulation of cell cycle-coupled mRNA levels.

In Vitro RNA–Protein Binding Assay

The 177-nucleotide (nt) cDNA sequence representing the putative protein-binding site (nt 4772–4948; Goswami et al.[11]) in the Topo II 3' UTR is amplified by PCR from plasmid DNA containing the entire Topo II 3' UTR (pTopUTR), using primer pairs 5' TAGTGACCATCTATGGG 3' and 5' CTGCTCTAGTTTTAGCT TAGTGG 3'.[11] PCR-amplified Topo II 3' UTR cDNA (177 nt) is cloned into transcription vector pGEM-T plasmid DNA (Promega, Madison, WI). β-Galactosidase transcript is used as nonspecific competitor in the RNA–protein binding assays. β-Galactosidase-coding sequence within the HpaI and ClaI restriction sites in pCMV plasmid (Clontech) is cloned into pBS II SK(+) and runoff transcript is generated by using T7 RNA polymerase. Riboprobes representing the sense strand of RNA from each plasmid are transcribed in vitro according to the protocol from Promega. Labeled riboprobes are transcribed by inclusion of [α-^{32}P]UTP (800 Ci/mmol; NEN Life Science Products, Boston, MA) in the transcription reaction. Reaction mixtures are treated with 1 U of DNase I and riboprobes are purified with the Promega Wizard PCR Prep DNA cleanup system. Bound transcripts are eluted with RNase-free water and stored at $-20°$ in the presence of RNasin (1 U/μl; Promega).

Radiolabeled riboprobe (1×10^5 cpm) is incubated with 15–20 μg of protein extract in buffer containing 12 mM HEPES (pH 7.9), 15 mM KCl, 5 mM MgCl$_2$, 2 mM dithiotreitol (DTT), 1 μg of tRNA, heparin (1 μg/μl), RNasin (1 U/μl), and

---

[17] J. Falck, P. Dagger, B. Jensen, and M. Sehested, J. Biol. Chem. **274**, 18753 (1999).

10% (v/v) glycerol in a total volume of 20 μl for 20 min at 25°. Total cellular protein extracts are prepared by repeated freeze–thawing in buffer containing 10 mM HEPES, 1.5 mM MgCl$_2$, 1 mM EDTA, 0.2 mM EGTA, 10 mM KCl, 1 mM DTT, 1 mM phenylmethylsulfonyl fluoride (PMSF), and 10% (v/v) glycerol. For competition experiments, the protein extract is first incubated for 10 min at 25° with unlabeled competitor RNA (specific or nonspecific) before the addition of the radiolabeled transcript. Control reactions without competitors are sham treated under identical conditions. RNA–protein binding reactions are treated with 5 U of RNase T1 for 15 min at 25° and separated by electrophoresis on a 4.5% (w/v) native polyacrylamide gel in 45 mM Tris, 45 mM boric acid, and 1.2 mM EDTA buffer, pH 7.4. Radioactive bands are visualized by exposing the dried gels to a PhosphorImager screen.

Results presented in Fig. 3 show that the mobility of the 177-nt Topo II 3' UTR transcript is retarded in the presence of protein extract compared with the control reaction without protein extract (compare lanes 1 and 2 in Fig. 3). The retardation

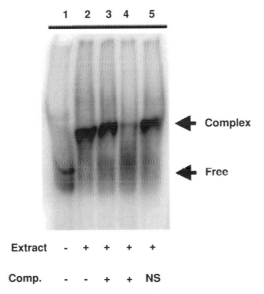

FIG. 3. *In vitro* RNA–protein binding assay. RNA–protein gel shift assay of [^{32}P]UTP-labeled 177-nucleotide Topo II 3' UTR riboprobe. Radiolabeled riboprobe (1 × 10^5 cpm) was incubated with (lane 2) or without (lane 1) 15 μg of protein extract prepared from asynchronously growing HeLa cells. For competition experiments, the protein extract was first incubated for 10 min at 25° with unlabeled specific [177-nucleotide Topo II 3' UTR, lane 3 (0.1 ng) and lane 4 (10 ng)], or nonspecific (10 ng of β-galactosidase, lane 5), competitor RNA before the addition of the radiolabeled transcript. Comp., Unlabeled competitor transcript; NS, nonspecific β-galactosidase transcript. Arrows represent mobility of the free and protein-bound riboprobes.

in mobility is due to an RNA–protein interaction, because proteinase K treatment of cellular extracts before the binding assay abolishes the retardation (data not shown). The specificity of the RNA–protein complex is determined by performing competition experiments in the presence of unlabeled 177-nt Topo II 3' UTR-specific competitor or nonspecific competitor RNA (β-galactosidase). Whereas addition of 0.1 ng (Fig. 3, lane 3) of unlabeled specific competitor has a minimal effect on protein binding, 10 ng (Fig. 3, lane 4) of the unlabeled transcript competes completely with the radiolabeled 177-nt Topo II 3' UTR RNA for protein binding. In contrast, 10 ng of the unlabeled nonspecific competitor RNA (β-galactosidase) does not compete with the radiolabeled transcript for protein binding (Fig. 3, lane 5). In control reactions without the protein extract (Fig. 3, lane 1), the radioactive bands above and below the complex could be due to RNase T1-resistant secondary structures of the transcript. These results show that protein binding to the 177-nt Topo II 3' UTR transcript is specific and demonstrate the applicability of the *in vitro* RNA–protein binding assay.

Redox Sensitivity of *in Vitro* RNA–Protein Binding Assay

The redox sensitivity of protein binding to the 177-nt Topo II 3' UTR is assayed by slight modifications of the method described above. Protein extract is prepared from asynchronously growing HeLa cells, using extraction buffer that lacks the reducing agent DTT. Similarly, DTT is omitted from the binding reaction buffer. Subsequently, binding reactions are performed in the presence of 2 mM DTT. The reaction mixture containing the protein extract is incubated with DTT for 10 min at 25° before the addition of the 177-nt riboprobe. The binding reaction is continued for 20 min after the addition of radiolabeled transcript and is analyzed by RNA gel shift assay as described above. Binding of cellular proteins to the Topo II 3' UTR increases 3- to 4-fold in protein extracts pretreated with the reducing agent DTT (compare lane 2 with lane 1 in Fig. 4). The assay is then repeated in extracts supplemented with 2 mM DTT and increasing concentrations of a thiol-oxidizing agent (diamide). A dose-dependent inhibition of protein binding is observed in extracts pretreated with the sulfhydryl-oxidizing agent diamide (0.1, 0.5, 1, 5, and 10 mM; Fig. 4, lanes 5–9). Whereas 0.1 mM diamide does not cause any significant change in protein binding (Fig. 4, lane 5), 0.5 and 1 mM diamide cause 30 and 50% decreases in protein binding (Fig. 4, lanes 6 and 7). Increasing the diamide concentration to 5 mM (Fig. 4, lane 8) and 10 mM (Fig. 4, lane 9) further inhibits RNA–protein complex formation. Taken together, these results indicate that reduced thiol residues in Topo II 3' UTR-binding proteins participate in the RNA–protein complex formation and that oxidation of these thiol residues to the disulfide form abolishes binding. These results show that the RNA–protein binding assay described here can be applied *in vitro* to determine the redox sensitivity of proteins binding to the 3' UTRs of cell cycle genes.

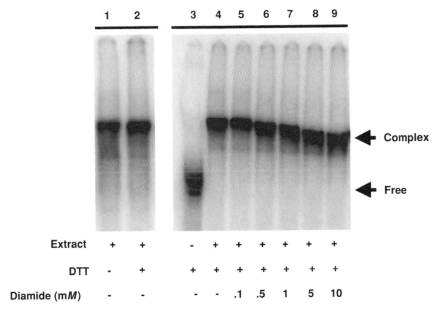

FIG. 4 Redox sensitivity of RNA–protein binding assay. RNA protein gel shift assay of the 177-nucleotide Topo II 3′ UTR riboprobe and HeLa cellular protein extracts pretreated with thiol-reducing or -oxidizing agents. (A) Protein extracts from asynchronously growing HeLa cells were prepared with extraction buffer without any DTT. RNA–protein binding reactions were carried out with protein extracts that were pretreated with (lane 2) or without (lane 1) 2 mM DTT. Lanes 4–9, protein extracts from asynchronously growing HeLa cells were prepared with regular extraction buffer containing 2 mM DTT and RNA–protein binding reactions were carried out in the presence of increasing concentrations of diamide (0.1, 0.5, 1, 5, and 10 mM). Lane 3 represents a control reaction without any protein extract.

In Vivo Manipulation of Redox State and in Vitro RNA–Protein Binding Assay

Asynchronously growing HeLa cells are treated with 20 mM N-acetyl-L-cysteine (NAC) and assayed for NAC uptake, reduced and oxidized glutathione content, RNA–protein binding, and Topo II mRNA levels. The pH of the NAC is adjusted to pH 7.0 with sodium bicarbonate. Intracellular reduced and oxidized glutathione as well as NAC levels are assayed according to previously published protocols.[18,19] Cell pellets are homogenized in 50 mM potassium phosphate buffer (pH 7.8) containing 1.34 mM diethylenetriaminepentaacetic acid. Total glutathione content is determined in sulfosalicylic acid [5% (w/v) SSA] extracts by the method

---

[18] R. V. Blackburn, D. R. Spitz, X. Liu, S. S. Galoforo, J. E. Sim, L. A. Ridnour, J. C. Chen, B. H. Davis, P. M. Corry, and Y. J. Lee, *Free Radic. Biol. Med.* **26,** 419 (1999).

[19] L. A. Ridnour, R. A. Winters, N. Ercal, and D. R. Spitz, *Methods Enzymol.* **299,** 258 (1999).

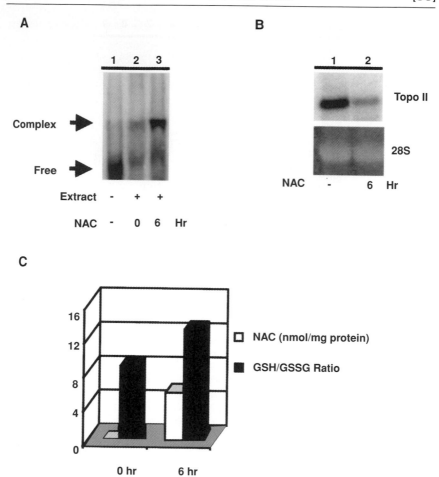

FIG. 5. Manipulation of intracellular redox state. Asynchronously growing HeLa cells were treated with 20 mM NAC (pH 7.0) and subjected to the RNA–protein gel mobility shift assay (A); Topo II mRNA levels were determined (B), as was NAC uptake (by HPLC) and glutathione content (by spectrophotometric recycling assay) (C). Changes in intracellular redox state were measured as the ratio of GSH (nmol/mg) to GSSG (2 × nmol/mg). Ethidium bromide-stained 28S ribosomal RNA levels were included for comparison of the Northern blot results.

of Anderson.[20] Reduced (GSH) and oxidized (GSSG) glutathione are distinguished by addition of 2 μl of a 1 : 1 mixture of 2-vinylpyridine and ethanol per 30 μl of sample followed by incubation at room temperature for 1.5 hr before addition of SSA. NAC levels in cells are measured after derivatization with N-(1-pyrenyl)

[20] M. E. Anderson, *in* "Handbook of Methods for Oxygen Radical Research" (R. A. Greenwald, ed.), p. 317. CRC Press, Boca Raton, FL, 1985.

maleimide, using a 15-cm C_{18} Reliasil column (Column Engineering, Ontario, CA) coupled with high-performance liquid chromatography with fluorescence detection.[19] All biochemical determinations are normalized to the protein content of whole cell homogenates, using the method of Lowry et al.[21]

Results presented in Fig. 5C show that treatment of cells with NAC alters the intracellular redox state to a more reducing environment. Thus, NAC levels are approximately 5.6 nmol/mg protein after 6 hr of treatment. Consistent with these results, the ratio of GSH to GSSG increases approximately 1.5- to 2.0-fold during this time frame. Total cellular protein extracts are prepared in the absence of DTT and in vitro RNA–protein binding assays are performed according to the method described above. These results show that protein binding to the Topo II 3′ UTR increases 3- to 4-fold in NAC-treated cells compared with untreated controls (compare lanes 2 and 3 in Fig. 5A). These results provide in vivo evidence that a shift to a more reducing environment enhances protein binding to the 177-nt Topo II 3′ UTR. Topo II mRNA levels are analyzed after a 6-hr treatment with NAC. An increase in intracellular reducing state induced by NAC treatment decreases Topo II mRNA levels by more than 90% (compare lanes 1 and 2 in Fig. 5B). Interestingly, the NAC-induced decrease in Topo II mRNA levels correlates with enhanced protein binding to the Topo II 3′ UTR (Fig. 5A). These results indicate that protein binding to the Topo II 3′ UTR is favored in a reducing environment, which appears to facilitate mRNA degradation. These results show the feasibility of manipulating the intracellular redox state and its subsequent effect on Topo II gene expression.

Conclusion

A growing body of literature suggests the importance of the intracellular redox state as the possible physiological regulator of cell proliferation. Although the molecular mechanisms are currently not fully understood, it is possible that at least some of the mechanisms could be at the level of cell cycle-coupled gene expression. The assays described here provide methods to study alterations in gene expression during the cell cycle and possible effects of alterations in the intracellular redox state on cell cycle-coupled variations in gene expression. Although the assays were optimized for the study of Topo II expression, similar approaches can be used to study the redox regulation of other cell cycle-coupled genes.

Acknowledgments

This work was supported by NIH Grants R29 CA-69593 (P.C.G.) and RO1 HL-51469 (D.R.S.).

---

[21] O. H. Lowry, N. J. Rosenbrough, A. L. Farr, and R. J. Randall, J. Biol. Chem. **193,** 265 (1951).

[39] Chemokine Expression in Transgenic Mice Overproducing Human Glutathione Peroxidases

By Nobuya Ishibashi and Oleg Mirochnitchenko

Introduction

Interest in the participation of reactive oxygen species (ROS) in diseases has grown rapidly and has been an area of active research. An issue that remains unresolved is the exact role of ROS in injury process. Although their harmful effects on lipids, proteins, and DNA are more or less understood, the ability of ROS and antioxidants to affect cellular signaling and in this way control gene expression needs additional investigation. As second messengers, ROS are thought to be involved in a number of inflammatory and/or immune-mediated disorders. Although many conditions can elicit inflammation, there are common pathways that contribute to the recruitment of inflammatory cells. A significant amount of data accumulated so far suggests that expression of chemokines, a large family of structurally related chemotactic cytokines produced by a variety of cells, is regulated by oxidative stress. Chemotactic factors not only induce leukocyte movement but also enhance endothelial and leukocyte adhesiveness and endothelial and extracellular matrix permeability. Activation of leukocyte receptors triggers these cells to secrete ROS and other mediators, providing a mechanism to amplify the inflammatory response. The chemokine system is characterized by apparent redundancies of ligands and receptors, which complicates investigation of specific chemokine-regulated events. Four chemokine subfamilies are now described (CXC, CC, CX3C, and C).[1] In general, CXC chemokines show chemoatractant activity for neutrophils, whereas CC chemokines are chemotactic for monocytes and lymphocytes, although there is some overlap. Chemokines interact with G protein-coupled seven-transmembrane domain receptors. Five human CXC receptors and nine human CC chemokine receptors have been reported.[2]

Although studies revealed that ROS are critical regulators of chemokine activation *in vitro,* few of them have addressed the potential role of these mediators *in vivo.* One of the ways to approach this issue is to use animals with a genetically changed level of specific antioxidant enzyme. This way it is possible to study chemokine behavior at the level of the whole organism in its natural environment. Another important aspect of this approach is that the level of oxidants/antioxidants changes under certain disease conditions, and the ability of these changes to influence the development of inflammation-related pathological conditions might

---

[1] A. Mantovani, *Immunol. Today* **20,** 254 (1999).
[2] L. M. Gale and S. R. McColl, *Bioassays* **21,** 17 (1999).

Copyright 2002, Elsevier Science (USA).
All rights reserved.
0076-6879/02 $35.00

be investigated as well. This chapter focuses on methods for studying chemokine expression in transgenic mice overexpressing human glutathione peroxidases. As a model of pathological conditions involving oxidative stress, kidney ischemia–reperfusion is used. Both animal studies and clinical data strongly implement ROS in ischemia–reperfusion injury, whereas the inflammatory response is crucial for amplification of kidney malfunction.

Oxidative Stress and Chemokine Activation

The best studied among chemokines in relation to oxidative stress-mediated activation is interleukin 8 (IL-8). In lipopolysaccharide (LPS)-stimulated human whole blood and cellular models, scavengers of oxygen and nitrogen intermediates inhibited IL-8 production.[3] Addition of exogenous ROS increased IL-8 expression in these cells. H_2O_2 induced IL-8 expression in epithelial cells, but not in endothelial cells.[4] The effect of reduced oxygen pressure in human glioblastoma and of oxidant tone in epithelium infected with respiratory syncytial virus on IL-8 activation was also mediated by ROS.[5,6] Intracellular glutathione redox status was an important modulator of the monocyte chemoattractant protein type 1 (MCP-1) level in a model of rat pulmonary granulomatous vasculitis.[7] Induction of MCP-1 protein in mesangial cells by tumor necrosis factor α (TNF-α) and IL-1β was sensitive to antioxidants.[8] Oxidative stress is also known to regulate macrophage inflammatory protein 1α (MIP-1α) mRNA expression in alveolar macrophages.[9]

Analysis of the pathways of activation of chemokines suggested at least two major mechanisms: one at the level of transcription activation and the other involving posttranscriptional stabilization of mRNA. 5′-Regulatory regions of chemokine genes contain several transcription elements, mostly combinations of NF-κB, AP-1, NF-IL-6, and C/EBP DNA-binding sites. These transcription factors are known to be sensitive to changes in the intracellular redox state.[10] Differential sensitivity of these transcriptional factors to oxidative stress establishes distinct patterns of cell-type specific and stimulus-specific gene expression. Ox/redox regulation of transcriptional factors occurs at the level of signal transduction pathways (including the activity of protein kinases and phosphatases) or redox status

[3] L. E. DeForge, J. C. Fantone, J. S. Kenney, and D. G. Remick, *J. Clin. Invest.* **90**, 2123 (1992).

[4] V. Lakshminarayanan, D. Beno, R. Costa, and K. Roebuck, *J. Biol. Chem.* **272**, 32910 (1997).

[5] J. G Mastronarde, M. M. Monick, and G. W. Hunninghake, *Am. J. Respir. Cell Mol. Biol.* **13**, 629 (1995).

[6] I. Desbaillets, A. Diserens, N. Tribolet, M. Hamou, and E. G. Meir, *Oncogene* **18**, 1447 (1999).

[7] A. Desai, X. Huang, and J. S. Warren, *Lab. Invest.* **79**, 837 (1999).

[8] J. A. Satriano, M. Shuldiner, K. Hora, Y. Xing, Z. Shan, and D. Schlondorff, *J. Clin. Invest.* **92**, 1564 (1993).

[9] M. Shi, J. Godleski, and J. Paulauskis, *J. Biol. Chem.* **271**, 5878 (1996).

[10] T. P. Dalton, H. G. Shertzer, and A. Puga, *Annu. Rev. Pharmacol. Toxicol.* **39**, 67 (1999).

of critical cysteine residues, influencing the interaction between subunits or with other factors or DNA.[11,12] Additional protein factors regulating the redox state of NF-κB and AP-1 were also described (thioredoxin and Ref-1).[13]

Transgenic Mice Overexpressing Antioxidant Enzymes as Model System to Study Role of Reactive Oxygen Species under Pathological Conditions

Free radical reactions and consequent lipid peroxidation are important regulators of cellular functions. Achieving the proper balance between the generation of ROS and the ability of different type of cells to detoxify or respond to ROS might be critical for a successful outcome under stress conditions. Previous animal studies addressing the role of free radicals and antioxidants have used mainly direct injection or infusion of antioxidants/oxidants. In most cases, this approach leads only to the modulation of the extracellular ox/redox balance, whereas the intracellular level of oxidative stress might be achieved by changing endogenous levels of antioxidant enzymes. When the increasing evidence of the role of ROS, not only as damaging agents, but also as intracellular signaling molecules is taken into account, it is clear that transgenic mice overexpressing antioxidant enzymes, or having these enzymes deleted, represent a new and promising tool for studying the mechanism of injury. For example, transgenic mice overexpressing intracellular and extracellular copper/zinc superoxide dismutase (Cu,ZnSOD), manganese SOD (MnSOD), and intracellular and extracellular glutathione peroxidases (GPs) were developed by us and others.[14–16] These animals show an increased resistance to a variety of pathological conditions, such as brain, heart, and kidney ischemia–reperfusion injury, hyperoxia, and doxorubicin (Adriamycin) and paraquat toxicity.[17–19] Nevertheless, GP mice showed increased sensitivity to hyperthermia and skin-induced

[11] Y. Sun and L. Oberley, *Free Radic. Biol. Med.* **21,** 335 (1996).

[12] Y. Suzuki, H. Forman, and A. Sevanian, *Free Radic. Biol. Med.* **22,** 269 (1996).

[13] K. Hirota, M. Matsui, S. Iwata, A. Nishiyama, K. Mori, and J. Yodoi, *Proc. Natl. Acad. Sci. U.S.A.* **94,** 3633 (1997).

[14] O. Mirochnitchenko, U. Palnitkar, M. Philbert, and M. Inouye, *Proc. Natl. Acad. Sci. U.S.A.* **92,** 8120 (1995).

[15] C. Epstein, K. Avraham, M. Lovett, S. Smith, O. Elroy-Stein, G. Rotman, C. Bry, and Y. Groner, *Proc. Natl. Acad. Sci. U.S.A.* **84,** 8044 (1987).

[16] M. E. Mirault, A. Tremblay, D. Furling, G. Trepanier, F. Dugre, J. Puymirat, and F. Pothier, *Ann. N.Y. Acad. Sci.* **738,** 104 (1994).

[17] M. Weisbrot-Lefkowitz, K. Reuhl, B. Perry, P. Chan, M. Inouye, and O. Mirochnitchenko, *Mol. Brain Res.* **53,** 333 (1998).

[18] Z. Chen, B. Siu, Y. Ho, R. Vincent, C. C. Chua, R. C. Hamdy, and B. H. Chua, *J. Mol. Cell. Cardiol.* **30,** 2281 (1998).

[19] N. Ishibashi, M. Weisbrot-Lefkowitz, K. Reuhl, M. Inouye, and O. Mirochnitchenko, *J. Immunol.* **163,** 5666 (1999).

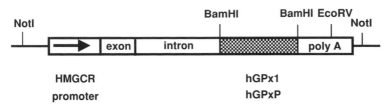

FIG. 1. The DNA fragments used for generation of transgenic mice overexpressing human GPx1 and PGxP. The cross-hatched box represents the coding region for GPx1 and GPxP. The promoter region (box with arrow), the first exon, and intron belong to the mouse HMGCR gene.

tumor promotion, whereas Cu,ZnSOD-overexpressing animals possess reduced microbicidal and fungicidal activity.[14,20,21] These results reflect the complex role of ROS generated under stress or disease conditions. In experiments described below, transgenic mice overproducing human extracellular and intracellular glutathione peroxidases (GPxP and GPx1, respectively) are used. GP is a critical antioxidant enzyme for the detoxification of peroxides. Its relatively low substrate specificity and particular kinetics make it an efficient reducer of peroxides, far more so than catalase, which is the other enzyme that detoxifies H_2O_2.

Construction of Transgenic Mice Overexpressing Human Glutathione Peroxidases

To create animals overexpressing antioxidant enzymes to study the effect of modulation of the level of oxidative stress on the development of pathological processes we use a non-tissue-specific promoter from the mouse hydroxymethyl-glutaryl-coenzyme A reductase gene (HMGCR) in order to avoid tissue-specific expression. It has been demonstrated that the bacterial *cat* reporter gene under the control of this promoter shows a ubiquitous pattern of expression in transgenic mice.[22] Fragments containing cDNA sequences of human GPx1[23] and GPxP[24] are inserted under the control of the HMGCR promoter (Fig. 1). Other groups reported construction of animals overproducing antioxidant enzymes with its own promoter[15] or with a tissue-specific promoter.[25]

*Not*I fragments are microinjected into the pronucleus of C57BL/6 × CBA/J hybrid eggs. Newborn babies are analyzed for the presence of transgene and

[20] Y.-P. Lu, Y. Lou, P. Yen, H. L. Newmark, O. I. Mirochnitchenko, M. Inouye, and M. T. Huang, *Cancer Res.* **57,** 1468 (1997).

[21] O. Mirochnitchenko and M. Inouye, *J. Immunol.* **156,** 1578 (1996).

[22] C. Gauthier, M. Methali, and S. Lathes, *Nucleic Acids Res.* **17,** 83 (1989).

[23] Y. Sukenaga, K. Ishida, T. Takeda, and K. Takagi, *Nucleic Acids Res.* **15,** 7178 (1987).

[24] K. Takahashi, M. Akasaka, Y. Yamamoto, C. Kobayashi, J. Mizoguchi, and J. Koyama, *J. Biochem.* **108,** 145 (1990).

[25] T. D. Oury, Y. Ho, C. A. Pantadosi, and J. D. Crapo, *Proc. Natl. Acad. Sci. U.S.A.* **89,** 9715 (1992).

expression of human glutathione peroxidases. In our experiments, GPx1 overexpression was detected in almost all tested tissues of several mouse lines,[14] whereas overexpression of GPxP was detected mostly in blood, as expected from its extracellular localization, as well as in kidney, which is known to be the major organ secreting GPxP in mice.[26] To obtain nontransgenic and heterozygous transgenic animals for experiments, transgenic founders are bred with (C57BL/6 × CBA/J)F$_1$ mice. Transgenic mice should be identified by Southern or slot-blot analysis of tail DNA, using labeled hGPxP and hGPx1 probes.[27] Nontransgenic littermates are used in experiments as wild-type controls.

Kidney Ischemia–Reperfusion Model

Ischemia–reperfusion injury of the kidney is a common problem with potentially catastrophic ramifications. Reactive oxygen species (ROS) and leukocytes recruited into the tissue have been implicated in the pathogenesis of this process. Insight concerning the mechanisms of leukocyte migration, as well as signal transduction pathways participating in cell activation, indicates that there is an important link between ROS generation and regulation of the inflammatory response. Pathways leading to oxidative stress-induced chemokine activation, including that mediated by NF-κB, are summarized in Fig. 2.

To test a hypothesis that a prooxidant state during kidney ischemia–reperfusion is a critical regulator of chemokine activation, unilateral renal artery occlusion model in mice overexpressing human GPs is used.[28] It is a reproducible, well-studied experimental procedure leading to uniform renal cell injury and severe renal impairment. Normal or transgenic males weighing 25–35 g are anesthetized with sodium pentobarbital (25 mg/kg) and xylazine (10 mg/kg) and administered heparin (300 USP units/kg of body weight) subcutaneously before surgery. Unilateral renal ischemia is induced by occluding the left side renal vein and artery with a microaneurysm clamp. Body temperature is maintained at 37° during the whole procedure. After 32 min of ischemia, the left kidney is reperfused by declamping the microaneurysm applicator and right nephrectomy is performed. Sham surgery consists of a surgical procedure that is identical except that the microaneurysm clamp is not applied. Mice are killed at various time points after surgery. Kidneys should be removed immediately after perfusion with cold phosphate-buffered saline (PBS) and rapidly frozen in liquid nitrogen to obtain extracts and perform RNA analysis.

[26] N. Avissar, J. C. Whitin, P. Z. Allen, D. D. Wagner, P. Liegey, and H. J. Cohen, *J. Biol. Chem.* **264,** 15850 (1989).

[27] B. Hogan, R. Beddington, F. Constantini, and E. Lacy, "Manipulating the Mouse Embryo: A Laboratory Manual." Cold Spring Harbor Laboratory Press, Cold Spring Harbor, NY, 1994.

[28] K. Kelly, W. Williams, R. Colvin, S. Meehan, T. Springer, J. Gutierrez-Ramos, and J. Bonventre, *J. Clin. Invest.* **97,** 1056 (1996).

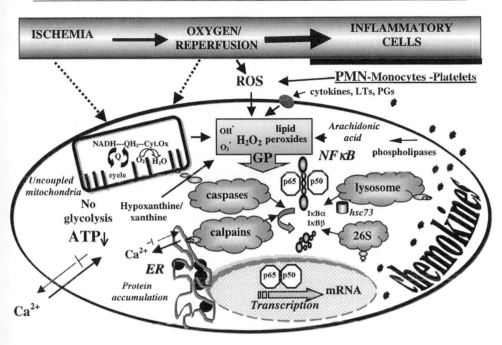

FIG. 2. Involvement of ROS in CXC chemokine activation during ischemia reperfusion. Among the primary sources of ROS are activated leukocytes and endogenous reactions such as xanthine oxidase-mediated conversion of xanthine to hypoxanthine, the mitochondrial electron transport chain, microsomal oxidation, and arachidonic acid metabolism. Cell activation by cytokines, leukotrienes (LTs), prostaglandins (PGs), and so on, also leads to the intracellular generation of ROS. NF-κB is shown as an example of an oxidative stress-sensitive transcription factor, leading to the induction of chemokine expression. Several proteolytic mechanisms (26S proteasome, caspases, calpains, and lysosomal proteases) able to degrade I-κB inhibitors in several circumstances are presented. The mechanism functional under ischemia reperfusion conditions is not known at present.

Assessment of Effect of Antioxidant Enzyme Overexpression on Chemokine Activation During Kidney Ischemia–Reperfusion

Several methods may be employed to analyze chemokine expression in kidneys of normal and transgenic mice after ischemia–reperfusion. Level of mRNA, depending on the abundance of the mRNA for a particular chemokine, is characterized by Northern blot analysis, RNase protection assay, or reverse transcription-polymerase chain reaction (RT-PCR). Two of those methods are described below. The use of RT-PCR for assessment of chemokine expression is described in detail elsewhere.[29] Primers for detection of several mouse chemokines are commercially available (e.g., from BioSource International, Camarillo, CA). Protein expression

[29] T. Standiford, S. L. Kunkel, and R. M. Strieter, *Methods Enzymol.* **288,** 220 (1997).

might be directly assessed by Western blot analysis or enzyme-linked immunosorbent assay (ELISA) in tissue extracts or by immunohistochemistry of the tissue sections, which is less quantitative but will give information regarding cell-specific localization of the protein.

Preparation of RNA and Northern Blot Analysis

Total RNA is isolated from kidneys, using TRIzol reagent (Life Technologies, Gaithersburg, MD) according to the manufacturer protocol. If necessary, poly $(A)^+$ RNA may be isolated with an oligo(dT) column. The RNA concentration is measured by a spectrophotometer. For Northern blot analysis, RNA samples [20 μg of total RNA or 3 μg of poly $(A)^+$ RNA] are denatured in $1\times$ morpholinepropanesulfonic acid (MOPS) electrophoresis buffer [0.04 M MOPS (pH 7.0), 0.01 M sodium acetate, 1 mM EDTA], 6.5% (v/v) formaldehyde, and 50% (v/v) formamide for 5 min at 65°. After cooling the samples on ice, a 1/10 volume of loading buffer containing 0.025% (w/v) bromphenol blue and 50% (w/v) glycerol is added and the mixture is loaded on a 1.4% (w/v) agarose gel with 2.2 M formaldehyde. After electrophoresis, RNA should be transferred from the gel to a nylon membrane (GeneScreen; New England Nuclear, Boston, MA). After transfer the membrane is fixed and prehybridized in 50% (v/v) formamide, $5\times$ SSPE (0.6 M NaCl, 0.04 M NaH$_2$PO$_4$, 5 mM EDTA, pH 7.4), $5\times$ Denhardt's solution [0.5% (w/v) polyvinylpyrrolidone, 0.5% (w/v) bovine serum albumin (BSA), 0.5% (w/v) Ficoll 400], 1% (w/v) sodium dodecyl sulfate (SDS), 10% (w/v) dextran sulfate, and carrier DNA (100 μg/ml) at 42° for at least 1 hr. As a probe DNA fragments labeled with a random primed DNA-labeling kit (Boehringer Mannheim, Indianapolis, IN) are used. Hybridization is performed overnight in prehybridization solution containing labeled probe at 42°. The membrane is washed once with $2\times$ SSPE solution for 15 min at room temperature, twice with $2\times$ SSPE and 2% (w/v) SDS for 30 min at 65°, and once with $0.1\times$ SSPE for 15 min at room temperature and autoradiographed. To compare RNA amounts in the bands, the films may be scanned by a densitometer and final values factored relative to the levels of control RNA, for example, β-actin. As shown in Fig. 3A, the level of KC mRNA assessed by Northern blot analysis was 10 and 4.5 times higher in normal mice compared with those in GPx1 and GPxP animals 6 hr after reperfusion, respectively.

RNase Protection Assay

The RNase protection assay is performed with a RiboQuant kit and an mCK-5 multiprobe template set (PharMingen, San Diego, CA) according to the manufacturer protocol. In brief, RNA probes for mouse chemokines [Ltn, RANTES (regulated on activation, normal T cell expressed and secreted), eotaxin, MIP-1β, MIP-1α, MIP-2, interferon-inducible protein 10 (IP-10), MCP-1, and TCA-3 (T-cell activation-3)] as well as for positive controls [L32 and glyceraldehyde-3-phosphate

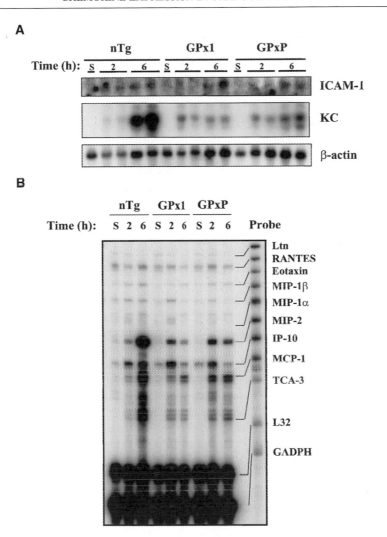

FIG. 3. Effect of renal ischemia on chemokine and ICAM-1 expression. (A) Twenty micrograms of total RNA was applied in each lane for the Northern blot analysis of KC expression. Data from two independent animals at 2- and 6-hr time points are shown. A significantly lower level of KC mRNA expression was detected in kidneys of GPx1 and GPxP mice in comparison with nontransgenic animals. For comparison, results of Northern analysis with ICAM-1-labeled probe, showing no difference between the same groups of animals, are also presented. As an internal control hybridization with β-actin was performed. (B) Five micrograms of total RNA was used for an RNase protection assay with the mCK-5 multiprobe template set (PharMingen). A probe set not treated with RNase is shown on the right. Corresponding RNase-protected fragments are highlighted. S, Sham-operated animals, which were killed 6 hr after operation. [Reprinted with permission from N. Ishibashi, M. Weisbrot-Lefkowitz, K. Reuhl, M. Inouye, and O. Mirochnitchenko, *J. Immunol.* **163,** 5666 (1999). Copyright © 1999 by The American Association of Immunologists.]

dehydrogenase (GAPDH)] are generated *in vitro* with a T7 polymerase transcription kit and [α-^{32}P]UTP as a label. Total RNA (5 μg) is incubated overnight at 56° with a mixture of labeled antisense probes (5 × 10^5 Cherenkov counts) in 80% (v/v) formamide, 1 mM EDTA, 400 mM NaCl, and 40 mM piperazine-N,N'-bis 2-ethanesulfonic acid (PIPES) (pH 6.7) for 12–16 hr. Samples are then treated with an RNase A–T1 mixture, following proteinase K digestion for 15 min at 37°. Samples are phenol extracted and precipitated by ethanol. The protected RNA is then dissolved in loading buffer [80% (v/v) formamide, 1 mM EDTA, 50 mM Tris–borate (pH 8.3), 0.05% (w/v) xylene cyanol, and 0.05% (w/v) bromphenol blue], denatured for 3 min at 90°, and electrophoresed on a 5% (w/v) polyacrylamide gel with urea. A standard curve is made by using undigested probes as markers and the identity of "RNase-protected" bands is established. To compare RNA amounts in the protected bands, the films are scanned with an imaging densitometer, and final values are factored relative to GAPDH levels. A similar approach may be used for any other chemokines not yet included in a kit, if analogous template plasmid with corresponding DNA fragment is created.

The results of a typical analysis by RNase protection assay of chemokine mRNA expression in kidneys from normal and GP transgenic mice subjected to ischemia–reperfusion are shown in Fig. 3B. Comparison of the level of activation of different chemokines indicates that they might be divided into several groups. Eotaxin was induced at 2 and 6 hr reperfusion in similar quantities in all three groups. MIP-1β, IP-10, MCP-1, and TCA-3 were equally activated at 2 hr in normal, GPx1, and GPxP mice, whereas at 6 hr a high level of expression persisted only in the first group. The most significant difference, however, was observed in the level of expression of MIP-2. This chemokine was highly inducible in normal mice. At 6 hr after ischemia, its level was 7.5 and 4.3 times higher in these animals in comparison with GPx1 and GPxP mice, respectively (Fig. 3B). Importantly, only a few polymorphonuclear cells (PMNs) might be detected in kidneys of normal mice by this time point, indicating that induction of this chemokine precedes massive migration of leukocytes. KC and MIP-2 belong to the family of CXC chemokines, which have been shown to cause PMN activation and mediate chemotaxis and induction of the respiratory burst.[30]

Immunostaining

Frozen tissue sections (10 μm) are used for immunostaining. Sections stored in a freezer at −70° are removed and immediately covered with ice-cold fixation solution [2% (v/v) formaldehyde in phosphate-buffered saline, pH 7.4]. Sections are incubated for 10 min at 4° and then washed three times for 3 min each with wash buffer–saponin solution, containing 0.1% (w/v) saponin (Sigma, St. Louis, MO)

[30] L. Feng, *Immunol. Res.* **21**, 203 (2000).

and Earl's buffered saline, pH 7.4 (EBSS; GIBCO-BRL, Gaithersburg, MD). Endogenous peroxidase activity is blocked by incubating sections in a blocking buffer [$3M$ NaN$_3$, 1% (v/v) H$_2$O$_2$, 0.1% (w/v) saponin in EBSS] for 60 min in the dark. Sections are then washed three times as described previously with wash buffer–saponin solution. Endogenous biotin activity is eliminated by washing with two blocking buffers (avidin–biotin blocking kit; Vector Laboratories, Burlingame, CA) sequentially, 1 hr each. Biotin solution is supplemented with 10% (v/v) donkey serum. After washing with wash buffer–saponin solution, sections are incubated overnight at room temperature with primary antibodies [goat anti-mouse antibodies (R&D Systems, Minneapolis, MN), 1 : 1000 to 1 : 2500 dilutions). Sections are then washed three times and incubated for 1 hr with biotin–donkey anti-goat IgG as a secondary antibody diluted 1 : 4000 (Jackson ImmunoResearch, West Grove, PA). Slides are washed three times in EBSS and incubated with Vectastain Elite ABC–peroxidase reagent (Vector Laboratories) supplemented with 0.1% (w/v) saponin, for 30 min in the dark. Samples are developed with 3,3′-diaminobenzidine tetrahydrochloride (DAB liquid substrate; Sigma) prepared according to the manufacturer instructions. After developing, sections are counterstained with hematoxylin, dehydrated, and mounted. Slides should be coded and interpreted by a pathologist who is blind to treatment. Each experiment is conducted several times and interpreted separately.

This method allowed us to demonstrate that MIP-2 and KC expression in a model of kidney ischemia–reperfusion injury is seen throughout the cortex, predominantly in tubules.[19] This conclusion is supported by results of experiments with cultured tubular and mesangial cell lines exposed to chemical anoxia–ATP repletion as a model of ischemia–reperfusion.[19] Mesangial cells play a key role in the inflammatory response in many pathological conditions, especially immune-mediated glomerular diseases. In contrast, in the ischemia–reperfusion model these cells produced small amounts of CXC chemokines and therefore tubular cells play a central role in their synthesis.

Preparation of Whole Cell Extract for Protein Analysis and Immunoprecipitations

Kidney tissues (100 mg) are homogenized in 10 volumes of radioimmunoprecipitation assay (RIPA) buffer [50 mM Tris-HCl (pH 7.6), 150 mM NaCl, 1% (v/v) Nonidet P-40, 0.5% (v/v) sodium deoxycholate, 0.1% (w/v) SDS, 1 mM dithiothreitol (DTT), 0.5 mM phenylmethylsulfonyl fluoride (PMSF), leupeptin, aprotinin, and pepstatin (10 μg/ml each), and the phosphatase inhibitors NaF (10 mM), NaVO$_4$ (1 mM), Na$_2$MoO$_4$ (1.5 mM), benzamidine (1 mM), glycerophosphate (20 mM), and p-nitrophenyl phosphate (20 mM)], using a Tissumizer (Tekmas, Cincinnati, OH). SDS–polyacrylamide gel electrophoresis sample buffer (5×) is added to a final concentration of 50 mM Tris-HCl (pH 6.8), 2% (w/v) SDS, 100 mM

DTT, 0.006% (w/v) bromphenol blue, and 10% (v/v) glycerol. Samples are boiled for 10 min and cooled on ice, and DNA is sheared by sonication. Lysates are finally centrifuged at 13,000g for 20 min at 4°, and supernatants are used for polyacrylamide gel electrophoresis and Western blotting. For immunoprecipitation analysis, RIPA homogenates are centrifuged at 13,000g for 20 min at 4°. The supernatants are transferred to new tubes. Anti-chemokine antibodies are added to 250 μg of protein extract and the mixture is incubated at 4° overnight. Thirty microliters of a 50 : 50 (w/v) slurry of protein A–Sepharose (Pharmacia, Piscataway, NJ) in RIPA buffer is then added to the samples. Immunopellets are collected and washed and protein is eluted by boiling for 1–2 min in SDS–polyacrylamide gel electrophoresis sample buffer. Samples are used for electrophoresis and Western blotting.

Involvement of Transcriptional and Posttranscriptional Mechanisms in Reactive Oxygen Species-Mediated Chemokine Activation

As noted in the Introduction, steady state levels of chemokine mRNAs may be modulated by transcriptional or posttranscriptional mechanisms. Transcriptional activation is mediated by increasing the activity of transcription factors. If transcription factors are induced in part by oxidative stress, in transgenic mice overexpressing antioxidant enzymes their level of activation should be significantly reduced. In this section we describe methods for analysis of activation of one of those factors, NF-κB, which plays one of the key roles in chemokine response in general, as well as during ischemia–reperfusion in particular.

Preparation of Nuclear Extracts and Electrophoretic Mobility Shift Assay

To prepare nuclear extracts, kidneys are homogenized in 4 volumes of buffer A [10 mM HEPES (pH 7.9), 0.25 M sucrose, 15 mM KCl, 5 mM EDTA, 1 mM EGTA, 0.15 mM spermine, 0.16 mM spermidine, 1 mM DTT, 0.2 mM PMSF, and leupeptin, aprotinin, and pepstatin (10 μg/ml each)], using eight strokes of a Dounce homogenizer (B-type pestle). After 10 min of incubation on ice, Nonidet P-40 is added to a final concentration 0.5% (v/v). Homogenates are centrifuged at 1000g for 10 min at 4° and the pellets are washed twice by suspension in buffer B (the same as buffer A, but without sucrose). Nuclear extracts are then extracted with 4 volumes of buffer C [100 mM HEPES (pH 7.9), 1.5 mM MgCl$_2$, 0.45 M KCl, 1 mM EDTA, 1 mM DTT, 10% (v/v) glycerol, 0.2 mM PMSF, and leupeptin, aprotinin, and pepstatin (10 μg/ml each)]. Mixtures are allowed to stay for 30 min at 4° and nuclear extracts are obtained by centrifugation at 14,000g for 20 min at 4°. Aliquots of the extract should be stored at −80°. Protein content is assayed with the Bio-Rad (Hercules, CA) protein reagent. Oligonucleotides used in the electrophoretic mobility shift assay (EMSA) are annealed, end labeled with polynucleotide kinase and $[\gamma\text{-}^{32}\text{P}]$ATP (New England Nuclear), and purified by polyacrylamide gel electrophoresis. For detection of NF-κB-binding activity the following oligonucleotides are used: 5′-AGCTCAGGGAATTTCCCTGGTCC-3′ (containing the mouse MIP-2 NF-κB-binding site) and 5′-GGCCAGGGAATTT CCCGGAGTA-3′ (containing the κB1 site of the mouse KC promoter region).[21] For EMSA, 5–10 μg of extract is incubated in a reaction mixture containing 20 mM HEPES (pH 7.9), 60 mM KCl, 5 mM MgCl$_2$, 1 mM DTT, 0.2 mM EDTA, 0.5% (v/v) Nonidet P-40, 1 μg of poly(dI/dC), and 8% (v/v) glycerol in a final volume of 20 μl for 20 min at 4°. After preincubation, 10^5 cpm of radiolabeled DNA probe is added and incubation is continued for 15 min at room temperature. The DNA–protein complexes are separated on native 5% (w/v) polyacrylamide gels in 0.25× Tris–borate–EDTA buffer. Supershift assays to prove the presence of specific proteins in shifted complexes are carried out after incubation of the nuclear extracts with 0.5 μg of rabbit polyclonal antibodies against p65 and p50 (sc-372 and sc-1192, respectively; Santa Cruz Biotechnology, Santa Cruz, CA) for 20 min at 4° followed by EMSA. For the competition assay, unlabeled double-stranded oligonucleotides should be added to the reaction mixtures at 10 or 25 molar excess over radiolabeled probes.

Results of a typical EMSA experiment with labeled MIP-2 NF-κB-binding site are shown in Fig. 4. At 6 hr of reperfusion NF-κB activity is significantly higher in kidneys from normal mice than from either GPx1 or GPxP transgenic animals. In both cases, binding was specific and the composition of binding complexes from p65/p50 proteins was proved by supershift assay. The presence of c-Rel was not observed in complexes. These data indicate that GP overexpression in kidneys of transgenic mice modulated DNA binding in the regulatory regions of at least two chemokines during ischemia–reperfusion, thereby affecting activation and migration of PMNs.

In the majority of cells, NF-κB exists in an inactive form in the cytoplasm by being bound to an inhibitory protein, such as I-κBα and I-κBβ.[31] Treatment of cells with various inducers, including ischemia–reperfusion, results in the degradation of I-κB proteins, thus releasing the bound NF-κB, which translocates to the nucleus and upregulate chemokine expression. To measure the presence of inhibitory proteins in kidney extracts and the dynamics of their degradation, Western blotting analysis is used.

Protein Electrophoresis and Western Blotting

After measuring the protein concentration, protein samples are denatured in a boiling bath of sample buffer containing 250 mM Tris-HCl (pH 6.8), 2% (w/v) SDS,

---

[31] S. Ghosh, M. May, and E. Kopp, *Annu. Rev. Immunol.* **16,** 225 (1998).

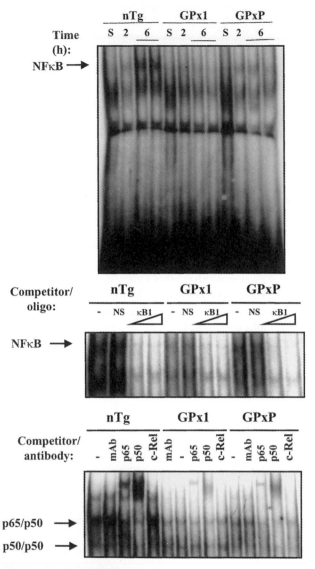

FIG. 4. Binding of proteins to the κB1 motif of the KC promoter region induced by ischemia–reperfusion. Kidney nuclear extracts from sham-operated animals (S) and mice after 32 min of ischemia and 2 or 6 hr of reperfusion were incubated with ^{32}P-labeled oligonucleotides, and then the EMSA was performed. Extracts from two independent animals are shown for the 6-hr time point. The specificity of the binding was verified in 6-hr extracts in the presence of a 25-fold excess of non-labeled noncompetitive oligonucleodites (NS) or a 10- and 25-fold excess of nonlabeled probe. The binding proteins were identified by incubating the reaction mixtures in the presence of p65, p50, or c-Rel antibodies as well as with normal serum (MAb). [Reprinted with permission from N. Ishibashi, M. Weisbrot-Lefkowitz, K. Reuhl, M. Inouye, and O. Mirochnitchenko, *J. Immunol.* **163,** 5666 (1999). Copyright © 1999 by The American Association of Immunologists.]

10% (v/v) glycerol, and 2% (v/v) 2-mercaptoethanol, and separated in minigels (7 cm; Bio-Rad) by SDS–15% (w/v) polyacrylamide gel electrophoresis. After electrophoresis, gels are electroblotted onto polyvinylidene difluoride (PVDF) membrane (Immobilon; Millipore, Bedford, MA). Membranes are blocked in a BLOTTO solution containing 1× TBS [10 mM Tris-HCl (pH 8.0), 150 mM NaCl], 5% (w/v) milk, and 0.1% (v/v) Tween 20 for 1 hr at room temperature. Membranes are then incubated with rabbit polyclonal antibodies against I-κBα and I-κBβ (sc-371 and sc-969, respectively; Santa Cruz Biotechnology) diluted in BLOTTO for 12 hr at 4°. After washings, membranes are incubated in BLOTTO with secondary antibodies (1 : 1000–1 : 20,000 dilution). After extensive washings in 1× TBS with 0.1% (v/v) Tween 20, proteins are detected with a Phototope-HRP Western blot detection kit and LumiGlo reagent (both from New England BioLabs, Beverly, MA). Autoradiographs are quantitated by densitometry.

As shown in Fig. 5, at 30 and 45 min after ischemia–reperfusion, fast degradation of I-κBα and I-κBβ was detected, although the level of depletion was significantly higher in normal mice. The degradation of I-κBα and I-κBβ correlates well with the level of nuclear p65. Therefore, overexpression of GPs seems to inhibit the influx of NF-κB into the nucleus by affecting degradation of its inhibitors.

To substantiate the transcriptional mechanism of CXC chemokine activation *in vivo*, the run-on transcriptional rate might be evaluated by using nuclei from kidneys isolated from sham-operated normal and transgenic mice as well as from

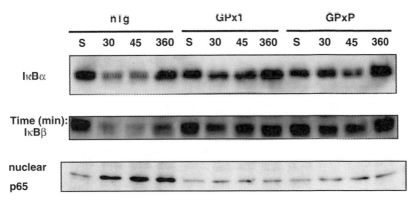

FIG. 5. Western blot analysis of degradation of I-κBα and I-κBβ induced by ischemia–reperfusion. Overexpression of GPx1 and GPxP in transgenic mice inhibits I-κBα and I-κBβ degradation in kidney extracts at 30 and 45 min of reperfusion in comparison with nontransgenic mice. S, Sham-operated animals. These data correlate with p65 protein accumulation in the nuclear extracts from the same animals. [Reprinted with permission from N. Ishibashi, M. Weisbrot-Lefkowitz, K. Reuhl, M. Inouye, and O. Mirochnitchenko, *J. Immunol.* **163**, 5666 (1999). Copyright © 1999 by The American Association of Immunologists.]

animals after ischemia–reperfusion. Another mechanism of posttranscriptional activation of chemokines involves an elevation in translational efficiency. To address this possibility, polysomal fractionation experiments are carried out in mouse kidneys after renal ischemia–reperfusion. The abundance of KC and MIP-2 transcripts is measured in the polysomal fractions, using RNA slot-blot analysis and chemokine-specific probes. These amounts should be adjusted by the amount of total mRNA in kidney homogenates. If increased amounts of polysomal bound mRNA are detected, it suggests an increased translatability for chemokines as one of the mechanisms for their activation during ischemia–reperfusion.

Nuclear Run-on Transcriptional Assay

To prepare nuclei from kidney cortex of mice after ischemia–reperfusion, ultracentrifugation in sucrose gradients is used as already described in detail.[32] Nuclei are resuspended in 50 mM HEPES (pH 8.0), 5 mM MgCl$_2$, 5 mM DTT, BSA (1 mg/ml), 25% (v/v) glycerol and stored in liquid nitrogen. On thawing, nuclei are diluted in reaction buffer [20 mM Tris-HCl (pH 7.9), 20% (v/v) glycerol, 140 mM KCl, 5 mM MgCl$_2$, 1 mM MnCl$_2$, 14 mM 2-mercaptoethanol, ATP, GTP, and CTP (1 mM each), 10 mM creatine phosphate, 20 units/ml creatine phosphokinase] with 1 mCi of [^{32}P]UTP (3000 Ci/mM) and incubated at 30° for 30 min, followed by DNase I treatment (150 units/ml). In pilot experiments, optimization of the incubation time may be needed. After proteinase K treatment, RNA is extracted and equal amounts of newly transcribed RNA are hybridized with 5 μg of denatured, linearized chemokine and β-actin cDNAs cross-linked to membranes. This method allows measurement of the rate of chemokine RNA transcription in kidney extracts from normal and transgenic mice after ischemia–reperfusion.

Quantification of Chemokine mRNA Bound to Polysomes

The polysome profile analysis is performed as follows: kidneys after ischemia-reperfusion are homogenized in 10 mM Tris-HCl (pH 7.4), 100 mM KCl, 10 mM MgCl$_2$, and 1 mM dithiothreitol. The homogenate is centrifuged at 13,000g for 30 min at 4° to remove the mitochondria. Ten absorbance units (OD$_{260}$) of the postmitochondrial supernatant is layered onto 12.5 ml of 10 to 45% (w/v) sucrose density gradient prepared in the above-described buffer. The gradients are centrifuged in a Beckman (Fullerton, CA) SW40 Ti rotor at 38,000 rpm for 100 min at 4°. Fractions (0.5 ml) are collected by upward displacement, and the OD$_{254}$ is measured and plotted. RNA isolated from the fractions is slot-blotted onto nylon

---

[32] D. L. Spector, R. D. Goldman, and L. A. Leinwand, "Cells: A Laboratory Manual." Cold Spring Harbor Laboratory Press, Cold Spring Harbor, NY, 1998.

membrane and hybridized with random primer ^{32}P-labeled chemokine cDNA probes and washed under stringent conditions.

Conclusion

Interaction between oxidative stress and inflammatory response is not a new concept. One of the best recognized sources of the ROS themselves is activated phagocytes, which under many pathological conditions can cause direct damage to the surrounding tissues. Several cytokines (such as TNF-α), released by the activated cells after interaction with target cells, are known to produce an oxidative stress response that might lead to dramatic changes in cellular fate (e.g., induce apoptosis). It has also become apparent that low levels of ROS play an important role in activation and recruitment of inflammatory cells through induction of adhesion receptors and regulation of the activity of lipoxygenases and cyclooxygenases, transcription factors, protein kinases, and other mediators. All this indicates that there are many ways in which ROS can influence the function of the immune system as well as its interaction with other tissues. In this chapter we have presented methods by which to use genetically engineered animal models, which are able to modulate the level of oxidative stress, to study the mechanisms of chemokine activation in pathological processes.

Overexpression of human GPs in transgenic mice significantly protects animals against kidney ischemia–reperfusion injury.[19] Both enzymes have similar substrate specificity, and are overproduced in the kidneys of our transgenic mice. The protective effect was accompanied by a significant decrease in chemokine expression (KC and MIP-2 in particular) and neutrophil migration. Although it might not be the only mechanism of protection in transgenic mice, the existence of compelling evidence for the role of neutrophils in renal ischemia–reperfusion indicates the ability of GPs to influence the outcome of ischemia reperfusion injury by modulation of early ROS production.

Although we[19] and others[33] have shown that several chemokines are induced during ischemia–reperfusion, the biological significance of the activation of different chemokines on the development of renal injury during ischemia–reperfusion has not been directly tested yet. Selective activation of certain chemokines and their cell type specificity warrants detailed analysis of the mechanisms of activation, including identification and study of the interaction of different transcription factors and posttranscriptional mechanisms. A series of transcription factors that participate in chemokine activation have been reported. Even though the most indispensable among them was NF-κB, synergistic participation of other transcription

[33] R. Safirstein, J. Megyesi, S. J. Saggi, P. M. Price, M. Poon, B. J Rollins, and M. B. Taubman, *Am. J. Physiol.* **261**, F1095 (1991).

factors also was shown. Importantly, various combinations of these factors mediate cell type-specific activation of a particular chemokine.

In this chapter we have focused mainly on analysis of the activation of different chemokines and characterization of factors directly affecting this activation. The next objective will be the study of upstream mediators, such as Rac proteins, stress- and mitogen-activated kinases, and ceramide and sphingomyelinase pathways. All of them are activated during ischemia–reperfusion and are sensitive to ox/redox regulation, and have been implicated in the activation of transcription factors inter-acting with chemokine promoters. Methods to study those mechanisms have been described.[34]

The possibility that modulation of chemoattractant activation via antioxidants can significantly affect tissue injury warrants further investigation and will lead to the development of new techniques.

[34] K. B. Bacon, *Methods Enzymol.* **288,** 340 (1997).

[40] Analysis of Promoter Methylation and Its Role in Silencing Metallothionein I Gene Expression in Tumor Cells

By KALPANA GHOSHAL, SARMILA MAJUMDER, and SAMSON T. JACOB

Introduction

Metallothioneins (MTs) are a group of evolutionarily conserved, cysteine-rich proteins implicated in a variety of biochemical reactions that include scavenging of oxygen-free radicals.[1] Cysteine moieties, mostly arranged in Cys-X-Cys clus-ters, are involved in the formation of metal–thiolate bonds that impart antioxidant properties to these molecules.[2] In fact, MT is the most potent scavenger of highly reactive hydroxyl radicals. Four different isoforms of MTs, encoded by different genes, have been identified in mammals.[3] Among these forms, MT-I and MT-II are ubiquitously expressed and highly induced in response to various environ-mental toxicants, various pathologic conditions, and stressors. MT-III is primarily expressed in the glutaminergic neurons of the brain, whereas the expression of MT-IV is confined to the stratified squamous epithelium of skin and tongue.

[1] J. A. S. Kagi, *Methods Enzymol.* **205,** 613 (1991).
[2] P. J. Thornalley and M. Vasak, *Biochim. Biophys. Acta* **827,** 36 (1985).
[3] C. J. Quaife, S. D. Findley, J. C. Erickson, G. J. Froelic, E. J. Kelly, B. P. Zambrowicz, and R. D. Palmiter, *Biochemistry* **33,** 7250 (1994).

Copyright 2002, Elsevier Science (USA).
All rights reserved.
0076-6879/02 $35.00

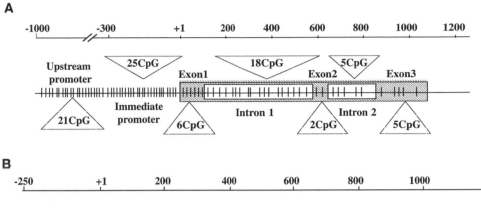

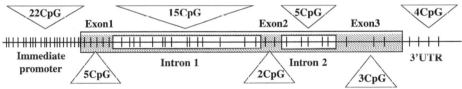

FIG. 1. (A) Schematic diagram of mouse *MT-I* gene, depicting the distribution of CpG dinucleotides within the promoter, exons, introns, and 3′-untranslated region (3′ UTR). (B) Schematic diagram of rat *MT-I* gene, denoting number of CpG base pairs at different regions. Vertical lines indicate positions of CpG dinucleotides. Numbers of CpG base pairs spanning different sites are denoted within triangles.

The *MT-I* gene is highly inducible by heavy metals in many different mammalian cell lines and in almost all tissues. Surprisingly, this gene is noninducible in some tumor cell lines[4,5] and in a solid tumor.[6] The lack of expression in these tumor cell lines is not due to the absence of functional *trans*-acting factors. Analysis of this gene promoter in mice and rats revealed that the *MT-I* proximal promoter region (which spans critical *cis* elements) harbors CpG islands (see Fig. 1A and B). CpG islands are usually located in the promoter and first exons of housekeeping genes and are characterized by high GC content, with the CpG-to-GpC ratio >0.6.[7] Analysis of the *MT-I* gene in these two species shows that CpG density is highest in the proximal promoter region (Fig. 1A and B).

Methylation at the 5-position of cytosine of complementary CpG base pairs is an epigenetic mechanism that plays a critical role in development, genomic imprinting, inactivation of the X chromosome, tissue-specific gene expression,

[4] S. Majumder, K. Ghoshal, Z. Li, Y. Bo, and S. T. Jacob, *Oncogene* **18,** 6287 (1999).
[5] R. D. Palmiter, *Experientia Suppl.* **52,** 63 (1987).
[6] K. Ghoshal, S. Majumder, Z. Li, X. Dong, and S. T. Jacob, *J. Biol. Chem.* **275,** 539 (2000).
[7] S. B. Baylin, *Adv. Cancer Res.* **72,** 141 (1998).

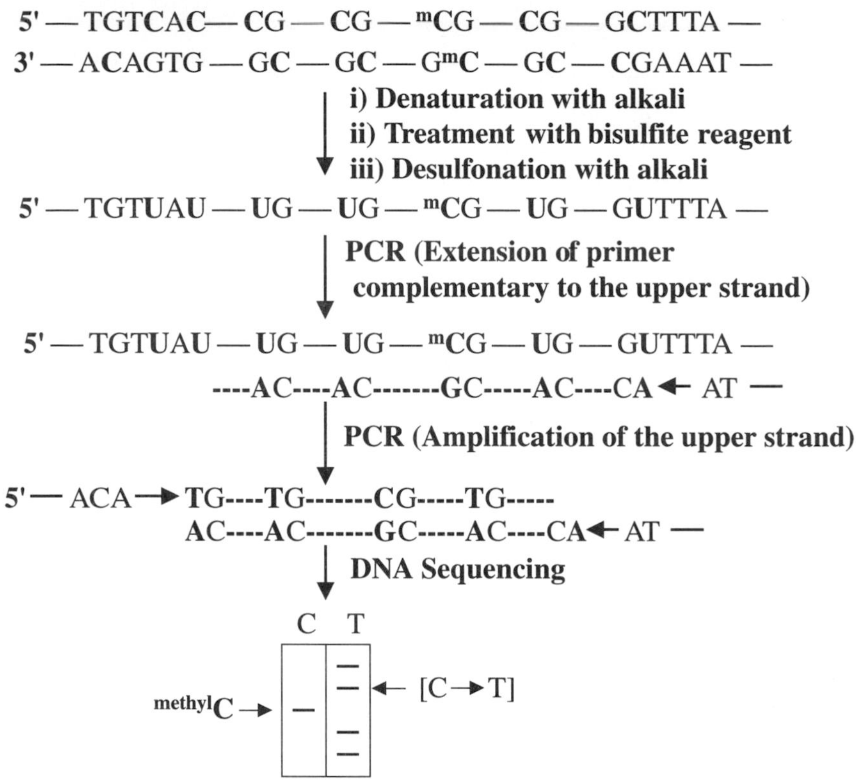

5' — TGTCAC— CG — CG — ^{m}CG — CG — GCTTTA —
3' — ACAGTG — GC — GC — G^{m}C — GC — CGAAAT —

> i) **Denaturation with alkali**
> ii) **Treatment with bisulfite reagent**
> iii) **Desulfonation with alkali**

5' — TGTUAU — UG — UG — ^{m}CG — UG — GUTTTA —

> **PCR (Extension of primer**
> **complementary to the upper strand)**

5' — TGTUAU — UG — UG — ^{m}CG — UG — GUTTTA —

----AC----AC-------GC-----AC----CA ◄ AT —

> **PCR (Amplification of the upper strand)**

5' — ACA ► TG----TG-------CG-----TG-----
       AC----AC-------GC-----AC----CA◄ AT —

> **DNA Sequencing**

FIG. 2. Schematic diagram representing steps involved in bisulfite genomic sequencing.

and carcinogenesis.[8] Aberrations in DNA methylation pattern characterized by global hypomethylation and regional hypermethylation occur in many human malignancies. In normal cells CpG islands in the promoter regions of most of the tumor suppressor genes, for example, those encoding p16, p15, retinoblastoma protein (Rb), and von Hippel-Lindau protein (VHL) are unmethylated and expressed. In cancer cells one or more of these tumor suppressor genes are rendered nonfunctional due to point mutations in their coding region. In contrast, some of the tumor suppressor genes in many human tumors are silenced by methylation of CpG islands in their promoters.[9] The signal that initiates alteration in methylation profile is yet to be understood.

A method exists to identify the methylated cytosines at specific sites within a gene. Bisulfite genomic sequencing, originally developed by Clark et al.,[10] is an innovative technique that can be used effectively to determine the methylation

[8] A. P. Bird and A. P. Wolffe, *Cell* **99**, 451 (1999).

[9] J.-P. Issa, *Curr. Topics Microbiol. Immunol.* **249**, 101 (1999).

[10] S. J. Clark, J. Harrison, C. L. Paul, and M. Frommer, *Nucleic Acids Res.* **22**, 2990 (1994).

status of specific CpG base pairs in a specific gene. Briefly, the procedure consists of denaturation, sulfonation, desulfonation, and polymerase chain reaction (PCR) amplification followed by restriction enzyme digestion and sequencing (see Fig. 2). Treatment of genomic DNA with bisulfite reagent under denaturing conditions converts cytosines in the DNA to uracils, amplified as thymine during subsequent PCR. Therefore, the few remaining cytosines in the sequencing gel are due to either 5-methylcytosine in the chromosomal DNA or incomplete bisulfite conversion, a common problem in this reaction. To overcome this artifact we have modified the original protocol. Denaturing DNA at 95° for 2 min after 30 min of desulfonation during this procedure facilitates completion of the reaction. Next, the amplified DNA is digested with several restriction enzymes before sequencing to check complete bisulfite conversion.

Methodology

Preparation of Genomic DNA from Cells or Tissues

Chromosomal DNA is isolated from mouse lymphosarcoma cells by lysis in digestion buffer [100 mM NaCl, 10 mM Tris-HCl (pH 8), 25 mM EDTA (pH 8), 0.5% (w/v) sodium dodecyl sulfate (SDS)] followed by overnight digestion with proteinase K (0.2 mg/ml) at 37°, phenol–chloroform extraction, and ethanol precipitation. The isolated DNA is allowed to dissolve overnight in TE [10 mM Tris-HCl (pH 8), 1 mM EDTA (pH 8)] buffer. To isolate DNA from tissues, pulverized frozen tissues are homogenized in the DNA digestion buffer before proteinase K digestion. Enough tissue or cells should be used to obtain at least 5–10 μg of DNA. For a detailed method for DNA isolation see Ausubel et al.[11]

Conversion of Unmethylated Cytosines of Genomic DNA to Uracils by Treatment with Bisulfite Reagent

The protocol for bisulfite genomic sequencing is shown in Fig. 3. Genomic DNA (5 μg) is denatured with 0.3 M NaOH (freshly prepared) at 37° for 30 min. PCR tubes (0.2 to 0.5 ml) are used for this reaction (see Fig. 3). The denatured DNA is treated with freshly prepared sodium metabisulfite (Sigma, St. Louis, MO) solution (final concentration, 2.35 M) containing hydroquinone (Sigma) (final concentration, 0.04 M) in a thermal cycler 20 times at 95° for 2 min and at 50° for 30 min. Thermal denaturation of DNA every 30 min at 95° prevents formation of double-stranded DNA (dsDNA), which is resistant to sulfonation by bisulfite reagent. This protocol results in complete conversion of cytosines to uracil. The bisulfite-treated DNA is purified with a Wizard DNA clean-up kit (Promega, Madison, WI), eluted in 50 μl of water, and treated with 50 mM NaOH (final concentration) in a volume of 100 μl at 37° for 30 min for desulfonation. DNA is then

---

[11] F. M. Ausubel, R. Brent, R. E. Kingston, and D. D. Moore (eds.), "Current Protocols in Molecular Biology," Section 2.2.1. John Wiley & Sons, New York, 1987.

Genomic DNA (5 μg in 10 μl) in 0.2 to 0.5 ml tubes
↓
Add 2 μl of freshly prepared 2 *M* NaOH
↓
Incubate for 30 min at 37°
↓
Add 100 μl of sodium bisulfite reagent

(5g Na metabisulfite + 50 mg hydroquinone + 600 μl 5M NaOH, add

water to make up the volume to 10 ml (final pH 5 automatically

adjusted) in a 15 ml tube, heat in a microwave oven for 2–3 sec

2-3X, for solubilization of hydroquinone)
↓
In a thermal cycler cycle at (95°/2 min, 50°/30 min) for 20 cycles,

store at 4° overnight.
↓
Clean up DNA using Wizard DNA Clean up kit, elute in 100 μl H₂O
↓
Add 6 μl of freshly prepared 5 *M* NaOH
↓
Incubate for 30 min at 37°
↓
Add 1 μg glycogen, 66 μl of 10 *M* ammonium acetate and 3 volume of

ethanol, incubate at –80° for an hr or overnight at –20°
↓
Spin down the DNA, wash with 70% ethanol, dry and dissolve in 20

μl of TE and use 2-5 μl (~100 –300 ng) for PCR

FIG. 3. Flow chart representing reaction conditions for bisulfite treatment and amplification of the converted DNA.

precipitated with ammonium acetate (2.5 *M,* final concentration) and 3 volumes of ethanol with 1 μg of glycogen as carrier at −80° for 1 hr or at −20° overnight. The DNA pellet is washed with 75% (v/v) ethanol, dried, dissolved in 20 μl of TE, and stored at −20° in small aliquots. As a control for bisulfite reaction, the promoter region of interest (in this case mouse or rat *MT-I* promoter) cloned into a bacterial plasmid DNA can be used. Because CpGs are not methylated in bacteria, complete conversion of cytosines to uracils is expected. Therefore, the cloned promoter DNA isolated from bacteria will serve as a negative control. The same plasmid DNA methylated with CpG methylase (New England BioLabs, Beverly, MA) *in vitro* before bisulfite treatment can be used as a positive control.

Amplification of MT-I Promoter from Bisulfite-Treated Genomic DNA by Nested
Polymerase Chain Reaction with Gene-Specific Primers

An aliquot of the converted DNA ($\sim$100 to 300 ng) is used for amplification of the *MT-I* promoter. We use nested PCR for amplification of the *MT-I* promoter from bisulfite-converted DNA. To amplify both methylated and unmethylated molecules we have designed primers from the regions of the promoter that do not harbor any CpG dinucleotides (see Fig. 4). Amplification of 200 to 500 bp is preferable, and it is difficult to amplify a region >550 bp as the size of chromosomal DNA after bisulfite treatment is considerably smaller. To amplify the upper strand of the bisulfite converted mouse *MT I* gene spanning the promoter region (Fig. 5) the following primers are used.[4]

For first round of PCR:
   mMTI-S1: 5′-TAGAGTAGATGGGTTAAGGTGAGTG
   mMTI-A1: 5′-ATCCCCACTTAATATTCTAAAAACC
For nested PCR:
   mMTI-S2: 5′-AGGAGTAGAGAATAATGTTGAGATGAGT
   mMTI-A2: 5′-CTTAAAAAACAACCTACCCTCTTTATAAT

Similarly, -304 bp to $+148$ bp of the rat *MT-I* promoter is amplified with two sets of primers from bisulfite-treated DNA derived from the liver and hepatoma. The following sets of primers are used to amplify the upper strand of rat *MT-I* promoter from bisulfite-treated DNA.[12]

For first round of PCR:
   rMTI-T1: 5′-GAAAGGAGAAGTTGAGGATAGTGTGTTATG
   rMTI-T2: 5′-TACCCCAAACCCCAACAAAAAACCATTC
For nested PCR:
   Hepa-A2: 5′-CCAAACCCCTACAACTAAATATTC
   Hepa-S2: 5′-GTATTGGATTAGTGATGGTTTGTAATAT

Restriction Enzyme Analysis of Polymerase Chain Reaction Product to Confirm
Completion of Bisulfite Treatment and Presence of Methyl-CpG Dinucleotides

Next, the PCR product is separated by agarose gel electrophoresis and the amplified DNA (524 bp for the mouse *MT-I* promoter and 452 bp for the rat *MT-I* promoter) is extracted from the gel slice and purified, using Qiaquick gel extraction kit (Qiagen). To test the efficiency of the bisulfite reaction, the amplified DNA is digested at 37° for 2 hr with the restriction enzymes *Apo*I (recognition site, G/C$^{\downarrow}$AATT$_{\uparrow}$A/T) and *Tsp*509I (cleavage site, $^{\downarrow}$AATT$_{\uparrow}$), which can cut only

[12] S. Majumder, K. Ghoshal, Z. Li, and S. T. Jacob, *J. Biol. Chem.* **274**, 28584 (1999).

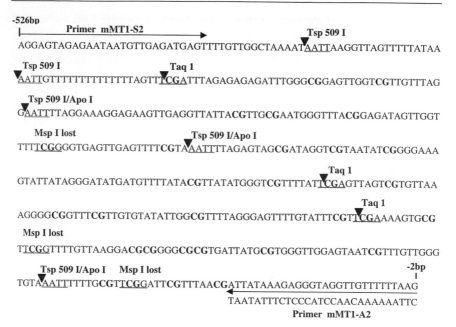

FIG. 4. Sequence of bisulfite-converted mouse *MT-I* promoter. The primers used for nested PCR of bisulfite-converted DNA and the restriction sites generated or lost are underlined.

the bisulfite-treated DNA (Fig. 4) but not the unconverted DNA (Fig. 5). The restriction sites for these two enzymes are generated only when cytosine residues are converted to thymine residues. The efficiency of bisulfite conversion is confirmed by a complete restriction cut of the amplified DNA (1 μg) with *Apo*I or *Tsp*509I. Partial cleavage indicates incomplete bisulfite conversion. The present

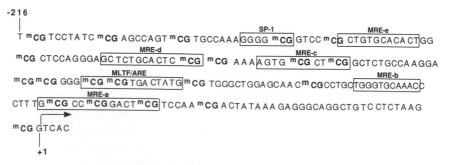

FIG. 5. Sequence of mouse *MT-I* proximal promoter region depicting the locations of *cis* elements and CpG dinucleotides. The *cis* elements, for example, MREs (metal response elements), MLTF/ARE (composite MLTF-binding site and antioxidant response element), and SP-1 sites are boxed. [Reproduced with permission from S. Majumder, K. Ghoshal, Z. Li, Y. Bo, and S. T. Jacob, *Oncogene* **18**, 6287 (1999).]

protocol always yields complete conversion of cytosines. To explore the presence of methylated CpG sequence in the chromosomal DNA isolated from a specific tissue or cell line the amplified *MT-I* promoter from mouse is digested with *Taq*I ($T^{\downarrow}$CGA) at 60° (for mouse *MT-I*) or with *Bst*UI ($CG^{\downarrow}$CG) at 65° (for rat *MT-I*). These sites are generated in the respective amplified DNA only if the CpG dinucleotides in the promoter region of chromosomal DNA are methylated.

The conditions for PCR amplification are the following. The reaction mixture for the first PCR contains 16.6 mM $(NH_4)_2SO_4$, 67 mM Tris-HCl (pH 8.8), 6.7 mM $MgCl_2$, and 10 mM 2-mercaptoethanol along with 0.2 mM dNTPs, 50 pmol of each primer in a volume of 50 μl. The PCR conditions used are as follows: 94°/2 min, 85°/2–4 min × 1 cycle, 94°/2 min, 60°/1 min, 72°/2 min × 35 cycles, and 72°/10 min × 1 cycle. Taq polymerase (2.5 units) was added when the reaction mixture was at 85° after initial denaturation at 94°. An aliquot of the first PCR reaction (1–3 μl) was then used for the nested PCR following the same reaction conditions with the exception of annealing temperature. The annealing temperature for nested PCR with both mouse and rat primers was 59°.

Sequencing of Polymerase Chain Reaction-Amplified Bisulfite-Converted MT-I Promoter

After completion of cytosine-to-thymine conversion is confirmed the PCR product is directly sequenced, using the Thermosequenase radiolabeled terminator cycle sequencing kit (United States Biochemical, Cleveland, OH) with the internal primer (5'-TTGGGGAAAGTATTATAGGGATATGATG) and the Hepa-S2 primer (annealing temperature of 50° for both) for the mouse and rat *MT-I* gene, respectively. Other PCR sequencing kits can also be used for PCR sequencing of the amplified product. Direct sequencing of the PCR product indicates whether an individual CpG dinucleotide exists in the unmethylated, methylated, or partially methylated state. Alternatively, the amplified product can be cloned into a plasmid vector such as TA cloning kit (Invitrogen) and individual clones can be sequenced.

Demethylation of MT-I Promoter with 5-Azacytidine

To determine whether *MT-I* promoter methylation is responsible for gene silencing, mouse lymphosarcoma P1798 cells are treated with 5-azacytidine (5-azaC), an inhibitor of DNA methyltransferase. These cells are grown in RPMI 1640 medium containing 25 mM HEPES (pH 7.2), 2 mM glutamine, 2% (w/v) sodium bicarbonate, 0.2 μM 2-mercaptoethanol, and 5% (v/v) fetal bovine serum. Cells at a density of 0.5×10^6/ml are treated with 2.5 μM 5-azacytidine or 5-deazacytidine (Sigma) for 72–90 hr until the cells divide at least once. The control or 5-azaC-treated cells at a density of 1×10^6/ml are treated with CdSO$_4$ (15 μM) or ZnSO$_4$ (50 μM) for 3 hr. Total RNA from these cells is subjected to Northern blot analysis with ^{32}P-labeled mouse MT-I cDNA as probe.

To explore the role of methylation in the expression of *MT-I* gene in the Morris hepatoma 3924A, a transplanted tumor, the hepatoma-bearing rats are injected with 5-azaC. Morris hepatoma 3924A is a dedifferentiated, rapidly growing tumor with a mean doubling time of 4–5 days.[13] It is grown by transplanting a thin slice (0.5 × 2–3 mm) of the solid tumor with a trochar in the hind leg of rats (ACI strain). The tumor on each leg grows to 15–20 g within 4–5 weeks. Most experiments are performed when the tumors attain this size. For heavy metal treatment, the tumor-bearing rats are injected intraperitoneally with $ZnSO_4$ (200 μmol/kg body weight) or the same volume of physiological saline.[6] After 4 hr, the animals are killed and the livers and hepatomas are immediately frozen in liquid N_2 for isolation of DNA and RNA. For 5-azaC treatment, rats bearing a hepatoma are injected intraperitoneally with the drug (5.0 mg/kg body weight) in 0.9% (w/v) NaCl or with saline alone (control) 15–20 days after tumor transplantation (when the tumor attains half of the maximal growth). 5-AzaC is injected on alternate days for 2 weeks. Animals are then injected intraperitoneally with $ZnSO_4$ (200 μmol/kg) or saline. After 4 hr the animals are killed to isolate RNA and DNA from the livers and hepatomas. The duration of 5-azaC treatment may vary from 10 to 15 days depending on the size of the tumor. The animals are killed when the tumors are reduced to 50 to 80% of the original size. This treatment regimen reduces the growth of the animals to some extent compared with the saline-injected controls without affecting the animals adversely.

Results

Responsiveness of MT-I Gene to Heavy Metals in Mouse Lymphosarcoma P1798 Cells and Morris Hepatoma 3924A Only after Treatment with 5-Azacytidine

As a first step to determine the role of methylation in preventing inducibility of the *MT-I* promoter in these tumor cells, we treated them with the demethylating agent 5-azaC, which forms a covalent complex with DNA methyltransferase–DNA adduct, inactivating the enzyme.[14] During subsequent DNA replication, methyl-CpG dinucleotides are unmethylated because of the lack of functional enzymes. After treatment with this agent MT-I mRNA is expressed in these cells on exposure to zinc (Fig. 6A and B). These results indicate that methylation is responsible for repression of the *MT-I* promoter in lymphosarcoma and hepatoma cells.

Methylation of MT-I Promoter at All CpG Dinucleotides in Mouse Lymphosarcoma Cells and Morris Hepatoma 3924A

To demonstrate that the *MT-I* promoter is indeed methylated in lymphosarcoma cells and Morris hepatoma, we performed bisulfite genomic sequencing.

[13] B. W. Duceman, K. M. Rose, and S. T. Jacob, *J. Biol. Chem.* **256,** 10755 (1981).
[14] D. V. Santi, C. E. Garrett, and P. J. Barrs, *Cell* **33,** 9 (1983).

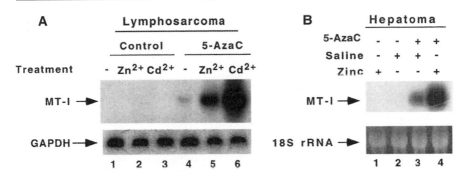

FIG. 6. Northern blot analysis of MT-I mRNA in lymphosarcoma cells and in Morris hepatoma before and after 5-azaC treatment. RNA (30 μg) isolated from cells or tissues is analyzed by Northern blotting first with random-primed, ^{32}P-labeled mouse MT-I cDNA and subsequently with rat glyceraldehyde-3-phosphate dehydrogenase (GAPDH) cDNA. [From S. Majumder, K. Ghoshal, Z. Li, Y. Bo, and S. T. Jacob, *Oncogene* **18**, 6287 (1999); and K. Ghoshal, S. Majumder, Z. Li, X. Dong, and S. T. Jacob, *J. Biol. Chem.* **275**, 539 (2000).]

Chromosomal DNA was isolated from mouse thymus and P1798 cells (with or without 5-azaC treatment for 72 hr), and then treated with bisulfite reagent. *MT-I* promoter (from -526 bp to -2 bp) is amplified with gene-specific primers. PCR-amplified DNAs from all three samples are fully cleaved by *Apo*I and *Tsp*509I, respectively, indicating complete bisulfite conversion (Fig. 7A). To detect the presence of CpG sites, if any, after bisulfite treatment, the amplified DNA is digested

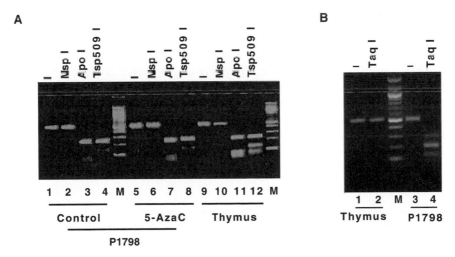

FIG. 7. (A and B) Restriction enzyme digestion profile of bisulfite-converted DNA from mouse thymus as well as control and 5-azaC-treated (72 hr) lymphosarcoma cells. Restriction enzyme digestions are performed according to the manufacturer protocol in buffers supplied by the company. [From S. Majumder, K. Ghoshal, Z. Li, Y. Bo, and S. T. Jacob, *Oncogene* **18**, 6287 (1999).]

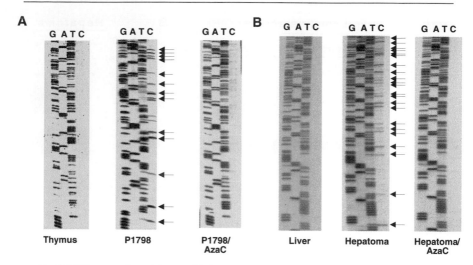

FIG. 8. PCR sequencing of the amplified *MT-I* promoter from bisulfite-converted mouse (A) and rat (B) chromosomal DNA. Arrows indicate methyl cytosines. [From S. Majumder, K. Ghoshal, Z. Li, Y. Bo, and S. T. Jacob, *Oncogene* **18,** 6287 (1999); and K. Ghoshal, S. Majumder, Z. Li, X. Dong, and S. T. Jacob, *J. Biol. Chem.* **275,** 539 (2000).]

with *Taq*I (T↓CGA). This restriction site is not present within the −526 bp to −2 bp region of mouse *MT-I* DNA, but is generated after C^{m}CGA (internal C methylation) is converted to TCGA on bisulfite conversion. Complete digestion of the amplified product from the lymphosarcoma DNA with *Taq*I and lack of digestion of that from the thymus with *Taq*I (Fig. 7B) demonstrate that *MT-I* promoter is methylated in the latter. Similarly, the PCR-amplified DNA from bisulfite-treated liver and hepatoma DNA is completely digested with *Apo*I and *Tsp*509I.

The amplified products are sequenced by the Femtomol sequencing method (Fig. 8A and B). All cytosines in mouse thymus and rat liver are converted to thymines, whereas only those cytosines that are followed by guanines remain intact in lymphosarcoma cells and hepatoma, indicating methylation *in vivo*. After treatment with 5-azaC only 20% of the CpGs are demethylated in P1798 cells, whereas in the hepatoma almost all the cytosines are demethylated.

Acknowledgments

The research performed in the authors' laboratory was supported, in part, by Grants ES 10874 and CA 81024 from the National Institute of Environmental Health Sciences and the National Cancer Institute, respectively.

[41] Functional Genomics: High-Density Oligonucleotide Arrays

By Sashwati Roy, Savita Khanna, Kimberly Bentley,
Phil Beffrey, and Chandan K. Sen

Introduction

The term *functional genomics* can be referred to as the "development and application of a global (genome-wide or system-wide) experimental approach to assess gene function by making use of the information and reagents provided by structural genomics."[1] It is characterized by high-throughput or large-scale experimental methodologies combined with statistical and computational analysis of the results. The fundamental strategy in a functional genomics approach is to expand the scope of biological investigation from studying single genes or proteins to studying all genes or proteins at once in a systematic fashion. Functional genomics promises to rapidly narrow the gap between sequence and function and to yield new insights into the behavior of biological systems.

As the Human Genome Project and related efforts identify and determine the DNA sequences of human genes, it is important that highly reliable and efficient mechanisms be found to assess individual genetic variation. Three methods for obtaining genome-wide mRNA expression data—oligonucleotide "chips,"[2] serial analysis of gene expression (SAGE),[3] and DNA microarrays[4,5]—are particularly powerful in the context of knowing the entire genome sequence (and thus all genes).[6]

Types of DNA Hybridization Arrays

Current array formats can be categorized into the following four groups.

1. *Macroarrays:* Macroarrays rely on robotically spotted probes that have been immobilized on a membrane-based matrix. The probe density is generally

[1] P. Hieter and M. Boguski, *Science* **278,** 601 (1997).

[2] S. P. Fodor, R. P. Rava, X. C. Huang, A. C. Pease, C. P. Holmes, and C. L. Adams, *Nature (London)* **364,** 555 (1993).

[3] V. E. Velculescu, L. Zhang, B. Vogelstein, and K. W. Kinzler, *Science* **270,** 484 (1995).

[4] M. Schena, D. Shalon, R. Heller, A. Chai, P. O. Brown, and R. W. Davis, *Proc. Natl. Acad. Sci. U.S.A.* **93,** 10614 (1996).

[5] M. Schena, D. Shalon, R. W. Davis, and P. O. Brown, *Science* **270,** 467 (1995).

[6] V. E. Velculescu, L. Zhang, W. Zhou, J. Vogelstein, M. A. Basrai, D. E. Bassett, Jr., P. Hieter, B. Vogelstein, and K. W. Kinzler, *Cell* **88,** 243 (1997).

Copyright 2002, Elsevier Science (USA).
All rights reserved.
0076-6879/02 $35.00

lower on these arrays compared with those of the other three groups. These arrays mostly use radioactive probe labeling. In some cases chemiluminescent labeling has also been described.

2. *Microarrays:* Microarrays use a glass or plastic slide as matrix. These arrays have a higher density of probes compared with macroarrays and use fluorescent labeling-based detection.

3. *High-density oligonucleotide arrays (gene chips):* The probe is generated *in situ* on the surface of the matrix. The leader in these arrays is Affymetrix (Santa Clara, CA) and their combinatorial synthesis method.

4. *Microelectronic arrays:* Microelectronic arrays represent one of the more recent formats of hybridization arrays currently under development by Nanogen (San Diego, CA). Instead of a membrane or a glass slide platform, these arrays consist of a set of electrodes covered by a thin layer of agarose coupled with affinity moiety (permitting biotin–avidin immobilization of probes). Selection and adjustment of proper physical parameters enable rapid DNA transport, site-selective concentration, and accelerated hybridization reactions to be carried out on active microelectronic arrays. These physical parameters include DC current, voltage, solution conductivity, and buffer species. Generally, at any given current and voltage level, the transport or mobility of DNA is inversely proportional to electrolyte or buffer conductivity. The incorporation of controllable electric fields gives a new degree of control over probe deposition and target hybridization.[7,8]

High-Density Oligonucleotide Arrays

The leading arrays in the category of high-density oligonucleotide arrays are manufactured by Affymetrix and utilize the combinatorial synthesis principle.[9] The arrays are designed by using a light-directed chemical synthesis process that creates a series of photolithographic masks to define chip exposure sites, followed by specific chemical synthesis steps. This process constructs high-density arrays of oligonucleotides. Approximately 20 different probe pairs represent each gene on a chip. Each probe pair consists of a perfect match (PM) oligonucleotide probe and a single-base mismatch (MM) oligonucleotide (Fig. 1). The arrays are designed for gene expression as well as single-nucleotide polymorphism (SNP) detection and they cover a large range of different species. The sequence data that Affymetrix uses to build the arrays are downloaded from public databases such as UniGene and GenBank.

[7] C. F. Edman, D. E. Raymond, D. J. Wu, E. Tu, R. G. Sosnowski, W. F. Butler, M. Nerenberg, and M. J. Heller, *Nucleic Acids Res.* **25,** 4907 (1997).

[8] W. M. Freeman, D. J. Robertson, and K. E. Vrana, *Biotechniques* **29,** 1042 (2000).

[9] G. C. Kennedy, *EXS* **89,** 1 (2000).

PM   99618_at (3vs1)
MM   **Intensity: 55 1738**

FIG. 1. Approximately 16–20 different probe pairs represent each gene on a chip. Each probe pair consists of a perfect match (PM) oligonucleotide probe and a single-base mismatch (MM) oligonucleotide.

For the chips to work properly, a sample must be prepared according to Affymetrix protocols. A brief description of the procedures involved in assessing a gene expression profile, using Affymetix GeneChip arrays, is provided.

Sample Preparation

The oligonucleotides on the chip or microarray are called the *probes* and the sample (total RNA or mRNA) that is put on to interrogate is called the *target*. The process is inverted from a traditional Northern analysis.

Total RNA Isolation

RNA is extracted from cells with an RNeasy total RNA isolation kit (Qiagen, Chatsworth, CA). For tissues, RNA is first extracted with TRIzol (Invitrogen, Carlsbad, CA) RNA extraction reagent and then cleaned up with an RNA isolation kit (Qiagen).

cDNA Synthesis

The first strand is synthesized by reverse transcribing the RNA, using the Superscript Choice system (Invitrogen) and oligo(dT)24-anchored T7 primer [high-performance liquid chromatography (HPLC) purified] at 42° for 60 min and then at 70° for 15 min. The second strand is synthesized by using the first-strand synthesis reaction, 5× second-strand buffer, *Escherichia coli* DNA polymerase, and T4 DNA polymerase. The cDNA is isolated according to the Phase Lock gel extraction (Eppendorf, Hamburg, Germany) procedure.

In Vitro Transcription, cRNA Clean-Up, and Fragmentation

Biotinylated RNA is synthesized with an RNA transcript labeling kit (BioArray HighYield: Enzo Diagnostics, Farmingdale, NY). A detailed protocol is provided with the kit.

In Vitro Transcription Clean-Up. Qiagen RNeasy minicolumns are used to clean up the *in vitro* transcription (IVT) cRNA. After the clean-up, cRNA is fragmented with 5× fragmentation buffer (200 mM Tris–acetate, pH 8.1; 500 mM potassium acetate; 150 mM magnesium acetate).

GeneChip: Hybridization, Washing, and Scanning

Further sample processing is mostly automated. The hybridization oven 640 automates the hybridization process for GeneChip probe arrays. The oven provides precise temperature control to ensure successful hybridization, and cartridge rotation to provide continuous mixing. Up to 64 arrays can be processed at one time. The GeneChip fluidics station automates the introduction of the nucleic acid target to the probe array cartridge and controls the delivery of reagents and the timing and temperature for hybridization of nucleic acid target to the probe array. Each fluidics station can independently process four arrays at one time. The probe array nucleic acid target is simply loaded on the fluidics station. Information about the type of array to be analyzed is punched in and the software automatically selects the appropriate protocol. Once processing is complete, messages displayed on the PC and the fluidics station indicate that the probe array is ready for scanning. The GeneArray scanner is from Agilent (Palo Alto, CA) and utilizes a charge-coupled device (CCD) camera and an argon ion laser to excite fluorescent molecules incorporated into the nucleic acid target to generate a quantitative hybridization signal (Fig. 2). With precise optics, the GeneArray scanner focuses the laser on 3-μm spots within each of the thousands of probe features contained on the GeneChip probe array. A high-resolution image of the probe array is displayed in real time during scanning, and fluorescence intensity data are automatically stored in a raw file.

A   **B**

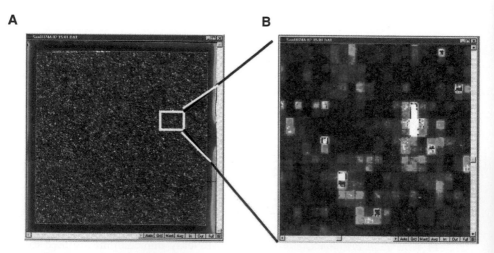

FIG. 2. A representative high-resolution image (A) and a zoomed area of the image (B) of hybridization signals generated by the GeneArray scanner.

Notes

Because chips can be used only once, the target must be tested to ensure that it is of high-enough quality to go onto the expression chip. The test chip serves as one control for the experiment, and the other controls are discussed below. It is critical that test chips be used before the target is put on the gene expression chip (sections below contain example data).

Test Chips: Target Labeling and Hybridization Efficiency Analysis

Once the two targets are prepared, they are each put on a test chip to determine whether they are of high-enough quality to go onto the expression chips. When looking at the data, the control gene names always begin with an AFFX prefix. These are AFFX-Murine BetaActin and AFFX-Murine GAPDH (glyceraldehyde-3-phosphate dehydrogenase) on the murine arrays. Ideally, when comparing $5'$ signal with $3'$ signal, a $1:1$ ratio will be seen. Empirically, targets having $5':3'$ ratios between 0 and 3 generally give good results, those with ratios between 3 and 4 give marginally good results, and those with ratios >4 give poor results. Thus test chips are used to determine target-labeling quality. Also included are control probe sets that interrogate phage sequences. BioB, BioC, BioD, and Cre are such sequences and are used as hybridization controls. These probe sets are designed to detect the prelabeled oligonucleotides that are contained in the eukaryotic hybridization control kit. This is to ensure that the hybridization, washing, staining, and scanning steps are capable of detecting a broad linear range of labeled cRNA with a high level of sensitivity. Each chip must first be analyzed, using the above-described controls as criteria, before the chips can be compared with one another. After basic analysis, scale factors must be examined between chips and should not vary by 3-fold.

Affymetrix has included controls on the chips so that the data can be quantified and also reproduced. The controls are also used to test different parts of the procedure so that troubleshooting can be performed if necessary. All the genes also act as their own control, with the perfect match and mismatch sequences that are used on the arrays, and the difference in hybridized signal between the probe sets is used to identify nonspecific hybridization and background signal. Affymetrix calls this value the *average difference* (see Data Analysis, below) and it is commonly used as the expression level for the probe set.

For the pilot study, RNA was isolated from normal and treated mice. The Affymetrix protocol was used to prepare the targets. The basic or absolute analysis of chip 1 had a value of 1.9 for the $5':3'$ ratio of GAPDH. Furthermore, the control probe sets (BioB, BioC, BioD, and Cre) were also present on the chip. The basic or absolute analysis of chip 2 (RNA from treated mouse) had a value of 2.0 for the $5':3'$ ratio of GAPDH. Control probe sets BioB, BioC, BioD, and Cre were also present on the chip. Both chips passed the specification of the controls, and the last criterion that they must pass is the fold change of the scale factor between them. This is a crucial point because if the criteria are met then the chips can be compared with one another and the differences in gene expression will be analyzed. For this study, the two chips had a fold change of 1.5 for the scale factor, which means that they can be compared with one another.

The second round of analysis is called *comparative analysis*. This analysis uses a control chip that is compared with an experimental chip. In the pilot study, the control chip would be chip 1 and the experimental chip would be chip 3. There are more than 12,000 genes and expressed sequence tags (ESTs) on one murine chip. The first step is to reduce the number of genes, so that the data show only those that have a significant change in either the control chip or the experimental chip. The software suite provided by Affymetrix does this analysis by using different algorithms that compare the two chips. It is important to note that the algorithms tend to be on the conservative side when determining whether a gene is present or not. In the pilot study, the reduced data, once run through the algorithms, showed that 283 genes had a significant difference between the control and experimental chips (Fig. 3). Each gene in a comparison analysis has five potential difference call outcomes: Increase, Marginally Increase, Decrease, Marginally Decrease, and No Change. To reduce the data, the No Change calls are removed, as are those with a fold change of less than 2. The fold change indicates the relative change in the expression levels between the experiment and control targets. Genes that indicate a marginal increase or decrease must be looked at individually to determine whether a definite call can be made. The reduced data can then be easily exported into a Microsoft Excel file for further manipulation.

Data Analysis

To analyze massive amounts of genome-wide data generated by microarray experiments is a challenging task. Gene expression data are useless unless biologically meaningful information can be extracted and presented in some readily comprehensible fashion.[10] The production of this information, involving many facets of image processing, statistical analyses, and data visualization, is possible only with computers powered by sophisticated software. The choice of data analysis strategy should be influenced by the purpose of the microarray experiment and the user's knowledge of the biology of the system under investigation.

Data Mining

The discovery of patterns in gene expression relationships is part of the realm of data mining. Known collectively as *clustering,* these multivariate statistical

---

[10] J. Quackenbush, *Nat. Rev. Genet.* **2,** 418 (2001).

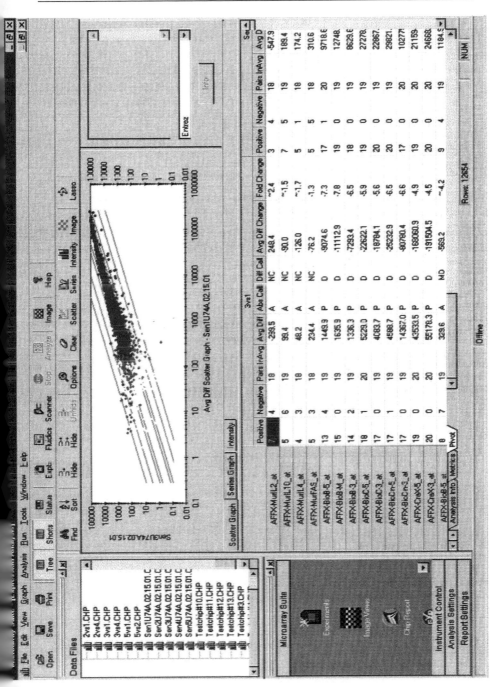

Fig. 3. A representative average difference scatter plot displayed by Microarray Suite (Expression analysis).

methods have become the essential tools for the elucidation of gene expression patterns in microarray data. A number of different methods, for example, k-means, self-organizing maps (SOMs), hierarchical clustering, support vector machines, and Bayesian statistics, are employed for clustering analysis.[10] Another useful data-mining method, *principal component analysis* (PCA), is a data reduction technique used to identify uniquely expressed genes. Bioinformatics and data storage are the culmination of the microarray analysis process. Many of the software analysis packages offer immediate access to many public or institutional genetic databases via the Internet.

For the present study, data were analyzed as described below, using Xalysis-Lite (XPROTEIN; Bioinformatics, San Rafael, CA). Three samples, each from a different control animal, were assigned to chips in control group A; likewise, three samples from different treated animals were assigned to chips in experimental group B. The goal to identify candidate genes that vary according to treatment should not be hindered by changes due to a single individual's traits. The analysis of individual probe sets and how they vary between the control group and the experimental group must take this into account. The approach described here is different compared with the "pooling" technique, in which samples are taken from a population of individuals, mixed, and analyzed on a single chip, thereby averaging their expression characteristics. In the present study, rather than mix the samples, a different chip is used for each. Assigning individual samples each to a different chip reduces biological variation because multiple different animals are used, and it reduces variation due to the measuring technology because multiple chips are used. As a result this technique can mask small, but real, expression differences because the source of variation is confounded. It may be coming from the individual animals, or the individual chips, or both. It is advisable, therefore, to design the overall experiment to use as many replicate chips measuring the same sample as time and budget allow.

After processing each chip as described earlier, the overall expression mean and standard deviation of each chip is calculated from each probe set's *average difference* value, one of the standard values output by the Affymetrix software. The highest 2% and lowest 2% of the values are considered outliers and not used in this calculation.

The statistics of experimental groups vary considerably, but because each comes from a different treated individual, all are accepted for further analysis (Table I). More stringent criteria for acceptance can always be adopted later.

Probe Set Analysis Protocol

The average difference value of each probe set is first clamped to zero and normalized by using the mean of its chip (i.e., from Table I). It is then scaled by using the mean of the group means. This calculation smooths out differences between individuals, but retains the overall expression level suggested by the group.

TABLE I
GROUP DATA

Group	Mean	Standard deviation
Control		
A1	29.16	61.92
A2	27.60	59.92
A3	29.74	63.91
Experimental		
B1	18.63	41.15
B2	38.66	81.23
B3	90.25	176.0

The consistency of each probe set's values across all experiments must be carefully evaluated before an effect can be considered real. To that end a two-sample independent t test between the control and experimental groups is applied to each probe set's values (after normalization and scaling) as follows:

$$ t = \frac{\bar{A} - \bar{B}}{\left[\frac{\sum A_i^2 - \frac{(\sum A_i)^2}{n_A} + \sum B_i^2 - \frac{(\sum B_i)^2}{n_B}}{(n_A - 1) + (n_B - 1)} \right]^{1/2} \left(\frac{1}{n_A} + \frac{1}{n_B} \right)} $$

In this equation, A_i refers to the individual values of one probe set from the control group and B_i refers to the individual values of the same probe set from the experimental group; n_A and n_B are the total number of experiments in each group; and $\bar{A}$ and $\bar{B}$ are the means. The t value can be used to determine the probability that values from group A belong to the same statistical distribution as values from group B. The probability value implied by t depends on the degrees of freedom of the data:

$$ df = (n_A - 1) + (n_B - 1) $$

Given t and df, the probability can be determined from a standard table of t values. If the probability is small, it is reasonable to assume that the treatment has altered the expression of the probe set in some way. Only those probe sets that pass this test are considered for further analysis (Table II). Regarding Table II, the first probe set illustrates data consistent among the experiments from each group; the second shows data that are highly inconsistent. Because the probability value of the first set is below the chosen threshold of 1% it is considered for future analysis; the second set, with a probability of more than 17%, is rejected.

It is reasonable to rely on this test to avoid the confounding problem because it rejects probe sets that vary in a nonspecific way between experimental groups, whether it be due to biological noise or noise from the technology. It is possible, however, that a real effect is being masked by technology noise and is, therefore,

TABLE II
STATISTICAL ANALYSIS

	Transcript A		Transcript B	
Probe set	Avg. diff.	Normalized	Avg. diff.	Normalized
A1	45.90	45.39	97.20	96.12
A2	39.00	40.75	58.60	61.22
A3	44.70	43.33	78.40	76.00
Mean of A		43.16		77.78
B1	50.20	132.50	37.60	99.24
B2	89.90	114.38	213.70	271.89
B3	207.40	113.02	237.20	129.25
Mean of B		119.96		166.80
t value		−11.96		−1.64
Probability		0.03		17.59
Fold change		2.80		2.1

Abbreviations/symbols: Avg. diff., Average difference.

rejected by this test. That is why it is advisable to run replicate experiments using the exact same sample when possible.

Scoring Candidate Probe Sets

Applying the above described test identified about 300 genes worthy of further examination. Which should be examined first? A quick way to decide is to score the candidate probe sets that passed the *t* test according to their fold change in expression weighted toward higher overall expression values. The mean value of the probe set within each experiment group is used from here on as its expression value within that group.

$$\text{Score} = \frac{\bar{B} + x}{\bar{A} + x} \quad \text{if } \bar{B} \text{ is greater than or equal to } \bar{A}, \text{ or}$$

$$\text{Score} = -\frac{\bar{A} + x}{\bar{B} + x} \quad \text{if } \bar{A} \text{ is greater than } \bar{B}$$

The value of variable x is set equal to the expression level deemed significant by the researcher. The list of candidates is then ordered according to this score. Probe sets near the top of the list are overexpressed in group B compared with group A; probe sets near the bottom of the list are underexpressed in group B compared with group A. Although the *t* value alone can be used to order the data in a similar fashion, the score calculated with this method is similar to the fold change value for the expression range of interest. Thus, the topmost and bottommost entries in this sorted list provide a convenient starting point for evaluation and further experimentation when examined by the critical eye of the researcher.

Acknowledgments

Supported by NIH GM 27345, the Surgery Wound Healing Research Program, and U.S. Surgical, Tyco Healthcare Group. The Laboratory of Molecular Medicine is the research wing of the Center for Minimally Invasive Surgery.

[42] Reporter Transgenes for Study of Oxidant Stress in *Caenorhabditis elegans*

By CHRISTOPHER D. LINK and CAROLYN J. JOHNSON

Introduction

For many studies of the effects of oxidant stress on cells it can be advantageous to visualize the transcriptional response of the cell *in vivo* in real time. In optically transparent model systems, gene expression can be directly visualized by the construction of reporter transgenes expressing green fluorescent protein (GFP), as originally demonstrated by Chalfie and colleagues.[1] We describe both the general considerations involved in the construction of GFP reporter transgenes responsive to oxidative stress and the specific details of constructing a representative transgenic reporter in the model nematode worm *Caenorhabditis elegans*. Although the details of the representative reporter transgene apply specifically to *C. elegans*, the general approach should be applicable to many model systems.

Identification of Oxidant Stress-Responsive Genes

Construction of oxidative stress-responsive reporter transgenes first requires identification of oxidative stress-responsive genes. Candidate responsive genes can be identified by extrapolation from studies of other systems [e.g., a *C. elegans* GFP reporter transgene based on the small *C. elegans* heat shock protein 16 (HSP16) was found to be responsive to oxidative stress,[2] an unsurprising result considering previous studies of mammalian small heat shock proteins[3]] or from direct gene

---

[1] M. Chalfie, Y. Tu, G. Euskirchen, W. W. Ward, and D. C. Prasher, *Science* **263,** 802 (1994).
[2] C. D. Link, J. R. Cypser, C. J. Johnson, and T. E. Johnson, *Cell Stress Chaperones* **4,** 235 (1999).
[3] X. Preville, H. Shultz, U. Knauf, M. Gaestel, and A. P. Arrigo, *J. Cell Biochem.* **69,** 436 (1998).

Copyright 2002, Elsevier Science (USA).
All rights reserved.
0076-6879/02 $35.00

expression data. An important consideration is the type of oxidative stress used in gene expression studies, as it is likely that not all oxidative challenges (e.g., exposure to redox quinones, H_2O_2, or hyperbaric oxygen) elicit the same spectrum of transcriptional responses. It is therefore preferable to choose candidate responsive genes on the basis of oxidative challenges similar to those planned for experiments with the reporter transgenes.

We have used the data of Tawe *et al.*[4] to construct oxidative stress-responsive reporter transgenes in *C. elegans*. These researchers used reverse transcriptase-polymerase chain reaction (RT-PCR) differential display to identify genes induced by the redox quinone paraquat, and then confirmed expression increases by Northern blot. In the future, it is likely that microarray-based gene expression studies[5,6] will identify a wealth of genes responsive to oxidative stress.

Design of Reporter Transgenes

Three major considerations concerning the design of a reporter transgene are as follows: (1) Should a translational or transcriptional fusion construct be made? (2) What extent of promoter region should be used? (3) What variant of GFP should be used for the chimeric construction? For translational fusions, the GFP-coding region is fused in frame with the coding region of the candidate responsive gene, usually at the extreme C or N terminus of the protein. An advantage of translational fusions is that they can reveal the subcellular localization of the candidate responsive gene, because GFP fusion proteins often retain the distribution (and function) of the natural protein.[7] This can be a disadvantage, however, for quantitative studies if, for example, the natural protein is secreted. In transcriptional fusions, the GFP-coding region is fused directly downstream of the candidate gene promoter. The major advantage of this approach is the relative ease of generating these constructs, which allows high-throughput production of reporter constructs by PCR-based techniques. A disadvantage of this approach is that important promoter elements (e.g., enhancers in introns) may be lost.

The appropriate extent of the promoter region used for the reporter constructs (i.e., how many kilobases of upstream sequence to use) is generally difficult to determine in the absence of experimental data. For organisms such as *C. elegans,* with relatively small, sequenced genomes, it is possible simply to use all the sequence between the candidate gene and the next upstream gene on the chromosome. (Note of caution for *C. elegans* researchers: This approach is obviously

[4] W. N. Tawe, M. L. Eschbach, R. D. Walter, and K. Henkle-Duhrsen, *Nucleic Acids Res.* **26,** 1621 (1998).

[5] S. Zou, S. Meadows, L. Sharp, L. Y. Jan, and Y. N. Jan, *Proc. Natl. Acad. Sci. U.S.A.* **97,** 13726 (2000).

[6] M. R. Volkert, N. A. Elliot, and D. E. Housman, *Proc. Natl. Acad. Sci. U.S.A.* **97,** 14530 (2000).

[7] H. Komatsu, I. Mori, J. S. Rhee, N. Akaike, and Y. Ohshima, *Neuron* **17,** 707 (1996).

not appropriate for genes that are downstream members of operons. It may be unwise to generate reporter constructs from any genes that are SL2-spliced.) As there has been no systematic analysis of *C. elegans* promoters to guide choices of upstream regions, we generally choose ∼3 kb upstream of the initiator ATG for our constructs when there are no other constraints on the promoter regions. Promoter sequences can be recovered by restriction enzyme digest of genomic clones, or by PCR from genomic DNA. Promoter recovery by PCR is significantly more flexible (e.g., it allows recovery of upstream sequences right up to the initiator ATG), although it entails the risk of PCR-induced mutations, and thus the resulting construct should be confirmed by DNA sequencing. As described below, we have had good success using PCR with high-fidelity polymerases to recover functional promoter fragments.

Many GFP sequence variants have been developed that vary in absorbance/emission spectra and stability.[8] *Caenorhabditis elegans* researchers are particularly fortunate because of the extensive series of GFP-based expression vectors constructed by Fire and colleagues, which have been made freely available to the research community (information available at http://ftp.ciwemb.edu/PNF:byName:/FireLabWeb/FireLabInfo/). The most common variants have substitutions (threonine or cysteine for serine) at residue 65, which leads to enhanced fluorescence and stability of the GFP. Constructs with Ser-65 substitutions can be readily visualized and quantified with epifluorescence microscopes equipped with standard fluorescein isothiocyanate (FITC) filter sets. For studies involving measurement of transient expression of responsive genes, it may be advantageous to use GFP variants with decreased stability.[9] GFP variants also exist with sufficiently different absorbance/emission spectra to allow distinction between coexpressed variants by using epifluorescence microscopes equipped with the appropriate filter sets.[10]

Construction of Representative *Caenorhabditis elegans* Oxidative Stress Reporter Transgene

Among the paraquat-inducible genes identified by Tawe *et al.* is a putative glutathione-*S*-transferase corresponding to predicted *C. elegans* gene K08F4.7. The predicted initiator ATG of this gene lies 727 bp upstream of another predicted gene (K08F4.6, also predicted to be a glutathione-*S*-transferase) transcribed from the opposite strand (see Fig. 1). This gene arrangement defines a short presumptive promoter region, making a good target for a PCR-based construction. Primers are

[8] A. B. Cubitt, R. Heim, S. R. Adams, A. E. Boyd, L. A. Boyd, and R. Y. Tsein, *Trends Biochem. Sci.* **20**, 448 (1995).

[9] J. B. Andersen, C. Sternberg, L. K. Poulsen, S. P. Bjorn, M. Givskov, and S. Molin, *Appl. Environ. Microbiol.* **64**, 2240 (1998).

[10] D. M. Miller, N. S. Desai, D. C. Hardin, D. W. Piston, G. H. Patterson, J. Fleenor, S. Xu, and A. Fire, *Biotechniques* **26**, 914 (1999).

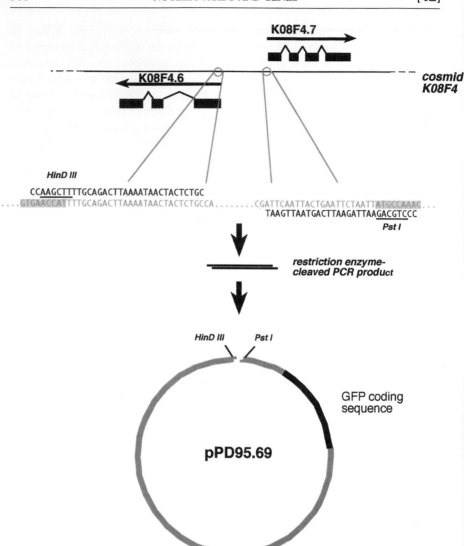

FIG. 1. Construction of an oxidative stress-inducible reporter transgene. The putative promoter region of the paraquat-inducible gene K08F4.7 was recovered by PCR, using cosmid K08F4 as a template and primers targeted immediately upstream of the K08F4.7 and K08F4.6 putative initiator ATG codons. Incorporated into the 5′ portion of these primers were restriction site sequences for *Hin*dIII and *Pst*I (underlined). After purification, this PCR fragment was cleaved with *Hin*dIII and *Pst*I and the "sticky end" product was recovered by gel purification, and ligated into a similarly purified *Hin*dIII/*Pst*I-linearized pPD95.69 vector fragment.

designed to amplify the sequence between the K08F4.6 and K08F4.7 ATG initiator codons, and to introduce 5' extensions containing convenient restriction enzyme sites. The promoter region is amplified from cosmid K08F4 by AmpliTaq Gold (PerkinElmer, Norwalk, CT) polymerase, as per the manufacturer protocol. [When possible, we amplify from cosmid clones to minimize amplification cycles and the likelihood of introducing PCR-generated mutations. We also more typically use high-fidelity polymerases such as *Pfu* (New England BioLabs, Beverly, MA).] The completed reaction is purified by passage through a PCR purification column (Qiagen, Chatsworth, CA), and then the PCR product is digested with *Hin*dIII and *Pst*I, and gel purified (Qiagen gel extraction kit). As a GFP expression vector, we have chosen pPD95.69 from the Fire collection. This is a promoterless expression vector that contains a GFP-coding region with the S65C substitution and five small introns. This vector is cleaved with *Hin*dIII and *Pst*I, gel purified, and ligated to the promoter fragment. After transformation, clones are recovered and checked by restriction digest, resulting in the identification of an appropriate clone, designated pAF15.

Recovery of Transgenic Lines

pAF15 is introduced into wild-type *C. elegans* animals by gonad microinjection, initially using the dominant morphological marker *rol-6(su1006)* encoded in coinjected plasmid pRF4[11] to identify transgenic animals independent of any GFP expression from the pAF15 reporter transgene. Microinjection solutions contain pAF15 (100 ng/μl) and pRF4 (100 ng/μl). [Gonad microinjection leads to the incorporation of injected DNA into germ cells and a fraction of the resulting progeny animals. These transformed F_1 progeny can be identified by the abnormal movement "Roller" phenotype they exhibit. A fraction of the transformed F_1 animals will contain heritable extrachromosomal arrays containing multiple copies of both injected plasmids, and thus will segregate transgenic Roller F_2 progeny, allowing the establishment of heritable (but mitotically and meiotically unstable) transgenic lines.]

After subsequently demonstrating oxidative stress-dependent induction of the pAF15 reporter transgene (see below), additional lines are generated by injecting only pAF15 and directly selecting transgenic animals on the basis of oxidative stress-induced GFP expression, using a Leica (Bensheim, Germany) MZ12 dissecting microscope equipped with epifluorescence optics to screen progeny of injected animals. One transmitting extrachromosomal line recovered in this manner, CL1166, has been used to generate completely stable chromosomally integrated lines. This is accomplished by exposing 20 transgene-bearing young adult animals to 7000 rad of γ rays from a cesium-66 source, and then cloning

---

[11] C. C. Mello, J. M. Kramer, D. Stinchcomb, and V. Ambros, *EMBO J.* **10**, 3959 (1991).

each exposed animal. In our work, 10 transgene-bearing first-generation progeny animals were subsequently cloned from each parental animal. After production of second-generation (F_2) animals, 200 plates are exposed to hyperbaric oxygen, and plates are screened for high proportions ($\geq 75\%$) of GFP-expressing F_2 animals, indicative of transgene stabilization via chromosomal insertion. F_2 animals are cloned from candidate integrated clones, and clones are identified in which all F_3 progeny express GFP in response to hyperbaric oxygen exposure. In our work, three completely stable lines were recovered and outcrossed to wild-type animals (six times for strain CL2166). Chromosomal integration of the transgene is confirmed by demonstrating appropriate Mendelian segregation of the transgene.

Hyperbaric Oxygen as Oxidative Stress

Redox quinones such as paraquat and juglone can generate intracellular superoxide, and have been routinely used in *C. elegans* to generate oxidative stress.[4,12,13] However, these compounds are both unstable and highly toxic, making them difficult to use to induce nonlethal oxidative stress. We therefore apply hyperbaric oxygen exposure to generate oxidative stress in *C. elegans,* using a cylindrical pressure vessel (16×17.5 cm external dimensions) custom machined from stainless steel (see Fig. 2). We have found that whereas longer exposures (e.g., 48 hr) to 40-psi oxygen (99.95% industrial grade) are lethal to *C. elegans,* shorter exposures (up to 8 hr) do not reduce the overall life span of wild-type animals, but can induce oxidative stress. We have therefore generally used hyperbaric oxygen to assay reporter responses.

Assaying Oxidative Stress Induction of Reporter Transgenes

Three independent chromosomally integrated transgenic lines carrying pAF15 (CL2166, CL3166, and CL4166) are exposed to a variety of oxidative stresses. All these lines show a low level of GFP expression in most tissues under normal growth conditions. (The basal GFP expression in these lines increases when animals are grown at higher temperatures, possibly reflecting increased endogenous oxidative stress in animals with higher metabolic rates.) This baseline level of GFP expression is strongly increased by a 4-hr exposure to hyperbaric oxygen (see Fig. 3A). To quantify transgene induction, cohorts of treated or untreated young adult animals are individually digitally imaged for GFP expression, using an Axioskop epifluorescence microscope (Zeiss, Thornwood, NY) equipped with a Cohu (San Diego, CA) monochrome camera, and the mean pixel density (corresponding to overall mean GFP expression in each animal) is determined with NIH

[12] N. Iishi, K. Takahashi, S. Tomita, T. Keino, S. Honda, K. Yoshino, and K. Suzuki, *Mutat. Res.* **237,** 165 (1990).
[13] J. R. Vanfleteren, *Biochem. J.* **292,** 605 (1993).

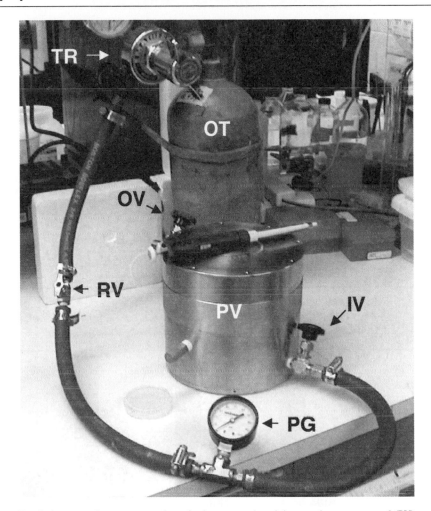

FIG. 2. Apparatus for exposure to hyperbaric oxygen. A stainless steel pressure vessel (PV) was connected to the regulator valve (TR) of a standard oxygen tank (OT) containing 99.95% pure oxygen (industrial grade) with vacuum tubing. Animals to be exposed to hyperbaric oxygen were propagated on solid agar medium petri plates in ambient air, and then placed inside the vessel. After sealing the vessel, air was purged with oxygen by opening the inlet (IV) and outlet (OV) valves, and allowing pure oxygen to flow through the vessel for 1 min. The outlet valve was then closed, and the subsequent increasing pressure was monitored with the in-line pressure gauge (PG). When the desired pressure was obtained (typically 40 psi), the tank regulator and intake valves were simultaneously closed. Line pressure was released by opening the line release valve (RV), and the pressure vessel was removed from the line and placed in a 20° incubator. At the conclusion of hyperbaric oxygen exposure, vessel pressure was released by partial opening of the output valve, which resulted in a relatively slow depressurization (i.e., >2 min) of the vessel.

A

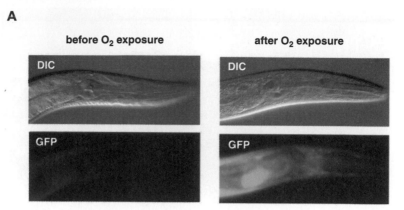

B

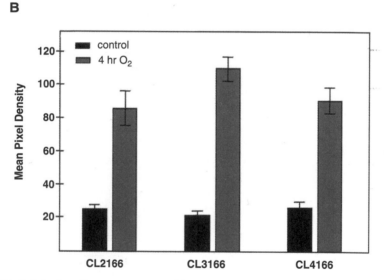

FIG. 3. Induction of reporter transgene with hyperbaric oxygen. (A) A young adult animal from reporter strain CL2166 (grown in ambient air at 20°) was mounted on a slide and imaged (*left*), and then returned to a solid agar medium plate and exposed to 40-psi hyperbaric oxygen in a pressure vessel for 4 hr. After a 3-hr recovery in ambient air, the animals were remounted and reimaged, using identical exposure conditions. Note the extensive increase in GFP expression in multiple (neuronal, intestinal, muscle, etc.) tissues. (B) Cohorts from three independent reporter lines (18–20 animals per strain per condition) were either exposed to 40-psi oxygen or maintained in ambient air as controls. GFP expression in each animal was then quantified with a digital camera and NIH Image software, and is expressed as mean pixel density.

Image software (NIH, Bethesda, MD).[14] Shown in Fig. 3B is the quantitation of GFP expression in response to a 4-hr exposure to 40-psi oxygen for these three transgenic lines. This level of oxidative stress produces a 4- to 5-fold increase in GFP expression in these animals. (Exposure to 40-psi air does not result in increased transgene GFP expression, indicating that this is clearly an oxidant stress effect, not a simple effect of pressure.) Similar analyses show that, as expected on the basis of the derivation of these reporters, exposure to juglone and paraquat also produces strong GFP induction in these animals.

Applications

Although GFP-based reporter constructs have been used previously to visualize oxidative stress-dependent relocalization of NF-κB[15] and protein kinase B,[16] few published works have described the use of GFP reporter constructs to specifically monitor the transcriptional response to oxidative stress. Our work with a limited set of reporter genes demonstrates the feasibility of this approach, and suggests a number of useful applications. For genetic model systems such as *C. elegans,* oxidative stress-dependent induction of GFP expression represents a phenotype that can be the basis of forward genetic screens. These screens can potentially identify genes involved in the induction of oxidative stress response genes (e.g., transcription factors and oxidative damage sensor proteins), or genes that can be mutated to preemptively block oxidative stress in the first place. More generally, oxidative stress-dependent GFP reporter transgenes can be used as surrogates to measure oxidative stress. For example, our initial rationale for developing these reporters was to create a tool to determine whether transgenic expression of the human β-amyloid peptide in *C. elegans*[17] resulted in oxidative stress. Given the well-established relationship between aging and oxidative stress,[18] these reporters may also be productively applied in studies of aging. Because GFP expression can be readily adapted to high-throughput assays, oxidative stress-dependent reporters can potentially be useful for screening chemicals for either prooxidant or antioxidant effects.

Acknowledgments

We thank Amy Fluet for construction of the pAF15 reporter plasmid, Edouard DeCastro for outcrossing of the CL2166 strain, and members of the Link and Johnson laboratories for advice and assistance.

[14] S. L. Shaw, E. D. Salmon, and R. S. Quatrano, *Biotechniques* **19,** 946 (1995).
[15] K. Tenjinbaru, T. Furuno, N. Hirashima, and M. Nakanishi, *FEBS Lett.* **444,** 1 (1999).
[16] A. Gray, J. Van der Kaay, and C. P. Downes, *Biochem. J.* **344,** 929 (1999).
[17] C. D. Link, *Proc. Natl. Acad. Sci. U.S.A.* **92,** 9368 (1995).
[18] R. S. Sohal, R. J. Mockett, and W. C. Orr, *Probl. Cell. Differ.* **29,** 45 (2000).

[43] Detection of DNA Base Mismatches Using DNA Intercalators

By ELIZABETH M. BOON, JENNIFER L. KISKO, and JACQUELINE K. BARTON

Introduction

The most prevalent mechanisms leading to mutations in DNA are direct mis-incorporation of bases during replication and sustained chemical damage. Under normal circumstances, the cell corrects these problems using DNA polymerase proofreading mechanisms as well as the complex repair machinery of the cell. In certain tissues that contain mismatch repair deficiencies, DNA mispairs may accumulate.[1-5] Even in healthy cells, however, mismatches and lesions can sometimes go unchecked, resulting in permanent alterations in the gene sequence for subsequent generations. Identification of genetic variations (single-nucleotide polymorphisms, SNPs) among individuals and populations has implications in understanding human disease and treatment, as well as the interaction of the environment and multiple genes during evolution.[6] Once these SNPs are identified and understood, rapid and reliable detection of them will be critical for the study, diagnosis, and treatment of genetically linked disease.

SNPs are detected as mismatches in heteroduplexes formed from a known copy of a gene and the test gene. Several procedures have been described to search duplex DNA for mismatches using chemical, enzymatic, and differential hybridization techniques.[7-15] None of these, however, are as selective, inexpensive,

[1] R. Kolodner, *Genes Dev.* **10**, 1433 (1996).

[2] P. Modrich, *Annu. Rev. Genet.* **25**, 229 (1991).

[3] P. Modrich, *Science* **266**, 1959 (1994).

[4] R. D. Kolodner, *Trends Biochem. Sci.* **20**, 397 (1995).

[5] B. A. Jackson and J. K. Barton, "Current Protocols in Nucleic Acid Chemistry," p. 6.2.1. John Wiley & Sons, New York, 2000.

[6] A. J. Brookes, *Gene* **234**, 177 (1999).

[7] D. H. Geschwind, R. Rhee, and S. F. Nelson, *Genet. Anal.* **13**, 105 (1996).

[8] J.-F. Xu, Q.-P. Yang, J. Y.-J. Chen, M. R. van Baalen, and I.-C. Hsu, *Carcinogenesis* **17**, 321 (1996).

[9] K. Yamana, Y. Ohashi, K. Nunota, and H. Nakano, *Tetrahedron* **153**, 4265 (1997).

[10] N. C. Nelson, P. Hammond, E. Matsuda, A. A. Goud, and M. Becker, *Nucleic Acids Res.* **24**, 4998 (1996).

[11] A. Ganguly and D. J. Prockop, *Electrophoresis* **16**, 1830 (1995).

[12] R. G. H. Cotton, N. R. Rodrigues, and R. D. Campbell, *Proc. Natl. Acad. Sci. U.S.A.* **85**, 4397 (1998).

[13] R. G. H. Cotton, H.-H. M. Dahl, S. Forrest, D. W. Howells, S. J. Ramus, R. E. Bishop, I. Dianzani, J. A. Saleeba, E. Palombo, M. J. Anderson, C. M. Milner, and R. D. Campbell, *DNA Cell Biol.* **12**, 945 (1993).

[14] D. H. Johnston, K. C. Glasgow, and H. H. Thorp, *J. Am. Chem. Soc.* **117**, 8933 (1995).

[15] G. Rowley, S. Saad, F. Giannelli, and P. M. Green, *Genomics* **30**, 574 (1995).

Copyright 2002, Elsevier Science (USA).
All rights reserved.
0076-6879/02 $35.00

and reliable as would be necessary for widespread application. Consequently, new strategies for mismatch detection are needed.

In this review, we describe approaches for mismatch discovery and diagnosis using intercalative DNA-binding molecules. Our approach to mismatch discovery utilizes a metal complex that binds by intercalation selectively to mismatches in duplex DNA due to the local thermodynamic and kinetic destabilization associated with these sites. On irradiation, this metal complex cleaves the DNA at its site of intercalation, the mismatched site. In this approach, the gene sequence does not have to be known, and we can search for mismatches on a genome-wide basis. During mutation diagnosis, we look for mismatches in short heteroduplexes formed from a known sequence and test oligonucleotides. In this assay, we take advantage of a distinctly different characteristic associated with mismatches, the perturbation of the electronic structure of DNA at the mismatched site. Figure 1 schematically illustrates mismatch discovery and diagnosis using these strategies.

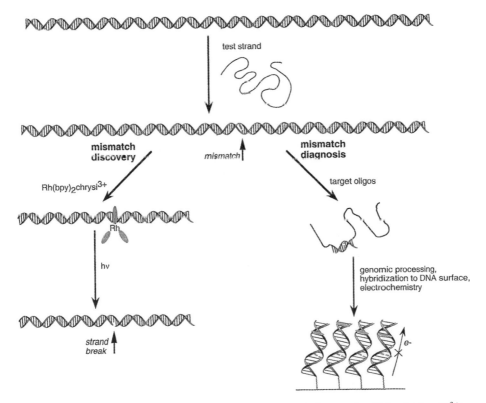

FIG. 1. Schematic representation of our approaches to mismatch discovery by [Rh(bpy)$_2$(chrysi)]$^{3+}$ photochemistry and mismatch diagnosis by electrochemistry. Both assays feature DNA intercalators as probe molecules.

Experimental Methods for Mismatch Discovery by a Designed Synthetic Intercalative Complex: Detection Based on Thermodynamic and Kinetic Destabilization

Metallointercalators of the type $[RhL_2(phi)]^{3+}$ have been shown to be effective in binding DNA via insertion of the 9,10-phenanthrene quinone diimine (phi) intercalator between base pairs in the DNA duplex.[16,17] For example, the complex Δ-α-$[Rh(R,R)$-$Me_2trien](phi)]^{3+}$ [$Me_2trien = (2R,9R)$-diamino-4,7-diazadecane] has been used to target the sequence 5'-TGCA-3', where functionalization of the ancillary ligands with methyl groups leads to specific van der Waals interactions.[18] High-resolution nuclear magnetic resonance (NMR) studies[19,20] and a crystal structure[21] of the metal complex bound to its target site have confirmed such sequence-specific interactions. The crystal structure, in which Δ-α-$[Rh(R,R)$-$Me_2trien](phi)]^{3+}$ is bound to a DNA octamer, shows the phi ligand to be deeply inserted and base stacked within the duplex from the major groove side.

There is widespread interest in site-specific recognition of DNA base pair mismatches. Strategies have exploited mismatch recognition proteins, differential chemical cleavage, hybridization of fluorescent conjugates, and DNA chip methodologies.[7-15] NMR studies have shown that, when a phi ligand intercalates between two base pairs, the phenanthrene ring stacks snugly into the base step and there is little unoccupied space left over (Fig. 2).[19,20] This suggested a mechanism for recognition in which a sterically demanding intercalating ligand, too large to fit readily into standard B-DNA, might recognize destabilized regions of the helical core. Our laboratory has reported the construction of a novel mismatch recognition agent, bis(2,2'-bipyridyl) (5,6-chrysenequinone diimine)rhodium(III), $[Rh(bpy)_2(chrysi)]^{3+}$, which has the differentiating characteristic of a four-, rather than three-, ringed intercalating ligand, chrysi.[22-25] The sterically bulky chrysi intercalating ligand is too wide to intercalate readily into B-form DNA, and therefore binding is restricted to destabilized regions at or near base pair mismatches. Thus, the complex is able to specifically target the duplex at or near mismatch sites and, on photoactivation, cleave the DNA backbone. Therefore, rather than

[16] K. E. Erkkila, D. T. Odom, and J. K. Barton, *Chem. Rev.* **99,** 2777 (1999).

[17] T. Johann and J. K. Barton, *Phil. Trans. R. Soc. Lond. A* **354,** 299 (1996).

[18] A. H. Krotz, B. P. Hudson, and J. K. Barton, *J. Am. Chem. Soc.* **115,** 12577 (1993).

[19] B. P. Hudson and J. K. Barton, *J. Am. Chem. Soc.* **120,** 6877 (1998).

[20] B. P. Hudson, C. M. Dupureur, and J. K. Barton, *J. Am. Chem. Soc.* **117,** 9379 (1995).

[21] C. L. Kielkopf, K. E. Erkkila, B. P. Hudson, J. K. Barton, and D. C. Rees, *Nat. Struct. Biol.* **7,** 117 (2000).

[22] H. Murner, B. A. Jackson, and J. K. Barton, *Inorg. Chem.* **37,** 3007 (1998).

[23] B. A. Jackson and J. K. Barton, *J. Am. Chem. Soc.* **119,** 12986 (1997).

[24] B. A. Jackson, V. Y. Alekseyev, and J. K. Barton, *Biochemistry* **38,** 4655 (1999).

[25] B. A. Jackson and J. K. Barton, *Biochemistry* **39,** 6176 (2000).

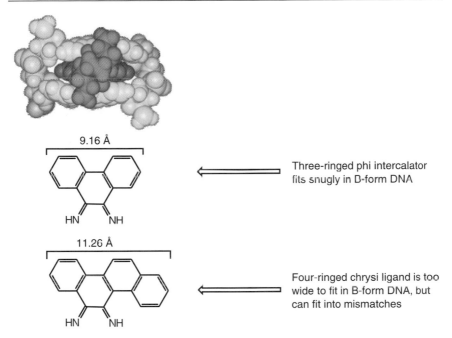

9.16 Å

Three-ringed phi intercalator
fits snugly in B-form DNA

11.26 Å

Four-ringed chrysi ligand is too
wide to fit in B-form DNA, but
can fit into mismatches

FIG. 2. Comparison of steric width of tricyclic phi and tetracyclic chrysene ligands reveals that the chrysene ligand would demand an additional 2 Å of space on intercalation into DNA. As shown in the space-filling model above the phi ligand, just enough room is afforded in normal DNA to allow phi intercalation. Thus, intercalation of the chrysene ligand preferentially targets DNA sites that are in some way distorted, or more easily distorted than normally base-paired sequences; simply put, the chrysene ligand is too sterically demanding for intercalation into normal DNA.

selecting for exposed functionalities of bases as with previously described metallo-intercalators, specificity for mismatch sites is enabled by selecting for the thermodynamic constraints associated with the mispair. $[Rh(bpy)_2chrysi]^{3+}$ was found to induce strand scission at approximately 80% of all mismatch sites in all possible surrounding sequence contexts and recognition correlated generally with the local helical destabilization associated with the mismatch.[25]

As Δ-$[Rh(bpy)_2(chrysi)]^{3+}$ has shown that a comparably sterically bulky intercalating ligand can enable binding specificity for mismatches on small, synthetic oligonucleotides, the next challenge was to test its ability to specifically recognize mispairs on long DNA polymers of biological origin.[24] Thus, an experiment using a mixture of two plasmids was performed; the plasmids differed in sequence at a single base position. Each plasmid was cut with *Sca*I to produce linear 2725-mers with the C/G variable site 975 base pairs from one end (Fig. 3). Equal amounts of the C- and G-containing plasmids were combined, resulting in a

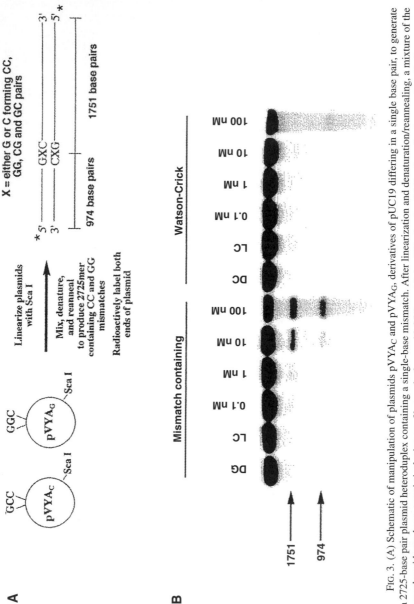

Fig. 3. (A) Schematic of manipulation of plasmids pVYA$_C$ and pVYA$_G$, derivatives of pUC19 differing in a single base pair, to generate a 2725-base pair plasmid heteroduplex containing a single-base mismatch. After linearization and denaturation/reannealing, a mixture of the two plasmids produces a statistical mixture of heteroduplexes with 25% containing CC mispairs. (B) Portion of an image of an alkaline agarose gel of the photolysis (440 nm, 10 min) products of the plasmid mixture described in (A) with increasing amounts of Δ-[Rh(bpy)$_2$(chrysi)]$^{3+}$. Conversion of the full-length plasmid to fragments of lengths consistent with specific cleavage at the mismatch site is observed as the concentration of Δ-[Rh(bpy)$_2$(chrysi)]$^{3+}$ is increased from 0.1 to 100 nM. Watson–Crick samples incubated with the metal complex in the absence of light (DC) or samples irradiated without metal complex (LC) show no cleavage.

mixture that recombined statistically to produce 50% correctly matched plasmids and 50% plasmids containing one (either CC or GG) mismatch. Each end of the 2725-mer was radioactively labeled and exposed to visible light in the presence of Δ-[Rh(bpy)$_2$(chrysi)]$^{3+}$. Remarkably, the specificity of [Rh(bpy)$_2$chrysi]$^{3+}$ for mismatches was shown to be sufficient to allow the discrimination of a single mispair in a 2725-base pair plasmid; irradiation of the plasmid samples without mismatches showed no specific cleavage. High-resolution polyacrylamide gel electrophoresis further revealed that the cleavage site directly neighbored that of the mismatch.

Cells that are mismatch repair (MMR) deficient are expected to result in a buildup of mismatches during the cell cycle.[1–5] As the complex [Rh(bpy)$_2$chrysi]$^{3+}$ has been shown to specifically intercalate at mismatched base pairs in DNA and photolytically promote strand scission on irradiation with visible light, we are now using this complex to test this hypothesis with DNA isolated from human tumor cells.[26] MMR-deficient cells treated with this complex result in a larger number of smaller sized fragments than DNA isolated from MMR-proficient cell lines. Furthermore, as this complex accumulates at mismatched sites, it promotes cytotoxicity in MMR-deficient cell lines at a level above that of cytotoxicity in MMR-proficient cell lines. Thus we are considering also the application of [Rh(bpy)$_2$chrysi]$^{3+}$ to probe for mismatched DNA *in vivo* and, perhaps, to develop such bulky intercalators for therapeutic strategies directed toward cancers associated with mismatch repair deficiency.

Experimental Methods

Synthesis of [Rh(bpy)$_2$chrysi]Cl, [bis(2,2'-bipyridine)5,6-chrysenequinone diimine rhodium(III) chloride].[22] Figure 4 outlines the complete synthesis of [Rh(bpy)$_2$chrysi]$^{3+}$. The [Rh(bpy)$_2$Cl$_2$]Cl intermediate is synthesized from RhCl$_3$ and 2,2'-bipyridine according to a procedure given in the literature.[27] To generate the diammine, [Rh(bpy)$_2$(Cl$_2$)]Cl is refluxed in ammonium hydroxide in a round-bottomed flask. Refluxing is continued until all yellow solid has dissolved (approximately 15–30 min). The solvent is removed from the reaction mixture on a rotary evaporator. The diammine product, [Rh(bpy)$_2$(NH$_3$)$_2$]Cl$_3$, which appears as a scaly, pale yellow powder in the flask, is typically recovered in 90–100% yield and should be characterized by ^{1}H NMR, UV–visible (UV–Vis), high-performance liquid chromatography (HPLC), and mass spectrometry.

The final diimine product is obtained via condensation with chrysenequinone. As chrysenequinone is not commercially available, its synthesis from chrysene follows.[28] Commercially obtained chrysene and sodium dichromate are added

[26] J. L. Kisko, D. T. Odom, O. Glebov, G. Kim, I. Kirsch, and J. K. Barton, unpublished results (2001).

[27] P. M. Gidney, R. D. Gillard, and B. T. Heaton, *J. Chem. Soc. Dalton Trans.* 2621 (1972).

[28] V. C. Greabe and F. Hönigsberger, *Ann. Chem.* **311,** 257 (1900).

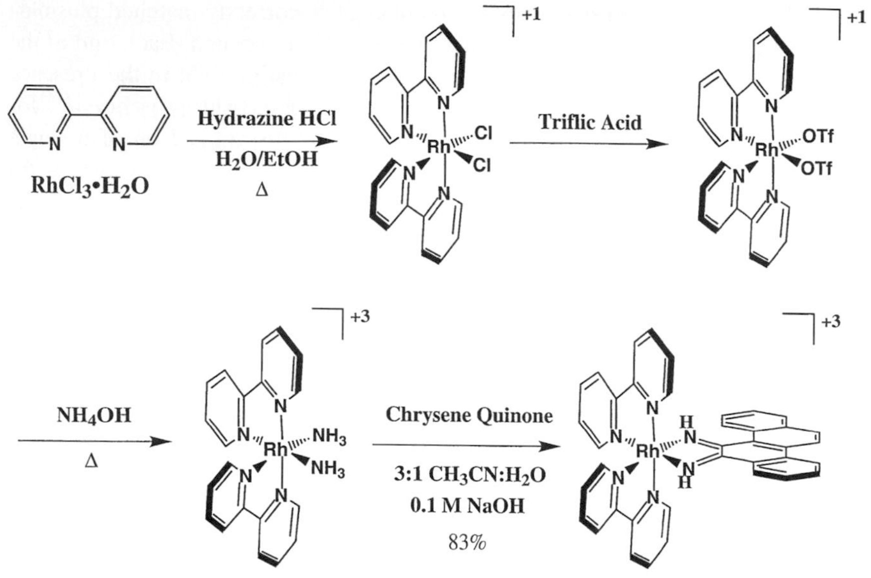

FIG. 4. Synthesis of novel mismatch recognition agent, bis(2,2′-bipyridyl)(5,6-chrysenequinone diimine)rhodium(III), [Rh(bpy)$_2$(chrysi)]$^{3+}$.

to a round-bottomed flask. Glacial acetic acid is then added and the solution is brought to reflux while stirring. As heating continues, the solution changes from the orange/yellow of the dichromate/chrysene to deep green with orange/red crystals of the chrysenequinone product. The mixture is refluxed for 9 hr. After the reaction mixture is removed from the heat, 75–100 ml of boiling water is immediately added. The product quickly precipitates while the green chromium by-products remain soluble. The solution is filtered to collect the product that is then thoroughly washed with water. Recoveries at this step are typically 80–90%. To purify the product away from any remaining chromium by-products or any present unreacted chrysene, it is recrystallized from ethanol. The product should be characterized by ^{1}H NMR, HPLC, and mass spectrometry.

Finally, chrysenequinone is dissolved in the acetonitrile portion of what will be a 3 : 1 (v/v) acetonitrile–0.1 M NaOH solvent system. The [Rh(bpy)$_2$(NH$_3$)$_2$]Cl$_3$ is dissolved in the 0.4 M NaOH aqueous portion and added to the chrysenequinone solution. The formation of the diimine product gradually changes the red/orange of the quinone solution to a dark brown. After the reaction is complete (approximately 8 hr), purification is achieved via cation-exchange chromatography, eluting with 0.15 M MgCl$_2$.[22,29] The product should be characterized by ^{1}H NMR, UV–Vis, HPLC, and mass spectrometry.

[29] I. P. Evans, G. W. Everett, and A. M. Sargeson, *J. Am. Chem. Soc.* **98**, 8041 (1976).

Probing Nucleic Acid Structure with Rhodium Reagents. The rich photochemistry of rhodium allows the straightforward identification of sites of binding by light-induced DNA cleavage. In the procedures described, end-labeled DNA is incubated with a transition metal reagent and then irradiated with visible light to promote cleavage of the nucleic acid. Local destabilizations in helix structure can thus be identified and analyzed.

The DNA samples used in these studies have been made synthetically using standard techniques or excised from plasmids using restriction endonucleases. All samples were labeled with [^{32}P]phosphate on either the 5' or 3' end by enzymatic methods.

Although the exact photolysis procedures that will be optimal under any given set of experimental conditions depend on the individual light source used, the following description includes many of the possible options commonly used in standard laboratories. The typical light source in our laboratory is a Thermo Oriel (Stratford, CT) mercury/xenon arc lamp equipped with an infrared (IR) filter, monochromator, and UV ($\lambda < 300$ nm) cutoff filter. Other sources used for these experiments have included a helium/cadmium laser (model 4200 NB, 442 nm, 22 mW; Linconix, Sunnyvale, CA) and a transilluminating light box (Spectroline model TR302 with broadband irradiation centered at 302 nm; Spectronics, Westbury, NY).

Equal volumes of the metal and DNA stock solutions are combined in a microcentrifuge tube, agitated, and centrifuged. Solutions are allowed to equilibrate for 5 to 15 min in the absence of light before irradiation. Total volumes are usually between 5 and 50 μl containing between 10,000 and 200,000 cpm of the end-labeled DNA. Total DNA concentrations are typically between 5 and 100 μM base pairs, depending on the binding constant of and its concentration relative to DNA for the chosen metal complex. Dark and light controls of each DNA of interest are run in parallel; the light control has no metal complex added and the dark control is not exposed to light.

Samples are irradiated by positioning the open tube in the output of the light source as close as possible to the focal point for between 5 and 30 min. The exact irradiation time depends on the strength of a given light source and the wavelength of irradiation. When a given source or wavelength is first used for an experiment, an irradiation time series should be performed to select the optimal irradiation time. In our experiments, the optimum wavelength for [Rh(bpy)$_2$chrysi]$^{3+}$ is 440 nm.

After irradiation, all samples are stored in the dark until the completion of the experiment. DNA is isolated from solution either by drying the samples or by ethanol precipitation. Samples are typically dried on a Speed-Vac concentrator (Savant, Hicksville, NY) or precipitated by addition of 1–5 μl of 9 mM calf thymus DNA, a 1/6 to 1/4 volume of 7.5 M ammonium acetate, and 4 volumes of absolute ethanol. The precipitates are mixed well and incubated on dry ice for 30 min before centrifuging at 14,000 rpm for 12 min on an Eppendorf centrifuge. The resulting pellets are rinsed with cold ethanol, dried *in vacuo,* counted on a scintillation counter, and resuspended in gel dye to a constant concentration of radioactivity before electrophoresis.

Standard DNA-sequencing electrophoretic techniques are used to identify cleavage sites on the DNA. Photocleavage samples are extracted into denaturing loading dye and electrophoresed in Tris–borate–EDTA (TBE) buffer on 8 to 20% (w/v) denaturing polyacrylamide gels. The samples are heated to 90° for 5 min before loading to promote denaturation of the DNA strands and dissociation of the metal complex probes. Sites of $[Rh(bpy)_2(chrysi)]^{3+}$ binding are indicated by cleavage bands in the experimental lanes. To identify the specific sites of nucleic acid cleavage, samples are run adjacent to samples of the same DNA strand that have been chemically sequenced using standard Maxam–Gilbert sequencing reactions. Gels are read using standard autoradiography and phosphorimagery techniques.

Single-Mismatch Detection in a Kilobase Linearized Plasmid. To test recognition of a single-base pair mismatch in nonsynthetic kilobase DNA samples, experiments with plasmids can be performed. In our experiments, each plasmid contained an oligonucleotide segment (57 bp) inserted between the *Sal*I and *Hin*dIII sites in the polycloning site of pUC19. The inserts differed in only one base pair to produce plasmids with either a cytosine (pVYA$_C$) or a guanine (pVYA$_G$) at position 467 (as numbered from the first base pair of pUC19).

In our experiments, samples of pVYA$_C$ and pVYA$_G$ were linearized with *Sca*I and dephosphorylated with shrimp alkaline phosphatase. After ethanol precipitation, samples of each plasmid were combined and denatured by addition of 10 M NaOH. After incubation for 5 min, the samples were neutralized with acetic acid, and sodium chloride and buffer (Tris, pH 8.5) were then added. The samples were then annealed via 30-min incubations at 65 and 37°. This mixture of heteroduplexes was desalted and extracted into 10 mM Tris buffer (pH 8.5) on a Microcon-10 spin filter (Millipore, Bedford, MA). Samples were then end labeled (on both ends of the plasmid) with $[^{32}P]ATP$ and polynucleotide kinase (PNK). Stocks of labeled plasmid for photocleavage were prepared from 2 μg of the labeled plasmid in 100 μl of 2× running buffer (100 mM Tris, 40 mM sodium acetate, 36 mM NaCl). Portions of the plasmid stocks (10 μl) were combined with equal volumes of solutions of Δ-$[Rh(bpy)_2(chrysi)]^{3+}$ (200, 20, 2, or 0.2 nM) at least 20 min before irradiation for 10 min at 440 nm. Samples were extracted into 1 volume of alkaline agarose loading dye [50 mM NaOH, 1 mM EDTA, 2.5% (v/v) Ficoll, 0.25% (w/v) bromocresol green] and run on an alkaline agarose gel [1% (w/v) agarose, 50 mM NaOH, 1 mM EDTA). The gel was fixed in 7% (v/v) trichloroacetic acid, dried *in vacuo,* and visualized by phosphorimagery.

To determine the cleavage sites of $[Rh(bpy)_2(chrysi)]^{3+}$ in pVYA$_{C/G}$ heteroduplexes at high resolution, a sample of pVYA$_C$ was digested with *Eco*RI, dephosphorylated with shrimp alkaline phosphatase, and 5′-end labeled using PNK and [^{32}P]ATP. The labeled plasmid was subsequently exposed to *Pvu*II, yielding three restriction fragments. The fragment of interest (270 base pairs) was run on a 6% (w/v) nondenaturing polyacrylamide gel and isolated by extraction into ammonium acetate buffer followed by ethanol precipitation. Samples of the labeled restriction fragments were combined with a large excess of either pVYA$_G$ or pVYA$_C$ that had been treated with both *Eco*RI and *Pvu*II. This mixture of labeled 270-mer and the three unlabeled fragments resulting from the double digests was denatured and reannealed as described above. Samples of these DNA solutions were diluted in 2× running buffer and mixed with an equal volume of Δ-[Rh(bpy)$_2$(chrysi)]$^{3+}$ solution 10 min before irradiation at 440 nm for 10 min. Samples were dried, extracted into formamide loading dye, and loaded on a 6% (w/v) denaturing polyacrylamide gel. After drying, the gel may be visualized by phosphorimagery.

Experimental Methods for Mismatch Diagnosis by DNA-Mediated Electron Transfer to DNA Intercalators: Detection Based on Electronic Destabilization

The study of DNA-mediated electron transfer has been of great interest. Although the finer kinetic and mechanistic issues surrounding DNA-mediated electron transfer are still being debated, it has become apparent that these reactions are extremely sensitive to DNA π-stacking.[30–38] The importance of the stacking of the electron donor and acceptor within the DNA base stack was highlighted in studies involving photoinduced electron transfer between modified bases in DNA.[31] DNA-mediated electron transfer is also highly dependent on the stacking of the base pairs intervening the donor and acceptor.[30–38] Base bulges,[36,37] flexible sequences,[39,40] and protein-induced distortions[41] all greatly influence the efficacy of long-range DNA-mediated electron transfer.

This sensitivity to base stacking provides the basis for mismatch discrimination by DNA-mediated charge transport. Mismatches are generally stacked in a DNA

[30] S. O. Kelley, R. E. Holmlin, E. D. A. Stemp, and J. K. Barton, *J. Am. Chem. Soc.* **119**, 9861 (1997).

[31] S. O. Kelley and J. K. Barton, *Science* **283**, 375 (1999).

[32] R. E. Holmlin, P. J. Dandliker, and J. K. Barton, *Angew. Chem. Int. Ed. Eng.* **36**, 2714 (1997).

[33] S. M. Gasper and G. B. Schuster, *J. Am. Chem. Soc.* **119**, 12762 (1997).

[34] C. J. Murphy, M. R. Arkin, Y. Jenkins, N. D. Ghatlia, S. Bossmann, N. J. Turro, and J. K. Barton, *Science* **262**, 1025 (1993).

[35] S. O. Kelley and J. K. Barton, *Chem. Biol.* **5**, 413 (1998).

[36] D. B. Hall and J. K. Barton, *J. Am. Chem. Soc.* **119**, 5045 (1997).

[37] D. B. Hall, R. E. Holmlin, and J. K. Barton, *Nature (London)* **382**, 731 (1996).

[38] D. T. Odom, E. A. Dill, and J. K. Barton, *Chem. Biol.* **7**, 475 (2000).

[39] T. T. Williams, D. T. Odom, and J. K. Barton, *J. Am. Chem. Soc.* **122**, 9048 (2000).

[40] M. E. Nunez, D. B. Hall, and J. K. Barton, *Chem. Biol.* **6**, 85 (1999).

[41] S. R. Rajski, S. R. Kumar, R. J. Roberts, and J. K. Barton, *J. Am. Chem. Soc.* **121**, 5615 (1999).

duplex but undergo somewhat greater dynamic motion than well-paired bases.[42–46] DNA charge transport detection, which depends on the electronic coupling within the base stack, is sensitive to these motions. For example, a single CA mismatch inserted into a DNA duplex between a covalently attached photoinduced electron donor (ethidium) and a covalently attached intercalating acceptor [Rh(phi)$_2$bpy^{3+}] significantly inhibits electron transfer as determined on the basis of the results of fluorescence quenching experiments.[30]

We have developed an electrochemical assay based on charge transfer through double-stranded DNA-modified gold electrodes to exploit the sensitivity of DNA-mediated charge transport in routine mismatch detection.[47–51] Modified electrodes are prepared by self-assembly of prehybridized duplexes (Fig. 5). These surfaces have been characterized by cyclic voltammetry, ellipsometry, radiolabeling of the duplexes, and atomic force microscopy (AFM).[47,48,52] We have found that densely packed monolayers with a 45° orientation of the helical axis with respect to the gold surface are formed when no potential is applied to the electrode surface. When the gold surface is charged, the morphology of the films responds electrostatically because of the highly negative charge of the DNA backbone. This voltage-induced morphology change is reversible and effectively constitutes a molecular switch.[48] The morphologies of our DNA surfaces are also greatly affected by the flexibility and orientation of the organic linker between the duplex and the gold surface.[52]

In our first electrochemical studies of mismatches in DNA films, the intercalator-binding site was fixed by site-selectively cross-linking daunomycin (DM) to guanine residues in the duplex.[49] In DNA surfaces that were completely Watson–Crick base paired, a well-resolved DM peak was observed in the cyclic voltammagram (CV). However, the presence of a single mispaired base (within the fully hybridized duplex) between the electrode and the site of intercalation switched off the electrochemical response entirely (Fig. 6). We have now established that in well-packed

[42] W. N. Hunter, G. A. Leonard, and T. Brown, *ACS Symp. Ser.* **682,** 77 (1998).

[43] B. A. Luxon and D. G. Gorenstein, *Methods Enzymol.* **261,** 45 (1995).

[44] J. E. Forman, I. D. Walton, D. Stern, R. P. Rava, and M. O. Trulson, *ACS Symp. Ser.* **682,** 206 (1998).

[45] N. Peyret, P. A. Seneviratne, H. T. Allawi, and J. SantaLucia, *Biochemistry* **38,** 3468 (1999).

[46] J. SantaLucia, *Proc. Natl. Acad. Sci. U.S.A.* **95,** 1460 (1998).

[47] S. O. Kelley, N. M. Jackson, J. K. Barton, and M. G. Hill, *Bioconjug. Chem.* **8,** 31 (1997).

[48] S. O. Kelley, J. K. Barton, N. M. Jackson, L. D. McPherson, A. B. Potter, E. M. Spain, M. J. Allen, and M. G. Hill, *Langmuir* **14,** 6781 (1998).

[49] S. O. Kelley, N. M. Jackson, M. G. Hill, and J. K. Barton, *Angew. Chem. Int. Ed. Eng.* **38,** 941 (1998).

[50] S. O. Kelley, E. M. Boon, N. M. Jackson, M. G. Hill, and J. K. Barton, *Nucleic Acids Res.* **27,** 4830 (1999).

[51] E. M. Boon, D. M. Ceres, T. G. Drummond, M. G. Hill, and J. K. Barton, *Nat. Biotechnol.* **18,** 1096 (2000).

[52] S. Mui, E. M. Boon, J. K. Barton, M. G. Hill, and E. M. Spain, unpublished results (2001).

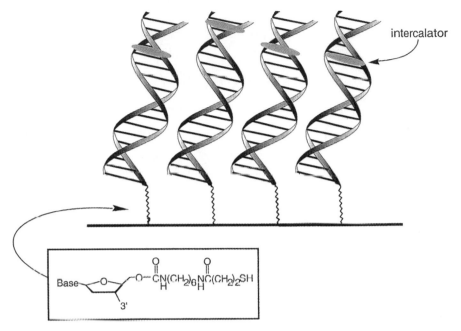

FIG. 5. Schematic representation of a DNA-modified gold electrode with a bound redox-active DNA intercalator for use in electrochemical assays.

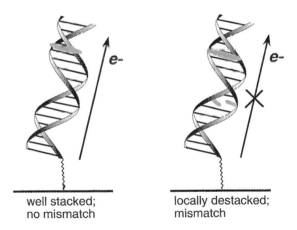

FIG. 6. Schematic representation of the efficacy of electron transfer in well-matched and mismatched duplexes. In mismatched DNA, base stacking is locally perturbed at the site of the mismatch and electron transfer is shut off.

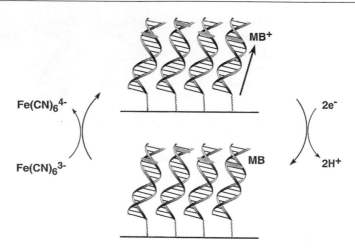

FIG. 7. Schematic representation of electrocatalytic reduction of $Fe(CN)_6^{3-}$ by MB^+ at a DNA-modified electrode. Electrons flow from the electrode surface to intercalated MB^+. Once reduced, MB can easily reduce $Fe(CN)_6^{3-}$ and regenerate MB^+ that can continue on in the catalytic cycle, and thus repeated interrogation of the DNA monolayer is achieved. MB^+ binding is primarily constrained to the top of the densely packed DNA monolayer, requiring charge transport through the DNA film, and electrostatic repulsion keeps $Fe(CN)_6^{3-}$ away from the interior of the anionic DNA film.

films of DNA duplexes, the DNA-mediated reaction can proceed using noncovalently bound intercalators such as DM or methylene blue (MB^+). Furthermore, these DNA films can be cycled by denaturation of the film leaving surface-bound single-stranded probes, followed by *in situ* hybridization with a single-stranded test sequence. Importantly, in order to detect a mutation, the electrochemical probe must be an intercalator, which binds to DNA by insertion between base pairs, effectively becoming a part of the base stack and thus electronically coupled into the base stack of DNA. This sensitivity to mismatches depends on electronic coupling within the base pair stack rather than on the thermodynamics of base pairing.[50]

By coupling the DNA-mediated electrochemical reaction to catalytic reoxidation of the intercalator in solution, we can now achieve greatly enhanced selectivity and sensitivity (Fig. 7).[50,51] In the electrocatalytic process, electrons flow from the electrode surface to intercalated methylene blue (MB^+) in a DNA-mediated reaction. The reduced form of MB^+, leukomethylene blue (MB), in turn reduces solution-borne ferricyanide, so that more electrons can flow to MB^+ and the catalytic cycle continues. In each experiment the surface-bound DNA is repeatedly interrogated. In duplexes containing mismatches, fewer MB^+ molecules are electrochemically reduced, so the concentration of active catalyst is lowered and the overall catalytic response is diminished. Electrocatalysis essentially amplifies the absolute MB^+ signal as well as enhancing the inhibitory affect of a mismatch. Furthermore, because of its catalytic nature, the measured charge increases with increased sampling times; longer integration times provide even greater absolute signals as well as increased differentiation between fully complementary and mismatched DNA (Fig. 8). Using this electrocatalytic assay, all single-base mismatches have been detected as well as several common DNA base damage products.[51] Furthermore, this technology is well suited to DNA chip-based formats and mismatch detection has been demonstrated on electrodes as small as 30 μm in diameter.[51]

Experimental Methods

Strategy for Construction of Thiol-Modified DNA. All reagents and solvents are purchased in their highest available purity and used without further purification. Millipore Milli-Q (18 MΩ cm) water is used in all experiments. All glassware and plasticware is DNase, RNase, and metal free.

Oligonucleotides are derivatized with alkane thiol linkers for self-assembly on gold surfaces (Fig. 5). The length of the oligonucleotide can be varied. The oligonucleotides are synthesized (trityl-off) at a 1-μmol scale on a DNA synthesizer using standard solid-phase phosphoramidite chemistry (1000-Å CPG). After the synthesis the DNA is still on the resin and fully protected with the exception of the 5'-OH terminus. This solid-phase DNA is transferred to a peptide reaction vessel (coarse frit). The 5'-OH is aminoacylated by reaction with carbonyldiimidazole (CDI) in dioxane (25 mg of CDI in 1 ml of dioxane for 45 min). After this activation reaction, a six-carbon amine-terminated linker is added by reaction with 1,6-hexanediamine in 9:1 dioxane–water (32 mg of linker in 1 ml of dioxane mixture for 25 min). This product is transferred to a 1.5-ml Eppendorf tube. The DNA-5'-NH$_2$ is cleaved from the CPG resin and all the bases are deprotected by incubation in 1 ml of concentrated NH$_4$OH at 55° for 8 hr. The DNA-5'-NH$_2$ product is then cooled, decanted, and evaporated to dryness *in vacuo.* This product is purified by reversed-phase HPLC on a C$_{18}$ 300-Å column with a gradient of 0–13% CH$_3$CN in 35 min, 13–50% CH$_3$CN in 50 min with ammonium acetate, pH 7, as the aqueous phase (monitored at 260 and 290 nm). This purified DNA-5'-NH$_2$ product is dried *in vacuo.*

The next step of the reaction is another aminoacylation to form the DNA-5'-SS product, which is then deprotected to form the final product, DNA-5'-SH. DNA-5'-NH$_2$ product is dissolved in 200 μl of 0.2 M, HEPES buffer (pH 8) and added to 15 mg of 3-(2-pyridyldithio)propionic acid N-hydroxysuccinimide ester in 100 μl of CH$_3$CN. This reaction proceeds at room temperature. After 1 hr the reaction is quenched by the addition of 700 μl of 5 mM phosphate, 50 mM NaCl (pH 7) buffer (hereafter referred to simply as PBS). On this addition the solution turns cloudy due to insoluble side products. This mixture is centrifuged for 5 min and the DNA solution is decanted from the solid side products and gel filtered on an NAP-10 column (Sephadex G-25, DNA grade) with PBS. The DNA-5'-SS eluant

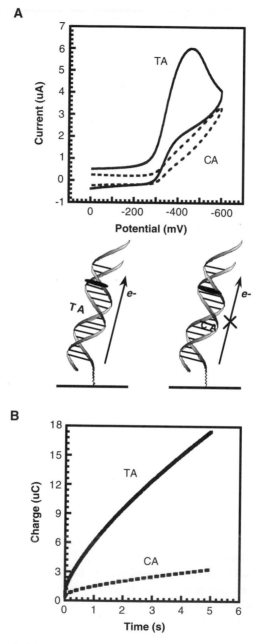

FIG. 8. Electrochemical detection of a single CA mismatch using both (A) cyclic voltammetry and (B) chronocoulometry.

is collected and the volume is reduced and purified by HPLC (analogously to DNA-5'-NH$_2$) and dried *in vacuo*. Next, the DNA-5'-SS product is resuspended in 200 μl of PBS. Dithiothreitol (DTT, 3 mg) is then added and the reduction reaction proceeds at ambient temperature for 40 min to form the final product, DNA-5'-SH. This is gel filtered on an NAP-5 column (Sephadex G-25, DNA grade) in PBS and then HPLC purified and dried *in vacuo*.

The thiol-modified single strand (DNA-5'-SH) can be tested for the presence of free thiol via HPLC by Ellman's test as follows. A small aliquot ($\sim$100 μl) of the DNA-5'-SH HPLC fraction is reinjected onto the HPLC for an analytical run followed by a second 100-μl aliquot to which 1 μl of 10 mM dithionitrobenzoic acid (DTNB; Ellman's reagent) had been added. Free thiol is monitored by a shift in the DNA-5'-SH chromatogram peak as well as new peaks in the UV–Vis spectrum at 330 and 410 nm.

The single-stranded oligonucleotide complementary to the thiol-modified strand is synthesized (trityl-off) at a 1-μmol scale on a DNA synthesizer using standard solid-phase phosphoramidite chemistry (1000-Å CPG). The CPG-bound DNA is then transferred to a 1.5-ml Eppendorf tube and mixed with 1 ml of concentrated NH$_4$OH and incubated at 55° for 8 hr to cleave the oligonucleotide from the CPG resin and deprotect all the bases. This product is then cooled, decanted, and evaporated to dryness *in vacuo*. This product is purified by reversed-phase HPLC.

The two purified, complementary single-stranded oligonucleotides are then quantitated and hybridized to make thiol-modified duplexes. Oligonucleotide stock solutions are prepared and quantitated by UV–Vis spectroscopy [$\lambda_{max} = 260$ nm, $\varepsilon\,(M^{-1}\,\mathrm{cm}^{-1})$]. The extinction coefficient of single strands is calculated by the sum of the extinction coefficients of the individual bases: $\varepsilon(dA) = 15,400$, $\varepsilon(dG) = 11,500$, $\varepsilon(dC) = 7400$, $\varepsilon(dT) = 8700$. Duplexes are formed by combining equimolar amounts of each strand in PBS for a final solution of 100 μM duplex. This solution is degassed and blanketed with argon, heated to 90° for 5 min, and then cooled slowly to room temperature (2 hr). Just before deposition on the clean gold electrode, 100 mM MgCl$_2$ is added to each sample.

Preparation of Electrode Surface. The gold electrodes are prepared by standard procedures. They are polished with 0.05-μm alumina, sonicated in distilled H$_2$O for $\sim$20 min, electrochemically etched in 1 M H$_2$SO$_4$, and rinsed well with distilled H$_2$O. The electrodes are then inverted and a 10-μl drop of the thiol-modified DNA duplex solution is deposited onto each electrode surface. The electrodes are kept in a moist environment at room temperature during the assembly process. To ensure maximum density of the monolayer, self-assembly is usually allowed to proceed overnight, but assembly does proceed much faster if the thiol and gold surface are perfectly clean. After assembly, the complementary DNA strand can be removed and replaced with test strands by *in situ* hybridization as follows. The DNA electrode is immersed in 90° PBS for 5 min and then rinsed thoroughly in PBS. Next, the electrode is immersed in a solution of 100 pM test strand oligonucleotide in

PBS with 100 mM MgCl$_2$ at 90° and allowed to cool to room temperature. Alternatively, if test strand is limited in quantity, a drop of the test strand in PBS with 100 mM MgCl$_2$ can be placed on the hot electrode surface and allowed to cool. During the cooling process, it is important to prevent evaporation of the solution, so drops of PBS should be placed on the surface as needed.

Electrochemical Assays for Single-Base Mismatch Detection at DNA-Modified Surfaces. Cyclic voltammetry (CV) is carried out in a two-compartment cell filled with PBS that is degassed and blanketed with argon. The DNA-modified gold working electrode and the platinum wire auxiliary electrode are separated from the saturated calomel reference electrode (SCE) by a modified Luggin capillary. Before electrochemical analysis, excess DNA is rinsed away with PBS and monolayer coverage is qualitatively checked with 2 mM Fe(CN)$_6^{3-}$ in PBS by scanning the potential from 0 to 600 mV and back at 100 mV/sec. If a CV signal is not observed, it is interpreted that the monolayer is so dense that Fe(CN)$_6^{3-}$ cannot diffuse to the gold surface and participate in redox chemistry. Thus in the following electrochemical studies of DNA intercalators, it is assumed that any CV signal observed is the result of charge transport through the DNA monolayer. Monolayers can be extensively characterized by AFM, ellipsometry, and radiolabeling of the duplexes, which is described elsewhere.[47,48]

If the electrode surface is sufficiently covered, a catalyst such as MB is added to the solution for a final concentration of 0.5 μM. Another catalyst can be used, but it must satisfy these requirements: it must bind to DNA by intercalation, it must not bind either too tightly or too loosely (the catalyst must be able to access the DNA base stack as well as the ferricyanide in solution), and its reduction potential must be about -200 to -600 versus SCE. This protocol assumes MB is used as the catalyst; if another intercalator is chosen, the electrochemistry conditions will need to be optimized. After addition of the catalyst and degassing with argon, the potential is scanned from 0 to -600 mV (vs. SCE) at 100 mV/sec. If the duplexes on the surface are well packed and contain no mismatches or other π-stack-disrupting lesions, a large catalytic wave is observed at about -350 mV. If there are mismatches, the background will start to rise at about -600 mV, but there will be no distinct peak (Fig. 8).

A more sensitive way to check for mismatches is to carry out chronocoulometry. The electrode potential is stepped to -350 mV versus SCE and allowed to integrate for 5 sec. If there are no base stack perturbations, the charge should steadily increase over the course of the experiment; the maximum amount of charge accumulated depends on the sequence, but typically the charge will reach 16 to 25 μC. If there is a mismatch, only 1 to 6 μC will accumulate, depending on the extent of perturbation and the sequence (Fig. 8).

[44] Deoxyguanosine Adducts of tert-4-Hydroxy-2-nonenal as Markers of Endogenous DNA Lesions

By Fung-Lung Chung and Lei Zhang

Introduction

Reactive oxygen species generated by normal cellular processes can directly or indirectly, through the production of reactive compounds, cause damage to protein and DNA.[1] A potentially important family of such compounds is α,β-unsaturated aldehydes or enals.[2] Enals are products of peroxidation of polyunsaturated fatty acids (PUFAs), and they are reactive bifunctional compounds that modify proteins and DNA via the Michael addition reaction.[3–7] The most commonly investigated enals occurring as products of lipid peroxidation are malondialdehyde (MDA), acrolein, and *t*-4-hydroxy-2-nonenal (HNE). Unlike other enals, HNE is a unique, long-chain enal product formed only by oxidation of ω-6 PUFAs.[2] Because of its reactivity, HNE has been investigated for its biological and physiological activities and been shown to display a wide range of activities.[8] Many laboratories investigated the chemical reactions of HNE with proteins as a basis for understanding its biological function. These studies have shown that HNE is capable of cross-linking proteins via thiol conjugation and Schiff's base formation.[9–13] The reaction products of DNA bases with enals, such as MDA and acrolein, were also studied more than a decade ago. However, not until more recent years have the reaction products of HNE with DNA been described. On reaction with deoxyguanosine (dG),

[1] B. Halliwell and J. M. Gutteridge, *Methods Enzymol.* **186,** 1 (1990).

[2] H. Esterbauer, H. Zollner, and R. J. Schaur, *in* "Membrane Lipid Oxidation" (C. Vigo-Pelfrey, ed.), p. 239. CRC Press, Boca Raton, FL, 1990.

[3] H. Esterbauer, R. J. Schaur, and H. Zollner, *Free Radic. Biol. Med.* **11,** 81 (1991).

[4] G. Witz, *Free Radic. Biol. Med.* **7,** 333 (1989).

[5] F.-L. Chung, R. Young, and S. S. Hecht, *Cancer Res.* **44,** 990 (1984).

[6] C. K. Winter, H. J. Segall, and W. F. Haddon, *Cancer Res.* **46,** 5682 (1986).

[7] P. Yi, D. Zhan, V. M. Samokyzyh, D. R. Doerge, and P. P. Fu, *Chem. Res. Toxicol.* **10,** 1259 (1997).

[8] R. J. Schaur, H. Zollner, and H. Esterbauer, *in* "Membrane Lipid Oxidation" (C. Vigo-Pelfrey, ed.), p. 141. CRC Press, Boca Raton, FL, 1991.

[9] K. Uchida and E. R. Stadtman, *J. Biol. Chem.* **268,** 6388 (1993).

[10] B. Friguet, E. R. Stadtman, and L. I. Seweda, *J. Biol. Chem.* **269,** 21639 (1994).

[11] L. M. Sayre, W. Sha, G. Xu, K. Kaur, D. Nadkarni, G. Subbanagounder, and R. G. Salomon, *Chem. Res. Toxicol.* **9,** 1194 (1996).

[12] B. A. Bruenner, A. D. Jones, and J. B. German, *Chem. Res. Toxicol.* **8,** 552 (1995).

[13] T. J. Montine, D. Y. Huang, W. M. Valentine, V. Amarnath, A. Saunders, K. H. Weisburger, D. G. Graham, and W. J. Strittmatter, *J. Neuropathol. Exp. Neurol.* **55,** 202 (1996).

Copyright 2002, Elsevier Science (USA).
All rights reserved.
0076-6879/02 $35.00

HNE yields $1\text{-}N^2$-propanodeoxyguanosine, an exocyclic adduct with an extended five-member ring homologous to those formed from acrolein and crotonaldehyde, as previously described.[5,7] The HNE-dG adducts, consisting of two pairs of diastereomers (HNE-dG 1,2 and 3,4), were isolated from the reaction mixture by reversed-phase high-performance liquid chromatography (HPLC). Other DNA bases, such as adenine and cytosine, can also form cyclic adducts with enals; however, their reactions with HNE have not yet been described.[14,15] Furthermore, HNE can be epoxidized, and the epoxy aldehyde has been shown to form exocyclic etheno adducts with guanine, adenine, and cytosine.[16]

As a protein- and DNA-reactive product generated by lipid peroxidation, HNE has been implicated in the pathogenesis of various human diseases that have been linked to chronic oxidative conditions. Immunochemical assays developed with antibodies against the HNE–protein adducts have shown that the HNE-bound proteins are present in tissue, and their levels are elevated in the neurodegenerative brain of Parkinson's and Alzheimer's patients.[17–20] It has also been proposed that HNE generated from low-density lipoproteins binds to apolipoproteins, and that this reaction contributes to the pathogenesis of atherosclerosis.[21–25] Because the promotion of certain cancers, such as colon, breast, and prostate, has been shown to be associated with dietary intake of ω-6 PUFAs, a unique source for HNE, questions also arise as to whether DNA is an *in vivo* target of HNE. To address this question, we have developed a highly sensitive method for the detection of HNE-dG adducts[24] based on the previously developed HPLC-based ^{32}P-postlabeling assay for the detection of $1\text{-}N^2$-propanodeoxyguanosine adducts derived from acrolein and crotonaldehyde, Acr-dG and Cro-dG, respectively.[25] The analysis of DNA isolated from liver and colon of untreated rats and humans by this

[14] R. S. Sodum and R. Shapiro, *Bioorg. Chem.* **16,** 272 (1988).

[15] C. K. Winter, H. J. Segall, and W. F. Haddon, *Cancer Res.* **46,** 5682 (1986).

[16] F.-L. Chung, H.-J. C. Chen, and R. G. Nath, *Carcinogenesis* **17,** 2105 (1996).

[17] K. Uchida, K. Itakura, S. Kawakishi, H. Hiai, S. Toyokuni, and E. R. Stadtman, *Arch. Biochem. Biophys.* **324,** 241 (1995).

[18] D. P. Hartley, D. J. Kroll, and D. R. Petersen, *Chem. Res. Toxicol.* **10,** 895 (1997).

[19] A. Yoritaka, N. Hattori, K. Uchida, M. Tanaka, E. R. Stadtman, and Y. Mizuno, *Proc. Natl. Acad. Sci. U.S.A.* **93,** 2696 (1996).

[20] K. S. Montine, S. J. Olson, V. Amarnath, W. O. Whetsell, Jr., D. G. Graham, and T. J. Montine, *Am. J. Pathol.* **150,** 437 (1997).

[21] M. S. Bolgar, C. Y. Yang, and S. J. Gaskell, *J. Biol. Chem.* **271,** 27999 (1996).

[22] K. Uchida, S. Toyokuni, K. Nishikawa, S. Kawakishi, H. Oda, H. Hiai, and E. R. Stadtman, *Biochemistry* **33,** 12487 (1994).

[23] R. G. Salomon, K. Kaur, E. Podrez, H. F. Hoff, A. V. Krushinsky, and L. M. Sayre, *Chem. Res. Toxicol.* **13,** 557 (2000).

[24] F.-L. Chung, R. G. Nath, J. E. Ocando, A. Nishikawa, and L. Zhang, *Cancer Res.* **60,** 1507 (2000).

[25] R. G. Nath, H.-J. C. Chen, A. Nishikawa, R. Young-Sciame, and F.-L. Chung, *Carcinogenesis* **15,** 979 (1994).

method has demonstrated that HNE-dG adducts, like Acro- and Cro-dG, are present in tissue DNA as background lesions. The levels of HNE-dG in tissue DNA examined so far are estimated to be in the range of 3–9 adducts per billion bases (3–9 nmol/mol guanine). Our studies further showed that the levels of HNE adducts were elevated in the livers of rats treated with carbon tetrachloride.[24] Additional evidence supporting the endogenous origin of this type of cyclic DNA adduct includes the following: (1) these adducts formed on incubation *in vitro* with PUFAs in the presence of ferrous sulfate; (2) depletion of glutathione in tissue resulted in a significant increase in adduct levels; and (3) adduct profiles appear to be tissue-specific.[26] The HNE-dG adducts in tissue, thus, could serve as a specific marker for the genetic damage caused by endogenous oxidation of ω-6 PUFAs.

Although the ^{32}P-postlabeling method has established that HNE causes endogenous DNA damage, the roles of these adducts in cancer or other human diseases are yet to be determined. Future challenges are not only to improve the method for better quantitation and efficiency and to develop methods for detecting new markers of DNA damage caused by other enals generated from lipid peroxidation, but also to study the mutational potential and repair of HNE adducts. These latter goals are pivotal for understanding the biological roles of the HNE adducts.

[32]P-Postlabeling/High-Performance Liquid Chromatography Analysis of tert-4-Hydroxy-2-nonenal-DNA Adducts in Tissues

Scheme 1 outlines the ^{32}P-postlabeling/HPLC method for the analysis of HNE-dG adducts in tissue DNA. This method was developed with synthetic standards and is specific for the detection of 1-N^2-propanodeoxyguanosine adducts of HNE. There are several notable features about this method. For example: (1) it offers remarkable sensitivity capable of detecting 0.06 fmol of HNE-dG adducts, or approximately 1 base modification per billion bases in DNA; (2) the incorporation of HPLC for purification and final analysis allows the detection of diastereomeric HNE adducts; and (3) the conversions of the HNE-dG 3′,5′-bisphosphates to the corresponding ring-opened products and 5′-monophosphates provide unequivocal proof of the identities of the adducts. Nevertheless, it should be mentioned that, although the ^{32}P-postlabeling/HPLC method described here is highly sensitive, the quantitation of this assay is somewhat compromised by the intrinsic variability due to its multistep nature and lack of internal standards.

Scheme 2 illustrates the chemical reactions involved in the formation of HNE-dG adducts in DNA and the detection and identification of these adducts. HNE produced by oxidation of ω-6 PUFAs binds to guanine presumably via

[26] F.-L. Chung, R. G. Nath, M. Nagao, A. Nishikawa, G.-D. Zhou, and K. Randerath, *Mutat. Res.* **424**, 71 (1999).

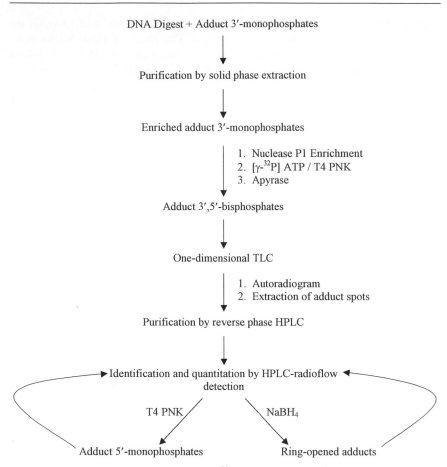

SCHEME 1. Outline of steps involved in the ^{32}P-postlabeling/HPLC method for the analysis of HNE–dG adducts in DNA. PNK, polynucleotide kinase.

an initial Michael addition by the exocyclic amino group, followed by ring closure at the N1 of guanine. Two pairs of diastereomers are formed as a result of the *trans* configuration between the hydroxyl and alkyl chains in the propano ring and the chiral carbon on the side chain (for simplicity only one stereoisomer is shown). Adducts are released as the 3′-monophosphates from DNA by enzymatic hydrolysis. After enrichment with a C_{18} solid-phase extraction cartridge, the fraction containing HNE adduct is labeled with $[\gamma\text{-}^{32}P]$ATP to yield the 3′,5′-bisphosphates of the adduct, which are purified on a polyethyleneimine cellulose plate, followed by two HPLC systems. The final analysis is carried out with an HPLC-UV detector system in conjunction with a radioflow detector, with

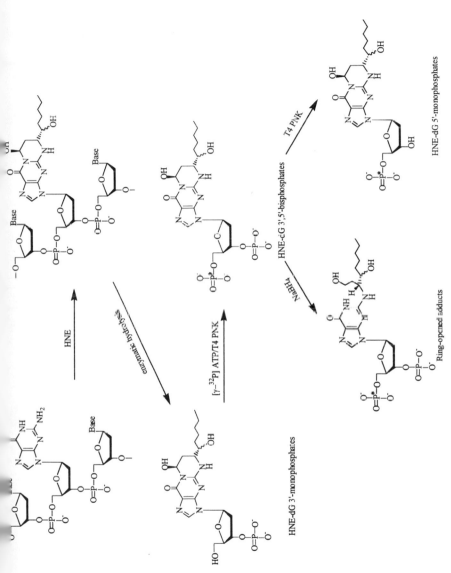

SCHEME 2. Formation of HNE–dG adducts from reaction of HNE with DNA and reactions involved in their detection and identification. Two pairs of diastereomers of HNE–dG 3'-monophosphates are formed as indicated by HPLC (Fig. 1). However, only one of the two pairs is shown here. [Reproduced with permission from F.-L. Chung, R. G. Nath, J. E. Ocando, A. Nishikawa, and L. Zhang, *Cancer Res.* **60**, 1507 (2000).]

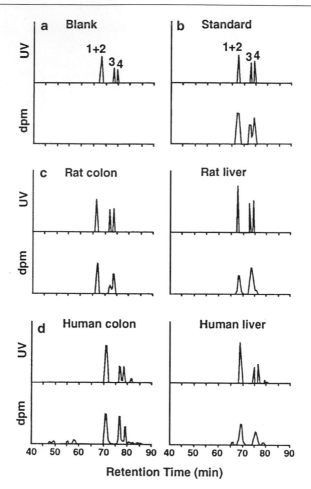

FIG. 1. Detection of HNE–dG adducts as 3′,5′-bisphosphates in rat and human tissues. These adducts are formed as two pairs of diastereomers 1,2 and 3,4 and isomers 1 and 2 at the 3′,5-bisphosphate level are not separated under the HPLC conditions used. The identities of the radioactive peaks obtained by ^{32}P-postlabeling are indicated by their HPLC comigration with the synthetic UV markers. (a) A blank sample of poly(dA-dC) : poly(dG-dT) is used to ensure that the assay is free of contamination; (b) external standards of HNE-dG adducts are used for quality assurance and quantitation; (c) rat tissues; (d) human tissues. [Reproduced with permission from F.-L. Chung, R. G. Nath, J. E. Ocando, A. Nishikawa, and L. Zhang, *Cancer Res.* **60,** 1507 (2000).]

the addition of standard adduct bisphosphates as UV markers. Figure 1 shows the final HPLC chromatograms demonstrating comigration of the HNE-dG adducts in rat and human liver and colon DNA with the UV markers. To confirm their identities, the adducts were converted to the 5′-monophosphates of HNE-dG with T4 polynucleotide kinase (T4 PNK) and to the ring-opened derivatives by sodium borohydride are then analyzed by HPLC as shown in Fig. 2.

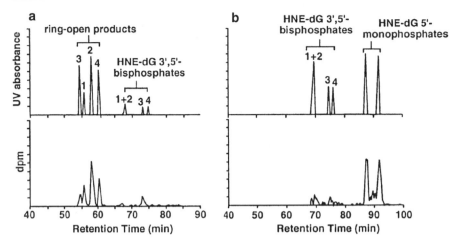

FIG. 2. Confirmation of HNE–dG adducts detected in rat colonic DNA. The radioactive peaks detected in tissue DNA, which comigrate with HNE–dG adducts in Fig. 1, are collected from the HPLC together with the UV standards. The concentrated fractions are then treated with $NaBH_4$ at pH 11, which opens the propano ring, and subsequently are reduced to N_2-[1-(hydroxyethyl)-2-hydroxyheptyl]-dG 3′,5′-bisphosphates (a) and with T4 PNK to yield the corresponding 5′-monophosphates (b). In both cases the products comigrated with their UV standards. HNE–dG adduct standards are added to the reaction mixture as references before analysis. Similar results are obtained with rat liver DNA and with human liver and colon DNA. [Reproduced with permission from F.-L. Chung, R. G. Nath, J. E. Ocando, A. Nishikawa, and L. Zhang, *Cancer Res.* **60**, 1507 (2000).]

Methods

Synthesis of tert-4-Hydroxy-2-nonenal-deoxyguanosine Adduct Standards as Ultraviolet Markers

HNE-dG adducts are prepared by reacting dG (20 mg, 70 μmol) in 4 ml of potassium bicarbonate buffer (20 mM, pH 9.4) with HNE (12 mg, 70 μmol) in 200 μl of methanol. The reaction mixture is stirred at 54° for 3 days. The products are isolated on a C_{18} reversed-phase semipreparative column (HPLC system 1) to afford 2 mg (6.4%) of HNE-dG. Two pairs of diastereoisomers, HNE-dG 1,2 and 3,4, are formed as previously reported.[7] The structure and purity of HNE-dG adducts are established by their UV, mass, and nuclear magnetic resonance (NMR) spectra and by HPLC. The 3′-monophosphates of HNE-dG are prepared by reacting HNE with dG 3′-monophosphate under the same conditions, and the products are then isolated by HPLC system 2. The collected fractions containing 3′-monophosphates of HNE-dG are combined and dried in a Speed-Vac (Savant, Hicksville, NY), reconstituted in H_2O, and rechromatographed using HPLC system 3 to remove the phosphate in the buffer. Amounts of adduct are quantified by HPLC by comparing the area of the adduct peaks with that of a previously standardized solution of 1-N^2-propano-dG adducts of crotonaldehyde (1 μg/μl). The 3′,5′-bisphosphates of HNE-dG are prepared by labeling its

3'-monophosphates of HNE-dG by the following procedure: A typical reaction mixture containing HNE-dG 3'-monophosphates ($\sim$30 μg) in 100 μl of H_2O, 100 μl of ATP (1 mg/ml in H_2O; Sigma, St. Louis, MO), 60 μl of 500 mM Tris base, and 50 μl of kinase buffer [a buffer made of 500 μl of 400 mM dithiothreitol (Sigma), 55 μl of 400 mM spermidine (Sigma), 500 μl of 400 mM $MgCl_2$, 500 μl of 800 mM Bicine (pH 9.8; Sigma), 445 μl of H_2O, and 7 μl of T4 PNK (30 units/μl; Amersham, Arlington Heights, IL)] is incubated at 37° for 60 min. The 3',5'-bisphosphates of HNE-dG formed are collected by HPLC system 4. The identities of the 3'-monophosphates and 3',5'-bisphosphates of HNE-dG are established by the characteristic UV spectra of 1-N^2-propano-dG and by conversion to the nucleosides with alkaline phosphatase. For dephosphorylation, 15–20 nmol of the bisphosphates is taken up in 30 μl of phosphate buffer (50 mM, pH 5.8) mixed with 30 μl of Tris-HCl (50 mM, pH 7.0) and 200 units of alkaline phosphatase (type V11 TA; Sigma). The mixture is incubated at 37° for 40 min. The comigration of the dephosphorylated products and their UV spectra, identical with the authentic HNE-dG adducts in HPLC system 5, serve to confirm their identities.

DNA Isolation

Tissue DNA is isolated by a modified Marmur's procedure.[27] Depending on the quantity of tissue available, the procedure may be modified. For example, in addition to reducing the volumes of the buffers proportionately, it is also possible to omit the interim precipitation step if only 100–300 mg of tissue is used for DNA isolation.

Tissue Homogenization. Typically, 2–5 g of tissue in a 100-ml beaker is thawed on ice, washed with a cold (4°) NaCl–citrate buffer (150–15 mM, pH 7.0) to remove blood, and minced to fine pieces with scissors. A mixture of cold NaCl–citrate buffer (8 ml/g of tissue) is then added. The tissue is homogenized with a Polytron Tissuemizer (Janke & Kunkel; IKA Works USA, Wilmington, NC) set at 70% of maximum speed for 60 sec. The homogenizer is stopped and the homogenate is cooled on ice at 30-sec intervals. The homogenized tissue is poured into a 40-ml plastic tube and centrifuged at 9500 rpm (12,000g) for 20 min at 4°, using a centrifuge (model J2-21M; Beckman, Fullerton, CA) with a JA-17 rotor. The pellet containing the nuclear fraction is saved and suspended in a cold Tris–EDTA–NaCl buffer (10 mM–1 mM–1 M, 6 ml/g of tissue).

Removal of Protein. To the above-described mixture containing the nuclear fraction is added 10% (w/v) sodium dodecyl sulfate (SDS; 0.1 ml/g of tissue). The mixture is vortexed and allowed to stand for 15 min at room temperature. Proteinase K (0.5 units/μl in H_2O, 5 units/g of tissue; Sigma) is then added, and

---

[27] J. Marmur, *J. Mol. Biol.* **3**, 208 (1961).

the mixture is incubated at 37° for 40 min. An equal volume of Sevag [chloroform–isoamyl alcohol, 24 : 1 (v/v); J. T. Baker, Phillipsburg, NJ) is added to the incubation mixture and the mixture is shaken vigorously on a wrist action shaker for 15 min. The mixture is then centrifuged at 9500 rpm for 15 min at 4°. The upper layer is pipetted into a Teflon flask and extracted again with an equal volume of Sevag to remove all the proteins. An equal volume of cold ethanol ($-20°$) is added to the flask containing the aqueous layer; the mixture is swirled gently and kept in a $-20°$ freezer for at least 60 min to facilitate the precipitation of DNA/RNA. The flask is then swirled gently and the precipitate is collected with a hooked glass rod or a pipette.

Removal of RNA. The DNA/RNA precipitate is dissolved in a Tris-HCl buffer (10 mM–1 mM, pH 7.0, 1 ml/g of tissue) in a 50-ml flask followed by vigorous vortexing. A solution of RNase A (9 units/g of tissue; Sigma) (see below) and RNase T1 [100,000 units in 200 μl, suspended in 3.2 M (NH$_4$)$_2$SO$_4$ solution, pH 6.0, 1600 units/g of tissue; Sigma] is added sequentially to the DNA/RNA solution. The mixture is incubated at 37° for 40 min, proteinase K (3 units/g of tissue) is then added, and the mixture is incubated for an additional 30 min to remove any protein contamination. The incubation mixture is extracted with Sevag (2 volumes). The mixture is shaken for 15 min and then centrifuged at 12,000 rpm (20,000g) for 20 min at 4°. The top layer is pipetted to a flask and extracted with 2 volumes of Sevag at least twice or until the UV absorbance of DNA in the aqueous layer shows a ratio of ∼1.8–2.0 at 260/280 nm and ∼0.40–0.45 at 230/260 nm. A solution containing NaCl (5 M, 0.1 ml/g of tissue) and an equal volume of cold ethanol ($-20°$) is added. After the solution is swirled gently, it is kept in a $-20°$ freezer for 60 min to precipitate DNA. The DNA precipitated is spooled out with a glass rod or a pipette, and washed with cold ethanol twice, a mixture of ethanol–ether (50 : 50, v/v) once, ether twice, and finally dried under a stream of N$_2$. The DNA is then dissolved in NaCl–citrate buffer (diluted 100-fold from a stock solution of NaCl–sodium citrate, 150 mM–15 mM) to make a final concentration of 2 μg/μl, and stored in $-80°$.

Preparation of RNase A Solution. The RNase A solution is prepared by dissolving 10 mg of RNase (78 units/mg; Sigma) in 1 ml of sodium acetate (10 mM, pH 5.0). DNase activity in the solution is destroyed by heating at 90° in a water bath for 10 min. The solution is then stored at $-20°$.

DNA Hydrolysis

DNA (80–100 μg, 2 μg/μl) in NaCl–citrate buffer (see above) is incubated for 3.5–4 hr at 37° with a freshly prepared solution containing micrococcal nuclease (0.3 units/μl, 15 μl/100 μg of DNA; Sigma), spleen phosphodiesterase (3 μg/μl, 15 μl/100 μg of DNA; Worthington, Freehold, NJ), sodium succinate buffer (300 mM, pH 6.0, 10 μl/100 μg of DNA), and CaCl$_2$ (100 mM, 10 μl/100 μg of DNA). The digested DNA is diluted to 1 ml with ammonium formate buffer (225 mM, pH 7.0) and the diluted DNA hydrolysate is taken up with a 1-ml tuberculin syringe and slowly filtered through an Acrodisc syringe filter (0.2 μm, 13-mm Minispike; Gelman, Ann Arbor, MI). One-tenth of the total hydrolysate is used to quantify dG 3'-monophosphate by HPLC (see section on quantification of HNE-dG adducts, below). The remaining hydrolysate is subjected to sample enrichment steps.

Enrichment of tert-4-Hydroxy-2-nonenal-deoxyguanosine Adducts

The large excess of normal nucleotides present in the DNA hydrolysate compared with HNE-dG adducts represents the main source of interference for labeling the adducts. Therefore, a method combining solid-phase extraction (SPE) and nuclease P1 was developed to enrich the adducts by removing normal nucleotides before ^{32}P labeling.

Solid-Phase Extraction. The DNA hydrolysate (80–100 μg of DNA) is applied to a preconditioned (washed with 15 ml of methanol and then with 15 ml of H_2O) C_{18} solid-phase extraction column (C_{18} Bond Elut, 3 ml/500 mg; Varian, Palo Alto, CA). The column is washed by gravity with 3 ml of 50 mM ammonium formate buffer (pH 7.0), followed by 3 ml of 10% (v/v) methanol in ammonium formate. The fraction containing the HNE-dG 3'-monophosphates is eluted with 6 ml of 50% (v/v) methanol in H_2O. The collected fraction is dried partially in a Speed-Vac and then transferred to a microcentrifuge tube.

Nuclease P1 Treatment. The fraction containing HNE adducts in a microcentrifuge tube is evaporated to dryness in a Speed-Vac. A premixed solution containing 12 μl of nuclease P1 (4 μg/μl in H_2O; Sigma), 2.5 μl of sodium acetate (1 M, pH 5.0), 5.5 μl of 1 mM zinc chloride, and 20 μl of H_2O is added to the adduct fraction. The mixture is thoroughly mixed by vortexing, spun in a microcentrifuge for 1 min, and then incubated at 37° for 60 min. At the end of incubation, a solution of 6.5 μl of 500 mM Tris base is added to the solvent before the mixture is evaporated in a Speed-Vac.

[γ-^{32}P]ATP Postlabeling and Purifications

A labeling mixture containing 4.5 μl of [γ-^{32}P]ATP (250 μCi in 25 μl; Amersham) and 0.35 μl of T4 PNK (30 units/μl; Amersham), 8 μl of H_2O, and 2.15 μl of kinase buffer (pH 9.8) is added to the enriched fraction. The mixture is incubated at 37° for 40 min, followed by treatment with 3 μl of apyrase (20 units/μl; Sigma) at 37° for 20 min. The total mixture is spotted slowly, using a pipette, onto a prewashed polyethyleneimine cellulose thin-layer chromatography (TLC) sheet (2.5 cm from the bottom, 1.5–2 cm apart for each DNA sample) [20 × 20 cm sheets (Machery Nagel, Duren, Germany) with a 5-cm wick (No. 2 filter paper; Whatman, Clifton, NJ)]. After the sheet is developed in 2.25 M NaH_2PO_4 buffer (pH 3.5) for 16–18 hr, it is air-dried and spotted with radioactive ink [200 μl of ^{99}Tc (0.5 mCi/ml)–800 μl of H$_2$O–1000 μl of black ink] and autoradiographed in steel cassettes on Eastman Kodak (Rochester, NY) XAR film for 20 min at room temperature. The film is then developed, fixed, and dried. Each adduct spot, identified by comparison with the simultaneously labeled 3′-monophosphates of HNE-dG standard, is excised and extracted with 1.5 ml of isopropanol–ammonia mixture [isopropanol–H$_2$O–ammonium hydroxide (15 N), 2.5 : 1.5 : 1 (v/v/v)] by shaking in a water bath for 12 min at 37°. The extract is filtered through an Acrodisc syringe filter into a 7-ml glass vial, and the filtrate is evaporated to dryness in a Speed-Vac and reconstituted in 750 μl of H$_2$O. The synthetic 3′,5′-bisphosphates of HNE-dG are added to the sample as UV markers before it is purified by reversed-phase HPLC systems 6 and 7 in sequence. The purified fraction containing the 3′,5′-bisphosphates of HNE-dG is again dried in a Speed-Vac and reconstituted in 500 μl of water.

High-Performance Liquid Chromatography Analysis and Confirmation of tert-4-Hydroxy-2-nonenal-deoxyguanosine Adducts

One-half of the purified fraction is analyzed with HPLC system 4. The identities of HNE-dG adducts are first established by their comigration with the UV standards (Fig. 1). Subsequently, they are verified by two conversion reactions; one involves enzymatic hydrolysis to the 5′-monophosphates of HNE-dG with T4 PNK, and the other involves treatment with sodium borohydride to yield the ring-opened products. The procedures for these conversion reactions are described below.

Chemical Ring-Opening Reduction. A quarter of the purified fraction is reacted with NaBH$_4$ (5 mg/ml) in 200 μl of K$_2$CO$_3$ (0.5 M, pH 12.0). The reaction mixture is stirred at room temperature for 20 min. The pH of the reaction is then adjusted to pH 7.0 with 1 N HCl. HPLC analysis (system 4) of the reaction mixture shows four new peaks eluting before the 3′,5′-bisphosphates of HNE-dG (Fig. 2a). Comigration with the ring-opened 3′,5′-bisphosphates of HNE-dG UV markers confirms their identities. It should be noted that the sequence of elution of the ring-opened adducts is different from that of the cyclic HNE adducts. The identities of the ring-opened 3′,5′-bisphosphates of HNE-dG UV standards are established by their UV spectra and by conversion to the corresponding nucleoside adducts with alkaline phosphatase. Comigration of the dephosphorylated products with the ring-opened HNE-dG (HPLC system 9), whose structures are characterized by UV, NMR and mass spectra,[7] is taken as a confirmation of the identities.

Enzymatic Hydrolysis. The remaining purified fraction is treated with 7 μl of T4 PNK (30 units/μl) in 200 μl of 1 M sodium acetate buffer (pH 5.0) with 60 μl of the kinase buffer [100 mM dithiothreitol, 500 mM MgCl$_2$, and 100 mM

Tris-HCl (pH 6.8)], and the mixture is incubated at 37° for 1 hr. After incubation, the mixture is filtered through an Acrodisc syringe filter and analyzed with HPLC system 8. Comigration of the hydrolyzed products with the UV markers of 5'-monophosphates of HNE-dG serves as confirmation of identities (Fig. 2b).

Quantification of tert-4-Hydroxy-2-nonenal-deoxyguanosine Adducts in DNA

Recovery of tert-4-Hydroxy-2-nonenal-deoxyguanosine Adducts. The recovery of adducts is determined with a sample of adduct standards. Five to 20 fmol of the 3'-monophosphates of HNE-dG are mixed with 15 μl of each of the four normal nucleotides (5 nmol/μl). This sample is analyzed as are other DNA samples, except that the DNA enzymatic hydrolysis step is omitted. The radioactivity of the comigrating peak in the final analysis is converted to femtomoles (assuming 100% labeling efficiency, 1 fmol = 10,000 dpm based on the specific activity of ATP, ~6000 Ci/mmol). The percentage of the recovery is calculated by dividing the amount of adduct obtained from the final analysis by the amount added at the beginning of the assay.

Levels of Adducts in DNA. The levels of adducts in DNA are determined by the radioactivity of adduct peaks after adjusting for decay and recovery, using the following equation, and are usually expressed as micromoles of adduct per mole of guanine (or femtomoles of adduct per nanomole of guanine). The amount of 3'-monophosphates of dG in DNA is quantified with 10% of DNA hydrolysate by comparison with the nucleotide standard, using HPLC system 10.

$$\text{Adduct level} = \frac{\text{dpm of adduct peaks} \times \left(\dfrac{100}{\% \text{ decay}} \right) \left(\dfrac{100}{\% \text{ recovery}} \right)}{1.0 \times 10^4 \text{ dpm/fmol} \times \text{dG (nmol)}}$$

Quality Assurance

To ensure the assay is free of contamination, a blank sample containing 50 μg of oligonucleotide [poly(dA · dC) : poly(dG · dT); Sigma] as a negative control is included for each set of DNA samples to be analyzed. In addition, it is advantageous to include a previously analyzed DNA sample of known adduct level as a positive control. We usually use rat liver DNA for this purpose.

High-Performance Liquid Chromatography Systems

The HPLC is performed on a Waters (Midford, MA) system, equipped with two model 510 pumps, a model 660 solvent programmer, and a Waters 994 photodiode array detector or a Waters 440 UV detector. For final analysis, a Waters 440 UV detector and a β-Ram flow through system (IN/US) are used. Linear gradients are used in all solvent programs. The Prodigy ODS 3 (Phenomenex, Torrance, CA), 5 μm, 4.6 × 250 mm C_{18} reversed-phase column is used for all the systems, except that a Prodigy ODS 3 (Phenomenex), 5 μm, 10 × 250 mm C_{18} reversed-phase column is used for system 1.

System 1: A, water; B, acetonitrile; 4 ml/min; 5% B, 5 min, 5–50% B, 55 min
System 2: A, 50 mM NaH_2PO_4 (pH 5.8); B, water–methanol (1 : 1, v/v); 1 ml/min; 0–100% B, 75 min
System 3: A, 25 mM sodium citrate; B, water–methanol (1 : 1, v/v); 0.6 ml/min; 50% B, 5 min, 50–100% B, 75 min
System 4: A, 50 mM NaH_2PO_4 (pH 5.8); B, water–methanol (1 : 1, v/v); 1 ml/min; 0–100% B, 100 min
System 5: A, water; B, acetonitrile; 1 ml/min; 5% B, 5 min, 5–50% B, 55 min
System 6: A, 50 mM NaH_2PO_4 (pH 5.0); B, water–methanol (1 : 1, v/v); 1 ml/min; 0–80% B, 80 min
System 7: A, 25 mM triethylamine phosphate; B, water–methanol (1 : 1, v/v); 0.6 ml/min; 50% B, 5 min, 50–100% B, 75 min
System 8: A, 50 mM NaH_2PO_4 (pH 5.8); B, water–methanol (1 : 1, v/v); 1 ml/min; 50% B, 5 min, 50–100% B, 75 min
System 9: A, water; B, acetonitrile; 1 ml/min; 5% B, 5 min, 5–30% B, 55 min
System 10: A, 50 mM NaH_2PO_4 (pH 5.8); B, water–methanol (1 : 1, v/v); 1 ml/min; 0–30% B, 0–10 min, 30–100% B, 10–13 min

Conclusion

The method described here is the only one currently available for the detection of isomeric HNE adducts in tissue DNA. Because it is a multiple-step method of superior sensitivity, caution must be taken to avoid contamination. Some of the precautions are as follows: (1) Use the designated HPLC unit pipettes and pipette tips for *in vivo* samples. Pipettes should be cleaned once a week by soaking in soapy water overnight, and rinsing thoroughly with water; (2) dispense the enzyme solutions to several microcentrifuge tubes to minimize the freeze–thaw process and possible cross-contamination; (3) store the kinase buffer at −80° for no more than 6 months; and (4) wash all TLC sheets by developing in water for 24 hr, air dry, and keep frozen at −20° before use. Although this method is routinely used in our laboratory, it still can be refined and improved. To perform this assay for the first time in the laboratory, it is highly desirable to initially use the synthetic HNE-dG standards before using DNA samples. The detection and identification capability of this assay for the isomeric HNE-dG adducts is clearly a unique strength. Using this assay, our studies have demonstrated its potential application for analyzing HNE-dG adducts in tissue as markers of endogenous DNA damage by lipid peroxidation. These studies will eventually help us understand the molecular mechanisms of chronic human diseases, including cancers, cardiovascular diseases, and neurodegenerative brain disorders.

Acknowledgments

This work is supported by NCI Grant CA 43159. We thank Dr. R. Nath for technical assistance and Drs. W. Davis and C. Conaway for critical reading of this manuscript and valuable suggestions.

[45] Transcription-Coupled Repair of 8-Oxoguanine in Human Cells

By Florence Le Page, Januario Cabral-Neto, Priscilla K. Cooper, and Alain Sarasin

Introduction

Oxidative damage to cellular DNA arises from attack by free radicals produced by cellular metabolism or by exogenous agents such as UV-A and ionizing radiation. Among the numerous types of base damage produced, 8-oxoguanine (8-oxoG or GO) is one of the most abundant. Because 8-oxoG readily mispairs with A, it is highly mutagenic, inducing G-to-T transversions in the absence of efficient repair. The base pair GO : C can be repaired in human cells by base excision repair (BER) initiated by the *hOGG1* gene product, a bifunctional DNA glycosylase with the dual function of lesion removal and incision of the DNA by AP (apurinic)-lyase activity.[1]

Although there are no known naturally occurring human mutations that specifically confer defects in BER, several human diseases are associated with deficiency in the nucleotide excision repair (NER) pathway, a versatile repair process that removes UV damage and other bulky lesions from DNA. These include the sun-sensitive disorders xeroderma pigmentosum (XP), which is characterized by skin changes and extreme predisposition to UV-induced skin cancer, and Cockayne syndrome (CS), a severe postnatal developmental disorder. Cells from CS individuals are specifically deficient in the preferential removal of lesions in template strands of genes transcribed by RNA polymerase II. Although this transcription-coupled repair (TCR) was initially presumed to be specific for DNA lesions

[1] S. Boiteux and J. P. Radicella, *Biochimie* **81,** 59 (1999).

Copyright 2002, Elsevier Science (USA).
All rights reserved.
0076-6879/02 $35.00

repaired by NER,[2] it is now clear that some lesions induced by ionizing radiation and other agents that generate free radicals are also removed by TCR. Although the full spectrum of oxidative lesions processed by TCR is not yet known, one of these is the oxidatively damaged base thymine glycol, which, like most other kinds of oxidative base damage, is primarily removed by BER.[3] We were interested in determining whether 8-oxoG could also be repaired by TCR. We addressed this question by the use of a single, site-specifically placed 8-oxoG paired with C in a defined location in a shuttle vector transfected into human cells, as previously described.[4,5]

Assay for Mutagenesis of 8-Oxoguanine-Containing Plasmids Replicated in Human Cells

A sequence containing a unique 8-oxoG was located in a replicating plasmid and inserted either into the 3'-untranslated region of the simian virus 40 (SV40) large T antigen (TAg) gene in the template strand transcribed from the early promoter of SV40 or into a nontranscribed corresponding position on the opposite (sense) strand of TAg. Use of these alternative constructs thus allowed comparison of cellular processing of 8-oxoG with and without transcription of the sequence in which it is located. The shuttle vector replicates autonomously in human cells transfected with the TAg of SV40 because it contains an SV40 origin of replication as well as itself encoding TAg, with essentially all transfected plasmids having replicated by 12 hr after transfection in one cell line studied.[4] Replication of the shuttle vector allows determination of the frequency of mutagenesis at 8-oxoG in human cells, which should reflect the efficiency of removal of the lesion (Fig. 1).

Construction of Closed Circular Replicating Monomodified Plasmids

Synthetic oligonucleotides used to construct the monomodified plasmids are from Genset (Paris, France) and are purified in a denaturing 20%(w/v) polyacrylamide gel. The modified 19-mer oligonucleotide carrying a unique 8-oxoG that we use contains a small fragment of the human Ha-*ras* gene from codons 10 to 14, with the lesion located on the second guanine of codon 12 (5'-GATC GGC GCC GGOC GGT GTG-3'). The 8-oxoG oligonucleotide is produced as described previously[6] by J. Cadet (CEA, Grenoble, France).

[2] P. C. Hanawalt, *Science* **266,** 1957 (1994).

[3] P. K. Cooper, T. Nouspikel, S. G. Clarkson, and S. A. Leadon, *Science* **275,** 990 (1997).

[4] F. Le Page, A. Guy, J. Cadet, A. Sarasin, and A. Gentil, *Nucleic Acids Res.* **26,** 1276 (1998).

[5] F. Le Page, E. E. Kwoh, A. Avrutskaya, A. Gentil, S. A. Leadon, A. Sarasin, and P. K. Cooper, *Cell* **101,** 159 (2000).

[6] V. Bodepudi, S. Shibutani, and F. Johnson, *Chem. Res. Toxicol.* **5,** 608 (1992).

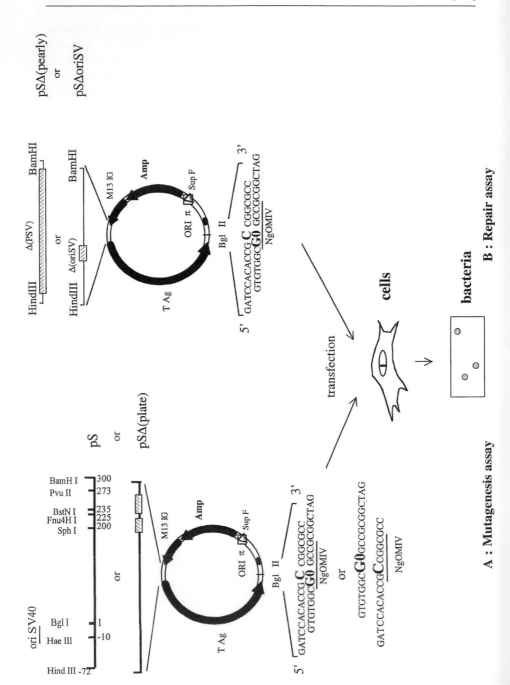

A : Mutagenesis assay

B : Repair assay

The shuttle vector used to construct the monomodified plasmids is derived from the pS189 shuttle vector and was given by M. Seidman (the National Institutes of Health, Bethesda, MD).

The replicative plasmids containing a unique 8-oxoG lesion paired with cytosine (GO : C) (Fig. 1) are constructed as described previously.[4] Briefly, a single-stranded modified plasmid pS(C) carrying a cytosine in the position opposite the lesion is constructed by inserting a 19-mer oligonucleotide complementary to the modified oligonucleotide carrying the 8-oxoG into the original plasmid DNA. The active M13 replication origin contained in pS189 is used to produce single-stranded DNA (ssDNA), using M13 helper phage (Pharmacia, Piscataway, NJ). A gapped double-stranded plasmid is then constructed by hybridizing the single-stranded pS(C) in 1× saline–sodium citrate (SSC) buffer with double-stranded DNA linearized at the BglII site. The mixture is heated at 98° for 10 min, cooled slowly to 65°, allowed to renature for 1 hr, and then ethanol precipitated. Twenty micrograms of gapped duplex plasmid DNA thus formed is hybridized overnight with 5 μg of 5′-^{32}P-phosphorylated 8-oxoG-containing oligonucleotide in the ligation buffer containing 1 mM ATP, and then ligated for 20 min with 400 units of T4 DNA ligase (Biolabs, Hertfordshire, UK) at 12°. Finally, covalently closed molecules are purified by isopycnic centrifugation on an ethidium bromide–cesium chloride gradient. Fractions collected from the gradient are analyzed on a 1% (w/v) agarose gel and those corresponding to the closed circular double-stranded DNA vector as detected by the ^{32}P label are pooled, dialyzed against TE [1 mM Tris (pH 7.8), 1 mM EDTA] and then ethanol precipitated (Fig. 2).

Plasmid containing an 8-oxoG in a nontranscribed sequence is constructed as described above, but beginning with a plasmid deleted for the late promoter sequences. The plasmid pSΔp$_{late}$ is constructed by a two-step protocol, first by digestion of single-stranded plasmid with SphI and Fnu4HI enzymes after hybridization of oligonucleotides at the restriction sites. The oligonucleotides are removed by heating and filtration through a Spun size S-400 column (Pharmacia) and the longest fragment is gel purified. The resulting single-stranded plasmid deleted of 25 bp is then digested with BamHI and BstNI restriction enzymes, using

---

FIG. 1. Experimental scheme and genetic maps of the various plasmids used. (A) Mutagenesis assay: The plasmid pS contains an 8-oxoG.C in a transcribed sequence under the control of the SV40 early promoter. The pSΔ(p$_{late}$) plasmid has been deleted in the SV40 late promoter sequence (hatched) and contains an 8-oxoG.C mispair in the opposite position in a nontranscribed sequence. After transfection and replication in human cells, plasmid DNA is recovered and amplified in recA$^-$ bacteria. Digestion by NgoMIV of clones allows the selection of mutants. (B) Repair assay: The plasmid pSΔoriSV has a deletion in the eukaryotic replication origin (hatched) and contains an 8-oxoG.C mispair in a transcribed sequence under the control of the SV40 early promoter. pSΔ(p$_{early}$) has a deletion of both the replication origin and the early promoter (hatched) that eliminates transcription of the 8-oxoG.C mispair. After transfection and repair in human cells, plasmid DNA was recovered and amplified in fpg$^-$/mutY$^-$ bacteria. Digestion by NgoMIV of plasmid DNA from individual colonies allows the selection of mutants.

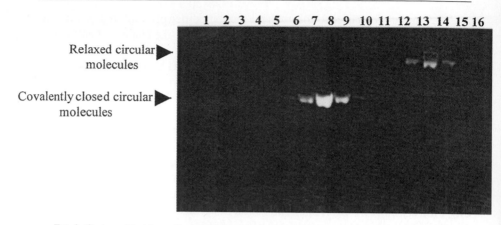

Relaxed circular molecules

Covalently closed circular molecules

FIG. 2. Cesium chloride gradient fractionation after plasmid construction allows the detection and purification of covalently closed circular molecules. Aliquots of about 20 fractions resulting from the isopycnic centrifugation of a GO : C construct on a cesium chloride gradient were loaded on 1% (w/v) agarose gels in 1 mM Tris, 20 mM sodium acetate, 2 mM EDTA (pH 7.8) adjusted with CH_3COOH. The fractions were collected from the bottom of the gradient. DNA markers for relaxed and covalently closed circular molecules are indicated. Closed circular monomodified plasmids were pooled from fractions 6–9 and used to transfect mammalian cells after dialysis.

the same procedure. The single-stranded DNA is recircularized by using 800 units of bacteriophage T4 DNA ligase for 4 hr at 16° to produce double-stranded plasmid DNA deleted of 90 bp. The SphI–Fnu4HI fragment corresponds to the sequence necessary for TAg-mediated transcriptional activation of the late promoter,[7] and the BamHI–BstNI fragment corresponds to a cluster of initiator-like elements. To allow early transcription to be active, the extent of deletion in the promoter region is limited to the minimum necessary to eliminate late transcription while still permitting efficient early transcription.

It is necessary to bear in mind that this kind of construction may produce certain false positives of closed circular plasmids due to the reconstitution of the double-stranded pS DNA. Such reconstituted double-stranded pS plasmid can be destroyed by treatment with the BglII restriction enzyme before CsCl gradient purification. However, a modified protocol introduced by Kodadek and Gamper[8] not only overcomes this particular problem but also increases the yield of the closed circular monomodified plasmids and the reproducibility of the construction. In this alternative procedure, the pS(C) plasmid is denatured and 10 μg of the single-stranded DNA is hybridized with 20 pmol of a 30-mer 5′-phosphorylated 8-oxoG-containing oligonucleotide in T4 DNA polymerase buffer (final volume, 40 μl). The mixture is heated at 70° for 10 min, and then cooled slowly to 25°.

[7] G. Gilinger and J. C. Alwine, J. Virol. 67, 6682 (1993).
[8] T. Kodadek and H. Gamper, Biochemistry 27, 3210 (1988).

To synthesize the complementary strand, using the 8-oxoG-containing oligonu-cleotide as a primer, the following components are added to the hybridized plas-mids (final volume, 100 μl) : T4 DNA polymerase buffer, 100 mM ATP, dNTPs (10 mM each), bovine serum albumin (BSA, 100 μg/ml), 10 U of T4 DNA li-gase, and 30 U of T4 DNA polymerase (both from New England BioLabs, Ozyme, Saint-Quentin, France). The reaction mixture is incubated for 1 hr and 30 min at 37° and then heated at 70° for 15 min. The resulting covalently closed molecules are purified as described above.[8]

Detection of Mutagenesis Induced by Unique 8-Oxoguanine

Eight hundred nanograms of closed circular double-stranded plasmid contain-ing a unique GO : C is transfected into different human cell lines (semiconfluent cells cultured in 10-cm^2 petri dishes). Monkey and transformed human cell lines are cultured in Dulbecco's modified Eagle's medium–Ham's F12 (3 : 1, v/v) supple-mented with 10% (v/v) fetal calf serum, amphotericin B (Fungizone) and penicillin (100 U/ml each), and streptomycin (100 μg/ml) (GIBCO-BRL, Life Technologies, Cergy Pontoise, France). To obtain repair-deficient cell lines corrected by expres-sion of the wild-type gene, cells are transfected with a plasmid containing the cDNA encoding the wild-type repair sequence[9] or transduced with a recombi-nant retrovirus as already described.[10] Transfectants are selected by growth in medium containing hygromycin (GIBCO-BRL) at increasing concentrations up to 375 μg/ml. Single clones are isolated after 10 days, propagated in 24-well plates, and analyzed for repair activity.

For the mutagenesis assay, the transfection method used is the cationic liposome 1,2-dioleoyl-3-trimethylammoniumpropane (DOTAP) procedure (Boehringer, Meylan, France). After transfection, cells are incubated for 72 hr and then col-lected. Extrachromosomal plasmid DNA is recovered by a small-scale alkaline lysis method.[11] Digestion with BglII restriction enzyme is performed to elimi-nate any progeny molecules arising from replication of double-stranded DNA that may have escaped the initial BglII digestion during the construction process and therefore may not have carried the modified oligonucleotide. This is followed by digestion with DpnI to eliminate any double-stranded molecules that have not replicated in the human cells.

The recovered plasmid DNA is then used to transform DH5αrecA$^-$ bacteria. For each transformation, one-tenth of the DNA recovered from mammalian cells is added to 40 μl of bacterial suspension. The mixture is transferred to a cold

[9] M. Carreau, E. Eveno, X. Quilliet, O. Chevalier-Lagente, A. Benoit, B. Tanganelli, M. Stefanini, W. Vermeulen, J. H. Hoeijmakers, A. Sarasin, *Carcinogenesis* **16**, 1003 (1995).
[10] M. Carreau, X. Quilliet, E. Eveno, A. Salvetti, O. Danos, J. M. Heard, M. Mezzina, and A. Sarasin, *Hum. Gene. Ther.* **6**, 1307 (1995).
[11] A. Stary, C. F. Menck, and A. Sarasin, *Mutat. Res.* **272**, 101 (1992).

Bio-Rad (Hercules, CA) Gene Pulser cuvette (0.2 cm) and electroporations are performed with a Sedd Cell Ject apparatus (Bio-Rad) under 40 μF, 192 Ω, and 2500 V. Colonies are selected by ampicillin resistance carried by the plasmid. DNA from individual bacterial colonies is prepared with the Jetsar genome miniplasmid purification system (Bioprobe Systems, Montreuil sous Bois, France). The frequency of mutations induced by replication of the lesion in the human cells is then determined by *Ngo*MIV digestion. The original shuttle vector plasmid pS189 has one *Ngo*MIV site (GCCGGC) and another is present in the inserted GO-containing oligonucleotide (pS189 GO : C). Therefore, the pS189 GO : C plasmid mutated at or around the GO site will be digested once outside the sequence of interest, observed as a linearized plasmid (form III). On the other hand, when the introduced lesion (8-oxoG) is correctly repaired, the plasmid is digested at both sites, giving rise to two fragments of 3407 and 1930 bp, respectively (Fig. 3). To determine the mutation spectra, linearized plasmids are sequenced by the chain termination method, using a Sequenase 2.0 kit (Amersham, Orsay, France). Three independent transformations are carried out for each experiment, and mutation frequencies are determined in each case from more than 100 individual plasmid clones.

Using this approach, we show that normal cells and cells with defects in NER produce a low mutation frequency on this plasmid (between 0 and 2%), whereas for other cells with defects in TCR the mutation frequency could reach 30–40%.[5]

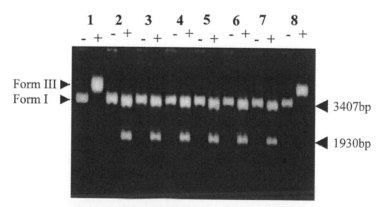

FIG. 3. Identification of 8-oxoG-induced mutations by restriction enzyme analysis. Form I DNA isolated from individual bacterial colonies was digested with *Ngo*MIV and the products were analyzed on 1% (w/v) agarose gels. The original shuttle vector plasmid pS189 has one *Ngo*MIV site and another is present in the inserted 8-oxoG-containing oligonucleotide. Thus the plasmids harboring a wild-type G : C sequence at the site of the original GO : C contain two restriction sites, giving rise to two DNA fragments of 3407 and 1930 bp, respectively (lanes 2–7), whereas plasmids mutated at the site originally containing the 8-oxoG are cut only once (lanes 1 and 8). The use of this enzyme therefore allows us to quantify the mutation frequency. All mutated plasmids were confirmed by DNA sequencing: (–) undigested DNA; (+) *Ngo*MIV-digested DNA plasmids.

Assay for Quantification of 8-Oxoguanine Removal from Plasmid DNA in Human Cells

A high mutation rate in some specific cells should correspond to a defect in removal of 8-oxoG in those cells. To examine this possibility, we constructed vectors having deletions of the SV40 origin of replication (Fig. 1), which therefore cannot replicate in human cells.[5]

Construction of Closed Circular Nonreplicating Plasmids Carrying Unique 8-Oxoguanine Lesion

Plasmid DNA deleted of the SV40 replication origin (pSΔSVori) is constructed as previously described.[4] Briefly, plasmid DNA is digested with *Pvu*II, and the *Pvu*II fragment is replaced with an identical fragment deleted of the SV40 origin of replication derived from the pLAS-wt plasmid, a generous gift from L. Daya-Grosjean (UPR 2169, CNRS, Villejuif, France).[12] The resulting plasmid, incapable of replicating, is able to transcribe the lesion by using the early promoter when transfected into human cells and is used to construct the monomodified plasmid DNA according to the protocol described above.

Plasmid deleted of its early region promoter (pSΔp$_{early}$) is constructed by digesting single-stranded plasmid with *Hae*III and *Pvu*II enzymes after hybridization of specific oligonucleotides able to hybridize and produce a short double-stranded DNA sequence at the restriction sites. The oligonucleotides are removed by heating at 90° for 2 min, followed by filtration through a Spun size S-400 column (Pharmacia). The longest restriction fragment is gel purified, using a Nucleotrap purification kit (Clontech, Montigny-le-Bretonneux, France) and recircularized, using 800 units of bacteriophage T4 DNA ligase for 4 hr at 16° to produce double-stranded pS189Δp$_{early}$GO : C deleted of 285 bp. This vector has therefore also lost its SV40 replication origin (Fig. 1) and can neither replicate nor transcribe either T antigen or the lesion-containing oligonucleotide inserted into its 3' untranslated region (UTR).

Quantification of 8-Oxoguanine Repair in Human Cells

Transfection of the nonreplicating GO : C plasmids with or without the early promoter deletion is performed as described for the mutagenesis assays. Cells are collected for recovery of extrachromosomal plasmid DNA after periods of incubation ranging from 2 to 72 hr. Elimination of any contaminating extracellular input DNA is performed by treatment of cell cultures with DNase I before extraction. Recovered plasmid molecules are individualized and amplified by transformation of the *Escherichia coli* PR195 *mutY*⁻/*fpg*⁻ strain (Δ*lac-pro* F' *pro lacI lacZ*

---

[12] L. Daya-Grosjean, M. R. James, C. Drougard, and A. Sarasin, *Mutat. Res.* **183**, 185 (1987).

mutY::kan fpg::kan Tn*10*), a generous gift from S. Boiteux (CEA, Fontenay-aux-Roses, France), to amplify the SV40 *ori*-deleted plasmid progeny and to avoid any repair of GO : C mispairs in the bacteria. Plasmid DNA from individual colonies is analyzed for repair of the lesion in human cells by digestion with *Ngo*MIV. Persistent 8-oxoG lesions in plasmids not repaired in human cells give rise to mutations in a fraction of the progeny in a given clone after replication in the repair-defective bacteria, thus giving a fraction of plasmid molecules in that clone that is resistant to digestion. In contrast, repair of the lesion in human cells yields clones after passage through the bacteria in which all the progeny DNA is digested by *Ngo*MIV. In good general agreement with estimates of the frequency with which A is inserted opposite 8-oxoG in *E. coli*,[13] approximately 20–30% of the plasmid molecules in the mixed clones is uncut. Repair of the lesion in human cells is calculated as a ratio of the number of colonies in which all progeny molecules are sensitive to *Ngo*MIV digestion to the total number of colonies analyzed. As a control for each experiment, this value is normalized to that found after direct transformation of bacteria with the original GO : C-containing vector. The result for each time point is quantified from three or four independent transformations, with approximately 100 colonies analyzed for each transformation.

Results for removal of 8-oxoG obtained using these plasmids agree very well with those obtained for mutagenesis frequency with replicating plasmids. Thus, the cells that induce high frequencies (30–40%) of G-to-T transversion at GO exhibit almost no repair of the lesion, whereas control cells with a low mutation rate repair almost 100% of GO in 12 hr.[5] Interestingly, all human cell lines known to be deficient in the transcription-coupled repair (TCR) of UV-induced DNA lesions, that is, Cockayne syndrome (CS) and combined CS and xeroderma pigmentosum (XP/CS) patients, are deficient in GO repair, but only when the lesion is located in a transcribed sequence.[5,14]

Detection of Transcription of Plasmid Sequences in Human Cells

To determine whether the sequence containing the 8-oxoG lesion is transcribed in each shuttle vector plasmid used, a standard protocol for reverse transcription-polymerase chain reaction (RT-PCR) is employed. Different monomodified plasmids (pSGO : C, pSΔoriGO : C, pSΔp$_{late}$C : GO, and pSΔp$_{early}$GO : C) are transfected into normal, nontransformed human cell strain 198VI, derived from a skin biopsy of a normal child, in which there is no TAg sequence expressed. After 12 hr, total RNA is prepared by the guanidium thiocyanate method[15] and treated

[13] J. Wagner, H. Kamiya, and R. P. Fuchs, *J. Mol. Biol.* **265,** 302 (1997).

[14] F. Le Page, A. Klungland, D. E. Barnes, A. Sarasin, and S. Boiteux, *Proc. Natl. Acad. Sci. U.S.A.* **97,** 8397 (2000).

[15] P. Chomczynski and N. Sacchi, *Anal. Biochem.* **162,** 156 (1987).

with DNase I to eliminate any residual plasmid DNA. Aliquots of isolated mRNA are analyzed for their quality by agarose gel electrophoresis. Five micrograms of total RNA is reverse transcribed by incubation for 30 min at 37° in a final vo ume of 50 μl containing 1000 U of Moloney murine leukemia virus (Mo-MuLV) reverse transcriptase (GIBCO-BRL), $1\times$ RT buffer [50 mM Tris-HCl (pH 8.3), 75 mM KCl, 3 mM MgCl$_2$], dNTPs (1 mM each; (Pharmacia), 4 mM dithiothreitol (DTT; GIBCO-BRL), RNase inhibitor (1 U/ml; Boehringer), and 1 μg of random hexamers (Pharmacia). The reaction is stopped by incubation for 5 min at 95°. PCR amplifications are performed with an automatic thermocycler (PerkinElmer, Courtaboeuf, France) in the presence of 2 U of *Taq* polymerase (AmpliTaq; PerkinElmer), 10 pmol of each sense and antisense primer of 25 bp, and PCR beads containing reaction buffer [10 mM Tris-HCl (pH 8.3), 5 mM KCl, 0.01% (w/v) gelatin], dNTPs (250 μM each), and 1 mM MgCl$_2$ (Pharmacia Biotech)]. The primers used, which amplify a 270-bp sequence, are complementary to the end of the coding sequence of the TAg gene and to the 3' end of the oligonucleotide carry-ing the lesion and are, respectively, 5'-AAAATGAATGCAATTGTTGT-3' and 5'-CTTGAGCGTCGATTTTTGTG-3'. After 1 cycle at 95° for 2 min, 30 cycles of denaturation, annealing, and extension are performed at 95° (1 min), 60° (1 min), and 72° (2 min), respectively. After 10 min at 72° to ensure that the final extension step is complete, aliquots of the products are analyzed by electrophoresis on a 1.5% (w/v) agarose gel and the DNA is extracted from the gel with a Nucleo-trap kit (Macherey Nagel, Düren, Germany). Sequencing of the amplified DNA fragments is carried out by the chain termination method, using a double-stranded DNA cycle sequencing system kit (GIBCO-BRL).

This approach has allowed us to verify that the absence of the SV40 early promoter prevents transcription of the oligonucleotide containing the GO lesion.

One possible explanation for the inability of cells that lack TCR to remove 8-oxoG in a transcribed sequence despite their capacity to repair 8-oxoG in a nontranscribed sequence is that access of repair enzymes to the lesion is prevented by a stalled RNA polymerase II (RNAP II). The classic model for TCR is that only lesions that block RNAP will be subject to preferential repair, and by extension that inability to perform TCR should result in an extended block to transcription. We therefore use RT-PCR to compare transcription 12 hr after transfection of the sequence containing the 8-oxoG in normal cells and TCR-defective cells, which do not remove the lesion. Transcription of the opposite strand does not interfere with the assay, because it occurs from the late promoter and hence later in time than transcription of TAg and thus of the lesion-containing sequence located in its 3' UTR.

To determine whether the lesion blocks transcription in cells defective in its repair, the RT-PCR analysis is performed on RNA extracted from diploid, nontransformed repair-defective cells isolated from a xeroderma pigmentosum group G patient also exhibiting Cockayne syndrome (XP-G/CS patient: XPCS1LV

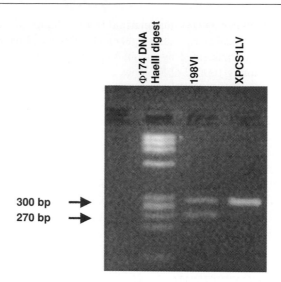

FIG. 4. Blockage of transcription *in vivo* at a plasmid sequence containing the 8-oxoG lesion. RT-PCR analysis of samples collected 12 hr after transfection of the nonreplicative shuttle vector pSΔoriGO : C into primary diploid cells isolated from a normal individual (198VI) or primary diploid cells isolated from a xeroderma pigmentosum group G patient also exhibiting Cockayne syndrome (XPCS1LV). Transcription of the TAg gene is indicated by the 300-bp product from primers located near the 5' end of the gene. Transcription of the sequence containing the 8-oxoG is revealed by the presence of a 270-bp product from primers spanning the site of the lesion. A 3-fold excess of product was loaded in the XP-G/CS cells to emphasize the absence of the 270-bp band. The left lane contains size markers. [Reproduced from F. Le Page, E. E. Kwoh, A. Avrutskaya, A. Gentil, S. A. Leadon, A. Sarasin, and P. K. Cooper, *Cell* **101,** 159 (2000), with permission.]

cell strain) in comparison with normal 198VI cells 12 hr after transfection of pSΔoriGO : C. The primers described above are used to amplify a 270-bp region containing the lesion. Transcription of sequences upstream of the lesion is also examined, as a positive control, by coamplification with another pair of primers (20 bp), 5'-GGAGGCTTCTGGGATGCAAC-3' and 5'-GAGCTTTAAATCTCTG TAGG-3', chosen to amplify a 300-bp coding region near the 5' end of the TAg gene. Transcription of the coding sequence of the TAg gene in the repair-deficient cells is compared with control. Whereas both cell strains transcribe TAg efficiently, only the normal cells transcribe the 8-oxoG-containing sequence (Fig. 4). No transcript that spans the lesion is detectable in repair-defective cells despite the occurrence of active transcription upstream of the lesion.

This technique allows us to correlate the state of transcription of the DNA sequences containing the lesion and the level of its repair on the transcribed strand. Our results strongly suggest that TCR of the oxidative lesion 8-oxoG is associated with *in vivo* blockage of RNA polymerase II at the lesion site.

Conclusion

Because there are no specific genotoxic agents able to produce only 8-oxoG in cellular DNA and because the antibodies raised against this lesion are not specific enough, we chose a shuttle vector assay allowing us to construct a plasmid DNA containing only one GO lesion at a given position. This assay is the only one at this time to allow the study of the repair and mutagenic potency of unique oxidative damage in human cells. This technology, which is equally applicable for other types of oxidative base damage, has been used in the past mostly for DNA lesions repaired by NER.[16] Results obtained with shuttle vectors in repair and mutagenesis analyses are highly reproducible and when used in NER studies have always been validated by genomic DNA data when available.[17]

Our results using a shuttle vector containing a single 8-oxoG showed for the first time the existence of transcription-coupled repair of this lesion in human cells.[5] The observed deficiency in this pathway in CS and XP/CS cells correlates with progressive neurological disorders observed in these patients. It is plausible that in nondividing brain cells the accumulation of unrepaired oxidative lesions on transcribed strands leads to apoptosis and therefore to progressive neurological deterioration. We have found that cells homozygous for mutations in the BRCA1 or BRCA2 gene, the major genes predisposing to breast and ovarian cancers, are also deficient in the TCR of 8-oxoG.[18] Several studies have reported a high level of oxidative stress in breast tissue that may be linked to estrogen metabolism and hormonal status. It is possible that the absence of repair of oxidative lesions in breast cells already mutant for the BRCA1 or BRCA2 gene will produce a high level of point mutations, thus contributing to malignancy.

The use of shuttle vectors carrying single lesions in defined locations as described in this chapter should allow monitoring of the status of repair of oxidative damage for cells isolated from numerous human diseases and eventually correlation of a given level of repair with a given type of malignancy.

Acknowledgments

The authors acknowledge the support of the Ligue Nationale contre le Cancer and the Association de Recherche sur le Cancer (Villejuif, France), and of the U.S. Public Health Service, National Cancer Institute. Januario Cabral Neto had a fellowship from the Radioprotection Department of Electricité de France (Paris, France).

The authors are grateful to Dr. J. Cadet (CEA, Grenoble, France) for providing the 8-oxoG-containing oligonucleotides and to Dr. S. Boiteux (CEA, Fontenay-aux-Roses, France) for the gift of the $fpg^-/mutY^-$ bacteria.

[16] A. Sarasin, J. Photochem. Photobiol. B 3, 143 (1989).
[17] C. F. Menck, C. Madzak, G. Renault, A. Margot, and A. Sarasin, Mutat. Res. 220, 101 (1989).
[18] F. Le Page, V. Randrianarison, D. Marot, J. Cabannes, M. Perricaudet, J. Feunteun, and A. Sarasin, Cancer Res. 60, 5548 (2000).

[46] Nucleobase and 5'-Terminal Probes for DNA Redox Chemistry

By Xi Hu, Stephen J. Lee, and Mark W. Grinstaff

Introduction

DNA, which possesses a unique helix structure formed from two complementary strands, plays a key role in genetic information storage and replication. Modifying DNA with a specific chemical functionality such as a radiolabel,[1] fluorescent dye,[2–6] antibody,[7–9] nanoprobe,[10] or redox probe[11–15] provides opportunities to study biological processes involving DNA. In addition, such labeled oligodeoxynucleotides can be used to detect medically relevant nucleic acid sequences or specific genetic markers of pathogens.

Redox probe-labeled oligodeoxynucleotides have attracted widespread interest for their applications in characterizing DNA-related redox reactions[16–30] and

[1] D. A. Melton, P. A. Krieg, M. R. Rebagliati, T. Manistis, K. Zinn, and M. R. Green, *Nucleic Acids Res.* **12**, 7035 (1984).

[2] D. M. Jameson and J. F. Eccleston, *Methods Enzymol.* **278**, 363 (1997).

[3] S. P. Lee and M. K. Han, *Methods Enzymol.* **278**, 343 (1997).

[4] K. M. Parkhurst and L. J. Parkhurst, *Biochemistry* **34**, 285 (1995).

[5] J. Li and Y. Lu, *J. Am. Chem. Soc.* **122**, 10366 (2000).

[6] J. B. Biggins, J. R. Prudent, D. J. Marshall, M. Ruppen, and J. S. Thorson, *Proc. Natl. Acad. Sci. U.S.A.* **97**, 13537 (2000).

[7] G. M. Barroso and S. Oehninger, *Hum. Reprod.* **15**, 1338 (2000).

[8] D. J. Hantowich, G. Mardirossian, M. Rusckowski, and P. Winnard, *Nucl. Med. Commun.* **17**, 66 (1996).

[9] D. Chen, A. E. Beuscher, R. C. Stevens, P. Wirsching, R. A. Lerner, and K. D. Janda, *J. Org. Chem.* **66**, 1725 (2001).

[10] T. A. Taton, R. C. Mucic, C. A. Mirkin, and R. L. Letsinger, *J. Am. Chem. Soc.* **122**, 6305 (2000).

[11] E. R. Holmlin, P. J. Dandliker, and J. K. Barton, *Bioconjug. Chem.* **10**, 1122 (1999).

[12] D. J. Hurley and Y. Tor, *J. Am. Chem. Soc.* **120**, 2194 (1998).

[13] S. I. Khan, A. E. Beilstein, and M. W. Grinstaff, *Inorg. Chem.* **38**, 418 (1999).

[14] S. I. Khan and M. W. Grinstaff, *J. Am. Chem. Soc.* **121**, 4704 (1999).

[15] X. Hu, G. D. Smith, M. Sykora, S. J. Lee, and M. W. Grinstaff, *Inorg. Chem.* **39**, 2500 (2000).

[16] P. J. Dandliker, R. E. Holmlin, and J. K. Barton, *Science* **275**, 1465 (1997).

[17] R. E. Holmlin, P. J. Dandliker, and J. K. Barton, *Angew. Chem. Int. Ed. Engl.* **36**, 2714 (1997).

[18] M. R. Arkin, E. D. A. Stemp, S. C. Pulver, and J. K. Barton, *Chem. Biol.* **4**, 389 (1997).

[19] M. R. Arkin, E. D. A. Stemp, R. E. Holmlin, J. K. Barton, A. Hormann, E. J. C. Olson, and P. F. Barbara, *Science* **273**, 475 (1996).

[20] T. Fiebig, C. Wan, S. O. Kelly, J. K. Barton, and A. H. Zewail, *Proc. Natl. Acad. Sci. U.S.A.* **96**, 1187 (1999).

[21] D. B. Hall, R. E. Holmlin, and J. K. Barton, *Nature (London)* **382**, 731 (1996).

[22] S. O. Kelley, R. E. Holmlin, E. D. A. Stemp, and J. K. Barton, *J. Am. Chem. Soc.* **119**, 9861 (1997).

METHODS IN ENZYMOLOGY, VOL. 353

Copyright 2002, Elsevier Science (USA).
All rights reserved.
0076-6879/02 $35.00

in electrochemically based diagnostic devices.[31–33] Highly reactive species, such as free radicals generated either by exogenous ultraviolet radiation or endogenous metabolic processes, are known to damage DNA via an oxidation reaction.[34–36] This oxidative damage to DNA, if left unrepaired, leads to mutations and finally to alteration in cell growth and promotion of tumorigenesis. To understand how oxidizing agents generated *in vivo* migrate to specific sites in DNA, and to determine the role of DNA in this biological charge-transfer reaction, a well-defined system in which DNA is modified with redox probes is desired.

Noncovalently bound intercalators were originally investigated to monitor DNA-mediated electrontransfer reactions, but the ill-defined structure of this system prevents the exact location of the redox probes on the DNA strand to be determined, thus limiting interpretation of the experimental data.[37,38] Currently, multiple groups are investigating DNA redox reactions with probes that are covalently attached to DNA at site-specific locations[13–15,27,39–44] or with probes that are bound and intercalated within the DNA base stacks at known positions.[11,24,26,29,45]

[23] M. E. Nunez, D. B. Hall, and J. K. Barton, *Chem. Biol.* **6,** 85 (1999).

[24] G. B. Schuster, *Acc. Chem. Res.* **33,** 253 (2000).

[25] P. T. Henderson, D. Jones, G. Hampikian, Y. Kan, and G. B. Schuster, *Proc. Natl. Acad. Sci. U.S.A.* **96,** 8353 (1999).

[26] S. M. Gasper and G. B. Schuster, *J. Am. Chem. Soc.* **119,** 12762 (1997).

[27] F. D. Lewis, R. Wu, Y. Zhang, R. L. Letsinger, S. R. Greenfield, and M. R. Wasielewski, *Science* **277,** 673 (1997).

[28] F. D. Lewis, R. L. Letsinger, and M. R. Wasielewski, *Acc. Chem. Res.* **34,** 159 (2001).

[29] K. Fukui, K. Tanaka, M. Fujitsuka, A. Watanabe, and O. Ito, *J. Photochem. Photobiol. B Biol.* **50,** 18 (1999).

[30] M. W. Grinstaff, *Angew. Chem. Int. Ed.* **38,** 3629 (1999).

[31] S. O. Kelley and J. K. Barton, *Bioconjug. Chem.* **8,** 31 (1997).

[32] S. O. Kelley, N. M. Jackson, M. G. Hill, and J. K. Barton, *Angew. Chem. Int. Ed.* **38,** 941 (1999).

[33] T. Ihara, M. Nakayama, M. Murata, K. Nakano, and M. Maeda, *Chem. Commun.* 1609 (1997).

[34] E. C. Friedberg, G. C. Walker, and W. Seide, "DNA Repair and Mutagenesis." ASM Press, Washington, D.C., 1995.

[35] J. Cadet, "DNA Damage Caused by Oxidation, Deamination, Ultraviolet Radiation and Photoexcited Psoralens." IARC Scientific Publications, Lyon, France, 1994.

[36] V. Machalik, *Int. J. Radiat. Biol.* **62,** 9 (1992).

[37] A. M. Brun and A. Harriman, *J. Am. Chem. Soc.* **114,** 3656 (1992).

[38] C. J. Murphy, M. R. Arkin, Y. Jenkins, N. D. Ghatlia, S. H. Bossmann, N. J. Turro, and J. K. Barton, *Science* **262,** 1025 (1993).

[39] F. D. Lewis, S. A. Helvoigt, and R. L. Letsinger, *Chem. Commun.* 327 (1999).

[40] T. J. Meade and J. F. Kayyem, *Angew. Chem. Int. Ed. Engl.* **34,** 352 (1995).

[41] A. E. Beilstein and M. W. Grinstaff, *Chem. Commun.* 509 (2000).

[42] Y. Kan and G. B. Schuster, *J. Am. Chem. Soc.* **121,** 11607 (1999).

[43] M. T. Tierney and M. W. Grinstaff, *J. Org. Chem.* **65,** 5355 (2000).

[44] M. T. Tierney and M. W. Grinstaff, *Org. Lett.* **2,** 3413 (2000).

[45] E. K. Erkkila, T. D. Odom, and J. K. Barton, *Chem. Rev.* **99,** 2777 (1999).

For site-specific covalent labeling of oligodeoxynucleotides with redox probes, three general synthetic strategies are used.[46,47] The first method introduces the probe phosphoramidite monomer during DNA synthesis. This phosphoramidite approach is amenable to derivatization of an oligodeoxynucleotide at the base 2′-deoxyribose, and at the phosphate. The second method, termed direct incorporation, involves labeling the solid support-bound oligodeoxynucleotide by palladium (Pd)-catalyzed cross-coupling chemistry. These two approaches are discussed below in detail. The third approach, not discussed, couples a redox probe to a previously modified oligodeoxynucleotide containing a terminal primary amine in solution. Table I[11–15,17,18,23,24,26,27,33,39–41,43–45,48–57] summarizes the redox probes that have been covalently attached to DNA.

Oligodeoxynucleotide Syntheses

Labeling Oligodeoxynucleotides via Phosphoramidite Synthetic Approach

A standard solid-phase DNA synthesis cycle is shown in Scheme 1.[58–60] This cycle includes the following reactions: detritylation, coupling, capping, and oxidation. The synthesis begins with detritylation of a solid-supported [typically control pore glass (CPG)] nucleoside, using trichloroacetic acid. The resulting 5′-free hydroxyl group is then coupled with a nucleoside phosphoramidite (A, C, G, and T) in the presence of tetrazole. Unreacted 5′-hydroxyl groups are subsequently converted to acetate esters by treatment with acetic anhydride/N-methylimidazole solution in the capping step. Finally, the 3′–5′ internucleotide phosphite linkage is oxidized to the more stable phosphotriester linkage with iodine. This cycle is repeated until the desired oligodeoxynucleotide sequence is assembled. At the end of the synthesis, the fully protected oligomer is cleaved from the solid support and

[46] G. H. Keller and M. M. Manak, "DNA Probes." Stockton Press, New York, 1993.

[47] A. E. Beilstein, M. T. Tierney, and M. W. Grinstaff, *Comments Inorg. Chem.* **22,** 105 (2000).

[48] K. Fukui and K. Tanaka, *Angew. Chem. Int. Ed.* **37,** 158 (1998).

[49] I. M. Abdou, T. L. Netzel, and L. Strekowski, *J. Heterocyclic Chem.* **4,** 387 (1998).

[50] D. Ly, Y. Kan, B. Armitage, and G. B. Schuster, *J. Am. Chem. Soc.* **118,** 8747 (1996).

[51] A. E. Beilstein and M. W. Grinstaff, *J. Organomet. Chem.* **637–639,** 398 (2001).

[52] R. C. Mucic, M. K. Herrlein, C. A. Mirkin, and L. Letsinger, *Chem. Commun.* 555 (1996).

[53] C. J. Yu, H. Yowanto, Y. Wan, T. J. Meade, Y. Chong, M. Strong, L. H. Donilon, J. F. Kayyem, M. Gozin, and G. F. Blacburn, *J. Am. Chem. Soc.* **122,** 6767 (2000).

[54] R. E. Holmlin, R. T. Tong, and J. K. Barton, *J. Am. Chem. Soc.* **120,** 9724 (1998).

[55] S. R. Rajski, S. Kumar, R. J. Roberts, and J. K. Barton, *J. Am. Chem. Soc.* **121,** 5615 (1999).

[56] S. I. Khan, A. E. Beilstein, M. Sykora, G. D. Smith, X. Hu, and M. W. Grinstaff, *Inorg. Chem.* **38,** 3922 (1999).

[57] J. F. Rack, E. S. Krider, and T. J. Meade, *J. Am. Chem. Soc.* **122,** 6287 (2000).

[58] M. J. Gait, "Oligonucleotide Synthesis: A Practical Approach." IRL Press, Washington, D.C., 1984.

[59] M. H. Caruthers, *Science* **230,** 281 (1985).

[60] M. H. Caruthers, *Acc. Chem. Res.* **24,** 278 (1991).

TABLE 1
REDOX PROBES INCORPORATED IN DNA

Redox probe	Research report	Ref.
Acridine	Fukui and Tanaka	48
Anthraquinone	Tierney and Grinstaff	43, 44
	Abdou et al.	49
	Schuster	24
	Gasper and Schuster	26
	Ly et al.	50
Ferrocene	Beilstein and Grinstaff	41, 51
	Ihara et al.	33
	Mucic et al.	52
	Yu et al.	53
Fluorenone	Tierney and Grinstaff	43
$M(diimine)_2(dppz)^{2+\,a}$	Holmlin et al.	17, 54
	Arkin et al.	18
	Erkkila et al.	45
$Rh(phi)_2(hpy)^{2+}$	Holmlin et al.	11
	Nunez et al.	23
	Rajki et al.	55
$Ru(bpy)_3^{2+}$	Khan et al.	13, 14, 56
	Hu et al.	15
	Lewis et al.	39
	Hurley and Tor	12
$Ru(bpy)_2(imi)^{3+}$	Meade and Kayyem	40
$Ru(impy)(bpy)_2^{2+}$	Rack et al.	57
Phenothiazine	Tierney and Grinstaff	43, 44
Stilbene	Lewis et al.	27

^{a} M = Ru, Os.

the phosphate and base-protecting groups are removed with concentrated NH_4OH at 55°.

Chemical functional groups can be introduced site specifically to oligodeoxynucleotides, using this phosphoramidite approach. This method allows for the incorporation of labeled nucleosides or nonnucleoside functionalities.[13,15,29,39,42,61] The modified phosphoramidite should be sufficiently soluble in and stable to the solvents and reagents used during DNA synthesis. The chemical and physical properties of the probe should also not alter after attachment to the oligodeoxynucleotide. The three nucleotide components (base, 2′-deoxyribose, and phosphate) are all sites where chemical functionality can be introduced in an oligodeoxynucleotide.

[61] K. M. Guckian, B. A. Schweitzer, R. X. F. Ren, C. J. Sheils, P. L. Paris, D. C. Tahmassebi, and E. T. Kool, *J. Am. Chem. Soc.* **118**, 8182 (1996).

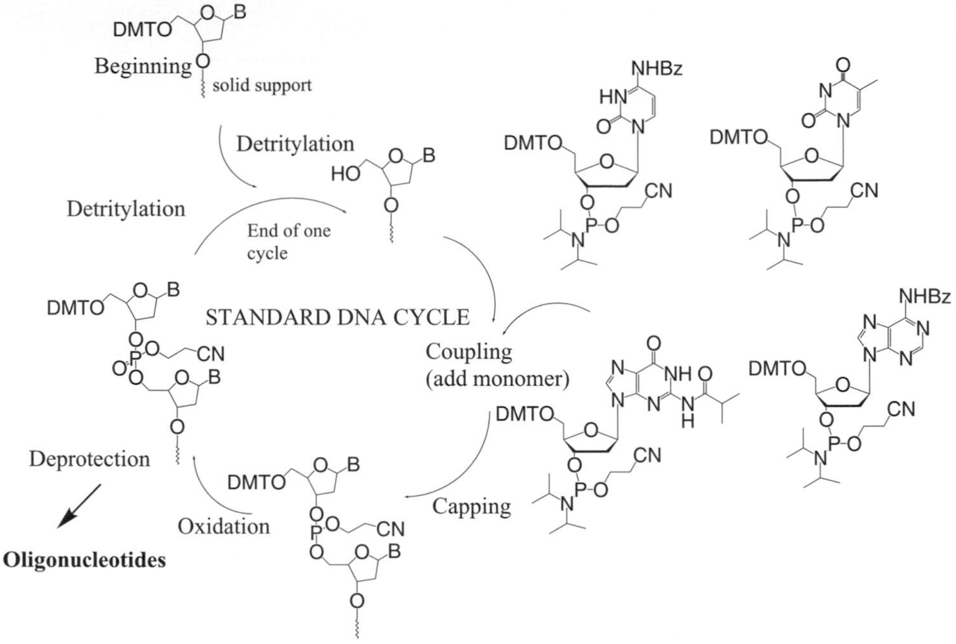

SCHEME 1. Standard automated DNA solid-phase synthesis.

Base Labeling

Base-labeled nucleosides are typically prepared by reacting a halide- or alkynyl-modified base with a desired redox probe, using a palladium-catalyzed cross-coupling reaction.[12–14,41,53] Probes such as metal complexes possess a number of favorable properties including tunable electronic structures, reversible electrochemical behavior, and emissive spectroscopic states. As shown in Scheme 2, a Pd(PPh$_3$)$_4$ cross-coupling reaction between the ruthenium complex, **3**, and 2′-deoxy-3′,5′-dibenzoyloxy-5-iodouridine, **5**, affords the metallo-nucleoside, **6**.[13,62] Benzoyl deprotection in methanolic ammonia, followed by reaction with 4,4′-dimethoxytrityl chloride and 2-cyanoethyl-*N*,*N*′-diisopropylchlorophosphoramidite, yields the ruthenium–nucleoside phosphoramidite, **9**, ready for automated solid-phase synthesis as shown in Scheme 1.

Compound 2: 4′-Methyl-2,2′-bipyridine-4-propargylamide. 4′-Methyl-2,2′-bipyridine-4-carboxylic acid[63] (1 mmol; **1**), propargylamine hydrochloride

[62] S. I. Khan, A. E. Beilstein, G. D. Smith, M. Sykora, and M. W. Grinstaff, *Inorg. Chem.* **38,** 2411 (1999).

[63] D. G. McCafferty, B. M. Bishop, C. G. Wall, S. G. Hughes, S. L. Mecklenberg, T. J. Meyer, and B. W. Erickson, *Tetrahedron* **51,** 1093 (1995).

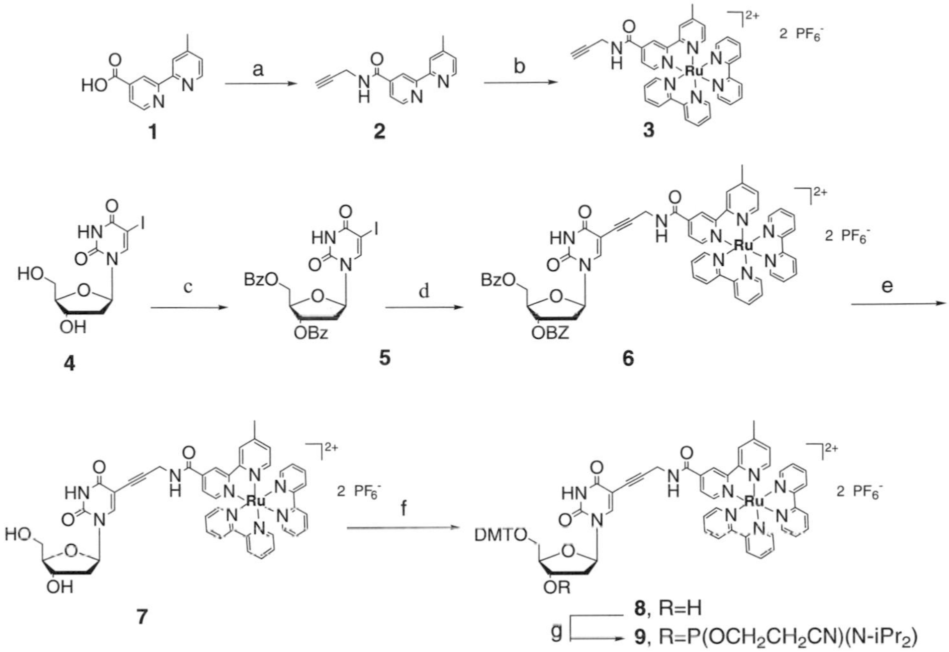

SCHEME 2. Synthesis of Ru(diimine)$_3^{2+}$-uridine phosphoramidite. Reagents and conditions: (a) propargyl amine · HCl, DCC, HOBt, DIPEA, DMF, 0°, 12 hr; (b) Ru(bpy)$_2$Cl$_2$, 70% aq. CH$_3$CH$_2$OH, reflux, 10 hr; (c) benzoyl chloride, C$_5$H$_5$N, 25°, 12hr; (d) **3**, Pd(PPh$_3$)$_4$, CuI, TEA, DMF, 25°, 8 hr; (e) NH$_3$/CH$_3$OH, 25°, 48 hr; (f) DMT-Cl, C$_5$H$_5$N, 25°, 12 hr; (g) ClP(iPr$_2$N)(OCH$_2$CH$_2$CN), DIPEA, CH$_3$CN, 25°, 2 hr.

(1 mmol), HOBt (1 mmol), and DIPEA (0.21 ml) are dissolved in dry DMF (15 ml) and cooled to 0°. DCC (1.2 mmol) is dissolved in DMF (3 ml) and added drop-wise to the reaction mixture. The mixture is stirred at room temperature overnight. The DCU is filtered off and the solvent is removed by vacuum distillation. The remaining solid compound is dissolved in ethyl acetate and washed with NaHCO$_3$ (5%), 0.5 N HCl, and brine, and dried over sodium sulfate. The solvent is removed by rotary evaporation and the compound is purified by column chromatography, using 2% methanol in chloroform as the solvent (76% yield). m.p. 146°; ^{1}H NMR (DMSO) δ 2.4 (s, 3H, CH$_3$); 2.5 (s, 1H, CH); 4 (s, 2H, CH$_2$); 7.25–8.8 (m, 6H, py); 9.4 (s, 1H, NH); FAB-MS calculated for C$_{15}$H$_{13}$N$_3$O [M]$^+$ 251.3; found [M + H]$^+$ 252.1. [Reprinted from S. I. Khan, A. E. Beilstein, G. D. Smith, M. Sykora, and M. W. Grinstaff, *Inorg. Chem.* **38**, 2411 (1999), with permission. Copyright 2000 ACS.]

Compound 3: Bis(2,2'-bipyridine) (4'-methyl-2,2'-bipyridine-4-propargyl-amide)-ruthenium(II) bis(hexafluorophosphate). Ru(bpy)$_2$Cl$_2$ (0.3 mmol) is added to a solution of 4'-methyl-2',2'-bipyridine-4-propargylamide (**2**, 0.3 mmol) in 70% ethanol/H_2O (25 ml) and refluxed for 10 hr. Next, the reaction mixture is cooled and ethanol is removed *in vacuo*. After the solution has stood for 4 hr at room temperature, it is filtered and the solid compound is washed with cold water. A saturated aqueous solution of NH_4PF_6 is added until no further precipitate is observed. The mixture is kept at room temperature for an additional 2 hr and then filtered, washed with cold water, ether and dried overnight to give an orange color product (82% yield). ^{1}H NMR (DMSO) δ 2.4 (s, 3H, CH_3); 2.5 (s, 1H, CH); 4.1 (s, 2H, CH_2); 7.4–9.2 (m, 22H, bpy); FAB-MS calculated for $C_{35}H_{29}N_7ORuP_2F_{12}$ [M-2 PF_6^-]$^+$ 664.7, [M-PF_6^-]$^+$ 809.7; found [M-2 PF_6^-]$^+$ 665.2, [M-PF_6^-]$^+$ 810.1.

Compound 5: 3', 5'-Dibenzoyloxy-2'-deoxy-5-iodouridine. 2'-Deoxy-5-iodo-uridine (**4**, 6 mmol) is dissolved in dry pyridine and cooled to 0°. Benzoyl chloride (36 mmol) is added slowly while the reaction is stirred for 12 hr at room temperature. The solvent is then removed and the crude material is dissolved in $CHCl_3$ (150 ml) and washed with 0.5 N HCl, water, and dried over Na_2SO_4. Silica gel column chromatography affords a white crystallinic solid in 80% yield. ^{1}H NMR (DMSO) δ 2.15 (t, 2H, C-2'); 3.6 (m, 2H, C-5'); 3.8 (m, 1H, C-3'); 4.2 (m, 1H, C-4'); 6.2 (t, 1H, C-1'); 7.2–7.9 (m, 10H, 2Ph); 8.2 (s, 1H, C-6); FAB-MS calculated for $C_{23}H_{19}N_2O_7$ [M]$^+$ 562.3; found [M + H]$^+$ 563.3.

Compound 6: (3',5'-Dibenzoyloxy-5-[(4'-methyl-2,2'-bipyridine-4-propargyl-amide) (bpy)_2Ru(II)]-2'-deoxyuridine bis(hexafluorophosphate). 3',5'-Dibenzoy-loxy-2'-deoxy-5-iodouridine (0.9 mmol), **5** (0.8 mmol), Pd(PPh_3)_4 (0.09 mmol), and CuI (0.2 mmol) are dissolved in dry DMF (15 ml) and degassed with N_2. Triethylamine (0.7 ml) is added and the reaction mixture is stirred for 8 hr. The solvent is then removed under reduced pressure. The crude product obtained is dissolved in acetonitrile and passed through a Sephadex column. Next, the solid is dissolved in acetonitrile and the addition of dry ether affords an orange-colored precipitate. The compound is filtered and dried to yield the ruthenium-modified 2'-deoxyuridine (79% yield). ^{1}H NMR (DMSO) δ 2.5 (s, 3H, CH_3); 2.8 (m, 2H, C-2'); 3.1 (m, 2H, C-5'); 4.25 (bs, 2H, CH_2); 4.6 (m, 1H, C-3'); 5.6 (m, 1H, C-4'); 6.2 (t, 1H, C-1'); 7.3–8.9 (m, 32H, Ph + bpy); 9.1 (s, 1H, C-6); FAB-MS cal-culated for $C_{58}H_{47}N_9O_8RuP_2F_{12}$ [M-2 PF_6^-]$^+$ 1099.1, [M-PF_6^-]$^+$ 1244.1; found [M-2 PF_6^-]$^+$ 1099.2, [M-PF_6^-]$^+$ 1244.1.

Compound 7: 5-[(4'-Methyl-2,2'-bipyridine-4-propargylamide) (bpy)_2Ru(II)]-2'-deoxyuridine bis(hexafluorophosphate). Compound **6** (1 g) is suspended in methanolic ammonia and left for 2 days at room temperature with occasional shaking. The solvent is removed by rotary evaporation. The compound is dissolved in a minimum volume of acetonitrile and precipitated with dry ether to yield a dark orange-colored compound (90% yield). ^{1}H NMR (DMSO) δ 2.5 (s, 3H, CH_3); 4.2 (s, 2H, CH_2); 6.15 (t, 1H, C-1'); 7.3–8.8 (m, 22H, bpy); 8.9 (s, 1H, C-6); FAB-MS calculated for $C_{44}H_{39}N_9O_6RuP_2F_{12}$ [M-2 PF_6^-]$^+$ 890.9, [M-PF_6^-]$^+$ 1035.8; found [M-2 PF_6^-]$^+$ 890.3, [M-PF_6^-]$^+$ 1035.2.

Compound 8: 5'-O-(4,4'-Dimethoxytrityl)-5-[(4'-methyl-2,2'-bipyridine-4-propargylamide) (bpy)$_2$Ru(II)]-2'-deoxyuridine bis(hexafluorophosphate). 5-[(4'-Methyl-2,2'-bipyridine-4-propargylamide) (bpy)$_2$Ru(II)]-2'-deoxyuridine bis(hexafluorophosphate), (**7**, 0.1 mmol) is dissolved in dry pyridine (5 ml) and the solvent is removed under high vacuum. This process is repeated twice. Next, the compound is dissolved in dry pyridine (15 ml) and dimethoxytrityl chloride (DMT-Cl) (2 mmol) is added to the flask in two portions. The mixture is stirred for 4 hr at room temperature. Methanol is then added to the reaction mixture to consume any excess DMT-Cl. Next, the solvent is removed and the compound is dissolved in CH_3CN and precipitated with ether to give an orange solid (81% yield). ^{1}H NMR (DMSO) δ 2.5 (s, 3H, CH_3); 3.8 (s, 6H, OCH_3); 4.2 (s, 2H, CH_2); 6.15 (t, 1H, C-1'); 7.2–8.8 (m, 33 H, bpy and ph); 8.9 (s, 1H, C-6); FAB-MS calculated for $C_{65}H_{57}N_9O_8RuP_2F_{12}$ [M-2 PF_6^-]$^+$ 1193.3, [M-PF_6^-]$^+$ 1338.3; found [M-2 PF_6^-]$^+$ 1193.3, [M-PF_6^-]$^+$ 1338.3.

Compound 9: 5'-O-(4,4'-Dimethoxytrityl)-3'-O-(2-cyanoethyl)-N,N'-diiso-propylphosphoramidite-5-[(4'-methyl-2,2'-bipyridine-4-propargylamide) (bpy)$_2$ Ru(II)]-2'-deoxyuridine bis(hexafluorophosphate). The DMT-protected metallonucleoside, **8** (0.1 mmol), is dissolved in dry CH_3CN (40 ml) and diisopropylethylamine (0.2 mmol) is added to the flask and cooled in an ice bath. Next, 2-cyanoethyl-N,N'-diisopropylchlorophosphoramidite (0.2 mmol) is slowly added under nitrogen. The reaction mixture is stirred for 2 hr under argon followed by the addition of CH_3CN (10 ml). The solvent is removed and the product is dissolved in CH_3CN and then precipitated with ether (90% yield). ^{31}P NMR ($CDCl_3$) 152 ppm (m). TLC: >95%. The compound is stored over KOH pellets under vacuum and subsequently used in an automated DNA synthesizer.

2'-Deoxyribose Labeling

The 2'-deoxyribose unit of the nucleoside is another site available for labeling.[15,40] One strategy explored is to substitute the 5'-hydroxyl group on the 2'-deoxyribose with an amine group, thus creating a more nucleophilic center for functionalization. Most ribose-labeling procedures rely on amide bond formation between a carboxylic acid-derivatized redox probe and the amino group on the ribose. For example, the redox probe, phenothiazine, can be covalently attached to a 5'-nucleoside as shown in Scheme 3. First, the 5'-position of thymidine is converted from a hydroxyl to an amine by treatment with methanesulfonyl chloride, followed by substitution with an azide, and finally reduction to the amine. The monocarboxylic acid-derivatized phenothiazine is then coupled to 5'-amino-5'-deoxythymidine, using CDI. The 5'-PTZ-labeled nucleoside, **14**, is reacted with

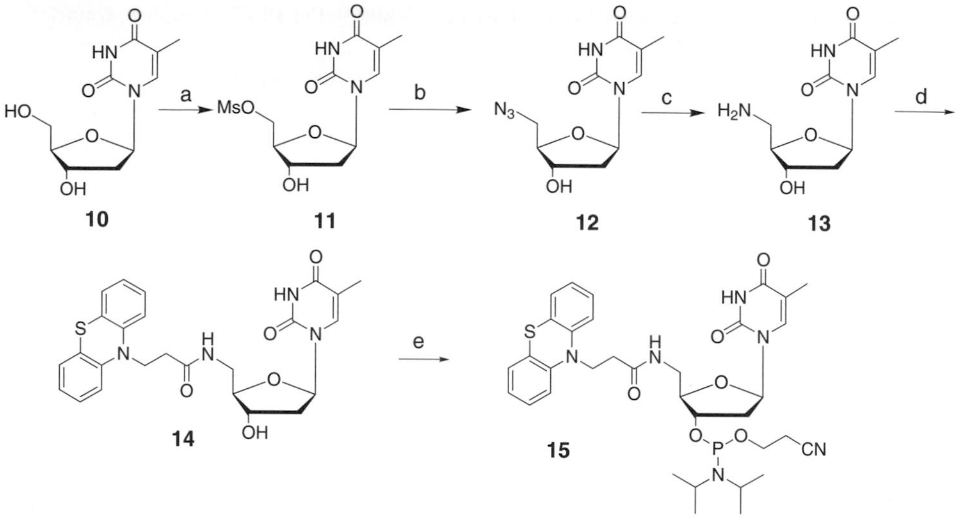

SCHEME 3. Synthesis of the phenothiazine-thymidine phosphoramidite. Reagents and conditions: (a) MsCl/pyridine, 0°, 12 hr; (b) LiN$_3$/DMF, 90°, 3 hr; (c) H$_2$, Pd/C, 50 psi, 25°, 5 hr; (d) PTZ-acid, CDI/DMF, LH-20/THF, 25°, 18 hr; (e) ClP(iPr$_2$N)(OCH$_2$CH$_2$CN), DIPEA, CH$_3$CN, 25°, 2 hr.

2-cyanoethyl-N,N'-diisopropylchlorophosphoramidite in dry acetonitrile to afford the PTZ-thymidine phosphoramidite, **15**.

Compound 11: 5'-O-Methanesulfonylthymidine. Thymidine (**10**, 12 mmol) is dissolved in 10 ml of dry pyridine and cooled to −10°. Next, mesyl chloride (14 mmol) is added dropwise over a period of 20 min. The reaction mixture is then kept at 0° for 12 hr. The next day, 10 ml of methanol is added to quench the reaction and the solvents are evaporated via high vacuum. The resulting crude product is checked by TLC and purified by column chromatography (silica gel; CH$_3$OH–CHCl$_3$, 1 : 9). A white powdered solid (**11**) is obtained (77% yield). ^{1}H NMR DMSO, δ 1.78 (s, 3H, 5-methyl), 2.08–2.22 (m, 2H, C2'), 3.22 (s, 3H, mesyl), 3.98 (q, 1H, C4'), 4.28 (m, 1H, C3'), 4.40 (m, 2H, C5'), 5.50 (s, 1H, 3'-OH), 6.22 (t, 1H, C1'), 7.48 (s, 1H, C6), 11.25 (s, 1H, N3). FAB-MS calculated for C$_{11}$H$_{16}$N$_2$O$_7$S [M]$^+$ 320; found (M + H)$^+$ 321.

Compound 12: 5'-Azido-5'-deoxythymidine. A solution of **11** (6 mmol) in 15 ml of DMF containing lithium azide (33 mmol) is stirred at 90° under nitrogen. After 3 hr, the reaction is stopped and the solvent is removed under vacuum. The resulting crude product is purified by column chromatography (silica gel; CH$_3$OH–CHCl$_3$, 1 : 9) to afford **12** (73% yield). ^{1}H NMR DMSO, δ 1.78 (s, 3H, 5-methyl), 2.08–2.22 (m, 2H, C2'), 3.57 (d, 2H, C5'), 3.85 (q, 1H, C4'), 4.20 (m, 1H, C3'), 5.50 (s, 1H, 3'-OH), 6.22 (t, 1H, C1'), 7.48 (s, 1H, C6), 11.25 (s, 1H, N3). FAB-MS calculated for C$_{10}$H$_{13}$N$_5$O$_4$ [M]$^+$ 267; found (M + H)$^+$ 268.

Compound 13: 5′-Amino-5′-deoxythymidine. A solution of 5′-azido-2′-deoxy-thymidine (**12**, 4 mmol) in 30 ml of methanol containing 10% Pd/C is shaken under 50 psi of hydrogen for 5 hr. The catalyst is removed by filtration and the filtrate is then concentrated to dryness via evaporation. The resulting crude product is purified by column chromatography (silica gel; CH$_3$OH–CHCl$_3$, 1 : 7) to yield **13** (90% yield). ^{1}H NMR DMSO, δ 1.78 (s, 3H, 5-methyl), 2.08–2.18 (m, 2H, C2′), 2.75 (s, 2H, amine), 3.37 (s, 2H, C5′), 3.75 (q, 1H, C4′), 4.20 (m, 1H, C3′), 5.20 (s, 1H, 3′-OH), 6.20 (t, 1H, C1′), 7.68 (s, 1H, C6). FAB-MS calculated for C$_{10}$H$_{15}$N$_3$O$_4$[M]$^+$ 241; found (M + H) 242.

Compound 14: N-(5′-Amino-5′-deoxythymidine)-3-phenothiazin-10-yl-propi-onamide. A solution of 3-phenothiazin-10-yl-propionic acid[44] (PTZ-acid, 1 mmol) and carbonyldiimidazole CDI (1.5 mmol) in 8 ml of dry DMF is stirred under nitrogen for 1 hr at 25°. The mixture is then diluted with 17 ml of dry THF, followed by addition of LH-20 resin (0.43 g, excess) to quench the excess CDI. One hour later, LH-20 is removed and **13** (1.0 mmol) is added to the reaction mixture. After stirring for 12 hr, the reaction is stopped and the solvents are removed. Column chromatagraphy (silica gel; CH$_3$OH CHCl$_3$, 1 : 9) yields a white solid (81% yield). ^{1}H NMR DMSO, δ 1.72 (s, 3H, 5-methyl), 2.03 (m, 2H, C2′), 2.55 (t, 2H, PTZ N-CH$_2$), 3.28 (m, 2H, C5′), 3.68 (m, 1H, C4′), 4.07 (m, 3H, carbonyl-CH$_2$ and C3′), 5.24 (d, 1H, 3′-OH), 6.08 (t, 2H, C1′), 6.87–7.16 (m, 8H, aromatic H of PTZ), 7.42 (s, 1H, C6), 8.08 (t, 1H, amide NH), 11.25 (s, 1H, C3). FAB-HRMS calculated for C$_{25}$H$_{26}$N$_4$O$_5$S 494.1618; found 494.1621.

Compound 15: N-(5′-Amino-3′-cyanoethoxydiisopropylaminophosphine-5′-deoxythymidine)-3-phenothiazin-10-yl-propionamide. 2-Cyanoethyl-N,N′-diiso-propylchlorophosphoramidite (0.6 mmol) is added to a solution of **14** (0.4 mmol) in 20 ml of dry CH$_3$CN containing diisopropylethylamine (0.31 ml). The reaction mixture is stirred under nitrogen for 2 hr. The solvent is then removed. The solid is rinsed with hexane and checked by ^{31}P-NMR (CDCl$_3$): δ 148.3 and 148.7 ppm. TLC: >95% yield.

Phosphate Labeling

The most common site for labeling an oligodeoxynucleotide is the 3′- or 5′-terminus.[11,42,43,64] Because an oligodeoxynucleotide terminus contains a free hydroxyl group, the phosphoramidite approach is well suited. As shown in Scheme 4, starting with commercially available 9,10-anthraquinone-2-carboxylic acid (**16**), the anthraquinone phosphoramidite is prepared in three steps.[43]

Compound 17: N-[2-(tert-Butyldiphenylsiloxy)-ethyl]-9,10-anthraquinone-2-carboxamide. 9,10-Anthraquinone-2-carboxylic acid (**16**, 3 mmol) is suspended in CH$_2$Cl$_2$ (20 ml). Oxalyl chloride (0.15 mmol) is added, followed by DMF (0.1 ml).

[64] S. M. Gasper, B. Armitage, X. Shui, G. G. Hu, G. B. Schuster, and L. D. Williams, *J. Am. Chem. Soc.* **120,** 12402 (1998).

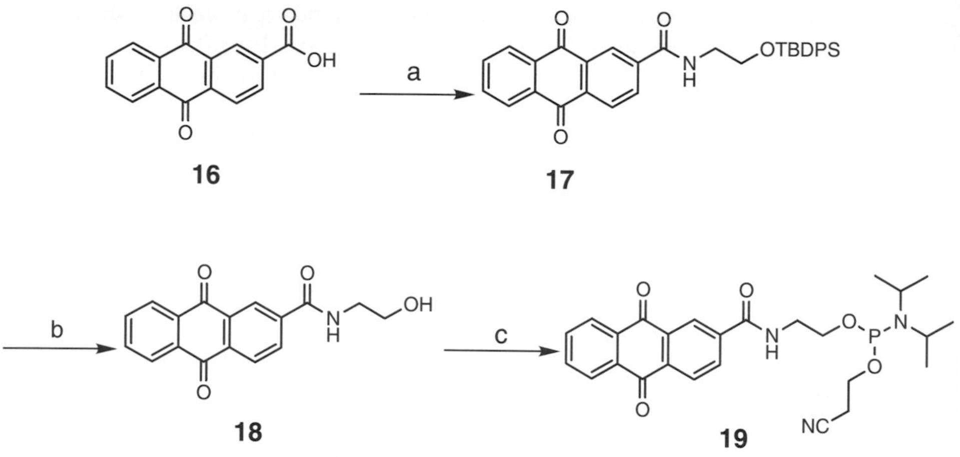

SCHEME 4. Synthesis of the anthraquinone phosphoramidite. Reagents and conditions: (a) i: (COCl)$_2$, DMF (cat.), CH$_2$Cl$_2$, 25°, 1 hr; ii: Cl · H$_3$NCH$_2$CH$_2$OTBDPS, CH$_2$Cl$_2$, DIPEA, −5 to 25°; (b) TBA$^+$F$^-$, THF; (c) ClP(iPr$_2$N)(OCH$_2$CH$_2$CN), DIPEA, CH$_2$Cl$_2$, 25°, 2 hr.

Immediate formation of gas is observed and the reaction is stirred until a yellow homogeneous solution is observed. The mixture is stirred for an additional 1 hr. Next, the volatiles are removed under high vacuum. The yellow residue is dissolved in CH$_2$Cl$_2$ (25 ml) and cooled to −5°. 2-(*tert*-Butyldiphenylsiloxy)aminoethane · HCl (3.5 mmol) is added followed by DIEA (10 mmol). The mixture is diluted with CH$_2$Cl$_2$ (50 ml) and washed with equal portions of water, 10% aqueous NaCO$_3$, water, and brine. Removal of the solvents followed by column chromatography (silica gel, 1% EtOH in CHCl$_3$) gives **17** as a yellow solid (87 % yield). ^{1}H NMR (CDCl$_3$) 8.56 (d, 1H), 8.35 (m, 3H), 8.15 (dd, 1H), 7.82 (m, 2H), 7.70 (m, 1H), 7.63 (m, 4H), 7.40 (m, 6H), 6.68 (bt, 1H), 3.89 (t, 2H), 3.64 (q, 2H), 1.08 (s, 9H). FAB-MS calculated for C$_{33}$H$_{31}$NO$_4$Si [M] 533; found [M]$^-$ 533.1. [Reprinted from M. T. Tierney and M. W. Grinstaff, *J. Org. Chem.* **65**, 5355 (2000), with permission. Copyright 2000 ACS.]

Compound **18**: *N-(2-Hydroxyethyl)-9,10-anthraquinone-2-carboxamide.* *N*-[2-(*tert*-butyldiphenylsiloxy)-ethyl]-9,10-anthraquinone-2-carboxamide (3 mmol), **17**, is dissolved in THF (50 ml) and tetrabutyl ammonium fluoride is added in small portions until TLC shows complete consumption of the starting material. The mixture is diluted with ether and the addition of a small amount of aqueous acetic acid (10 ml) is followed by crystallization of a yellow solid. This solid is filtered, washed with water and isopropyl alcohol, and extensively dried under vacuum (65% yield). **18** ^{1}H NMR (DMSO-d$_6$–CDCl$_3$, 1 : 1) 8.87 (bt, 1H), 8.57 (s, 1H), 8.28 (m, 1H), 8.18 (m, 3H), 7.88 (m, 2H), 4.76 (t, 1H), 3.51 (q, 2H), 3.34 (q, 2H). HR-FAB-MS calculated for C$_{17}$H$_{14}$NO$_4$ *m/z* 296.0923 (M + H)$^+$; found 296.0919.

Compound 19: N-[2-(β-Cyanoethyl-N',N''-diisopropylphosphino)ethyl]-9,10-anthraquinone-2-carboxamide. 2-Cyanoethyl-*N*,*N*'-diisopropylcholophosphoramidite (0.6 mmol) is added to a solution of **18** (0.4 mmol) in 20 ml of dry CH$_3$CN containing diisopropylethylamine (0.31 ml). The reaction mixture is stirred under nitrogen for 2 hr. The solvent is then removed. The solid is rinsed with hexane and checked by ^{31}P NMR (CDCl$_3$): ^{31}P NMR 148.7. TLC: >95% yield.

Labeling Oligodeoxynucleotides via Direct Incorporation Synthetic Approach

In the previous section, the redox probes are incorporated in oligodeoxynucleotides using redox-labeled phosphoramidites during DNA synthesis. Although this approach is widely used, when large, chemically complex, or redox-active phosphoramidites are incorporated the synthetic reactions and isolation procedures can be challenging and thus lead to low yields. The incorporation of a redox probe can also be accomplished by labeling the solid support-bound oligodeoxynucleotide.[14] Scheme 5 shows the direct (on-column) incorporation approach for labeling an oligodeoxynucleotide with ferrocene.[47] Standard DNA synthesis is performed until incorporation of the 5'-DMT-3'-cyanoethyl-*N*,*N*'-diisopropylphosphoramidite-2'-deoxy-5-iodouridine. Next, the synthesis is stopped without deprotecting the 5'-hydroxyl or cleaving the oligodeoxynucleotides from the resin. The column is removed from the synthesizer, attached to a syringe, and the alkyne-derivatized ferrocene, Pd(Ph$_3$P)$_4$, CuI, and 150 μl of dry DMF–Et$_3$N (3.5 : 1.5) are added. After 3 hr, the column is rinsed with 10 ml of THF–Et$_3$N (9 : 1) and 40 ml of acetonitrile, dried under N$_2$ for 0.5 hr, and reinstalled on the synthesizer. DNA synthesis is resumed until the final oligodeoxynucleotide product is prepared. Depending on the organic (biotin; amines, PTZ, AQ) or inorganic [Fc, Ru(diimine)$_3$$^{2+}$] probes used, the yields vary from 40 to 95%.[14,41,51,65]

Purification

After synthesis, the oligodeoxynucleotides are deprotected at 55° in NH$_4$OH for 12 hr. Purification of these synthetic oligodeoxynucleotides is accomplished by either high-performance liquid chromatography (HPLC) or polyacrylamide gel electrophoresis (PAGE).

High-Performance Liquid Chromatography

Reverse-phase chromatography is generally used to isolate and identify a particular synthetic oligodeoxynucleotide. In reversed-phase HPLC, the crude

[65] S. I. Khan, A. E. Beilstein, M. T. Tierney, M. Sykora, and M. W. Grinstaff, *Inorg. Chem.* **38**, 5999 (1999).

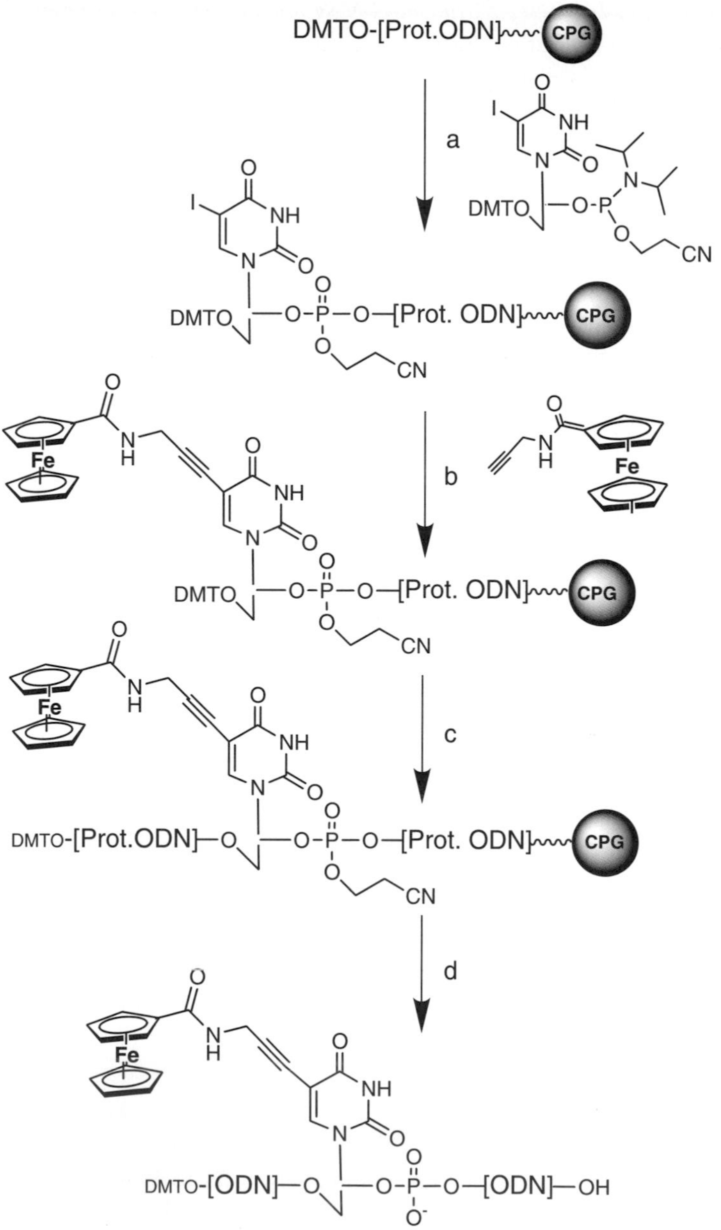

SCHEME 5. On-column synthesis of ferrocene-labeled oligodeoxynucleotides. Reagents and conditions: (a) DNA synthesis; (b) off-synthesizer, Pd(Ph₃P)₄ CuI, TEA, DMF, 3 hr; (c) resume DNA synthesis; (d) 30% NH₄OH, 55°, 16 hr; ODA, Oligodeoxynucleotide; Prot., protected.

material undergoes adsorption and desorption processes between the nonpolar stationary phase and polar mobile phase. The hydrophobic interactions between the oligodeoxynucleotides and the stationary or mobile phase determine the retention time. In most cases, functionalized oligodeoxynucleotides are more hydrophobic and thus have a longer retention time than unlabeled oligodeoxynucleotides. Besides purification, HPLC also provides a means to determine the yield.

A typical purification procedure for a labeled oligodeoxynucleotide involves the following steps.

1. Deprotect the synthetic oligodeoxynucleotide (1 μmol) after DNA synthesis with NH$_4$OH at 55° for 12 hr.

2. Lyophilize the sample and redissolve the residue in 100 μl of deionized water.

3. Equilibrate the reversed-phase C$_{18}$ column (semipreparative) with 95% TEAA buffer and 5% acetonitrile.

4. Inject 20–30 μl of sample and elute with a gradient of 5–50% acetonitrile over 30–40 min.

5. Collect the desired product as it elutes from the column by monitoring the oligodeoxynucleotide absorbance at 254 nm.

Polyacrylamide Gel Electrophoresis

Polyacrylamide gel electrophoresis (PAGE) is routinely used to separate and purify synthetic oligodeoxynucleotides.[66] DNA possesses a negative phosphate backbone and consequently orients and migrates in an electric field. The electrophoretic mobility of single-stranded or double-stranded DNA is closely related to its chain length. The larger the DNA, the slower it migrates. The redox-labeled oligodeoxynucleotide typically moves more slowly compared with an unlabeled oligodeoxynucleotide. The effective range of separation depends on the different concentrations of polyacrylamide gel. For example, a denaturing 20% acrylamide gel containing 7–8 M urea is able to separate oligodeoxynucleotides whose lengths differ by as little as 0.2% (1 in 500 bases).

A typical procedure is outlined below.

1. Deprotect the synthetic oligodeoxynucleotide (1 μmol) after DNA synthesis with NH$_4$OH at 55° for 12 hr.

2. Lyophilize the sample and redissolve the residue in a solution of 20 μl of deionized water and 40 μl of loading buffer (95% formamide, 20 mM EDTA, 0.25% xylene cyanol).

[66] J. Sambrook, E. F. Fritsch, and T. Maniatis (eds.), "Molecular Cloning: A Laboratory Manual," 2nd Ed. Cold Spring Harbor Laboratory Press, Cold Spring Harbor, NY, 1989.

3. Prepare a 20% denaturing polyacrylamide gel (15 ml of 40% acrylamide–bisacrylamide, 3 ml of 10× TBE, 14.5 g of urea).

4. Apply 5 μl of the sample to each well of the gel and run the gel for 2–3 hr.

5. Visualize the bands under UV light by laying the gel (covered with Saran Wrap) on TLC plates.

6. Slice the desired band, and crush and elute the desired oligodeoxynucleotide with sodium acetate buffer.

Characterization

Matrix-Assisted Laser Desorption Ionization Mass Analysis

Purified synthetic oligodeoxynucleotides are often analyzed by negative matrix-assisted laser desorption ionization (MALDI) mass spectrometry to confirm the modification.[67,68]

Spectroscopic Analysis

Electronic absorption can be used to determine DNA concentration. All the nucleobases (A, G, C, and T) possess extensive π electron systems, which give rise to intense $\pi-\pi^*$ transitions that are observed in the electronic absorption spectra at ~250 nm. Because of base interactions, the absorbance of an oligodeoxynucleotide is not the direct sum of the individual base absorbances. In fact, the absorption spectrum for an oligodeoxynucleotide is a consequence of base composition and base interactions. To obtain an accurate extinction coefficient of an oligodeoxynucleotide, nuclease P1 (*Penicillium citrinum*) is used to cleave the entire oligodeoxynucleotide to its nucleotide components and thus eliminate all the A, C, G, and T π-stacking interactions.[46] Once the oligodeoxynucleotide extinction coefficient is known, the sample concentration can be determined from Beer's Law.

The procedure for enzyme digestion and concentration determination is as follows.

1. Redissolve the pellet of HPLC-purified oligodeoxynucleotide in 100 μl of deionized water.

2. Remove 5 μl of the oligodeoxynucleotide solution and add this sample to a microcentrifuge tube containing 1 μl of a nuclease P1 solution (1 mg/ml, 20 mM NaOAc, pH 5.5) and 44 μl of sodium acetate buffer (20 mM, pH 5.5).

3. Incubate the sample at 60° for 0.5 hr.

4. Cool the sample to room temperature, dilute 10 μl of the sample to 1 ml with water, and measure the UV–visible (UV–Vis) absorbance.

[67] K. J. Wu, A. Steding, and C. H. Becker, *Rapid Commun. Mass Spectrom.* **7**, 142 (1993).
[68] M. L. Gross and L. K. Zhang, *J. Am. Soc. Mass Spectrom.* **11**, 854 (2000).

5. Determine the concentration [$A = \varepsilon bc$, where c is the concentration, A is the absorbance, ε is the sum of the absorption coefficients ($A = 15,300 \text{ cm}^2 \text{ mol}^{-1}$, $C = 7400 \text{ cm}^2 \text{ mol}^{-1}$, $G = 11,800 \text{ cm}^2 \text{ mol}^{-1}$, and $T = 9300 \text{ cm}^2 \text{ mol}^{-1}$), and b is the path length].

Melting Temperature

Electronic absorption spectroscopy can be used to monitor the denaturation and renaturation of double-stranded DNA. On denaturation, double-stranded DNA separates into single-stranded DNAs, with an absorption increase of approximately 40%. This increase in absorbance occurs over a narrow temperature range, with a sigmoidal curve indicating that the denaturation process is a cooperative event in which the collapse of one section of the structure destabilizes the remainder. The melting temperature (T_m) is defined as the temperature at which the maximum absorbance change (dA/dT) occurs. Longer DNA duplexes have higher melting temperatures and more stable secondary structures. The DNA sequence also influences the melting temperature, as a duplex possessing a large number of GC pairs will have a higher melting temperature compared with a duplex rich in AT pairs. The ionic strength of the solution also is known to affect the melting temperature. The melting temperature of labeled oligodeoxynucleotide duplexes is also dependent on the location of the redox probes on DNA and the interaction between the probe and DNA structure. The T_m can be, for example, higher if the probe is intercalated or lower if it is tethered to the base.

To form a duplex from two complementary strands, combine equal amounts of DNA strands in buffer solution (5 mM NaH$_2$PO$_4$, 50 mM NaCl, pH 7) and heat the resulting solution at 80° for 2–3 min, and then slowly cool the sample to room temperature. The melting curve profile is obtained with the following parameters on a Hewlett-Packard UV–Vis spectrophotometer: monitoring wavelength, 254 nm; temperature range, 20–70°; rate of temperature change, 0.5°/min.

Circular Dichroism Spectroscopy

Dichroism is the phenomenon in which light absorption changes for different directions of polarization. Circular dichroism (CD) refers to the absorption of the two different types of circularly polarized light.[69] Nucleic acids have an intrinsic asymmetry because a chiral sugar is present within the structure. The formation of double-stranded oligodeoxynucleotides into a helical structure presents an asymmetry, which gives rise to strong interactions between the chromophore bases, generating an intense CD spectrum. As a result, CD spectra provide information about DNA secondary structure as A-form and B-form DNA (10.4 and 10.2), and Z-form DNA possess unique CD signatures. For example, the CD spectrum of

[69] W. C. Johnson, "CD of Nucleic Acids." VCH, New York, 1994.

B-form (10.4) DNA contains positive CD features at 280 nm and 190 nm, and a negative feature at 240 nm.

Differential Scanning Calorimetry

Differential scanning calorimetry (DSC) is another valuable technique to characterize the DNA melting transition. During the melting process, DNA strands change from a fully duplexed state to a fully separated single-strand state, and heat is absorbed as a result of this transition. Likewise, during hybridization, heat is generated. This absolute amount of heat change presents a direct way to determine the thermal stability of a labeled DNA duplex, and allows for the comparison of this DSC model-independent enthalpy determination with the model-dependent optical experiment method.[70,71] In conventional DSC, the difference in heat flow between a sample and an inert reference is measured as a function of time and temperature. During the experiment, the change in heat is recorded as the temperature is increased and the area under this curve corresponds to the transition enthalpy.

A Microcal DSC instrument with a cell size of 0.51 ml is used for the DNA thermal analysis experiments. In a typical experiment, both the sample and reference cells are filled with buffer (5 mM NaH$_2$PO$_4$, 50 mM NaCl, pH 7) first and the cells are scanned from 10 to 80° repeatedly at 0.5°/min. Once reproducible up-and-down baseline scans are obtained, the double-stranded oligodeoxynucleotide solution (35 μM concentration) is loaded in the sample cell, leaving the reference cell unchanged. Again, the sample and control chambers are scanned from low to high temperature at a constant rate of 0.5°/min. The resulting curve is subtracted from the baseline and then normalized to concentration to afford the corrected heat capacity-versus-temperature curve. Finally, the area underneath this curve is calculated, and this enthalpy change can be used to compare the stability of labeled versus unlabeled DNA duplexes.

DNA Redox Reactions

Methods to study oxidative damage in DNA may lead to a better understanding of DNA damage *in vivo* as well as provide insights about the mechanism(s) of charge transfer through different media. Currently, the effects of distance, sequence, structure, and energetics on reaction rate and product(s) formation are being investigated. One method to initiate a redox reaction in DNA and study

[70] L. A. Marky and K. J. Breslauer, *Biopolymers* **26,** 1601 (1987).
[71] C. A. Gelfand, G. E. Plum, A. P. Grollman, F. Johnson, and K. J. Breslauer, *Biochemistry* **37,** 7321 (1998).

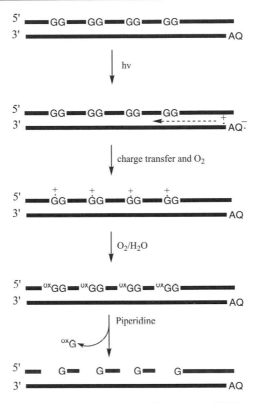

SCHEME 6. Photoinduced oxidative damage of DNA.

guanine oxidation is to use a photoinduced trigger. A number of organic[72-76] and inorganic[18,21,23,55,77] triggers covalently attached or intercalated in DNA are being explored.

Henderson *et al.* describe the use of an anthraquinone (AQ)-linked DNA duplex containing four separate GG blocks to probe long-range guanine oxidation.[25] As illustrated in Scheme 6, irradiation of the tethered anthraquinone at 350 nm

[72] I. Saito, T. Nakamura, K. Nakatani, Y. Yoshioka, K. Yamaguchi, and H. Sugiyama, *J. Am. Chem. Soc.* **120,** 12686 (1998).

[73] D. B. Hall, S. O. Kelly, and J. K. Barton, *Biochemistry* **37,** 15933 (1998).

[74] E. Meggers, M. E. Michel-Beyerle, and B. Giese, *J. Am. Chem. Soc.* **120,** 12950 (1998).

[75] E. Meggers, D. Kusch, M. Spichty, U. Wille, and B. Giese, *Angew. Chem. Int. Ed.* **37,** 460 (1998).

[76] B. Giese, S. Wessely, M. Spormann, U. Lindermann, E. Meggers, and M. E. Michel-Beyerle, *Angew. Chem. Int. Ed.* **38,** 996 (1999).

[77] D. B. Hall and J. K. Barton, *J. Am. Chem. Soc.* **119,** 5045 (1997).

generates anthraquinone radical anion and an adjacent base radical cation.[25,78,79] This radical cation then migrates through DNA until it is trapped by the 5′-guanine of a GG block to yield an 8-oxoguanine in the presence of O_2/H_2O. Subsequent hot piperidine treatment cleaves DNA at the oxidized guanine site and the resulting shorter DNA strands are analyzed by polyacrylamide gel electrophoresis to compare GG cleavage efficiency with different distances. Long-range charge transfer can occur over distances of approximately 200 Å.

Conclusions

Using the procedures described above, redox probes can be covalently attached to the nucleobase, ribose, or phosphate moiety of DNA. The synthetic procedures are amenable to both inorganic and organic redox probes. The site-specifically labeled ssDNA and dsDNA can be characterized by a number of techniques including mass spectroscopy, gel electrophoresis, HPLC, melting temperature, CD, and DSC to obtain information about identity, purity, duplex solution structure, and duplex thermodynamics. Redox probe-labeled oligodeoxynucleotides provide a unique means to systematically study charge-transfer reactions in DNA.

Acknowledgments

This work was supported in part by the Army Office of Research. M.W.G. also thanks the Pew Foundation for a Pew Scholarship in the Biomedical Sciences, the Alfred P. Sloan Foundation for a Research Fellowship, and the Dreyfus Foundation for a Camille Dreyfus Teacher Scholarship.

[78] D. T. Breslin and G. B. Schuster, *J. Am. Chem. Soc.* **118**, 2311 (1996).
[79] B. Armitage, *Chem. Rev.* **98**, 1171 (1998).

[47] Expressed Sequence Tag Database Screening for Identification of Human Genes

By AGNÈS RÖTIG, ARNOLD MUNNICH, and PIERRE RUSTIN

Introduction

Because the identification of human genes makes possible a better understanding of physiological and pathological processes, it represents one of the main goals of human geneticists. It has long been true that working at the bench was the only way to reach this goal, but more recently the enormous amount of data resulting from the human sequence project and the development of computing capabilities

Copyright 2002, Elsevier Science (USA).
All rights reserved.
0076-6879/02 $35.00

have revolutionized the field. New approaches and new tools are now routinely used for gene identification. A first useful resource is the development of the Expressed Sequence Tags database (EST; as of January 20, 2001 there were more than 2.9×10^6 ESTs), generated from a large variety of human tissues and representing a large number of human genes. On the other hand, the availability of the complete human genome sequence will presumably boost the identification of a large number of human genes. Indeed, a huge amount of information about the human genome sequence is now available in several databases and can be readily used for the identification of human genes. Finally, the increasing number of online computing programs allows these sequence databases to be easily exploited. Nevertheless, it is worth remembering that a nucleotide sequence picked up in a human sequence database is generally not sufficient per se to determine a cDNA or gene sequence and that further experimental work must be carried out for each particular gene.

When possible, the quickest approach to identify a human gene is often to use cross-species comparison by computer analysis. Because a large number of biochemical functions or metabolic pathways have been conserved during evolution, protein and/or nucleotide sequences are often, in part, highly conserved. Thus the knowledge acquired about model organisms such as the yeast *Saccharomyces cerevisiae*, the fly *Drosophila melanogaster*, or the nematode *Caenorhabditis elegans*, is of great help in identifying genes in other species, especially in humans. Such *in silico* cloning, allowing the identification of a yet unknown human gene, takes only a few hours compared with tedious and day-consuming library screening or functional complementation. After identifying part of the searched sequence, experimental work is still to be done to check this *in silico* cloning and reconstitute the complete sequence.

We describe below the different steps involved in *in silico* cloning, using several databases and programs. A standard personal computer connected to the Internet is the only required material. This strategy is facilitated by the use of a molecular biology server such as the Deambulum server[1] which allows rapid connection to several databases and sequence analysis programs (Fig. 1). Almost all the databases and programs mentioned below can be easily accessed through this server. Alternatively, each database can be accessed by entering their particular location (Web address indicated as footnotes), using an Internet navigator.

Identification of Human Gene by Sequence Homology

The step-by-step procedure described below allows identification of the human gene encoding the counterpart of a protein already known in another organism. We have chosen the yeast *MRS4* gene, encoding a splicing protein, a member

[1] Deambulum: http://www.infobiogen.fr/services/deambulum/fr/

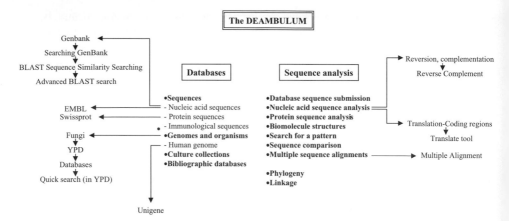

FIG. 1. Presentation of the Deambulum server and its links with various molecular biology servers referred to in this chapter.

of the mitochondrial carrier family (MCF) that suppresses mitochondrial splicing defects. The human counterpart of this gene is unknown as of January 2001.

Because the amino acid sequences have been better conserved than nucleotide sequences through evolution, it is more efficient to start with the yeast protein sequence to identify the human nucleotide sequence. The yeast protein sequence encoded by *MRS4* can be retrieved from various databases, for example, GenBank,[2] EMBL,[3] SWISS-PROT,[4] and Yeast Proteome Database (YPD).[5] Using this last resource, the amino acid sequence is retrieved by (1) clicking on *Quick search* for YPD (*Saccharomyces cerevisiae*), (2) entering the acronym MRS4 in the *Gene names* section and clicking on *Submit,* and (3) selecting on the next screen the retrieved result, MRS4 (in hypertext format), which leads to the general description of the gene and the protein. Then, in the *Sequence* section of the scrolling screen, select *See protein sequence* (hypertext format). The sequence displayed on the next screen is selected (no need to eliminate the amino acid numbering) and copied [Ctrl-C], and this particular window is closed.

This amino acid sequence is used to identify human ESTs that could represent part of the human cDNA encoding the MRS4 human protein homolog (Fig. 2). This is performed by using the Basic Local Alignment Search Tool (BLAST), that is, BLAST sequence similarity searching.[6] Different kinds of analyses can be performed with this program. The one to be used for our purpose is the Translated BLAST Searches and the program to be chosen is tblastn (click on

---

[2] GenBank: http://www.ncbi.nlm.nih.gov/Genbank/GenbankOverview.html

[3] EMBL: http://www.ebi.ac.uk/embl/index.html

[4] SWISS-PROT: http://www.expasy.ch/sprot/

[5] YPD: http://www.proteome.com/databases/index.html

[6] BLAST: http://www.ncbi.nlm.nih.gov/BLAST/

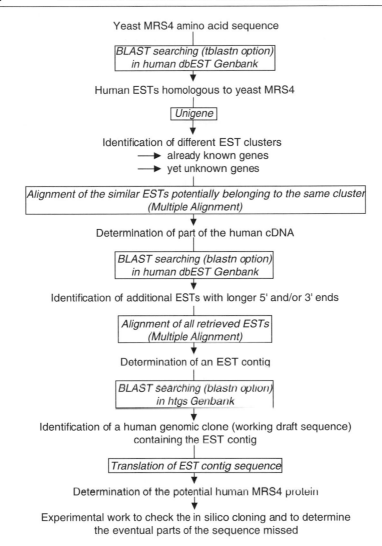

Yeast MRS4 amino acid sequence

BLAST searching (tblastn option)
in human dbEST Genbank

Human ESTs homologous to yeast MRS4

Unigene

Identification of different EST clusters
→ already known genes
→ yet unknown genes

Alignment of the similar ESTs potentially belonging to the same cluster
(Multiple Alignment)

Determination of part of the human cDNA

BLAST searching (blastn option)
in human dbEST Genbank

Identification of additional ESTs with longer 5' and/or 3' ends

Alignment of all retrieved ESTs
(Multiple Alignment)

Determination of an EST contig

BLAST searching (blastn option)
in htgs Genbank

Identification of a human genomic clone (working draft sequence)
containing the EST contig

Translation of EST contig sequence

Determination of the potential human MRS4 protein

Experimental work to check the in silico cloning and to determine
the eventual parts of the sequence missed

FIG. 2. Schematic of the method used for human MRS4 sequence retrieval through homology with yeast MRS4 protein.

Protein query—Translated db [tblastn] in hypertext format) which compares a protein query sequence against a nucleotide sequence database dynamically translated in all reading frames. The memorized amino acid sequence previously retrieved from the *YPD* database should be copied and pasted [CtrlV] in the query window. It would be also possible to only enter the accession number or GI of the protein sequence. The database to be used is *est_human*. The result of the search is obtained by clicking on *Blast!* and then on *Format!* on the next screen. It appears as a scheme showing the various retrieved human ESTs with lines of different colors, each color corresponding to a homology score (red for score ≥200, pink for score between 80 and 200, green for score between 50 and 80, etc.). The list of the human ESTs is then presented with the score and the *E* value given for each alignment. The meaning of all these variables is explained in the introductory pages of the BLAST program. The alignments of the query sequence with all translated ESTs are shown in order of decreasing homology scores. Several indications help determine the likelihood of having really retrieved ESTs corresponding to the human counterpart of the protein of interest (in our case MRS4 protein): the score, the *E* value, the identities, and the gaps between the query sequence and the retrieved EST (Fig. 3). Frame = +1 indicates that the EST sequence has been translated in the first 5′ frame. As EST sequences are "single-pass" sequences, they are prone to contain errors. Alignment of the yeast protein sequence with the human protein sequence obtained using the six possible reading frames allows the identification of homology despite potential errors. Thus, in the chosen example, MRS4, two blocks of homology are identified for two reading frames. The sequences presenting the

```
AV704087 ADB Homo sapiens cDNA clone ADBAFE05 5'.
          Length = 681

 Score =  104 bits (259), Expect(2) = 7e-39
 Identities = 53/114 (46%), Positives = 69/114 (60%), Gaps = 2/114 (1%)
 Frame = +1

Query: 120  PMKTALSGTIATIAADALMNPFDTVKQRLQLDTNL--RVWNVTKQIYQNEGFAAFYYSYP 177
            P    + +G +AT+  DA MNP + VKQR+Q+  +     RV  +  + ++QNEG  AFY SY
Sbjct: 334  PATKSAAGCVATLLHDAAMNPAEVVKQRMQMYNSPYHRVTDCVRAVWQNEGAGAFYRSYT 513

Query: 178  TTLAMNIPFAAFNFMIYESASKFFNPQNSYNPLIHCLCGGISGATCAALTTPLD 231
            T L MN+PF A +FM YE    +  NPQ  YNP  H L G  +GA  A  TTPLD
Sbjct: 514  TQLTMNVPFQAIHFMTYEFLQEHXNPQRLYNPSSHVLSGASAGAVAARATTPLD 675


 Score = 79.7 bits (195), Expect(2) = 7e-39
 Identities = 41/94 (43%), Positives = 60/94 (63%), Gaps = 1/94 (1%)
 Frame = +3

Query: 13   DYEALPSHAPLHSQLLAGAFAGIMEHSLMFPIDALKTRVQAAGLNKAAS-TGMISQISKI 71
            DYEALP+ A + +  ++AGA AGI+EH +M+PID +KTR+Q+   + AA   ++   + +I
Sbjct: 69   DYEALPAGATVTTHMVAGAEAGILEHCVMYPIDCVKTRMQSLQPDPAARYRNVLEALWRI 248

Query: 72   STMEGSMALWKGVQSVILGAGPAHAVYFGTYEFC 105    ◀── yeast MRS4
             EG    +G+     GAGPAHA+YF  YE C          ◀── conserved amino acids
Sbjct: 249  IRTEGLWRPMRGLNVTATGAGPAHALYFACYEKC 350    ◀── human retrieved EST
```

FIG. 3. Result of a BLAST search with the yeast MRS4 amino acid sequence. Alignment of the yeast and human sequences identified two blocks of homology corresponding to two different reading frames (amino acids 120–231 for frame +1 and amino acids 13–105 for frame +3). Dashes indicate the gaps between the query and the retrieved sequence.

highest E values, with high identity and a low percentage of gaps, are obviously the most likely candidates. It should be kept in mind, however, that high E values can be associated with genes encoding proteins of similar function, the highest score not always corresponding to the gene of interest.

A simple way to discriminate between two or more eventual genes encoding different proteins of the same family is to check the origin of each EST. This can be done with the UniGene server,[7] which consists of a collection of human sequences, defined as clusters representing a unique gene with its map location and the corresponding ESTs. In our example, we search UniGene with the AV704087 EST, presenting the highest homology score ($E = 7 \times 10^{-39}$) with the yeast MRS4 protein sequence. The corresponding UniGene cluster (Hs.326104) corresponds to MRS3/4 putative mitochondrial solute carrier, whereas the UniGene cluster Hs.300496 (mitochondrial solute carrier), which can be retrieved with the AI133696 EST, with a similar high homology score ($E = 1 \times 10^{-33}$), does correspond to another gene encoding a mitochondrial solute carrier. As this last EST corresponds to an already identified gene that is not the human MRS4 counterpart, AV704087 seems to be a better candidate to represent the human MRS4 counterpart. A similar search should be done for all retrieved EST sequences in order to identify their origin. In our example, 12 of the 20 first ESTs correspond to Hs.326104 and 8 to Hs.300496 UniGene clusters.

Alignments of ESTs similar to AV704087 and belonging to the Hs.326104 Unigene cluster, using the Multiple Alignment program,[8] are used to confirm that they all correspond to the same gene. In our example, EST AV704087 is one of the largest ESTs (681 bp) but represents only part of the human cDNA. To possibly identify the 5' and 3' parts of this sequence, AV704087 is used as a template to identify additional human ESTs. Such identification can be done with the BLAST Search program,[6] using the *blastn* option, which compares a nucleotide query sequence against a nucleotide sequence database, the database to be used being est_human. Several additional ESTs are thus identified that share high homology with the 3'-half sequence of AV704087. The first 15 ESTs overlap and are highly similar, some of them having a longer 3' end. Some of these ESTs are in the opposite direction (indicated by plus/minus). In this case, the sequence can be reverted by using the Reversion Complementation program from BCM Search Launcher.[9] Finally, the alignment of all the sequences, using the Multiple Alignment program,[8] allows construction of an EST contig. In our example, one of the ESTs (AA743110) appears to present a poly(A) tail, indicating that the complete 3' part of the cDNA should be included in this 1255-bp-long contig.

[7] UniGene: http://www.ncbi.nlm.nih.gov/UniGene/index.html
[8] Multiple alignment: http://www.genebee.msu.su/services/malign_reduced.html
[9] Reverse complement: http://dot.imgen.bcm.tmc.edu:9331/seq-util/seq-util.html

An EST contig possibly contains several errors, as the EST sequences are only "single-pass" cDNA sequences. These sequence errors can often be corrected after comparing the EST contig with genomic sequences provided by the complete sequence of the human genome that is available in GenBank. Searching for a human genomic clone containing an EST contig can be performed by a BLAST search in the High-Throughput Genome Sequence (htgs) database. In our example, this identifies the RP11-85A1 clone (GenBank accession number AC007643, a working draft sequence) as containing the totality of the EST contig. In addition, this BLAST search also indicates the location in the genomic clone of the different exons of the EST contig. Mismatches or gaps in the alignment of the two sequences permit restoration of the most probable sequence.

The nucleotide sequence of the EST contig can now be translated into an amino acid sequence according to the three possible 5′ and three possible 3′ frames using the Translate Tool program.[10] As in our example, the first BLAST search with the yeast MRS4 protein indicates that the AV704087 EST is 5′–3′ oriented (Fig. 3), and thus only the three translations in the 5′ frame must be considered. One of these amino acid sequences should contain the small amino acid sequences homologous to the yeast protein initially identified by the BLAST search (Fig. 3). In our example, the alignment of this incomplete amino acid sequence with the yeast MRS4 protein, using the Multiple Alignment program, reveals a 45.8% homology between the two sequences, with few gaps and obvious consensus sequence blocks in several regions of the proteins, including the C-terminal part (Fig. 4). The total time required to obtain this amino acid sequence is less than 10 hr of computing. However, in our example, the 5′ part of the EST contig and consequently the N-terminal part of the protein sequence are missing. At this point, experimental work needs to be started to identify the 5′ part of the human MRS4 cDNA. The most efficient way would be to perform 5′ RACE (rapid amplification of cDNA ends) on poly(A)$^+$ RNA. This should result in the identification of the ATG translation initiation codon. However, this search was done in December 2000; other ESTs can probably be retrieved when reading this chapter, some of them possibly containing the 5′ part of the cDNA.

This procedure, which allows us to identify this EST contig, consists of performing a BLAST search in the human EST database. Another approach could be to perform this BLAST search in the nr database, instead of Human ESTs, which contains well-defined sequences, that is, complete gene, cDNA, or protein sequences from GenBank but no ESTs. This has been systematically performed for all yeast protein sequences reported in YPD.[5] Indeed, information found in YPD about MRS4 indicated, in the Related Proteins section, that the yeast MRS4 presents several related proteins from different species, including human. Details of these human related proteins show that the first three human related proteins

---

[10] Translate tool: http://www.expasy.ch/tools/dna.html

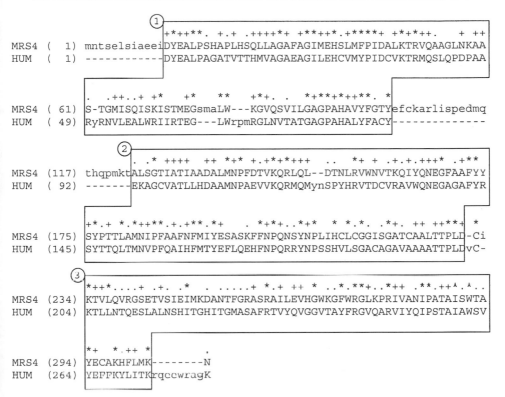

```
                           +*++**. +.+ .++++*+ +**++*.+*****+ +*+*++.    + ++
MRS4  (   1) mntselsiaeeiDYEALPSHAPLHSQLLAGAFAGIMEHSLMFPIDALKTRVQAAGLNKAA
HUM   (   1) ------------DYEALPAGATVTTHMVAGAEAGILEHCVMYPIDCVKTRMQSLQPDPAA

             .  .++..+ +*    +*    **    +*+. .  *+***+*++**. *
MRS4  (  61) S-TGMISQISKISTMEGsmaLW---KGVQSVILGAGPAHAVYFGTYefckarlispedmq
HUM   (  49) RyRNVLEALWRIIRTEG---LWrpmRGLNVTATGAGPAHALYFACY-------------

             .  .* ++++  ++ *+* +.+*+*+++  ..   *+ + .+.+.+++* .+**
MRS4  ( 117) thqpmktALSGTIATIAADALMNPFDTVKQRLQL--DTNLRVWNVTKQIYQNEGFAAFYY
HUM   (  92) -------EKAGCVATLLHDAAMNPAEVVKQRMQMynSPYHRVTDCVRAVWQNEGAGAFYR

             +*.+ *.*++**.+.+**.*+   . *+*+..*+* * *.*. .*+. ++ ++**+ *
MRS4  ( 175) SYPTTLAMNIPFAAFNFMIYESASKFFNPQNSYNPLIHCLCGGISGATCAALTTPLD-Ci
HUM   ( 145) SYTTQLTMNVPFQAIHFMTYEFLQEHFNPQRRYNPSSHVLSGACAGAVAAAATTPLDvC-

             *++*....+ .+. .*  . .....+ *.+ ++ * ..*.**+..*++ .**.++^.^..
MRS4  ( 234) KTVLQVRGSETVSIEIMKDANTFGRASRAILEVHGWKGFWRGLKPRIVANIPATAISWTA
HUM   ( 204) KTLLNTQESLALNSHITGHITGMASAFRTVYQVGGVTAYFRGVQARVIYQIPSTAIAWSV

             *+  *.++ *          .
MRS4  ( 294) YECAKHFLMK--------N
HUM   ( 264) YEFFKYLITKrqccwragK
```

FIG. 4. Comparison of the Multiple Alignment program of the translated human EST contig sequence (HUM) and the yeast MRS4 protein sequence. Boxes indicate the blocks of homology between human and yeast sequences. The two blocks previously identified by the BLAST search (see Fig. 2) are framed again. A third block of homology is also shown. The homology between the human and the whole yeast sequences is 45.8%; it corresponds to 55.6, 53.1, and 43.5% in blocks 1, 2, and 3, respectively.

(retrieved by BLAST searching, using the yeast MRS4 protein) are the HT015 protein (accession number NP_057696), the hypothetical protein FLJ10618 (accession number NP_060625), and the CGI-69 protein (accession number NP_057100), which present, respectively, 60, 41, and 46% similarity and 40, 25, and 27% identity with the yeast MRS4 sequence. The sequence alignments presented in the scrolling screen show a stretch of homology between the yeast MRS4 protein and the human HT015 protein, but this is not observed for the hypothetical FLJ10618 protein or the CGI-69 protein, which should lead one to be cautious with this homology. Nevertheless, these three proteins are mitochondrial carrier proteins, as indicated after clicking in the corresponding GenBank number. Looking for information about the HT015 protein reveals that this protein is the mitochondrial solute carrier (UniGene Hs.300496) that we previously noted in the first step of the BLAST search. Alignment of the yeast MRS4 protein and the human HT015 protein reveals 42.1% homology (not shown), which is slightly less than the homology between the yeast MRS4 and our EST-deduced amino acid sequence. At this step, no other available data allow us to know which of the EST-derived amino acid sequences, or which of the HT015 proteins, is the real MRS4 homolog; only complementation studies in yeast might answer this question. Since December 2000, the complete sequence of the human homolog of yeast MRS4 has been identified (UniGene Cluster Hs.326104, Genbank AAK49519) and matches the partial protein sequence retrieved by the method described here.

Bench work is then absolutely required to check the *in silico* cloning before claiming to have identified a human cDNA. Oligonucleotide primers must be designed from the genomic or EST contig sequence to perform PCR or RT-PCR followed by sequencing in order to correct eventual mismatches and gaps.

In the selected case of MRS4, BLAST searching retrieves several ESTs showing significant homology scores. However, in some cases, no human homologous sequences can be identified. As most ESTs have been obtained by sequencing the 3' part of mRNAs, the 5' part of large mRNAs is often lacking. Consequently, this identification strategy may be inefficient if the C-terminal regions of the known gene from the model organism and its human counterpart present low levels of homology. This can result either from interspecies differences in protein equipment or from loss of particular functions of one given protein during evolution. Another explanation may stem from a low degree of sequence conservation between species. Finally, the absence of a human ortholog in the EST database of a gene identified in a model organism may also be due to a low transcription level of the corresponding gene, as the EST database is a qualitative and quantitative picture of the genome transcriptome.

Identification of Human Gene by Expressed Sequence Tag Database Searching

The EST database can also be exploited when positional cloning does not disclose one or more obvious candidate genes mapping in a genetic interval identified by linkage analysis. However, this approach is not so straightforward as for identification of a human gene by homology sequence with an as yet unknown sequence, as described above. Indeed, the researcher must continuously ply between computing and experimental work (YAC or PAC contig constitution, segregation analysis of polymorphic markers in families, search for transcribed sequences, etc.), which cannot be linearly described.

Author Index

Numbers in parentheses are footnote reference numbers and indicate that an author's work is referred to although the name is not cited in the text.

A

Aaron, C. S., 446
Abadie, C., 322
Abboud, F. M., 209
Abdou, I. M., 550, 551(49)
Abe, H., 3
Abe, K., 371
Abe, Y., 107, 108(26)
Abkevich, V. I., 16
Abola, E. E., 12
Abraham, R. T., 71, 72, 77–80
Abramson, J. J., 241, 244, 244(2), 246(12), 248, 251(12)
Abreu, I. A., 141, 148(19), 149(19)
Abu-Soud, H. M., 119, 120
Adams, C. L., 487
Adams, J. A., 279
Adams, M. W., 70
Adams, M. W. W., 141, 148(16), 149(16), 151(16)
Adams, S. R., 216, 499
Adhya, S., 446
Adjadj, E., 36
Adlington, R. M., 301
Aebersold, R., 177
Aebi, H., 394
Affara, N., 347
Aggarwal, S. K., 213
Agius, D. R., 398, 408(3)
Agopyan, N., 366
Agrawal, R. S., 342, 344(24)
Agrawal, S., 425, 426(27)
Ahamed, B., 216
Ahmed, S. A., 49, 50(16)
Ahn, B. W., 261
Aidoo, A., 446
Aitken, J. F., 187
Aizawa, S., 81, 83, 84(6), 85(6), 86(6)
Akaike, N., 498

Akasaka, M., 463
Alahari, S., 426
Alam, J., 164, 353, 360(12), 375
Alam, M. R., 233, 235(5), 237(5), 238(5)
Albee, L. D., 449, 452(11), 454(11)
Alben, J. O., 187(54), 189
Albers, R., 24
Alefounder, P. R., 123
Alekseyev, V. Y., 508, 509(25)
Alexander, M. Y., 322
Alexander, R. W., 220, 225(1), 263
Alexov, E., 187, 206(14)
Alfsen, A., 187, 190(10), 206(10)
Alhava, E., 342, 344(24)
Ali, S. F., 399
Alkayed, N. J., 375
Allavena, G., 23
Allawi, H. T., 516
Allen, M. H., 58, 59(20)
Allen, M. J., 516, 522(48)
Allen, P. Z., 464
Allen, R. G., 56
Allis, C. D., 288
Allison, W. S., 46
Allocatelli, C. T., 187(34), 188, 206(34)
Almli, L. M., 389, 395
Aloy, P., 10, 16(2)
Altman, R. B., 15
Altschul, S. F., 21
Alwine, J. C., 540
Amarnath, V., 523, 524
Ambros, V., 501
Ames, G. F.-L., 122
Amici, A., 261, 282
Amiconi, G., 187(39; 42; 43), 188
Andersen, J. B., 499
Andersen, J. F., 44
Anderson, C. M., 397
Anderson, J. L., 190, 194(87)
Anderson, M. E., 104, 105, 109(25), 458

575

D

H

P

Z

Subject Index

A

Aconitase, superoxide assay for NADPH oxidase detection, 232–233

Amine oxidase, *see* Copper-containing amine oxidase

Antisense oligodeoxyribonucleotide, *see* NADPH oxidase; Protein kinase C

Ascorbic acid, assay in manganese superoxide dismutase transgenic mice, 405–406

B

Basic Local Alignment Search Tool, identification of human genes, 569, 572–574

B cell

BALB/c mouse plasmacytoma model
genetic susceptibility, 435
induction, 434–435
pristane granuloma role, 435–436
transgenic shuttle vector assays of oxidative mutagenesis in plasmacytomagenesis
comparison of assays, 445–447
λLIZ assay
magnetic cell sorting, 440–441
oxidative mutagenesis in lymphoid tissues and purified B cells, 439–441
principles, 437, 439
pristane treatment, 439–440
overview, 436–437
pUR288 assay
phagocyte-mediated oxidative mutagenesis, 444–445
principles, 442–444

BH4, *see* Tetrahydrobiopterin

Biotinylated glutathione ethyl ester
affinity purification of labeled proteins, 111, 113
metabolism in cells, 113
protein incorporation assay
cell loading, 109

lysate preparation, 110
materials, 107, 109
Western blot, 110
rationale for protein labeling, 103–104
structure, 104
synthesis
free sulfhydryl determination, 106
materials, 106
overview, 104–105
purification, 106–107
reaction mixture, 106

BLAST, *see* Basic Local Alignment Search Tool

C

Caenorhabditis elegans oxidant stress-responsive genes
green fluorescent protein reporters and applications, 497–499, 505
oxidative challenge, 497–498
paraquat-inducible glutathione S-transferase promoter
green fluorescent protein detection, 502, 505
hyperbaric oxygen as inducer, 502
plasmid design, 499, 501
transgenic line recovery, 501–502
reporter transgene design, 498–499

Calcineurin
calcium/calmodulin activation, 71
oxidative stress reporter assay
advantages and limitations, 80–81
buffers and media, 74–75
controls, 77–78
hydrogen peroxide-induced stress, 71, 78–80
luciferase reporter assay, 76
plasmids, 74, 79
principles, 73–74, 76–77
rationale, 70
T cell culture, 75
transfection, 75–76
structure, 71

F

O